T0186895

DEVELOPMENTS IN GEOTECHNICAL ASPECTS OF EMBANKMENTS,
EXCAVATIONS AND BURIED STRUCTURES

SYMPOSIUM ON UNDERGROUND EXCAVATIONS IN SOILS AND ROCKS, INCLUDING
EARTH PRESSURE THEORIES, BURIED STRUCTURES AND TUNNELS / BANGKOK / 1989
SYMPOSIUM ON DEVELOPMENTS IN LABORATORY AND FIELD TESTS IN
GEOTECHNICAL ENGINEERING PRACTICE / BANGKOK / 1990

Developments in Geotechnical Aspects of Embankments, Excavations and Buried Structures

Edited by
A.S. BALASUBRAMANIAM
Y. HONJO / D.T. BERGADO / N. PHIEN-WEJ / B. INDRARATNA & PRINYA NUTALAYA
Asian Institute of Technology, Bangkok, Thailand

A.A. BALKEMA / ROTTERDAM / BROOKFIELD / 1991

CIP-DATA KONINKLIJKE BIBLIOTHEEK, DEN HAAG

Developments

Developments in geotechnical aspects of embankments, excavations and buried structures / ed. by A.S. Balasubramaniam...(et al.). – Rotterdam (etc.): Balkema. – [1].
Symposium on underground excavations in soils and rocks, including earth pressure theories, buried structures and tunnels, Bangkok, 1989. – Symposium on developments in laboratory and field tests in geotechnical engineering practice, Bangkok, 1990. – With index.
ISBN 90 5410 019 2 bound
Subject heading: geotechnology.

The texts of the various papers in this volume were set individually by typists under the supervision of each of the authors concerned.

Published by
A.A. Balkema, P.O. Box 1675, 3000 BR Rotterdam, Netherlands
A.A. Balkema Publishers, Old Post Road, Brookfield, VT 05036, USA

ISBN 90 5410 019 2

Developments in Geotechnical Aspects of Embankments, Excavations and Buried Structures, Balasubramaniam et al. (eds)
© 1991 Balkema, Rotterdam. ISBN 90 5410 019 2

Table of contents

Preface IX

1. *Laboratory and field tests*

An introduction to centrifuge model testing 3
Naotoshi Takada

Centrifugal modeling of soil-structure interaction 17
James C.Ni & Chwen Song Yeh

Centrifugal model tests – Recent projects 23
L.D.Fuglsang

Influence of a non-random fabric on macroscopic properties of geomaterials 29
S.Pietruszczak

CPT correlations for sands from calibration chambers 41
J.H.A.Crooks

Testing of very soft clay 55
Y.Watari & H.Shinsha

Full scale laboratory and field test for development of gravel drains system 67
Y.Okita & K.Ito

X-ray diffraction of the five component clay shales 79
Sanga Tangchawal

Correlation of sampling effect in Hong Kong 87
K.W.Chau

Mechanics of wave pulse in rock fragmentation 103
Sanga Tangchawal

Filter design for tropical residual soils 113
B.Indraratna & E.L.G.Dilema

Strength and displacements of model infilled rock joints 123
N.Phien-wej, U.B.Shrestha & Ching-Yuan Ho

Large scale model loading tests performed in a pneumatic caisson 133
Shunta Shiraishi

Performance and monitoring of a deep, large diameter shaft 143
Kenneth R.Peaker & Shaheen A.Ahmad

Pile driving measurements for assessment of hammer performance and pile adequacy 151
P.Somehsa & R.F.Stevens

Pile diagnostics by FPDS 167
Chandra Prakash, R.K.Bhandari, A.K.Sharma & P.C.Rastogi

The study of the swelling properties of the altered granite by means of large scale field tests 183
for underground excavations of the largest pumped storage power station in China
Ge Xiurun

2. *Embankments, excavations, slopes, and earth retaining structures*

Excavation of submarine ground in the world's largest pneumatic caisson 199
Shunta Shiraishi

Full-scale slope failure at Selborne, UK 209
D.J.Petley, E.N.Bromhead, M.R.Cooper & D.I.Grant

Study on the mechanisms of slope failure due to heavy rainfall by using laboratory 225
and numerical models
Naftali S.Mshana, Atsumi Suzuki & Yoshito Kitazono

Time-dependent deformation of anchored excavation in soft clay 241
S.L.Lee, U.Parnploy, K.Y.Yong & F.H.Lee

Evaluation of earth pressure on retaining walls 247
P.J.Moore

Groundwater recharge to prevent subsidence of adjacent structure in a subway excavation 261
Yusake Honjo & Akira Morishima

Model tests of slope failure due to seepage water 269
S.Murata & H.Shibuya

Model test on sheet-pile countermeasures for clay foundation under embankment 277
H.Ochiai, S.Hayashi, T.Umezaki & J.Otani

3. *Tunnels and buried structures*

Behavior and simulation of sandy ground tunnels 291
Toshihisa Adachi, Atsushi Yashima & Keisuke Kojima

Earth pressure acting on buried pipes 331
J.Tohda

Large pipes buried in soft soil – Prefabricated pipes in river Nile sediments 347
H.-U.Werner & F.Eichstädt

Analyses of the longitudinal bending strain of pipelines subjected to vehicle loads   357
Nobuo Takagi

TBM tunneling and supports in ravelling and squeezing rock   367
Noppadol Phien-wej & Edward J.Cording

Behavior of a buried flexible steel pipe under groundwater level   377
Masami Fukuoka, Yoshinori Imamura & Takaki Omori

Shotcrete mix design, process of shotcreting for tunnel support (Narmada Power House   393
Tunnel)
A.V.Shroff & M.J.Pandya

Geotechnical experiences from construction of large diameter underground water pipe in soft   401
Bangkok clay
Noppadol Phien-wej, A.S.Balasubramaniam & Wichai Suthipongkiat

4. *Soil and rock improvements*

Consolidation of clayey soils using combined dewatering method   411
H.Kotera, Y.Wakame, T.Sakemi & T.Matsui

Some examples of field tests for soil improvement methods in Japan   425
M.Aoyama, M.Nakamura, M.Kuwahara & M.Nozu

Soil improvement for tank foundation on sand deposit   443
Y.Ozawa, S.Sunami & M.Kosaka

Behavior of the improved ground by the Deep Mixing Method under embankment loading   449
*Yusuke Honjo, Chung-Ho Chen, Lin Der Guey, D.T.Bergado, A.S.Balasubramaniam &
Ryosuke Okumura*

Mechanically stabilized earth (MSE) and other ground improvement techniques   461
for infrastructure constructions on soft and subsiding ground
D.T.Bergado, C.L.Sampaco, M.C.Alfaro & R.Shivashankar & A.S.Balasubramaniam

Interaction of steel geogrids and low-quality, cohesive-frictional backfill and behavior   483
of mechanically stabilized earth (MSE) wall on soft ground
D.T.Bergado, A.S.Balasubramaniam, K.H.Lo, R.Shivashankar, C.L.Sampaco & M.C.Alfaro

A computational method for the design of passive bolts   499
B.Indraratna

5. *Site investigation and selected topics*

Geotechnical aspects of the construction of the Singapore MRT   507
J.N.Shirlaw & D.F.Stewart

The geotechnical aspects of Malaysia's North-South Expressway   525
R.A.Nicholls

Pile-tip cap driving method for precast concrete pile   537
Tatsuo Ishikawa

Stochastic three dimensional joint geometry modelling including a verification to an area 545
in Stripa Mine, Sweden
Pinnaduwa H.S.W.Kulatilake & Deepa N.Wathugala

Geotechnical characteristics of soft soils in Indonesia 557
J.S.Younger

Field and laboratory tests of tropical residual soils as materials for dam 571
Didiek Djarwadi

Preliminary report on the sand liquefaction induced by the 16 July earthquake in Dagupan 583
City (Pangasinan Province, Philippines)
G.Rantucci

Author Index 597

Developments in Geotechnical Aspects of Embankments, Excavations and Buried Structures, Balasubramaniam et al. (eds)
© 1991 Balkema, Rotterdam. ISBN 90 5410 019 2

Preface

This volume on Developments in Geotechnical Aspects of Embankments, Excavations and Buried Structures is ninth in the series of Balkema Publications on the last two Annual Geotechnical Symposia sponsored by the Asian Institute of Technology, Canadian International Development Agency and the Southeast Asian Geotechnical Society. The previous eight ones are on Geotechnical Problems and Practices of Dam Engineering (1982), Geotechnical Aspects of Coastal and Offshore Structures (1983), Recent Developments in Ground Improvement Techniques (1985), Recent Developments in Laboratory and Field Tests and Analysis of Geotechnical Problems (1986). Geotechnical Aspects of Mass and Material Transportation (1987), Environmental Geotechnics and Problematic Soils and Rocks (1988), Computer and Physical Modeling in Geotechnical Engineering (1989), and Restoration Works on Infrastructures and Monuments (1990).

For ease of reference, the contents of this volume are presented in five sections. Seventeen papers are contained in Section 1 which deals with laboratory and field tests. The majority of the papers in this Section deal with 1 g and centrifugal model tests, and they refer to a wide range of geotechnical problems. The second Section contains eight papers under the title embankments, excavations, slopes and earth retaining structures. Here again some of the papers refer to Laboratory Model tests and Full Scale tests.

The third Section is on tunnels and buried structures. There are eight papers included in this Section and they cover a wide selection of topics on buried pipelines and tunnels. The Section four is titled as soil and rock improvements. Except for one paper on rockbolts all the others, relate mainly to soft soil improvements. The last Section contains five papers on site investigation as well as other selected topics.

This volume is a combined volume of contributed papers on Laboratory and Field Tests as well as Embankments, Excavations, Slopes and Buried Structures.

The editors like to express their sincere thanks for the help and support provided by several sponsoring organizations and individuals in releasing this volume. Sincere gratitude is expressed to the General Committee Members of the Southeast Asian Geotechnical Society: Dr Chin Der Ou, Prof. Seng Lip Lee, Dr Za-Chieh Moh, Dr E.W. Brand, Dr Ting Wen Hui, Dr Tan Swan Beng, Dr Ooi Teik Aun, Dr Surachat Sambhandharaksa, Dr Victor Choa, Dr R.A. Fraser, Dr K.Y. Yong, and Dr Clive Franks.

At the Asian Institute of Technology, sincere thanks are expressed among others to the President Prof. Alastair M. North, Dr Yordphol Tanaboriboon, Dr A. Fukuda and Dr H. Mori. The assistance provided by Mr Manoj Kumar Panda, Mr Shahadat Hossain, Mr S. M. Mainuddin Chowdhury and Mr Rey S. Ofren in checking and proofreading of the manuscripts have made it possible to have this volume in its present form.

A. S. Balasubramaniam
Y. Honjo
D. T. Bergado
N. Phien-wej
B. Indraratna
Prinya Nutalaya

1. Laboratory and field tests

Developments in Geotechnical Aspects of Embankments, Excavations and Buried Structures, Balasubramaniam et al. (eds)
© 1991 Balkema, Rotterdam. ISBN 90 5410 019 2

An introduction to centrifuge model testing

Naotoshi Takada
Osaka City University, Japan

ABSTRACT A centrifuge provide a scale model with prototype stress conditions, which is most important in geotechnical model testing in simulating the prototype behavior where mechanical properties are greatly stress-dependent. The usefulness of centrifuge model test has become a common acceptance in the geotechnical engineering field, and the number of centrifuges is rapidly increasing at present. This paper deals with the principle, examples of application and problems of centrifuge model testing, and the centrifuge facilities, on the basis of the author's experience.

Introduction

The purpose of model tests in the engineering field is to predict the prototype behavior, to investigate the validity of a theory and a method of analysis, to evaluate engineering properties of materials, to find new engineering mechanism, and also for educational purposes. In the geotechnical engineering field, stresses induced by the body force of the soil usually occupy a predominant part in the total working stresses, in addition the mechanical properties are strongly stress dependent. Thus, model tests using small size models in the earth's 1-g gravity field can not reproduce the prototype behavior because the stress level due to its selfweight is much lower than in the prototype. A centrifuge enables a scale model to properly generate the same stress level as the prototype, therefore we can expect at a certain extent the prototype behavior in the model.

ISSMFE Technical Committee on Centrifuge Testing established in 1982 concentrated on gathering and providing information of new model testing technique. International symposia on centrifuge modelling were held consecutively in Tokyo, Manchester and Davis in 1984, and in Paris in 1988. Discussion sessions were conducted at the ICSMFE in San Fransisco in 1985, and in Rio de Janeiro in 1989. Next symposium is to be held in Colorado in 1991. The news letter "Geotechnical Centrifuge" is circulated quarterly since 1988 to parties and individuals involved with centrifuge modelling. Under these circumstances recognition of usefulness of centrifuge modelling by researchers in the geotechnical engineering field has developed a

rapid increase of the number of centrifuges during these 10 years.

In Japan, the centrifuge model test in the geotechnical engineering was initiated in Osaka City University in 1965 for selfweight consolidation of very soft clays using a small size centrifuge with a rotor radius of 1m. Since then, centrifuges were installed in Tokyo Institute of Technology in 1969 (Kimura, 1985), in Port and Harbour Research Institute in 1980 (Terashi, 1985), and so on. At present, 18 centrifuges are in operation for research and practical work, four of which belong to laboratories of private companies. The trend of introducing centrifuge is now booming and the number of centrifuge will exceed 20 in near future. All, except one in Port and Harbour Research Institute, are middle size centrifuges. This tendency is somewhat different from the move in Europe and USA where large machines are being constructed.

The number of reports dealing with centrifuge model tests and of data produced by centrifuge model tests has rapidly increased, and the centrifuge test has become a common acceptance in Japan. Tsuchi-To-Kiso, the monthly journal of Japan Society of Soil Mechanics and Foundation Engineering, in the issues from December, 1987, through August, 1988, provided the articles relating to the centrifuge modelling to the society members for their understanding the principle, roles, examples of test results, the method of model testing and the centrifuge facilities, together with the limitation and problems in application.

In Southeast Asia, except Japan, China is a leading country followed by Singapore in the centrifuge model testing. Though the number of centrifuge is small at present, the number of researchers who have been involved with the centrifuge model test outside their countries is not small. They are, the author thinks, willing to introduce the centrifuge facilities to their laboratories in near future.

This paper deals with the principle , examples of application and problems in centrifuge modelling, and the centrifuge facilities on the basis of my 25 years' experience since 1965 when the first Japanese geotechnical centrifuge was introduced in author's laboratory. Although there is repetitive of other papers relating to the centrifuge modelling, an attempt is made to benefit in the profession of those who are interested in but not much familiar to the centrifuge model test, rather than of those who are involved with it, in being able to better understand the centrifuge model test.

1. Principle and Role of Centrifuge Test

1.1 To generate deformation and failure in the model.

The earth structure scale models which are formed with the same materials as those of the prototype are not always led to failure in a 1-g gravity field. Fig.1 is an example, where we consider the failure of a clay slope with a circular slip surface under undrained conditions (Mikasa & Takada, 1973). The safety factor of the prototype failure is calculated by

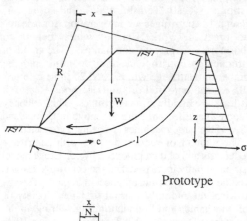

Prototype

N-th scale model

Fig.1 Stability of C-material slope

$$F_p = \frac{Rcl}{Wx} \qquad ----- (1)$$

where R is the radius of the slip circle, c is the undrained shear strength, l is the length of the circular arc, W is the weight of the soil mass, x is the horizontal distance from the origin of the slip circle to the center of gravity of the soil mass. The safety factor of an N-th scale model (i.e. 1/N size of the prototype) formed exactly with the same material as the prototype is calculated by :

$$F_m = (\frac{R}{N} c \frac{l}{N}) / (\frac{W}{N^2} \cdot \frac{x}{N}) = \frac{NRcl}{Wx} \qquad ----- (2)$$

This equation indicates that this model has a safety factor N times as large as the prototype. If this model is put in the increased acceleration field N times the earth's gravity, the selfweight of the soil mass, W, increases to NW, and F_m corresponds to F_p. Thus, the model will fail in the same manner as the prototype.

1.2 To generate prototype stress-strain relations in the model.

Since stress-strain relations and strength of soils are generally stress dependent, the reproduction of prototype stress conditions in the scale model is all important to produce the prototype behavior in the model. The following are typical examples of stress dependent engineering properties.

(1) Stress-Strain Relations

If the strain in every soil element remains within a low level, the compressive and shear behaviors of the soil mass are regarded as being elastic, and can be simulated by a model with an elastic material such as gelatine. However, the stress-strain relation of soil is generally nonlinear, particularly in the range close to failure. In addition, the dilatancy, which is peculiar to the soils and is strongly dependent on the stress level, causes the behavior of small models in the 1-g gravity field to differ from that of the prototype; negative dilatancy in shear behavior of the prototype will turn to positive under a low stress level in the scale model.

(2) Compressibility

Volume compressibility changes drastically when the consolidation pressure exceeds the consolidation yield stress (equal to the preconsolidation pressure if the clay is normally consolidated). In the over-consolidation range, volume compression is usually proportional to the consolidation pressure, whereas in the normally consolidation range it is proportional to log of pressure. In the case of selfweight consolidation of a very soft clay, for example, the final water content distribution of a model in the 1-g gravity field corresponds to the only upper part of the prototype as shown in Fig.2. If this model is put in the centrifugal acceleration field N times the earth's gravity, the water content will distribute just in the similar manner to the prototype.

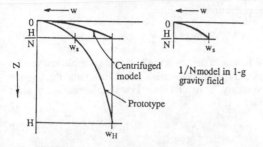

Fig.2 Distribution of final water content in selfweight consolidation

(3) Strength Characteristics

Generally the shear strength envelope of a clay is strongly dependent on its stress history as shown in Fig.3. Even for granular materials, the strength envelope is not always straight lines but likely to decrease their slope with increasing stress level due to particle crushing and/or other effects.

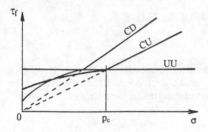

Fig.3 Shear strength envelopes for clay

2. Similarity Rule

(1) Stress and Strain

The basic principle of centrifuge modelling is to test geometrically N-th scale model formed with the prototype material in the acceleration field N times the earth's gravity. Under these conditions, following relations exist:

Length	$l_m = l_p/N$	----- (3)
Acceleration	$\alpha_m = \alpha_p$	----- (4)
Mass density	$\rho_m = \rho_p$	----- (5)

Where subscripts m and p designate the model and prototype, respectively.
From Eq.(3):

| Area | $A_m = A_p/N^2$ | ----- (6) |
| Volume | $V_m = V_p/N^3$ | ----- (7) |

From Eq.(4), (5), (6):

| Force | $F_m = F_p/N^2$ | ----- (8) |
| (body force and external force) | | |

| Stress | $\sigma_m = \sigma_p$ | ----- (9) |

Since the model is assumed to be formed with the same material as the prototype, not only vertical stress but also all other stresses of the model are identical to those of prototype. Similarly, the strains between the two are identical.

| Strain | $\varepsilon_m = \varepsilon_p$ | ----- (10) |

Displacement is the integration of the strain with length, thus

| Displacement | $S_m = S_p/N$ | ----- (11) |

(2) Flow of pore water

Let t, v and Δl be the time, the velocity of pore water, and the seepage length, respectively, then

$$t_m = \Delta l_m/v_m$$
$$t_p = \Delta l_p/v_p \qquad ----- (12)$$

Assuming that the model and prototype have the same permeability and the Darcy's law is valid:

$$v_m = k\,i_m = k\,(\Delta h_m/\Delta l_m)$$
$$v_p = k\,i_p = k\,(\Delta h_p/\Delta l_p) \qquad ----- (13)$$

Where i is the hydraulic gradient and Δh the hydraulic head difference. Since the hydraulic heads in respective locations in the model and prototype are the same,

$$\Delta h_m = \Delta h_p \qquad ----- (14)$$

Therefore,

$$i_m = N\,i_p \qquad ----- (15)$$

From Eqs.(13) and (15)

$$v_m = N\,v_p \qquad ----- (16)$$

5

Using Eqs.(3), (12) and (16), we obtain

$$t_m = t_p/N^2 \qquad \text{----- (17)}$$

Eq.(17) indicates that in the model the hydraulic behavior, such as time-consolidation of clays and seepage flow, progresses N^2 times faster than in the prototype.

(3) Capillarity

In the capillary zone above a water table, capillary pressure causes the increase of the intergranular pressure (effective pressure), which leads to the shear strength increase (Fig.4). The capillary rise, H, in the N-th scale model formed with the prototype material in the acceleration field N times the earth's gravity is

$$H_m = H_p/N \qquad \text{----- (18)}$$

In the capillary zone, the degree of saturation generally decreases with increasing height. However, the capillary zone can be simulated in the centrifuge model to a certain degree in respect with the pore pressure distribution. In the partially saturated zone above the capillary zone, the pore water forms a meniscus, which also causes the increase of the intergranular pressure. For a model of sandy soils, however, we experienced the downward movement of the pore water when the model was put in the centrifugal acceleration field. This pore water movement lasts until the capillary water in a form known as contact moisture comes to a equilibrium conditions with the applied centrifugal acceleration. Thus the soil in the centrifuge is as if being in the "pF test". In such a situation, the capillary pressure in the model differs from that in the prototype. Therefore, we can not simulate the prototype shear behavior in the centrifuge model where the pore water movement above the capillary zone would be expected. We have not enough

knowledge of the scale effect on the centrifuge modelling of partially saturated soils.

(4) Dynamic motion

The effects of earthquake, for instance, is a major concern to the geotechnical profession. Suppose a simple model where sinusoidal motion expressed by Eq.(19) is applied to the horizontal base of the model perpendicular to the direction of centrifugal acceleration.

$$x = a \sin \omega t \qquad \text{----- (19)}$$

Where x, a and ω is the horizontal displacement, the maximum displacement (amplitude) and the angular velocity, respectively. For displacement, a scale factor is N:

$$x_m = x_p/N \qquad \text{----- (20)}$$

The inertial horizontal body force in the model must be increased by a factor N corresponding to the increased vertical body force (Fig.5). The horizontal acceleration must satisfy the following relation:

$$\alpha_m = N\alpha_p \qquad \text{----- (21)}$$

The horizontal velocity and acceleration are derived from Eq.(19) as:

$$\frac{dx}{dy} = -a\omega \cos \omega t \qquad \text{----- (22)}$$

$$\frac{d^2x}{dt^2} = a\omega^2 \sin \omega t = \alpha \qquad \text{----- (23)}$$

From Eqs.(19), (20), (21), (23), we obtain

$$\omega_m = N\omega_p \qquad \text{----- (24)}$$

Thus the time is reduced by a factor N. From Eq.(22), it can be derived that the velocity of the model is the same as that of the prototype.

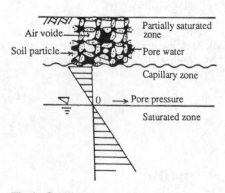

Fig.4 Capillary zone

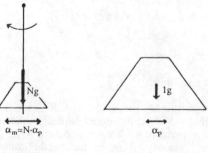

Model   Prototype

Fig.5 Scale relations in dynamic motion

Model shaking in the centrifuge has been performed by the spring actuated shaking system; the explosive method; the bumpy road method; the servo-controlled hydraulic actuator system; the piezoelectric shaking system. Recently the servo-controlled hydraulic actuator system is a majority. However, dynamic model test in a high gravity field with a high frequency together with a small amplitude needs a large cost in installation of the shaking system and the data acquisition system. As an alternative, the author's laboratory employed the "tilting method" which gives the model a lateral body force statically, which is specified in the ordinary design criteria for earthquakes.

There are numerous difficulties in dynamic modelling, one of which, for instance, is container confinement producing undesired reflection of stress waves into the soil mass.

(5) Scaling problems

The scaling relations relating to the geotechnical centrifuge model test are summarized in Table 1 (Ko, 1988). As can be seen in the table, some scaling relations for time conflict to each other. For instance, in simulating sand liquefaction under dynamic loads in the centrifuge, there are conflicts between the scale factors of dynamic motion $1/N$ and that of permeation of pore water $1/N^2$. In this case, as a countermeasure, the pore water is replaced by a viscous liquid, such as oil, glycerine-water mixture or methylcellulose-water mixture so as to have pore fluid permeability $1/N$ compared with water.

Another conflicts appear in the deformation process when the strain develops and concentrates to form a slip surface. During the period when the stress level is not close to failure, the strains in the model distribute in the same manner as in the prototype. After a very thin shear band or a slip surface is formed, the value of scale factor of relative displacement between soil masses deviates from N and rapidly becomes much smaller than N. Such a deviation may appear between the soil and the smooth surface of other structures. This point of view is very important in considering the similarity in the models at failure and must be investigated.

4. Scale Effect

4.1 Concept of Modelling of Models

Reliability of the scale model test results which extrapolate to the specified or imaginary prototype behavior have to be verified. It is best to do this by comparing the model with the prototype. However, it is often impractical and almost impossible except small and simple structures. As an alternative, a technique called "the modelling of models" is used to validate the internal consistency of the test programme and the scaling relations. This concept is graphically presented in Fig.6. The horizontal axis represents the representative length, L, of the model in a log scale, and the vertical axis represents the acceleration ratio, N, which equals to the model scale, also in a log scale. In the diagram, all the models tested at 1-g gravity field are defined as prototypes, and models specified on the diagonal lines with the slope of 45° correspond to the same prototype, since $L_m \cdot N = L_p$ constant. It is important that these models are also the models of each other. Comparing these models with one another on the basis of the scaling relations, the validity of scale relations and the accuracy of the models can be

Table 1 Scaling Relations

Quantity	Prototype	Model
Length	N	1
Area	N^2	1
Volume	N^3	1
Velocity	1	1
Acceleration	1	N
Mass	N^3	1
Force	N^2	1
Energy	N^3	1
Stress	1	1
Strain	1	1
Mass Density	1	1
Energy Density	1	1
Time (Dynamic)	N	1
Time (Diffusion)	N^2	1
Time (Creep)	1	1
Frequency	1	N

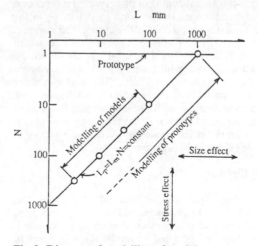

Fig.6 Diagram of modelling of models

evaluated, before their performance is extrapolated to the prototype.

Reviewing past papers, we find a lot of research work using centrifuge model tests covers almost all the subjects with which geotechnical engineers and researchers have been concerned. However, papers demonstrating the repeatability of model tests and validity of scale relations by means of modelling of models are limited. Demonstrating these is very important not only to develop the centrifuge technique but also to evaluate the possibility and limitation of centrifuge modelling. Fuglsang and Ovesen collected and reviewed these sorts of model tests from past papers (Fuglsang and Ovesen, 1988).

4.2 Examples of Modelling of Models

(1) Circular footing

Fig.7 shows an example of evaluating validity of "modelling of models" tests of a circular footing with a rough bottom surface on the dry sand (Fuglsang and Ovesen, 1988). All the models represent the same prototype having a diameter $D_p=1m$. The model grounds were formed with a uniform sand, whose particle sizes range from 0.3 to 0.6mm. The dimensionless peak-values of the loads on the model footings together with the standard deviations of the values are presented in Fig.7(1). The dimensionless load-displacement curves are also shown in Fig.7(2) (Ovesen, 1979). No scale effect is observed not only on peak-values but also load-displacement curves. In this case, ratios of the model footing diameter to the average particle size ranging from 35 to 200.

(2) Square anchor plate

Fig.8 shows "modelling of prototype" test results on a square anchor plate where a model in 1-g is, by definition, the prototype (Fuglsang and Ovesen, 1988). The prototype anchor with a width $B_p=0.75m$ was buried at an angle $\theta=45°$ to the horizontal plane with a well-graded sand having an average particle size of 0.8mm. The sand with water content of 4.7% was compacted by a vibrator. The model anchor having $B_m=20mm$ wide was buried in the same sand as used in the prototype, except that over 2mm particles were removed. Fig.8 shows no scale effect on the peak force.

(3) Clay Slope

The clay used was an alluvial clay ($w_L=101$, $w_P=31\%$), which was remolded into a slurry with $w=140\%$ and was consolidated under 0.5kgf/cm². The slope is 1 : 0.5 and the model shape is shown in Fig.9. The slope height of models varies from 1cm

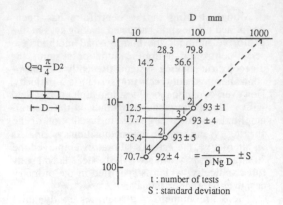

(1) Maximum loads

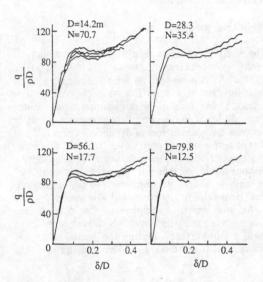

(2) Load-displacement curves

Fig.7 Circular footing

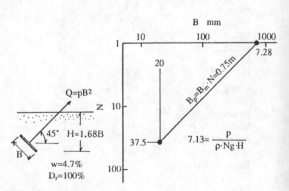

Fig.8 Square anchor

8

to 18cm. For models with slope heights of 4-18cm, OCU Centrifuge Mark 5 with a rotor radius of 256cm (Fig.17) was used, and for models with slope heights of 1-4cm, a small centrifuge with a rotor radius of 25cm (Fig.18), which was deviced for educational purposes, was used. Since the model slopes were led to failure rapidly by increasing acceleration, failure is assumed to be under undrained conditions.

Fig.9 shows the test results ; the model height is converted to the prototype height by a relation $H_p = H_m \cdot N$ where N is the acceleration ratio at failure, and plotted against the model slope height (Takada and Kusakabe, 1987). Every model shows the similar prototype height at failure, though a slightly smaller height is observed in small models. Fig.10 shows the examples of failure patterns of models. The larger models show a clear circular slip surface, whereas the smaller ones show a ductile failure. This tendency suggests a minimum requirement in the model size in respect with the width of the shear band.

(4) Selfweight consolidation of slurry clay

The clay used was an alluvial clay ($w_L=98\%$, $w_P=40\%$), which was remolded into a slurry with a water content of 160%. The model was setup in an acrylic cylinder. The model heights were 10, 5, 2.5cm and models were singly and doubly drained. Fig.11 shows the test results in terms of the prototype degree of consolidation against the prototype elapsed time (Mikasa et al, 1980). They coincide to each other indicating the validity of scale relations. This test was conducted by using OCU Centrifuge Mark-1.

(5) Heavy Tamping

The prototype ground was assumed as an axi-symmetrical cylindrical column, half of which was simulated in a semi-cylindrical column, 30cm in diameter and 20cm in depth. The vertical cross-section of the model ground including its center axis was supported by a glass plate, through which the ground behavior can be observed. Ram blows were applied to the ground surface through a short semi-cylindrical wooden penetration rod to produce ground deformation clearly. The soil used was a sandy soil containing several percents of fines, which was compacted to the model ground with a relative density of 75% and a water content of 8%. The model scales were N=65, 75, 100 and the respective combinations of blow effort for models were: [ram mass; ram base diameter (penetration rod diameter); ram drop height] = [36.4grams; 3.5cm; 34.7cm], [23.7; 3.0; 29.8] and [20; 2.3;21.9]. These combinations represent the same prototype blow effort of [20t; 2.3m; 20m]. In the centrifuge in

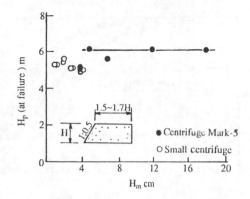

Fig.9 Clay slope

h_m=18cm

h_m=4cm

h_m=2cm

Fig.10 Failure patterns of clay slopes

9

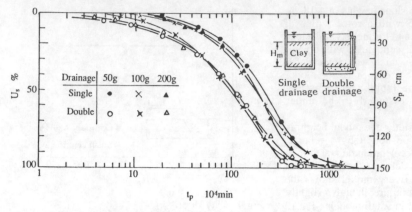

Fig.11 Selfweight consolidation of slurry clay

flight, ram must be guided to the radial direction against "Coriolis force". The ram drop height in the above brackets are corrected so as to have the prototype ram velocity at blow with consideration of the difference between the accelerations at radii of ram release and ram blow.

Fig.12 shows test results in terms of prototype ram penetration into the ground against the number of ram blows (Takada et al, 1987). Three curves show the validity of scale relations.

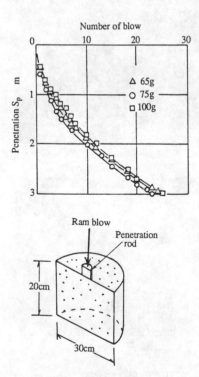

Fig.12 Heavy tamping

5. Problems in simulating prototype behavior

All importance in the centrifuge model is to induce the prototype stress conditions in the model. There are numerous problems arising in modelling the prototype as follows. Some are common with ordinary models in the 1-g gravity field and others are peculiar to the centrifuge models.

(1) Similarity of Soil Conditions

1) Particle size limitation

If the prototype materials contain a percentage of large particles, such as dam materials, the model materials must be chosen by some method with consideration of the particle size distribution. To use such materials, they must be sieved to remove over size particles because of model size limitation. Since these model materials are regarded as alternatives, the accurate prototype behavior cannot be expected in the model. This problem is common with tests for mechanical properties of soils, such as the compaction test and the shear test.

2) Stress history

A normally consolidated young clay stratum can be replicated by consolidating a slurry clay in a certain centrifugal acceleration either under selfweight or selfweight plus a surcharge load. The alluvial clay stratum having a over-consolidated crust due to deccication and/or other effects is replicated in the model by means of a partial consolidation method (Davies, 1981). However, these model clay grounds will exhibit lower sensitivity than the real clay ground because of the short aging period; the deformation of the model is more ductile than that of the prototype in the field.

The ground which consists of overconsolidated

10

aged clayey stratum or of strata with different stress histories cannot be replicated in the model any more, even if the undisturbed soil sample from each stratum is available, because the model ground assembled with samples taken discretely from different depths do not reproduce the same continuous mechanical and physical properties along the depth as in the field. An alternative approach is to use the undisturbed soil block at the key part of the model structure (Mikasa et al, 1981).

(2) Setback in reduction of model time

It is of great benefit that the centrifuge model reduces greatly the time spent for test. However, the reduced test period makes difficult to accurately control the test conditions in some cases, such as undrained condition during shear deformation, and the simplified assumptions, which are important in the idealized model tests, cannot be assured. This difficulty is likely to occur in the models with a low plastic silty clay and a clay layer with a small thickness.

(3) Size effect

As described before, the width of shear band is not proportional to the model size but decreasing relatively to the model size with increasing model size. Therefore, the local deformation may differ between in models with different sizes. The width of the shear band is considered to be chiefly dependent on the sensitivity of soils. Therefore, the small size model using a laboratory-consolidated low sensitive clay is likely to deform ductilely with a wide shear band, thus the minimum requirement of model size must be taken into consideration.

As to the effect of the particle size, it is recommended that the minimum ratio of the model footing/average particle size should be 30 (Fuglsang and Ovesen, 1988). Under this condition, the soil is regarded as a continuous medium.

(4) Segregation of coarse particles in slurry clay

Segregation of sand particles in a slurry clay occurs when the model is put in the high acceleration field (Takada & Mikasa, 1985). This phenomenon, of course, usually occurs in a mudpond into which a slurry clay dredged from the sea bed and mixed with water by a suction pump is poured. The high acceleration field magnifies this because the centrifuged slurry clay with high water content can not hold sand particles with increased weight. Segregation of sand particles depend on the water content and the activity of clay and the centrifugal acceleration. In order to avoid excessive segregation of sand particles in the selfweight consolidation tests,

the lower acceleration field should be chosen.

(5) Errors in the centrifugal acceleration field

The centrifugal acceleration works radially in direction and is proportional to the radius in magnitude as illustrated in Fig.13. The water surface in the model container shows a circular arc and the horizontal plane of the model corresponds to a convex curved surface in connection with the direction of body force. In addition the vertical stress does not distribute linearly along the model depth. Provided that the "total weight" is in accordance with the scale relations described before, the mean radius at which the nominal centrifugal acceleration is defined is the distance from the rotor center to the point of 2-3rd model depth. The stresses due to the body force in the upper part and the lower part from this depth are smaller and larger, respectively, than those in the constant acceleration field.

Another error in the centrifugal acceleration field is caused by the "Coriolis force", which affects dynamic events. In the case of the heavy tamping

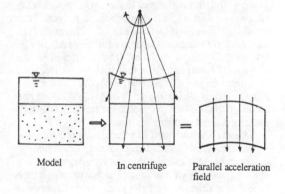

Model   In centrifuge   Parallel acceleration field

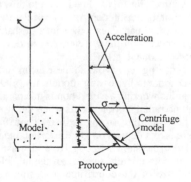

Fig.13 Errors in centrifuge: direction and magnitude of acceleration

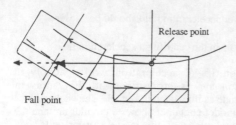

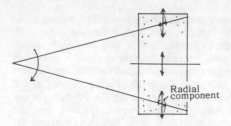

(1) Coriolis force

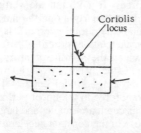

Fig.14 "Free fall" in centrifuge

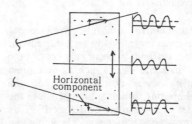

(2) Horizontal acceleration

Fig.15 Error sources in dynamic motion in centrifuge

described before, the ram released at a certain radius in the centrifuge in flight flies at a constant speed along a straight line tangent to the circle on which the ram was whirling, till it reaches, if it could, the model ground as shown in Fig.14. Therefore, the ram does not increase speed, nor fall perpendicular to the model ground surface. The deviation in the horizontal distance of the ram drop point from the ram release point increases with decreasing ram release radius and with increasing ram drop height. In order to simulate the prototype free fall, the ram must be guided to the radial direction.

The mass movement is affected by the Coriolis force when it has a radial component. In simulating earthquake motions in the centrifuge, even if a shaking motion of the model is horizontal in the model container, each model part, except the central part, has a radial component resulting in generating the Coriolis force, Fig.15(1). In addition to this, the horizontal component of centrifugal acceleration, except at the central part, causes the center of shaking motion in respect with the acceleration to deviate from the nutral value, Fig.15(2).

A large centrifuge can avoid all these errors in the case of 2-D models by testing models in a radial plane so that the working acceleration is in a parallel manner and no radial motion occurs when the model is shaken horizontally. In the ordinary centrifuge design, these errors depend on the "distortion index" (Mikasa, 1984) as shown in Fig.16, and must be considered in determining the size of the model as well as the size of the centrifuge in designing a centrifuge.

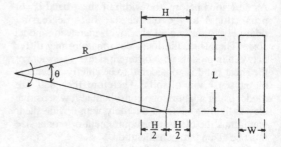

Fig.16 Distortion indexes: θ and H/R

6. OCU Centrifuges

6.1 History

Table 2 shows the specifications of past centrifuge, Mark-1, 2, 3, 4, and the present one, Mark-5. Geotechnical Engineering Laboratory of Osaka City University began with a very simple centrifuge Mark-1 to conduct the selfweight consolidation test of slurry clays. The rotor of this centrifuge weighed only 30kg. Two steel vessels were hinged at the both ends of the rotor, and acrylic cylindrical model containers were set in them. During flight, measurement was made for surface settlement visually by using a stroboflash. Several series of model tests showed the validity of the modelling of models, and revealed the peculiar behavior of

Table 2 Specifications of centrifuges in Osaka City University

Centrifuge	Date of completion	Max. acceleration	Nominal radius	Motor capacity	Model container L·H·W(cm)	Container inclination	Distortion index θ	H/R	Max. capacity
Mark-1	1, 1965	200g	100cm	2.2KW	φ10·H20	——	6°	0.2	0.63g·ton
Mark-2	8, 1966	100	100	2.2	35·18·7.5	——	20	0.18	0.95
Mark-3	8, 1968	200	100	3.7	50·30·10	16.7°	29	0.30	6
Mark-4	8, 1975	200	155	11	50·30·16.7	16.7	19	0.19	10
Mark-5	9, 1983	200	256	22	50·30·16.7	15.1	12	0.12	10

selfweight consolidation of very soft clays, which was evaluated by Mikasa's consolidation theory. This centrifuge also provided the author much experience of both the centrifuge and centrifuge model testing.

Centrifuge Mark-2 was designed for bearing test of footings. The loading force was supplied from a electric motor mounted on the center shaft of the rotor through a set of bevel gears to suit the swing type model container, and was measured electrically. Some series of bearing tests showed the validity of modelling of models and provided the peculiar role of the dilatancy in the shear behavior of sand beneath the footing (Mikasa, et al, 1973).

Centrifuge Mark-3 was designed for stability tests of a rockfill dam, especially for earthquakes and the rapid drawdown of the reservoir (Mikasa, et al, 1969). For these purposes, Mark-3 was equipped with a middle size model container, a container tilting system as described in section 2.(4), and a system controlling the water level in the model container. This centrifuge was used thereafter for various purposes.

Centrifuge Mark-4 was an enlarged version of Mark-3 to ease the distortion indexes, only a rotor together with a housing shell and a electric motor being replaced.

Centrifuge Mark-5, still in principle, is an enlarged version of Mark-3. Its rotor radius was further elongated to 256cm and the rotor was installed in an underground pit for both safety and laboratory space. This machine inherits all the functions that Mark-3 possessed and the size of model container remains the same which enable two persons to carry it with the specimen and test instruments inside. The main part of Mark-5 is illustrated in Fig.17.

A very small hand-made centrifuge for educational purposes described before is also shown in Fig.18. This centrifuge consists of a 25cm radius rotor and a wooden housing. The rotor is of disc shape to reduce aerodynamic resistance and is driven by a 60W variable speed induction motor to the maximum acceleration of 560g.

6.2 Centrifuge Mark-5

(1) Rotor

The rotor consist of a pair of H-shaped beams. At both ends of the beams, links are hinged by means of universal ball joints, on which model container and a counter weight container (or another model container) are pinned. Two beams are supported on the center shaft in the manner that enable them to move in the parallel and opposite direction to each other by a screw jack system powered by a 0.4 KW motor installed inside of the center shaft. This movement will tilt the model container in the centrifugal acceleration plane providing the model with a lateral force corresponding to the converted static earthquake force. The load in designing the rotor and the center shaft was determined under the conditions of loosing one model container in flight with the model inside. The rotor was also designed to have a small height (9.3cm) with the least requirement of rigidity so as to minimize the aerodynamic resistance which greatly contributes the power consumption and the temperature rises of air inside housing. For dynamic test, usually a high rigid rotor is employed. However, the CALTEC centrifuge that has a very slender light weight rotor manages the dynamic test by means of the dynamic counter weight system (Scott, 1985). This slender rotor can carry a shaking equipment in the same manner as the CALTEC centrifuge, though we have no intent to conduct dynamic tests at present.

(2) Motor

A three-phase induction motor with a relatively small capacity was resulted from the reduction of the aerodynamic resistance. The speed is controlled by the frequency control by means of an electronic invertor. The driving force from the motor is conducted to the center shaft through a set of bevel gears and a V-belt. The employment of the V-belt is for safety; it can be easily broken in case of emergency.

(3) Model Container

The model container of standard size is directly hinged to the links. The swing platform shown in Fig.19 is also used, on which a model container inevitably smaller than that of standard size is mounted. The front frame of every model container to which a thick laminated glass plate is fixed is removable for model preparation. The glass plate must be often exchanged because of abrasion by sand particles.

(4) Housing

The rotor is installed in an underground reinforced concrete pit for safety and laboratory space, though we have not experienced any serious danger. In order to reduce the aerodynamic resistance, the inner height of the pit is minimized keeping a least requirement for working space and the inner surface is made as smooth as possible. The model behavior can be visually observed by means of stroboflash through an observation window which is located in a

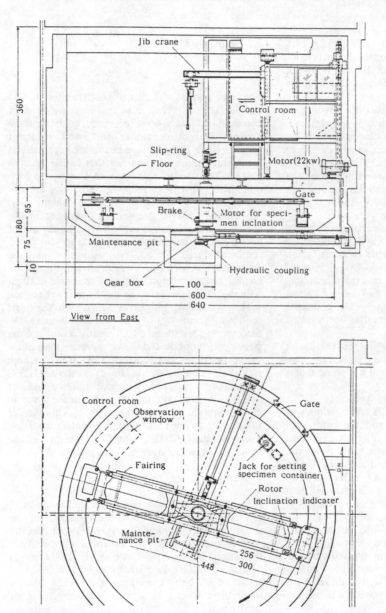

Fig.17 OCU Centrifuge Mark-5

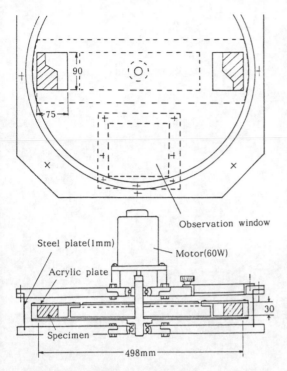

Fig.18 Hand-made small centrifuge

Fig.19 Swing platform and model container containing a mud-trench model inside

dark room, where the control of test process and measurement are made.

For safety measure, two sets of thin steel wire are tensioned vertically a few centimeter apart from the vertical wall of the pit, and also two pair of electrode plates are arranged on the inside surface. If, during centrifuge operation, some hard body detached from the rotor or model container cuts the wire or some amount of water splashed from the model container shorts circuit, the power will shut-off and the pneumatic brake works automatically to stop rotation.

(5) Electric and Hydraulic Joints

Electric joints are installed on the center shaft; nine are for power supply to the motor for model container tilting and test devices on the model container; sixty are for electric measuring devices. Recent development of electronic devices, such as a switching box, can reduce the necessary number of electric joints for measurement.

Five hydraulic joints are installed at the lower end of the center shaft for pressurized oil or air supply. They are designed to have a small diameter of the rotary axle (about 2.8cm) to reduce wear of the seals.

7. Conclusive Remarks

The centrifuge provides us with various purposes of model tests such as the modelling of prototypes, modelling of models, parametric experimental studies, evaluation of numerical methods, fact finding studies, etc. The benefit of these model tests is great in the geotechnical engineering area because of the distinct stress dependency of the engineering properties of soil materials. There are many interesting subjects to be experimentally examined by this method.

The centrifuge model also has an advantage for educational purposes; introducing students and young engineers to realistic behaviors of models replicating field events that they seldom observe enhances their interest and understanding. For this purposes, some companies are developing and standardizing very small and small centrifuges with low cost.

Finally, the author thinks following factors are important and should be taken into consideration when introducing the centrifuge:

1. Research projects (sizes and kinds of prototype structures)
2. Sizes and volumes of models
3. Rotor radius of centrifuge (distortion indexes)
4. Maximum acceleration
5. Driving power
6. Cost (construction, operation, maintenance)
7. Personnel
8. Laboratory space
9. Number of models per year
10. Volume of soil materials

References

Davies, M.C.R. (1981). Centrifugal modelling of embankments on clay foundations, Ph. D. Thesis, Cambridge University.

Fuglsang, L. D. & Ovesen, N.K. (1988). The application of the theory of modelling to centrifuge studies, Centrifuge in Soil Mechanics, pp.119-138, Balkema, Rotterdam.

Kimura, T. (1985). Geotechnical centrifuge model tests at the Tokyo Institute of Technology, Proc. of Int. Symposium on Geotechnical Centrifuge Model Testing, Tokyo, pp.59-79.

Ko, H.Y. (1988). Summary of the state-of-the-art in centrifuge model testing, Centrifuges in Soil Mechanics, pp.11-18, Balkema, Rotterdam.

Mikasa, M., Takada, N. and Yamada, K. (1969). Centrifugal model test of rockfill dam, Proc. of 7th ICSMFE, vol.2, pp.325-333

Mikasa, M. and Takada, N. (1973). Significance of centrifugal model test, Proc. of 8th ICSMFE, vol.1, pp.273-278

Mikasa, M., Takada, N. and Mochizuki, A. (1980). Centrifuge model test of earth structures, Tsuchi-To-Kiso, Journal of JSSMFE, 28-5, pp.15-23, (in Japanese).

Mikasa, M., Mochizuki, M. and Matsumoto, T. (1981). Stability test of a rockfill dam by centrifuge, Proc. of 10th ICSMFE, vol.1.3, pp.475-478.

Mikasa, M. (1985). Two decades of centrifugal testing in Osaka City University, Proc. of Int. Symposium on Geotechnical Centrifuge Model Testing, Tokyo, pp.43-49.

Ovesen, N.K. (1979). The scaling law relationship-Panel discussion, Proc. 7th European Conf. on SMFE, vol.4, pp.319-323.

Takada, N. & Kusakabe, O. (1987). Principle of centrifuge model, Tsuchi-To-Kiso, Journal of JSSMFE, 35-12, pp.89-94, (in Japanese).

Takada, N., Takeuchi, I., Mikasa, M. and Ikeda, M. (1987). Centrifuge model test of heavy tamping (1st report), Proc. of annual convention JSCE, III, pp.16-17.

Takada, N. and Mikasa, M. (1985). Determination of consolidation parameters by selfweight consolidation test in centrifuge, STP 892, Consolidation behavior of soils, pp.548-556, ASTM.

Terashi, M., Kitazume, M. and Tanaka, H. (1985). Application of PHRI geotechnical centrifuge, Proc. of Int. Symposium on Geotechnical Centrifuge Model Testing, Tokyo, pp.164-171.

Scott, R.F. (1985). Centrifuge model testing at Caltech, Proc. of Int. Symposium on Geotechnical Centrifuge Model Testing, Tokyo, pp.103-119.

Centrifugal modeling of soil-structure interaction

James C. Ni
Civil Engineering Department, Tamkang University, Taiwan

Chwen Song Yeh
Taiwan Provincial Water Conservancy Bureau, Taiwan

ABSTRACT: The experimental program described in this paper was initiated for the purpose of gaining an understanding of the response of a horizontal shelter buried in soil. In order to verify the validity of these calculations, the experiment was conducted to measure the moduli of subgrade reaction to be compared with the values calculated from the analytical solutions and the values calculated from analytical formulae and conventional triaxial testing results.

INTRODUCTION

Models are used in many engineering fields to reduce the time and cost of solving technical problems. Geotechnical centrifugal modeling is particularly important to the soil-structure interaction problems, such as buried structure and earth dam etc. Experiments can be carried out in a centrifuge where an artificial gravity was created by the centrifugal acceleration to simulate the body forces generated by the self-weight of the material in the prototype. Such simulation of the gravity-induced stresses is important due to the fact that the soil stiffnesses, and hence the soil-structure interaction phenomena, are governed by the level of confining pressure acting in the soil [1,2].

Length can be chosen as first fundamental quantity, fixing the scale relation between prototype and model as $\lambda = L/L'$, and specific force (force per unit area) as second fundamental quantity fixing the relative scale as $\xi = \sigma / \sigma'$ [11,12,13,14,15,16,17]. In practice, modeling of static geotechnical problems can be divided into two categories : (1)Models in which both the scale of the length λ and the scale of the specific force ξ are greater than 1. This type of modeling requires that the soil materials used should meet the following failure and stress-strain, conditions:

$$\sigma = \xi \sigma', \quad \tau = \xi \tau', \quad \varepsilon = \varepsilon'$$

as shown in Figures 1 and 2. In earth structures, however, it is extremely difficult, if not impossible, to satisfy these two critical conditions imposed on the model materials.

(2) Models in which only the scale of the length $\lambda > 1$, while that of stress $\xi = 1$. In this case, the material of the proto type can be used for the model provided that body forces can be ignored, compared with boundary forces. Body force is, however, almost invariably significant in soil mechanics prototype problems and the investigator is then faced with the requirement of the similarity of body force, i.e. $\gamma' = \gamma * \lambda$. Without centrifugal model testing schemes, it is obvious that satisfying the above condition is impossible. The centrifuge used is shown as in Figure 3, and its specification is listed in Table 1.

ANALYTICAL MODELING

The horizontal underground shelter to be simulated is circular cylinder and is made of reinforced concrete with a length of 1980 inches, an outside diameter of 216 inches and a wall thickness of 21 inches approximately. The behaviour of shelter under the internal loading is a complex soil-structure interaction problem [4,5,8,9,10]. These interactions can be replaced by a series of radial (K_r), tangential (K_θ) and longitudinal (K_x) springs at 22.5° spacing circumferentially and at a longitudinal spacing of 100 inches as shown in Figures 4 and 5.

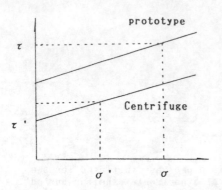

Figure 1 – Failure Envelope in Prototype and Centrifugal Modeling

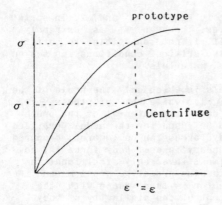

Figure 2 – Stress-Strain Relation in Prototype and Centrifugal Modeling

Table 1 – University of Colorado Centrifuge Specifications

Manufacturer	Genisco
Model	1230-5
G-Range	Variable 1 to 262 g at 42 in nominal radius
Driving System	25 HP hydraulic
Working Radii	42.0 in - center to basket hinge 11.5 in - hinge to basket floor
RPM Range	0-470 RPM
Payload Capacity	35,000 g lbs (350 lb at 100 g)
Test Package	18 in x 18 in x 18 in
Electrical Rings	56 slip rings
Fluid Transfer	2 hydraulic slip rings
Test Recording	Closed circuit TV 35 mm SLR camera

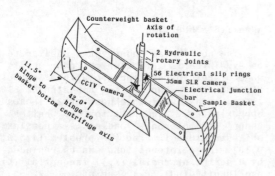

Figure 3 – Schematic of the Centrifuge

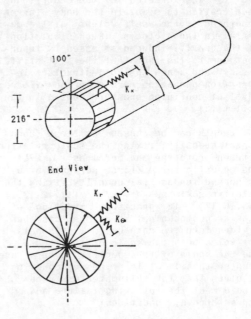

Figure 4 – Cylinder Representation

18

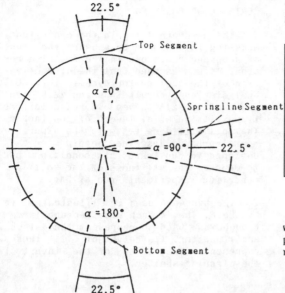

Figure 5 - Segmentation of Cylinder

Table 2 - Elastic Solutions to Evaluate Moduli of Subgrade Reaction

Segment	K	K	K
Top (α=0)	0.042E	0.046E	0.041E
Springline (α-90)	0.048E	0.043E	0.059E
Bottom (α=180)	0.050E	0.056E	0.059E

The soils used to backfill the shelter will be compacted in layers. The index properties and standard Proctor compaction results are listed in Table 3.

Table 3 - Soil Engineering Properties

INDEX PROPERTIES			COMPACTION RESULTS	
LL	PL	PI	MAXIMUM DRY DENSITY	OPTIMUM MOISTURE CONTENT
(%)	(%)	(%)	(pcf)	(%)
16.1	12.2	3.9	125.9	10.8

GRADATION RESULTS						
FINES	C	C	D	D	D	Gs
(%)			(mm)	(mm)	(mm)	
<5	3.1	1.0	0.09	0.17	0.32	2.66

The longitudinal springs K_x can be assessed in a manner similar to axial loading of a pile in which axial tangential resistance is generated. The tangential and circumferential springs K_θ can be assessed similarly to a pile under torsional loading. The radial springs K_r can be envisioned to provide resistance to the expansion of a cylindrical cavity in the soil. The literature is scarce on the subjects of axial and torsional loadings of horizontal piles, and of expansion of horizontal cylindrical cavities. The matter is complicated by the horizontal position of the shelter located at a shallow depth in a material whose stiffness increases with depth. The elastic solutions for embedded areas in a semi-infinite medium can be utilized to calculate the moduli of subgrade reaction. These solutions, representing different variations of Mindlin's problem[3,4].

The top and bottom segments can be approximated as flat horizontal rectangular areas, while the segment at the spring line can be approximated as a flat vertical area. For these segments, the elastic solutions to be used for the calculation of the various moduli of subgrade reactions can be summarized in Table 2, where E is the Young's modulus of the soil, taken to be the tangent modulus since small deformations are expected in dynamic soil-structure interaction computations.

In addition, a series of conventional triaxial compression tests were conducted again under various confining pressures in spite of triaxial testing conducted before [6,7] , because of the significant discrepancy. Four different backfill conditions were simulated as shown in Table 4. The values of E for soils in Table 2 depending on the confining pressure and therefore depth are listed in Table 5 for four soil conditions where a Poisson ratio of 0.2 is assumed. The values of K_r calculated at springline and bottom segments shown in Table 6 are used to compare with the results from centifugal modeling.

19

Table 4 - Backfill Conditions

Backfill Condition	Dry Density	Moisture Content
1	95% of maximum γ	OMC
2	95% of maximum γ	dried to 2%
3	90% of maximum γ	OMC
4	85% of maximum γ	OMC

Table 5 - Triaxial Testing Results

Backfill Condition	Confining Pressure (psi)	Initial Moduli (ksf)
1	5	576
	20	1037
	40	1440
2	5	1066
	20	1080
	40	1138
3	5	540
	20	720
	40	1044
4	5	518
	20	672
	40	1010

Table 6 - Analytical K Values (psi/in)

Backfill Condition Segment	1	2	3	4
Springline	330	440	258	244
Bottom	425	440	295	275

CENTRIFUGAL MODELING

The hardware setup in the centrifugal modeling is designed to measure the moduli of subgrade reactions in the radial direction. It was assumed that the discretization of the shelter for dynamic interaction analysis would result in using 22.5° segments of the 216 inches o.d. circular section with an axial length of 100 inches. The soil stiffness tests in the experimental program described herein were designed to duplicate this condition, and to measure the stiffness of the soil in centrifuge experiments run at 54g.

The hardware used is illustrated in Figure 6. The 4-inch wide rectangular trough was used to represent the 4-inch model shelter. Its height of 5.111 inches represents the depth of the invert of the circular shelter.

Dimensions of Openings

A 2.778" * 1.178"
B 1.852" * 0.785"
C 1.389" * 0.589"

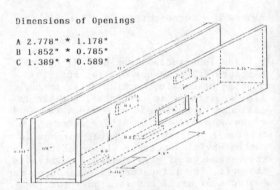

Figure 6 - Hardware Arrangement in Soil Stiffness Testing

A depth of 6.389 inches soil was compacted below the rectangular trough as well as both sides. Five windows were made in the trough through which stamps could be pushed into the soil to perform a loading test. Three sizes of stamps were employed, A, B, and C. The middle size of B was a 1/54-scaled model of the rectangular area formed by the projection of the 22.5° segment of a 216 inches o.d. pipe and 100-inch long, while sizes A and C are respectively 1.5 and 0.75 times those of B. A jack rod and LVDT rod arrangements are used for loading the stamps to determine the soil stiffness. The same four cases of soil condition as used in the laboratory triaxial compression testing were employed in this centrifugal model testing.

The testing results are illustrated in Figures 7 through 11. The radial springs K_r at springline and invert under four soil conditions estimated from these figures are listed in Table 7. The centrifugal testing results are repeatable with a reasonable accuracy, therefore the results in Table 7 can be used to justify the validation of analytcial solutions in Table 6. After comparing these data, positive conclusions were drawn and are illustrated in the next section.

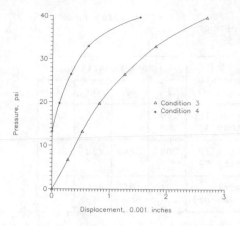

Figure 9 – Centrifugal Testing Results
       of Stamp B2

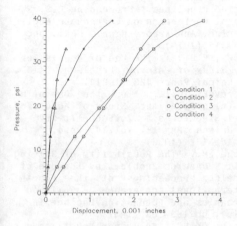

Figure 7 – Centrifugal Testing Results
       of Stamp B1

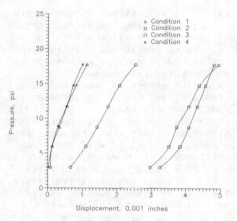

Figure 10 –Centrifugal Testing Results
       of Stamp A

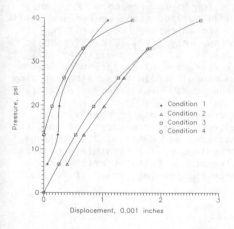

Figure 8 – Centrifugal Testing Results
       of Stamp B3

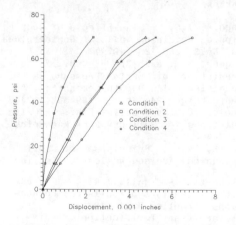

Figure 11 –Centrifugal Testing Results
       of Stamp C

21

Table 7 - K Values (psi/in) from
Centrifugal Testing

Segments	Soil Condition			
	1	2	3	4
Springline (Stamp A)	220	--	180	220
Springline (Stamp C)	300	2240	230	300
Invert (Stamp B2)	694	1040	496	--

CONCLUSIONS

1. K_r values at springline obtained from centrifugal testing under four soil conditions are smaller than the corresponding analytical solutions. This is reflected from the anisotropy of soils compacted in layers.
2. The sizes of stamps or the rigidities of stamps affect the K_r values measured in centrifugal testing.
3. K_r values at invert from centrifugal testing are higher than the analytical solutions. This is likely caused by the difference of boundary condition at base, i.e. a rigid base within finite distance versus a semi-infinite medium.
4. Centrifugal modeling testing is very valuable to calibrate the parameters used in the dynamic analysis of soil-structure interaction problem.

REFERENCE

1. Ingold, T.S. , " The Effects of Compaction on Retaining Walls, " Geotechnique 29, No.3, 265-283 (1979).
2. Aggour, M.S. and Brown, C.B., " The Prediction of Earth Pressure on retaining Walls Due to Compaction, " Geotechnique 24, No.4, 489-502(1974).
3. Poulos, H.G. and Davis, E.H., Elastic Solutions for Soil and Rock Mechanics (1974).
4. Scott, R.F., Principles of Soil Mechanics, Addison Wesley Publication (1963).
5. Richart, F.E., Hall, J.R. and Woods, R.D., Vibrations of Soils and Foundations, Prentice Hall (1970).
6. "Nevada-Utah Verification Studies, FY79-Geotechnical Data, " Fugro, report FM-TR-27 (1979).
7. "Laboratory Test Results, MX Program, Nevada-Utah," Report No.79-266,(1979).
8. Whitman, R.V. and Richart, F.E., " Design Procedures for Dynamically Loaded Foundations, " Journal of Soil Mechanics and Foundations Division, ASCE, 93, SM6, pp.169-193 (1967).
9. Hardin, B.O., " The Nature of Stress-Strain Behaviour for Soils, "Earthquake Engineering and Soil Dynamics Specialty Conference, ASCE, Pasadena, California, pp.3-90 (1978).
10. Parmelee, R.A. and Ludtke, C.A., Seismic Soil-Structure Interaction of Buried Pipelines," Proceeding U.S. National Conference on Earthquake Engineering, Ann Arbor, Mich., EERI, pp. 406-415 (1975).
11. Bucky, P.B., " The Use of Models for the Study of Mining Problems, "AIME Technical Paper 425 (1931).
12. Pokrovsky, G.I., " On the Ose of a Centrifuge in the Study of Models of Soil Structures," Zeitschrift fur Technische Physik, Vol.14, No.4, pp.160-162 (1933).
13. Rocha, M., "The Possibility of Solving Soil Mechanics Problems by the Use of Models," Proceedings of the Fourth International Conference on Soil Mechanics and Foundation Engineering, Vol.1, pp.183-188 (1957).
14. Roscoe, K.H., "Soils and Model Tests," Journal of Strain Analysis, Vol.3, pp.57-64 (1968).
15. Roscoe, K.H. and Poorooshasb, "A Fundamental Principle of Similarity in Model Tests for Earth Pressure Problems," Proceedings 2nd Asian Conference on Soil Mechanics and Foundation Engineering, Tokyo, Vol.1, p.134 (1963).
16. Pokrovsky, G. and Boulytchev, V., " Soil Pressure Investigation on Sewers by Means of Models," Technical Physics of the USSR, Vol.I,No.2, pp.121-123 (1934).
17. Pokrovsky, G. and Fyodorov, I.S., " Studies of Soil Pressures and Soil Deformations by Means of a Centrifuge," Proceedings of the First International Conference on Soil Mechanics and Foundation Engineering, Vol.1, p.70 (1936).

Developments in Geotechnical Aspects of Embankments, Excavations and Buried Structures, Balasubramaniam et al. (eds)
© 1991 Balkema, Rotterdam. ISBN 90 5410 019 2

Centrifugal model tests – Recent projects

L. D. Fuglsang
Danish Engineering Academy, Lyngby, Denmark

ABSTRACT: A brief introduction to centrifugal principles and model laws is presented. Four categories of centrifuge model tests are described, and two projects illustrate the capability of the centrifuge in model testing.

INTRODUCTION TO CENTRIFUGE MODELLING

Centrifuge modelling techniques have become well established in geotechnical engineering in the last twenty years. A thorough state of the art was presented in a book edited by Craig et al (1988). Authors from ten countries contributed.

The essence of centrifuge modelling is the ability to reproduce soil self-weight stress distributions that have significant influence in most geotechnical structures. When a model at 1:n scale is subjected to n times gravity in the centrifuge, the stress distribution is the same in the model and the full scale prototype. Besides, all effects related to flow of pore water, for example, consolidation, are speeded up by a factor of n^2 if the prototype soil is used in the model.

In the past the effect of soil self-weight in the prototype was often extrapolated from laboratory model tests performed at different scales. Sometimes the use of appropriate boundary forces can provide an adequate replacement for self-weight of the soil in specific problems. Also a hydraulic gradient can be used to generate effective stresses. However, in general, it is necessary to turn to the centrifuge if self-weight is to be properly represented in geotechnical model testing.

THEORY OF MODELLING

Before model tests are carried out, similarity conditions and subsequent model laws should be derived either through dimensional analysis or from the differential equations that describe the phenomenon. Langhaar (1951) has described the principles of dimensional analysis, while for instance Roscoe (1968) used differential equations to derive scaling relations for soil properties.

Fuglsang & Krebs Ovesen (1988) illustrate the use of the two methods in a simple geotechnical problem. They also establish the similarity requirements and scaling factors for basic soil parameters when modelling in the centrifuge is intended, see Table 1. The dimensionless numbers, also called π-products, directly give the model requirements. Similarity between model and prototype is attained, when each dimensionless number has the same value in the model and the prototype. If the scaling factor N of a parameter is defined as the model value over the prototype value of that parameter, each dimensionless number leads to a relation between the scaling factors of the parameters in that number. These relations also express the similarity conditions.

Fuglsang & Krebs Ovesen conclude that distorted test results may appear in granular soils due to particle size-effects, and also in flow problems due to time-effects. It is necessary to perform model tests at different scales to check this, see Figure 1.

Figure 1 illustrates the principle of modelling of models. The model size is plotted logarithmically against the gravity level of the test. A 1000 mm prototype can be modelled by a 100 mm model at 10 times gravity, or by a 10 mm model at 100 gravities. The two models are models of the same prototype, but they are also models of each other. Therefore they can be compared with one another and their behaviour should both

Table 1. Scaling factors in centrifuge tests.

parameter	symbol	dim.less number	similarity requirement	scaling factor
acceleration	a		$N_a =$	n
model length	l		$N_l =$	$\frac{1}{n}$
soil density	ρ		$N_\rho =$	1
particle size	d	$\frac{d}{l}$	$N_d =$	1
void ratio	e	e	$N_e =$	1
saturation	S_r	S_r	$N_S =$	1
liquid density	ρ_l	$\frac{\rho_l}{\rho}$	$N_{\rho l} = N_\rho =$	1
surface tension	σ_t	$\frac{\sigma_t}{\rho_l a d l}$	$N_\sigma = N_\rho N_a N_d N_l =$	1
capillarity	h_c	$\frac{h_c \rho_l a d}{\sigma_t}$	$N_h = N_\sigma N_\rho^{-1} N_a^{-1} N_d^{-1} =$	$\frac{1}{n}$
viscosity	η	$\frac{\eta}{\rho_l d \sqrt{al}}$	$N_\eta = N_\rho N_a N_d N_a^{\frac{1}{2}} N_l^{\frac{1}{2}} =$	1
permeability	k	$\frac{k\eta}{d^2 \rho_l a}$	$N_k = N_d^2 N_\rho N_a N_\eta^{-1} =$	n
particle friction	φ	φ	$N_\varphi =$	1
particle strength	σ_c	$\frac{\sigma_c}{\rho a l}$	$N_\sigma = N_\rho N_a N_l =$	1
cohesion	c	$\frac{c}{\rho a l}$	$N_c = N_\rho N_a N_l =$	1
compressibility	E	$\frac{E}{\rho a l}$	$N_E = N_\rho N_a N_l =$	1
time:				
inertia	t_1	$t\sqrt{\frac{a}{l}}$	$N_t = N_l^{\frac{1}{2}} N_a^{-\frac{1}{2}} =$	$\frac{1}{n}$
lam. flow	t_2	$t\frac{k}{l}$	$N_t = N_l N_k^{-1} =$	$\frac{1}{n^2}$
creep	t_3			1

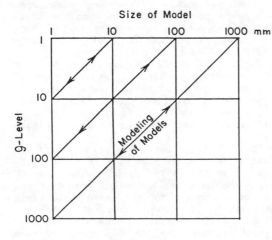

Figure 1. Modelling of models.

extrapolate to the same prototype performance by means of the scaling relations.

By modelling of models it is then possible to detect the model scale where particle size-effect or the conflict in time scaling become significant. Such model test series have already been conducted on various geotechnical structures, and examples were collected by Fuglsang & Krebs Ovesen.

Other research workers have written about scaling factors in centrifuge modelling, see for example Goodings (1984), Scott (1988).

USAGE OF THE CENTRIFUGE

Centrifuge testing has been used for at least four different purposes: Modelling of prototypes, studies of new phenomena, parametric studies, validation of numerical methods. They are described by Ko (1988). A summary is given below, and examples of typical projects are outlined in subsequent paragraphs.

Modelling of prototypes seems an obvious application of the centrifuge technique. A model made of the prototype material at a comprehensive model scale should replicate the prototype behaviour. However, it often becomes extremely difficult to build all details that control the behaviour of the prototype. Therefore, this type of usage is absolutely questionable for complicated structures.

The study of a phenomenon, which is not fully understood, can be undertaken in the centrifuge. Observations in such experiments can provide an insight on the structural behaviour and form the basis for the formulation of a theory.

Parametric studies can generate data for formulating design charts for specific situations that cannot be dealt with properly by theoretical means. The bearing capacity of footings situated in slopes and the stability of clay slopes under partially drained conditions belong to this group of problems. Such studies have only been conducted in a few cases.

The validation of numerical methods can be secured. Centrifuge experiments can be run close to prototype conditions and compared with the results of the numerical methods. This procedure is particularly useful for finite element analyses because every aspect of the test, including the boundary conditions, can be accounted for in such analyses. Also the sensitivity of the numerical model can be checked by comparison with centrifuge test results.

The validation of model laws by modelling of models mentioned before could be added as another test purpose, although it is a prerequisite for the utilization of the centrifuge.

All the approaches mentioned above are valid reasons for using the centrifuge for model testing. It is worth noting that usual laboratory model tests at much reduced scales have been carried out in the past to fulfill the same objectives. Because of the lack of similarity the results are often doubtful. Scott (1988) concludes that the majority of papers on this work ignore scaling or even reference to extrapolation to prototype behaviour.

TEST FACILITIES

A number of centrifugal machines have been built in recent years for specific use in geotechnical engineering. The facilities vary according to factors such as money available, the particular work envisaged by the designer, the existing equipment.

Ko (1988) gave the capacity of the existing geotechnical centrifuges. He plotted the payload size versus the maximum g-level. A few more are described in the proceedings of the conference CENTRIFUGE 88, edited by Corté (1988).

SURFACE FOOTINGS ON A SAND WITH A CAPILLARY ZONE

Bagge et al (1989) have reported tests with surface footings on a sand with a capillary zone. A field test on a circular footing was performed, and three centrifuge laboratories (in Denmark, England and France) modelled the full-scale conditions in the centrifuge using their own choice of techniques.

The investigation served two purposes, to evaluate the variability in centrifuge test results among the three teams, and to show whether it was possible to model the full-scale prototype in this case. The latter purpose makes this investigation belongs to the category, modelling of prototypes, mentioned above.

The prototype test was carried out on a prepared test site, so the conditions were perhaps more ideal than usual for a prototype structure. The sand was compacted in 0.3 m layers, and then saturated from beneath through a system of drains. The diameter of the footing was 1.6 m, and the height of the capillary zone was 0.4 m.

Given the site and loading conditions, centrifuge model tests were conducted with 57 mm footings at 28 gravities (57 mm × 28 = 1600 mm) on the prototype sand.

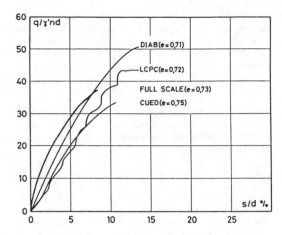

Figure 2. Full-scale test and model test results.

Figure 2 shows the load-deflection curves of the full-scale test as well as the model tests. Note that dimensionless quantities are plotted in the figure, where q is the applied pressure, $n\gamma'$ is the submerged unit weight of the soil in the centrifuge, and d is the diameter of the footing. s is the vertical displacement.

Bearing in mind that the model test results refer to slightly different void ratios, the loads agree reasonably at large deflections. However, the initial inclination of the curves show some discrepancy.

Tests were also performed at different void ratios and without the capillary zone. The load at 10% relative deflection is

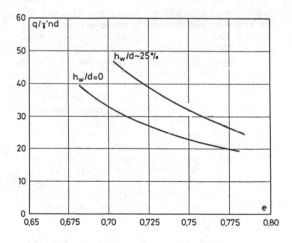

Figure 3. Load at 10% deflection.

plotted versus the void ratio in Figure 3. The results stress the dependency of the bearing capacity on void ratio e and depth of the water table h_w. This is also seen in the classical bearing capacity formula :

$$q = \tfrac{1}{2}\,\gamma'd\,N_\gamma + (\gamma's + \gamma_w h_w)\,N_q$$

That part of the project is mentioned here to illustrate the capability of the centrifuge to perform parametric studies in a controlled manner.

Finally it should be emphasized that all the model tests were carried out in the centrifuge because the behaviour of sand is highly dependent on the stress level. The self-weight stresses due to both sand and capillary water were modelled in similarity with the prototype stresses.

SUCTION PILES IN A SEA BED OF CLAY

The breakout resistance of steel tubes closed at the top end has been investigated by Fuglsang & Steensen-Bach (1991). The undrained tensile capacity of the piles was measured in centrifugal model tests, and the results were compared with a suggested analytical design method.

It is likely that suction piles will be installed under lightweight steel jackets in marginal oil and gas fields in the North Sea. The jacket is supposed to rest on three piles, 5 m in diameter and 10-20 m deep. When the pile is subjected to uplift load, the unloading of the soil results in a reduction of the pore pressure inside and beneath the pile, if the soil permeability is low. The concept is that the tensile capacity of the pile is increased by the reduction of the pore pressure. Suction piles seem beneficial both during installation and under extreme horizontal loads on the jacket, i.e. wave loading and ship impact.

The model piles were 65 mm and 80 mm in diameter and two diameters long. They were installed in consolidated beds of kaolin clay and covered with 100-200 mm water in order to simulate the sea depth. The piles were then loaded vertically in a strain controlled manner at 40 times gravity in the centrifuge. Besides the main test programme with piles closed at the top, tests were also performed with open piles to measure skin friction.

The most significant purpose of the experimental investigation was to evaluate the nature of the failure in the clay beneath the pile base. It was asked whether the designer should expect a sudden fracture across the base or would a plastic

Table 2. Results of centrifuge tests with suction piles.

Pile diameter	65 mm		80 mm
v (mm/s)	1.0	0.1	1.0
c_u (kPa)	8.5	7.0	17.5
N	8.1	8.5	6.5

failure occur? Thus it was an investigation belonging to the category of unknown phenomena.

The test results showed that the clay failed in plastic flow. After the test a shallow crater with a diameter of about three pile diameters was observed around the pile at the clay surface.

The failure seems to correspond to the pattern of plastic deformation in a reversed bearing capacity failure, i.e. the shearing of the soil is similar to a failure beneath an embedded footing, but the direction is reversed. Then the breakout capacity of the pile is:

$$F = W_p + W_s + W_w + \alpha c_u A_e + (N c_u - q)A_b$$

including the weights of pile, soil, and water above the pile, the external skin friction and the reversed undrained bearing capacity of the clay beneath the pile q is the total stress in the clay beside the pile base.

The bearing capacity factor N may be derived from the maximum breakout force in each test. The result is given in Table 2 together with the loading rate and the clay strength. It has not been possible to explain why the value was lower in the stronger clay. However, it should be noted that the value of N is highly affected by the uncertainty of c_u. The clay strength was measured by means of a laboratory vane.

The empirical solutions in literature give N between 6.3 and 9 depending on the depth of penetration.

It was important to conduct the testing in the centrifuge because the overburden of soil and water at the pile base should be similar to the prototype value. Thereby, positive absolute pore pressures were maintained in the clay, and the suction capacity before pore water cavitation was as high as in the field. This was emphasized by a number of normal laboratory model tests (1 g-tests), where tensile fracturing took place in the clay at the pile base. The analysis indicated tensile

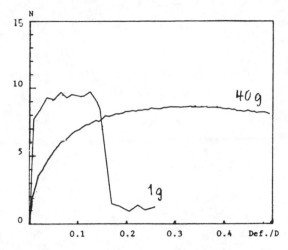

Figure 4. Dimensionless load-deflection curves.

stresses up to 100 kPa at the base in these tests, while there were compressive stresses at the pile base in the centrifuge tests.

Typical load-deflection curves from tests at 40 g and 1 g are given in Figure 4. Instead of the load F, the bearing capacity factor N is plotted versus relative deflection. It is believed that the two curves reflect the two failure mechanisms already mentioned.

CONCLUDING REMARKS

Centrifugal model testing has come a long way in the last 20 years. At present new test facilities have been built around the world for specific use in geotechnical engineering.

The technique is not equally powerful in all the categories of application mentioned in this paper, but it can be used for all these purposes. Ko (1988) outlines the refinements that are needed in the test technique. He mentions the development of miniature transducers, in-flight soil characterisation, detailed modelling of stress and loading history.

It is most likely that centrifugal model tests will play an increasing role in soil mechanics research in the future. Schofield (1988) foresees new areas that probably will involve centrifugal testing. Seepage flow with pollutant migration, interaction of ice and embedded structures in the cold oceans and failures in earthquakes are mentioned.

Fuglsang & Krebs Ovesen (1988) give recommendations in order to improve the reliability of results: Repeat more tests than seen hitherto, build models as large as possible, and use the modelling of models method to detect possible scale-effects.

REFERENCES

Bagge et al (1909): Surface footings on a sand with a capillary zone. 12th ICSMFE, Rio de Janeiro.
Corté (1988): Centrifuge 88. Balkema, Rotterdam.
Craig (1984): The application of centrifuge modelling to geotechnical design. Balkema, Rotterdam.
Craig et al (1988): Centrifuges in soil mechanics. Balkema, Rotterdam.
Fuglsang & Krebs Ovesen (1988): The application of the theory of modelling to centrifuge studies. In Craig et al (1988).
Fuglsang & Steensen-Bach (1991): Breakout resistance of suction piles in clay. Centrifuge 1991. Balkema (forthcoming).
Goodings (1984): Relationships for modelling water flow in geotechnical centrifuge models. In Craig (1984).
Ko (1988): Summary of the state-of-the-art in centrifuge model testing. In Craig et al (1988).
Langhaar (1951): Dimensional analysis and the theory of models. John Wiley, New York.
Roscoe (1968): Soils and model tests. Journal of Strain Analysis.
Schofield (1988): An introduction to centrifuge modelling. In Craig et al (1988).
Scott (1988): Physical and numerical models. In Craig et al (1988).

Developments in Geotechnical Aspects of Embankments, Excavations and Buried Structures, Balasubramaniam et al. (eds)
© 1991 Balkema, Rotterdam. ISBN 90 5410 019 2

Influence of a non-random fabric on macroscopic properties of geomaterials

S. Pietruszczak
McMaster University, Hamilton, Ont., Canada

ABSTRACT: This paper is concerned with the mathematical description of the
deformation process in geomaterials. In particular, the response as observed on the
macroscale is correlated with some implicit measures of the material microstructure.
First, a plasticity formulation is reviewed which attributes the anisotropy in
mechanical behaviour to a bias in the spatial distribution of voids. Subsequently,
the case of a spontaneous loss of material homogeneity, resulting from formation of a
shear band, is addressed. A new mathematical formulation for the description of
localized deformation is derived, based on considering the material as a two-phase
composite. The resulting homogenized constitutive relation can represent the unstable
material response, i.e. strain softening.

1.   INTRODUCTION

Most of the engineering materials contain
some imperfections, such as dislocations,
pores, fissures, etc., which are
considered to be the source of
irreversible deformations. In the
materials with compact structure, such as
metals, it is believed that the
dislocation movement and its growth is a
predominant mechanism of irreversible
deformations. On the other hand, in
materials with open structure, such as
soils, the deformation process is largely
influenced by the nucleation of fissures
and the evolution of voids, which results
in splitting of grains and clusters. The
permanent deformations are generated
mainly by relative sliding, rolling and
lifting of particles ([5],[12]), since
the interparticle forces may be broken
much easier than the interatomic forces
of mineral crystals. In general, the
inherent material fabric (loosely defined
as a space composition of solid
particles) and its progressive evolution
during the deformation process, influence
to a large extent the mechanical response
as observed on the macroscale. In
particular, a non-random fabric often
results in the directional dependence
(i.e., anisotropy) of the mechanical
properties.

Anisotropy effects in natural soil
deposits are already well documented in
the literature (e.g., [6],[7]) and are
inevitably associated with nonhomogeneity
of the internal structure. An alignment
of soil particles has been detected in
various natural clay and sand deposits
(e.g., [2],[1]) and it appears to be
induced by both depositional and
environmental conditions. Microscopic
evidence suggests that the preferred
orientation of soil particles can also be
achieved during one-dimensional K_o-
consolidation process. Thus, the question
of anisotropy, as observed on the
macroscale, entails identification of
initial soil structure/fabric as linked
to inherent anisotropy and its
progressive evolution resulting in
induced anisotropy.

A large number of experimental
investigations, on both the micro (using
electron microscopy) and macroscopic
level, have been performed. The main
problem in correlating the results of
these experiments is the lack of a
commonly accepted measure of the internal
structure. Based on the existing
evidence, it seems rational to relate the
directional dependence of soil properties
to the distribution of pores and
fissures. Thus, the spatial distribution
of voids, or a derivative quantity
(porosity/void ratio) can be accepted as

an implicit measure of soil fabric. Such a concept was put forward and defined in mathematical terms by the present author and his colleague in ref. [10].

This paper deals with the local description of the deformation process, from its early stages when the material is treated as a homogeneous one (on the macroscale) up to an advance stage when the deformation becomes localized along a preferred orientation. The latter mechanism, associated with a strongly oriented fabric perceived as nonhomogeneity of the material, results in an unstable response, often referred to as strain softening. In the following section, a plasticity formulation is reviewed (after ref. [10]) that incorporates some tensorial measures of the evolution of material fabric derived from the directional distribution of voids. In subsequent section, a new mathematical formulation is proposed for describing the unstable deformation process resulting from the inception of a shear band and a spontaneous loss of homogeneity of the sample.

2. PLASTICITY FORMULATION INCORPORATING SOME MEASURE(S) OF MATERIAL FABRIC

In the geotechnical design the external load must often be limited to the value constrained by the allowable deformation rather than the bearing capacity of soil. The assessment of the deformation field can be obtained by the analysis of the proficiently defined boundary value problem of continuum mechanics. Such an analysis, regardless of the numerical technique employed, requires an appropriate constitutive relation, which is then deterministic in the context of reliability of the obtained solution.

It has been shown [11] that constitutive laws built within the classical framework of the theory of plasticity do not adequately model the anisotropy induced by the plastic flow. A procedure to enhance the plasticity based models has been proposed in ref. [10]. It consists of representing the soil anisotropy in terms of anisotropy in its phase distribution. As an implicit measure of soil fabric, a 'directional porosity' has been chosen, which could be identified as a generalized, direction-dependent counterpart of 'porosity', i.e., the scalar valued measure of void space in soil mass.

In order to define the 'directional porosity', consider a unit sphere (S), which encloses a representative volume of the material. Select now a test line of length $L^* = 2R$ ($R = 1$ being the radius of the sphere) and the orientation $v = \{v_1, v_2, v_3\}$ with respect to the fixed Cartesian coordinate system. The fraction of L^* occupied by voids can be defined as

$$L(v) = l(v) / L^* \qquad (1)$$

where $l(v)$ represents the total length of interceptions of this line with soil pores. For uniformly distributed test lines, the mean value of the quantity L, averaged over the domain S, is

$$L_{av} = \frac{1}{4\pi} \int_S L(v) \, dS \qquad (2)$$

It has been proven in ref. [10] that L_{av} can be identified with the average porosity of the material, n_0, whereas the lineal fraction occupied by pores $L(v)$ is an unbiased estimator of $n(v)$, i.e., volume fraction of voids in the direction v.

The scalar valued function $n(v)$, defined over the unit sphere S, can be described in terms of symmetric traceless tensors Ω, $[\Omega]$, ...

$$n(v) \approx n_0 (1 + \Omega^T m + m^T [\Omega] m + ...) \qquad (3)$$

where the second order tensor Ω has a vector representation
$\Omega = \{\Omega_{11}, \Omega_{22}, \Omega_{33}, \Omega_{12}, \Omega_{13}, \Omega_{23}\}^T$
and
$m = \{v_1^2, v_2^2, v_3^2, v_1 v_2, v_1 v_3, v_2 v_3\}^T$.
The higher rank tensors $[\Omega]$, ... relate to the higher order fluctuations in void space distribution. Thus, in order to describe a smooth orthogonal anisotropy it is sufficient to employ an approximation based on the first two terms of the expansion (3).

Incorporation of the concept of 'directional porosity', eq. (3), into plasticity framework requires the formulation of an appropriate evolution law for the components of Ω. The latter should refer the rate of change of Ω to the deformation history. In particular, the components of $\dot{\Omega}$ may be related to the strain rate deviator $\dot{e}$, through an isotropic tensor valued function

$$\dot{\Omega} = \dot{\Omega}(\Omega, \dot{e}, n_0) \qquad (4)$$

The details concerning the specification of function (4) are discussed in ref. [10].

30

It is postulated now that a strong inherent anisotropy, present in materials with large grain size (e.g. gravel with flat grains, sand with elongated grains, etc.), can be described by means of a fabric tensor α, which is defined as a tensor valued function of Ω, i.e, $\alpha = \alpha(\Omega)$. Such an anisotropy will inevitably affect both the elastic properties as well as the functional form of the failure criterion. Thus, for this class of materials

$$\sigma = \sigma(\varepsilon, \alpha) \qquad \text{in elastic range}$$
$$F = F(\sigma, \alpha) = 0 \tag{5}$$

The proper mathematical representation of both the elasticity tensor and the failure criterion in terms of fabric tensor is provided in ref. [10].

In materials with small grain size the inherent anisotropy can be considered as being relatively weak, so that $\alpha \approx \{1, 1, 1, 0, 0, 0\}^T$. On the other hand, the anisotropy induced by the deformation process may develop, resulting in directional dependence of certain mechanical characteristics. The latter form of anisotropy can be described in terms of progressive evolution of the 'directional porosity'. One way of formulating the problem is to assume that the yield criterion f=0 is an isotropic function of σ and Ω,

$$f = f(\sigma, \Omega, \kappa) = 0 ; \qquad \dot{\varepsilon}^P = \dot{\lambda} \frac{\partial f}{\partial \sigma} \tag{6}$$

i.e., it depends on the deformation history κ, and on ten (in general) functionally independent invariants of both tensors.

An alternative, and perhaps simpler, formulation can be derived by expressing the flow rule in the form

$$\dot{\varepsilon}^P = \dot{\lambda} \frac{\partial f}{\partial \sigma} + \dot{\mu} G; \quad f = f(\sigma, \kappa) = 0 \tag{7}$$

where G is a symmetric second order tensor whose components are function of Ω and the deformation history κ, i.e., $G = G(\Omega, \kappa)$. Various general representations for G are considered in ref.[10]. The simplest one has been obtained by assuming coaxiality of G and Ω, i.e.,

$$G = h\Omega; \qquad \dot{\mu} = \dot{\lambda} \tag{8}$$

which results in

$$\dot{\varepsilon}^P = \dot{\lambda} \left(\frac{\partial f}{\partial \sigma} + h\Omega \right) \tag{9}$$

where h is a scalar-valued function of the deformation history. The above functional form is analogous to a non-associated flow rule. The deviation from the normality is attributed to a bias in the spatial distribution of voids induced by the deformation process.

The numerical performance of the above reviewed mathematical framework has been examined in references cited in this section. In order to provide an illustration, consider, for example, a specimen of a normally consolidated Weald clay [14] subjected to a series of hypothetical undrained deformation histories as depicted in figures below.

Fig. 1 shows the predicted effective stress trajectories corresponding to the chosen program. The initial K_0-consolidation process, which is simulated very closely, results in an induced cross-anisotropy. This anisotropy manifests itself in a bias in the directional distribution of void ratio, Fig. 3. The K_0-consolidation is followed by an undrained uniaxial compression. First, the response of specimens trimmed at different orientations relative to the direction of the major consolidating stress is examined. The effective stress paths, for both vertical and horizontal samples, are virtually the same (Fig. 1), however the vertical sample displays a stiffer response than the horizontal one (Fig. 2). These results are in qualitative agreement with the

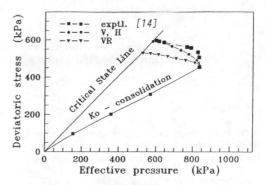

Fig. 1 Effective stress paths for K_0-consolidation followed by undrained uniaxial compression; (V - vertical sample, H - horizontal sample, VR - vertical sample under a continuous rotation of principal stress directions at 0.4°/kPa)

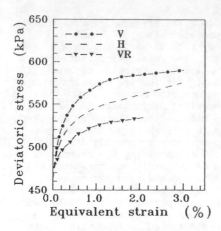

Fig. 2   Deviatoric characteristics corresponding to undrained compression.

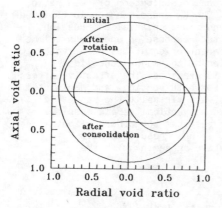

Fig. 3   Evolution of spatial distribution of void ratio for the considered loading history.

experimental data available in the literature (e.g., [6]). Next, the undrained response of the vertical sample under a continuous rotation of principal stress axes is examined. It is evident that in this case both the effective path and the deviatoric characteristic are affected. In particular, the undrained shear strength is reduced as compared to the former loading history (i.e., with no rotation).

The details, pertaining to the specification and the selection of material parameters, are provided in ref. [11]. The results, although very fragmentary, illustrate the ability of the formulation to account for directional dependence of material properties as well as their sensitivity to the rotation of the principal stress directions.

3.   DESCRIPTION OF LOCALIZED DEFORMATION

At advanced stages of the deformation process, the fabric of the material may undergo a spontaneous restructuring resulting in nonhomogeneity on the macroscale. A typical example is the localized deformation, i.e, formation of a shear band along a certain preferred orientation.

The phenomenon of localization may be considered as a bifurcation problem, i.e., an instability in the constitutive description of homogeneous deformation. Such an instability may trigger a nonhomogeneous mode involving the inception of a planar band under the conditions of continuing equilibrium. The theoretical framework for strain localization was provided in refs. [3] and [13]. The necessary condition consists of a path-dependent (in general) criterion governing the inception of localization. The criterion, if met, furnishes the corresponding orientation of the shear band [9].

If the strain localization takes place, the constitutive law describing homogeneous deformation is no longer applicable. A rigorous solution to the problem requires the analysis of an appropriately defined boundary value problem. The approach proposed in this paper is based on considering the material as a two-phase composite. The elastoplastic matrix, representing the intact material, is assumed to be intercepted by a planar band, with distinct mechanical properties, formed along a predetermined orientation. The macroscopic behaviour is described in terms of the mechanical properties of both constituents, their volume fraction and the geometrical arrangement. The mathematical formulation is based on Hill's mixture rule and the spatial averaging procedure commonly adopted in the mechanics of composite materials. The idea is similar to that put forward in ref. [8] (a 'smeared' shear band approach). The present mathematical framework however, is more rigorous.

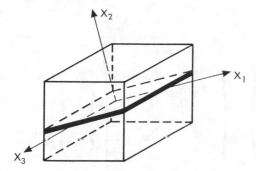

Fig. 4   Geometry of the problem

$$\dot{\varepsilon}_{11} = \dot{\varepsilon}_{11}^m = \dot{\varepsilon}_{11}^s \; ; \qquad \dot{\sigma}_{22} = \dot{\sigma}_{22}^m = \dot{\sigma}_{22}^s$$

$$\dot{\varepsilon}_{33} = \dot{\varepsilon}_{33}^m = \dot{\varepsilon}_{33}^s \; ; \qquad \dot{\sigma}_{23} = \dot{\sigma}_{23}^m = \dot{\sigma}_{23}^s \qquad (12)$$

$$\dot{Y}_{13} = \dot{Y}_{13}^m = \dot{Y}_{13}^s \; ; \qquad \dot{\sigma}_{12} = \dot{\sigma}_{12}^m = \dot{\sigma}_{12}^s$$

Assume now that both constituents are elastoplastic and that their mechanical properties are described by constitutive relations

$$\dot{\boldsymbol{\sigma}}^m = [D]\dot{\boldsymbol{\varepsilon}}^m \; ; \qquad (13a)$$

$$\dot{\boldsymbol{\sigma}}^s = [K]\dot{\boldsymbol{\varepsilon}}^s \qquad (13b)$$

Substitution of the above relations into constraints (12) leads, after some algebraic manipulations, to a set of three simultaneous equations

$$[\delta]\dot{\boldsymbol{\varepsilon}}^m = [A]\dot{\boldsymbol{\varepsilon}} + [B][\delta]\dot{\boldsymbol{\varepsilon}}^s \qquad (14a)$$

where

$$[\delta] = \begin{bmatrix} 0 & 1 & 0 & 0 & 0 & 0 \\ 0 & 0 & 0 & 1 & 0 & 0 \\ 0 & 0 & 0 & 0 & 0 & 1 \end{bmatrix} ;$$

$$[A] = \begin{bmatrix} a_{11} & \cdots & a_{16} \\ \cdot & & \cdot \\ a_{33} & \cdots & a_{36} \end{bmatrix} ;$$

$$(14b)$$

$$[B] = \begin{bmatrix} 0 & b_{12} & b_{13} \\ b_{21} & 0 & b_{23} \\ b_{31} & b_{32} & 0 \end{bmatrix}$$

and the coefficients of both matrices [A] and [B] are specified in the Appendix. The equations (14) can now be combined with the relation (10a) expressed in the form

$$[\delta]\dot{\boldsymbol{\varepsilon}} = \eta_s[\delta]\dot{\boldsymbol{\varepsilon}}^s + \eta_m[\delta]\dot{\boldsymbol{\varepsilon}}^m \qquad (15)$$

Thus, after some transformations, the following set of equations is obtained

$$[\delta]\dot{\boldsymbol{\varepsilon}}^m = \left([I] + \frac{\eta_m}{\eta_s}[B]\right)^{-1}\left([A] + \frac{1}{\eta_s}[B][\delta]\right)\dot{\boldsymbol{\varepsilon}} = [\overline{S}^m]\dot{\boldsymbol{\varepsilon}}$$

$$(16)$$

where [I] is a 3x3 unit matrix.

Finally, taking into account the identities (12), the strain rate averages $\dot{\boldsymbol{\varepsilon}}^m$ can be related to the overall

(i)   General three-dimensional formulation

Consider a volume of the material in the form of a parallelepiped intercepted by a planar band. Choose the Cartesian coordinate system as shown in Fig. 4, i.e., with x_2-axis normal to the shear band. The analyzed composite consists of two continuous constituents, the matrix (m) and the shear band material (s). Both constituents are assumed to exist simultaneously and be perfectly bonded. Let the macroscopic behaviour be described in terms of macrostress/strain tensors σ and ε, understood as respective averages over the elementary volume of the composite. Then, according to mixture theories [4]

$$\dot{\boldsymbol{\varepsilon}} = \eta_s\dot{\boldsymbol{\varepsilon}}^s + \eta_m\dot{\boldsymbol{\varepsilon}}^m \qquad (10a)$$

$$\dot{\boldsymbol{\sigma}} = \eta_s\dot{\boldsymbol{\sigma}}^s + \eta_m\dot{\boldsymbol{\sigma}}^m \qquad (10b)$$

where η's denote the volume fractions of both constituents and

$$\dot{\boldsymbol{\varepsilon}}^i = \frac{1}{V_i}\int_{V_i} (\dot{\boldsymbol{\varepsilon}}^i)' \, dV_i \; ; \qquad i = m, s \qquad (11a)$$

$$\dot{\boldsymbol{\sigma}}^i = \frac{1}{V_i}\int_{V_i} (\dot{\boldsymbol{\sigma}}^i)' \, dV_i \qquad (11b)$$

i.e, $\dot{\boldsymbol{\varepsilon}}^i$, $\dot{\boldsymbol{\sigma}}^i$ (i=m, s) are the body averages of the actual strain/stress rates (..)' taken over the corresponding volumes V_i.

The set of equations (10) should now be supplemented by kinematic compatibility and equilibrium constraints, which with reference to the chosen coordinate system, Fig. 4, take the form

33

macroscopic measure $\dot{\varepsilon}$, i.e.,

$$\dot{\boldsymbol{\varepsilon}}^m = [S_m]\dot{\boldsymbol{\varepsilon}} \qquad (17a)$$

where the concentration factors $[S_m]$ are defined as

$$[S_m] = \begin{bmatrix} 1 & 0 & 0 & 0 & 0 & 0 \\ \bar{S}_{11} & \bar{S}_{12} & \bar{S}_{13} & \bar{S}_{14} & \bar{S}_{15} & \bar{S}_{16} \\ 0 & 0 & 1 & 0 & 0 & 0 \\ \bar{S}_{21} & \bar{S}_{22} & \bar{S}_{23} & \bar{S}_{24} & \bar{S}_{25} & \bar{S}_{26} \\ 0 & 0 & 0 & 0 & 1 & 0 \\ \bar{S}_{31} & \bar{S}_{32} & \bar{S}_{33} & \bar{S}_{34} & \bar{S}_{35} & \bar{S}_{36} \end{bmatrix} \qquad (17b)$$

The matrix $[S_S]$, required to derive the corresponding strain rate averages for the shear band material, can easily be obtained from the identity

$$\eta_s[S_s] + \eta_m[S_m] = [I] \qquad (18)$$

which is implied by eq. (10a). Thus,

$$\dot{\boldsymbol{\varepsilon}}^s = [S_s]\dot{\boldsymbol{\varepsilon}}; \quad [S_s] = \frac{1}{\eta_s}([I] - \eta_m[S_m]) \qquad (19)$$

Given both concentration factors $[S_m]$ and $[S_S]$, the macroscopic stress rates are completely defined by the relation (10b).

(ii) Plane strain case

Let us examine now, in more detail, the case of a planar deformation, which is the most common in geotechnical applications. Select the coordinate system as shown in Fig. 5. Under the plane strain regime,

$$\dot{\varepsilon}_{33} = \dot{\varepsilon}^s_{33} = \dot{\varepsilon}^m_{33} = 0; \quad \dot{Y}_{13} = \dot{Y}^s_{13} = \dot{Y}^m_{13} = 0;$$

$$\dot{\sigma}_{23} = \dot{\sigma}^s_{23} = \dot{\sigma}^m_{23} = 0 \qquad (20)$$

the compatibility and equilibrium constraints (12) reduce to

$$\dot{\varepsilon}_{11} = \dot{\varepsilon}^m_{11} = \dot{\varepsilon}^s_{11}; \quad \dot{\sigma}_{22} = \dot{\sigma}^m_{22} = \dot{\sigma}^s_{22};$$

$$\dot{\sigma}_{12} = \dot{\sigma}^m_{12} = \dot{\sigma}^s_{12} \qquad (21)$$

Assume that the properties of the material outside the shear band are described by

$$\dot{\boldsymbol{\sigma}}^m = [D]\dot{\boldsymbol{\varepsilon}}^m \qquad (22a)$$

or in an explicit form,

$$\begin{Bmatrix} \dot{\sigma}^m_{11} \\ \dot{\sigma}^m_{22} \\ \dot{\sigma}^m_{33} \\ \dot{\sigma}^m_{12} \end{Bmatrix} = \begin{bmatrix} D_{11} & D_{12} & D_{14} \\ D_{21} & D_{22} & D_{24} \\ D_{31} & D_{32} & D_{34} \\ D_{41} & D_{42} & D_{44} \end{bmatrix} \begin{Bmatrix} \dot{\varepsilon}^m_{11} \\ \dot{\varepsilon}^m_{22} \\ \dot{Y}^m_{12} \end{Bmatrix} \qquad (22b)$$

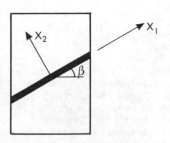

Fig. 5   Geometry of the problem

Since the thickness of the shear band is small as compared to other dimensions of the elementary volume, it may be convenient to express the constitutive law (13b) in terms of velocity discontinuities at the interface rather than explicit strain rate measures. Such an approach is advantageous since those quantities are directly measurable during a typical experiment (say, a simple shear test). In the elastic range, the constitutive relation assumes the form

$$\begin{Bmatrix} \dot{\sigma}^s_{22} \\ \dot{\sigma}^s_{12} \end{Bmatrix} = \begin{bmatrix} K_N & 0 \\ 0 & K_T \end{bmatrix} \begin{Bmatrix} \dot{g}_N \\ \dot{g}_T \end{Bmatrix}; \qquad (23)$$

$$\dot{\sigma}^s_{33} = \dot{\sigma}^s_{11} = (K_N - 2K_T)\dot{g}_N$$

where g_N, g_T are the discontinuities in the normal and tangential components of the velocity and K_N and K_T are the corresponding elastic stiffnesses. The above representation can easily be derived from the Hooke's law by imposing plane strain conditions and assuming that the normal strain rates in the direction

x_1 are negligible as compared to the remaining components.

In the elastoplastic range, the plastic (i.e., irreversible) part of velocity discontinuities can be derived by, for example, employing a yield function

$$f = f(\sigma^s_{22}, \sigma^s_{12}, \kappa) = 0 \; ; \quad \kappa = \kappa(g^p) \qquad (24)$$

and a flow rule

$$\dot{g}^p = \dot{\lambda} \frac{\partial Q}{\partial \sigma^s} \; ; \quad Q = Q(\sigma^s_{22}, \sigma^s_{12}) = \text{const} \qquad (25a)$$

where

$$\frac{\partial Q}{\partial \sigma^s} = \left\{ \frac{\partial Q}{\partial \sigma^s_{22}}, \frac{\partial Q}{\partial \sigma^s_{12}} \right\}^T \qquad (25b)$$

Following the standard plasticity procedure, i.e., satisfying the consistency condition and utilizing the Hooke's law (23), one obtains

$$\left\{ \begin{array}{c} \dot{\sigma}^s_{22} \\ \dot{\sigma}^s_{12} \end{array} \right\} = \left[\begin{array}{cc} K_{11} & K_{12} \\ K_{21} & K_{22} \end{array} \right] \left\{ \begin{array}{c} \dot{g}_N \\ \dot{g}_T \end{array} \right\} \qquad (26a)$$

where

$$[K] = [K^e] - \frac{1}{H} [K^e] \frac{\partial Q}{\partial \sigma^s} \left(\frac{\partial f}{\partial \sigma^s} \right)^T [K^e] \qquad (26b)$$

$$H = \left(\frac{\partial f}{\partial \sigma^s} \right)^T [K^e] \frac{\partial Q}{\partial \sigma^s} - \frac{\partial f}{\partial \kappa} \left(\frac{\partial \kappa}{\partial g^p} \right)^T \frac{\partial Q}{\partial \sigma^s}$$

and $[K^e]$ represents the elastic stiffness, viz. eq.(23).

In particular, if the yield and plastic potential functions are selected according to a simple Coulomb friction law, i.e.,

$$f = \sigma^s_{12} - \mu \sigma^s_{22} = 0 \; ; \quad \mu = \mu(\kappa) \; ; \quad \kappa = g^p_T \qquad (27)$$

$$Q = \sigma^s_{12} - \mu' \sigma^s_{22} = \text{const.} \; ; \quad \mu' = \text{const} \qquad (28)$$

then the constitutive matrix [K], eq. (26), has the representation

$$[K] = \left[\begin{array}{cc} K_N - \frac{1}{H} K^2_N \mu \mu' \; ; & \frac{1}{H} K_N K_T \mu' \\ \\ \frac{1}{H} K_N K_T \mu \; ; & K_T - \frac{1}{H} K^2_T \end{array} \right] \qquad (29a)$$

where

$$H = \sigma^s_{22} \frac{\partial \mu}{\partial g^p_T} + K_N \mu \mu' + K_T \qquad (29b)$$

Other, more sophisticated forms of the friction law (27) may be implemented, taking into consideration the specific media-media interaction problem and the microstructure of the actual contact area. Within the present framework, eqs. (23) and (26), the constitutive law for the interface material takes the form

$$\dot{\sigma}^s = [\overline{K}] \dot{g} \qquad (30a)$$

or in explicit terms

$$\left\{ \begin{array}{c} \dot{\sigma}^s_{11} \\ \dot{\sigma}^s_{22} \\ \dot{\sigma}^s_{33} \\ \dot{\sigma}^s_{12} \end{array} \right\} = \left[\begin{array}{cc} \left(1 - 2\dfrac{K_T}{K_N}\right)K_{11} & \left(1 - 2\dfrac{K_T}{K_N}\right)K_{12} \\ K_{11} & K_{12} \\ \left(1 - 2\dfrac{K_T}{K_N}\right)K_{11} & \left(1 - 2\dfrac{K_T}{K_N}\right)K_{12} \\ K_{21} & K_{22} \end{array} \right] \left\{ \begin{array}{c} \dot{g}_N \\ \dot{g}_T \end{array} \right\}$$
$$(30b)$$

The mathematical formulation can now be completed following a similar procedure to that outlined in previous section (i). Substitution of both constitutive relations (22) and (30) into compatibility and equilibrium constraints (21) results in

$$[\delta] \dot{\varepsilon}^m = [A] \dot{\varepsilon} + [B] \dot{g} \qquad (31a)$$

where

$$[\delta] = \left[\begin{array}{ccc} 0 & 1 & 0 \\ 0 & 0 & 1 \end{array} \right] \; ; \quad [A] = \left[\begin{array}{ccc} a_{11} & a_{12} & a_{13} \\ a_{21} & a_{22} & a_{23} \end{array} \right]$$

$$[B] = \left[\begin{array}{cc} 0 & b_{12} \\ b_{21} & 0 \end{array} \right] \qquad (31b)$$

and the coefficients a's and b's are specified in the Appendix. This set of equations can now be supplemented by relations (10a). It should be noted that since $\eta_m \to 1$, eqs. (10a) may be expressed in the form

$$[\delta] \dot{\varepsilon} \approx [\delta] \dot{\varepsilon}^m + \eta \dot{g} \qquad (32)$$

where $\eta = \eta_s/t$ and 't' is the

35

corresponding thickness of the shear band. Assigning some arbitrary dimensions, say b x h, to the elementary volume shown in Fig. 5, it is evident that $\eta = (h \cos\beta)^{-1}$, i.e., the value of η is in fact independent of the interface thickness, t. Solving the system of equations (31) and (32), one obtains

$$[\delta]\dot{\varepsilon}^m = ([I]\eta + [B])^{-1} \ ([A]\eta + [B][\delta])\dot{\varepsilon} \qquad (33)$$

or alternatively,

$$[\delta]\dot{\varepsilon}^m = [\bar{S}_m]\dot{\varepsilon} \qquad (34a)$$

where, in view of eqs. (31b), the components of the matrix $[\bar{S}_m]$ are defined as

$$[\bar{S}_m] = \begin{bmatrix} \bar{S}_{11} & \bar{S}_{12} & \bar{S}_{13} \\ \bar{S}_{21} & \bar{S}_{22} & \bar{S}_{23} \end{bmatrix}$$

$$\bar{S}_{11} = (\eta^2 a_{11} - \eta \, a_{21} b_{12})/c;$$

$$\bar{S}_{12} = (\eta^2 a_{12} - \eta \, a_{22} b_{12} - b_{12} b_{21})/c;$$

$$\bar{S}_{13} = (\eta^2 a_{13} + \eta b_{12} - \eta a_{23} b_{12})/c;$$

$$\bar{S}_{21} = (-\eta a_{11} b_{21} + \eta^2 a_{21})/c;$$

$$\bar{S}_{22} = (-\eta a_{12} b_{21} + \eta^2 a_{22} + \eta b_{21})/c;$$

$$\bar{S}_{23} = (-\eta a_{13} b_{21} - b_{21} b_{12} + \eta^2 a_{23})/c;$$

$$c = \eta^2 - b_{12} b_{21}$$

$$(34b)$$

Thus, taking into account the identity $\dot{\varepsilon}_{11}^m = \dot{\varepsilon}_{11}$, eq. (21), the strain rate averages $\dot{\varepsilon}^m$ are related to the macroscopic measure $\dot{\varepsilon}$ by

$$\dot{\varepsilon}^m = [S_m]\dot{\varepsilon} \qquad (35a)$$

where

$$[S_m] = \begin{bmatrix} 1 & 0 & 0 \\ \bar{S}_{11} & \bar{S}_{12} & \bar{S}_{13} \\ \bar{S}_{21} & \bar{S}_{22} & \bar{S}_{23} \end{bmatrix} \qquad (35b)$$

The concentration factors $[S_s]$ can be derived from equation (32). Assuming that

$$\dot{g} = [S_s]\dot{\varepsilon} \qquad (36)$$

the set of equations (32) and (34a) yields

$$[\delta] = [\bar{S}_m] + \eta[S_s] \qquad (37)$$

so that

$$[S_s] = \frac{1}{\eta}([\delta] - [\bar{S}_m]) \qquad (38)$$

It should be emphasized again that, the thickness of the shear band does not enter the formulation as both concentration factors depend on the properties of constituents and the geometrical factor η, which is invariant with respect to 't'.

Finally, the macroscopic stress rates $\dot{\sigma}$ can be derived from the decomposition (10b). Since $\eta_s \ll \eta_m$, it is likely that the following approximation may be employed

$$\dot{\sigma} \approx [D][S_m]\dot{\varepsilon} \qquad (39)$$

which represents the overall homogenized constitutive relation.

It appears that the strain softening response, as observed on the macroscale, will be the result of instability (in Drucker's sense) of σ^s-ε^s characteristics for the interface material. This conclusion is however speculative and needs to be verified numerically. The mathematical framework derived above is now at the stage of numerical implementation. A comprehensive study in that respect will be presented shortly.

4.   FINAL REMARKS

The engineering design can be characterized as the process of progressive data reduction; the data from both the field and laboratory investigations are processed with the help of continuum mechanics and applied mathematics. In such a process the understanding of physical behaviour of soil under the anticipated loading conditions is essential. Thus, the fundamental investigations, aimed at mathematical description of soil deformation, are very meritorious. On one hand, they serve the purpose of an improved and economized design, on the other one they eventually allow to approach complex structural problems, which cannot be solved by conventional design procedures.

In this paper, a mathematical framework has been outlined for the description of the deformation process of geomaterials. The framework correlates the macroscopic mechanical behaviour with the internal structure of the material and its progressive evolution. For the homogeneous deformation mode, some implicit measures of material fabric have been cast within the phenomenological plasticity formulation. Such a an enhanced framework can depict, at least in a qualitative sense, various manifestations of soil anisotropy. The failure mechanism, associated with a spontaneous loss of homogeneity, has been described within the framework of the mechanics of composite materials. The proposed homogenized constitutive relation is rigorous and may be adequate for representing the unstable material response, i.e., strain softening.

REFERENCES

1. Delage, P. & Lefebvre, G. 1984. Study of the structure of a sensitive Champlain clay and of its evolution during consolidation. Canadian Geotechnical Journal. 21: 21-35.

2. Duncan, J.M. & Seed, H.B. 1966. Anisotropy and strength reorientation in clays, Journal of Geotechnical Engineering Division, ASCE. 92: 21-50.

3. Hill, R. 1962. Accoleration waves in solids. Journ. Mech. Phys. Solids. 10: 1-16.

4. Hill, R. 1963. Elastic properties of reinforced solids; some theoretical principles. Jour. Mech. Phys. Solids. 11: 357-372.

5. Mitchell, J.K. 1970. Fundamentals of soil behaviour, John Wiley & Sons, Inc.,.

6. Mitchell, R.J. 1972. Some deviations from isotropy in a lightly overconsolidated clay, Geotechnique. 22: 459-467.

7. Ochiai, H. & Lade, P.V. 1983. Three-dimensional behaviour of sand with anisotropic fabric. Journal of Geotechnical Engineering Division, ASCE. 109: 1313-1328.

8. Pietruszczak S. & Mroz, Z. 1981. Finite element analysis of deformation of strain softening materials. Int. Journ. Num. Meth. Eng. 17: 327-334.

9. Pietruszczak, S. & Stolle, D.F. 1987. Deformation of strain softening materials, part II: Modelling of strain softening response. Computers & Geotechnics. 4: 109-123.

10. Pietruszczak, S. & Krucinski, S. 1989. Description of anisotropic response of clays using a tensorial measure of structural disorder. Mechanics of Materials. 8: 237-249.

11. Pietruszczak, S. & Krucinski, S. 1989. Considerations on soil response to the rotation of principal stress directions. Computers & Geotechnics. 8: 89-110.

12. Rowe, P.W. 1962. The stress-dilatancy relation for static equilibrium of an assembly of particles in contact. Proceedings of the Royal Society. A269: 500-527.

13. Rudnicki, J.W. & Rice, J.R. 1975. Conditions for the localization of deformation in pressure-sensitive dilatant materials. Journ. Mech. Phys. Solids. 23: 371-394.

14. Skempton, A.W. & Sowa, V.A. 1963. The behaviour of saturated clays during sampling and testing. Geotechnique. 13: 269-290.

Appendix

Derivation of strain concentration factors

(i) Three-dimensional formulation

Consider first identity

$$\dot{\sigma}_{22} = \dot{\sigma}_{22}^m = \dot{\sigma}_{22}^s \qquad \text{(A-1)}$$

as specified within the set of static constraints (12). Substituting the constitutive relations (13), the equation (A-1) becomes

$$D_{21}\dot{\varepsilon}_{11}^m + D_{22}\dot{\varepsilon}_{22}^m + D_{23}\dot{\varepsilon}_{33}^m + D_{24}\dot{\gamma}_{12}^m + D_{25}\dot{\gamma}_{13}^m + D_{26}\dot{\gamma}_{23}^m =$$

$$K_{21}\dot{\varepsilon}_{11}^s + K_{22}\dot{\varepsilon}_{22}^s + K_{23}\dot{\varepsilon}_{33}^s + K_{24}\dot{\gamma}_{12}^s + K_{25}\dot{\gamma}_{13}^s + K_{26}\dot{\gamma}_{23}^s \qquad \text{(A-2)}$$

Now, taking into account the representation (10a) together with the kinematic constriants (12), the above equation can be expressed in the form

$$\dot{\varepsilon}_{22}^m = a_{11}\dot{\varepsilon}_{11} + a_{12}\dot{\varepsilon}_{22} + a_{13}\dot{\varepsilon}_{33} + a_{14}\dot{\gamma}_{12} + a_{15}\dot{\gamma}_{13}$$
$$+ a_{16}\dot{\gamma}_{23} + b_{12}\dot{\gamma}_{12}^s + b_{13}\dot{\gamma}_{23}^s \qquad \text{(A-3)}$$

where

$$a_{11} = (K_{21} - D_{21})/C_1 \; ; \qquad a_{12} = K_{22}/(\eta_s C_1);$$

$$a_{13} = (K_{23} - D_{23})/C_1; \quad a_{14} = -D_{24}/(\eta_m C_1) \; ;$$

$$a_{15} = (K_{25} - D_{25})/C_1; \quad a_{16} = -D_{26}/(\eta_m C_1) \; ;$$

$$b_{12} = (K_{24} + \frac{\eta_s}{\eta_m} D_{24})/C_1 \; ; \quad b_{13} = (K_{26} + \frac{\eta_s}{\eta_m} D_{26})/C_1 \qquad \text{(A.4)}$$

and

$$C_1 = D_{22} + \frac{\eta_m}{\eta_s} K_{22} \qquad \text{(A-5)}$$

Taking the next static constraint (12), i.e.,

$$\dot{\sigma}_{12} = \dot{\sigma}_{12}^m = \dot{\sigma}_{12}^s \qquad \text{(A-6)}$$

and following the same procedure to that outlined above, one obtains

$$\dot{\gamma}_{12}^m = a_{21}\dot{\varepsilon}_{11} + a_{22}\dot{\varepsilon}_{22} + a_{23}\dot{\varepsilon}_{33} + a_{24}\dot{\gamma}_{12}$$
$$+ a_{25}\dot{\gamma}_{13} + a_{26}\dot{\gamma}_{23} + b_{21}\dot{\varepsilon}_{22}^s + b_{23}\dot{\gamma}_{23}^s \qquad \text{(A-7)}$$

where

$$a_{21} = (K_{41} - D_{41})/C_2 \; ; \quad a_{22} = -D_{42}/(\eta_m C_2);$$

$$a_{23} = (K_{43} - D_{43})/C_2; \quad a_{24} = K_{44}/(\eta_s C_2) \; ;$$

$$a_{25} = (K_{45} - D_{45})/C_2; \quad a_{26} = -D_{46}/(\eta_m C_2) \; ;$$

$$b_{21} = (K_{42} + \frac{\eta_s}{\eta_m} D_{42})/C_2 \; ; \quad b_{23} = (K_{46} + \frac{\eta_s}{\eta_m} D_{46})/C_2 \qquad \text{(A-8)}$$

and

$$C_2 = D_{44} + \frac{\eta_m}{\eta_s} K_{44} \qquad \text{(A-9)}$$

Finally, the last identity

$$\dot{\sigma}_{23} = \dot{\sigma}_{23}^m = \dot{\sigma}_{23}^s \qquad \text{(A-10)}$$

results in

$$a_{31} = (K_{61} - D_{61})/C_3 \; ; \qquad a_{32} = -D_{62}/(\eta_m C_3);$$

$$a_{33} = (K_{63} - D_{63})/C_3 \; ; \qquad a_{34} = -D_{64}/(\eta_m C_3) \; ;$$

$$a_{35} = (K_{65} - D_{65})/C_3; \qquad a_{36} = -K_{66}/(\eta_s C_3) \; ;$$

$$b_{31} = (K_{62} + \frac{\eta_s}{\eta_m} D_{62})/C_3 \; ; \quad b_{32} = (K_{64} + \frac{\eta_s}{\eta_m} D_{64})/C_3 \qquad \text{(A-11)}$$

where

$$C_3 = D_{66} + \frac{\eta_m}{\eta_s} K_{66} \qquad \text{(A-12)}$$

The above relations define completely the matrices [A] and [B] in eq. (14).

(ii) Plane strain case

Consider the second identity in eq. (21), i.e.,

$$\dot{\sigma}_{22} = \dot{\sigma}_{22}^m = \dot{\sigma}_{22}^s \qquad \text{(A-13)}$$

Upon substitution of the constitution relations (22b) and (30b), the above equation becomes

$$D_{21}\dot{\varepsilon}_{11}^m + D_{22}\dot{\varepsilon}_{22}^m + D_{24}\dot{\gamma}_{12}^m = K_{11}\dot{g}_N + K_{12}\dot{g}_T \qquad \text{(A-14)}$$

Now taking into account the relation (32), together with the constraint $\dot{\varepsilon}_{11} = \dot{\varepsilon}_{11}^m = \dot{\varepsilon}_{11}^s$, e.q. (21), one obtains

$$\dot{\varepsilon}_{22}^{m} = a_{11}\dot{\varepsilon}_{11} + a_{12}\dot{\varepsilon}_{22} + a_{13}\dot{\gamma}_{12} + b_{12}\dot{g}_T \qquad \text{(A-15)}$$

where

$$a_{11} = -D_{21}/C_1 \; ; \; a_{12} = K_{11}/(\eta\,C_1) \; ; \; a_{13} = -D_{24}/C_1$$

$$b_{12} = (K_{12} + \eta\,D_{24})/C_1$$

$$\text{(A-16)}$$

and

$$C_1 = D_{22} + \frac{1}{\eta}\,K_{11} \qquad \text{(A-17)}$$

Finally, taking

$$\dot{\sigma}_{12} = \dot{\sigma}_{12}^{m} = \dot{\sigma}_{12}^{s} \qquad \text{(A-18)}$$

according to eq. (21), and following the same procedure, one obtains

$$a_{21} = -D_{41}/C_2 \; ; \; a_{22} = -D_{42}/C_2 \; ; \; a_{23} = K_{22}/(\eta\,C_2)$$

$$b_{21} = (K_{21} + \eta\,D_{42})/C_2$$

$$\text{(A-19)}$$

where

$$C_2 = D_{44} + \frac{1}{\eta}\,K_{22} \qquad \text{(A-20)}$$

The relations (A-16) together with (A-19) define both matrices [A] and [B] in eq. (31).

Developments in Geotechnical Aspects of Embankments, Excavations and Buried Structures, Balasubramaniam et al. (eds)
© 1991 Balkema, Rotterdam. ISBN 90 5410 019 2

CPT correlations for sands from calibration chambers

J.H.A.Crooks
Golder Associates, Calgary, Alb., Canada

SUMMARY

* The CPT is a robust test which provides reliable and repeatable measurements. However, estimating mechanical properties of soils based on empirical correlations is problematic in the case of sands; the common use of relative density to characterize sand behaviour exacerbates this situation.

* Soil behaviour can be conveniently normalized in terms of the state of the material which is quantified in terms of the distance between current state and a reference state on a void ratio - stress plot. For clays, OCR is commonly used to define the current state of the material while for sands, state parameter (ψ) provides a similar function. Consistent relationships have been developed relating the results of laboratory tests on sands and clays, and their respective state descriptions.

* Interpretation of the available calibration chamber test results for sands in terms of state parameter provides a comprehensive and accurate approach in determining the state of sands from CPT data.

* The compressibility of the sand is an important factor in the relationship between Qc and ψ. This effect can be quantified using the slope of the SSL to represent compressibility.

* It is estimated that the accuracy of the Qc - ψ relationships for sands is within +/- 0.03 in terms of void ratio. This is a significant improvement over conventional interpretations and is sufficiently precise for most engineering purposes.

* Based on reasonable assumptions, it is possible to relate the state parameters for sands (ψ) and clays (OCR-I). The Qc - state relationships for clays and sands which result from this process are consistent and provide a unified interpretation for both materials.

1 INTRODUCTION

The cone penetration test is now routinely used in geotechnical practice for stratigraphic logging, estimating soil properties and construction control. Over the past decade, there have been extensive developments from the original Dutch mechanical friction cone including electronic measurement of data, acoustic transmission of signals and mud injection systems to facilitate penetration. Measurement of porewater pressure response during penetration is now common and adds much to the quantity and quality of data obtained by the CPT. Further improvements can be anticipated through increased measurement capability together with standardization of equipment (eg cone design, location of porewater pressure sensor etc), test procedures and data reduction methods (eg end area corrections). Notwithstanding future improvements, the cone, even in its current state, is a sophisticated yet robust instrument which is capable of providing highly repeatable information.

The usefulness of the CPT for logging stratigraphy has been well established for some time and has been much enhanced by the measurement of porewater pressure response during penetration. The CPT is capable of detecting relatively thin soil layers. However in most cases, a layer thickness of about 0.5 m is required to ensure full development of tip resistance and therefore allow estimation of soil properties. For qualitative profiling, as opposed to estimation of material properties, this is not a serious drawback.

Determination of soil properties based on CPT data is more problematic. There are a number of basic approaches which can be adopted as follows:

- Direct correlation of a CPT measurement with a design parameter (eg bearing capacity with cone tip resistance). This type of correlation is of limited value since it is specific in terms of material type and problem application. However, it is useful if good local calibration is available.
- Theoretical solution of what is in effect a complex boundary value problem based on sophisticated constitutive models and numerical solutions. This approach is at an early stage of development although good insight into the problem can be obtained.
- Empirical relationships between quantities measured in the CPT and mechanical properties measured in other tests. The following discussion will focus on this approach, mainly in relation to granular materials.

For clays, it is relatively simple to directly establish a relationship between a CPT-measured quantity and a property measured in another test (eg Q_c vs S_u from field vane tests or laboratory tests on undisturbed samples). For sands, this procedure is significantly more difficult because it is not possible to routinely obtain undisturbed samples for laboratory testing. Thus, an indirect approach is used whereby Q_c - density relationships are developed based on calibration chamber tests and then combined with density - property correlations from laboratory strength tests to obtain generalised Q_c - property correlations.

There is a basic problem with the traditional implementation of this approach in that to account for different material characteristics, it is necessary to express density in a dimensionless manner. The common description used in this regard is relative density, a measurement which is plagued by well documented difficulties. Of equal importance is the fact that different sands do not behave in the same manner even if they exist at the same relative density and stress level. Nor does the same sand behave in the same manner even if the density is the same but the stress level is different.

To overcome these problems, use is made of the recently developed "state parameter" concept instead of relative density for both characterizing sand behaviour and interpreting CPT data. This paper describes the use of "state parameter" to interpret the results of calibration chamber tests and to derive general correlations between CPT measurements and mechanical properties of sands.

2 CHARACTERIZING SOIL BEHAVIOUR

2.1 The state of soils

The development of a "universal", comprehensive constitutive model for soils together with usable numerical methods to solve practical boundary value problems in geotechnical engineering, is still at a formative stage. In its absence, it has been common practice to represent soil behaviour in terms of mechanical properties measured in tests which purport to reflect at least the important factors affecting the soil behaviour. For example, laboratory tests are carried out on "undisturbed" samples of clay to measure undrained strength. The samples are reconsolidated to the in situ effective stress state, the tests are carried out with no drainage and stress paths are chosen to reflect the field problem and to recognize inherent anisotropy. Thus, the test is viewed as a prototype of the field problem and the measured property is applied directly in analysis. It is evident that the "property" measured in a "prototype" test of this nature reflects the soil behaviour only under very specific conditions. To better understand the limits of applicability of the measured property and the behaviour of the soil in general, consideration must be given to the basic factors which control behaviour. If these basic factors can be isolated and formulated in terms of (say) a single quantity, then this quantity would be a useful "signature" of the material and normalizing parameter for comparison of properties etc.

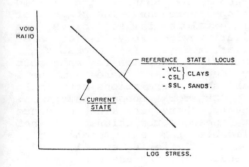

Fig.1   State of soils

At the risk of appearing simple, it can be stated that the behaviour of a soil is controlled, at least in a first order sense, by its void ratio and effective stress level which together define the state of the soil (Figure 1). To use state as a normalizing parameter, it is necessary that it is quantified.

For clays, the virgin consolidation line (VCL) has traditionally been used, either implicitly or explicitly, for this purpose. The VCL is the locus of states following (primary) consolidation of the clay. (It is noted that the VCL should only include primary consolidation strains; the inclusion of arbitrary quantities of secondary compression depending on the time of loading and material characteristics, is not logical). An alternative reference developed by workers at Cambridge is the critical state line (CSL) which is the locus of states following undrained shearing at large strains when the effective stresses in the soil are constant. It is not important which is used as a reference for current state; what is important is that the reference line represents the result of a relevant physical process and that it can be repeatably defined.

An example of the explicit use of the VCL as a reference for current state is the SHANSEP concept (Ladd and Foott, 1974). In this approach, the current state is quantified in terms of the ratio of the maximum past vertical effective stress to the current vertical effective stress. The maximum past vertical effective stress is the intersection of the reconsolidation line and the VCL. This ratio is referred to as the over-consolidation ratio (OCR) and many publications describe the relationships between OCR and behavioural properties.

The SHANSEP approach is now well established in practice largely because it provides a robust link between the state of the soil and its mechanical properties. This allows comparison of the results of different test types on the same soil to define the differences which are the result of the test boundary conditions (Figure 2). It also allows comparison of the behavior of different soil types (Figure 3).

It should be appreciated that there are many similarities between the SHANSEP approach and other concepts for characterizing clay behaviour. For example, it has long been recognised that the maximum past effective stress which the clay has experienced, is a major

43

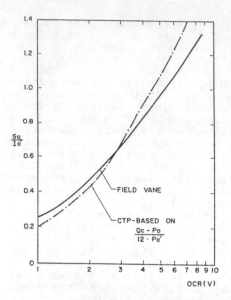

Fig.2 Normalized strength of clays from different tests

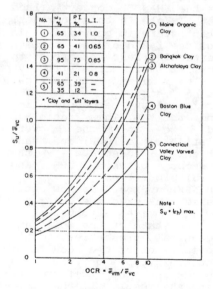

Fig.3 Normalized DSS strengths for different clays (Ladd & Foott,1974)

factor controlling its undrained strength. Thus, it is common practice to express undrained strength as a ratio in terms of maximum past effective stress (ie preconsolidation pressure). The result is similar to that obtained

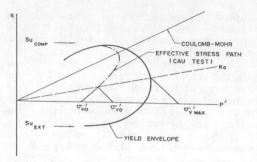

Fig.4 Yield envelope model for clays

from a more formal SHANSEP approach at least for lightly over-consolidated clays.

There are also many similarities between the SHANSEP approach and the qualitative "yield envelope" model for representing the behaviour of soft clays (Crooks, 1981). As shown on Figure 4, the undrained shear behaviour of a clay consolidated to its in situ effective stress state, is a function of the size and shape of the yield envelope. Size is basically defined by the maximum past effective stress (ie OCR) and this is the basis of similarity between the SHANSEP and yield envelope approaches. The shapes of yield envelopes for soft clays are similar provided Ko values are also similar. This is probably a reasonable assumption for many soft clay deposits and there are well established correlations between Ko and OCR for these materials. However, there is also evidence that these correlations are not applicable for all recent clay deposits (Jefferies et al, 1987) and indiscriminate use can lead to incomplete and/or incorrect understanding of behaviour.

Further, the shape of the yield envelope depends on whether or not the clay is highly structured or cemented. If it is, then the normal SHANSEP procedure of consolidating samples well past their maximum past effective stresses before allowing swelling to create specific OCR values, can destroy the natural soil structure. As a result, the subsequent normalized behaviour will not reflect the

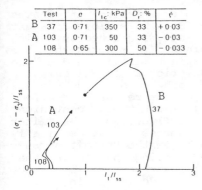

Test	e	I_{1c}: kPa	D_r: %	ψ
B 37	0.71	350	33	+0.03
A 103	0.71	50	33	−0.03
108	0.65	300	50	−0.033

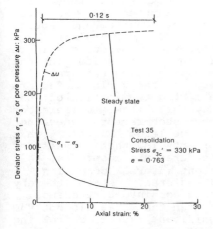

Fig.5 Dependence of sand behaviour on stress level (Been & Jefferies, 1985)

behaviour of the in situ material.

Finally, because there is a consistent physical relationship between virgin consolidation behaviour and the critical state of a clay, the SHANSEP approach could also be viewed as a simplification of critical state soil mechanics (CSSM). Thus CSSM is a full stress-strain-strength model while SHANSEP is based on a single parameter characterization of the soil.

2.2 The state parameter concept for sands

The purpose of the above discussion on characterizing the behaviour of clays is to provide a general framework for a similar discussion on sands. It is convenient to describe general concepts in terms of clay behaviour because these are better developed than is the case for sands as a direct consequence of the relative ease with which clays can be sampled and tested.

Traditionally, relative density has been used as a normalising parameter for sands. As noted previously, there are two serious disadvantages to using relative density. Firstly, it is notoriously difficult to measure maximum and minimum densities repeatably and as a result, relative density is an imprecise quantity. Secondly, relative density does not account for the important effect that stress level exerts on sand behaviour. For example, Figure 5 shows the stress paths for two samples of the same sand at essentially the same density and subjected to undrained triaxial compression. The only difference is the consolidation stress level. Sample A at the lower stress level shows strongly dilatent behaviour while Sample B at the higher stress level displays a contractive response. This basic difference in behaviour is not reflected by relative density. Thus it is necessary for a normalizing parameter for sands to include a description of stress level.

Referring to the basic concept for quantifying the state of soils described previously, the current state of a sand should be related to a reference line on the state diagram (ie void ratio - stress plot). For sands, it is not possible to define a unique virgin consolidation line and this cannot be used as a reference. Further, the critical state line is not readily determined for sands using typical laboratory testing techniques. However, the steady state of sand is a repeatably measureable behaviour and provides a suitable reference. The steady state line (SSL) is the locus of states at large strain following undrained shearing of sands. Thus it is similar to the critical state line although there are differences of opinion regarding the relationship between the CSL and SSL. The steady state condition is determined in an undrained load controlled triaxial compression test on a contractive sample (Figure 6). At large strains the effective stresses in the sample

45

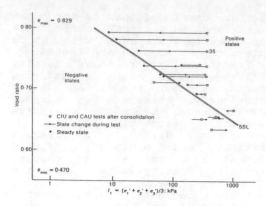

Fig.6 Steady state line determination (Been & Jefferies, 1985)

are constant; these stresses together with the known initial void ratio define the end state of the test. The locus of end states for a series of tests define the SSL for a given sand (Figure 6).

Using the SSL as a reference, Been and Jefferies (1985) quantified the state of a sand as the void ratio difference between current void ratio and the void ratio at steady state for the same stress level (Figure 7). They termed this quantity the state parameter of the sample (ψ). It is conceptually similar to OCR except that instead of using a stress difference as is the case for clays, a void ratio difference is used. Samples which exist in a state above the SSL will exhibit contractive behaviour and are defined as having +ve ψ values.

Those samples existing in states below the SSL will exhibit dilatent behaviour and are defined as having -ve ψ values. It is noted that the first invariant (I1) is used as the stress descriptor for defining ψ because both major and minor principal stresses affect behaviour. Thus, it is convenient that the influence of the minor principal stress is automatically incorporated in the definition of ψ. Extensive testing has shown that the initial stress ratio does not affect the SSL defined using I1.

Provided consistent test procedures are adopted such as those described by Castro and Poulos (1977), the SSL for a given sand can be repeatably measured. There is continuing discussion in the literature regarding the relative importance of the various factors which affect determination of the SSL. However, the differences caused by such factors does not appear to be large. Further, what is important is that consistent test procedures are adopted so that a consistent material condition is achieved at steady state thereby providing a consistent reference.

It is known that state is not the only significant factor controlling sand behaviour; the initial fabric of the sand (ie the physical arrangement of particles) is also important. The major problem in this regard is that there is as yet no convenient way to quantify initial fabric. The effect of initial fabric is not likely to be important in relation to large strain behaviour of sand such as effective friction angle. Further, steady state behaviour is unaffected by differences in sample preparation methods which would be expected to induce different initial fabrics. However, initial fabric will affect how the stress-strain behaviour develops. Further, it will have a significant effect on sand behaviours associated with small strain (eg porewater presure response to rotation of principal stresses). In these cases, ψ alone may not be an adequate description of the material.

Constitutive models which describe the stress-strain-strength behaviour of sands are currently being developed and offer the promise of significant insight into

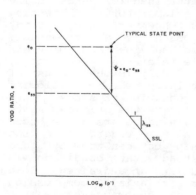

Fig.7 Definition of state parameter (Been & Jefferies, 1985)

46

sand behaviour. The particulate mechanics approach presented by Rothenburg and Bathhurst (1989) is particularly interesting because it accounts for both the initial fabric and state of the sand; the latter is incorporated in terms of state parameter. Modelling the response of the sand during shear is mainly based on changes in fabric during the shearing process which in turn depend on the relative location of the transient state in relation to the steady state. Development of this model benefitted initially from the state parameter concept and in turn has provided considerable insight for the further development of the concept. The relationship between the state parameter concept and Rothenburg's particulate mechanics model is essentially the same as the relationship between SHANSEP and CSSM as discussed above.

2.3 Sand property - state parameter correlations

The usefulness of the approach in terms of understanding sand behaviour can be illustrated by reference to the example shown on Figure 5. As noted previously, these two samples of the same material had the same void ratio but were consolidated to different stress levels. Sample B at the higher stress level lies above the SSL and therefore is in a contractive state while Sample A lies below the SSL and is in a dilatent state. The inclusion of stress in the state parameter characterization allows for this difference to be readily explained. This is not the case if relative density is used.

As stated above, the SSL for a given material can be reliably determined using consistent testing procedures. However, the SSLs for different sands are quite unique as indicated on Figure 8. These differences are due to variations in mineralogy, fines content, compressibility, grain size, gradation etc. By describing the state of a sand using its unique SSL, the effects of different material characteristics are included in the state parameter description.

A number of correlations between

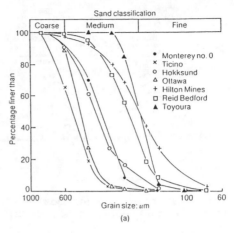

(a)

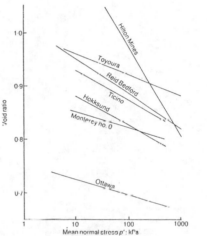

Fig.8 Steady state lines for different sands (Been et al, 1987a)

the mechanical properties of sands and state parameter have been developed. For example, angles of internal friction measured in triaxial tests on samples of 20 sands are plotted against on Figure 9. Despite the wide variation in material type and test conditions, there is a well defined relationship between ϕ' and within a band represented by +/- 2 degs. In contrast, Figure 10 shows that there is a much greater scatter in the data using relative density as the normalizing parameter.

Further examples of correlations between the properties of sands obtained in drained and undrained

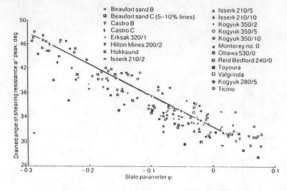

Fig.9 Ø' - State parameter correlation for sands (Been et al,1987a)

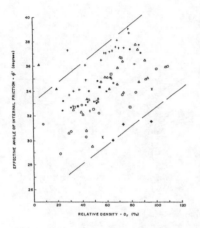

Fig.10 Ø' - Relative density correlaiton for sands

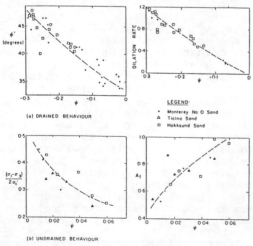

Fig.11 Dependence of sand behaviour on state parameter

triaxial tests are shown on Figure 11. These correlations provide further evidence of the benefit of using Ψ to represent sand behaviour.

3 CPT INTERPRETATION IN TERMS OF STATE PARAMETER

3.1 Interpretation

Calibration chambers are used to determine correlations between cone penetration resistance (Qc), density and stress level. Basically, a calibration chamber test involves placing sand at a known void ratio in a test chamber, imposing known axial and radial stresses on the sample (these need not be equal), and carrying out a cone penetration test through the prepared sample to determine the characteristic tip resistance. The test conditions can be varied in terms of the void ratio of the sample and the imposed axial and radial stresses. Details of the test procedures and data processing to account for differences in chamber size/ design are discussed in the next section of this paper.

Traditionally, chamber test results have been presented in terms of relative density-axial stress-tip resistance correlations as shown on Figure 12. Different

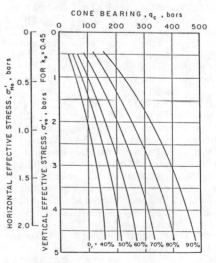

Fig.12 Qc - Dr - Stress correlation

axial/radial stress ratios require additional plots. As stated previously, relative density is not a good normalising parameter for sand behaviour measured in laboratory strength tests and the same criticisms apply to its use for representing the results of chamber tests. Further, as shown on Figure 13, the traditional presentation of Qc - stress - relative density relationships indicates differences which are reportedly dependent on material compressibility. However, the traditional method of interpretation provides no indication regarding how to take this obviously important factor into account in practical applications.

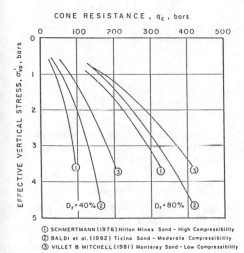

CONE RESISTANCE, q_c, bars

① SCHMERTMANN (1976) Hilton Mines Sand - High Compressibility
② BALDI et al. (1982) Ticino Sand - Moderate Compressibility
③ VILLET & MITCHELL (1981) Monterey Sand - Low Compressibility

Fig.13 Comparison of Qc - Dr - stress correlations for different sands

Since the basic variables associated with chamber tests (ie void ratio and stress level) also define the state of the sand, it is logical to interpret the chamber test results in terms of state parameter (Been et al, 1986, 1987a). This interpretation was based on work carried out in two stages:- 1) based on available published information, and 2) additional testing in a newly constructed chamber.

3.2 Existing data

Interpretation of existing published information involved the following steps:
- The published results of over 400 chamber tests on a total of 6 sands were summarized in terms of void ratio, imposed stress conditions and measured tip resistance. Only results from tests using right cylindrical cones were included to avoid inconsistencies in the data set as a result of using different cone types.
- Samples of the sands used in the chamber testing programs were obtained and their SSLs determined. These SSLs are identified on Figure 8.
- The value of Ψ for each chamber test was calculated based on the known state of the chamber sample and the SSL for the material.
- The data were plotted for each material in terms of :

$$\frac{Qc - P}{P'} = f(\Psi)$$

The average of the axial and radial stresses (P, P') was used to normalize Qc in order to account for different stress ratios.

The correlations determined in this manner are presented on Figure 14. Because of the method of interpretation, only one correlation is required for each material. It is clear that well defined correlations were obtained for all of the sands.

3.3 Additional data

A new calibration chamber was constructed by Golder Associates to allow Qc - density - stress relationships to be developed for other materials. The test chamber is described in detail by Been et al (1987b) and the first material tested was Erksak sand from the Beaufort Sea off Canada's northern coast. It is this material that is most commonly used for the construction of artifical islands for offshore hydrocarbon exploration. The major features of the chamber are:

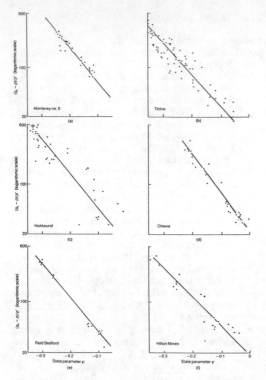

Fig.14 State parameter interpretation for chamber test for different sands (Been et al, 1987a)

- Height = 1m, diameter = 1.4m. The cone:chamber diameter ratio is 38 which requires minimal correction for chamber size effects for loose - medium samples.
- Rigid base plate which better represents field conditions than does a "soft" base plate.
- Stress controlled lateral boundary.
- Different axial and radial stresses can be imposed, each with an upper limit of 1 mPa.
- Back pressure can be applied to the sample.
- 10 sq cm, 60 deg, right cylindrical electric cone with porewater pressure and horizontal stress measurement.

The Erksak test data is summarized on Figure 15 and like the other data sets, the relationship between normalized Qc and ψ is well defined.

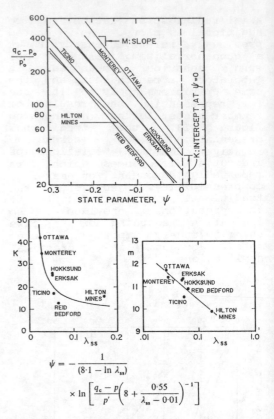

$$\psi = -\frac{1}{(8 \cdot 1 - \ln \lambda_{ss})}$$

$$\times \ln\left[\frac{q_c - p}{p'}\left(8 + \frac{0 \cdot 55}{\lambda_{ss} - 0 \cdot 01}\right)^{-1}\right]$$

Fig.15 Generalized state parameter interpretation for CPT in sand (Been et al, 1987a)

3.4 Generalized Qc - state parameter relationship for sands

The Qc - state parameter relationships for the individual sands used in the chamber testing programs are plotted together on Figure 15. The normalized Qc - relationships can be represented by the equation:

$$\frac{Qc - P}{P'} = k \cdot \exp(-m)$$

As indicated on Figure 15, there are significant differences in the relationships for different sands. It is noted that comparison of the scatter in the individual data sets with the difference between the relationships indicates that differences in Qc - ψ relationships as a function of sand type are real.

The apparent dependence on compressibility indicated by the traditional interpretation of chamber test data suggests that this property could be used to quantify dependence of the Qc – ψ relationship on material type. The question is – how should compressibility be defined? It is noted that the slope of the SSL (λss) is dependent on compressibility in that it reflects the porewater pressure generated during undrained shear. Further, during cone penetration, there is an increase in stress level within the sample. As the stress level increases and the state of the material approaches the SSL, the rate at which it approaches will depend on the slope of the SSL (λss).

Based on the above, the parameters "k" and "m" which define the location of the Qc – ψ relationships, were plotted against ss as shown on Figure 15. The effect of compressibility represented in this manner, is evident and consistent. Thus, based on state parameter interpretation of the available chamber test data, it is possible to represent all of the data in terms of the plots and equation shown on Figure 15.

The degree to which the test data agree with the general expression describing the Qc – ψ relationship for sand is shown on Figure 16. Thus for 80% of the data base, this expression predicts to an accuracy of within 0.035. This accuracy represents about 10% of the range in likely to be encountered in the field since values of are rarely encountered outside the the range between -0.3 and +0.05 for the broad class of subangular to subrounded sands. This level of accuracy is acceptable for most practical engineering purposes.

3.5 Factors affecting chamber test data

The obvious question arises regarding how well chamber tests represent field conditions. It is tempting to expect that because the sand sample tested in a calibration chamber is large (ie in the order of 2 tons) that the test should be relatively free of boundary effects. In fact, this is not the case because of the extent of displacements caused by cone insertion into a relatively incompressible material. There are effects of both size and the nature of the lateral boundary as indicated on Figure 17. However, this data can also be used to "correct" the raw chamber data to standard conditions. It is noted that there is little effect for loose to medium sands.

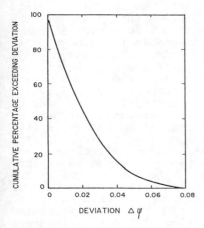

Fig.16 Accuracy of generalized relationship on Fig.15 (Been et al, 1987a)

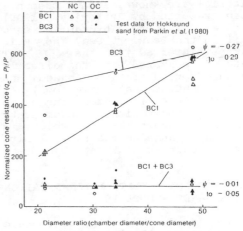

BC1 Lateral stress control
BC3 Lateral strain control

Fig.17 Chamber test boundary effects (Been et al, 1986)

Other factors which could affect the accuracy of interpretation of chamber tests are discussed by Been et al (1988a) and include:

- Errors in density measurement.
- Non-uniform distribution of stresses within the sample.
- Degree of saturation (not a factor for "clean sands).

Other potential sources of error such as incorrect cone calibration, temperature sensitivity, zero drift, errors in applied pressure, membrane leaks etc can be discounted. They are unlikely to cause systematic shifts in the data particularly since a number of organisations were involved in carrying out the tests.

In their evaluation of possible errors which could be associated with chamber tests, Been et al (1988a) indicate a potential maximum error of +/- 0.06 assuming that all errors are cumulative and operate in the same sense. Actual errors are likely to be significantly lower.

The most persuasive approach to evaluating the potential errors associated with Qc - correlations developed from chamber tests, is to carry out a cone penetration test in material in which samples have been obtained for direct void ratio measurement. It is also necessary that λss and Ko for the in situ material are known. Such a case has been reported by Jefferies (1988) for an offshore hydraulic sand fill in the Beaufort Sea. The void ratios inferred from the CPT data are compared with the directly measured void ratios (based on water content determinations) on Figure 18. As indicated there is good agreement with a random data scatter of about +/- 0.03 which is the same range as the 80 % limits for the chamber test data base.

4 UNIFIED Qc - STATE INTERPRETATION FOR SANDS AND CLAYS

As discussed previously, there are conceptual similarities between the state parameter approach for sands and the SHANSEP approach for clays. OCR and Ψ are both single parameter descriptions of state and have been shown to provide a consistent basis

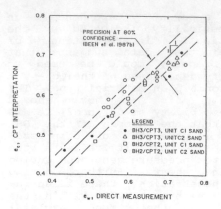

Fig.18 Comparison of void ratios inferred from CPT and measured directly (Jefferies, 1988)

for normalizing the behaviour of the respective materials. Thus, in the same way that CPT data for sands can be interpreted in terms of Ψ, it is also possible to interpret CPT data for clays in terms of OCR.

This interpretation was presented by Crooks et al (1988) and is reproduced on Figure 19. The Qc - OCR correlations are presented in terms of both the conventional definition of OCR (ie vertical stresses) and also the first stress invariant (I1). The reason for the latter interpretation is that it requires the actual value of Ko to be used to define the in situ stresses; it also includes the effect of horizontal yield stresses. Given the nature of the displacement field around a cone, it is reasonable to expect that horizontal in situ and yield stresses will be at least as important as the equivalent vertical stresses. By describing stress in terms of I1, this factor is taken into account. Further by including actual Ko values in the description of in situ stress, the cases where Ko is different from the traditionally expected value, can be included in the interpretation framework. (It is beyond the scope of this paper to discuss the problems associated with measuring Ko and the inherent bias against such measurement. Suffice it to say that simply because it is a difficult condition

52

to determine does not mean that it should be ignored. Since knowledge of the in situ stress state is important for interpretation of in situ tests, the issue should be addressed despite the obvious problems involved).

As indicated on Figure 19, consistent relationships are obtained for both OCR definitions for a wide range of materials and OCR values. Thus interpretation of CPT data for both sands and clays in terms of the initial state of the material appears to be reasonable.

A further step in the process of state interpretation of CPT data can be realized if it is assumed that the VCL and the SSL are parallel and uniquely related. As discussed by Been et al (1988b), this condition is approximately satisified for clays. Assuming that:
- the SSL and the CSL are in fact the same for clays, and
- the relationship between the CSL and the VCL is as described by Roscoe and Burland (1968),

then a relationship between Ψ and OCR defined in terms of the first stress invariant, can be determined. On this basis, it is possible convert the Qc - OCR(I) relationship for clays to Qc - Ψ relationships. The combined relationships for sand and clay are shown on Figure 20. Bearing in mind

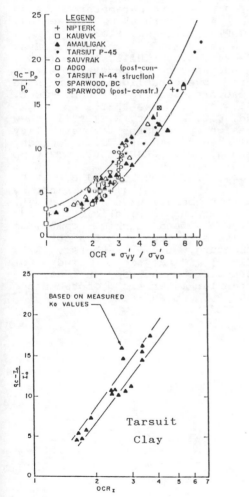

Fig.19 State interpretation of CPT in clays (Crooks et al, 1988; Been et al, 1988b)

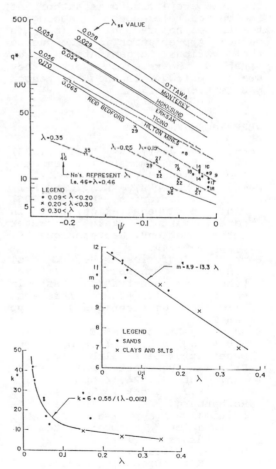

Fig.20 Unified state interpretation for CPT in clays and sands (Been et al, 1988b)

the compressibility dependence of the Qc - Ψ relationships for sands, values of compression index for the clay data set are also included on Figure 20. The "m" and "k" values which define the location of the Qc - state relationships are consistent for both the clay and sand data sets. While it is appreciated that the data base used to develop this unified CPT interpretation is limited, the approach does appear to hold considerable promise.

5 ACKNOWLEDGEMENTS

It is evident that this paper is based on research by a number of people over a lengthy period of sustained effort. Most of those involved are included as authors of the quoted references; I apologise to those not recognised in this manner. Appreciation is also due to the various organisations who financed the research work; they are acknowledged in the referenced publications.

REFERENCES

Been, K. & Jefferies, M.G. (1985) "A state parameter for sands". Geotechnique 35, 2, pp 99-112.

Been, K., Crooks, J.H.A., Becker, D.E. & Jefferies, M.G. (1986) "The cone penetration test in sands: Part I, state parameter interpretation". Geotechnique 36, 2, pp 239-249.

Been, K., Jefferies, M.G., Crooks, J.H.A. & Rothenburg, L.(1987a) "The cone penetration test in sands: Part II, general inference of state". Geotechnique 37, 3, pp 285-299.

Been, K., Lingnau, B.E., Crooks, J.H.A. & Leach, B. (1987b) "Cone penetration test calibration for Erksak (Beaufort Sea) sand". Can. Geot. J., 24, 4, pp 601-610.

Been, K., Crooks, J.H.A. & Rothenburg, L. (1988a) "A critical appraisal of CPT calibration chamber tests". Proceedings ISOPT-I, Orlanda, Florida.

Been, K., Crooks, J.H.A. & Jefferies, M.G.(1988b) "Interpretation of material state from the CPT in sands and clays". Proceedings "Penetration Testing in the UK" published by Thomas Telford, London.

Castro,G. & Poulos, S.J. (1977) "Factors affecting liquefaction and cyclic mobility". J. Goetech. Engng. Div., ASCE 103, GT 6, pp501- 505.

Crooks, J.H.A. (1981) "A qualitative stress-strain (time) model for soft clays" ASTM Symposium on Laboratory Shear Strength of Soil, STP 740, Chicago, Illinois.

Crooks, J.H.A., Becker, D.E., Been, K., & Jefferies, M.G. (1988) "CPT interpretation in clays". Proceedings ISOPT-I, Orlanda, Florida.

Jefferies, M.G. (1988) "Verification of Qc - state parameter function in sand strata". Proceedings ISOPT-I, Orlanda, Florida.

Jefferies, M.G., Crooks, J.H.A, Becker, D.E. & Hill, P.R. (1987) "Independence of geostatic stress from overconsolidation in some Beaufort Sea clays". Can. Geot. J., 24, 3, pp342-356.

Ladd, C.C. & Foott, R. (1974) "New design procedures for stability of soft clays". J. Geotech. Engng. Div., ASCE 100, GT7, pp 763- 786.

Roscoe, K.H. & Burland, J.B. (1968) "On the generalized stress strain behaviour of wet clay". Engineering Plasticity, ed. Heyman and Leckie, Cambridge University Press, pp 535-609.

Rothenburg, L. & Bathurst, R.J. (1989) "Analytical study of induced anisotropy in idealized granular materials". Geotechnique 39, 4, pp 601-614.

Developments in Geotechnical Aspects of Embankments, Excavations and Buried Structures, Balasubramaniam et al. (eds)
© 1991 Balkema, Rotterdam. ISBN 90 5410 019 2

Testing of very soft clay

Y. Watari & H. Shinsha
Engineering Research Institute, Penta-Ocean Construction Co., Ltd, Tokyo, Japan

ABSTRACT: Because nearly 90% of its land being mountainous in Japan, the land necessary for industrial and agricultural expansions was built by reclamation and draining of shallow water areas since the ancient ages.

Using the soil dredged from the sea bottom with a pump dredger is the most economical method, and this method has been used in most of the large scale land reclamation projects.

Today, land reclamation is done not to make more land but to dispose the soil generated by dredging to open up navigational channel. In such a project, the volume of the dredged soil poured into the reclamation site becomes larger than the original volume of the sea bottom soil due to mixing of a large amount of sea water with dredged soil. Therefore, we must be able to quantify the volumetric change of soil in order to determine how much space is necessary to dispose the dredged soil.

The authors conducted a series of research work to solve this problem, and this paper describes the tests that were conducted relating to very soft clays and the results of these tests.

Keywords: Reclamation; very soft ground; sedimentation; consolidation; vane shear test

1 INTRODUCTION

In a project in which clayey soil is dredged from the sea bottom using a pump dredger and dumped in a reclamation site, the soil particles in the dredged soil is mixed with a large amount of sea water for transportation to the reclamation site in the dredging pump and pipeline.

At the reclamation site, the soil particles are initially suspended in sea water in the slurry state, but they gradually sink to the bottom and accumulated, eventually forming a ground.

The first question that must be answered is from what state we call the reclaimed ground.

In the early part of the sedimentation process, soil particles settle near the sea bottom and the density of the slurry near the sea bottom increases. At this stage, a large amount of sea water is still present in the voids between soil particles.

The soil particles that reached and become accumulated at the sea bottom will eventually begin to form a ground by drain out the pore water by their own weight.

Next comes the self-weight consolidation process which can be clearly observed.

In this process, from what point can be define the consolidating soil as having formed a ground?

In order to calculate the volumetric change of a reclaimed ground, we must perform a self-weight consolidation calculation. Before we can perform a self-weight consolidation calculation, we must know the void ratio of the soil when consolidation begins. In order to know the void ratio when a ground is formed, we must define the virgin state ground.

At this point, we need a test procedure for obtaining the coefficient of consolidation to be used in the self-weight consolidation calculation.

Next, for us to use a very soft ground, we need data on ground strength. However, it is difficult to accurately obtain the strength of a ground that is close as liquid state using the conventional soil test and exploration procedures.

For the purpose of solving these problems, for those who are in the position to actually perform reclamation work, we have devised a sedimentation test, a consolidation test and a vane shear test.

2 SETTLEMENT AND SEDIMENTATION TESTS

When sea water is added to clayey soil collected from the sea bottom and the mixture is poured into a cylinder, observation of the soil's sedimentation in one-dimensional form will progress over time as shown in Fig. 1.[1]

Specifically, sedimentation of soil particles in the suspension begins from the starting condition (a) in Fig. 1 with the larger grain size particles reaching the bottom first and forming a sedimentation layer.

In this sedimentation layer, separate depositions of the coarse grain particles such as sands and the fine grain soil such as silt, clay can be clearly seen.

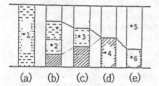

				*5	

*1 Suspension
*2 Deposition surface
*3 Settling surface
*4 Deposited layer
*5 Clear phase
*6 Consolidated layer

(a)　(b)　(c)　(d)　(e)

Fig.1　Settlement and self-weight consolidation process

This phenomenon is clearly predicted by the Stokes' Law.

The suspension will change to two layers, specifically a clear water layer at the top and a suspension layer with increased particle density at the bottom.

The interface of these two layers is called the settling surface, and dropping of the settling surface and rising of the deposition surface eventually eliminates the suspension layer.

In the sedimentation layer during this process, self-weight consolidation is advancing from the lower portion and the height of the sedimentation layer is decreasing. However, new soil particles continue to fall onto the sedimentation layer so that the overall thickness of the sedimentation layer is increasing.

At an actual reclamation site, the conditions (b) and (c) in Fig. 1 exist during the reclamation work. When pouring of dredged soil is completed, the condition shown as (d) in Fig. 1 occurs. Then, self-weight consolidation continues and the condition of the reclamation site reaches that which is shown as (e) in Fig. 1.

2.1 One-dimensional Sedimentation Tests

The one-dimensional sedimentation tests are performed as described below.

The measuring cylinder that is used for settlement test should have a diameter of at least 15 cm. If the diameter is so small compare the height, frictions between soil and the cylinder surface can affect the ground settlement caused by self-weight consolidation, resulting in testing errors.

A slurry sample obtained by mixing the clayey portion of the soil sample collected from the sea bottom with the sea water collected form the same location (dredged site) in such a way that the water content is same as that of the slurry poured from a pump dredger is poured into the measuring cylinder. The sea water from the same area as the soil sample was collected should be used if at all possible. This is because the mineral ions present in sea water affects flocking of soil particles and different type and volume of minerals cause soil particles' settlement speed to be different.

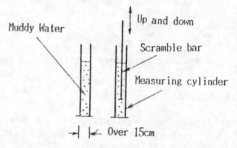

Fig.2 One-dimensional sedimentation test

Next, many holes are punched at the circular disc whose diameter is just a shade smaller than the inner diameter of the measuring cylinder and a rod is attached to the disc. Then, this scramble bar is inserted into the measuring cylinder and moved up and down to stir the suspension. The suspension that is poured into measuring cylinder will begin to flock immediately, and this flocking must be held to minimum until the test begins.

The data to be collected from the one-dimensional sedimentation test are the heights of settling surface and deposition surface at various elapsed times.

The one-dimensional sedimentation test can be done in two ways, specifically by repeating the test several times by pouring different amounts of suspension into the measuring cylinder and by dividing the total amount of suspension to be poured into several batches and pouring each batch at a certain time interval. The latter method is called the multi-step sedimentation test. (See Fig. 5 for illustration of the sedimentation test methods.)

2.2 Results of One-dimensional Sedimentation Test and Their Use

Fig. 3 shows sample results from one-dimensional sedimentation test. H_0 is the initial slurry height, t_0 is the starting time of self-weight consolidation which is defined as the time at which the top of the sedimentation layer (which rises as the soil particles in the slurry settles) is at the same height as the boundary of suspension and clear water (which drops as the soil particles in the slurry settles), and t_{100} is the self-weight consolidation ending time which is defined as the point where the slope of settlement curve changes abruptly (see Fig. 3). The settlement that occurs beyond t_{100} is considered to be a creep process.[2]

The height of the soil sample at time t_0 is defined as H_{t0}, and the height of the soil sample at time t_{100} is defined as H_t100. The phenomenon that causes the soil samples height to change from H_{t0} to H_t100 as shown in Fig. 3 occurs only when the soil particles in the suspension are in coagulating free settlement. When the suspension's water content falls to 300% and below, such phenomenon no longer occurs because the soil particles interfere with each other and they can no longer be in

free settlement condition. In general, when reclaiming a land with soil from the sea bottom that has been dredged by a pump dredger, the water content of slurry is in the 1,000% to 2,000% range, so that the points H_{t0} and H_t100 can be found.

Fig. 4 is a plotting of the data from a series of one-dimensional sedimentation tests with different slurry pouring heights. If, on this curve, H_{t0} is A and H_t100 is B, then points A_1, A_2 and A_3 are on a straight line and points B_1, B_2 and B_3 are also on a straight line. Therefore, the self-weight consolidation starting and ending times for any given depth of reclamation using dredged sea bottom soil can be estimated by performing one-dimensional sedimentation test with three or more different slurry pouring heights.

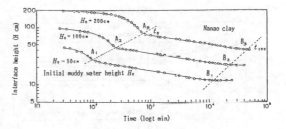

Fig.4 Sedimentation and consolidation curve of plural step test

At an actual reclamation site, slurry is not poured all at once. In fact, slurry pouring continues for several months or, in some cases, several years. Because self-weight consolidation advances on the clay ground that is formed early, the estimates obtained by one-dimensional sedimentation testing do not match the condition of the actual reclamation site exactly. For this reason, multi-step sedimentation test is also performed. In a multi-step sedimentation test, suspension is poured as in one-dimensional sedimentation test but in several batches. Fig. 5 illustrates the multi-step sedimentation test procedure.

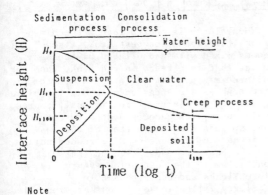

Note

*t_0 : Starting time of consolidation sedimentation
*t_{100} : Ending time of self-weigh consolidation
*H_{t0} : Interface height when t_0
*Ht_{100}: Interface height when t_{100}

Fig.3 State of sedimentation and consolidation process

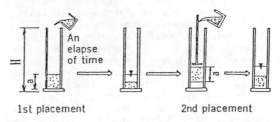

1st placement   2nd placement

Fig.5 Placing method of plural step test

Fig. 6 shows the results of pouring, at different speeds, dredged soils in the suspension condition at total pouring height of 100 cm with the measurement cylinder's maximum pouring height (which corresponds to an actual reclamation site's revetment crest height) assumed to be 40 cm.[3] These curves show that, under the stated conditions, slurry pouring speed must be no more than 0.42 cm/hr to prevent overflowing of the dredged soils at the actual reclamation site. This means that the speed of slurry pouring and the amount of soil that can be poured into a reclamation site are closely interrelated, and the amount of dredged soil that can be poured into a reclamation site without causing overflow can be estimated by performing multi-step one-dimensional sedimentation test.

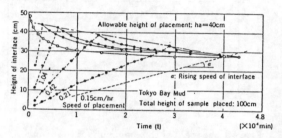

Fig.6 Family of sedimentation curves with different speed of placement

2.3 Two-dimensional Sedimentation Tests

As stated in an earlier section, sedimentation of soil particles in the slurry that has been poured into a reclamation site have different settlement speeds depending on their grain sizes, and this causes separate sedimentation layers. This phenomenon can be more clearly understood when a two-dimensional sedimentation test is performed.[4]

Settlement and sedimentation at the actual reclamation site are a three-dimensional process. However, because the flow of poured slurry has a certain width and the direction of flow change is two-dimensional, three-dimensional sedimentation test does not have a reproducibility. Therefore, we built a two-dimensional water tank as shown in Fig. 7 which has a slurry flow width that has been estimated from measurements that were taken at an actual reclamation site. The dimensions of the water tank are 30 cm (width), 400 cm (length) and 65 cm (depth). Sea water was poured into this tank at depth of 50 cm to simulate the condition of the reclamation

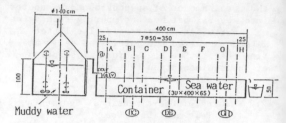

Fig.7 Two dimensional sedimentation test apparatus

site before pouring of slurry begins.

On one end of the water tank is installed a spill way, and slurry that has been adjusted to the same initial water content as in the actual reclamation site is poured. The amount of slurry pouring is decided based on the pour load rate which is defined as the ratio of the pouring amount and the pouring surface area at the actual reclamation site. Table 1 summarizes the test conditions for a number of cases tested.

Tab.1 A sample of test condition

Case	Sample	W_o [%]	d [mm]	q [ml/min]	U [m/sec]	$\square$ [min]	Pouring condition
S−I	Clay	1260	3	150	0.35	720	Continuous
S−II	Clay :6 +Sand :4	1000	4	120	0.16	720	″
L−I	Clay	1000	4	250	0.33	240	″
L−II	Clay	1000	6	500	0.29	150	″
L−III	Clay	1000	8	1000	0.33	120	″
L−IV	Clay :6 +Sand :4	1000	6	712	0.42	1320	″
L−V	Clay :6 +Sand :4	1000	6	700	0.42	1380	Separate

When slurry is poured into a two dimensional sedimentation test tank, it becomes mixed with and diluted by the sea water that is already in the tank, and the diluted mixture settles near the bottom of the tank as a suspension with high specific gravity. As slurry is continuously poured, this suspension becomes a density current and spreads over the bottom of the tank. Thickness of suspension differs depending on the pouring load rate, but the soil particles in the suspension eventually form flocks and begin to settle.

Slurry collection equipments are installed at points BC, DE and GH in Fig. 7. Also, scales are installed at points A through H to measure the heights of the slurry suspension surface and the soil particle deposition surface.

Excess water that is drained from the spill way is collected and its turbidity is measured.

At the same time as pouring of slurry begins, water content of the samples that are collected by the slurry collection equipment are measured. The height of the suspension surface and the height of the soil particle deposition surface are also measured.

Even after pouring of slurry is finished, settlement of the soil particle deposition surface is measured until self-weight consolidation ends. When self-weight consolidation ends, the deposited soil samples are collected from a number of depth levels and measurements are taken to obtain water content and grain size distributions.

2.4 Results of Two-dimensional Sedimentation Test and Their Use

Fig. 8 shows the water content distribution curves at points BC, DE and GH of the two-dimensional settlement and sedimentation test tank for the slurry that is in the process of settlement and the clayey soil that has deposited at the bottom. These curves show that the water contents are in the approximately 20,000% or higher range which is very high. From the observations made, settlement of the dredged soils can be classified as coagulating free settlement.

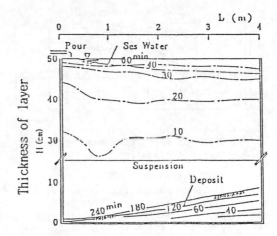

Fig.9 Sedimentation and deposition state of cohesive soil

sedimentation. Sedimentation surface near the slurry inlet is higher than at the outlet because the tank is relatively short.

If the same test is performed using a slurry sample which is a mixture of clayey soil and sandy soil, coarse grain soil particles become completely separated from fine grain soil particles and they are deposited near the slurry inlet. (See Fig. 10)

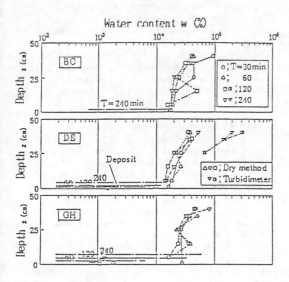

Fig.8 Distribution of water content in deposition container

Fig. 9 shows how clayey soils become deposited. The dashed line shows the rise of the slurry surface as slurry is poured, and the solid line shows the progress of

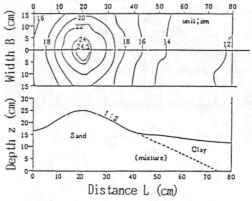

Fig.10 Deposition state of coarse grain soil

Because of this phenomenon, deposited soil samples were collected after pouring test is finished at points that are at different distances from the slurry inlet to test the soil characterists. Water content distribution is shown in Fig. 11, and grain size distribution is shown in Fig. 12.

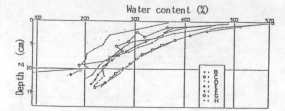

Fig.11 Percentage of water content
distribution of deposition soil after
self-weight consolidation

Fig. 11 shows that water content rises
as the distance from the slurry inlet
increases, and Fig. 12 shows that larger
numbers of finer grain soil particles are
deposited at points that are farther away
from the slurry inlet.

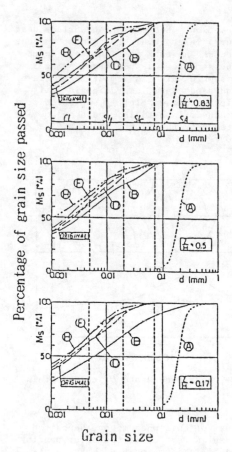

Fig.12 Grain size distribution curve of
deposited soil

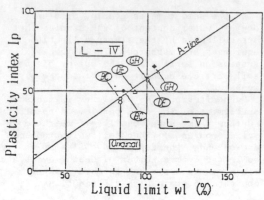

Fig.13 Consistency characteristics

Fig. 13 shows the consistency characteristics at various points, and it clearly
shows that the soils poured into a reclamation site become separated and the soils
characteristics are not same at   different points.

The one-dimensional and two-dimensional
sedimentation tests only enable measurement of the relationship between void ratio
e and load p in the self-weight consolidation range. In short, these tests provide
the e-p relation only in a range of extremely small stress.

In order to calculate the amount of
settlement due to consolidation on an
actual ground, tests must be conducted for
the stress levels up to the standard consolidation test. Therefore, the seepage
consolidation test was developed to fill
the e-p curve between the sedimentation
test and the standard consolidation test.

3 SEEPAGE CONSOLIDATION TEST

The seepage consolidation test is designed
to obtain the consolidation time and the
settlement amount on high water content
fine grain soil in the region where the
stress loaded on soil is extremely small by
passing pressurized water through the soil
sample in the axial direction of the cylindrical container.

The seepage consolidation test apparatus
consists of a consolidation test cylinder,
an osmotic pressure loading equipment and
a device that measures the amount of water
that is transmitted through the soil
sample.[5]

Fig. 14 shows the makeup of the consolidation test cylinder. The osmotic pressure
that acts on the soil sample is loaded by
creating a pressure difference between
the pressurized water that is injected from

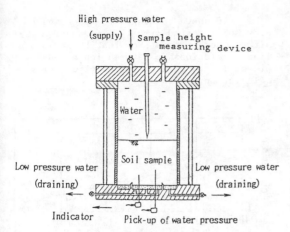

High pressure water

(supply)

Sample height measuring device

Water

Soil sample

Low pressure water
(draining)

Low pressure water
(draining)

Indicator   Pick-up of water pressure

Fig.14   Seepage consolidation test cylinder

the supply inlet at the top and the pressurized water that is drained from the
drain port at the bottom. In other words,
the osmotic pressure that flows through
the soil sample acts as the consolidation
load, so that the consolidation stress can
be measured by measuring the pore water
pressure by inserting needle shaped pressure pickup into the soil samples from the
bottom of the consolidation test cylinder.
At the same time, the amount of settlement
by consolidation is measured using a sample
height measuring device at the upper portion of the test cylinder.

When consolidation test is finished, the
entire sample is cut into a number of thin
layers, and the water content levels are
measured continuously on the samples collected from these layers.

The soil sample is prepared by directly
pouring slurry into the test cylinder, as
was done in the one-dimensional sedimentation test, and letting the soil particles
settle and deposit.

This test produces a consolidation coefficient (c_v) curve which is calculated for
each pair of permeability coefficient k and
volumetric compressibility coefficient m_v.

It must be noted that correct water content figures cannot be obtained for high
water content clay soils by using the furnace drying method of measurement because
salt and other substances dissolved in sea
water become crystallized and measured as
part of solid. Therefore, the result obtained must be corrected using the following formula, where ω_m is the water content
obtained by furnace drying method and β is
the ratio of the masses of salt and pure
water in the sea water that is passed
through the consolidation test cylinder:

$$\omega = \frac{1 + \beta}{1 - \beta \cdot \omega_m} \cdot \omega_m$$

ω_m; Water content by heat drying
β ; Proportion of quality between salt
    water and plain water

4 THE e-p CURVE IN THE LOW STRESS PART

Fig. 15 shows the e-log p curves obtained
by performing settlement and sedimentation
test, seepage consolidation test and standard consolidation test on same soil sample.

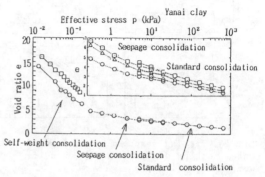

Fig.15   e-log p curve

In Fig. 15, the low consolidation stress
part was obtained by two-dimensional sedimentation test, the medium value part was
obtained by seepage consolidation test, and
the high value part was obtained by standard consolidation test, and the three
curves thus obtained can be connected nearly without a break. This means that these
three tests can provide the consolidation
characteristics of very soft ground in a
wide broad range of stress levels from low
level to the normal level.

To perform an actual consolidation calculation, the stress at the start of consolidation and the stress at the end of
consolidation must be known. In the normal
ground stress condition, these stress
values are known. In the sedimentation
condition, the stress at the time soil
particles settle through slurry and reach
the deposition surface is thought to be
the consolidation starting time stress.
The void ratio at this point defined the
limit void ratio $e_i{}^*$, and its value was
obtained by the following method:[6]

(1) It was assumed that e and log p have
a linear relationship, and a linear equation that approximates the data from consolidation test was obtained.

(2) Next, it was assumed that the soil

initially has uniform e_i^* value, and the distribution of water contents at the end of self-consolidation process was obtained by varying the value of e_i^*.

(3) The calculated and actually measured water content distribution curves were compared to obtain e_i and e_i^* that found the best matching.

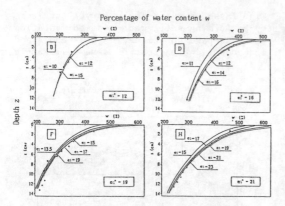

Fig.16 Fitting state of percentage of water content distribution after self-weight consolidation

Fig. 16 compares the calculated and actually measured water content distribution curves. Because of large fluctuation of the actual measurement values, the curves could not be fitted very well, but we selected the value shown in Fig. 17 for e_i^*.

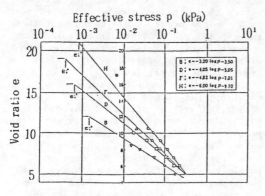

Fig.17 e_i^*-log p relation diagram

Fig. 17 shows the e_i^* value determined from Fig. 16 on the e-log p curves. It appears that a void ratio that corresponds roughly to $p = 10^{-3}$ kPa should be used as e_i^*.

Fig. 18 shows the log e-log p curves on clay soils tested from two reclamation sites. Both curves show that there is a linear relationship between log e and log p, and it can be used in consolidation calculation for this ground.

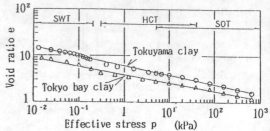

SWT;Self-weight consolidation test
HCT;Seepage consolidation test
SOT;Standard consolidation test

Fig.18 Log e-log p curve at field

5 CALCULATION TO ESTIMATE RECLAMATION VOLUME

When clayey soil sedimentation at the sea bottom are dredged using a pump dredger and poured into a reclamation site, settlement, sedimentation and self-weight consolidation occur simultaneously. The extent of these three processes differ depending on the amount of slurry that is poured and the elapsed time, but this type of land reclamation using dredged sea bottom soils require a procedure that takes into consideration the amount of clay soil sedimentation and the amount of settlement due to self-weight consolidation. Mikasa has pointed out that consolidation analysis on very soft clay ground must take the following items into consideration, and proposed a consolidation formula that meet this requirement:

(1) Effect of self-weight;
(2) Changing of layer thickness;
(3) Non-linearity of stress and strain, and stress dependence of permeability coefficient.

On the other hand, the rapid advancement of computer technology in the recent years have made it possible to easily solve the consolidation problem using an FEM analysis technique. Therefore, we have conducted a one-dimensional consolidation analysis on clay soil using FEM analysis of elastic method as the basis and taking into consideration the items listed above.

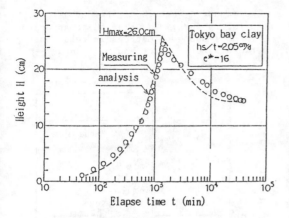

Fig.19 Comparison between measuring and analysis of deposition

Fig. 19 compares predicted increase in deposition soil height while slurry is being poured and subsequent settlement process to the values obtained by the tests that produced the data for Fig. 8 and Fig. 9. There are good matches where the deposition height is increasing as well as where settlement is occurring after slurry pouring is finished.

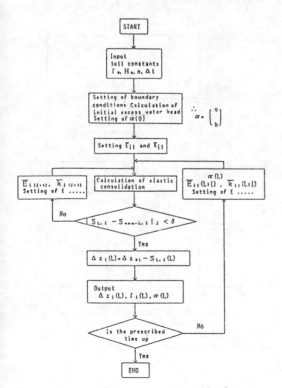

Fig.20 Flow diagram of consolidation analysis

Therefore, the analysis used here is adequate for analyzing the actual reclamation site. Fig. 20 shows the flow of this analytical calculation.

To predict the amount of soil deposition that occurs at an actual reclamation site, we must classify and predict the characteristics of the soils that become deposited at various sections and calculate the appropriate amount of consolidation that occurs for each soil type. In our experiment, this soil classification was made using the coefficient of plasticity (I_p), and a table of soil deposition heights at different deposition speeds for different I_p values. Then, this table was used to predict the deposition height at the reclamation site. Fig. 21 shows the predicted and actually measured values at the field. The amount of slurry that is poured

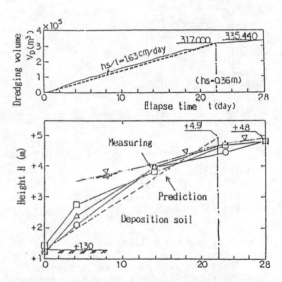

Fig.21 Prediction and measuring height of deposition

at an actual reclamation site changes daily, and the deposition height that is predicted based on the average daily pouring rate does not necessarily match the actual deposition height, but the method described above appears more than adequate to be of use.

6 VANE SHEAR STRENGTH OF VERY SOFT CLAY

Ground strength is one of the most important data, along with consolidation data, for geotechnical engineers who work with soft grounds.

On a very soft clay ground, the entire

stages from   where the ground is in a near liquid condition through that were the ground has begun to exhibit supporting force must be handled. However, there are a number of difficult problems in dealing with these conditions by the soil survey and test techniques available today. Therefore, we developed laboratory use and field use vane shear testing apparatus for application on very soft grounds.

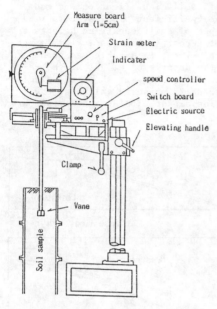

Fig.22 Vane shear test apparatus in laboratory

Fig. 22 shows the laboratory use vane shear test apparatus, and Fig. 23 shows the field use vane shear test apparatus. The laboratory use vane shear test apparatus can test a wide range of soils from very soft clay soil to the generally found alluvial clay soil by changing the range of measurement strength for different soil strengths. The field use vane shear test apparatus is designed to measure the vane shear resistance directly above the vane to avoid the weight of the boring rod affecting the measurement value.

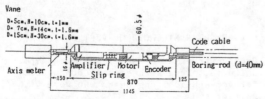

Fig.23 Vane shear test apparatus at field

We measured vane shear strengths in laboratory and field using these apparatus and found good matches in relationship between water content and vane shear strength as shown in Fig. 24.

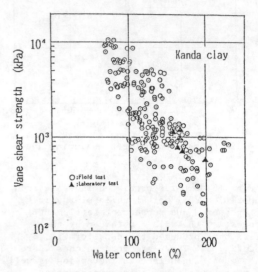

Fig.24 Comparison of vane shear strength between laboratory and field

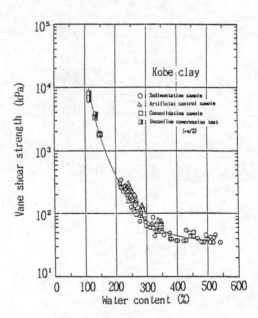

Fig.25 Vane shear strength in variable laboratory sample

Fig. 25 shows the results of vane shear testing in laboratory using differently premaking method of soil samples. The

64

results match closely, indicating that the relationship between vane shear strength and water content for high water content soils is not affected by how the samples are prepared. It also shows that vane shear strength approaches a constant when water content rises over 300%.

7 CONCLUSION

The following conclusion can be made on the tests that must be performed to investigate a ground that has been built by filling the sea with clayey soils dredged from the sea bottom using a pump dredger:

(1) From settlement and sedimentation test
- Settlement of dredged clay soil particles is the coagulating free settlement type.
- Dredged clay soil particles become deposited in separate groups. Therefore, soil sample that has been sorted to nearly uniform grain size by pre-settlement, etc., must be used in a one-dimensional sedimetation test.
- One-dimensional sedimentation test enables easy understanding of geotechnical phenomena in reclamation site.
- Two-dimensional sedimentation test can closely duplicate the settlement and sedimentation processes that take place at reclamation site.

(2) From seepage consolidation test
- Seepage consolidation test provides the various coefficient relating to the consolidation relationship between the low stress part of self-consolidation condition and the standard consolidation test part.

(3) From reclamation volume estimation
- Continuous e-log p curve from low to high stress region obtained by the reclamation volume estimating calculation enables analysis of consolidation to calculate self-weight consolidation that occurs when load is applied continuously and thereby calculating the volume of soil necessary to reclaim a given area.

(4) From vane shear test
- The vane shear strength values obtained by the newly developed laboratory use vane shear test apparatus and the field use vane shear test apparatus on soil samples with identical water content values are in good match. Nearly identical vane shear strength values can be obtained on soil samples reproduced in laboratory as long as the water content is same regardless of how the soil samples are prepared.

REFERENCES

1) Yano, K; Imai, G; Tsuruya, K.: Settlement and self-weight consolidation of soil dredged by pump dredger, Penta-Ocean Construction Co., Ltd. Annual Report of Engineering Research Institute, Vol. 6. pp. 1-5, 1977 (in Japanese)
2) Yano, K; Tsuruya, K; Sedimentation and deposition of suspended very fine soil particles. The 17th Japan national conference on soil mechanics and foundation engineering pp. 185-188. 1982 (in Japanese)
3) Yano, K; Imai, G; Tsuruya, K: The study on volume change of reclamated cohesive soil, Penta-Ocean Construction Co., Ltd. Annual Report of Engineering Research Institute. Vol. 8. pp. 65-68, 1979 (in Japanese)
4) Watari, Y; Shinsha, H.; Hayashi, K.; Aboshi, H.: Settlement and separate deposition of dredged cohesive soil; Research report of Engineering Faculty, Hiroshima University, Vol. 34. No. 2 pp. 175-187, 1986 (in Japanese)
5) Watari, Y.; Yano, K.; Tsuruya, K.: The seepage consolidation test method of extremely soft fine soil, Penta Ocean Construction Co., Ltd. Annual Report of Engineering Research Institute. Vol. 10. pp. 109-113, 1981 (in Japanese)
6) Watari, Y.; Shinsha, H.; Hayashi, K., Aboshi, H.: Dredge soil volume pre-estimation based on self-weight consolidation analysis of cohesive reclamation soil, Research report of Engineering Faculty, Hiroshima University. Vol. 34, No.2, pp. 159-173, 1986 (in Japanese)

Developments in Geotechnical Aspects of Embankments, Excavations and Buried Structures, Balasubramaniam et al. (eds)
© 1991 Balkema, Rotterdam. ISBN 90 5410 019 2

Full scale laboratory and field test for development of gravel drains system

Y. Okita & K. Ito
Konoike Construction Co., Ltd, Osaka, Japan

ABSTRACT: This paper deals with various full scale laboratory and field tests which are carried out for development and improvement of gravel drains system. Although these tests are expensive, they are indispensable to develop the new improvement technique. And also they bring extra merit such as quality and reliability on the new method.

1 INTRODUCTION

Since Niigata earthquake in 1964 as a turning point, it has been widely recognized that the ground liquefaction during earthquakes is caused by the pore water pressure of loose sand deposits. Many countermeasures against liquefaction have been proposed such as sand compaction piles, vibro-floatation and so on. Although the representational measure of them is the sand compaction piles, its applications are often subject to restriction as it spreads too much vibration and noise hazards with acompanying ground displacement.

Then the concept of gravel drains was presented by Seed in 1977. The measure is founded on the new principle, which prevents the liquefaction with quick dissipation of pore water pressure during earthquakes in the ground. As its execution is not accompanied with vibration or noise, it can be applied in urban area and still near by existing structures. Co-operating with NKK Corporation, we have developed an exclusive gravel drain pile driver with the flat tip tamping rod at the center of casing auger in 1982, and has been driven more than one million meters of the gravel drain pile lengths in Japan for recent eight years. During this period, we performed various laboratory and field tests on a full scale, in order to authorize the design method, select the gravel drain materials, verify the effect of preventing liquefaction, and improve the execution machine.

In this paper the concept and aspect of gravel drain are introduced. Also, the various full scale laboratory and field tests for development and improvement of gravel drains system are reported.

2 CONCEPT AND ASPECT OF GRAVEL DRAINS

In 1977, Seed proposed a new countermeasure to prevent the liquefaction which is to install a system of gravel drains in loose sandy ground, as shown in Fig.1, so that pore water pressure generated by cyclic loadings of an earthquake may be dissipated almost as fast as they are generated.

The basic equation of the one-dimensional theory for dispersion of the excess pore water pressure is shown below.

$$\frac{k_h}{m_v \gamma_w}\left(\frac{\partial^2 U}{\partial r^2} + \frac{1}{r}\frac{\partial U}{\partial r}\right) + \frac{k_v}{m_v \gamma_w}\frac{\partial^2 U}{\partial z^2} = \frac{\partial U}{\partial t} - \frac{\partial U_g}{\partial t}$$

Where, m_v : coefficient of volume compressibility, γ_w: unit volumetric weight of water, k_h : coefficient of permeability in the horizontal direction, U: excess pore water pressure in sand layer, r: radius, t: time, z: vertical coordinate and $\partial U_g/\partial t$: rate of production of excess pore pressure by repeated shearing in non-improved ground.

In 1978, a new countermeasure against liquefaction was taken in the area behind the east shore wall of the Ohgishima reclaimed island. Then, the gravel drain piles were installed by Benoto type boring or doughnut auger. Then in 1982, we developed the exclusive gravel drain pile driver with flat tip tamping rod at the center of casing auger, because the first developed machine has low execution speed and high cost.

Since then, the gravel pile driver

equiped with flat tip tamping rod installed more than one million meters of the gravel drain pile lengths. In 1989, a fullscale verification test for improvement of compaction effect was carried out again. This test brought us an effective compaction method which tends to stabilize ground with cone-shaped tip tamping rod. N-value of liquefable ground is increased 50% by using the cone-shaped tip rod. This result was certified in a field test while noise and vibration remained as they were.

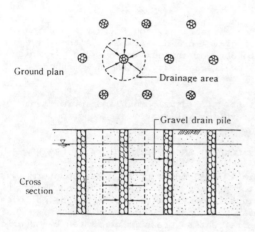

Fig.1 Arrangement of Grave Drain Piles

3 FIRST FIELD TEST FOR VERIFYING PERFORMANCE OF GRAVEL DRAIN AND PILE DRIVER

The vibration source of the performance test is four vibration rods used for improving a loose sand ground. According to the performance tests (Fig.2), the gravel drains system and the pile driver have the following feature and effects.

3.1 Effect of Gravel Drain
(Generation and Dissipation of pore-water pressure)

Fig.3 shows the test result of the generation and dissipation of pore-water pressure, where u/c represented as pore-pressure ratio (u: excess pore water pressure   c: effective confining pressure). It means the liquefaction state when the pore pressure ratio becomes 1.0. No difference in pore-pressure generation between improved ground and original ground exists on the stage of generation, because of the high power level of vibration. But the difference exists on the stage of dissipation. The quick dispersion of pore-water pressure means maintaining the ground in a safe condition and minimizing the damage due to liquefaction. In actual condition, there exists the ground densification effect of the

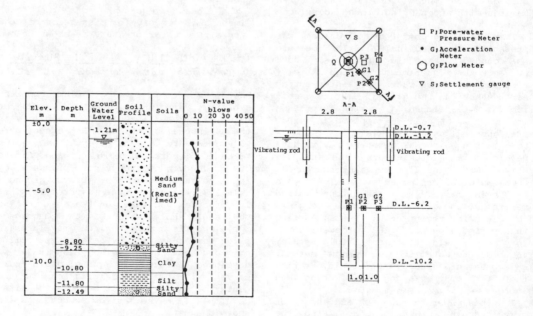

Fig.2 Performance Test of Gravel Drain and its Pile Driver

gravel drain installed with the tamping rod, so the perfectly prevention from ground liquefaction is expectable.

3.2 Low Vibration and Low Noise Levels in Operation

Fig.4 shows damping characteristics of vibration according to distance, of the gravel drains methods in comparison with vibrocompaction methods like the sand compaction method and others. The source

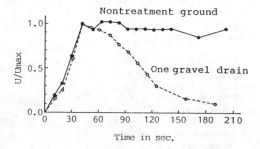

Fig.3 Generation and Dissipation of Pore-water Pressure Ratio

of vibration in the sand compaction method is the vibration machine itself, and the vibration level is not so low even when a countermeasure like adoption of high frequency vibro is taken. Contrarily, the vibration level of the gravel drains method is similar to background vibration level as no vibrators are used, less than 75dB the value prescribed in the Construction Work Control Standard of Japan. Also, the noise level of the gravel drains method is low as shown in Fig. 5, less than 85dB which is the noise control value in town areas.

3.3 Effect of Preventing Ground Displacement during Construction

Fig.6 shows the results of underground displacement measured with inclinometer when gravel drains were compacted by (a) Benoto method, (b) compressed air method (casing auger method which does not use compaction rod), and (c) special execution machine, in sandy soil with N-value of ranging from 7 to 12. The maximum horizontal displacement was 6 cm by Benoto Method, 3 cm by compressed air method, and

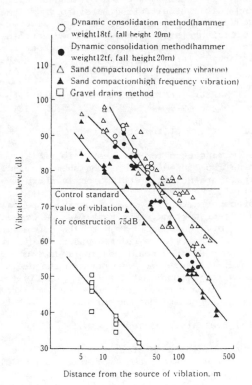

Fig.4 Damping Characteristics of Vibration according to Distance

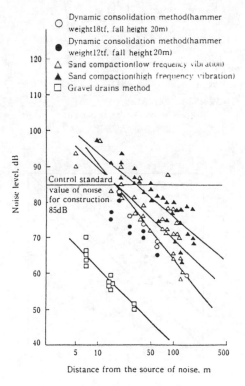

Fig.5 Damping Characteristics of Noise according to Distance

less than 4 mm by the special execution machine which is almost negligible. Thus, it was confirmed that the use of compaction rod offers greater effect on preventing ground displacement.

3.4 Increase in N-value after the Execution

From a design point of view, this method was expected to disperse excess pore pressure quickly in sandy soil, but according to the study results of the ground where gravel drain piles were actually driven, comparison of N-value before and after execution indicates an increase in the N-value as shown in Fig.7. The figure means that compaction affect ground can be convinced. However, since the N-value shown is based on soil type in the Niigata area, the proportion of fine grained soil is low. In the future

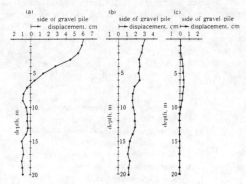

Fig.6 Measurement Results of
Underground Displacements

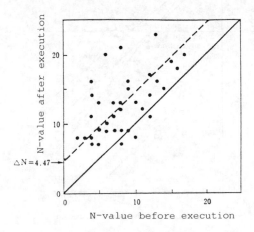

Fig.7 Change of N-value
Before and After Execution

it will be necessary to accumulate data in soils with a higher proportion of fine grains, to study quantitatively the relationship between the increase in N-value and particle size.

4 LABORATORY TEST FOR DESIGNING PARAMETER

In designing gravel drains, diameter, spacing and material size (particle size), they are determined so as to keep the value of the excess pore water pressure generated by an earthquake below the value needed to maintain the stability of the structure, by using the above equation. In this case, an evalution of the following items are important:

(1) Evaluating the generated characteristics of the excess pore water Ug in non-improved ground during earthquake.
(2) Coefficients of permeability and volume compressibility of the sandy ground.
(3) Selection of drain materials which are characterized with a high permeability and free-choke.

Several attempts were carried out for item (1) which is studied by Seed et al., and Civil Engineering Laboratory of the Construction Ministry of Japan. However, sufficient studies have not been yet made for items (2) and(3), although these are important factors in design.

4.1 Measurement of Volume Compressibility of Sandy Soil

(1) Outline of the test

Specimens used for the test were reclaimed sand collected from the Ohgishima island. Their grain size distribution is shown in Fig 8. A cyclic triaxial apparatus of electro-pneumatic stress control type was used for testing. The test was performed by applying repeated load to the specimen, isotropically consolidated under $1.0 kgf/cm^2$ in an undrained condition to produce porewater pressure and then opening the drain cock to reconsolidate, and the coefficient of volume compressibility m_v was obtained from the consolidation curves.

(2) Evaluation of Test Results

With definitions of static m_v for the coefficient calculated from the tangential inclination of the virgin consolidation curve, dynamic m_v for the coefficient obtained from the tangential inclination of the re-consolidation curve and the average dynamic $\overline{m_v}$ for that obtained from the linear inclination between starting and ending points of the examination of the relation between U/C for pore water pressure ratio (u : excess pore water pressure, c: effective confining pressure) and m_v for the coefficient of volume compressibility is shown in Fig. 9. It is known from Fig. 3 that the range of test result is constant within the value of U/σc <0.5 regardless of the difference in the definition of m_v.

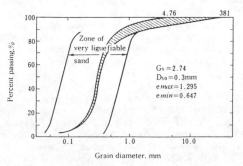

Fig.8 Grain Size Distribution Curve

4.2 Permeability Test of Gravels

(1) Outline of the Test

It is necessary for gravels to be used as drain materials because of their comparatively high permeability in respect to in-situ soil. Since no test methods have been established yet regarding the evaluation of coefficient of permeability of gravels and their characteristics. We carried out large scale constant head permeability tests. The test apparatus is shown in Fig. 10. The selected gravels were a single grade gravel for road construction (JIS A 5001).

(2) Results and Discussion

The relationship between the coefficient of permeability of gravels and hydraulic gradient is shown in Fig.11. It can be seen that the coefficient of permeability decreases with the increase in hydraulic gradient. According to the result of field vibration tests, it is known that when the

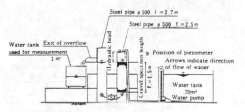

Fig.10 Apparatus of Constant Head Permeability Test

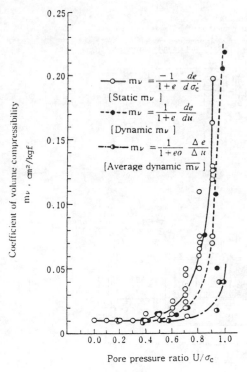

Fig.9 Coefficient of Compressibility versus Pore Pressure Ratio Relationship

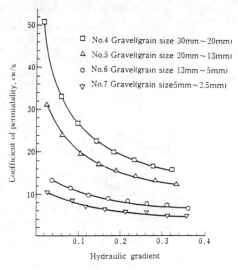

Fig.11 Relationship between Permeability and Hydraulic Gradient

soil surrounding the drains liquefies, a hydraulic gradient of approx. 0.3 is produced in the gravel drains. In the case where soil liquefaction is prevented by gravel drain piles, the hydraulic gradient in the drain is generally estimated to be less than 0.3. Based on the above results, it is clear that, although the design coefficient value of permeability of gravels varies with particle size, a value of 5 to 15 cm/sec corresponding to a hydraulic gradient of 0.3 is considered as the adequate general rule.

4.3 Filter Test of Gravels

(1) Outline of the test

Fig.12 shows detail and dimensions of the filter test apparatus of gravel. The density under water with a compacting rod, and base sands are filled uniformly in a saturated condition. Then, the effective confining pressure is given by perforated plates, and under constant head water flowed through the base sand and gravel filter. Filter zone is observed through Acrylic wall and the seepage velocity is measured with a flow meter. At the same time, the density distribution of specimens is measured with γ-Ray density meter(Fig. 13). This apparatus consists of a γ-Ray source and counter have 10mm silt of lead plate, so γ-Ray transparent zone the specimen. The Decreased γ-Ray from the mass of the specimen is counted by scintillator, then it is converted into the density distribution.

(2) Test results

The density distributions of the specimen in the various conditions are shown in Fig 14,15,16 and 17. The base sand of the test is silica sand No.6 for ceramics of JIS. As the filter zone is mixed with sand and gravel, its density becomes higher than the other part.

The test results show that the clogging state depends on the effective grain size ratio D_{g15}/D_{s85}, where D refers to the grain size, and the subscript (15,85) denotes the percent which is smaller. Fig.14 shows that there is no base loss when D_{g15}/D_{s85} is smaller than 7. As $7 \leq D_{g15}/D_{s85} < 9$ the base material flows into gravel, which forms the stable filter zone

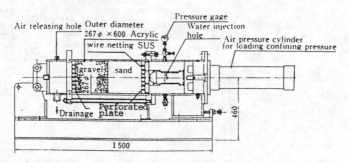

Fig.12 Filter Test Apparatus

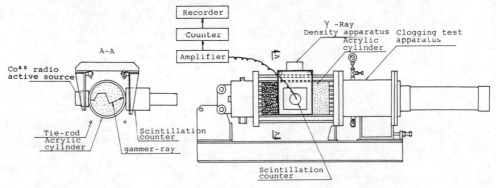

Fig.13 Measuring the Density Distribution with γ-Ray Density Meter

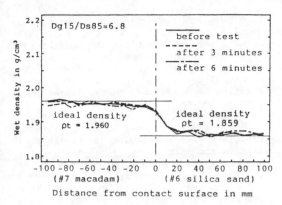

Fig.14 No Clogging State of Filter
(Dg15/Ds85 < 7)

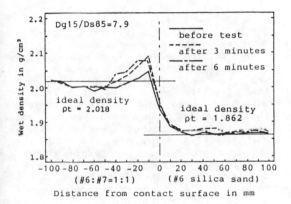

Fig.15 Formation of Stable Filter Zone
(7 ≤ Dg15/Ds85 < 9)

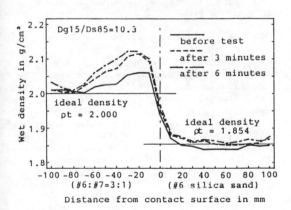

Fig.16 Progress of Unstable Filter Zone
(9 ≤ Dg15/Ds85)

(Fig.15). This zone keeps the permeability constant. Beyond the clogging criterion (9 ≤ Dg15/Ds85) the filter zone grows up as shown in Fig.16, while the permeability

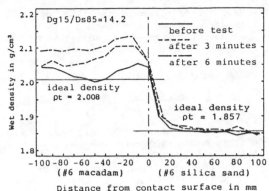

Fig.17 Clogging State of Filter
(9 ≤ Dg15/Ds85)

decreases gradually. In the end, the base permeability drops extremely resultantly (Fig.17). As a result , it has become clear that the criterion of clogging with uniformly graded soil can be obtained by the following condition:

Dg15/Ds85 < 9

4.4 Selection of Standard Conditions for Drain Materials

To enhance the draining effect of the gravel drains, drain materials of larger particle size should be selected. It must be noticed that, if the particle size is much larger than the particle size of in-situ soil, sand may cause choking, clogging and then lowering the permeability. Materials of smaller particle size, however, may be good for preventing choking but cannot be used as a drain materials because their permeability becomes inadequate. Therefore, the selection of a particle size standard is needed to select gravels as drain materials having filter effect without clogging. Conventionally, Terzaghi's standard has been generally employed as a selection standard for filter materials, which is described in the Road and Earth Work Guidance or others. However, since this standard contains excessive safety factors, we pursued an accurate boundary of clogging, using the above test apparatus. Further, if a value equivalent to more than 400 times the coefficient of permeability of ground as the limit of drain materials, the equation becomes as below, in which the coefficient of permeability is proportional to square of 15% particle diameter.

Dg15/Ds15 < 20

73

where, Ds15 represents 15% particle diameter of sand.

As a result, the following equation has been obtained as the particle size selection standard for gravels used for drains:

$$20Ds15 < Dg15 < 9Ds85$$

5. FULL SCALE LABORATORY TEST AND FIELD TEST FOR IMPROVING

5.1 Full Scale Gravel Drain Pile Driving Model for Improving the method of Gravel Drain

In order to improve the compaction effect, much effort and experience were necessary for improving the machine in field, because of the nonuniformity of the test ground. The exclusive facilities of the same scale as prototype gravel drain pile driver as well as the ground for improving machine.

Fig.18 shows the general view of the exclusive facilities used for taking data to improve the gravel drain machine. The facilities consist of a tower crane for lift, a compacting device upon the platform, a soil vessel for testing container and instrumentations for measurement. The tamping device upon the elevation platform was diverted the real execution machine for gravel drain, and the soil vessel was a pile of 6 rings of 2 meters in diameter and 1.8 meter in height. The casing tube was suspended from the platform and raised up together with the platform. The improved ground strength was measured by the portable hydraulic cone penetrometer. The soil vessel was covered with the lid to give a confining pressure by hydraulic cylinder. The sample is the fine sand of Mt.Sengen in Chiba Prefecture while a single grade No.7 gravel of JIS for road produced for drain.

5.2 Factors Affecting Compaction

From preliminary test it is clear that the main factors affecting compaction is the tip shape of tamping rod, diameter of tamping rod and the stroke of tamping. Among the factors affecting compaction, the main test was especially focused both on the tip shape and the rate of tamping. Table 1 shows the test cases to improve the gravel drain pile driver. The stroke of tamping is 200mm , the same as existing execution, and the diameter of tamping rod is 150mm.

Fig.19 shows the result of test B as a typical example. The cone penetration resistance was measured at intervals of 10 cm from the surface of gravel drain before and after the installation. The cone resistance increases linearly according to the depth, that is, the effective confining pressure. The solid and broken lines represent the ground strength after and before installation. The difference between the solid and broken line represents the improvement by the tamping accompanied with the installation.

Fig.21 shows the deterioration of the cone resistance increment according to the distance from gravel drain center. The test result are the average value at the depth between 60 cm and 120 cm . Although the effective zone of flat tip rod is ranging between 70-80 cm from the center, the improved zone of cone-shaped tip rod is over 100 cm, that is the radius of the soil vessel. Also, the flat tip rod requires larger numbers of tamping to the same extent of compaction as the cone-shaped tip rod. Moreover, the higher is the rate of tamping , the more increasing the cone penetrating resistance.

According to the previous result, it is obvious that, the cone shaped tip is more effective on compacting than the flat tip of tamping rod.

5.3 Verification by Field Test in Niigata

Niigata City is located to the north of Tokyo and looks out on the Japan Sea. The verification field test was carried out in the liquefied areas due to Niigata earthquake in 1964.

The soil layers of test site are the soft ground filled with fine sand from Shinano river. Fig.22 shows the soil profile of the test site. Test upper layer to 4 to 5 meter consists of finer sand or

Table 1   Case of Improvement Test

	Shape of Tip	Number of Tampings (blows/m)	Stroke of tamping (mm)	Diameter of tamping rod (mm)
A	Conical	150	200	150
B	Conical	100	200	150
C	Flat	200	200	150
D	Flat	100	200	150

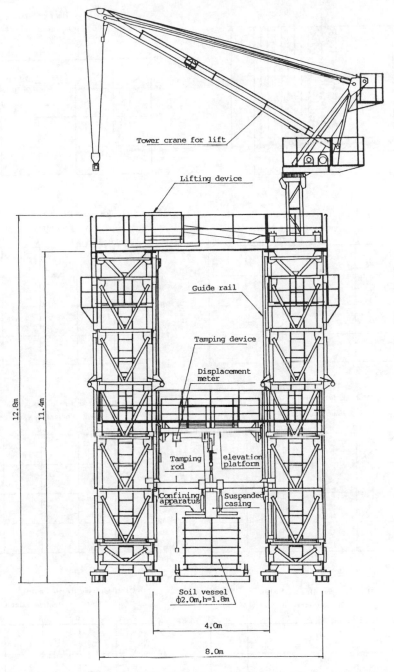

Fig.18 General View of Exclusive Facilities for Improvement Test

sandy silt containing more than 20% fines (passing # 200), and the lower consists of pure and uniform fine sand containing about 7% fines.

The gravel drains were installed to 10 m in depth with the new machine which has cone-shaped tip of tamping rod . The rod is 150 mm in diameter , the stroke of tamping is 200 mm and the rate of tamping was 75 blows per   minute.

The soil strength was measured with Ram-sounding and converted into normalized N-

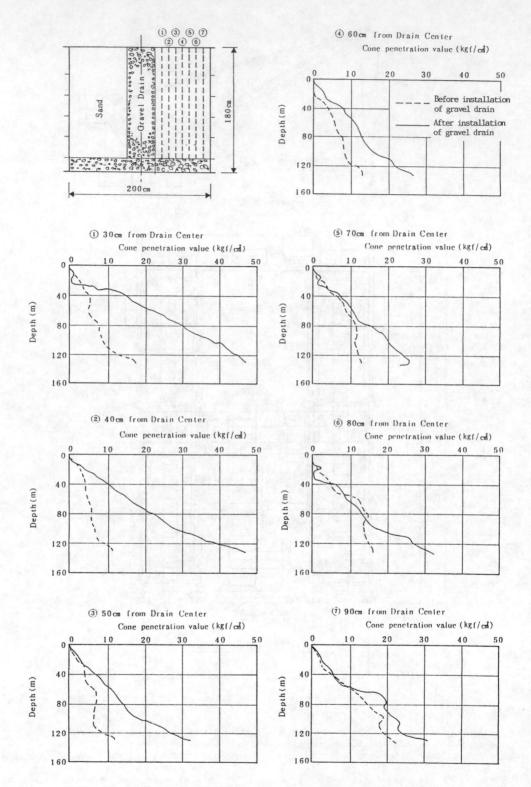

Fig.19   Test Result of Improvement by Gravel Drain Installation

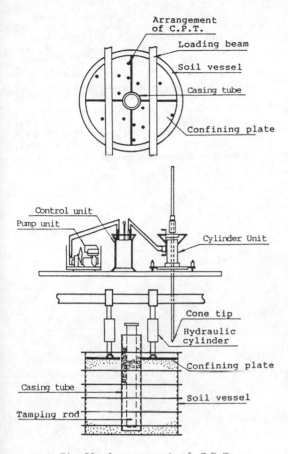

Fig.20 Arrangement of C.P.T.

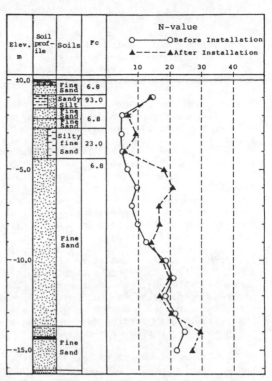

Fig.22 Profile of Test Site and Results

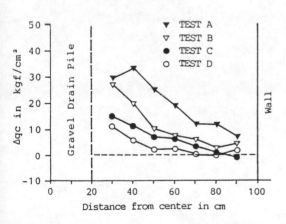

Fig.21 Deterioration of Cone Resistance

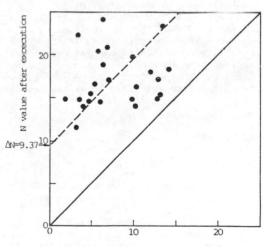

Fig.23 Strength Increment with
Cone-shaped Tip Rod

value before and after the installation. Within the layer from 4 to 10 meters in depth , which contains about 7% fines, the increment of normalized N-value is 10 blows. When the fine content was greater than 20% or the N-value was greater than 15 blows, no remarkable increase in penetration resistance of Ram-sounding was observed.

77

6.CONCLUSIONS

This paper aims to introduce the field and full scale laboratory test for development and improvement of the countermeasure against liquefaction during earthquake. These tests are taken to confirm the effect of a new method and they bring us the proper design method as well as the suitable execution.

Based on the results of field and full scale laboratory tests, the following conclusion can be derived :

(1) The full scale laboratory test brought us a new type of gravel drain pile driver equipped with cone-shaped tip rod which produces a dense gravel drain as well as compact the ground in the vicinity of the drain efficiently.

(2) the effective grain size ratio $Dg15/Ds85$ prescribes the state of clogging. At $7 < Dg15/Ds85 < 9$, the stable filter zone is existed , which prevents the progress of clogging. At $Dg15/Ds85 < 7$, no clogging occurs. And at $9 < Dg15/Ds85$, the permeability decreases drastically owing to the progress of clogging.

The Gravel Drain System has many advantages such as low vibration, low noise levels, negligible ground displacement, ease of execution, low cost and so on.

Measures against ground liquefaction are important in the advancement of the earthquake-proof characteristic of civil engineering structures and facilities. It is expected that the applications of this method will further increase in the future.

Finally, we wish to express our deepest appreciation to Prof. Ishihara of University of Tokyo, Pro. Yoshimi of Tokyo Technical University,and other fellows of NKK.

REFERENCES

Saito,A., Ohno,Y., Yoshida, Nagayama and Yokoyama. 1982. Experimental study on effect of measure for liquefaction by gravel drains method. Nippon Kokan Technical Report, No.92 (in Japanese).

Seed,H.B. and J.R.Booker. 1977. Stabilization of potentially liquefiable sand deposits using gravel drains. J. GED ASCE, Vol.103, No.GT7.

Ohno,Y.,Ito,K. and Okita,Y. 1983. Coefficient of volume compressibility for gravel drain design. Annual Meeting of JSSMFE, No.18 (in Japanese).

Nakajima,Y.,Ito,K.,Okita,Y. and Shimaoka,H. 1985. Permeability and clogging limit of gravel drain. Annual Meeting of JSSMFE, No.20 (in Japanese).

Ohno,Y., Ito,K., Minamikawa,Y. and Okita,Y. 1984.Short term clogging limit of gravel drain. Annual Meeting of JSSMFE, No.19 (in Japanese).

Okita,Y.,Miki,S.,Ito,K. and Kumahara.1989. Difference of density distribution from preparation of sample. The 44th Annual Conference of JSCE (in Japanese).

Okita,Y.,Miki,S., and Ito,K. 1990. Full scale model test of gravel drain pile and compaction(Part2). The 45th Conf. of JSCE (in Japanese).

Developments in Geotechnical Aspects of Embankments, Excavations and Buried Structures, Balasubramaniam et al. (eds)
© 1991 Balkema, Rotterdam. ISBN 90 5410 019 2

X-ray diffraction of the five component clay shales

Sanga Tangchawal
Department of Mining Engineering and Mining Geology, Faculty of Engineering, Chulalongkorn University, Thailand

ABSTRACT: A full lengthy X-ray diffraction technique for clayey rock sampling and preparation, clay mineral recognition and interpretation is exhibited. The procedure used illite as an internal standard for clay mineral interpretation. The proposed formula for different clay mineral types, adopted and modified from various publications, are set for a high clay content sedimentary rock. Problems involved in this semi-quantitative method are discussed. The application of proposed formula gives satisfactory results and is not unreasonably time consuming.

1 INTRODUCTION

This paper describes the specimen sampling and preparation for using the X-ray diffraction technique, and the semi-quantitative interpretation for clay mineralogy of the five component rock. As often observed, a shale when direct exposed exhibits properties related to a rock but with time these properties undergo some internal properties changes to those more closely identified with a soil.

The terms "clay shales" are defined as "fine grained, argillaceous, over-consolidated materials." When materials are completely disaggregated into individual particles, they have the properties of clay. Materials included in this study may be massive, fissile, laminated, or slickensided.

2 SPECIMEN SAMPLING AND PREPARATION

Samples can be obtained from cores, fresh surface and underground excavations. All samples are sealed tightly to prevent moisture migration. No out-crop or otherwise appreciably weathered samples are included in order to avoid possible weathering effects on clay mineral composition.

Specimen preparation for clay mineralogy study is tedious and time consuming. The following suggested technique is aimed to avoid grinding and crushing as much as possible. The clay mineralogy specimen, about 500 gm each, is placed in a 500-ml beaker and covered with distilled water for a few days.

To disaggregate clay particles from other non-clay particles, crushing is preferable to grinding. A high speed (up to 12,000 rpm) rotary blender should be used for a minimum time (10 minutes) and likewise ultrasonic vibration. Good dispersion requires removal of soluble salts which, if present, cause flocculation of the clay particles. Chemical aids to dispersion and vigorous shaking in water followed by centrifugation may be necessary.

Separation of particle fractions larger than 2 microns (up to 5 microns for highly clay mineral materials) by settling in water under gravity or by centrifugation often removes or greatly decreases the non-clay minerals such as quartz, feldspar, coarse particle micas, and carbonate minerals.

The smearing technique is used to perpare and oriented, air-dry specimen for the X-ray diffraction test. The fine size clay slurry is lightly smeared uniformly on a petrographic slide with a spatula. It is then left to air dry at room temperature for 24 hours prior to X-ray scanning.

Conventional clay interpretation uses the air-dry condition as a basic technique to obtain information of different clay mineral constituents. Recent publications especially those of the Mineralogical Society Monograph No.5 (p.305-325) and the author's previous work confirmed that some clay mineral types need to be activated under controlled atmosphere of ethylene glycol or glycerol.

A limitation of preliminary examination is that mixed-layer clay mineral types are scarcely be recognized. In most cases, they are characterized by non-integral diffractions and the diffraction peaks are broad and often have asymmetrical profiles.

Mixed-layer clays often contain swelling layers (montmorillonite) combined with non-swelling layers (mostly illite). Ethylene glycol saturation impresses a standard condition on the swelling component. The optimum condition for glycol saturation is to expose the specimen slide to glycol vapor for three days, at 50-60 degrees C to increase the vapor pressure of the glycol.

To distinguish the normal illite mineral from the mixed-layer minerals, the heat-treated condition is suggested. The specimen slide is placed in an electric temperature-controlled furnace and heated at 350 degrees C for three hours. Diffraction patterns from oriented specimens in the air-dry condition, after ethylene treatment and after heat treatment often allows recognition of the different clay mineral types.

3 SEMI-QUANTITATIVE INTERPRETATION OF CLAY MINERALS

Clay shale rocks for this study have their major amounts of clay minerals as illite, kaolinite, chorite, mixed-layer minerals, and their minor amounts of montmorillonite. The identification for clay minerals was carried out on the specimen slide, clamped in place in the path of a collimated X-ray beam. The semi-quantitative evaluation for clay minerals was based on the characteristic basal reflections of the clay minerals and the effects of ethylene glycol and heat treatments upon the position, shape, and peak intensity of the reflections.

The identification methods above seem to be reliable for the low angle 2θ range. The main idea is that the clay percentage of each mineral components is calculated based upon their basal peak areas in order to make direct comparisons. The sum of the weighted peak areas is 100 percent of clay specimen mineralogy.

Semi-quantitative interpretation procedures of clay minerals are complicated. An example of clay mineral interpretation is given in Figures 1, 2, and 3. There are five types of clay minerals encountered in the identification process and the following subsections is the explanation on how each mineral type is detected and evaluated.

3.1 Illite

The presence of illite is detected from the glycol-treated specimen with a regular 001 basal reflection at 10 angstroms(equal to 8.8 degrees 2θ using CuKα X-radiation) and a 002 basal at 5 angstroms. The 003 basal reflection coincides with the strongest reflection of quartz at 3.34 angstroms; therefore, it is not suitable for comparison with the other close coincidence peaks (a 002 kaolinite and a 004 chlorite reflection).

For the clay mineral calculation shown in Figure 2, the relative content of illite is always assumed to be equal to 1.0 since it has been used as the internal standard mineral. The semi-quantitative equation is:

$$I_1^g / I_1^g = 1.0 \qquad (1)$$

where I_1^g = Illite first-order peak area for glycol-treated specimen, cps

3.2 Mixed-layer Minerals

X-rays diffracted by a heterogeneous mixed-layer material, included random stratifications of hydratable and non-hydratable clay particles, do not resolve into distinct maxima after treatment with ethylene glycol. Instead, the diffraction trace is distributed over a broad angular range. This region of broad diffuse reflections in random positions commonly starts on a 2 degrees 2θ and ranges up to the low angle side of the 001 illite reflection.

X-RAY DIFFRACTION DATA SHEET

1. General Information

Project Mine C
Dirll Hole No.4
Size Range 2 Microns

Date 02/12/87
Depth 1000.2 ft
Air-Dry 24 hours
  Treatment Period
X-ray GE XRD-700
  Diffraction Unit

Performed By Tangchawal
Rock Type Grey Mudstone
Glycol Treatment 72 hours
  Period
Monochromatic CuKα
  X-radiation

Heat Treatment 3 hours
  Period

2. Semi-Quantitative Interpretation on Clay Minerals

Type of Specimen Treatment	Position of Peak degrees 2θ	Height of Peak, cps	Peak Width at Half Height	Area of Peak, cps	Description	Symbol
Air-Dry	17.5	85	22	1870	Illite Second Order	I_2^a
	8.7	100	12	1200	Illite First Order	I_1^g
Glycol	17.5	80	12	960	Illite Second Order	I_2^g
	–	–	–	–	Montmorillonite First Order	M_1^g
	8.7	280	8	2240	Illite First Order+ Mixed-Layer Minerals	M_{xl}^h
	18.4	45	12	540	Chlorite Third Order	C_3^h
Heat	25.0	60	10	600	Chlorite Forth Order	C_4^h
	–	–	–	–	Kaolinite First Order	K_1^h
	24.7	150	10	1500	Kaolinite First Order	K_2^h

Figure 1 X-ray diffraction data sheet for the identification of each clay mineral in the total clay mineral fraction

3. Content Calculation

Clay Content Calculation	Parts in Ten	Percent in Clay Mineral
I = Illite in clay fraction $= I_1^g / I_1^g = 1.0$	3.2	32 %
M = Montmorillonite in clay fraction $= M_1^g / ((6)I_1^g) = 0$	–	–
CIC= Correction due to incomplete collapse of mixed-layer minerals = The difference in baseline intensity of the glycol and heat treatment specimen = 0		
M_x = Mixed-layer or expandable minerals in clay fraction * if no discrete montmorillonite $= ((M_{x1}) - (I_1) + CIC) / I_1^g$ $= (2240 - 1200 + 0) / 1200 = 0.87$ * if discrete montmorillonite present $= (M_{x1}^h) - (M_1^g /6) - (I_1^g) + CIC) / I_1^g =$	2.8	28 %
C = Chlorite in clay fraction $= C_3^h / I_2^g = 540/960 = 0.56$	1.8	18 %
K = Kaolinite in clay fraction * if no discrete chlorite $= K_1^h / I_2^g =$ * if discrete chlorite present $= (K_1^h /(2)C_4^h) \cdot (C) = (1500/(2)600) \cdot (0.56)$	2.2	22 %
Total = 3.13	10	100 %

Figure 2 **X-ray diffraction** data sheet for the content calculation of each clay mineral in the total clay mineral fraction

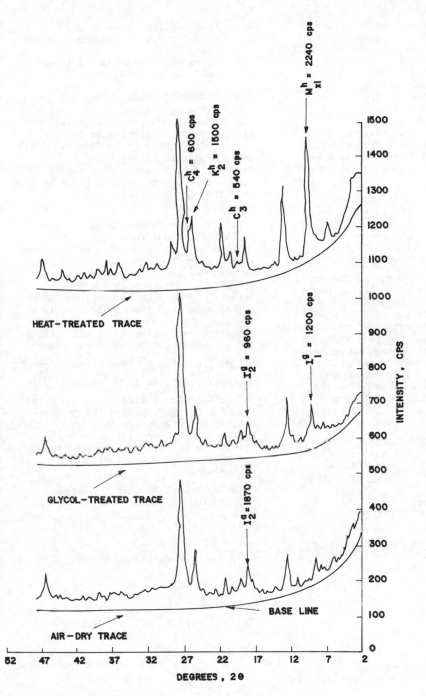

Figure 3 The comparison for X-ray diffraction profile of various treatment
types on the same specimen and showing the area of each important
peak

Interlayer water and ethylene glycol can be removed from the mixed-layer material by heat treatment at 350 degrees C or higher. The amount of mixed-layer material is equal to the difference in peak intensity areas between the heat-treated specimen and the glycol-treated specimen at a 10 angstrom peak (plus the correction value due to incomplete collapse of mixed-layer material, in some case). The peak intensity difference is then compared in a 1:1 ratio to illite.

The correction process for incomplete collapse follows that indicated in Figures 2 and 3. At the 10 angstrom peak, the glycol-treated trace is compared with the heat-treated trace (Johns, Grim, and Bradley, 1954; Odom, 1963; Schultz, 1964; Tangchawal,1988).

A straight line is drawn on the low angle side from the apex through the base of the 10 angstrom peak on the glycol trace and on the heat-treated trace. This line is projected to a common base line and the peak width is determined. A straight forward percent correction can be made. An example to obtain the fraction for incomplete collapse of the mixed-layer minerals and the correction value (CIC) is given on the data sheet calculation in Figure 2.

The empirical formula for the mixed-layer or expandable minerals in clay content fraction M_x, is as follows:

$$M_x = \frac{[(M_{x1}^h) - (I_1^g) + CIC]}{I_1^g} \qquad (2)$$

if there is no discrete montmorillonite and

$$M_x = \frac{[(M_{x1}^h) - (M_1^g/6.0) - (I_1^g) + CIC]}{I_1^g} \qquad (3)$$

if there is the discrete montmorillonite
where :

M_x = Mixed-layer or expandable mineral peak area for glycol-treated specimen, cps

M_{x1}^h = Illite first-order and mixed-layer peak area for heat-treated specimen, cps

I_1^g = Illite first- order peak area for glycol-treated specimen, cps

M_1^g = Montmorillonite first-order peak area for glycol-treated specimen, cps

CIC = Correction value due to incomplete collapse of mixed-layer minerals

3.3 Montmorillonite

Most high clay content rocks contain small amounts of montmorillonite. The discrete montmorillonite is detected by the broad humps at low angular range and its peak at 17 angstroms (5.2 degrees 2θ). Some non-swelling is not produced, however, a minor trace of the mixed-layer material may be detected instead.

The relative intensity of the montmorillonite basal reflection after ethylene-glycol treatment has been converted in proportion with the illite first-order basal reflection for an ethylene-glycol treated specimen. Brindley (1980) quoted a suggestion by Reynolds in the Mineralogical Society Monograph No.5 (p. 434), published by Mineralogical Society of London that a factor value of 6.0 should be used to divide the ratio M_1^g (for glycol-treated montmorillonite at 17 angstroms)/I_1^g (glycol-treated illite at 10 angstroms). This factor accounts for the relative scattering intensities of glycol-montmorillonite and illite with the allowance for unit cell compositions and the Lorentz and polarization parameters.

The empirical formula for montmorillonite in the clay fraction, M, will be:

$$M = \frac{M_1^g}{6(I_1^g)} \qquad (4)$$

where:

M_1^g = Montmorillonite first-order peak area for glycol-treated specimen, cps

I_1^g = Illite forst-order peak area for glycol-treated specimen, cps

3.4 Chlorite

Chlorite is identified by its 003 basal reflection at 4.7 angstroms (18.9 degrees 2θ) and its 004 basal reflection at 3.54 angstroms (25.1 degrees 2θ). The 001 and 002 basal reflections of chlorite are not used since the 001 chlorite basal reflection coincides with the 001 vermiculite basal reflection and the 004 chlorite reflection coincides with the 002 kaolinite basal reflection.

In some scanning specimens the amount of chlorite is small and the 004 chlorite reflection is only a shoulder on the high angle side of the 002 kaolinite peak; on the other hand, if the amount of kaolinite is small, the 002 kaolinite reflection is only a shoulder on the low angle side of the 004 chlorite peak.

The ratio fo chlorite to illite (Huang 1981, Tangchawal, 1988) is computed by dividing the heat-treated peak area intensity of 003 chlorite reflection (C_3^h) by the glycol-treated peak area intensity of 002 illite reflection (I_2^g).

The empirical formula for the chlorite in the clay fraction, C, will be:

$$C = \frac{C_3^h}{I_2^g} \qquad (5)$$

where

C_3^h = Chlorite third-order peak area for heat-treated specimen, cps

I_2^g = Illite second-order peak area for glycol-treated specimen, cps

3.5 Kaolinite

As often happens, the recognition of kaolinite in the presence of chlorite has a problem. The 002 and 004 reflections of chlorite (d = 7.1 and 3.54 angstroms) overlap the 001 and 002 reflections of kaolinite (d = 7.15 and 3.58 angstroms).

If no discrete chlorite is present, the 001 basal reflection of heat-treated kaolinite (K_1^h) is compared directly to the 001 basal reflection of glycol-treated illite (I_1^g) in a 1:1 ratio basis. The empirical formula for kaolinite in the clay fraction K, in the absence of chlorite, is as follows:

$$K = \frac{K_1^h}{I_1^g} \qquad (6)$$

where

K_1^h = Kaolinite first-order peak area for heat-treated specimen, cps

I_1^g = Illite first-order peak area for glycol-treated specimen, cps

If chlorite is present, the above empirical formula is no longer valid. The amount of kaolinite in the clay fraction is then calculated in two steps.

First, the 002 reflection of heat-treated kaolinite (K_2^h) and the 004 reflection of heat-treated chlorite (C_4^h) are to compare as a ratio of their peak areas. It is necessary to multiply the 004 chlorite-reflection peak area by 2.0 to compensate for the form factor (Johns, Grim and Brindley, 1954). This will allow comparing the reflection in a 1:1 ratio with the 002 kaolinite-reflection peak area.

Secondly, the resulting ratio of to chlorite has to be multiplied by the ratio of chlorite to illite (symbol as "C") in order to relate the abundance of kaolinite to illite where illite is used as the internal standard mineral. This final result is in accordance with the requirement that the basal reflection from which ratios are derived should have similar angular functions.

The empirical formula for kaolinite in the clay fraction K, in the presence of chlorite, appears as follows:

$$K = \frac{K_2^h \; (C)}{(2) \; C_4^h} \qquad (7)$$

where

K_2^h = Kaolinite second-order peak area for heat-treated specimen, cps

C_4^h = Chlorite forth-order peak area for heated-treated specimen, cps

C = Percent of chlorite in the clay fraction, percent

Additional interpretation steps for the semi-quantitative analysis of clay mineral assemblages need to be made before the final evaluation of each clay mineral percentage can be made. These additional steps are to calculate the content ratio from the weighted-peak areas and convert to a parts-in-ten basis so that the percentages for each component clay mineral in the clay mineral fraction can be obtained.

Ratio of clay mineral types as previously mentioned are calculated as "the content ratio." For example, illite has the content ratio of 1.0 and other clay minerals will have their own content ratios (see Column 1 in Figure 2). These ratios are added together and result in "the total content value" of the clay mineral assemblage.

Each mineral content ratio, when divided by the content value and multiplied by 10, will be "the parts-in-ten value" of clay minerals (Column 2 in Figure 2). The sum of all clay mineral percentages is 100 percent. Multiplying parts-in-ten values by 10 gives "the resulting percentage value" of each clay mineral assembled in the total clay mineral fraction (Column 3, Figure 2).

4. CONCLUSION

One should realize that the method in this paper is useful for a high clay content rock and is only semi-quantitative as several approximations are made during the interpretation and calculation procedures. It would be worthwhile to prove that the proposed method is reliable for a rock in particular stratigraphic units. Some duplicate specimens of the same rock type and location should be X-ray scanned and the results compared. If the results compare well, chemical wet analysis should be followed.

Further studies can proceed with non-clay minerals (quartz, feldspar, calcite, etc.) and organic matter in the whole rock sample. These would be helpful in gaining a better understanding of the effects of compositions on the engineering properties of shales (Tangchawal, 1988). The determination for the total amounts of non-clay minerals can be made by using the empirical formula and applied the correction factor during the calculation procedure. The organic matter interpretation can be indirectly investigated by using the application of polarizing microscope to compare

the organic matter with the whole mineral content.

ACKNOWLEDGEMENTS

The author wishes to express his deep appreciation to those previous contributors as referred in the references and to those who assisted in the preparation of this paper.

REFERENCES

Brindley, G.W. 1980. Quantitative X-ray mineral analysis of clays, Chapter 7 in Monograph No. 5. Mineralogical Society of London: 411-438

Brindley, G.W. and Brown G. 1980. Crystal structures of clay minerals and their X-ray identification. Monograph No. 5. Mineralogical Society of London.

Johns, W.D., Grim R.E., and Bradley W.F. 1954. Quantitative estimations of clay minerals by diffraction methods. Sedimentary Petrology 24: 242-251.

Huang, S.L. 1981. Swelling behavior of Illinois coal shales. Unpublished Ph.D. Dissertation. University of Missouri-Rolla.

Odom, I.E. 1963. Clay mineralogy and clay mineral orientation of shales and claystones overlying coal seams in Illinois. Unpublished Ph.D. Dissertation. University of Illinois.

Schultz, L.G. 1964. Quantitative interpretation of mineralogical composition from X-ray and chemical data for the Pierre shale. Geological Survey Professional Paper. No. 391-C. U. S. Government Printing Office.

Tangchawal, S. 1988. Engineering properties and clay mineralogy of Illinois coal mine roof shales and underclays. Unpublished Ph.D. Dissertation. University of Missouri-Rolla.

Developments in Geotechnical Aspects of Embankments, Excavations and Buried Structures, Balasubramaniam et al. (eds)
© 1991 Balkema, Rotterdam. ISBN 90 5410 019 2

Correlation of sampling effect in Hong Kong

K.W.Chau
Hong Kong Polytechnic, Hong Kong

ABSTRACT: Laboratory simulation of the field sampling process in the current practice and the investigation on the effect of sampling disturbance on the soils in terms of the dry densities employing two sampling tubes of different wall thickness are presented. Two types of residual soils most commonly found in Hong Kong, namely, decomposed granite and decomposed volcanics, were tested in the remoulded state under a wide range of moisture contents. The disturbed soil densities from the driven sampling tubes were compared with the soil densities prepared by different compaction operations. The results demonstrate that the sampling process yields very good samples of medium dense soils, but tends to densify loose soils and loosen dense soils. Correlation of dry densities were derived under different moisture contents, thicknesses of the sampling tube and degrees of compaction, which are useful in evaluating site investigations to truly reflect the subsurface soil properties.

1 INTRODUCTION

In order to investigate geotechnical considerations that have enormous impacts on the feasibility, design, cost, performance, and risks of any construction projects, a detailed site investigation is necessitated prior to the construction activites. It is of principal significance that they be adequate in terms of thoroughness, suitability of methods used, and quality of execution of the work to assure that all imperative conditions have been detected and reliably evaluated. One of the main tasks in carrying out site investigation is to obtain as far as practicable undisturbed samples truly representing the properties of the subsurface materials.

There is a demand for high quality undisturbed samples of cohesionless soils, which has been highlighted in recent years by increasing awareness of the need to evaluate the seismic stability of soils in the foundations of important structures such as nuclear power plant and earth dams. In order to evaluate the dynamic response and seismic stability of soils, high quality undisturbed soil samples are required for the laboratory determination of cyclic strength and large-strain elastic moduli and for accurate determination of in-situ density.

Consequently, it is necessary to preserve as well as possible both the in-situ density and the in-situ soil structure, being two separate and distinct properties of the soil, including grain-to-grain contacts, with a minimum of disturbance.

In fact, there is hardly any truly undisturbed sample, for two reasons:

1. A sampling tube displaces a certain amount of soils, which inevitably produces strain and some disturbance of the sample;

2. Even in "perfect sampling", an imaginary process that eliminates disturbance due to soil displacement, the state of stress in the soil sample undergoes a complex, and to some degree indeterminate, history of change during the sampling, handling, shipping, storage, extrusion, specimen preparation, and laboratory set-up processes. Moreover, the in-situ state of stress, stress history, and state of stress in the sample are seldom known except by crude approximation. These shortcomings are widely recognized, and the term "undisturbed" sample is conventionally used to mean a sample that is obtained and handled by methods designed to minimize these effects (Hvorslev, 1980).

In this laboratory simulation, the disturbance effect in terms of the dry density is evaluated. Laboratory works are carried out to simulate the to-date sampling process of the subsurface soils in the fields. The soil density from the driven sampling tubes are compared with the original soil density using remoulded residual soils. A range of original soil densities can be obtained by using different compaction procedure, i.e. different compaction machine and different frequency of compaction.

The two most commonly encountered residual soils in Hong Kong, namely, decomposed granite and decomposed volcanics rocks, are analyzed in this investigation. Basically, they are formed by weathering, either physical or chemical, of the respective parent rocks. Their properties may differ considerably due to various factors and an attempt is made to compare the two soils in view of the sampling disturbance characteristics of their remoulded samples.

It is generally expected that two sampling tubes of same diameter but different wall thickness will give different sampling disturbance effect. It is valuable to investigate the density change for these two cases and to compare whether or not their results are compatible with the stresses caused as calculated from the theoretical background.

Besides, the studies of soil engineers indicated that many properties of soils are significantly influenced by the amount of water present in the soil. Therefore, investigation is made to study the effect of sampling disturbance over the normal range of moisture content.

With tests performed in this direction, it is possible to derive the correlations between the undisturbed and the disturbed densities of any soil. This information will be useful as, based on it, appropriate adjustments can be effected in order to ensure that the site investigation is evaluated reliably.

2 ANALYSIS OF SOILS CONSIDERED

The types of soil used in the investigation are decomposed granite (from Pokfulam Road) and decomposed volcanics (from Shatin) in Hong Kong.

Residual soils formed by the decomposition of rock covers a large extent of the earth's surface, particularly in the tropics. In Hong Kong, most of the soils have been formed by the in-situ decomposition of acid igneous rocks. Furthermore, as a result of the sub-tropical climate with high temperatures and heavy rainfall in the summer, the decomposition processes are very active and have produced great depths of residual soil. The decomposition process may be physical, as in the case of erosion by the forces of wind, water or glacier, and disintegration by alternate freezing and thawing. The other way is by chemical decomposition which results in changes in the mineral contents of the rock.

The main soil types encountered in Hong Kong are the residual soils, decomposed granite and decomposed volcanics; the marine deposits, soft normally consolidated clays and loose to dense sands; and the older alluvium, firm to stiff preconsolidated clays and clayey sands occasionally found between the residual soils and the marine deposits. Detailed information regarding Hong Kong residual soils can be found in Lumb (1962 & 1965).

While there are considerable differences between the residual soils — decomposed granite and decomposed volcanics — from a geological point of view, they are very similar in mineral content and hence no resistance to weathering or chemical decomposition. Indeed, the only significant difference between the two rocks (granite and volcanics) is that of grain-size, the granite minerals being of coarse sand size and the volcanic minerals in the ground-mass of medium to fine silt size. The rate of decomposition is primarily governed by the grain-size of the original rock and the jointing pattern, which control the ease of penetration of ground water.

2.1 *Decomposed granite*

As a consequence of the wide joint spacing, what occurs is a series of decomposition in which a large block of fresh rock is gradually reduced to a spheroidal boulder which itself will eventually decompose. Owing to the coarse grain-size of the granite, the individual mineral grains become separated at a very early stage but since no relative movement occurs between grains the original texture of the granite is preserved in the final soil. At the surface of the decomposed granite an aerated zone develops in which the iron-containing minerals are oxidized, the resulting ferric oxide giving a characteristic red colour to the soil. This soil, which is the final completely decomposed phase of the sequence, is very much different in nature and properties from the underlying incompletely decomposed phase. As the decomposition proceeds from fresh rock to the surface red earth the feldspar grains become progressively smaller, so the grading of the soil changes from coarse sand, when the grains first separate, through sand to clayey sand. The unoxidized coarse sand and silty sand is referred to as decomposed granite.

2.2 *Decomposed volcanics*

The same general pattern also applies to decomposed volcanics, but with modifications due to the differences in grain-size and joint spacing. In the early stages the boulders present are small, with a size of 300 mm or 600 mm, and these boulders decompose at a slower rate than the granite boulders, owing to their lower permeability to water. As the boulders decrease in size they produce first gravel and then sand, but even the sand grains are still composite and not individual quartz or feldspar grains, as is the case for the granite. Eventually all the sand fraction should decompose and in the final completely decomposed phase the coarsest grains present would be the silt-size quartz grains. Oxidation at the surface is not as pronounced as with the decomposed granite and it is not necessary to distinguish between oxidized and unoxidized decomposed volcanics. The soil changes from well graded sandy gravel in the early stages to clayey silt in the final stage, and the original volcanic texture and joint planes are preserved.

Furthermore, the depth of decomposition of volcanic rock varies considerably even over short distances, owing to the inherent variations in grain-size, mineral content, jointing pattern, and drainage conditions, but generally the decomposed granite is thicker than the decomposed volcanics. Also, the change from fresh to decomposed granite is always sharp but the basal surface generally bears no relation to the surface topography, while with the volcanic, there is a more gradual transition from fresh to decomposed and zones of soft rock are often encountered.

2.3 *Batch of soils tested*

There are results showing that the residual soils of Hong Kong derived from igneous rocks fall into groups having significantly different properties. For the properties of the soils under consideration, it has been found that they can be classified by the shape of the grading curve. Also, they have

appreciable angles of shearing resistance when fully drained, the average angle being about 35° for the decomposed granite, and 30° for the decomposed volcanics. For individual samples the angle of shearing resistance can vary considerably from the average but the variations are statistically independent of voids ratio, grading, or degree of saturation, and can be considered as random variations.

When unsaturated, all soils have a large cohesive component of strength but on complete saturation the drained cohesion drops to zero or a very small value. This drop in strength is particularly noticeable with the decomposed volcanics, and cuttings in these soils which are quite stable in the dry season often collapse during the rainy season after infiltration of rainwater and consequent softening.

The compressibility of the residual soils is generally low, except for loose decomposed granite, and is not affected largely by the degree of saturation. Besides, the soils cannot be regarded as linear elastic materials since the stress-strain relationship is strongly influenced by the confining pressure. However, for small pressure increments, the modulus of elasticity can be approximated to increase linearly with the subsurface depth.

3 SAMPLING DISTURBANCE

It is impossible to obtain a completely undisturbed sample, since the act of sampling inevitably disturb the soil to some extent.

3.1 *Causes of sample disturbance*

The principal causes of sample disturbance are (Marcuson & Franklin, 1980):
1. the boring process
2. driving the sample tool
3. withdrawing the sampling tool, and
4. the relief of stress in the soil.

Fig. 1 and 2 illustrate respectively typical forces acting during the driving and withdrawal operation of the sampling process.

The disturbance caused in driving the sampling tool depends on the thickness of the tube and the manner in which it is driven. A thin-walled sampler causes much less disturbance, but is easily damaged. The least disturbance is caused when the sampler is driven in by steady pressure but for stiff soils this requires a provision of a secure anchorage. For routine investigations, the tool is usually driven down by blows from a monkey.

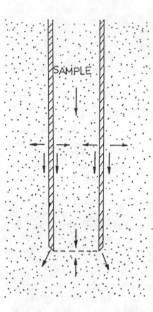

Fig. 1 Forces acting during the driving operation of sampling

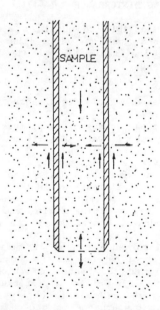

Fig. 2 Forces acting during the withdrawal operation of sampling

3.2 *Drive sampling — force and deformation*

In principle a drive sampler is a tube which is forced into the soil without any rotation or chopping action and without removing the soil displaced by the walls of the sampler. This soil is simply pushed aside with consequent severe stress changes and plastic deformations in the surrounding soil. A simplified and rough analysis of the forces and deformations during the drive sampling operation and during the removal operation is classified as:

1. Forces during driving
2. Entrance of excess soil
3. Influence of the inside wall friction
4. Deflection and failure of soil below the sampler
5. Influence of pressure or vacuum over the sample
6. Influence of the outside wall friction
7. Forces during the withdrawal
8. Disturbance during the withdrawal.

The pressure on top of the sample and the inside wall friction tend to compress and distort the soil layers and increase the pressure on the circular area directly below the sampler. The pressure on the surrounding annular area is very high on account of the edge resistance and a part of the soil below this area must be displaced as the sampler is forced into the soil. Outside the annular area the pressure is governed by the weight of overlying soil and the outside wall friction.

As the sampler advances, a part of the soil under the annular area is displaced by the walls of the sampler, pushed aside and some may be forced into the sampler. It will thereby increase the thickness and cause convex distortions of the soil layers in the upper part of the sample. Figure 3 depicts typical distortions on the soil due to entrance of excess soil.

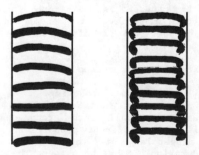

Fig.3 Distortion by entrance of excess soil

The inside wall friction and a positive pressure on top of the sample increase the pressure below the sampler and thereby tend to compact the soil before it enters the sampler, and moreover, tend to produce a convex curvature of the soil layers. The distortion is generally small in the central part and increases sharply and may be confined entirely to a zone of drag close to the surface of the sample but in some cases large and nearly parabolic convex distortions may be produced, as shown in Figure 4.

The total inside wall friction and thereby also the pressure on the soil below the sampler increases with the penetration of the sampler and the length of the sample. The first effect of an increase in the pressure is to decrease and later prevent entrance of excess soil, but with increasing penetration well will finally exceed the bearing capacity of the soil. From then on the soil layers below the sampler will be deflected downward, stretched, and reduced in thickness before they enter the sampler. (Figure 5) At a still greater penetration of the sampler and deflection of the oil layers, the latter will assume a distinct concave curvature. (Figure 6) This curvature, especially near the surface of the sample, may be reversed after the soil enters the sampler and is exposed to direct action by the inside wall friction.

The other influencing factors include the pressure or vacuum over the sample and the outside wall friction.

3.3 Forces and disturbance during the withdrawal

The direction of the wall friction is reversed during the withdrawal. An examination of the various forces acting on the sample during withdrawal shows that difficulties in retaining the sample generally decrease with increasing length and increasing diameter of the sample.

When the sample is separated from the subsoil by a direct pull or rotation, the necessary tensile or torsional forces must be transmitted from the walls of the sampling tube to the soil over a certain length of the lower part of the sample. These forces will generally cause a partial disturbance of the lower part of the sample, and this disturbance may be increased by swelling on account of contact with free water in the bore hole and/or internal migration of water from the undisturbed to the partially disturbed sections, or vice versa.

3.4 Principal dimensions of samplers

The principal dimensions of a drive sampler and certain ratios between these dimensions are defined in Fig. 7. Their optimum values depend on the area ratio C_a, which is approximately equal to the ratio between the volume of displaced soil and the volume of the sample, and the inside clearance C_i, which is the ratio between the difference from inside diameter of the sampling tube and the diameter of the cutting edge.

Fig.4 Distortion due to inside wall friction

Fig.5 Downward deflection of soil layers below sampler

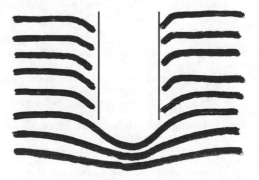

Fig.6 Formation of soil cone below sampler

91

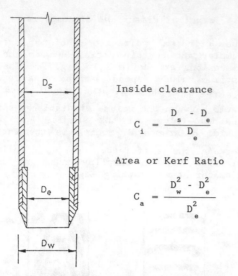

Inside clearance

$$C_i = \frac{D_s - D_e}{D_e}$$

Area or Kerf Ratio

$$C_a = \frac{D_w^2 - D_e^2}{D_e^2}$$

Fig.7 Characteristic dimensions of sampler

Two samplers of different wall thickness were used with principal dimensions as tabulated in Table 1.

Table 1. The principal dimensions of the soil samplers.

	sampling tube 1	sampling tube 2
thickness (mm)	2.7	1.8
internal dia. (mm)	38	38
weight (g)	374.05	283.3
area ratio (%)	15	10

3.5 Disturbance of soil samples

The disturbances to which soil samples may be subjected can be classified in the following basic types, proceeding from relatively slight and common disturbances to grave and usually avoidable disturbances:

1. change in stress conditions
2. change in water content and void ratio
3. disturbance of the soil structure
4. chemical changes, and
5. mixing and segregation of soil constituents.

The influence of these disturbances on the results of laboratory tests depends not only on the type and degree of disturbance but also on the character of the soil and the testing technique and is subject to extreme variations.

4 LABORATORY WORKS

The investigation basically comprises two parts: compaction and sampling. Consequently the experimental works involved in the study are also divided under the two above headings. A description on the details of the laboratory works involved in these respective sections is given below.

4.1 Compaction process

The soil is compacted in a specified manner from very loose to standard compaction over a range of moisture content from about 12 percent to 21 percent, the range normally encountered. In order to effect different compaction efforts, different modes of compaction were performed by varying the number of blows from a constant height of the standard 2.5kg rammer, similarly to that prescribed by B.S. 1377:1975.

The apparatus involved in the compaction process are: (1) A cylindrical metal CBR (California Bearing Ratio) mould having an internal diameter of 152mm and an internal effective height of 179mm, with a detachable baseplate and a collar 50mm deep as shown in Figure 8, (2) a metal rammer having a 50mm diameter circular face, weighing 2.5kg and being equipped with a suitable arrangement for controlling the height of drop to 300mm, as shown in Figure 9 or a mechanical CBR/proctor compactor, as shown in Figure 10, (3) other ancillary equipment including a balance, a palette knife, a straight-edge, a 20mm BS test sieve, a receiver, a 600mmx500mmx80mm deep metal tray and the apparatus for moisture content determination.

A 7kg sample of air-dried soil passing the 20mm BS test sieve was taken and mixed thoroughly with a suitable amount of water depending on the soil type. The mould, with baseplate attached, was placed on a solid base. The moist soil was then compacted into the mould, with the extension attached, in five layers of approximately equal mass, each layer being given a certain number of blows from the rammer dropped from a height of 300mm above the soil. It was ensured that the blows were distributed uniformly over the surface of each layer and that the tube of the rammer was kept clear of soil so that the rammer always fell freely. Moreover, it was ensured that the amount of soil used was sufficient to fill the mould, leaving not more than 6mm to be struck off when the extension was removed. The

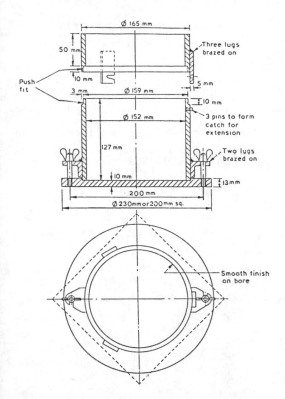

Fig.8 A standard CBR Mould

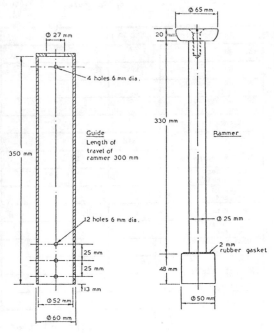

Fig.9 A standard 2.5kg rammer

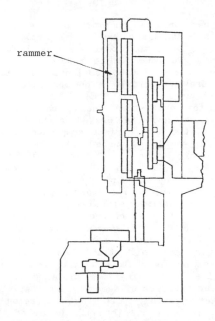

Fig.10 The schematic elevation of a CBR/ proctor compactor

extension was then removed and the compacted soil was levelled off carefully to the top of the mould by means of a straight-edge.

4.2 *Sampling process*

After the soil sample had been prepared at a certain bulk density by a certain compaction, they would be tested with either one of the two sizes of sampling tube (one with thicker wall thickness and another with thinner wall thickness).

The apparatus involved in the sampling process comprise: (1) An open drive metal sampling tube having internal diameter 38mm, thickness 2.7mm and length of 179mm with some diametrically opposite openings of diameter 6mm at one end and sharpened edge at the other end of the sampler, (2) another open drive metal sampling tube similar to the above except with thickness 1.8mm and length of 178mm, (3) a metallic cap having a diameter about same as that of the sampling tubes to protect the samplers from damages during the driving process, (4) other ancillary tools such as a balance, a 2kg hammer to drive the sampler and cap, a 4mm diameter and 250mm long metal rod, a steel rule, a pair of caliper and apparatus for extracting the whole soil specimen from the sampling tube without great disturbance.

Either one of the two soil samplers (having different wall thickness) was taken. The sample tool was usually well oiled inside and out to reduce friction, and was clear from any soil particles left from previous trial. The metal cap was placed on the blunt end of the sampling tube for protection during the driving process. They were lowered on the central part of the soil surface compacted inside the CBR mould and pushed into the soil a little bit by hand. The tool was forced into the soil either by blows from a hammer, or by jacking. Although jacking is preferable, satisfactory results would still be obtained by careful driving and serious remoulding of the sample would not be caused. The distance which the tool was driven should be checked as, if driven too far, the soil will be compressed in the sampler by the cap. Attention should be given to maximum distance driven without the risk of damaging the sample.

After driving, with the cap removed, metal rod was put into the holes on the blunt end. They were rotated to break off the core and the sampler was steadily withdrawn. The soil sample was removed from the tube by means of the extrusion machine with great care. The length and weight of the sample were then measured. The soil left inside the CBR mould and the soil sample taken out were both excavated out to a tray whose weight had been determined beforehand. The tray containing the soil mass was determined again. A representative sample of the specimen was taken and its moisture content was determined. The remainder of the soil specimen was broken up, rubbed through the 20mm BS sieve, and then mixed with the remainder of the original sample, suitable increments of water was added successively and mixed into the sample, and the above procedure including compaction was repeated for each increment of water added.

5 ANALYSIS OF EXPERIMENTAL DATA

5.1 *Computations of densities*

The undisturbed bulk density, D_{b1} in Mg/m^3, of each compacted specimen was calculated from the equation:

$$D_{b1} = \frac{m_2 - m_1}{3248} \qquad (1)$$

where m_1 is the mass of tray in g and m_2 is the mass of soil and tray in g.

The undisturbed dry density, D_{d1} in Mg/m^3, was calculated from the equation:

$$D_{d1} = \frac{100 \, D_{b1}}{100 + w} \qquad (2)$$

where w is the moisture content of the soil in %.

The disturbed bulk density, D_{b2} in Mg/m^3, of each soil samples obtained was calculated from the equation:

$$D_{b2} = \frac{m_4 - m_3}{1.134L} \qquad (3)$$

where m_3 is the mass of the empty sampling tube in g and m_4 is the mass of the soil sample and the sampling tube in g.

The disturbed dry density, D_{d2} in Mg/m^3, was calculated from the equation:

$$D_{d2} = \frac{100 \, D_{b2}}{100 + w} \qquad (4)$$

Soil Type & Source :-

Characteristics of Sampling tube :-

a. Type (thick/thin) :   b. Internal diameter =   mm

c. Weight   = . g d. Thickness   = . mm

MOISTURE CONTENT			
Tin No.			
Wt. of tin (g)			
Wt. of wet soil + tin (g)			
Wt. of dry soil + tin (g)			
Moisture Content (%)			

SAMPLING			
Determination No.	1	2	3
Wt. of sampling tube + soil (g)			
Length of sample obtained (mm)			
Wt. of tray (g)			
Wt of tray + soil (g)			
Undisturbed Bulk Density (Mg/m³)			
Undisturbed Dry Density (Mg/m³)			
Disturbed Bulk Density (Mg/m³)			
Disturbed Dry Density (Mg/m³)			

Fig.11 Blank sheet for filling of test results

Soil Type & Source :- decomposed granite (Pokfulam Road)

Characteristics of Sampling tube :-

a. Type (thick/thin) : thick   b. Internal diameter = 38 mm

c. Weight   = 374.05g   d. Thickness   = 2.7mm

MOISTURE CONTENT			
Tin No.	40	6	43
Wt. of tin (g)	29.83	29.56	29.83
Wt. of wet soil + tin (g)	82.51	106.48	110.36
Wt. of dry soil + tin (g)	75.41	96.73	100.59
Moisture Content (%)	15.5	14.5	14.8
SAMPLING			
Determination No.	1	2	3
Wt. of sampling tube + soil (g)	534.85	563.5	569.3
Length of sample obtained (mm)	122	137	134
Wt. of tray (g)	465	465	460
Wt of tray + soil (g)	4333	4749	4835
Undisturbed Bulk Density (Mg/m^3)	1.19	1.32	1.35
Undisturbed Dry Density (Mg/m^3)	1.03	1.15	1.18
Disturbed Bulk Density (Mg/m^3)	1.16	1.29	1.28
Disturbed Dry Density (Mg/m^3)	1.01	1.13	1.13

(a)

Soil Type & Source :- decomposed volcanics (Shatin)

Characteristics of Sampling tube :-

a. Type (thick/thin) : thick   b. Internal diameter = 38 mm

c. Weight   = 374.05g   d. Thickness   = 2.7mm

MOISTURE CONTENT			
Tin No.	5	32	14
Wt. of tin (g)	29.84	29.75	29.885
Wt. of wet soil + tin (g)	114.14	73.59	107.34
Wt. of dry soil + tin (g)	100.16	66.29	94.695
Moisture Content (%)	17.9	18.0	18.2
SAMPLING			
Determination No.	1	2	3
Wt. of sampling tube + soil (g)	614.85	580.15	562.8
Length of sample obtained (mm)	113	95	87
Wt. of tray (g)	466	463	465
Wt of tray + soil (g)	6718	6842	6550
Undisturbed Bulk Density (Mg/m^3)	1.93	1.96	1.87
Undisturbed Dry Density (Mg/m^3)	1.61	1.64	1.57
Disturbed Bulk Density (Mg/m^3)	1.88	1.91	1.91
Disturbed Dry Density (Mg/m^3)	1.57	1.59	1.60

(a)

Soil Type & Source :- decomposed granite (Pokfulam Road)

Characteristics of Sampling tube :-

a. Type (thick/thin) : thin   b. Internal diameter = 38 mm

c. Weight   = 283.3 g   d. Thickness   = 1.8mm

MOISTURE CONTENT			
Tin No.	15	4	47
Wt. of tin (g)	29.80	29.78	29.45
Wt. of wet soil + tin (g)	99.32	84.365	98.78
Wt. of dry soil + tin (g)	87.185	74.86	86.74
Moisture Content (%)	21.1	21.1	21.0
SAMPLING			
Determination No.	1	2	3
Wt. of sampling tube + soil (g)	524.9	524.0	460.9
Length of sample obtained (mm)	137	122	125
Wt. of tray (g)	467	466	460
Wt of tray + soil (g)	5332	6221	4211
Undisturbed Bulk Density (Mg/m^3)	1.50	1.77	1.16
Undisturbed Dry Density (Mg/m^3)	1.24	1.46	0.95
Disturbed Bulk Density (Mg/m^3)	1.56	1.74	1.25
Disturbed Dry Density (Mg/m^3)	1.28	1.44	1.04

(b)

Fig.12 Examples of test result sheet for decomposed granite

Soil Type & Source :- decomposed volcanics (Shatin)

Characteristics of Sampling tube :-

a. Type (thick/thin) : thin   b. Internal diameter = 38 mm

c. Weight   = 283.3 g   d. Thickness   = 1.8mm

MOISTURE CONTENT			
Tin No.	10	50	19
Wt. of tin (g)	29.665	29.535	29.75
Wt. of wet soil + tin (g)	81.775	102.18	99.265
Wt. of dry soil + tin (g)	72.32	88.57	86.72
Moisture Content (%)	21.2	21.0	21.0
SAMPLING			
Determination No.	1	2	3
Wt. of sampling tube + soil (g)	419.1	496.2	487.1
Length of sample obtained (mm)	83	103	92
Wt. of tray (g)	463	462	463
Wt of tray + soil (g)	3760	5582	6264
Undisturbed Bulk Density (Mg/m^3)	1.02	1.58	1.79
Undisturbed Dry Density (Mg/m^3)	0.83	1.28	1.46
Disturbed Bulk Density (Mg/m^3)	1.44	1.82	1.95
Disturbed Dry Density (Mg/m^3)	1.18	1.48	1.60

(b)

Fig.13 Examples of test result sheet for decomposed volcanics

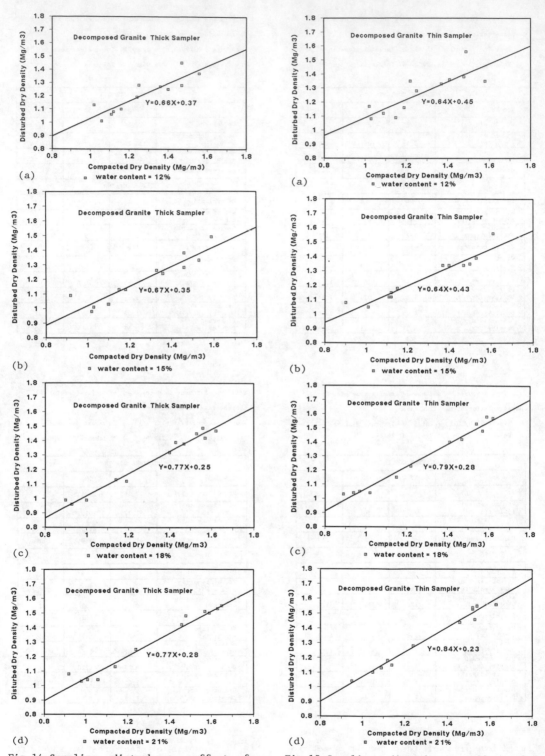

Fig.14 Sampling disturbance effect for remoulded decomposed granite using 2.7mm thick sampler at various water contents

Fig.15 Sampling disturbance effect for remoulded decomposed granite using 1.8mm thick sampler at various water content

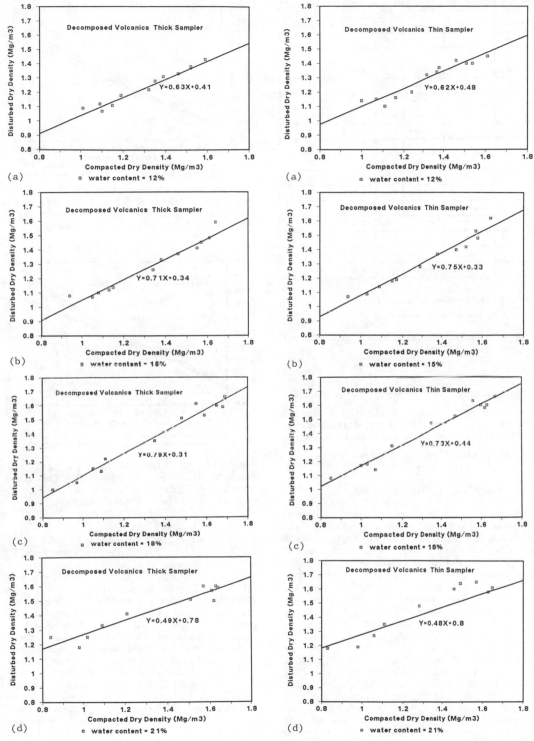

Fig.16 Sampling disturbance effect for remoulded decomposed volcanics using 2.7mm thick sampler at various water contents

Fig.17 Sampling disturbance effect for remoulded decomposed volcanics using 1.8mm thick sampler at various water content

Soil type	moisture content	type of tube	y = ax + b		regression coefficient r
			a	b	
D.G.	12	thicker	0.66	0.37	0.89
D.G.	15	thicker	0.67	0.35	0.93
D.G.	18	thicker	0.77	0.25	0.99
D.G.	21	thicker	0.77	0.28	0.98
D.G.	12	thinner	0.64	0.45	0.84
D.G.	15	thinner	0.64	0.43	0.96
D.G.	18	thinner	0.79	0.28	0.99
D.G.	21	thinner	0.84	0.23	0.99
D.V.	12	thicker	0.63	0.41	0.98
D.V.	15	thicker	0.71	0.34	0.98
D.V.	18	thicker	0.79	0.31	0.99
D.V.	21	thicker	0.49	0.78	0.96
D.V.	12	thinner	0.62	0.48	0.96
D.V.	15	thinner	0.75	0.33	0.98
D.V.	18	thinner	0.73	0.44	0.99
D.V.	21	thinner	0.48	0.80	0.76

N.B.   D.G. denotes decomposed granite &

D.V. denotes decomposed volcanics

Fig.18 Tabulated results on linear regression analysis

The disturbed dry density, D_{d2}, obtained in a series of determinations was plotted against the corresponding undisturbed dry density, D_{d1} under the specified moisture content, soil type and sampler type.

5.2 Linear regression analysis

The test results are entered in the standard form as shown in Figure 11. Figures 12 and 13 is a typical test result sheet duly completed. Figures 14 to 17 depict typical graphical representations of sampling disturbance effect for remoulded decomposed granite and remoulded decomposed volcanics by using samplers with different thicknesses at various water contents.

Linear regression analysis is used to analyze the results obtained and the results are shown on Figure 18. Figure 19 to Fig. 22 depict the general sampling disturbance effect at different water contents, by using the results of the linear regression analysis, for the respective combinations of decomposed granite/decomposed volcanics with thicker/thinner sampler tubes.

From the graphs of undisturbed dry density and disturbed dry density shown

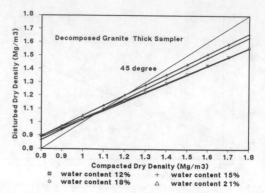

Fig.19 Linear regression analysis of sampling disturbance effect for remoulded decomposed granite using 2.7mm thick sampler at different water contents

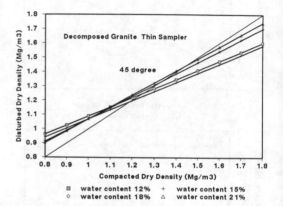

Fig.20 Linear regression analysis of sampling disturbance effect for remoulded decomposed granite using 1.8mm thick sampler at different water contents

above for various types of soil with different wall thickness of sampler, it can be generally concluded that the sampling process tends to densify loose soils and loosen dense soils.

For both types of soil used, decomposed granite and decomposed volcanics, it is generally true for the thinner-walled samplers to give less disturbance effect than that for the thicker-walled samplers.

For decomposed granite, the results seem to be of similar pattern for all the four moisture contents used in the experiment. In other words, the correlation between undisturbed and disturbed dry densities for decomposed granite only changes slightly under different water contents.

On the contrary, the results for decomposed volcanics at a high water

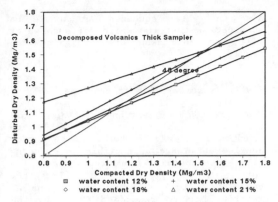

Fig.21 Linear regression analysis of sampling disturbance effect for remoulded decomposed volcanics using 2.7mm thick sampler at different water contents

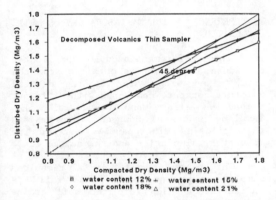

Fig.22 Linear regression analysis of sampling disturbance effect for remoulded decomposed volcanics using 1.8mm thick sampler at different water contents

content, i.e. 21%, show a rather distinct discrepancy from the above although the tendency was still to densify loose soil and to loosen dense soil. The point for no change in density after sampling shifts up to a higher values of dry density, in the range of 1.4Mg/m^3 to 1.8Mg/m^3. Large disturbance effect is observed for moderately loose soil and it is densified enormously.

The above results are reasonable and can be interpreted by the fact that the decrease of cohesion strength with water content is more significant for decomposed volcanics than for decomposed granite. As has been mentioned earlier, decomposed volcanics has smaller grain size when compared with decomposed granites. In high water content of 21%, the cohesion strength of decomposed volcanics drops to

such a low level that the loose soil is easily compacted even by the sampling action and as such large disturbance effect is observed.

For normal range of moisture content of decomposed volcanics, nevertheless, the behaviour accords with the general pattern similar to its counterpart of the decomposed granite.

6 POSSIBLE SOURCES OF ERRORS AND WAYS TO ALLEVIATE

There are a number of sources of error affecting the accuracy of the experimental results.

6.1 *Non-uniform compaction*

Although the remoulded sample is compacted in five separate layers, the effect of non-uniform compaction cannot be eliminated but only reduced to a minimum.

6.2 *Loss of water*

The moisture content quoted are that of soil after thorough mixing. However, water may be lost in a number of ways:-
1. Water adhering to the plastic bag
2. Water adhering to the hand during mixing
3. Water squeezed out from bottom if heavy compaction is necessary to achieve the pre-determined density
4. Water loss due to air-conditioning.

6.3 *Non-uniform moisture content*

As the soil mass used in each sampling procedure is 6kg or so, it is difficult to ensure the uniformity of moisture content within the soil mass of enormous size in spite of the tremendous effort to attain this target. The sample taken for determination of moisture content may not be representing that of the soil mass in the CBR mould. It is important that the water is mixed thoroughly and adequately with the soil when samples are being prepared, since inadequate mixing gives rise to variable test results. This is particularly important with cohesive soils when adding a substantial quantity of water to the air-dried soil. With clays of high plasticity, or where hand mixing is employed as in the tests performed in this study, it may be difficult to distribute the water uniformly through the air-dried soil by mixing alone, and it may be necessary to store the mixed sample in a sealed container for a minimum period of 16 hours or so before continuing the experiment.

6.4 Attachment of soil on rammer

The weight of the rammer may change during compaction due to the soil attached to it. To minimize this error, it is essential to remove the soil attached to the rammer before each layer is compacted.

6.5 Non-uniform sample surfaces

The soil sample frequently contains some holes on the top after it is levelled with a straight edge after compaction has completed due to replacement of relatively large particles. These holes must be refilled with fine soil and then compacted by hand and the surface re-levelled before weighing.

6.6 Deviation from the designed drop

The first blow is usually not equal to the designed drop, i.e. 300 mm for the compaction machine. This error can be eliminated by lowering the hammer by hand slowly until it rests on the soil and then setting the counter before switching on power.

6.7 Control of variation of density

If the amount of soil struck off after extension is too large, the variation of density cannot be kept under control corresponding to the compactive effort used. Thus it is necessary to control the total volume of soil sample. One difficulty arises as the amount of soil for each soil type required is not uniform. A solution is to perform a few preliminary tests for each compactive effort in advance so as to grasp an idea of the approximate amount of soil to be used.

6.8 Variation of water content in mould

For wetter samples, water will escape through the bottom of the mould causing a variation of water content within the mould. Thus it is necessary to take soil samples from different parts within the mould for water content determination.

Also the water squeezed out should be absorbed so that the weight of such water will not be taken when weighing the sample.

6.9 Soils susceptible to crushing

The procedures mentioned before only applies to soils not susceptible to crushing. If the soils under consideration are susceptible to crushing (i.e. they contain granular materials of a soft nature like soft limestone, sandstone, etc.), the contained granular material may be reduced in size by the action of the rammer. In such cases another procedure in compaction given in Test 12 of B.S. 1377:1975 has to be followed.

7 DIFFERENCE BETWEEN LABORATORY & SITE CONDITIONS

In the aforementioned experimental procedure, samples were obtained from the site, mixed with water to the desired moisture content, compacted with different compaction effort and tested with the sampling process. However, there exists the following discrepancies between the laboratory simulation and the actual field conditions:

1. The tested sample is a remoulded sample while on site the soil is undisturbed. Therefore the soil grains arrangement are quite different for two cases. For comparison purpose, a graph of undisturbed dry density against disturbed dry density is plotted for natural soil as shown in Figure 23. The

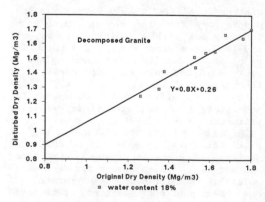

Fig.23 Sampling disturbance effect for natural residual soil (decomposed granite with 18% water content using 1.8mm thick sampler) for comparison purpose

soil type used is decomposed granite and the correlation is found to be of the similar type and pattern as the results found for remoulded soil. Hence it can be concluded that there will be little effect between natural and remoulded Hong Kong residual soils with respect to the sampling disturbance effect on soil density.

2. In the field condition the soil properties adjacent to the soil sample will usually be similar to certain distance but for the existence of a hard rock layer. However, in the laboratory sampling process, the boundary effect from both the sides and the bottom of the CBR mould cannot be ignored. The stress field within the soil mass in this case may be completely different from that occurred in the sampling process in the field. In the experiments, soil mass of larger volume could not be used due to the limitation imposed by the availability of apparatus and the required operations. The CBR mould is the largest possible for the sampling procedure to be carried out smoothly in order to avoid the difficulties that would have been encountered in manipulating a large soil mass. Nevertheless, it can be envisaged that the effects of bottom and sides can be minimized to certain extent by reducing the length of the soil sample. Due consideration should also be given to the loss of sample during the withdrawal of the samplers, which dictates the minimum length of the soil sample.

3. The dimensions of driving samplers used in the field will usually be of larger size than the ones used in the laboratory. It is reasonably suggested that there will be less disturbance effect on the soil mass for larger size of samplers and thus the laboratory results are on the conservative side.

8 APPLICATION OF TEST RESULTS ON SITE

A knowledge of density is essential, as indicated in B.S. 5930:1981, in all engineering problems where the body weight of the strata is an important factor, e.g. (i) in the stability of slopes and of earth dam, (ii) in determining the earth pressure on retaining walls, tunnel linings, and the timbering of excavations. Dry density is used as an essential indication of an adequate state of compaction in earth dams and embankments and in stabilized soil roads. Usually sample is taken out by sampling tube from the filled embankment and its density is measured to indicate the state of compaction acquired. The laboratory

results accrued from the study can act as a helpful tool here as the filled soil is also in the remoulded state despite the fact that the sampler used on site is usually of larger diameter, e.g. 100mm.

9 CONCLUSIONS

Based on the captioned laboratory simulation to the field sampling process of the subsurface materials in the current practice, we can conclude that the process yields very good samples of medium dense soils, but tends to densify loose soils and loosen dense soils. For both Hong Kong decomposed granite and decomposed volcanics tested, it is found that thinner-walled samplers give less disturbance effect as compared with thicker-walled samplers. Moreover, the correlation between undisturbed and disturbed dry densities varies slightly with water content and follows the same general pattern except for decomposed volcanics at high moisture content, which deviate from this trend.

The works contribute significantly to the better understanding of the properties of the tested residual soils in Hong Kong — decomposed granite and decomposed volcanics. Besides, the correlations will be extremely useful during the site investigation stage of any project as, based on it, appropriate adjustments can be effected to evaluate the field sampling results acquired during the process. Secondly, the true soil density attained in geotechnical engineering project such as a filled embankment can also be determined reliably.

Nevertheless, the investigation was limited only to two types of residual soils in Hong Kong, decomposed granite and decomposed volcanics. As such, it is strongly recommended that the above laboratory simulation studies can be extended to other types of soil in Hong Kong as well as to soils in other parts of the world so that the full scenario of the effect on sampling disturbance for soils with different properties can be gleaned.

REFERENCES

British Standard Institution. 1975. B.S. 1377: *Method of test for soils for Civil Engineering Purpose*.
British Standard Institution. 1981. B.S.5930: *Code of Practice for Site Investigations*.
Hvorslev, M.J. 1980. Subsurface explorat-ion and sampling of soils for Civil

Engineering purposes. *Report on a research project of the Committee on sampling and testing*, Soil Mechanics and Foundations Division, ASCE.

Lumb, P. 1962. The properties of decomposed granite. *Geotechnique*, 12 :226-243.

Lumb, P. 1962. General nature of the soils of Hong Kong. *Symposium on Hong Kong Soils*, Hong Kong Joint Group of the Institutions of Civil, Mechanical and Electrical Engineers: 19-31.

Lumb, P. 1965. The residual soils of Hong Kong. *Geotechnique*, 15: 180-194.

Marcuson, W.F. & A.G. Franklin 1980. State of the art of undisturbed sampling of cohesionless soils. *Geotechnical Engineering*, Journal of Southeast Asian Society of Soil Engineering, 11, No. 1 :31-53.

Scott, C.R. 1980. *An Introduction to Soil Mechanics and Foundations*.

Terzaghi, K. & Peck, R.B. 1967. *Soil Mechanics in Engineering Practice*, Wiley.

Developments in Geotechnical Aspects of Embankments, Excavations and Buried Structures, Balasubramaniam et al. (eds)
© 1991 Balkema, Rotterdam. ISBN 90 5410 019 2

Mechanics of wave pulse in rock fragmentation

Sanga Tangchawal
Department of Mining Engineering and Mining Geology, Faculty of Engineering, Chulalongkorn University, Thailand

ABSTRACT: An impulse load causing rock fragmentation is in the form of transient stress pulse. Four basic types of these stress wave pulses are rod or bar, plane, spherical, and cylindrical. The derivation of elastic wave equations and examples of wave mechanics for impulsive loading in drilling and conventional blasting were demonstrated. These would gain a better understanding of mechanisms of energy transfer and its application to the wave propagation and to the fractures of rock caused by impacts and explosives.

1 INTRODUCTION

Principles of rock fragmentation involve in the dynamic application of energy. Most rock fragmentation is accomplished by mechanical drilling and blasting with low to high order detonation explosives. A better understanding of the energy transfer, particularly to both soil and rock, resulting from wave propagation and the effects of these wave on rock structures, is critical to the solutions of problems.

The impulse load from a source is in the form of transient stress (or strain) wave pluse. Transient characteristics of wave pluse are of relatively short duration (10^{-6} to 10^{-5} seconds) and have different shapes. Geometrically, they are approximately plane, spherical, or cylindrical in shape. These types serve to illustrate the basic mechanics of wave propagation and consequent fracturing of the rock.

For the sake of simplicity, the wave pulses are considered in theoretical analyses only to longitudinal pulses and having symmetrical forms. The medium rock also assumes to be elastic, isotropic, homogeneous, and continuous medium.

2 ANALYSIS OF WAVE PULSES

When an elastic body is disturbed by impact pressure, the equations of motion of the disturbance can be obtained by considering the variations in stress occurring across the infinitesimal element. Some forms of wave equations are easily derived and can be readily evaluated. However, some wave equations are complicated and the differential wave equations can only be solved by use of transform calculus of other numerical methods.

An approach to the analysis of wave pulses, obtaining practical solutions, is to divide various characteristics of wave pulses into two main categories: planar and nonplanar wave pulses. The transient planar wave pulses are the bar (or rod) wave and the plane (dilatation) wave. These two planar pulses propagate at a great distance from the region of initiation, the radius of curvature of the wave front will become infinite so that the front can be treated as plane. Other transient waves behave differently, their propagation characteristics are of nonplanar waves and the wave front is assumed curved, such as spherically and cylindrically expanding elastic waves. These are the types often produced by explosions and highly localized impacts.

One major difference between nonplanar and planar wave pulses is that both spherical and cylindrical pulses change shape as they advance, altering markedly the distribution of stress and particle velocity within the pulse. In both spherical and cylindrical waves, the normal stresses on an element are compressive in the radial direction and usually tensile in the tangential direction.

2.1 Bar (rod) wave pulse

This type of wave pulse propagates in a bar or rod of small diameter in which the Poisson's ratio effects due to lateral extension caused by a longitudinal stress wave pulse are negligible.

The wave equation for a bar wave developed by considering a small longitudinal section of an elastic solid bar subject to a stress wave pulse moves in accordance with Newton's law of motion (Figure 1). The plane wave equation for a wave pulse in a bar is

$$\frac{\partial^2 u}{\partial x^2} = \frac{\rho}{E} \cdot \frac{\partial^2 u}{\partial t^2} \qquad (1)$$

where
- u = displacement of particle
- ρ = density of the medium
- E = Young's modulus

If substitute in terms of the bar velocity of longitudinal wave c_b, equation 1 becomes

$$\frac{\partial^2 u}{\partial x^2} = \frac{1}{c_b^2} \cdot \frac{\partial^2 u}{\partial t^2} \qquad (2)$$

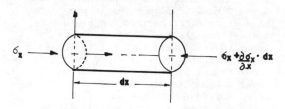

Figure 1 Compressive stresses on an infinitesimal section of the bar wave

2.2 Plane (dilatation) wave pulse

An assumption for a plane (dilatation) wave pulse is that a pulse traveling in the x direction in an infinite medium.

The material is constrained in the y and z directions so that the displacement and strain in these directions is zero (not for the stress). Consider compressive stresses on the infinitesimal cubic element which is shown in Figure 2 below.

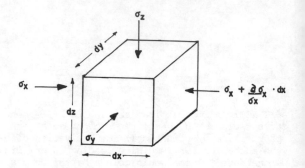

Figure 2 Compressive stresses on the infinitesimal cubic element with no strain produced in the plane perpendicular to x direction

The equation of equilibrium of a plane wave pulse leads to

$$\frac{\partial \sigma_x}{\partial x} = \rho \cdot \frac{\partial^2 u}{\partial t^2} \qquad (3)$$

From theory of elasticity, for an infinite medium, the stress-strain relationship is given by

$$\sigma_x = \lambda e + 2G\varepsilon_x \qquad (4)$$

where
λ = Lame's constant

$$= \frac{\nu E}{(1 + \nu)(1 - 2\nu)}$$

ε = strain

e = dilatation (volume expansion)

$\quad = \varepsilon_x + \varepsilon_x + \varepsilon_x$

G = shear modulus = $\dfrac{E}{2(1 + \nu)}$

ν = Poisson's ratio

To obtain the equation of motion for a plane wave pulse from the equilibrium equation, and substitute all constraint variants

$$(\lambda + 2G) \frac{\partial^2 u}{\partial x^2} = \rho \cdot \frac{\partial^2 u}{\partial t^2} \qquad (5)$$

or, in terms of a plane wave velocity c, the equation gives

$$\frac{\partial^2 u}{\partial x^2} = \frac{1}{c^2} \cdot \frac{\partial^2 u}{\partial t^2} \qquad (6)$$

in which
c = longitudinal (dilatation) wave velocity

$$= \sqrt{\frac{\lambda + 2G}{\rho}}$$

2.3 Spherical wave pulse

Analysis of spherical wave pulse can be derived from the equation of equilibrium for a small spherical element in polar coordinates. Figure 3 shows the compressive stresses on a small symmetrical spherical element in which the center of the spherical coordinate coincides with the polar coordinate system.

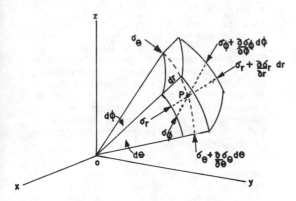

Figure 3 Compressive stress on a symmetrical spherical element where there is no rotation

Summation of forces on the point P in the r direction and in accordance with Newton's law of motion, F = mA.

$$\sigma_r(rd\theta)(rd\phi) - (\sigma_r + \frac{\partial \sigma_r}{\partial r}dr)(r + dr) \cdot$$

$$d\theta (r + dr) d\phi + 2\sigma_\theta \sin \frac{d\theta}{2} (r + dr) \cdot$$

$$d\theta dr + 2\sigma_\phi \sin \frac{d\phi}{2} (r + dr) d\phi dr$$

$$= mA \qquad (7)$$

As the dimensions of the element are made smaller and smaller, the angles $(d\theta/2)$ and $(d\phi/2)$ can substitute for the values of $\sin(d\theta/2)$ and $\sin(d\phi/2)$, respectively. Second order terms are neglected, so equation 7 is simplified to yield

$$\frac{\partial \sigma_r}{\partial r} + 2\frac{\sigma_r}{r} - \frac{\sigma_\theta}{r} - \frac{\sigma_\phi}{r} = \rho \frac{\partial^2 u}{\partial t^2} \qquad (8)$$

For a dilatation wave there is no rotation of element, hence

$$\frac{\partial \sigma_r}{\partial r} + \frac{2(\sigma_r - \sigma_\theta)}{r} = \rho \cdot \frac{\partial^2 u}{\partial t^2} \qquad (9)$$

The equation of motion for a spherical wave in terms of displacement and elasticity constants can be obtained by substituting all constraint variants.

$$(\lambda + 2G) \frac{\partial^2 u}{\partial r^2} + \frac{2}{r} \frac{\partial u}{\partial r} - \frac{2u}{r^2} \qquad (10)$$

$$= \rho \cdot \frac{\partial^2 u}{\partial t^2}$$

Equation 10 can also rewrite in terms of displacement and wave velocity.

$$\frac{\partial^2 u}{\partial r^2} + \frac{2}{r} \frac{\partial u}{\partial r} - \frac{2u}{r^2} = \frac{1}{c^2} \cdot \frac{\partial^2 u}{\partial t^2} \qquad (11)$$

The displacement u in terms of the displacement potential is

$$u = \frac{\partial \phi}{\partial r} \qquad (12)$$

By substituting equation 11 with the displacement potential value and factoring, the general equation for a symmetrical spherical wave pulse of the one dimensional wave equation is

$$\frac{\partial^2 (r\phi)}{\partial r^2} = \frac{1}{c^2} \cdot \frac{\partial^2 (r\phi)}{\partial t^2} \qquad (13)$$

2.4 Cylindrical wave pulse

In the case of a cylindrical cavity embedded in a homogeneous, elastic medium, infinite in extent, which is subjected to an impulsive internal pressure, the displacement takes place in the direction of the radius vector. The problem can be solved by postulating that compressive stresses act on a small element of a body cut out by two radii and two circular arcs, Figure 4. The position of a point at the center of this element, in two dimensions, is given by r and θ.

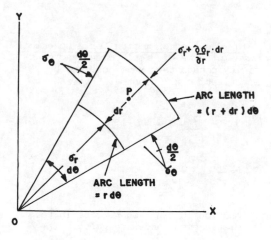

Figure 4   Compressive stresses on a small symmetrical cylindrical element

For an axisymmetrical problem, the shear stresses on the element surface are zero and the tangential stresses are independent of the angle θ. Summation of forces in the radial direction for zero body force gives the symmetrical

cylindrical wave equation.

$$\sigma_r (rd\theta) - (\sigma_r + \frac{\partial \sigma_r}{\partial r} \cdot dr)(r + dr)d\theta$$

$$+ 2\sigma_\theta \sin \frac{d\theta}{2}(dr) = mA \qquad (14)$$

As a small cylindrical element, neglecting some terms will give this equation close to that of a spherical wave equation.

$$\frac{\partial \sigma_r}{\partial r} + (\frac{\sigma_r - \sigma_\theta}{r}) = \rho \cdot \frac{\partial^2 u}{\partial t^2} \qquad (15)$$

General equation of a cylindrical wave in one dimensional case is

$$\frac{\partial^2 \psi}{\partial r^2} + (\frac{1}{r}) \cdot \frac{\partial \psi}{\partial r} = \frac{1}{c^2} \cdot \frac{\partial^2 \psi}{\partial t^2} \qquad (16)$$

$$\psi = (\frac{1}{r}) \cdot \frac{\partial (ru)}{\partial r} \qquad (17)$$

3 DISCUSSION

An inception of productive applications of modern processes of impulsive loading on how stress transients behave in solids and their use has been essential to the knowledge of rock fragmentation. Basic understanding of energy transfer in rock fragmentation will recently be made in increments, but these small advances may be critically important for the rational design and controlling of impacts and explosions.

Examples of wave mechanics for impulsive loading in this paper are twofold: impact drilling and conventional blasting. Each type of application serves to illustrate the basic mechanics of wave propagation and to the fractures of rock caused by impacts and explosives.

3.1 Stress wave in drill steel

When a piston in a percussion drill strikes the end of a drill rod, the impact generates a wave in both the piston and the rod. To achieve an understanding of the mechanism of energy transfer, one must have a thorough knowledge of the initiation and propagation of the plane elastic stress wave (Clark, 1979).

To illustrate the stress wave in drill steel, a piston and an infinite bar (rod) of the same diameter are chosen. A compressive pulse travels in a bar when it meets a change in cross section where part of the wave will likewise be reflected and part transmitted, Figure 5a. Continuity of forces at the change in cross section requires that

$$F_i + F_r = F_t \qquad (18)$$

where F_i = incident force

F_r = reflected force, and

F_t = transmitted force

The application of impulse concept and change of momentum gives

$$\sigma = \rho c_b v \qquad (19)$$

c_b represents the wave velocity in the bar, and

v represents the particle velocity in the bar

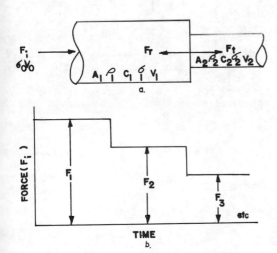

Figure 5   a) Forces within a bar for area change
b) Force-time diagram which results as the unit step wave form

Continuity of motion also requires that

$$v_o = v_p + v_b \qquad (20)$$

Assuming the piston and the bar are made of the same material and

$$q = A_b / A_p \qquad (21)$$

Substitute all values from equation 19 to equation 21 into equation 18, one gets the reflected force for the first impact is

$$F_r = \frac{(q-1)}{(q+1)} \cdot F_i \qquad (22)$$

or

$$F_r = R_c (F_i) \qquad (23)$$

Similarly, the transmitted force for the first impact is

$$F_t = \frac{2q}{(q+1)} (F_i) \qquad (24)$$

or

$$F_t = T_c (F_i) \qquad (25)$$

R_c is defined as the reflection coefficient, and T_c is the transmission coefficient. Thus, for each successive pulse reflection, the new incident force in the infinite bar by an amount equal T_c multiplied by the last incident wave in the piston resulting in a decreasing step wave form, as shown in Figure 5b.

Equations 23 and 25 can also rewrite in terms of stress as

$$\sigma_r = R_c(\sigma_i) \qquad (26)$$

and

$$\sigma_t = t_c(\sigma_i) \qquad (27)$$

The stress wave analysis as illustrated is only basic processes of one dimensional propagation of waves for the simple system. In a percussion drill, it would also be necessary to consider the effects of waves reflected from the end of the drill rod, the effects of the drill chuck, and other factors that were not taken into account in the above.

3.2 Stress wave in conventional blasting

The problem of the generation of elastic waves by explosion pressures is usually chosen to state the problem, by ignoring any effects in actual initiation process as follows: given a spherical cavity of

radius "a" within a homogeneous, ideally

elastic infinite medium of density "ρ." Also assuming that the point of explosive initiation is at the center of circular and the propagated wave pulse has the compressional dilatation wave velocity

"c" and to find the elastic wave motion parameters which results from application of an arbitrary pressure P(t) at point P. Figure 6 illustrates the statement of the problem.

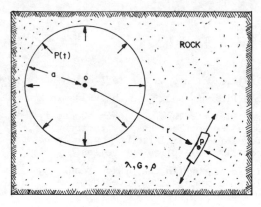

Figure 6 Illustration of the stated model for the elastic wave pulses generated by explosion pressures

To solve the problem which has been formally stated above, one needs to find the elastic wave displacement which is a solution of the equation of motion for

a wave pulse, vanishes at infinity, and satisfies the given boundary conditions. The equation of motion of a spherical elastic wave is

$$\frac{\partial^2 (r\phi)}{\partial r^2} = \frac{1}{c^2} \cdot \frac{\partial^2 (r\phi)}{\partial t^2} \quad (28)$$

where $\quad u = \dfrac{\partial \phi}{\partial r}$

The Dirac delta function can simulate the unit applied pressure P(t). Hence, the radial stress at the interior surface of the cavity.

$$- \left[(\lambda + 2G) \frac{\partial u}{\partial r} + \frac{2\lambda u}{r} \right]_{r=a} = P(t) \quad (29)$$

At this point it is convenient to assume $\lambda = G$ or the Poisson's ratio is 0.25. For the periodic function, the applied pressure can be represented by a Fourier integral. The equation that satisfies all boundary conditions becomes

$$- \rho c^2 \left[\partial u / \partial r + 2u / 3r \right]_{r=a} \quad (30)$$

$$= 1/2\pi \int_{-\infty}^{\infty} \int_{-\infty}^{\infty} P(\gamma) e^{in(\gamma - t)} \, dn \, d\gamma$$

Sharpe (p. 146-150) suggested a solution of the motion equation (equation 28) which represents a diverging pressure pulse from the surface of the spherical cavity is

$$\phi \qquad - \frac{(1)}{r} e^{-in\tau} \quad (31)$$

where

$$\tau \qquad = t - \frac{(r-a)}{c}$$

If multiply any functions which is independent of r and t, the right hand side of equation 31 is still a solution.

108

$$\phi = (1/2\pi r)A(n)P(\gamma)e^{in(\gamma-\tau)} \qquad (32)$$

where

$$A(n) = \text{arbitary function}$$

To sum up such solution in equation 32, one has a solution.

$$\phi = 1/2\pi r \int_{-\infty}^{\infty} \int_{-\infty}^{\infty} A(n)P(\gamma)e^{in\ (\gamma-\tau)} \, dn \, d\gamma \qquad (33)$$

Forming $u = (\partial\phi/\partial r)$ and $(\partial u/\partial r)$ from equation 32, and substituting the resulting expressions into the equation 30 gives

$$-\rho c^2/2\pi a \int_{-\infty}^{\infty} \int_{-\infty}^{\infty} A(n)(4/3a^2-4in/3ac-n^2c^2).$$
$$P(\gamma)e^{in(\gamma-\tau)} \quad dn \quad d\gamma$$
$$= 1/2\pi \int_{-\infty}^{\infty} \int_{-\infty}^{\infty} P(\gamma)e^{in(\gamma-\tau)} \quad dn \ d\gamma \quad (34)$$

The above equation (equation 34) will satisfy identically provides that the arbitary function $A(n)$ has the value

$$A(n) = (a/\rho)(n^2 + 4inc/3a - 4c^2/3a^2)^{-1} \qquad (35)$$

To achieve a formal solution of the wave pulse motion produced when a pressure of arbitary form is applied to the interior surface of a spherical cavity, one substitutes the expression for $A(n)$ in equation 32. The displacement potential of equation below satisfies all boundary conditions, that is

$$\phi = \frac{a}{2\pi\rho r} \int_{-\infty}^{\infty} \int_{-\infty}^{\infty} \frac{P(\gamma)e^{in\ (\gamma-\tau)}}{n^2 + 4inc/3a - 4c^2/3a^2} \, dn \, d\gamma$$
$$(36)$$

Since $P(t)$ can have a Fourier integral representation, the applied pressure is described by a single exponential pressuse pulse.

$$P(t) = P_o e^{-\alpha t} \quad \text{for} \ t \geqq 0 \qquad (37)$$
$$= 0 \qquad \text{for} \ t < 0$$

The pressure P_o is a constant pressure pulse and represents the initial and highest pressure attained (detonation pressure). The α symbol is a positive time decay constant.

Using this form of pressure in equation 36, carrying out the γ integration, and factoring the denominator, one has

$$\phi = \frac{iaP_o}{2\pi\rho r} \cdot \qquad (38)$$

$$\int_{-\infty}^{\infty} \frac{e^{-in\tau}(dn)}{\left[n+i\alpha\right]\left[n+2c(1+\sqrt{2})/3a\right]\left[n+2c(i-\sqrt{2})/3a\right]}$$

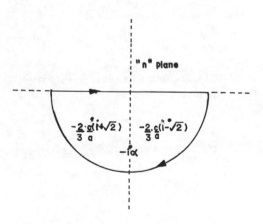

Figure 7 Contour used in evaluation (theory of residues) of the integral of equation 38, showing the location of the integrand

By using the path of integration, from the theory of residues, as the contour shown in Figure 7, the integral in equation 38 can be evaluated. The result for the general solution of the displacment potential is

$$\phi = \frac{aP_o/\rho r}{(\omega/\sqrt{2}-\alpha)^2+\omega^2} \left[-e^{-\alpha\tau} +e^{-\omega\tau/\sqrt{2}}\{(1/\sqrt{2}-\alpha/\omega)\right.$$

$$\left.\cdot(\sin \omega\tau)+ \cos \omega\tau\}\right] \text{ for } \tau \geq 0$$

$$= 0 \text{ for } \tau < 0 \qquad (39)$$

where ω = angular frequency for circular oscillation

$$= \frac{2\sqrt{2}c}{3a}$$

and λ = G, or the Poisson's ratio is 0.25

To let the applied pressure has a unit function form, one permits the time decay constant (α) to approach zero.

$$P(t) = P_o \qquad \text{for } t \geq 0 \qquad (40)$$

$$P(t) = 0 \qquad \text{for } t < 0$$

The displacement potential solution for α approaches zero is

$$\phi =$$

$$\frac{2aP_o}{3\rho\omega^2 r}\left[-1+\sqrt{3/2} \ e^{-\omega\tau\sqrt{2}} \ \sin(\omega\tau+\tan^{-1}\sqrt{2})\right]$$

$$\text{for } \tau \geq 0$$

$$= 0 \text{ for } \tau < 0 \qquad (41)$$

From equation 41 the displacement of particle u in the rock medium corresponding to the application of unit function pressure to the interior

surface of the cavity. The general solution of the particle displacement is

$$u = \frac{\partial\phi}{\partial r} = \frac{aP_o}{4G}\left[(a/r)^2-\sqrt{3/2} \ (a/r)^2\right.$$

$$e^{-\omega\tau/\sqrt{2}} \ \sin(\omega\tau+\tan^{-1}\sqrt{2})+\sqrt{2}(a/r)e^{-\omega\tau/\sqrt{2}}.$$

$$\left.\sin\omega\tau\right] \qquad \text{for } \tau \geq 0$$

$$u = o \qquad \text{for } \tau < 0 \qquad (42)$$

where λ = G, or the Poisson's ratio

is 0.25

The above equation is the partial derivative of equation 41 with respect to r and has reintroduced the definitions of ω and G into the coefficient outside the bracket.

Another solution which suggested by Duvall (p. 311-312) using the Poisson's ratio as 0.25, the displacement potential resulting from application of a pressure pulse of the form,

$$P(t) = P (e^{-\alpha t}- e^{-\beta t}) \text{ for } \beta>\alpha$$

where α and β are constants of pressure pulse parameters $\qquad (43)$

4 CONCLUSION

Stress wave pulse in drill steel as discussed (Section 1 of the Discussion Part) is only applied to the simple piston case. A digital computer may be used to determine the shape of the stress wave pulse produced by a percussion drill piston of complex geometry.

The force-time diagram (stress waveform) displays the energy time relationship. This provides a basis for the determination at which energy is transferred from piston to drill rod. A better understanding of the integrated process as a whole is to investigate the characteristics of energy transfer at all points in the system.

Section 2 of the discussion is the application of stress wave pulse in conventional blasting. The discussion was concentrated primarily on having the solution of the radial wave pulse parameters: displacement potential and displacement of particle. Furthermore, the foregoing analysis can obtain other successive wave parameters: particle velocity, particle acceleration, stress and strain by a similar solving method.

The solutions defined by Hornsey and Clark (p.116-119) during the Tenth Rock Mechanics Sympsosium are more general than those given by Sharpe and Duvall, with no restriction on the value of Poisson's ratio. This latter developed method is more accurate and may exhibit better significant features of real pressure pulses.

Comparison of the experiment between observed waveforms and numerical calculations of wave equations usually be made using granite as the medium rock. Curves of various wave parameters, especially those of displacement and decay preesure pulse forms were plotted as functions of time and radial distance. The details of comparison are out of the scope of this paper. However, it can be concluded that the displacement

on the cavity radius "a" and increase with the cavity radius value. The decay rate of pressure pulse (from the initial peak value) becomes inversely proportional to radial distance as rise time of the pressure pulse decreases. It also approaches the inverse first power of the distance as the cavity radius and radial distance become large.

At best it appears that shock wave pulses and pulses from large underground explosions, which transverse the rock beyond the fracture and plastic zones, do not behave as that predicted by elastic wave equations. At small to intermediate distances, the attenuation rates of the peak values of displacement and other wave pulses parameters are generally greater than those predicted by an elastic wave theory. Only at large distances that the elastic wave theory may be used as a first approximation for real waves provided appropriate values of radius

cavity "a" and initial constant pressure P_o could be found.

In view of the competent hard rocks (granite, sandstone, limestone), there are relatively small effect of changes in Poisson's ratio and the fact that most competent rocks do not differ too significantly in density, an onward analysis should be made on the basis of average values of wave pulse parameters. Changes in elastic wave pulses from one rock medium to another are functions of only the wave velocity of the rock medium.

ACKNOWLEDGEMENTS

The author wishes to express his deep appreciation to those previous contributors as referred in the references and to those who assisted in the preparation of this paper.

REFERENCES

Clark, G.B. 1979. Wave transmission in drill steel. In quarterly of Colorado school of mines "principles of rock drilling"74: 7-11.

Clark, G.B. 1987. Principles of rock fragmentation. John Wiley and Sons.

Dutta, P.K. 1968. The determination of stress wave forms produced by percussive drill pistons of various geometrics design. Int. J. Rock Mech. Min. Sci. & Geomech. Abstr. 5: 501-518.

Duvall, W.I. 1953. Strain wave shapes in rock near explosions. Geophysics 18: 310-323.

Hornsey, E.E. and Clark, G.B. 1968. Comparison of spherical elastic, Voigt, and observed wave forms for large underground explosions. Proceedings U.S. Tenth Sysmposium on Rock Mechanics, Austin, Texas: 115-148.

Selberg, H.L. 1952. Transient compression wave from spherical and cylindrical cavities. Arkiv för Fysik 5: 97-108.

Sharpe, J.A. 1942. The production of elastic waves by explosion pressure. I. Theory and empirical field observations, Geophysics 7: 144-154.

Developments in Geotechnical Aspects of Embankments, Excavations and Buried Structures, Balasubramaniam et al. (eds)
© 1991 Balkema, Rotterdam. ISBN 90 5410 019 2

Filter design for tropical residual soils

B. Indraratna & E. L. G. Dilema
Division of Geotechnical & Transportation Engineering, Asian Institute of Technology, Bangkok, Thailand

Abstract: In this paper, design criteria for cohesionless granular filters are proposed to minimize the erosion of residual soil under conditions of concentrated leaks. Earth dam engineering had increasingly emphasized the detrimental effects of concentrated leaks developed through a cracked core, responsible for considerable erosion of fine materials, mainly clays and silts. Sealing of cracks depends on the capability of the downstream filter to retain such fine particles. Eroded particles rarely exist in dispersed conditions due to the presence of calcium and other multivalent cations in the reservoir water that encourage flocculation. Therefore, economical filter design should attempt to consider the retention of clay flocs rather than individual dispersed particles. Design guidelines based on filter permeability and retained equivalent particle sizes are discussed in this paper. From the experimental results, a distinct demarcation between effective and non-effective filters for a given equivalent floc size has been identified. Such a relationship established for these fine erodible particles can be used in design for the protection of dam cores composed of similar residual soils.

1 Introduction

The filter design techniques adopted for Balderhead Dam conforms to the filter rules for non-cohesive soils which are based on the piping ratio d_{85}/D_{15} by Bertram (1940) and Sherard et al (1963). These design criteria had also been employed successfully for the filter drains of the Selset Dam (Bishop and Vaughan, 1962). However, when the reservoir of the Balderhead Dam had been impounded, a large 'sink-hole' had appeared in its crest. Investigations had shown that the core had suffered cracking by hydraulic fracture before impounding had been completed. Internal erosion of the core material had taken place slowly for a period of 14 months (Vaughan et al, 1970).

The occurrence of concentrated leaks is related to embankment cracking which poses detrimental effects on earth dam engineering. When these leaks occur followed by erosion of the walls of the leakage channels, the erodible particles may be broken down into its fine constituents which are carried in suspension through the leakage channel towards the filter. For an impervious core of non-dispersive soils, the smallest particle that arise during erosion and segregation generally exists in its flocculated state, influenced by the mineralogy of the base material, chemistry (salinity) of the reservoir water and the soil concentration in suspension. Therefore, the economical design of filters should consider the retention of flocculated clay particles, instead of adopting stringent criteria to prevent the erosion of individual dispersed particles. The downstream filter adjacent to the dam core must not only be capable of retaining the eroded particles, but must also sustain stability at increased pressure gradients imposed by the eroding fluid forces.

The scope of this paper is to propose a filter criterion for effective retaining of erodible clay flocs from a residual soil. A typical lateritic residual soil from Saraburi Province in Thailand has been selected for this particular study, since very similar soils are being considered as embankment fill and core construction material by the Royal Irrigation Department of Thailand, through which this research program has been motivated. For this study, granular (cohesionless) soils were considered as the filter material, whereas the base material consisted of the fine fraction sieved from the lateritic residual soil.

2 Design of granular filters for lateritic residual base soils

2.1 Filter as a crack control measure

Filters should not only serve as a measure of dissipating excess pore pressures and draining seepage water, but must also serve as a barrier in resisting erosion of fine particles through a cracked core. For a filter to be effective, it must be non-cohesive, otherwise, it may sustain an open crack, hence fail to protect the damaged core. If a crack forms in the cohesive core, the filter should be able to arrest the eroded fine particles transported by the seepage water, and subsequently reduce the rate of leakage through the crack. An effective filter in this manner may provide sufficient time for the crack to seal by gradual swelling of the clay, due to wetting of the core in the vicinity of the crack.

Under circumstances of large differential settlements, multiple cracks can occur within the core leading to excessive seepage. Even

if the filter is well designed, effective retention depends on the size of fine particles that are carried by the seepage water. In order to evaluate the effectiveness of filters under such situations, the degree of flocculation of the clay particles needs to be examined (Vaughan, 1978; Sherard et al, 1984) with respect to the filter permeability. The filter design method used for the Cow Green Dam and the Empingham Dam has been based on the above concept, where the compacted core material is composed of boulder clay and upper lias clay, respectively (Vaughan and Soares, 1982). These design principles have been extended in this study, in order to design granular filters for tropical residual soils that are commonly used as a core construction material in the Asia-Pacific region.

2.2 Filter material and base soil preparation

The granular filters consisted of sub-angular quartz river sands which are brownish yellow in colour and are locally available. Rinsing of sand was necessary to remove fine particles (less than 75 μm) which can add cohesion to the otherwise cohesionless filter. Mechanical sieving was employed to separate the material into different ranges of grain sizes. Both uniform and graded filters were prepared. In the case of graded filters, sieved material of pre-determined sizes were blended in various proportions, and their uniformity coefficients were computed. Uniform filters consisted of an array of carefully sieved fine, medium and coarse sand. Variations in uniformity coefficients and grain sizes were evaluated to compare and contrast the efficiency of different filters.

Soil fraction passing sieve #200 was considered to be within the size range of erodible particles. The equivalent particle sizes of the fine fractions were determined by hydrometer sedimentation analysis. The purpose of conducting sedimentation tests was not to determine the exact individual particle sizes, but the equivalent particle sizes associated with the specific soil-water system. The clay suspension simulating a concentrated leak consisted of distilled water, 25 g/l of fine soil fraction and CaCl$_2$ solution at a concentration of 3.5 me/l.

2.3 Filtration apparatus

The filtration apparatus was designed to simulate a concentrated leak through a crack in the dam core, discharging into a downstream filter as illustrated in Figure 1a. It consists of two transparent cylinders connected by a central disk (Figure 1b). The lower cylinder contains the filter material with appropriate tapping points for head loss measurements and the upper cylinder contains the soil slurry. The apparatus was connected to a constant head of about 2.0 m, and a hydraulic gradient in the order of 8 – 10 was maintained throughout the test.

2.4 Influence of water chemistry

The degree of flocculation of fine soil particles is a direct function of the water chemistry (Arulanandan et al, 1975 and Soares, 1980). If the reservoir water is rich in multivalent cations such as calcium, the eroded soil particles reaching the downstream filter will be flocculated. Therefore, the design of the filter does not require stringent criteria to retain the smallest possible, dispersed individual clay particles. Instead, the clay flocs can be considered as the size of the base particles required to be arrested by the filter.

Figure 2 illustrates the variations of the mean equivalent size of clay flocs as a function of the calcium concentration in the slurry. The sizes of clay flocs were estimated by hydrometer sedimentation tests. It was observed that the mean particle size of flocs increases with [Ca^{++}], but attains a maximum at a concentration of about 10 me/l. A similar trend has been observed by Soares (1980) for the flocculation of boulder clay used for the Cow Green Dam core. Cation concentrations of typical reservoir waters are also indicated in Figure 2 with several selected dams in

Thailand, from which reservoir water samples were analyzed by the atomic absorption apparatus.

2.5 Influence of soil concentration

Investigation of base soil grain sizes with varying soil concentration indicated a significant reduction in settling velocities for a concentration of 50 g/l. In other words, at high soil concentrations, the equivalent particle sizes are underestimated. Allen (1975) has compared the particle sizes determined by sedimentation analysis to those examined by microscopic techniques. According to the findings of this study, a soil particle concentration of 1% by volume has been suggested, which corresponds to 27 g/l approximately. Although, Allen (1975) has not been concerned with the influence of cations on fine particle flocculation, it has been verified that an increase of the clay concentration in suspension affects the settling velocity of the particles. It may be postulated that even if the settling velocity of one particle could accelerate another particle along its path, the quantity of settling volume would be counteracted by an upward flow of equal volume, giving a net result of reduced settlement. In this study, a base soil quantity of 25 g/l (dry soil) has been considered as appropriate.

3 Laboratory simulation

3.1 Experimental methodology

The filtration test procedure is summarized as follows:

(1) The filter material was compacted in a conventional permeameter, in several lightly tamped layers, above a 1" thick uniform fine gravel. The filter was subsequently saturated and the coefficient of permeability was measured.

(2) The soil suspension was prepared. Sedimentation analysis was carried out prior to each filtration test for the determination of equivalent particle sizes.

115

(3) The upper cylinder of the apparatus was then connected to the lower cylinder containing the granular filter.

(4) The soil slurry was transferred into the upper cylinder above the filter material with the use of a deflector to prevent any disturbance of the interface.

(5) The upper end plate was attached and the water intake valve was opened simultaneously with the discharge valve. The constant water head was then applied, which maintained a hydraulic gradient between 8 and 10.

(6) Flow rates and head losses were measured at 5 minute intervals to evaluate the filter behaviour. Effluents were sampled for turbidity and colour comparison tests at various time intervals.

3.2 Filter effectiveness

The filtration tests were classified according to certain factors that influence the efficiency of filters in retaining floc-size particles. The following factors were taken into consideration in determining the effectiveness of filters:

Effective filters - The filter is able to retain the fine particles by establishing a thin skin of base material initially trapped at the interface. A noticeable drop of the flow rate and permeability due to this self-filtering process is initially observed, which attains a constant level with time. Turbidity measurements of the effluent must be within acceptable limits (less than 25 NTU), throughout the test.

Ineffective filters - Appreciable amount of fine particles either clog the filter completely or get totally washed out in a muddy flow giving rise to highly turbid effluents. Flow rate and permeability measurements are not indicative of effective self-filtering process. In fact on several occasions, the flow rate and permeability suddenly increased, as the thin skin which attempts to create a self-filtering layer could not withstand the sudden enhanced hydraulic gradient, thereby undergoing interface "blow out".

Clogging filters - This phenomenon could be detected by rapidly decreasing flow rates, which do not stabilize to constant levels with time, but instead diminish monotonically to practically insignificant values.

3.3 Filtration tests and data interpretation

The first series of sand filters tested consisted of uniform fine, medium and coarse sand having almost the same uniformity coefficients. The fine and medium sand filters effectively control erosion by retaining the base soil at the filter-slurry interface. Therefore, the effluent turbidities corresponding to these filters are insignificant. As a result of the relatively small permeability of these effective filters, the associated effluent flow rates are also very low. The initial drop of the effluent flow rates is associated with the self-filtering phenomenon which is fully effective within 30 minutes, beyond which constant flow rates are measured. The effluent flow rate corresponding to the coarse filter does not indicate significant development of self-filtering, and consequently becomes unable to retain the flocculated clay particles. This is attributed to the insufficient packing associated with coarse uniform grains. Consequently, the coarse sand filter does not effectively retain the base soil, as indicated by the excessively high turbidity , where severe erosion of the soil occurs through the filter during the first few minutes. Although an increase of the mean uniformity coefficient from 1.35 to 2.3 , indicates a significant reduction in erosion, its permeability is still too excessive to establish a stable self-filter interface .

The next series of tests was conducted on fine - medium sand filters, where the D_5 , D_{10} or D_{15}

was kept common for each group of filters, but with varied uniformity coefficients. For a specific grain size to be held constant, the coarse fraction of the filter material should be increased giving rise to a higher C_u. It was observed that irrespective of the common grain size, all three sets of filters revealed similar trends with regard to permeability, effluent flow rate and turbidity values. These filters can be regarded as effective due to the low effluent turbidity measurements, with an average time of about 20 minutes for self-filtering to be fully effective, except for one filter ($C_u = 3.8$) which becomes clogged with time, demonstrating a continuously decreasing flow rate, without attaining a constant value. The permeability of the above filter could not be measured properly, due to the very small and unsteady head between tapping points. In contrast to D_5, D_{10} and D_{15} filter groups, a group of filters with common mid-range grain sizes was tested. Observations clearly indicated that the mid range grain sizes such as D_{50} do not seem to be directly related to the filter permeability. However, it was observed that the filter permeability systematically decreases with increasing C_u, and that a number of filters having a wide gradation ($C_u > 4$) generally indicate clogging.

In general, it can be observed that for this lateritic soil of residual nature, uniform coarse filters are ineffective in comparison with fine and medium filters of similar uniformity coefficients. On the other hand, the risk of clogging is enhanced at elevated values of C_u. In order to evaluate further the performance of coarse filters, another series of filter tests with fixed D_{100} was conducted for different gradations (C_u @ 2 - 4). In this manner, the largest size of the filter particles was controlled to provide greater flexibility in the assortment of lower size fractions.

Two distinct types of ineffective coarse filters were observed, namely clogging and interface instability. The filter with the widest gradation ($C_u = 3.4$) indicates gradual clogging, producing a monotonically decreasing flow rate. The permeability of this filter could not be determined accurately after the inception of clogging , as a result of the difficulties associated with small and unsteady head between the tapping points. However, its permeability is expected to be the lowest due to the greatest degree of packing. The other filters (ineffective) initially indicate a diminishing flow rate, but experience an abrupt increase in flow after several minutes . This is attributed to the initial self-filtering process, where a thin skin of fine materials temporarily seals or bridges the pores of the slurry-filter interface. Subsequently, the hydraulic gradients rapidly increase at the vicinity of the interface, causing instability (piping) and destroying the activated layer, thereby increasing the flow rate again. In contrast to the two ineffective filters, clogging behaviour is characterized by very low effluent turbidity from the beginning .

Considering the complete series of filters that have been investigated with regard to erosion control, the following aspects can be summarized:

Fine to medium grain size filters are generally effective, and rarely undergo clogging, unless the uniformity coefficient approaches four.

Uniform coarse sand filters are always ineffective, unless blended with some proportions of fine and medium sand. However, these graded filters increase the risk of clogging.

3.4 Design guidelines

(a) Filter permeability - grain size relationships

For granular soils of similar grading and grain shapes, permeability is a function of grain size and porosity. It has been verified earlier that the porosity is mainly influenced by the smaller size fractions of the filter (USACE,

117

1953; Honjo and Veneziano, 1989). Figure 3 summarizes the relationship obtained for permeability coefficient against D_5, D_{10} and D_{15}, respectively based on the results of this study. It is clearly seen that these grain fractions could be closely linked to the coefficient of filter permeability (k). The following expressions have been obtained by best fit analysis:

(i) $k = a D_5{}^b$;
lower bound: a = 0.41, b = 1.4
upper bound: a = 0.57, b = 1.79

(ii) $k = m D_{10}{}^n$;
lower bound: m = 0.39, n = 1.53
upper bound: m = 0.63, n = 1.80

(iii) $k = p D_{15}{}^q$;
lower bound: p = 0.38, q = 1.76
upper bound: p = 1.24, q = 1.83

The difference between the upper and lower bounds of D_{15} is somewhat greater than that of D_{10} and D_5. For grain sizes significantly larger than D_{15}, the scatter of data points becomes even more pronounced, hence, a realistic correlation between such grain sizes and the corresponding filter permeabilities cannot be properly defined. In fact, these findings seem to be in agreement with Kenney and Ofoegbu (1984) who have declared that permeability is primarily influenced by grain sizes less than D_{10}.

(b) Filter permeability – particle floc size relationship

The permeability of the filter can be regarded as a unique engineering property which characterizes the filter behaviour during internal seepage. The results of the above tests characterize the effectiveness of a filter to its permeability with regard to the gradation curve of the base soil. A specific particle size (δ_{85}) is selected for the base soil gradation curve to correlate with the filter permeability. This specific size, δ_{85} does not change significantly with moderate changes in cation concentration of the water (Dilema, 1990), which may be attributed to the flocculated particles that have attained their largest possible

size, where the particle weight is greater than the unbalanced force existing at its surface. Therefore, 15% of the largest fraction (i.e. δ_{85}) is more than adequate to create a self-filtering layer as also resolved from previous studies by Vaughan (1978) and Soares (1980).

The findings of this study, together with the results of previous studies conducted by Vaughan and Soares (1982) have been plotted in Figure 4. There is no doubt that a distinct boundary can be identified which separates the effective filters from the ineffective ones. The boundary which corresponds to the residual soils used in this study is not in agreement with that proposed for well-graded tills in the United Kingdom. This is expected as a result of the different mineralogical and aspects between these soils, which affect the extent of flocculation.

The boundary between the effective and ineffective filter zones for the lateritic residual soil can be quantified by the best fit regression as given below:

$k = 6.5 \times 10^{-4} \delta_{85}{}^{1.25}$ for k in cm/s and δ_{85} in µm

In contrast, the expression proposed by Vaughan and Soares (1982) for boulder tills is given by:

$k = 6.7 \times 10^{-4} \delta_{85}{}^{1.52}$, for k in cm/s and δ_{85} in µm.

It is important to note that in this study, clogging of filters has been considered as ineffective, because, ultimate plugging of a part of the filter generates rapid increase of hydraulic gradients in other portions of the filter, which can subsequently induce localized piping of filter grains in the vicinity.

4 Practical Implications

The permeability of the granular filter required for a particular dam project can be determined, by considering the possible sizes of the fine particles that are susceptible to erosion by seepage water, under cracked core conditions. The filter permeability can then be related to its particle distribution curve by specifying the quantities D_5, D_{10} and D_{15}, if the granular filter material is graded. Although previous investigations such as Vaughan and Soares (1982) and Sherard et al (1984) have stressed the role of filter permeability as a basis for filter design, the current practice is still greatly influenced by the use of grading ratios. Grading ratios which had been developed initially for cohesionless base particles (Bertram, 1940) have been extended by USACE (1955) for cohesive soils with the following major criteria given by: $D_{15}/\delta_{85} < 5$ and $D_{15} < 0.4$ mm. In order to realize the implications of the above guidelines, values of D_{15} against δ_{85} obtained in this study have been plotted in Figure 5, together with the USACE criteria. On the basis of the experimental evidence, it seems that for this tropical residual soil, $D_{15} = 0.3$ mm is a more realistic upper bound for the filter grains. On the other hand, the USACE grading ratio of $D_{15}/\delta_{85} < 5$ (originally documented by Bertram, 1940) is definitely acceptable. Considering the total range of base particles (45 - 90 µm), a slight refinement to the above grading ratio may be applied for this residual soil as indicated by the proposed demarcation (hatched) line:

(i) δ_{85} 45 to 60 µm ; D_{15}/δ_{85} 5.5 to 5 and (ii) δ_{85} 60 to 90 µm ; D_{15}/δ_{85} 5 to 4.

The quantification of an exact boundary between effective and ineffective filters for a given soil, requires a large number of test data in the vicinity of $D_{15} = 0.30$ mm, for the complete range of flocculated clay particles.

Experience in the North and Northeast Thailand has shown that the erosion of residual soils has always been a major concern. If the USACE recommendation of $D_{15} < 0.4$ mm is adopted for this particular lateritic soil, Figure 3 indicates an effective permeability coefficient in the order of 0.10 - 0.15 cm/sec. Consequently, the size of retained particles would be in

the range of 60 - 100 µm (δ_{85}), according to the relationship presented in Figure 4. On the basis of this investigation, the USACE criterion seems to be overestimated, hence, as proposed earlier, a value of D_{15} less than 0.3 mm seems to be more realistic for the residual soils in this region. It is hoped that these developments would be beneficial for future construction of dam cores with lateritic residual soils.

5 Conclusions

As in the case of most dam core materials, compacted lateritic residual soils can sustain open cracks due to their cohesive properties. Therefore, precautions must be taken to arrest the erosion of particles under cracked core conditions. The filter design criterion developed for this lateritic residual soil proposes an empirical relationship between the size of retained base particles (δ_{85}) and the effective filter permeability. It identifies a distinct demarcation between effective and non-effective filters corresponding to the flocculated particle sizes, and is quantified by the following equation:

$k = 6.5 \times 10^{-4} \; \delta_{85}^{1.25}$ where, k in cm/s and δ_{85} in µm.

Although the proposed relationship is developed for a lateritic, residual soil from the Saraburi province, Thailand, it is anticipated that it can be extended to other residual soils in the region with similar mineralogy and weathering aspects. For instance, certain core materials considered at the moment by the Royal Irrigation Department of Thailand in the provinces of Udon Thani, Suphanburi and Chonburi are very similar in nature. However, a definite limitation of the above proposed equation is that, it may not be used in confidence for any type of fine grained soil, such as one which is either relatively recent (unweathered) or contains a considerably higher clay content.

The filter rationale proposed in this study for the lateritic residual soils is not formulated to arrest the 'dispersed' size clay particles. It takes advantage of some degree of 'self-filtering', where larger size particles are first expected to bridge the interface pores during initial seepage, subsequently activating the filter to retain relatively smaller size particles without clogging. Therefore, an acceptable erosion level of fine particles (effluent turbidity < 25 NTU) is tolerated at the beginning.

Furthermore, the filter permeability is directly dependent on the smaller grain fractions of the filter material. The proposed relationships between the permeability and D_5 or D_{10} can be considered as particularly reliable due to convincing statistical correlations. These relationships can be used as important guidelines, at least in the preliminary design stages. Considering the complete array of filter tests conducted in this study, it can be concluded in general, that fine to medium sand filters are effective and rarely experience clogging, unless the uniformity coefficient is close to four. Uniform coarse sand filters are incapable of forming a self-filtering interface, and if excessively graded, the risk of clogging is introduced.

Acknowledgements

The authors gratefully appreciate the financial support provided by the Canadian International Development Agency (CIDA). Sincere thanks are also due to Laboratory Technicians for their assistance during the extensive experimental phase. Thanks are also extended to Mr. Golam Rasul (Research Engineer) for his assistance during preparation of the manuscript.

Bibliography

Allen, T., Particle Size Measurements, 2nd Edition, Chapman & Hall, London; 1975.

Arulanandan, K., Loganathan, P. and Krone, R.B., Pore and Eroding Fluid Influences on the Surface Erosion of a Soil, Journalof

Bertram, G.E., An Experimental Investigation of Protective Filters, Harvard Graduate School of Engineering, Publication No. 267, Vol.6; 1940.

Bishop, A.W. and Vaughan, P.R., Selset Reservoir: Design & Performance of the Embankment, Proceedings of Institution of Civil Engineers, Vol. 23, London, pp. 305-346; 1962.

Dilema, E.L.G., Development of permeability-floc size criterion for granular filter design, M.Sc. Thesis No. GT-89-4, Asian Institute of Technology; 1990.

Honjo, Y. and Veneziano, D., Improved filter criterion for cohesionless soils, Journal of Geotechnical Engineering, ASCE, Vol. 115, No. 1, pp. 75-94, 1989.

Kenney, T.C. and G.I. Ofoegbu, Permeability of Compacted Granular Materials, Canadian Geotechnical Journal, Vol. 21, pp.726-729; 1984.

Krishnayya, A.V.G., Eisenstein, Z. and Morgenstern, N.R., Behaviour of Compacted Soil in Tension, Journal of Geotechnical Engineering Division, Proceedings of the ASCE, Vol. 100, GT-9, pp. 1051-1061; 1974.

Sherard, J.L., Emabankment Dam Cracking, Embankment Dam Engineering, Casagrande Volume, Hirschfeld, R.C. and Poulos, S.J. Editors, John Wiley & Sons, pp. 271-353; 1972.

Sherard, J.L., Dunnigan, L.P., and Talbot, J.R., Filters for Silts and Clays, Journal of the Geotechnical Engineering Division, Proceedings of the ASCE, Vol. 110, no. 6, pp. 701-718; 1984.

Soares, H.F., Experiments on the Retention of Soils by Filters, Thesis, submitted to the University of London, England;1980.

USACE, Filter Experiments and Design Criteria, Technical Memorandum no. 3 - 360, U.S. Waterways Experiment Station, Vicksburg, Mississippi; 1953.

Vaughan, P.R., Kluth, D.J., Leonard, M.W., and Pradoura, H.M., Cracking and Erosion of the Rolled Clay Core of Balderhead Dam and the Remedial Works adopted for its Repair, 10th International Congress on Large Dams, Montreal, Vol. 3, pp. 73-93; 1970.

Vaughan, P.R., Design of Filters for the Protection of Cracked Dam Cores Against Internal Erosion, Journal of the Geotechnical Engineering Division, Proceedings of the ASCE Convention and Exposition, pp. 1-22; 1978.

Vaughan, P.R. and Soares, H.F., Design of Filters for Clay Cores of Dams, Journal of the Geotechnical Engineering Division, Proceedings of the ASCE, vol. 108 pp. 17-31; 1982.

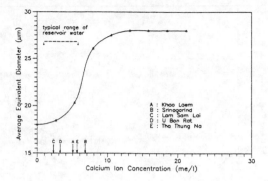

Figure 2 Flocculation behaviour of fine particles from lateritic residual soil

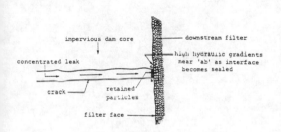

Figure 1a Concentrated leak through a dam core discharging into the downstream filter

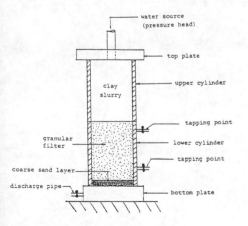

Figure 1b Schematic diagram of the filter test apparatus developed at AIT

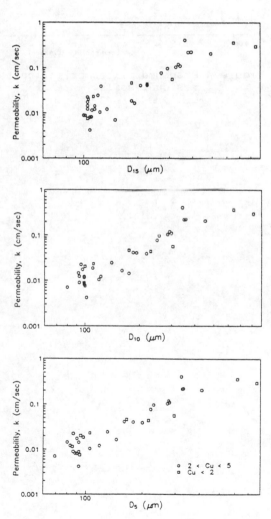

Figure 3 Influence of specific grain sizes on filter permeability

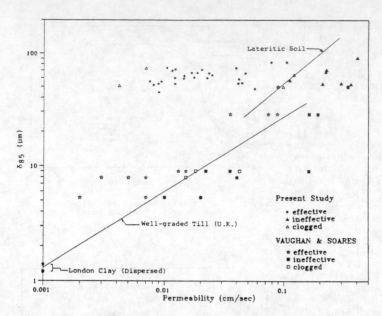

Figure 4 Proposed filter criterion for lateritic
residual soils

Strength and displacements of model infilled rock joints

N. Phien-wej, U. B. Shrestha & Ching-Yuan Ho
Asian Institute of Technology, Bangkok, Thailand

ABSTRACT: An experimental study on model infilled rock joints made with gypsum/celite revealed that the strength and displacement characteristics were influenced by the type of infill material, thickness of infill in relation to amplitude of joint roughness, roughness geometry and normal stress. The presence of infill results in a reduction in the interaction between joint walls during shear and thus a reduction in dilation, shear stiffness and peak strength of the joint. It was found that an infill thickness sufficient to eliminate all the joint wall interaction was smaller for a weaker infill material and for a smoother joint. The strength of idealized saw-toothed joints with stiff clay infill was reduced to the strength of the infill material when the infill thickness reached about 0.5 to 1 time the roughness amplitude, varying with joint roughness angle (i). The greater the roughness angle the greater the required infill thickness. For joints with dry-bentonite infill, whose friction angle was almost the same as that of the smooth joint wall, the threshold thickness was around twice the roughness amplitude. The infill thickness required to prevent joint dilation was smaller than the thickness required for full reduction of the joint strength and was smaller for a weaker infill.

1. INTRODUCTION

Many instability problems in rock engineering involve sliding along joints with weak infill materials. An example is the catastrophic Vaoint Slides, whose sliding of a huge rock mass was thought to occur along a joint with thick clay infill (Hendron and Patton, 1986). All of us realize that the presence of an infill material will result in a reduction in strength of a rock joint. And when an infill becomes relatively thick, strength of a rock joint will be solely controlled by the infill rather than the joint wall (Barton, 1974). However, as long as the thickness of an infill has not yet reached that thickness, the influence of the infill and the roughness of the joint will both have roles in the joint properties. In such a case, besides infill thickness the rock joint behavior should be also affected by the type of infill, the roughness of joint wall and normal stress. Previous works to understand the behavior of infilled rock joint are very limited, due to difficulties in performing tests and obtaining rock joint samples that represent a broad spectrum of rock joint roughness. Therefore, quantitative data on the influence of infill on strength of rock joints are meager.

Previous studies on the influences of infill thickness on strength of artificial rock joints are such as those of Goodman (1970); Lama (1978) etc. However, these studies were only superficial since they only investigated one factor, i.e. infill thickness. Another study was made by Neito (1974) who looked into details of the interaction between natural rough joint surfaces under a variation in thickness of clay infill. He categorized the interaction into three modes, namely, interlocking, interfering and non-interfering. Interlocking condition occurs when there is contact between rock surfaces during shear. Non-interfering condition occurs when interaction between joint wall roughness is totally eliminated and the behavior is solely controlled by the infill material. Interfering condition is the transition between the two modes and no joint wall contact occurs. A similar characterization of response of infilled rock joints was made by Barton (1974), who divided the response into 4 categories and described tendency of rock joint dilation and compression in each category.

In order to obtain a more detailed knowledge on the behavior and properties of infill rock joints, a research program has been set up at the Asian Institute of Technology. Large direct shear tests were adopted as a means to study model rock joints and natural joints. The model rock joints were used in the first phase of the study in order to provide a means to control and cover a complete range of influencing factors (i.e. joint roughness and type and thickness of infill), so that the fundamentals of shearing of infilled joint can be established. This paper presents the results of the first phase of the work which involved an experimental study on model rock joints with two types of infill material (high and low strength infills).

2. MODEL MATERIALS

Gypsum mixed with celite was used to make an idealized soft rock joint. This model material has been used by previous investigators (e.g. Einstein, et al 1969; Goodman, 1970; Yudhbir et al (1983); etc.). It was found to give the best combination of brittle behavior, acceptable strength, lack of bleeding, smooth surface characteristics and minimum bubble holes.

In the preparation, gypsum and celite (celite 1.72% by weight) was dry mixed and then mixed with water (water-gypsum ratio = 0.55:1). The mix was poured in the mold and cured in a humid room for at least 72 hours. Twenty four hours prior to testing, the specimens were forced to dry out in an oven to achieve a water content within a range of 18 to 21 %. At this water content, the model material was found to be a good representation of soft rock behavior such as limestone, shale and tuff because it satisfies the similitude requirements (Table 1). Two types of infill materials were used, i.e. dry bentonite and stiff clay.

Dry commercial bentonite was used to simulate a higher strength infill material. The bentonite was oven dried at 50° – 60°c for at least 2 days prior to preparation on the joint surface to obtain a desired thickness. In order to ensure the consistency, the water content of the infill was checked after the tests and if exceeded 3%, the test was discarded.

The stiff Bangkok Clay was used for the clay infill. The saturated clay was collected from boreholes drilled at AIT campus at a depth of 10-12 m. The clay had water content = 20-30% PI = 20-50 and LI = 0.2-0.4.

Direct shear tests showed that the strength envelopes of the two infill materials, the smooth planar rock joints and the clay-rock interface were linear with zero cohesion intercept. The friction angles of these materials are summarized in Table 2.

Table 1 Similitudes of model material (gypsum/celite) and soft rocks

n Factor	Model Rock	Shale	Tuff	Limestone
σ_c/σ_t	4.25	3-10	5	13-29
E_t/σ_c	594	160-300	33	400-1000
τ_o/E_t	$4.76*10^{-4}$	$4*10^{-4}$	$8*10^{-4}$	$6*10^{-4}$
ν	0.2	0.2	0.29	0.1-0.2
ϕ°	32°	32°	30°	

Property	Values
σ_c (MPa)	10.6
σ_t (MPa)	2.5
E_t (GPa)	6.3
γ, (kN/m³)	13.2

Table 2 Friction angles of model rock and infill materials

	Friction Angle ϕ = Degree
Infill Material	
Dry Bentonite	31.6
Stiff Clay	15.1
A = 15 cm x 15 cm	
Rock/Rock	
I = 0°	
at peak	35.6
at residual	32.0
Rock/Clay	
Joint	15.7

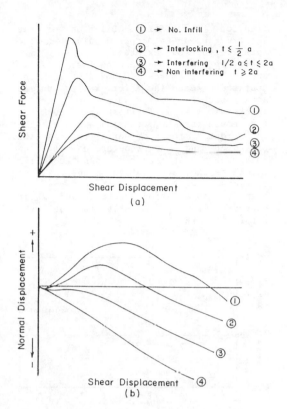

Fig. 1 Generalized relationship between shear force, shear displacement and normal displacement of dry-bentonite infilled saw-toothed joints

3. TEST PROGRAM

Large direct shear tests were performed on the model rock joints with various infill thicknesses. For tests on dry bentonite infilled joints, the shear surface of the specimens were 150 mm x 150 mm. For clay infill joints, a larger specimens of 300 mm x 300 mm were also used in addition to the 150 mm x 150 mm specimens. The joint roughness geometries used in the study were planar, regular saw-toothed and corrugated. The influence of joint roughness was investigated by varying roughness angle (i) of the saw-toothed roughness, from 0°, 7.5°, 15° and 30°. Only one corrugated surface was tested in this early part of the study program.

The roughness amplitude (a) was kept at 10 mm for the non-planar joints. The tests for each joint roughness and infill thickness were carried out at normal stresses of 0.3, 0.5, 0.75 and 1 MPa, which encompass the usual range mostly encountered in rock engineering problems (Barton, 1974). All the tests were run at a shear rate of 1 mm/min and continued up to a displacement of 30 mm.

In the tests of clay infilled joints, the joint surfaces were coated with three layers of a waterproof coating (methyl ancohol based) material in order to prevent moisture loss of the thin saturated clay infill to the gypsum.

4. BEHAVIOR OF DRY-BENTONITE INFILLED JOINTS

The dry bentonite which was in the form of compacted powder acted like a granular infill. It had almost the same friction angle as the smooth surface of model rock joint. Thus it simulated a high-strength infill. The test results showed that the shear force-displacements relationship of the non-planar rock joints can be categorized with reference to the range of the infill thickness as shown in Fig. 1. This generalized responses are similar throughout the range of normal stress investigated in the test program.

The shear force-displacement behavior was distinguished on the basis of the interaction between the joint surfaces (Neito, 1974). Non-interfering condition existed when the infill thickness exceeded about twice the roughness amplitude, in which condition the shear force-shear displacement relationship and normal displacement-shear displacement relationship of the infilled joint are practically controlled by the infill material. With a decrease in the infill thickness, the interaction between the joint surfaces increases, resulting in an increase in shear strength and stiffness, and gradual change in normal displacement from compression to dilation. Dilation of the joint during shear occurred only when the interlocking mode existed.

For the two roughness angles, i = 15° and 30°, the tests showed that the contact between the joint walls during shear occurred only when the infill thickness was smaller than about 60 % of the roughness amplitude; and for the investigated range of normal stress the rising-up and shearing-through of the joint teeth both occurred. When the infill thickness exceeded that value there was no contact between the joint surfaces and therefore no dilation occurred; and shear ruptures developed mostly through the infill material.

Fig. 2 summarizes dilation-compression (expressed as tan i = dh/dv at peak stress) of the joints. Dilation only occurred when the infill thickness was small enough to promote the interlocking condition. For flat joint, no dilation occurred even with a very thin infill. The result also showed that compression occurred earlier for a rougher joint; and at a higher normal force the amount of dilation was higher for a rougher joint and for a higher normal force.

The effect of infill thickness on the peak shear strength of the model joints are shown in Fig. 3 in the form of the ratio of peak shear strength to normal stress (τ_p / τ_n) versus infill thickness

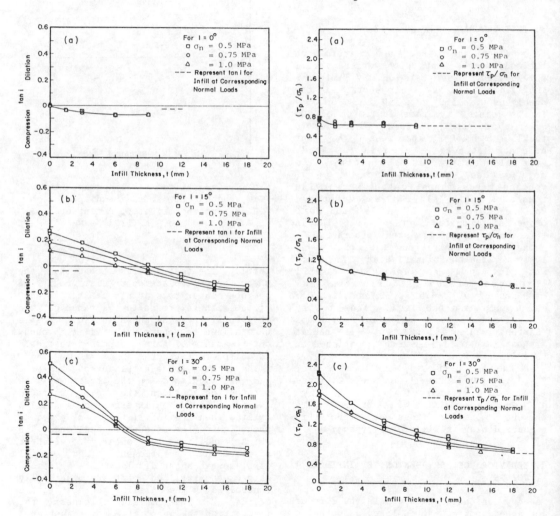

Fig. 2 Effect of infill thickness on normal displacement of dry-bentonite infilled joints

Fig. 3 Effect of infill thickness on peak strength of dry-bentonite infilled joints

(t). For the planar infilled joint τ_p/τ_n was equal to that of the infill material even with a very thin infill (1.5 mm). Only the joint with no infill showed slightly higher strength due to the effect of micro roughness of this planar joint. For the non-planar joints, the strength gradually dropped from the values of the non-infilled joints to that of the infill material as the infill thickness increased. The strength of the joints dropped to the strength of the infill when the infill thickness reached the magnitude of twice the roughness amplitude, regardless of normal stress. This is somewhat similar to a previous report by Goodman (1970) who did a test on and idealized saw-tooth rock surface ($i = 30^\circ$) cast with a plaster-celite material with crushed mica infill. The test which was done at only one normal stress showed that the thickness of filling needed to be at least 50 percent greater than the roughness amplitude so that the strength of the joint will be as low as the strength of the infill.

Fig. 4 shows peak shear strength envelopes for various thickness of infill of each joint roughness. For the joint with roughness angle of 15, the envelope for each infill thickness is approximately linear, except for the non-infilled joint. However, for the rougher joint ($i = 30^\circ$), the test results showed bilinear envelopes. The bi-linearity of the envelope becomes less pronounced as the infill thickness increases. The envelope becomes linear when the infill thickness approaches 1.5 times the roughness amplitude. The linear strength envelopes of the infilled rock joints with roughness angle of 15, result in a constant ratio of τ_p/τ_n for all levels of normal stress. This is not the case for the rougher joint ($i = 30^\circ$), whose τ_p/τ_n decreased with an increase in normal stress when the infill was relatively thin.

For shear behavior of a rock joint, there are two other important parameters:- the amount of shear displacements (dh) required to reach peak strength and the shear stiffness (peak shear strength/shear displace-ment at peak) of the joint (Kss). Even though the test data were scattered, there was a definite trend that for a given non-planar joint surface, the required shear displacement for the development of peak shear strength increased with an increased in infill thickness and then leveled off. At a small infill thickness, the peak strength was reached when the shear displacement

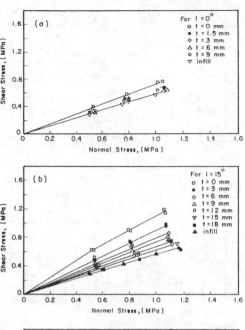

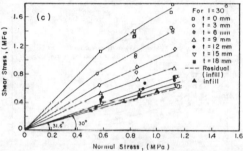

Fig. 4 Peak strength envelopes of dry-bentonite infill joints

reached approximately the value of the infill thickness. For a thick infill, the required shear displacement for peak strength was smaller at a higher normal stress.

On the shear displacement (du) required to reach the minimum strength of the infilled joints, it was found that for the planar joints the minimum strength was reached when du reached about 8–12 mm, the same amount required when testing infill material alone. For the roughest joint ($i = 30^\circ$) with infill thickness less than 1.5 roughness amplitude, the minimum shear strength was reached after at least 20 mm of shear displacement. Above that infill thickness, the du dropped down and approached the amount required by the infill material alone.

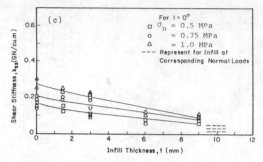

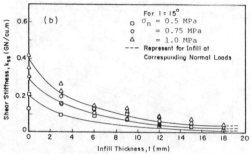

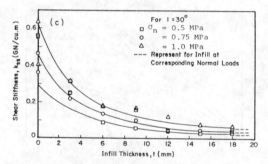

Fig. 5 Effect of infill thickness on secant shear stiffness of dry-bentonite infilled joints

The effect of the infill thickness on the shear stiffness of the joints is summarized in Fig. 5. It can be seen that even for the planar joint surface the presence of the infill had an effect in reducing the shear stiffness of the joint. However, the effect was more pronounced for rougher joints. The shear stiffness of the two rough joints decreased to the value of the infill material alone when the infill thickness reached about twice the roughness amplitude, similar to the effect on strength.

5. BEHAVIOR OF STIFF CLAY INFILLED JOINT

The tests were conducted on joints of planar surface and saw-toothed surfaces with three roughness angle, i = 7.5°, 15° and 30°. Larger size specimens (300 mm x 300 mm) were only used for joints with i = 7.5° and 15°. The stiff clay infill had drained friction angle of 15°, about one half of the value of the model rock joint wall. Therefore, it simulated rock joints with a weak infill.

The test results showed generalized shear force-displacements relationships similar to those of the dry-bentonite infilled joints. However, the threshold infill thicknesses are slightly different. It was found that the interlocking mode could only occur when the infill thickness was still smaller than about one half of the roughness amplitude, the same as the dry-bentonite infilled joints. However the non-interfering mode of the two rough joints (i = 15° and 30°) could be reached when the infill thickness increased to about the magnitude of the roughness amplitude. For the least rough joints (i = 7.5°), the non-interfering mode was reached when the infill thickness was only about one half of the roughness amplitude.

The dilation compression behavior of the joints in each of the three interaction modes were found to be similar to that of the dry-bentonite infilled joints. Dilation of the joints for all normal stresses (0.3-1.0 MPa) occurred only in the interlocking mode (i.e. $t/a > 1/2$), in which the joint dilated after an initial compression. The joints compressed when the infill thickness was greater than the roughness amplitude. For thickness between this two values, the joints were observe to dilate at a very low normal stress.

The influence of the stiff clay infill on the peak strength of the joints are summarized in Fig. 6. The clay infill reduced drastically the strength of the clean rock joints down to the value of the clay infill when the infill thickness reached about 0.8-1.0 times the roughness amplitude for the two rougher joints (i = 15° and 30°). For the flattest non-planar (i = 7.5°), the strength dropped to the infill strength when the infill thickness reached only about 1/2 the roughness amplitude. This may be attributed to a less compaction of the clay infill in between the approaching teeth walls of the joint with a smaller roughness angle. For the planar joint, only the infill thickness of 3 mm was sufficient to eliminate the effect of the micro roughness of the joint walls and the strength dropped to the infill's value.

The test results showed that the effect

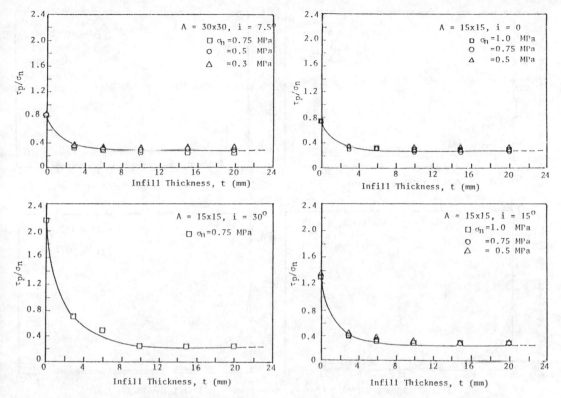

Fig. 6 Effect of infill thickness on peak strength of stiff clay infilled joints

of the thickness of the stiff clay infill in reducing the peak shear strength envelopes and the shear stiffness of the non-planar rock joints are practically in a similar manner to that of the dry-bentonite infill. The variation in shear stiffness of the clay infilled joints with the infill thickness are shown in Fig. 7. Note that for a given infill thickness, the secant shear stiffness was smaller for the saw-toothed joint than that of a corresponding planar joint. However, tangent shear stiffness were practically of no difference for the four joint roughnesses.

6. INFLUENCING FACTORS

The most significant factors controlling strength and displacement behavior of infilled rock joints is the thickness of the infill in relation to the amplitude of the joint roughness. Peak strength and shear stiffness of a non-planar joint will be reduced to the values corresponding to those of the infill as the thickness reached over a certain threshold value.

This is attributed to adverse effect of infill in reducing a portion of shear resistance of a joint which is derived from the undulation of the joint wall. It was found that the threshold infill thickness at which a joint strength and stiffness reduced to the minimum values (the infill's values), varied with the strength of the infill (i.e. type of infill). The threshold thickness is smaller for a weaker infill. The test results showed that the peak strength and shear stiffness of the saw-tooth joints were reduced to the values corresponding to those of the infill material when the infill thickness reached about twice and one time, the roughness amplitude for the dry bentonite and stiff clay infill, respectively. In direct shear, the dry bentonite infill had almost the same friction angle as that of the smooth gypsum/celite planar surface ($\emptyset = 31^\circ$) while that of the stiff clay infill was only about one half ($\emptyset = 15^\circ$). The threshold thickness for a given infill material seemed to be unaffected by the degree of joint except when the joint roughness become relatively flat. The

129

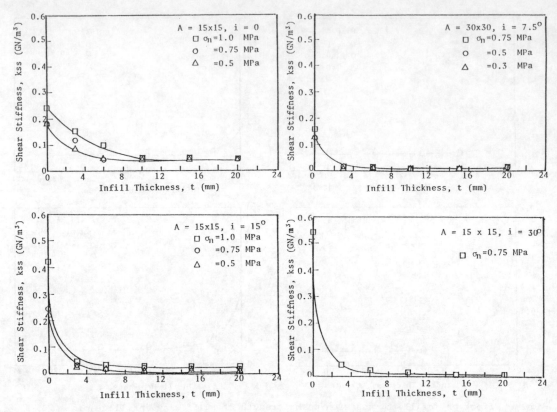

Fig. 7 Effect of infill on secant shear stiffness of stiff clay infilled joints

tests on the clay infilled joints showed that the threshold t/a ratio was around 1.0 for rougher joints ($i = 15°$, and $30°$), but it was reduced to 0.5 for the flatter joint ($i = 7.5°$). For a planar joint, the threshold thickness was less than 3 mm.

On the effect of normal stress, it was found that the normal stress did not show any significant effect on the threshold thickness even though the test data seemed to suggest that a higher normal stress may result in a slightly larger threshold thickness. Normal stress had an effect on the infill thickness at which the normal displacement of rock joints transitioned from dilation to compression and on the amount of dilation angle (dv/dh) of joints with a thin infill as shown in Fig. 2. A higher normal stress resulted in a smaller dilation angle at peak strength and a smaller transitional infill thickness. Except for the roughest joints ($i = 30°$) with a relatively thin infill, other joints showed a linear peak shear strength envelopes throughout the range of normal stress. The strength envelope of the roughest joints having a relatively thin

infill could be approximated as bi-linear envelope as suggested by Patton (1966).

On the size effect, it was preliminarily found that the stiff clay infilled joints tested using two specimen sizes (150 mm x 150 mm vs. 300 mm x 300 mm) did not show any obvious differences in their peak shear strength and shear stiffness. Additional tests would be required before any conclusive remarks can be made on the effect of specimen size.

With regard to the effect of joint roughness geometry, a small number of tests was conducted on the stiff clay infilled joints with corrugated surfaces. The joint had an amplitude of 10 mm, the same as of the saw-toothed joint. However, its average joint roughness angle was 38°, As shown in Fig. 8, the test data showed that the peak strength of the joints dropped to that of the clay infill at a greater infill thickness ($t/a = 1.5$) than that of the saw-toothed joints ($t/a = 1.0$). This was solely attributed to the curved surface of the corrugated joints and may be to the higher inclination angle of the undulation.

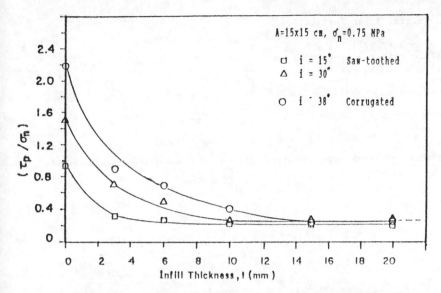

Fig. 8 Variation of peak shear strength with infill thickness of saw-toothed
and corrugated joint surfaces

CONCLUSIONS

On the basis of the results of the ongoing
experimental study on the non-planar
idealized infilled rock joints so far
completed, it can be concluded that the
most important parameter that affects the
strength and displacement characteristics
of the joints is the thickness of the
infill. As the thickness of the infill
increases, the peak strength and shear
stiffness of the joint decreases from the
high values of the non-infilled joints to
the infill's values. The presence of an
infill tends to reduce the effect of
roughness (undulation) on the dilation and
shear resistance of joints. An infill
thickness required to eliminate all the
roughness effect varied with the roughness
amplitude as well as the type of infill
and, to a lesser degree, on the geometry
and inclination angle of joint roughness.
Peak shear strength and shear stiffness of
an infilled joint would drop to minimum
values corresponding to the infill's when
the infill thickness reached about twice
the roughness amplitude for the high
frictional strength infill. However, an
infill thickness of only 1/2 to 1 time the
roughness amplitude was necessary for the
clay infilled joints. The required infill
thickness decreased with decreasing
inclination angle of joint roughness,
especially when the inclination angle was
low.

REFERENCES

Barton, N. 1974. A review of the shear
strength of filled discontinuities in
rock. NGI Publication 105, Oslo, 1-38.
Einstein, H.H., Nelson, R.A., Ruhn, R.W.
and Hirschfeld, R.C. 1969. Model study
of jointed rock behavior, 11th Symp.
Rock Mechanics, Berkeley, 83- 104.
Goodman, R.E. 1970. The deformability of
joints. Determination of the insitu
modulus of deformation of rock. Symp.
ASTM, 1969, Special Technical Publica-
tion 477, 174-196.
Hendron, A. J. and Patton, F.D. 1987. The
Vaoint Slide - A geotechnical analysis
based on new geologic observations of
the failure surface Engineering Geology,
No. 24, 475-491.
Ho, Ching-Yuan, 1990. Experimental study
of strength of clay infilled rock joints.
M.Sc. Thesis, Asian Institute of Techno-
logy.
Lama, R.D. 978. Influence of clay filling
on shear behavior of joints, Proc. 3rd
Cong. of Int. Asso. of Eng. Geology,
Madrid, Vol. 2, 27-34.
Neito, A.S. 1974. Experimental study of
the shear stress-strain behavior of clay
seams in rock masses, Ph.D. Thesis,
University of Illinois.
Patton, F.D. 1966. Multiple modes of shear
failure on rock. Proc. 1st Cong. ISRM,
Lisbon Vol. 1, 503-519.
Shrestha, U.B. 1989. Effect of infill
thickness on shear behavior of rock

joints. M. Eng. Thesis, Asian Institute
of Technology.
Yudhbir, Lemanza, W. and Prinzl, F. 1983.
An emperical failure criterion for rock
mass. Int. Cong. Rock Mechanics, Vol.1b,
Melbourne.

Developments in Geotechnical Aspects of Embankments, Excavations and Buried Structures, Balasubramaniam et al. (eds)
© 1991 Balkema, Rotterdam. ISBN 90 5410 019 2

Large scale model loading tests performed in a pneumatic caisson

Shunta Shiraishi
Shiraishi Corporation, Tokyo, Japan

ABSTRACT: A series of large-scale in-situ loading tests up to 1 950 tf of load were performed in a pneumatic caisson on deep-laying volcanic sand(scoria) for the purpose of assessing the ultimate bearing capacity of large-size foundations. It was found that, since the crushing of scoria grains was very pronounced in the highly stressed zone of the bearing ground, the failure occurred progressively resembling the failure in a cohesive soil of high compressibility and that the intensity of failure load did not varied widely with respect to the size of the models.

1 INTRODUCTION

The variation in the bearing capacity factor $N\gamma$ of dense sands was studied by reviewing the results of the model loading tests performed in the past and the following equation was proposed to define the variation in the value of $N\gamma$ in the prototype range of foundations (Shiraishi 1990).

$$N\gamma' = N\gamma p / B^{0.2} \quad \cdots\cdots\cdots\cdots (1)$$

Where $N\gamma'$: Varied value of $N\gamma$.
$N\gamma p$: Prandtl's value of $N\gamma$ for the foundation with a smooth base.
$N\gamma$: Terzaghi's value of $N\gamma$ for the foundation with a rough base.
B : Breadth of foundation base(m).

According to the Eq.(1), the value of $N\gamma'$ decreases as B increase. The value of $N\gamma'$ is equal to $N\gamma p$ or $0.631N\gamma p$ respectively when B is equal to 1 m or 10 m.

The ultimate bearing capacity formula, prescribed in the Road Bridge Specifications compiled by the Japan Road and Highway Association and used in other publicly authorized design specifications, define the theoretical values of $N\gamma p$ which are much greater than the true values of prototype foundations.

Consequently, the values of ultimate bearing capacity computed by the specified formula are considered to be very large.

Therefore, a majority of bridge foundations are designed not on the basis of the specified formula but on the basis of the tables of empirical allowable bearing capacities which may give very low values too much on the safe side.

This kind of discrepancy arises simply because the true values of $N\gamma'$ are not verified with an ample proof of theories and experiments. The Eq.(1) has been based on the results of a small number of model tests performed mostly in laboratories using small test models. It has not been verified by proving extensive experimental data on large models.

In order to supplement the credibility of the Eq.(1), the following large scale model loaing tests were performed in a pneumatic caisson under construction.

2 LOADING SETUPS IN PNEUMATIC CAISSON

The Eq.(1) suggests that the value of $N\gamma'$ may be smaller than $N\gamma p$ in the range of B being larger than 1 m.

There are many small model test data where B is smaller than 0.3 m. However, no well controlled test data of ultimate bearing capacity of firm ground is found where the value of B exceeding 0.3 m (De Beer 1965).

On the dense sand where the value of $N\gamma p$ exceeds 200, more than several hundred tf of test load may be required to produce the failure state in the bearing ground supporting a large test model. Such a heavy load may not be possible in a laboratory. For setting up a temporary heavy loading facility on the ground, a considerable amount of money, probably more than US$ 1 million, may be needed to build it.

Therefore, the series of the large scale loading tests up to the loading force of 1 950 tf were performed in the working chamber of a pneumatic caisson under construction.

This caisson is the foundation of the pier No.5 of the Ashigara Bridge in the widening project of Tokyo-Nagoya Highway at the place about 22km east of Mt.Fuji. Those loading tests were performed in combination with the construction work of the caisson as a part of the contract awarded by the Japan Highway Public Corporation to the joint venture of Shiraishi Corporation and Wakachiku Construction Company.

It had been estimated that the maximum loading force of 1 950 tf might be required to produce a failure state in the bearing ground supporting the largest test block of 1.3 m square. The reaction of this very heavy load was arranged to be held by the huge weight of the caisson body.

The surface of the ground dug out in the working chamber of the caisson was graded carefully to form a smooth horizontal plane. As shown in Figs.1 and 2, the test footing block was placed on the leveled surface of the ground in tandem with the hydraulic jacks for exerting the loading force. On the top of those jacks, the shim blocks and the grid girders were solidly placed for transmitting the reaction of the loading force upwardly to the reinforced-concrete ceiling slab covering the working chamber.

For the test block larger than 0.49 m² in base area, three 650-tf hydraulic jacks were employed.

The settlements of the four corners of the test block were measured with the micro gages attached to the reference beams which had been installed free from the influence of the loading force.

Both the loading operation as well as the settlement-measurement were remotely controlled and monitored electronically from the monitoring room installed on the ground near the caisson(Fig.3).

The loading force readings were converted electronically from the oil pressure in the hydraulic jacks. The loading

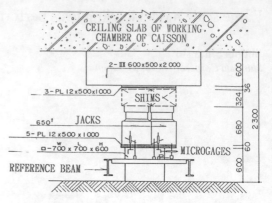

Fig.1 Loading setups

Fig.2 Loading setups(Photo)

Fig.3 Monitoring room

force of every jack had been calibrated beforehand with respect to the oil pressure.

The load-settlement relations in graphic forms were displayed on the monitoring TV as shown in Fig.4.

Fig. 4 Graphic display on TV

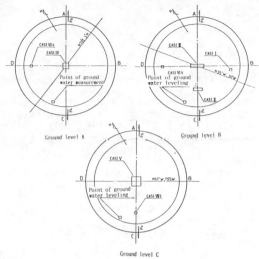

Ground level A

Ground level B

Ground level C

Fig. 6 Positions of test points

As shown in Fig. 5, the loading tests were performed on the three different ground levels. The ground level A, B and C were at 11m, 14m, and 18m respectively below the ground surface or at 13m, 16m and 20m respectively below the ground water table.

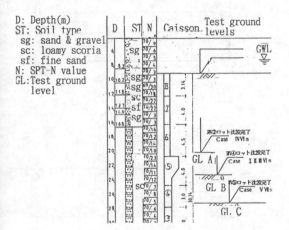

D: Depth(m)
ST: Soil type
  sg: sand & gravel
  sc: loamy scoria
  sf: fine sand
N: SPT-N value
GL:Test ground
  level

Fig. 5 Ground levels

Fig. 6 shows the positions of the test points on each of those ground levels.

TABLE I shows the size of the test blocks and the loading capacity for each case of the loading tests.

3 GEOLOGY OF GROUND

The ground tested underneath the caisson was composed of very dense, dark brown loamy scoria (volcanic sand) which

TABLE I Loading test cases

Ground level	Test case	Size of test block B(m) × L(m)	Loading capacity (tf × number of jacks) (tf)
A	VIIs	0.3 × 0.3	350 × 1 = 350
	IV	0.7 × 0.7	650 × 3 = 1 950
B	VIs	0.3 × 0.3	350 × 1 = 350
	I	0.4 × 0.4	650 × 1 = 650
	II	0.4 × 1.2	650 × 2 = 1 300
	III	0.4 × 2.0	650 × 3 = 1 950
C	VIIIs	0.3 × 0.3	350 × 1 = 350
	V	1.3 × 1.3	650 × 3 = 1 950

belongs to the eruptions of the Old Hakone Volcano. The STP-N value of this scoria is over 70 and estimated to be in the range between 100 and 500. The scoria is depositted in an irregular pattern being its dip variable between 5° and 30°.

Although the scoria is well grained and very densely compacted, its grains are porous and fragile. The crushing strength of the scoria grains is around 100 kgf/cm² (9 810 kPa).

The predominant crushing of the coarse scoria grains under the very high intensity of the test load stress seems to be obscured the shear failure pattern of the bearing ground at some of the tests.

4 LOADING TESTS

The test loadings were made in 3 cycles with 9 uniform loading steps in a cycle as a rule.

The time interval between the loading steps was 15 minutes at the virgin loading and at the zero load, and 2 minutes between the unloading steps or the reloading steps. The uniform loading rate of about 6- 9 tf/㎡/min(59- 88 kPa/min) was chosen for producing a drained condition in the nearly saturated bearing ground.

When the load-settlement relation of the uniform-rate loading test was plotted on a section paper of bi-logarithmic scales, the yield point appeared clearly at the point of refraction between the two curves as shown in Fig.7.

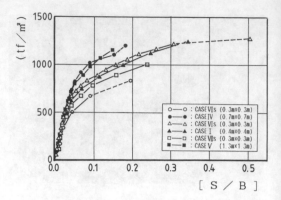

Fig.8 Q - S/B curves(Square models)

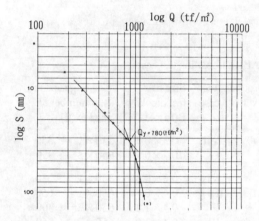

Fig.7 Log Q - log S curve

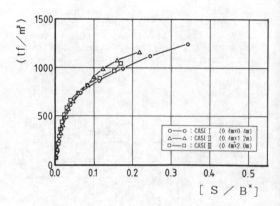

Fig.9 Q - S/B curves(Rectangular models)

The load-settlement curves of some of the tests are shown in Figs.8, 9, 10, 11 and 12.

The results of all the loading tests are shown in TABLE II.

The size effect which may be expressed by the Eq.(1) could not be sorted out from the values of Q_f shown in TABLE II. It is partly because, the strength of the bearing ground at each test case varied randomly and partly because, the crushing of the coarse scoria grains obscured the size effect.

The values of Q_f on the ground level B are generally higher than Q_{fs} in the levels A and C. It suggests that the crushing strength of the ground at the level B may be the highest, the second highest at the level C and the lowest at the level A.

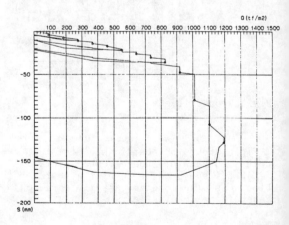

Fig.10 Q - S curve(Case IV)

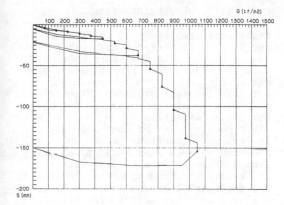

Fig. 11 Q - S curve(Case Ⅲ)

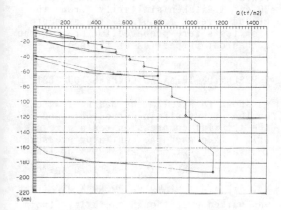

Fig. 12 Q - S curve(Case V)

TABLE Ⅱ Results of loading tests

Ground level	Test case	Size of test block(m)	Yield stress Qy (tf/m²)	Ultimate stress Qf (tf/m²)	Qy/Qf
A	Ⅵs	0.3x0.3	430	833	1.94
	Ⅳ	0.7x0.7	760	1 194	1.53
B	Ⅵs	0.3x0.3	600	1 278	2.13
	Ⅰ	0.4x0.4	790	1 250	1.62
	Ⅱ	0.4x1.2	550	1 167	2.12
	Ⅲ	0.4x2.0	680	1 050	1.54
C	Ⅷs	0.3x0.3	630	1 000	1.59
	V	1.3x1.3	533	1 166	2.19

5 PHYSICAL PROPERTY OF SCORIA

The undisturbed soil samples taken up from the test ground adjacent to the load-testing points were tested in the laboratory and the following physical properties of the scoria were measured. Although the specific gravity Gs is almost uniform at about 2.8 regardless of the depth of the ground levels and the position on each of the ground levels, the unit weight γ, the void ratio e and the degree of saturation sr vary in a wide range. The values of γ fluctuate between 1.64 t/m3 and 1.82 t/m3, the values of e between 1.01 and 1.43 and the values of sr between 77.9 % and 90.4 %, respectively. The values of γ, e and sr vary widely even on the same depth of ground level.

TABLE Ⅲ shows the average values of γ, e and sr on each of the three ground levels.

TABLE Ⅲ Average values of γ, e and sr

Ground level	γ (t/m3）	e	sr(%)
A	1.75	1.19	87.8
B	1.79	1.06	81.2
C	1.70	1.28	85.3

The values of γ, e and sr shown in TABLE Ⅲ suggest that the strength of the ground may be highest on the ground level B, second highest on the level C and lowest on the level A.

The grain-size distribution of the scoria varies very widely and randomly with little coherence with the depths or positions. The grain-size distribution also changed from the natural state due to the high stress caused either by the triaxial compression tests or by the in-situ loading tests. Those changes in grain-size distribution were caused by the crushing of the coarse grains of scoria under the very high test stress.

The grain-size distributions of the test samples before and after the triaxial compression tests(TCT) are shown in TABLE Ⅳ. In TABLE Ⅳ, "gr" denotes the gravel-size component of grain diameter d $\geq$ 2mm, "sa" the sand-size component of d = 0.074 mm - 2.0 mm and "si" the silt-size component of d = 0.005mm - 0.074mm.

The grain-size distributions of the test samples before and after the loading tests (LT) are shown in TABLE V.

TABLE IV Change in grain-size distribution

Ground level	Before TCT			After TCT		
	gr(%)	sa(%)	si(%)	gr(%)	sa(%)	si(%)
A	63	31	6	48	44	8
B	55	41	4	36	57	7
C	65	29	6	45	46	9
Average	61	33.7	5.3	43	49	8
Change				-18.0	+15.3	+2.7

TABLE V Change in grain-size distribution

Ground level & test case		Before LT			After LT		
		gr(%)	sa(%)	si(%)	gr(%)	sa(%)	si(%)
A	VIIs	44	49	7	36	49	15
	IV	59	38	3	20	55	25
B	VIs	43	55	2	29	60	11
	I	33	64	3	18	62	20
	II	31	66	3	19	68	13
	III	24	74	2	18	72	10
C	VIIIs	76	14	10	29	52	19
	V	38	60	2	33	64	3
Average		43.5	52.5	4.0	25.3	60.3	14.5
Change					-18.3	+7.8	+10.5

While the gravel-size component decreased by about 18 % in average both through TCTs and LTs, the sum of the components of sand-size and silt-size grains increased by the same percentage. This change was resulted from the crushing of the coarse grains in the gravel-size component into pieces of sand-size grains and of silt- size grains.

The maximum crush-down ratio occurred at the Case VIIIs of loading test where 47 % of the gravel-size component was crushed into pieces so as to increase the sand-size component by 38 % and the silt-size component by 9 %. The radical change in grain-size distribution was accompanied by a considerable decrease in the volume of the ground rated to be more than 10 % in volumetric strain.

The decrease in volume thus occurred caused remarkable change in the values of γ and e, increase in γ and decrease in e, although those values were not measured after the TCTs or the LTs.

6 TRIAXIAL COMPRESSION TESTS

Eight core samples of 10 cm in diameter and 20 cm high were cut out vertically (δ = 90 °) with a diamond core cutter from the undisturbed frozen square blocks of 0.3m x 0.3m x0.4m which was chiseled out from the test grounds. Five other core samples of same size cut out in horizontal direction (δ = 0°) and six other core samples of same size cut out in inclined directions (δ = 30° , 60° and - 30°) were also made from other frozen square blocks of same size taken out similarily from the test grounds.

Altogether, 3 ground levels x 19 pcs = 57 pcs of core samples for triaxial compression tests , two core samples taken from the ground level A for unconfined compression tests and sixteen more rectanguler samples of 8 cm x 10 cm x 20 cm size taken similarily from the ground level B for plane-strain compression tests were prepared for proving the mechanical properties of the scoria.

7 RESULT OF TRIAXIAL COMPRESSION TEST

As shown in Fig.13, the peaks of stress on the stress-strain curves obtained by the triaxial compression tests were clearly marked when the confining pressure σ_3 was 3.0 kgf/cm² (294 kPa) or less.

However, when σ_3 was 15.0 kgf/cm² (1 472 kPa) or more, no peak of stress was marked even though the axial strain ε grew as high as 15 % as shown in Fig.14. The volumetric strain ε_v during the compression tests when σ_3 = 7 kgf/cm² (686.7 kPa) or more was consistently in contraction, nearly the same magnitude as the axial strain ε.

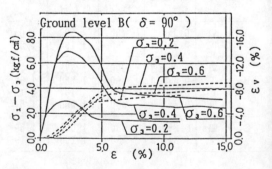

Fig.13 Stress-strain curves
(σ_3 = 0.2 - 0.6 kgf/cm²)

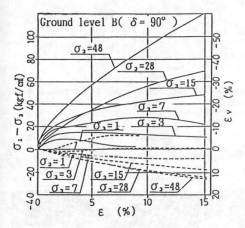

Fig.14 Stress-strain curve
(σ_3 = 1.0 - 48 kgf/cm²)

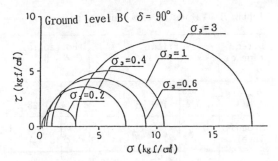

Fig.16 Mohr´s circles
(σ_3 = 0.2 - 3.0 kgf/cm²)

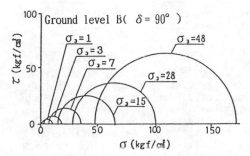

Fig.17 Mohr´s circles
(σ_3 = 1.0 - 48 kgf/cm²)

It is presumed that the trend of dilatancy, volume increase or negative volumetric strain, was offset by the overwhelming influence of volume decrease due to the crushing of the coarse grains. The failure of the bearing ground was mainly due to the crushing of grains which could not be defined to be any shear failure of regular pattern.

As shown in Fig.15, no clear rupture plane appeared in the samples tested at the confining pressure σ_3 of 15 kgf/cm² (1472 kPa) or more.

TCT	(UU, CU, CU, CD)				Sheet		11
					Date		
(90°) (0.00 m~0.00 m)					Name		
						***** 0000	kgf
0.20% /min .			kgf/cm²/min				℃
	No 1	No 2	No 3	No 4	No 5		
σ_3 kgf/cm²	1.00	3.00	7.00	15.00	28.00		
$(\sigma_1 - \sigma_3)_f$ kgf/cm²	9.557	15.190	28.471	49.449	71.527		
ε_f %	2.35	5.53	9.62	15.01	15.03		
u_f kgf/cm²							
A_f							
e_f							
ε_{vf} %	-1.596	0.191	3.409	7.962	11.411		

Fig.15 Rupture plane

The shape of failure envelope covering the Mohr´s circles was upwardly convex as shown in Figs.16 and 17. It suggests not only that the angle of internal friction ϕ was stress dependant but also that the influence of the crushing of coarse grains was seriously involved.

The degree of stress dependancy of ϕ was extremely high. At the initial tangent of the failure envelope, the originating value of ϕ is nearly equal to 63° where σ_3 = 0.4 kgf/cm² (39 kPa) while the average value of ϕ in the range of σ_3 = 1 - 28 kgf/cm² is 33.4° .

TABLE VI shows the values of ϕ and the cohesion c determined from the diagram of Mohr´s circles. The highest values of the axial stress σ_1 occurred in the triaxial compression tests were roughly equal to the ultimate vertical stresses developed by the loading tests.

Because of the influence of the crushing of coarse grains was very remarkable, the scoria behaved like a cohesive soil of high compressibility under the high confining pressure σ_3 exceeding 7 kgf/cm² (687kPa) while the scoria behaved like a cohensionless soil under the very low confining pressure of 0.4 kgf/cm²(39 kPa)

TABLE VI Values of c, ϕ and $_{max}\sigma_1$

Direction of σ_1	Vertical ($\delta = 90°$)			Horizontal ($\delta = 0°$)	
σ_3 (kgf/cm²)	0-0.4	1 - 28		0-0.4	1-28
Ground level	c　ϕ	c　ϕ	$_{max}\sigma_1$	c　ϕ	c　ϕ
A	0 63.4	1.93 34.0	106.3	- -	2.5 32.8
B	0 63.0	1.94 33.4	99.5	- -	2.4 35.0
C	0 63.0	1.59 34.4	110.7	- -	2.0 36.2
Average	0 63.1	1.82 33.9	105.5	- -	2.3 34.7

Remarks: c and ϕ are in kgf/cm² and in degree respectively.

or less. This kind of change in the mode of behavior may be explained as follows.

The scoria which contains many porous grains having numerous micro voids enclosed in the grains themselves resembles highly sensitive cohesive soils with very unstable cellular or honeycomb grain structures. The latters may not be highly compressible under low stress but may become highly compressible once the unstable grain skeletons collapse.

Likewise, the scoria is not very compressible before its porous grains are crushed but as the crushing of the porous grains proceeds it may become highly compressible and weak like a remolded sensitive clay.

8 MODE OF FAILURE

The mode of failure in the scoria ground which had undergone the test loads was observed on the vertical sections of the ground cut out through the center of the loaded area after the loading tests had been finished.

In the example of the Case IV (0.7 m x 0.7 m) test, the crushing of the scoria grains immediately beneath the test footing block was very pronounced as is shown in Fig. 18 and the color of this portion changed lighter than it had been before and the failure mode by shearing rapture was obscured by the grain crushing.

Because the white vertical pigment strips were distorted in multiple layers, the outward shear rupture seems to be developed progressively after the grain crushing occurred extensively. The gro-

und surface adjacent to the loaded area begun to move up only after the loading step 17 which corresponds to the load intensity of 1 010 tf/m² exceeding the yield-point load of 780 tf/m² (7 652 kPa). This delayed occurrence of the ground movement suggests that the outward shear rupture begun to develop only at the last steps of loading when the test block settled 80 mm or 0.114B (B; width of test block) mainly due to the crushing of scoria grains.

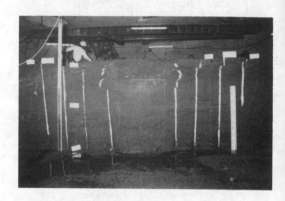

Fig. 18 Mode of failure(Case IV test)

In contrary to the Case IV test, little trace of grain crushing appeared in the Case III test (0.4 m x 2.0 m) where the gravel-size component was only 18 % and decreased only by 6 % due to the crushing of grains. In the Case III test, the rupture bands by shear failure developed toward right side were clearly marked as shown in Fig. 19.

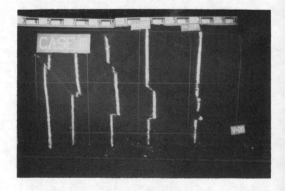

Fig. 19 Mode of failure (Case III test)

In the Case Ⅲ test where the strain-gaged pipe meters has been burried in the adjacent ground for measuring the horizontal movement of the ground at various depths, a measurable displacement begun to take place even from the first loading step of 75 tf/㎡ (736 kPa). The maximum outward displacement of the ground surface at the failure load intensity of 1 050 tf/㎡(10 300 kPa) was 9.2 mm at the point 50 cm apart from the edge of the loaded area as is shown in Fig.20.

As a matter of course, the direction of the horizontal displacements shown in Fig.20 were in the same direction as the outward movement of the ground shown in Fig.19.

The mode of failure at the other Cases of Ⅰ, Ⅱ, Ⅴ, Ⅵs, Ⅶs and Ⅷs were of the patterns intermediate in between the two extreme cases of Ⅳ and Ⅲ.

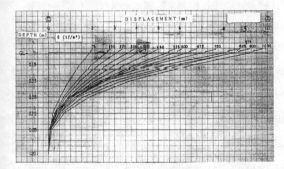

Fig.20 Horizontal displacement of ground

9 INTERPRETATION OF TEST RESULTS

The results of the afore mentioned loading tests are featured in the following points.

(1) The settlement of the test block was magnified due to the crushing of porous grains.

(2) The scoria behaved like a cohesive soil of high compressibility when it was placed under the very heavy test stress while it essentially is a cohesionless soil.

(3) The cohesive behavior of the scoria was pronounced in that the ultimate stress Qf of the test Case I of the square model was highest among the Qfs of the test Cases Ⅰ, Ⅱ and Ⅲ. To the

contrary, in a cohesionless soil the Qf of the most oblong model of the Case Ⅲ should be highest and the Qf of the Case I lowest.

(4) The shapes of the rupture bands differ from the authentic shape according to the Terzaghi´s theory for the foundation with a rough base on cohesionless ground. The shape of the rupture band of the Case Ⅲ test resembles more to the hypothetical shape according to the Prandtl´s theory for the foundation with a smooth base on cohesionless soil despite the fact that the test model had a rough base.

The size effect which may be expressed by the Eq.(1) could not be sorted out of the results of the loading tests.

If the values of $N\gamma'$ and $N\gamma p$ are counted back from the values of Qf obtained by the loading tests by the following equations, the values of $N\gamma'$ scatter randomly in a wide range as are shown in TABLE Ⅶ.

$$N\gamma' = 2Qf/\beta\gamma_b B \cdots\cdots (2)$$
$$N\gamma p = N\gamma' B^{0.2} \cdots\cdots (3)$$

Where β (shape factor) = 1 - 0.4B/L
γ_b (submerged unit weight) = 0.8 t/m3

Judging from the values of c and ϕ shown in TABLE Ⅵ which do not vary so widely scattered as the values of $N\gamma p$ in TABLE Ⅶ do, the application of the Eqs.2 and 3 for interpreting the test results may not be appropriate.

TABLE Ⅶ Values of $N\gamma'$ and $N\gamma p$

Test case	Size (BxL)(m)	Qf (tf/㎡)	$N\gamma'$	$B^{0.2}$	$N\gamma p$
Ⅰ	0.4x 0.4	1 250	12 900	0.833	10 840
Ⅱ	0.4x 1.2	1 167	8 413	0.833	7 004
Ⅲ	0.4x 2.0	1 050	7 133	0.833	5 939
Ⅳ	0.7x 0.7	1 194	7 107	0.931	6 617
Ⅴ	1.3x 1.3	1 166	3 738	1.054	3 940
Ⅵs	0.3x 0.3	1 278	17 747	0.786	13 949
Ⅶs	0.3x 0.3	833	11 574	0.786	9 097
Ⅷs	0.3x 0.3	1 000	13 889	0.786	10 917

When ϕ = 34°, γ_b = 0.8 t/m3 and B = 0.3 m - 1.3 m, the $N\gamma$ term of the bearing capacity formula $\beta N\gamma' B/2$

$= \beta\ N\gamma\ p\beta^{0.8}\ /2$ may be in the range
from 1.7tf /㎡ to 9.9 tf/㎡ and negle-
gibly small as compared to the high
value of Qf of around 1 000 tf/㎡.
   Thence, Qf may be computed by the following approximate equation.

$$Qf\ =\ \alpha\ cNcp\ \cdots\cdots\cdots\cdots (4)$$

   Where α (shape factor) = 1 + 0.3B/L
           Ncp: Cohesion term of bearing
                capacity factor, Prandtl´s
                value for the foundation
                with a smooth base.

   The values of Qf computed by the
Eq.(4) are shown in TABLE Ⅷ together
with the test values of ultimate bearing
capacity Qdt.

TABLE Ⅷ Values of Qf and Qdt

Ground level	Test case	Size (BxL)(m)	c (tf/㎡)	Ncp	Qf (tf/㎡)	Qdt (tf/㎡)
A	Ⅶs	0.3x0.3	19.3	44	1 104	833
	Ⅳ	0.7x0.7	19.3	44	1 104	1 194
B	Ⅰ	0.4x0.4	19.4	42	1 050	1 250
	Ⅱ	0.4x1.2	19.4	42	896	1 167
	Ⅲ	0.4x2.0	19.4	42	864	1 050
	Ⅵs	0.3x0.3	19.4	42	1 059	1 278
C	Ⅴ	1.3x1.3	15.9	45	930	1 166
	Ⅷs	0.3x0.3	15.9	45	930	1 000

   The computed values of Qf are rather
conservative values ranging between
0.768 Qdt and 0.925 Qdt except for the
Case Ⅶs test where Qf is 1.325 Qdt.
The ground where the Case Ⅶs test was
made might be exceptionally weak.

10 CONCLUSION

Summerizing the preceeding discussions,
it may be concluded as follows.
   The strength of the ground of scoria
with porous, fragile grains was seriously
affected by the crushing of the grains.
If the grains were solid and strong, the
test values of ultimate bearing capacity
Qdt would have been much higher than
the values shown herein before.
   It was the first experience in the history of model loading tests to observe
the phenomena of grain crushing.
   Some crushing of soild, strong gravelly sand was reported elsewhere in the

load testing of deeply embedded large-
diameter piles. It was estimated that
the crushing strength of the solid strong sandy grains might be around 2 000
tf/㎡ or more.
   If the solid sandy grains were infini-
tively strong, the ultimate bearing
capacity might have been much in excess
of 2 000 tf/㎡.
   Thus, the ultimate bearing capacity of
cohesionless soil is more or less associated with grain crushing in ultra-high
stress zone.
   The crushing strength of the scoria
mentioned herei before was around 1 000
tf/㎡ varying in the range of - 17 % to
+ 28 %.
   The calcarious sand grains prevailing
in the tropical regions made mostly of
fossil of coral may have the crushing
strength of the order similar to the
scoria.
   Where the crushing strength dominates
the ultimate bearing capacity, the failure of the ground may occur progressively resembling the failure in a cohesive
soil of high compressibility even if the
ground is under drained condition like
in the afore-mentioned loading tests.

ACKNOWLEDGEMENTS

The loading tests and the soil tests
were made under the assignment let by
the Japan Highway Public Corporation.
The writer is highly grateful to the
esteemed personnel of the Public Corporation, particularly to Mr. Y. Maeda who
was in charge of those tests, for their
support and courtesy offerred to the
writer.
   The writer is also grateful to Associate Professor O. Kusakabe, University of
Utsunomiya and other advisors of academic circles for their valuable advise
and cooperation.

REFERENCES

Shiraihi, S. 1990. Variation in bearing
   capacity factors of dense sand assessed by model loading tests, Soils and
   Foundations, Japanese Society of Soil
   Mechanics and Foundation Engineering,
   Vol.30, No.1, pp 17 - 26.
De Beer, E.E. 1965. Bearing capacity and
   settlement of shallow foundation on
   sand, Proceedings, Symposium on Bearing Capacity and Settlement of Foundations, Duke University, pp 15 - 33.

Developments in Geotechnical Aspects of Embankments, Excavations and Buried Structures, Balasubramaniam et al. (eds)
© 1991 Balkema, Rotterdam. ISBN 90 5410 019 2

Performance and monitoring of a deep, large diameter shaft

Kenneth R. Peaker & Shaheen A. Ahmad
Trow Geotechnical Ltd, Brampton, Ont., Canada

ABSTRACT: A large circular shaft 33 m diameter by 30 m deep was constructed through the overburden for the Metropolitan Toronto raw water intake at the Easterly Filtration Plant. To facilitate construction, geotechnical recommendations included a rectangular earth pressure diagram and an earth pressure coefficient $\emptyset$ of 0.2 for a shoring system consisting of 914 mm soldier piles and timber lagging. Overburden to shale bedrock consists of dense silts and sands with a relatively high groundwater level. Dewatering using eductors proved reasonably successful. The designers specified no detectable vertical movement adjacent to the deep well, since buildings supported by spread footings at the top of the well were to be unaffected.

Monitoring of the construction showed that the earth pressure assumptions were reasonable, however, a trapezoidal distribution is likely to be more applicable. Inward movement of the shoring was monitored at less than 15 mm. No vertical movement was noted.

INTRODUCTION

Metropolitan Toronto (Ontario, Canada) pumps raw water from Lake Ontario for treatment at the Easterly Filtration Plant by way of a deep large diameter pumping station. Immediately surrounding the deep pump station are the ancillary structures housing the related treatment facilities. Figure 1 shows a schematic cross-section of the construction. Original ground level was near Elevation 103.6 m, while the general basement level of all structures is Elevation 97 m. The deep pumping station continues through the overburden, a further 30 m below the general basement level. The pumping station was constructed as a circular building 33 m in diameter. This paper discusses the geotechnical aspects of planning and construction of this deep structure.

The configuration of the deep well building was decided as circular by the designers James F. MacLaren Limited. The designers also specified that the shoring system through the 30 m of overburden (the structure continues for an additional 30 m through bedrock), prevents all loss of ground into the excavation to guarantee

stability of the spread footings founded above the base of the pump station.

SUBSOIL DATA

The site[1] is located at the top of the Scarborough Bluffs, a well known local phenomenon. The borehole data indicated that below a thin veneer of topsoil, the subsoil comprises an initial 2.5 m to 6 m of dense and very dense silt and fine sand with the odd small lens or layer of slightly coarser sand with occasional small gravel inclusions. This is believed to be the surficial Lake Iroquois beach deposits. This dense silt contains wet sand seams, generally below a depth of 3 m.

Underlying the surficial Lake Iroquois beach deposits, a zone of very dense silt or silty sand with a till-like structure was encountered down to depths of up to 9 m to 12 m below original grade. This deposit is believed to be the Leaside Till deposits of the early Wisconsin era. The horizon between this silt till and the overlying very dense silt (often containing small gravel inclusions) is difficult to define accurately, therefore

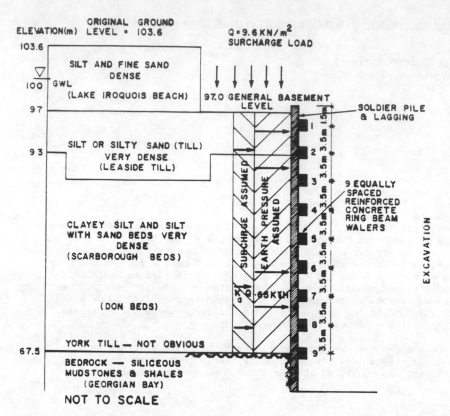

ORIGINAL GROUND
ELEVATION(m) LEVEL = 103.6

$Q = 9.6 KN/m^2$
SURCHARGE LOAD

103.6

SILT AND FINE SAND
DENSE
(LAKE IROQUOIS BEACH)

100 GWL

97.0 GENERAL BASEMENT
LEVEL

97

SILT OR SILTY SAND (TILL)
VERY DENSE
(LEASIDE TILL)

93

CLAYEY SILT AND SILT
WITH SAND BEDS VERY
DENSE
(SCARBOROUGH BEDS)

SURCHARGE ASSUMED
EARTH PRESSURE ASSUMED

$K_a Q = 50 KTH$

(DON BEDS)

YORK TILL — NOT OBVIOUS

67.5

BEDROCK — SILICEOUS
MUDSTONES & SHALES
(GEORGIAN BAY)

SOLDIER PILE
& LAGGING

9 EQUALLY
SPACED
REINFORCED
CONCRETE
RING BEAM
WALERS

EXCAVATION

1.5m 3.5m 3.5m 3.5m 3.5m 3.5m 3.5m 3.5m 3.5m

1 2 3 4 5 6 7 8 9

NOT TO SCALE

Figure 1. General Detail

the thickness of the Leaside Till has been estimated approximately.

Below depths of approximately 12 m, deposits of very dense stratified clayey silt and silt grading gradually into massive beds of very dense wet silts and sands were encountered to depths of up to 21 m to 27.5 m below existing ground level. These are believed to be the Scarborough Beds of the early Sangamonian Interglacial sediments.

The Scarborough Beds, in turn, are underlain by stratified clayey silt and/or sand with layers and seams of clay, i.e., the Don Beds of the late Sangamonian Interglacial sediments.

In the lower depths, the Illinoian deposits of either York Varves or York Till appear to be either absent or very thin.

The deep borings penetrated into bedrock at a sufficient depth to distinguish both the initial rock contact and the upper

limit of sound rock beneath the site. The bedrock down to approximately Elevation 42 m comprises a series of repeatedly interbedded siliceous grey mudstones and shales, with less frequent beds of current bedded fine sandstones and slightly calcareous siltstones. The beds form the lowermost unit of the Georgian Bay of the Meaford-Dundas Formation and pass transitionally downward into the Collingwood shales below Elevation 43 m approximately. The Collingwood shales are bituminous and are marked by a significantly lower degree of fracturing, jointing and bedding than the overlying Dundas shales.

Gas is present in minor but persistent quantities in the Collingwood shales.

Extensive fracture zones, and less extensive rubble zones or thin clay gouges (seams) were noted within the Georgian Bay. Only one or two major bedding planes with thin clay gouges (indicating bedding plane slip in recent geological times) were indicated within the Collingwood shales.

GROUNDWATER OBSERVATIONS

From the observation recorded during the fieldwork program, groundwater was recorded in wet silt and sand seams at random depths throughout the overburden below an elevation of approximately 100 m, i.e., groundwater was encountered below ground surface at a depth of about 3 m to 5 m. Generally this groundwater was observed at very slow seepages into the open boreholes from the dense wet silt and sand seams. The quantity of groundwater entering the boreholes, therefore, from the very dense fine-grained soils, was small. In many instances, the open boreholes remained open to well below the water table for a number of days after the completion of the fieldwork before partially collapsing and caving below the water level.

Long-term water levels recorded after the completion of the fieldwork and in a number of piezometers installed at various depths in the overburden, indicated that the hydrostatic level of the groundwater within the overburden was at approximate Elevation 100 m.

THE SHORING SYSTEM

The systems considered for shoring were soldier pile and lagging, sheeting, or drop caisson. The sheeting alternative was not considered seriously because of the difficulty in driving the sheeting; a drop caisson was rejected because of a lack of local contracting experience and a fear of ground loss adjacent to the caisson. The soldier pile and lagging alternative was considered the most likely to be successful in retaining the ground provided the groundwater could be lowered - this system was elected as a construction procedure.

The soldier pile and lagging system consisted of 36 steel thin walled concrete filled tube soldier piles 914 mm in diameter. The soldier piles were located at 10° intervals in a circle with a 34 m diameter. The soldier piles utilized steel angles welded to the tube to support the timber lagging. The bracing system for the soldier piles was designed as nine reinforced concrete ring beams spaced at 3.5 m centres. The first ring beam was poured very close to the top of the excavation, while the ninth was located just below the surface of the bedrock. The reinforced concrete ring beams or walers were designed to be incorporated

into the permanent wall and provided support for the form work of the permanent wall.

GEOTECHNICAL CONSIDERATIONS

The geotechnical terms of reference were to provide a system that resulted in no vertical settlement of footings founded adjacent to and 30 m above the base of the overburden excavation. In addition, the design should be realistic in terms of costs, i.e., the earth pressure coefficients should be appropriate. Peck [2] in his state-of-the-art concludes that for a dense cohesive granular soil, little information is available, however, surface settlements in the neighbourhood of 0.5 percent at 0.5 distance from the wall are probably realistic. For this project, 0.5 percent of 30 m would result in predicted settlements of 15 mm. While it is not possible to directly relate settlement and inward movement of the shoring, it is not unrealistic to assume inward movements of at least 15 mm should be anticipated.

Earth pressures that may be generated are conventionally recommended by calculating/estimating values of k from typical relationships such as $\tan^2$ (45-$\emptyset$/2). For this project, values of $\emptyset$ were not measured, but estimated to be in the region of $\emptyset = 40^\circ$. If we consider $\emptyset = 40^\circ$, an appropriate k value may be 0.2.

Based on local experience [4] with the dense cohesive granular soils present at this site, a k value of 0.2 was recommended as being the highest value likely to occur during construction. A rectangular earth pressure distribution for the multi-strutted structure was recommended such that the maximum pressure would approach 0.65 k H. For long-term conditions, the water table will rise back to its original level near the top of the pump station, hence, a triangular distribution where k = 0.35 and the submerged weight of soil plus water pressure was suggested.

The designers requirement of no settlement adjacent to the surface of the pump station was a difficult criteria to meet. In fact, zero movement would be an impossible criteria and their requirement was interpreted as 'no detectable movement adjacent to the structure'. With this criteria, it was decided that if construction movements inward of the shoring system could be limited to less than 15 mm, that no movement would be detected at ground surface. The designer was therefore advised that with good construction practice, no surface movement would be noted with the proposed soldier pile and lagging system provided the site was dewatered to bedrock level around the pumphouse and provided good construction practice was applied.

STRUCTURAL DESIGN [3]

Based on the recommended soil parameters, the temporary structure was designed using a surcharge load of 9.6 kN/m[2] for the entire area, plus a uniformly distributed loading of 76.6 kN/m[2]. Timber lagging was designed as a simple supported beam and utilized 150 mm thick lagging boards. Considering the soldier piles and the nine reinforced concrete ring walers, the soil loading was calculated to be 36 identical, equally spaced, radial loads of 800 kN. For the ideal ring conditions of uniform soil pressure and a perfect circular member bending moments varied from a maximum of 190 kN.m at the load points to 85 kN.m at the mid points. The calculated force in the waler was 4,450 kN.

Any non-uniformity of loading would increase the bending moments greatly above the ideal values calculated. The designer elected to incorporate a possible 50 mm deflection of the waler and 75 mm out of round condition. These factors increased the calculated bending moment to 1350 kN.m.

INSTRUMENTATION

The instrumentation selected for the project consisted of: piezometers to ensure the groundwater level was below excavation level; inclinometers to estimate the inward movement of the shoring; and strain gauges in the waler ring beams to estimate the force and moment in these beams.

a) Piezometers:

Four piezometers were installed around the perimeter of the pump station and one piezometer was located in the centre of the circular structure. The piezometers were of the standpipe-type placed in augered holes backfilled with filter sand and sealed to prevent entry of surface water. A decision was made to leave the piezometers open to all horizons despite the fact that various water levels

existed; the piezometers therefore were to act as standpipes recording the overall effect of groundwater and dewatering.

b) Inclinometers:

Three inclinometers 120° apart were located in the soldier piles. The inclinometers terminated below the base of the soldier piles approximately 5 m into sound rock.

c) Strain Gauges:

Walers' No. 1, 4 and 7 of the nine reinforced concrete ring walers were selected for instrumentation. On each of the three selected walers, nine equally spaced stations were selected and two "Allteck" electrical resistance weldable strain gauges were installed. One gauge at each station was welded to the inside face reinforcing steel, the other on the outside. All 54 gauges were read from a central read out station.

DEWATERING

Prior to construction a dewatering system was installed by a specialist subcontractor. The contractor elected to use an eductor system to control the groundwater. Figure 2 indicates a typical grading range for the material to be dewatered, however, it is worth noting that the subsoil contained numerous seams and pockets of more permeable waterbearing sands. Little continuity of the sand seams or pockets was expected, hence, the actual flow from the system was expected to be small.

The eductor system consisted of 10 deep eductors continuing to the surface of the bedrock and 34 eductors that terminated 7 m above bedrock level. The eductors were evenly spaced around the perimeter of the pump station. Holes 350 mm in diameter were augered and then backfilled with soil and water. A 300 mm casing was then flushed into place and the eductor placed inside, the casing backfilled with filter sand and the casing removed. Some problems with silt eroding the eductor nozzles occurred and it was necessary to introduce clean water into the circulating system.

Site construction began from Elevation 95+m and eductors were installed from May 1976 to July 1976 start up. The initial groundwater level was about 95 m. This level dropped steadily to Elevation 79.0

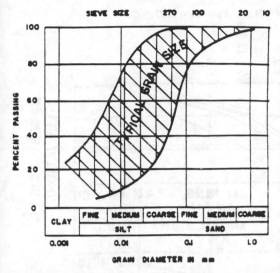

SIEVE SIZE   270   100   20 10

TYPICAL GRAIN SIZE

| CLAY | FINE | MEDIUM | COARSE | FINE | MEDIUM | COARSE |
| | SILT | | | SAND | | |

GRAIN DIAMETER IN mm

Figure 2. Typical Grain Size Distribution over Entire Depth

by the end of August 1976, and then slowly dropped to Elevation 73 m, when the excavation level and water level coincided in March 1977. The water level was never lowered below Elevation 69.5, some 2.5 m above rock surface, however, this did not cause major contractural problems. The well points were removed in December 1977, when the permanent wall was in place. Figure 3 shows a history of groundwater level and construction.

CONSTRUCTION

The installation of the soldier piles was a relatively simple construction operation, considering the very dense nature of the subsoil. The top of soldier pile was above Elevation 97 m and the base was located approximately 0.6 m into rock at Elevation 67 m. The soldier piles were installed in pre-drilled oversize auger holes that were grouted in place. Some problems with caving ground did occur and water entry from the lower level was a problem despite the dewatering. Specifications called for plumbness to be within 2 percent and all soldier piles, except one, met this specification. Most piles were installed with an actual out-of-plumb of less than 0.5 percent. The one pile that did not meet spec was removed, the hole backfilled with lean concrete and then satisfactorily reaugered.

PERFORMANCE

(i) Dewatering -

Figure 3 shows that the dewatering was marginally satisfactory. The groundwater level was not lowered to the bedrock surface, nor was the water level near this level when the soldier piles were installed. Some minor sloughing and seepage problems occurred, but, nothing that was of sufficient magnitude so as to delay the job. To provide for total dewatering, it would have been necessary to install the system well in advance of augering the soldier pile holes and it would be necessary to extend the system well into bedrock to take care of the upward gradient in the bedrock. From a cost/performance consideration the dewatering system was adequate.

(ii) Inclinometers -

Figure 4 shows the performance with time of the shoring when considering inward movement. The maximum inward movement measured was 15 mm on one inclinometer station, with the other two stations showing less than 10 mm inward movement. This performance was felt to be excellent and followed the predicted performance estimated prior to construction. Much of this performance can be credited to the design and contractor team who understood the importance of this aspect of the project.

(iii) Strain Measurements -

The designers J.F. MacLaren were particularly concerned with the possibility of eccentric loading of the ring walers. To confirm that the eccentricity was within the design values, the strain measuring system was installed. Figure 5 shows the results of typical stations on walers 4 and 7. This figure indicates the axial force as related to the sum of the strains from both faces, as well as the estimated bending moment as measured by the difference in strains measured at the inside and outside faces of the waler.

The strain results from waler No. 1 located 2.5 m below top of soldier pile were not clear. They were both erratic and small. It is concluded that little load reached waler No. 1.

Waler No. 4 was located 13 m below the top of caisson, after a few days of erratic fluctuation, the measured bending

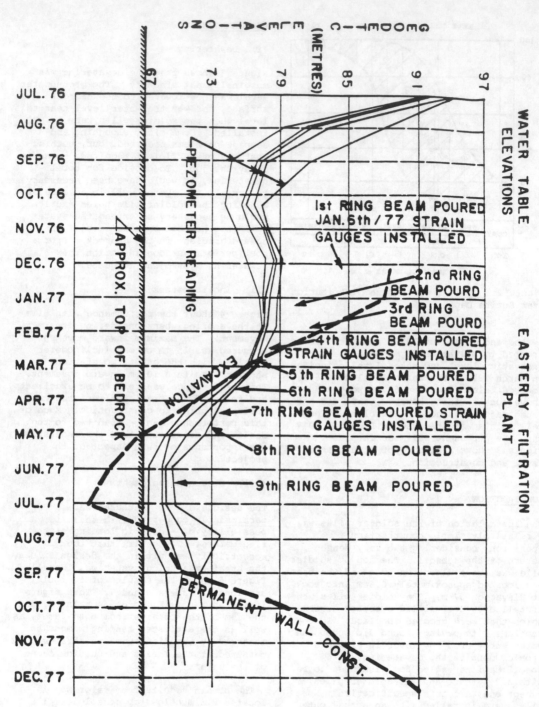

Figure 3. Construction Sequence

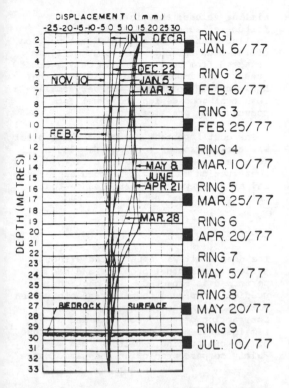

Figure 4. Maximum Horizontal Displacement at Inclinometer 12

moment remained constant. The total load in the ring increased as the excavation proceeded and levelled off near 6,400 kN at rock excavation level.

Waler No. 7 repeated the reading history of ring No. 4 with very small bending moments and a total load of 4,700 kN.

The designers estimated that because of creep of the concrete under load there will be a transfer of a portion of the internal forces from concrete to steel. This creep when related to time produced a creep factor to modify the total load measured from strain in the steel, such that the total load in ring 4 was 4,270 kN and ring 7, was 3,100 kN. The predicted total load in each ring was 4,450 kN.

(iv) Surface Settlement –

Construction of adjacent buildings was underway during the construction of this pumping station making it difficult to provide accurate surface measurement to detect vertical movement. The criteria was that a surface or vertical deflection would not be detectable. After nearly 10 years of performance of the pump station and adjacent buildings, it can now be safely stated that there was no detectable vertical displacement.

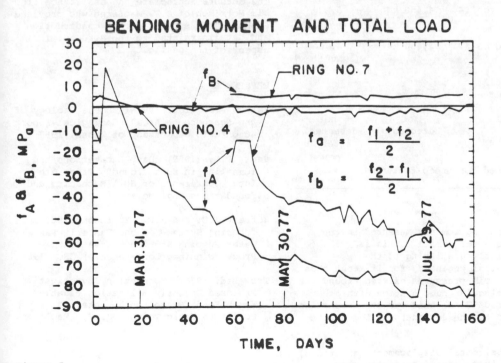

Figure 5.

149

COMPARISON OF PREDICTION VS MEASUREMENTS

Table No. 1 summarizes the data. The rectangular soil pressure diagram tended to over estimate the soil pressure at the top and base of the wall. A trapezoidal distribution would probably be more realistic considering the measurements. The maximum soil loading in the central region as indicated by ring 4 is close to the 0.65 k H value assuming k = 0.2, but it is likely that an even lower value is more realistic considering that virtually no load was recorded in ring 1 and only 70 percent of the estimated value was present in ring 7.

Table 1. Prediction Vs Measurements

	Predicted	Measured
Bending Moment #1	190 - 85 kN.m	negligible
Bending Moment #4	190 - 85 kN.m	negligible
Bending Moment #7	190 - 85 kN.m	negligible
Total Force #1	4,450 kN	very small
Total Force #4	4,450 kN	6,430 (4,270)* kN
Total Force #7	4,450 kN	4,700 (3,100)* kN
Horizontal Displacement	15 mm	15 mm
Vertical	not detectable	not detectable after 10 years

*corrected for creep factor

The bending moments measured in the rings were very small. This is particularly gratifying to the geotechnical engineer, for it meant that uniform soil pressures existed around the entire circular pump station with a diameter of 33 m; this was the assumption provided to the designers at the design stage.

The horizontal displacement measured by the inclinometer and predicted show similar values; this is somewhat fortunate. Errors and drift in the inclinometer can be a significant portion of the measurement. The value of 15 mm selected from Peck's work is only a crude guess. The important point is that the horizontal displacement was very small - a result of good design and excellent construction practice.

The most important aspect of the project was the control of vertical displacement. No accurate measurements were taken during construction so that it is much easier to say after 10 years of performance that this criteria was satisfied.

CONCLUSION

The deep shaft was constructed without serious contractual problems and has performed as expected. The geotechnical predictions proved to be satisfactory when compared to measured values. The interaction of soil, concrete and steel remains extremely complicated and because of the use of fiddle factors, can we fairly compare prediction and performance?

ACKNOWLEDGEMENTS

The authors appreciate the assistance of Jim Morrison of J.F. MacLaren who provided enthusiastic support during construction and details related to the structural aspect of the design.

REFERENCES

Karrow P.F. (1967) "Pleistocene Geology of the Scarborough Area", Geological Report 46 Ontario Department of Mines, 1967.

Peck, P.B. (1969) "Deep Excavations and Tunneling in Soft Ground", Proc. 7th Int. Conference of Soil Mechanics and Foundations Engineering

Jofreit J.C. and J.F. Morrison, 1980 "Shoring System for the Easterly Raw Water Pumping Station Excavation". Proc. Winnipeg Conference of CSCE 1980.

Trow Ltd. 1976, "Foundation Investigation Proposed Easterly Filtration Plant, Manse Road, Scarborough, Ontario" A report to Metro Toronto/J. F. MacLaren.

Developments in Geotechnical Aspects of Embankments, Excavations and Buried Structures, Balasubramaniam et al. (eds)
© 1991 Balkema, Rotterdam. ISBN 90 5410 019 2

Pile driving measurements for assessment of hammer performance and pile adequacy

P. Somehsa & R. F. Stevens
Fugro-McClelland, Houston, Tex., USA

ABSTRACT : This paper outlines the application of dynamic pile tests by Fugro-McClelland for the evaluation of the adequacy of offshore piles. Dynamic pile testing, also often referred to as pile monitoring during driving, has now become an industry accepted procedure for evaluating hammer performance, driving stress and soil resistance during driving (SRD). An overview of the experience gained from a variety of installation conditions in this part of the world, including marine clays, (normally and overconsolidated) sands and rocks (eg limestone) is presented. The final section presents the author's assessment of future direction in the understanding of SRD in clay soils. It is hoped this "industry" contribution would stimulate further academic focus in the observations and problems encountered.

INTRODUCTION

For onshore piling projects, it is normal to perform static load tests to verify pile adequacy and thereby eliminate unnecessary conservatism in design. However, such tests are prohibitively expensive for offshore application due to the logistics and magnitude of loads involved. An economic alternative for assessment of pile adequacy is to monitor piles during driving. Pile monitoring is carried out by monitoring strain and acceleration near the pile top during hammer impact. The measurements are compared with theoretical simulations using the stress wave procedure to evaluate hammer performance and pile adequacy.

Field monitoring of hammer performance and pile adequacy has been pioneered by our company through research and development work since the early 1970's. To-date, we have monitored pile installation at over 120 offshore platform sites throughout the world. Our field experience has provided the basis of expansion of our technical support capabilities. For example, in 1978 we developed a graphical method for estimating cushion properties and hammer efficiency from characteristic parameters of pile force (Appendix 1). This method provides our clients with detailed information about the performance of driving system components. Experience with field assessments of pile adequacy led to the development of a new acceptance logic flowchart in 1982 (Appendix 2). This guideline establishes a recommended procedure for reconciliation of design assumptions with actual driving experience to reduce the need for subjective judgements in these field assessments. Recent developments include a statistical analysis of all the data collected to-date, procedures for controlled hard driving and the estimation of maximum stress along the pile length.

Our field operations are tailored to provide hammer performance and pile adequacy assessment on-site, during the progress of driving. Specifically our objectives are to:

(1) Help the contractor optimise the performance of his hammers;

(2) Control pile driving stresses;

(3) Evaluate pile acceptance, in case of unexpected driving conditions develop a reconciliation procedure so as to minimise, if not eliminate, remedial action.

Pile monitoring has provided cost and time savings during platform installation that far outweigh the cost for the monitoring service itself.

HAMMER PERFORMANCE AND CUSHION CHARACTERISTICS

Hammer performance parameters and cushion characteristics obtained with pile monitoring include:

(1) Energy transmitted to the pile and hence <u>driving system efficiency</u>;

(2) Estimated kinetic energy of ram at impact and hence <u>hammer efficiency</u>;

(3) Cushion properties including stiffness and coefficient of restitution.

Procedures for obtaining the above are outlined in Appendix 1.

Hammer Performance

The rated energy of air/steam hammers commonly used offshore is usually stated as the potential energy of a ram weight free falling through its nominal stroke height. The actual energy available for advancing the pile may be significantly less. The various forms of energy losses are identified as follows:

(1) Reduced ram stroke height due to insufficient steam pressure, pile batter and friction in guides;

(2) Ram descent impeded by steam trapped in hammer prior to impact;

(3) Losses in cushion and pile cap;

(4) Losses near pile top due to imperfect pile cap-to-pile contact and/or fatigue (bulging) in driving head;

(5) Losses due to flexural strain resulting from excessive stick-up length ("whip" phenomenon described by Poskitt et al) (1984);

(6) Losses due to unmatched or

152

imperfect follower-to-pile connection, if present.

Losses due to items (1) and (2) are recognised by the hammer efficiency (HE) parameter, while losses due to items (1) through (4) are accounted for by the driving system efficiency (SE). Losses due to (5) and (6) are not explicitly evaluated, their occurrence, however, is relatively infrequent and can often be deduced by observation of driving. "Normal" ranges of hammer and driving system efficiency, as obtained for our measurements are as follows:

HE :   60 to 80 percent
SE :   30 to 50 percent

From a pile drivability standpoint, the driving system efficiency is our <u>preferred</u> parameter for describing hammer performance since it better describes the energy available for driving and is also more accurately determined from our measurements. Hammer efficiency on the other hand appears to be the more frequent industry reference. Knowledge of both parameters, however, is useful for identifying the location of energy losses, if present.

Cushion Characteristics

When driving is commenced with a new cushion, the stiffness gradually increases over the first few hundred blows. It then stabilises for most of its life which may, depending on its make-up, vary from 3,000 to 10,000 blows.

Towards the end of its life, the stiffness can either deteriorate or increase again, depending on the materials used. Increasing the cushion stiffness improves drivability, while reducing stiffness reduces drivability. Increasing the stiffness also increases driving stress and, if excessive can result in hammer damage. By monitoring cushion stiffness during driving, it is possible to provide guidance to the contractor on optimisation of the cushion make-up as well as to indicate when a cushion change should be made. Average values of the cushion stiffness as obtained from our measurements would depend on the cushion material used and typically varies from 30,000 to 100,000 kips/in. for hammers having a rated energy of about 300,000 ft-lbs. The coefficient of restitution helps to monitor losses in the cushion and pile cap and typically varies between 0.6 and 0.9 with an average for air/steam hammers of about 0.75.

DRIVING STRESSES

Stresses when Driving in Soils

When driving in cohesive or granular soils, the driving stress is generally not critical so long as the pile has been designed adequately for driving. To ensure adequate wall thickness and steel grade for driving, wave equation analyses should be performed for an upper bound set of parameters, such as:

153

Hammer efficiency : 90 percent

Coefficient of restitution: 0.9

Cushion stiffness : Upper range for given hammer and cushion type

End bearing : Depending on soil type, typically about 50 to 90 percent

Stresses during driving are measured directly at the gage location. The force and velocity measurements as well as their reflections from the pile tip are used in the expression below, presented by Stevens et al (1984), to estimate the maximum stress over the entire pile length.

$$S_{max} = \frac{1}{2}\frac{1}{A}(1(F(t_1) + F(t_2)) + \frac{E}{c}(V(t_1) + {}^*V(t_2)))$$

Where

  A = minimum pile cross sectional area;

  E = pile modulus of elasticity;

  c = velocity of wave propagation;

  t_1 = time at which maximum velocity occurs;

  $t_2 = t_1 + 2Lc$, and

  L = length of pile below gages

Generally $V(t_1) + V(t_2)$ gives the maximum stress while $v(t_1) - v(t_2)$ would apply when most of the soil resistance is mobilised at the pile tip.

Stresses when Driving in Rocks

While the above procedures for checking stresses are adequate for piling in soils, extreme caution is needed when driving in rock layers since the theoretical maximum driving stress during reflection at the tip can be as much twice the incident stress, and much more if only partial contact occurs between pile tip and the rock. Partial contact can result when the rock surface is non–uniform or from oblique angle contact, for example when a battered pile encounters a horizontal rock surface. When planning to drive piles into rock or through rock layers, the required steel grade and wall thickness, especially at pile tip, should be evaluated from wave equation analyses assuming various levels of possible resistance at the pile tip. It should be noted that the API (RP 2A) criteria for pile refusal blowcounts may not be appropriate for driving in rock and can result in pile tip damage. Refusal criteria should be based on wave equation pre-analyses and pile monitoring so as to control driving stresses. When special care is taken in the planning of the driving system and pile, it is feasible to drive piles in certain rock types (eg gypsum) without significant risk of tip damage by carefully observing pile monitoring data, especially the form of the stress wave reflected from the pile tip (Stevens et al) (1984).

154

For limestone layers of limited thickness, pile drivability, or its rather susceptibility to tip damage, has been noted to be influenced by the relationship of the pile size to the layer thickness, the pile wall thickness and the hammer size used. For instance a large diameter pile may drive relatively easily through a 10-ft thick shallow limestone layer while a smaller diameter conductor may encounter driving difficulty and/or tip damage. Some characteristics of pile tip damage have been recorded with pile monitoring and may be useful for planning future installations in similar material. However since damage can only be identified after the occurrence, careful consideration should be given to the risk of having to repair or replace piles which are damaged at the tip. It is also suggested that when planning pile installation in limestone or other rocks where the surface is suspected to be non-uniform, geophysical surveys should be performed (in addition to the detailed geotechnical investigation) so as to evaluate the extent of non-uniformity of the rock surface.

Allowable Driving Stresses

Based on extensive observations as well as the accuracy of our measurements of driving stress, we normally recommend an allowable driving stress of 0.9 times the yield stress of the pile when driving in soils. A similar value is also applicable to rock, provided there is sufficient confidence that no stress concentrations are present (such as the partial contact situations discussed above). If stress concentrations are considered likely, then the allowable stress should be reduced accordingly. Consideration of other stresses, such as due to pile handling, hammer placement or bending stresses (eg curved conductors), is beyond the scope of this paper but should also be checked.

SOIL RESISTANCE

Due to the extensive scope of this subject and the limitation of this presentation, this section is confined to a discussion of the various soil resistance concepts in cohesive soils such as those found in Southeast Asia.

Blowcounts

Pile driving contractors usually regard blowcount as the primary measure of soil resistance. This is reasonable when the hammer performance is known and remains constant throughout driving. In reality, however, this is not the case. It is necessary to remove uncertainties due to changes in hammer performance when evaluating SRD. Typical characteristics of hammer blowcounts when driving in clay are illustrated on Figure 1 and are discussed below.

When a pile is first lowered, it will initially penetrate some distance (depending on the clay strength) under its own self weight, momentum and hammer weight. Thereafter when splices (new sections) are welded and driving is commenced, blowcounts typically start very high for about 2 to 5 ft, followed by a reduction to some intermediate value. This high blowcount is attributed partly to lower drivability resulting from a softer (new) cushion, lower hammer efficiency (warm-up period) and also to an increase in soil resistance occurring when a pile is allowed to rest for a while. This phenomenon of increased soil resistance in clay is referred to as "set-up". Thereafter, the hammer efficiency and cushion stiffness stabilizes and the blowcount gradually decreases with penetration for about 20 to 50 ft. If the pile is driven further continuously, blowcounts would remain constant or increase very slightly with penetration. The intermediate to gradual reduction in blowcount is attributed to the gradual reduction of SRD to a stable, fully "remolded" value, with stabilised hammer efficiency. The constant or slight increase in blowcount thereafter is attributed to the gradual increase in pile penetration, again at a stabilised efficiency. The envelope of minimum

blowcounts is described as **"continuous driving"** blowcounts while the envelope of high blowcounts can be described as **"after set-up"** blowcounts.

When piling is stopped for example to weld a new splice, the pattern of blowcounts is repeated. If the delays for welding are excessive or the hammer of insufficient driving capability, then premature refusal due to excessive set-up can occur. The above observations are for situations of "normal" driving with no hammer or pile anomalies that could cause energy losses and result in a different blowcount pattern.

Soil Resistance to Driving (SRD)

The Soil Resistance to Driving (SRD) or the "static" component of total driving resistance is probably the most relevant soil resistance concept in the evaluation of pile drivability, and thereafter the assessment of pile adequacy. Procedures for "measuring" and "predicting" SRD are discussed below.

Measured SRD. An approximate "measured" SRD can be obtained in the field using simplified closed form solutions (CASE methods) if there is sufficient confidence regarding the soil type. Alternatively, the measured hammer properties and blowcount can be entered into wave equation analyses to obtain a more correct estimate of

SRD. SRD's obtained in this way are illustrated on Figure 3. SRD analyses may also be performed using the measured force and velocity records directly in a wave equation program (eg CAPWAP).

Despite the availability of hammer performance measurements and the removal of uncertainties relative to the hammer, the evaluation of SRD using the wave equation procedure is still somewhat uncertain. The main uncertainties lie in the accuracy of the modelling of the soil elements both inside and outside the pile and the choice of quake and damping factors appropriate to the chosen model. In our current practice, we remain confident in the use of the conventional Smith (1960) model together with the quake and damping factors recommended by Roussel (1979). A discussion of these is presented below.

Soil **quake** and **damping** parameters required in the wave equation analyses are not intrinsic soil properties, but rather correlation coefficients that incorporate all that is not clearly understood about the process of driving a pile. Some of the data on which previous correlation studies have been based are for small diameter (12 to 18-in) closed end piles, and may not be applicable to the large diameter open end piles used offshore. Other studies performed for easy driving may not be appropriate for hard driving because blowcounts determined by wave equation analysis are insensitive to variations of soil and hammer properties at low blowcounts. Soares, Matos and de Mello (1984) and Swann and Abbs (1984) have summarised soil quake and damping parameters, indicating that a wide range of parameters give satisfactory results. It is difficult to say which parameters are incorrect, and which parameters are the accepted industry values.

Prior to the early 1980's, the damping parameters proposed by Smith (1960) for sand and by Coyle, Bartoskewitz, and Berger (1973) for clay were used. Later the Roussel (1979) parameters, which were determined from a comprehensive correlation study performed for large diameter offshore piles, were adopted. In that study, driving records of 58 piles at 15 offshore sites in the Gulf of Mexico were analyzed. The side and point quake are assumed equal, with a magnitude of 0.10 in. for stiff to hard clay, silt and sand. Side damping in clay decreases with increasing shear strength, which is in agreement with the laboratory test results of Coyle and Gibson (1968) and Heerema (1979). Our pile monitoring case histories, some of which have been presented by Stevens, Wiltsie and Turton (1982), generally support these parameters.

Predicted SRD

The SRD expected to be mobilised during driving can be predicted using various procedures described in the literature (e.g. Wright et al, Semple et al, Stevens et al). In these procedures, SRD's measured during previous pile installations were used to develop empirical relationships involving various soil properties, such as remolding, state of consolidation, cone penetration resistances etc. Unit resistances obtained from these relationships are used in a computation procedure analogous to static pile capacity analyses to obtain the predicted SRD's. For the marine clays of Southeast Asia, SRD predictions based on a simple proportion of the static capacity appear to be reasonable, see Figure 1.

The applicable embedded pile surface areas to be used for the analyses depend on whether the pile **plugs** or **cores** during driving. When the soil mass inside the pile (or plug) moves with the pile as a combined unit, the pile is described as being plugged and the SRD is computed as external side resistance plus end bearing resistance acting over the entire pile tip area including the pile wall and soil plug. When the soil plug does not move downwards with the pile, the pile is described as coring. The SRD, in this case, is computed as external and internal side resistances plus end bearing on the pile wall area only.

The applicable unit resistances to be used depend on whether the pile is being driven "continuously" or "after set-up". The available data on marine clays suggest that piles generally core during continuous driving. For the "after set-up" situation, there is some evidence of partial (or temporary) plugged behaviour (see below). The SRD for continuous driving in marine clays is frequently represented by 30 to 50 percent of static skin friction, taken to act on the external pile area only, plus end bearing on the pile wall. Internal side resistance is neglected. The measured SRD's after a 4 to 8 hour delay exceeds the "continuous" driving resistance by a set-up factor ranging from 1.5 to 2.0. For extended delays, the SRD's generally equal or exceed static pile capacity. Both measured and predicted SRD's are illustrated on Figure 1.

Soil Plug Behaviour

Stevens (1988) discussed the inertial force acting on the soil plug, and presented data that showed large diameter piles driving unplugged to depths of 350 ft in clay. The tendency to plug or core was demonstrated to be dependent on the inertial acceleration and strength of the soil mass within the pile. Piles would core above a certain threshold acceleration which depended on the pile size and soil properties. Wright et al discuss soil plug behaviour and suggest (from pile monitoring data) that plugging may occur. Piles generally do not plug in clay during continuous driving. After a short delay, however, we have observed (from pile monitoring data) piles driving in a plugged mode for several feet in clay.

In the paper by Paikowsky et al (1989) titled "A New Look at the Phenomenon of Offshore Pile Plugging" the authors conclude that pile plugging during driving is a frequent occurrence and is of greater significance than previously recognised. For the Gulf of Mexico, pile plugging occurs for piles driven to depths greater than 75 times the pile diameter. The authors estimated that 15 percent of the piles driven in the Gulf of Mexico are plugged and a further 25 percent are most likely plugged.

Although the behaviour of the soil plug during driving is governed by its inertial acceleration, plug behaviour during static loading would be based on static calculations. In general for long offshore piles in clay, internal static skin friction exceeds end bearing, thus the 'plugged' capacity would be applicable.

Driving Shoe

For most offshore installations in cohesive soils, a larger wall thickness, internal driving shoe is incorporated in the pile to expedite the installation by reducing internal resistance (and to promote coring behaviour). When a driving shoe is omitted the SRD mobilised inside the pile can be significant and there is an increased tendency for plugging.

Discussion of SRD during Continuous Driving in Clay Soils

The comments in this section are based on the author's own experience with pile monitoring data although much of the direction is based on the work of Heerema (1979).

According to Heerema, the mobilised soil resistance is a function of normal stress and the idea of soil remolding or the role of moisture between pile and soil is largely discounted. During a hammer blow, a compression wave, having the shape of a bulge (Poisson effect), would push soil outwards from the pile, exerting normal stresses as it passes the soil. (Additional normal stresses may also be present due to bending effects discussed by Poskitt et al). If it can be visualised that the clay soil, once pushed outwards, takes a longer time to close back on the pile than the small time interval between hammer blows, then contact, and therefore normal stress, would diminish over the upper region of the pile.

159

Based on the foregoing, it is proposed that a contributing factor to the apparent decrease in resistance during "continuous" driving as compared to "after set-up" is that the **area mobilised** during continuous driving may be significantly less than after set-up. A typical calculation of this area is as follows:

$$LW = T \times C$$
$$SRD = LW \times A \times Su$$

where

LW = effective length of compression wave

T = duration of compression wave (approximately 2 to 3 times the rise time, see Appendix 3)

C = wave speed

A = outer pile surface area /unit length

Su = shear strength over lowest LW of pile penetration

SRD predictions using the above expression gave generally similar results as calculations based on the current, widely used concept of remolding in clay soils.

For granular (sand) soils, the concept of mobilised area as presented above would be equally applicable if it can be visualised that the time taken for sand to close back on the pile is shorter than the time interval between blows during continuous driving. Therefore, the area mobilised and therefore soil resistance would be similar during continuous driving as well as driving after delays. It is noted however that in granular soils, the effects of soil plug behaviour as well as densification (or loosening) can result in differences in expected SRD and require further study.

For clay soils nevertheless, it remains to be seen whether mobilised area, soil remolding, or some combination of both, will come to be regarded as the appropriate concept for evaluating of SRD.

CONCLUSIONS

Criteria for hammer performance evaluation and typical ranges of measured efficiency, stiffness and coefficient of restitution are presented. Pile monitoring can be used for controlling hammer performance and where necessary help expedite the installation without expensive remedial action or subjecting the pile driving equipment to unnecessary risk.

Procedures for controlling stresses in the pile, especially when driving in rock are discussed. Direct application of the current API RP 2A guidelines for refusal blowcounts **may not be safe for driving in rock**.

The evaluation of soil resistance to driving (SRD) in clay soils is discussed together with uncertainties and involved.

REFERENCES

Coyle H M, Bartoskewitz R E, and Berg W J (1973), "Bearing Capacity Prediction by Wave Equation Analysis - State-of-the-Art", Texas Transportation Institute, Texas A&M University, Research Report 125-8.

Coyle H M, and Gibson G C (1970), "Empirical Damping Constant for Sands and Clays", Journal, Soil Mechanics and Foundations Division, ASCE, New York, Vol 96, SM3, pp 949-965.

Heerema E P (1979), "Relationships Between Wall Friction, Displacement Velocity and Horizontal Stress in Clays and in Sand for Pile Drivability Analysis", Ground Engineering, Foundation Publications Ltd, Brentwood, Essex, Vol 12, No. 1, pp 55-65.

Lee S L, Y K Chow, G P Karunaratne and K Y Wong (1988), "Rational Wave Equation Model for Pile Driving Analysis", Journal of Geotechnical Engineering, ASCE, Volume 114, No 3, March.

Paikowsky S G, Whitman R V and Baligh M M (1989), "A New Look at the Phenomenon of Offshore Pile Plugging", Marine Geotechnology.

Poskitt T J and Ward G, (1984), "The Effect of Stick-Up on Pile Drivability" Seminar on Foundations for Offshore Installations, Society for Underwater Technology, Volume 10, No. 1.

Randolph M F, (1987), "Modelling of the Soil Plug Response During Pile Driving" 9th Southeast Asian Geotechnical Conference, Bangkok.

Roussel H J (1979), "Pile Driving Analysis of Large Diameter High Capacity Offshore Pipe Piles", Ph.D thesis, Dept of Civil Engineering, Tulane University, New Orleans.

Semple R M and Gemeinhardt J P (1981) "Stress History Approach to Analysis of Soil Resistance to Pile Driving", Proceedings, 13th Offshore Technology Conference, Houston, Texas, Vol 1, pp 165-172.

Soares M M, Matos S F D and de Mello J R C (1984), "Pile Drivability Studies, Pile Driving Measurements", Proceeding, 2nd International Conference on the Application of Stress Wave Theory on Piles, Stockholm, pp 64-71.

Smith E A L (1960), "Pile Driving Analysis by the Wave Equation", Journal, Soil Mechanic and Foundations Division, ASCE, New York, Vol 86, SM4, pp 35-61.

Stevens R F, Wiltsie E A and Turton, T H (1982), "Evaluating Pile Drivability for Hard Clay, Very Dense Sand and Rock", Proceedings, 14th Annual Offshore Technology Conference, Houston, Vol 1, pp 465-481.

Stevens R F (1988), "The Effect of a Soil Plug on Pile Drivability in Clay", Proceedings, 3rd International Conference on the Application of Stress Wave Theory to Piles, Ottawa, pp 861-868.

Stevens R F, Wiltsie E A, and Middlebrooks J R (1984), "Controlled Hard Driving", Proceedings, 2nd International Conference on the Application of Stress Wave Theory on Piles, Stockholm, pp 162-169.

Swann L H and Abbs A F (1984), "The Use of Wave Equation in Calcareous Soils and Rocks", Proceedings, 2nd International Conference on the Application of Stress Wave Theory on Piles, Stockholm, pp 421-434.

Wright N D, Tamboezer A J, Windle D, van Hooydonk, W R and Ims, B (1982), "Pile Instrumentation and Monitoring During Pile Driving Offshore N W Borneo", Proceedings, Offshore Technology Conference, Paper No. 4204, Houston, Texas.

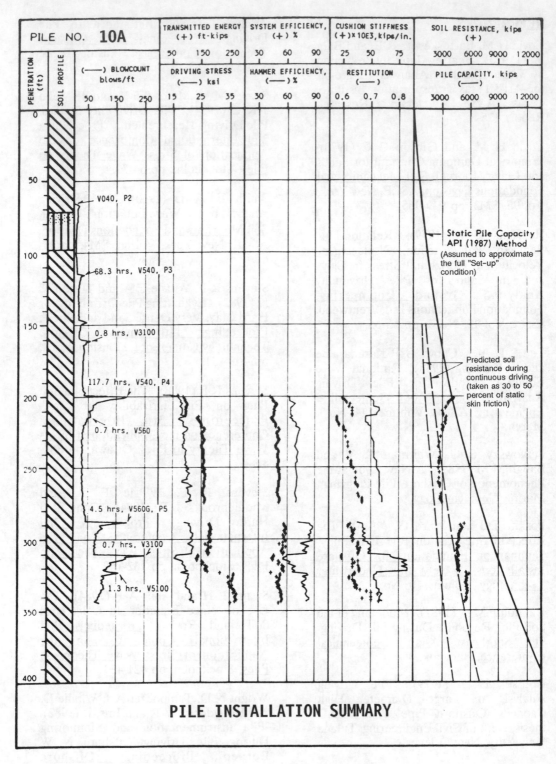

Fig. 1

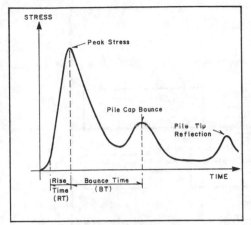

Fig.A Typical Stress Time Record

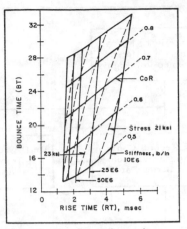

Fig.B Cushion Properties and
Hammer Efficiency

Cushion Properties

The rise time (RT) is a function of cushion stiffness while the pile cap bounce time (BT) is a function of the coefficient of restitution (CoR). Both are relatively independent of efficiency. From wave equation predictions for a range of stiffness and CoR, a nomograph of rise times and bounce times is plotted (see Figure B). Measured rise and bounce times are entered into Figure B to estimate cushion properties.

Hammer Efficiency

Also shown in Figure B are predicted peak stresses at the gage location for 100% efficiency. Since the peak stress is proportional to the ram impact velocity, the following relationship is used to compute hammer efficiency (HE) for the measured stiffness and coefficient of restitution:

$$HE = (V/V_{100})^2 = (S/S_{100})^2$$

where V is the estimated ram impact velocity corresponding to the measured initial peak stress, S. The subcript 100 is for the respective computed values at 100 percent efficiency.

Driving System Efficiency

The energy actually transmitted to the pile is calculated by numerical integration of the recorded data using the equation:

$$ENTHRU = \Sigma FV \, dt$$

where ENTHRU = maximum energy transmitted to the pile;

F   = recorded pile force;

V   = recorded pile velocity; and

dt   = time interval between data points.

The energy transmitted to the pile, expressed as a ratio of the rated energy of the hammer, gives the **driving system efficiency.**

EVALUATION OF HAMMER PERFORMANCE FROM PILE MONITORING DATA

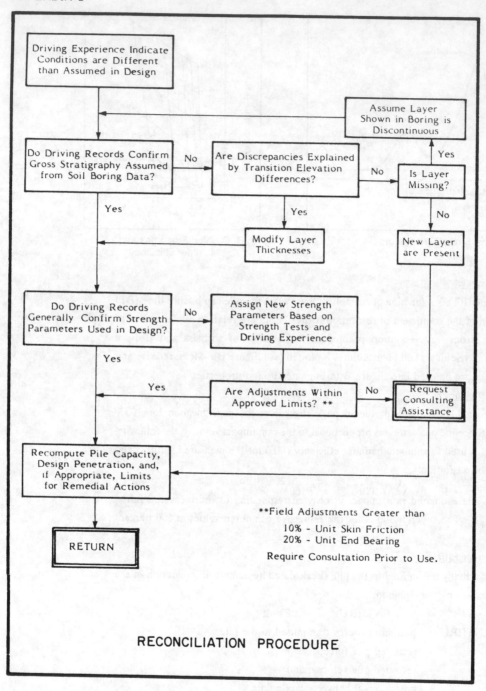

RECONCILIATION PROCEDURE

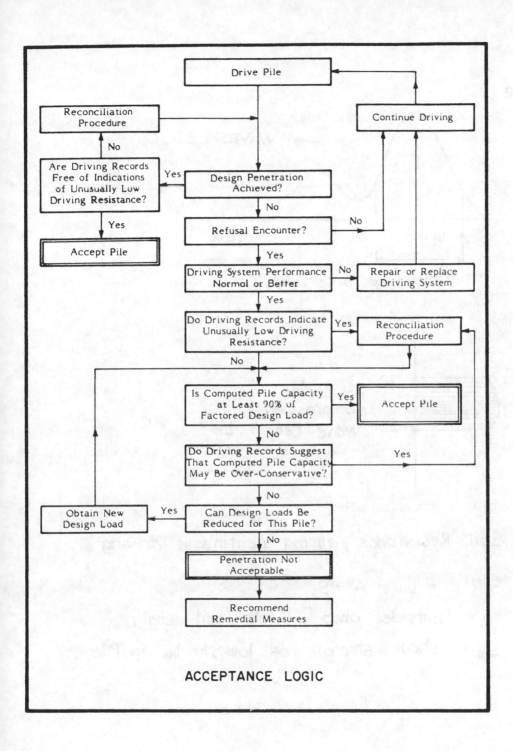

ACCEPTANCE LOGIC

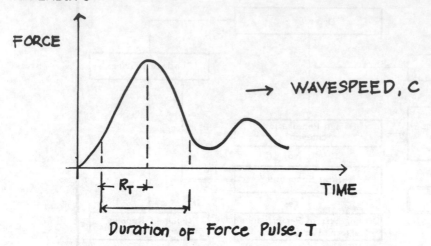

FORCE

→ WAVESPEED, C

R_T

TIME

Duration of Force Pulse, T

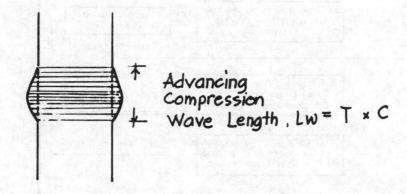

Advancing
Compression
Wave Length, $L_W = T \times C$

Soil Resistance during continuous driving

$SRD = L_W \times A \times S_u$

A = outside area of Pile / unit length

S_u = shear strength of lowest L_W of Pile

Developments in Geotechnical Aspects of Embankments, Excavations and Buried Structures, Balasubramaniam et al. (eds)
© *1991 Balkema, Rotterdam. ISBN 90 5410 019 2*

Pile diagnostics by FPDS

Chandra Prakash, R. K. Bhandari, A. K. Sharma & P. C. Rastogi
Central Building Research Institute, Roorkee, India

ABSTRACT   Low strain integrity testing technique based on one dimensional stress wave analysis applicable to piles, has found wide application in various parts of the world to evaluate suspected defects in piles. In India, integrity testing has been done by Foundation Pile Diagnostic System (FPDS-2) which utilises the solutions of one dimensional wave propagation based on 'method of characteristics' and digital data processing technique for monitoring velocity reflectograms. The principle and testing methodology alongwith some of the results of integrity tests conducted on various type of concrete piles at different sites of the country and few examples of signal matching for quantitative estimate of discontinuities are presented. The method of integrity testing has been found quick and reliable, fit for use in the field shortly after pile construction without interfering much with site activity. The signal matching technique using linear soil model may yield the quantitative determination of defects in suspected piles to a reasonable degree of accuracy.

1 INTRODUCTION

The formation of a sound pile of the required shape and dimensions is one of the important aspect of piling technology. Sometimes, however, constructions do end up with piles of questionable integrity (Bhandari et.al. 1982). This is the reason that quality of piles has always been a matter of great concern. Integrity testing to evaluate suspected defects in piles shortly after their construction using a definitive and economical non-destructive procedure has, therefore, captured imagination of professionals from time to time (Mohan 1981, Palmer & Levy 1981, Sliwinski and Fleming 1984). Of the various available methods, low strain sonic integrity testing method based on one dimensional stress wave analysis applicable to piles has been used successfully in various parts of the world (Middendorp and Van Brederode 1983, Reiding et.al. 1984, Seitz 1986, Rausche et.al 1988). The method basically involves recording of pile head acceleration caused by blow of a hand held hammer on pile top. In few cases, where low strain method had not provided sufficient information, high strain method was employed by using higher impact energies and additional measurements of the pile top acceleration and strain (Rausche and

Seitz 1983). In India, the work on integrity testing of piles has been started recently by Foundation Pile Diagnostic System (FPDS-2), TNO make (Prakash et. al. 1989, Bhandari et.al. 1989). The testing methodology, principle and some of the results of tests conducted on various type of concrete piles at different sites are discussed. A few examples of signal matching for quantitative estimate of discontinuities are also highlighted.

2 PRINCIPLE OF INTEGRITY TESTING

The principle of integrity testing method is time-domain reflectometry of stress wave propagation through pile material acting as a one dimensional medium. The wave is generated by a short hammer blow impact on pile head which travels down the length of pile in axial direction with the speed of sound. The particle velocity V at any level is dependent on force, F and impedance of pile, Z at that level (V = F/Z). The impedance is directly proportional to the area of cross-section of pile, A ($Z = EA/C = A\sqrt{E\rho}$, C = stress wave velocity, E = Young's modulus and ρ= density of pile material) & any change in it due to irregularities or defects present in the pile stem, causes variation in the particle velocity. As a result of these changes, a part of the compression wave is reflected from the location of variations in impedance of pile shaft. If there is no variation in impedance throughout the length of pile, the compression wave will reflect from pile toe only. Monitoring and analysis of these reflections form the basis of integrity testing. Fig.1(a) through Fig.1(c) represent the velocity-time reflectograms for a continuous pile, a pile with necking and for a pile with increase in cross-section respectively. Difference in the type of reflections in a pile with necking (Fig.1(b)) and a pile with increased cross-section (Fig.1(c)) helps in distinguishing the shape of pile in two cases.

3 METHOD OF INTEGRITY TESTING

The diagnosis of a pile is based on two equally important aspects, (a) monitoring of stress wave using either analogue or digital data processing technique, and (b) analysis & interpretation of velocity reflectogram. The Foundation Pile Diagnostic System incorporates the digital data processing technique for monitoring of stress wave which provides more accurate results in comparison to analogue technique. The set-up for integrity test by the FPDS is shown in Fig.2. The test is conducted by striking pile head by a small hand held hammer, struck in such a way that a blow with a short rise time is achieved. The reflections are picked up by an accelerometer pressed on pile top close to the location of hammer blow. The observed signal is amplified by the computer controlled amplifier. It is then numerically integrated to convert accelerations into velocity through signal conditioning subsystem, fitted to the back of FPDS computer. Some low pass filtering is also done to improve the signal to noise ratio. The signal conditioning sub-system also controls the triggering of signal for digitising the same by a 12 bit analogue to digital converter, fitted to the computer. The whole process is controlled by the computer and after processing of signals, the results are displayed on computer screen in the form of particle velocity versus length records providing the information about defects, if any and approximate pile length. The results are also stored on the hard disk of FPDS computer for subsequent analaysis. The generated compression wave experiences damping effect due to soil friction acting along pile shaft. However, increase in gain

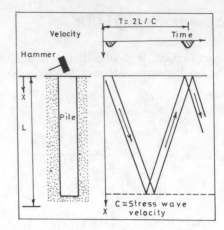

(a) Pile with uniform section

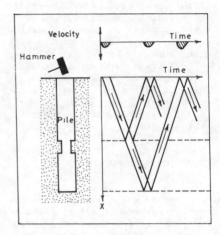

(b) Pile with decrease in section

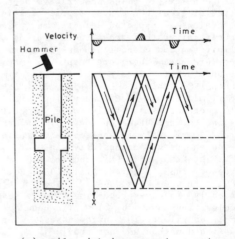

(c) Pile with increase in section

Fig.1 – Stress wave pattern of reflections

Fig.2 - Set-up for integrity test by FPDS

with time compensates for signal loss due to soil friction and to obtain a clear toe reflex, signal can be amplified upto desired level by selecting suitable gain value.

The system concept of integrity testing procedure alongwith a typical record obtained for structurally sound bored cast-insitu concrete pile is depicted in Fig.3.

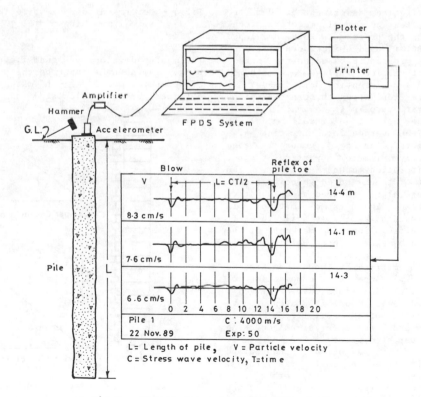

Fig.3 - System concept of integrity test

The reflectograms obtained for each pile tested are used to evaluate the structural integrity of piles. Since each type of discontinuity, for example, changes in cross-section, cracks, joints in pile, soil inclusions, pile toe, change in soil stiffness and reinforcement overlapping in case of heavily reinforced piles, provides unique type of reflections, the first hand information about the type of discontinuities and their approximate locations can be had directly by reflectograms. The effect of sudden variation in soil stiffness on the observed signal can be ascertained using sub-soil data. A change from stiff layer to a soft one yields same type of signal as a decrease in cross-section while in case of change from soft soil layer to a stiffer one, the trend of signal is same to that of increased section. However, the reflections are not so sharp as in the case of variations in pile cross-section. This helps in identifying whether the variation in signal is due to discrepancy in pile shaft or due to change in ground conditions.

4 QUANTITATIVE ESTIMATE OF DISCONTINUITIES

For quantitative assessment of defects, integrity testing signal matching technique using the program TNOWAVE, developed by TNO-IBBC, is used (Middendorp & Reiding 1988). The program is based on the 'method of characteristics' applied to a mathematical model representing one dimensional stress wave propagation through the pile.

The first step in signal matching involves the derivation of soil model by matching the average integrity test signal, selected for a structurally good representative pile for a particular site, by using iteration method, Fig.4. The first part of the measured velocity signal i.e. impact blow, is used as an input to the pile head. The first estimate of soil model is based on available soil investigation results. Since the movement of the pile caused by a small hand held hammer is very small (<< 0.1 mm), the shaft resistance is schematized by linear dampers and the toe resistance by a combination of linear spring and a linear damper, Fig.5. The derived soil model is taken to be representative of the soil for a particular area and is used for signal matching of other suspected piles. The model, if required, can be adjusted to incorporate any local change in soil stratigraphy, density or hardness for any other pile location in the area. For signal matching of suspected piles also, the first part of

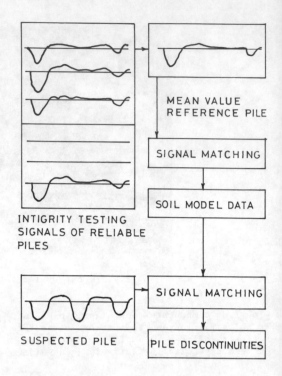

Fig.4 – Determination of pile defects by signal matching

the measured velocity signal i.e., impact blow, is taken as input to the pile head of computer model. A good match is then obtained between calculated and measured

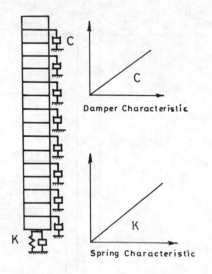

Fig.5 – Model of integrity testing simulation

signals by varying cross-section of computer model pile at location of defects through a number of iterative steps.

5 EXAMPLES OF FIELD TEST

The integrity tests were conducted on (i) especially cast bored concrete defective piles, (ii) bored cast-in-situ concrete production piles, (iii) driven cast-in-situ concrete production piles, and (iv) precast concrete production piles, at various sites covering different type of soil deposits, Table 1.

5.1 Especially Cast Defective Piles

5.1.1 CBRI Site

The integrity test results for three piles, cast at this site (Fig.6) are given in Fig.7. The velocity trace of pile (1) is like a free end pile as expected in such deposits. The signal is repeated four times, each time the dip i.e. toe reflex is reduced due to dissipation of energy with time. The trace also shows upward rise at about 1.5m which is probably due to increased soil stiffness at this level. The velocity trace of pile (2) shows a sharp reflection of reduction in pile impedance at about 2.1 m, similar to pile toe reflex. The strong repetition of signal indicates that the pile is of only about 2.1 m length, thus confirming the discontinuity created in the pile. In the velocity trace for pile (3), a reduction in pile impedance at about 1.2 m followed by increase in impedance, clearly reflects reduction in pile cross-section. The toe reflection is not so clear. Also the equivalent built in length of pile corresponds to a reduced wave velocity of 3000 m/s against 3600 m/s, indicating concrete of inferior quality at any section in this pile in comparison to piles (1) and (2). The observed velocity reflectogram thus appears to be in accordance with the discontinuity created in the pile.

5.1.2 Vivek Vihar Site

The integrity test results of four piles, cast with different type of discontinuities at this site (Fig.8), are given in Fig.9. In piles (2), (3) and (4), the enlargement of pile section in different shapes was made by hand. To create discontinuity in pile (3) at lower level, the concreting was done upto 3.0 m depth and

then the tremie pipe was taken out. The soil was next poured in the bore hole from its top. Obviously, during pouring the soil got mixed with the drilling mud and finally with the laitence available at top of green concrete poured in the bottom portion of pile. After two hours, the tremie pipe was again lowered and concreting completed upto the top. In pile (4), the discontinuity was created by lowering soil, packed in a polythene bag in cylindrical form of about 100 mm dia. and 500 mm length, through tremie pipe during concreting. It is obvious that creation of discontinuity was not well controlled.

The velocity reflectogram for pile (1) at this site is also identical to the velocity trace of a free end pile with low side friction as expected in such deposits. The local increase in pile impedance at about one metre followed by decrease in impedance, reflected by the velocity trace of pile (2), is in accordance with the built in shape of pile. The observed velocity trace for pile (3), shows little local increase in impedance at about one metre and considerable reduction in impedance at about 3.7 m followed by some increase in impedance. Since there is no repetition in signal and a clear toe reflection is observed, the decrease in impedance indicates reduction in pile section at this level and not a complete break in the continuity. Further, the built in length of pile could be obtained corresponding to the reduced stress wave velocity of 3000 m/s against 3600 m/s in case of piles (1) & (2). This also suggests the presence of inferior quality of concrete in any portion of the pile in comparison to the concrete of piles (1) and (2). The local variations in impedance, reflected in top portion of the velocity trace of pile (4),are in accordance with the built in cross-section of pile. The presence of soil inclusion at about 6 m is reflected by reduction in impedance at this level. However, the reflection is not so good since the discontinuity created at this level was not well controlled as pointed out earlier. Further, in this case also the built in length of pile corresponds to stress wave velocity of 3000 m/s, indicating presence of inferior quality of concrete in any portion of the pile.

5.2 Bored Cast-In-Situ Production Piles

5.2.1 Paschim Vihar Site

The observed velocity reflectograms of two typical bored compaction underreamed piles

171

Table 1 - Summary of integrity tests conducted by FPDS

Sl.No.	Site Details			Pile Details			
	Location	Soil Type	W.T. (m)	Designation	Type	Dia. (mm)	Depth (m)
1.	C.B.R.I., Roorkee	Silty sand (SM)	4.25	Especially cast	Bored concrete with known discontinuities	300	4.0
2.	Vivek Vihar, Delhi	Clayey silt (ML) followed by silty sand (SM)	1.50	Especially cast	Bored concrete with known discontinuities	375	7.0
				Production	Bored cast-in-situ concrete	300	5.7
3.	Paschim Vihar, Delhi	Clayey silt (ML)	5.50	Especially cast	Bored concrete with known discontinuities	375	5.0
				Production	Bored cast-in-situ concrete	375	5.0- 6.0
4.	Anand Vihar, Delhi	Silty sand/poorly graded sand (SM/SP)	3.50	Production	Bored cast-in-situ concrete	400, 500	10.0-15.0
5.	N.F.L., Delhi	7-10m Clayey & sandy silt with kankar followed by rock	1.50	Production	Bored cast-in-situ concrete	350, 400, 500	8.5-11.0
6.	NCPP, Vidyut Nagar, Ghaziabad	Silty clay (CL)/ clayey silt (MI) followed by silty sand (SM)	2.00	Production	Bored cast-in-situ concrete	400, 500, 760	13.0-18.0
7.	Delhi Cantt.	Sandy silt/silty sand (SM)	2.00	Production	Driven cast-in-situ concrete	400, 530	17.0-18.0
8.	NFCL, Kakinada	Dense sand (SP) 6 m followed by 8 m soft clay overlying stiff clay	1.00	Production	Precast concrete	400 mm x 400 mm	22.7

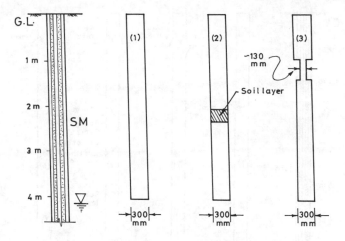

Fig.6 - Especially cast bored piles at CBRI

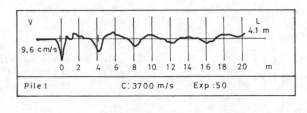

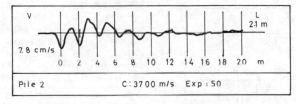

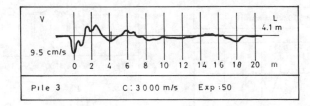

Fig.7 - IT results of especially cast bored piles, CBRI (Fig.6)

at this site alongwith the expected design shape of piles are shown in Fig.10. The construction of pile involves, boring, underreaming, pouring of concrete in the usual manner and finally driving a rein-forcement cage assembly alongwith a steel pipe through freshly laid concrete to compact both the concrete and soil around simultaneously.

The velocity reflectograms of both the piles correspond to same settings of stress wave velocity, 3800 m/s and expo-nential gain, 5. The length of piles indicated by the integrity tests is very close to that constructed. The reflection in the form of increase in impedance at about 4.5 m depth in case of velocity reflectogram of pile (1), confirms the bulb formation as expected in the built in shape of pile. However, the velocity reflectogram of pile (2) does not show such increase in impedance at this depth. But there is an increase in impedance at about 2.5 m depth, indicating increase in

173

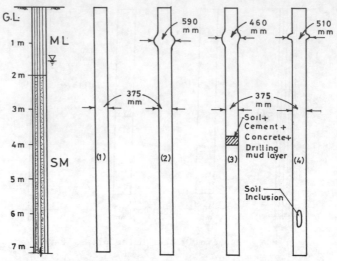

Fig.8 - Especially cast bored piles at Vivek Vihar

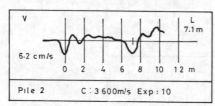

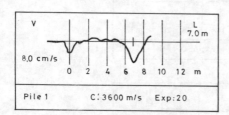

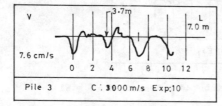

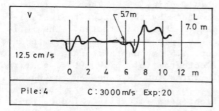

Fig.9 - IT results of especially cast
bored piles, Vivek Vihar (Fig.8)

pile cross-section at this level. Since
the piles were not excavated, these reflectograms could not be verified with the
actual built in shape of piles.

5.2.2 NPL Site

The velocity reflectograms of four 500 mm
diameter bored cast-in-situ concrete
piles, taken down to rest on rock, at this
site are given in Fig.11. The reflectogram
of pile (1) is clearly indicative of a

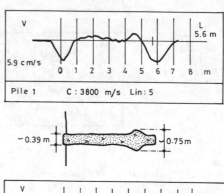

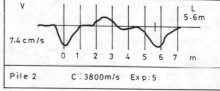

Fig.10 - IT results of two typical production piles, Paschim Vihar

174

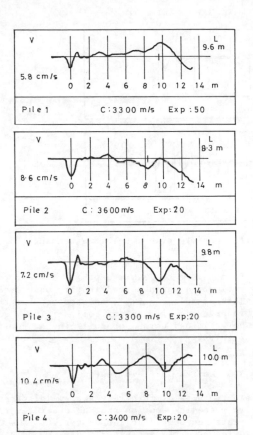

Fig.11 - IT results of bored cast-in-situ
concrete piles, NPL

structurally good pile with toe resting on
rock as reflection of toe is in opposite
direction to that of hammer blow. In case
of pile (2) the toe reflection is not so
sharp as in case of pile (1) and it is
somewhat indicative of the presence of
very thin layer of muck between the toe
and rock. The toe reflection in case of
pile (3) is in the same direction to that
of hammer blow, clearly indicating the
presence of muck at toe level. The reflectogram of pile (4) shows reduction in
cross-section at about 4.75 m and clear
toe reflection indicative of the presence
of muck at toe level. The lengths recorded
are very close to those constructed 9.6 m
in both the cases for pile (1), 8.3 m
against 8.6 m for pile (2), 9.8 m against
10.0 m for pile (3) and 10.0 m against
10.6 m for pile (4).

5.2.3 NCPP Site

At this site a large number of integrity
tests were performed on bored cast-in-situ

concrete piles, constructed to support
various structures of a Thermal Power
Plant. Fig.12 shows velocity reflectograms
of two typical piles of 500 mm diameter
and 13.6 m depth below cut-off level (cut-
off level 2.0 m below ground level). At
the time of integrity tests on these
piles, lean concrete base was already laid
to cast the cap and the pile heads were
about one metre above the lean concrete.
Both the reflectograms show sharp increase
in impedance at about one metre followed
by little decrease in impedance, confirming the presence of lean concrete at this
level. In case of pile (1), an increasing
trend with minor variations in impedance
upto clear toe reflection at 14.7 m, indicates pile of good structural integrity.
On the other hand in case of pile (2),
large variations in impedance can be seen
beyond 8 m depth and it is difficult to
establish clear toe reflex. Also 14.7 m
length corresponds to stress wave velocity
of 3600 m/s against 4000 m/s in case of
pile (1). These observations suggest that
the pile (2) has irregular shape beyond 8
m and also the quality of concrete in this
pile is not so good as in pile (1).

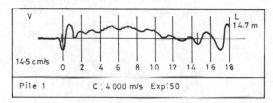

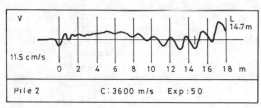

Fig.12 - IT results of bored cast-in-situ
concrete piles, NCPP

5.3 Driven Cast-In-Situ Production Piles

The integrity tests were conducted on 400
mm and 530 mm diameter driven cast-in-situ
concrete piles of 17 m to 18 m depth at
Delhi Contonment site. The velocity reflectograms of two typical 400 mm diameter
piles at this site are given in Fig.13.
Both the reflectograms correspond to same
stress wave velocity of 3800 m/s and exponential gain of 100. While the reflectogram of pile (1) is clearly indicative
of structurally good pile of proper shape

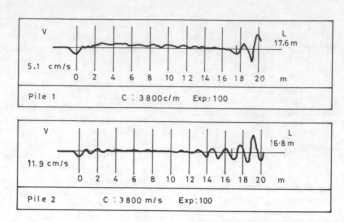

Fig.13 - IT results of driven cast-in-situ concrete piles, Delhi Cantonment

upto full depth, the variations in impedance beyond 12 m in case of pile (2)
indicate some what irregular shape beyond
this depth.

5.4 Precast Concrete Production Piles

Integrity tests were conducted on a large
number of 400 mm x 400 mm square cross-
section and 22.7 m long precast concrete
piles, driven to support various structures of a fertilizer plant at Kakinada.
The piles were penetrated by water jetting
through top 6.0 m dense silty sand layer
and then, these were driven with 4 to 5 t
drop weight keeping the height of fall
1.25 to 1.0 metre.
   Integrity test result of a typical pile
at the site is given in Fig.14 alongwith
the static cone penetration resistance.
The variations observed in the velocity
trace can be very well explained in the
light of cone penetration resistance and
driving process. The pile belongs to a
group for which excavation for laying cap
was completed and also the lean concrete
base at about 1.2 m below the pile head
level was laid prior to the test. The
first small sharp increase in impedance at
about 1.2 m is due to the presence of lean
concrete at this level. The almost smooth
trace upto about 11.0 m shows low friction
on the pile which is because the pile was
penetrated by water jetting upto about 6.0
m and beyond that soft clay layer (static
cone resistance 0.2 kg/cm^2) extends upto
11.0 m depth. The gradual increase in
impedance beyond 11.0 m is due to friction
offered by stiff to very stiff clay layer
(static cone resistance 2 to 10 kg/cm^2 for
11.0 m to 12.0 m depth and 15 to 25 kg/cm^2

from 13.0 m to 23.0 m depth). A clear toe
reflection was observed in this case also.
This example clearly demonstrates that the
interference effect of soil stiffness can
be explained using sub-soil investigation
results.

5.5 Remarks on Field Tests

The results of integrity tests, discussed
in preceding section and the other several
tests conducted so far on various type of
piles at different sites clearly indicate
that the low strain integrity testing
based on one dimensional stress wave approach as conducted by the FPDS is an
efficient technique for assessing the
integrity of concrete piles. The method
is found suitable to both cast-in-situ as
well as to precast concrete piles. For
proper interpretation of integrity test
data, a good understanding of sub-soil
investigation results, method of pile
construction, sequence followed during
casting of piles and quality of concrete
used, is essential. The variations in
velocity of hammer blow do not effect much
the consistency of measured signal. But
the condition of the pile head particularly the quality of concrete at that level
and the quality of hammer blow do effect
the quality of integrity test reflectogram. Further, the presence of inferior
quality of concrete in any portion of the
pile reduces the stress wave velocity and
the probable built-in length of pile corresponds to a reduced stress wave velocity. In case of bored cast-in-situ concrete
piles particularly in soil deposits with
high water table where stabilization of
bore hole is done by drilling mud and

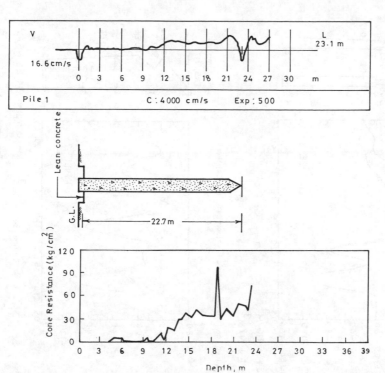

Fig.14 - IT result of precast concrete pile and static cone penetration resistance, NFCL, Kakinada

concreting is done by tremie method, large variations in the stress wave velocity for piles, cast using the same grade of concrete, were observed.

6 EXAMPLES OF INTEGRITY TESTING SIGNAL MATCHING

6.1 Especially Cast Defective Piles

The best fit signal matches for three piles, constructed at site I (Fig.6) are shown in Fig.15. The soil model, used to match the signals of piles (2) and (3) has been derived by matching the signal of pile (1). In pile (2) the best fit match shows a complete break at the level of discontinuity in the model. Since the compactness of loose soil at the level of discontinuity, is different than the compactness of undisturbed soil available at pile toe, little reduced values of linear damper and spring at the level of discontinuity provided a better match between the measured & calculated signals as shown in Fig.12(d). In case of pile (3), model shows a compatible cross-section with the expected built in shape of pile though the exact dimensions are slightly different. The pile could not be excavat-

ed due to some reasons to verify these results. Both of the signal matches at this site show that the technique provides a reasonably good quantitative estimate of discontinuities.

Fig.16 shows the best fit signal matches for the piles cast at Site II (Fig.8). Soil model in this case also was derived by matching the signal of pile (1). The modelled piles in general conform to the expected built in discontinuities. But it is difficult to pin point exact difference between the discontinuities formed in the actual built in piles and those predicted by models as the piles were not exhumed in this case also. The location of discontinuities is almost the same to that can be expected in built in piles. The maximum lateral dimensions of the discontinuity in the top portion of modelled piles are upto 15 per cent less than those expected in built in piles. The discontinuity created in pile (2) at lower level is reflected by reduced cross-section in model. The soil inclusion in pile (3) is reflected by a minor reduction in model pile.

6.2 Production Piles at Paschim Vihar Site

Since there was no straight shafted production pile and also no exhumed pile was

177

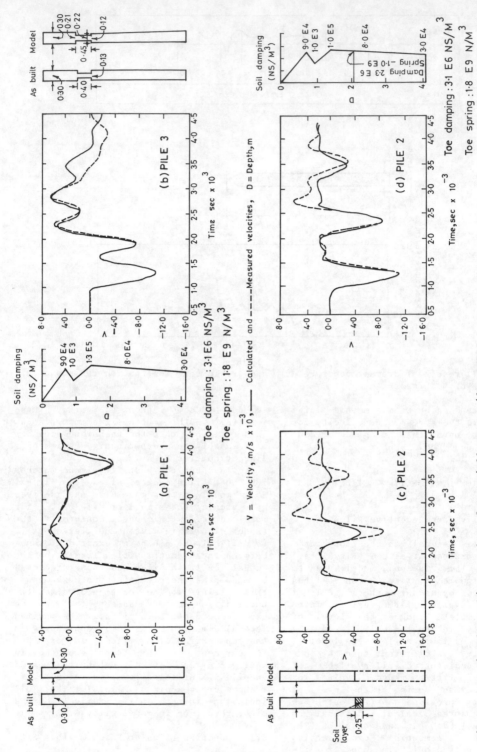

Fig.15 · IT signal matching results of especially cast piles, CBRI

178

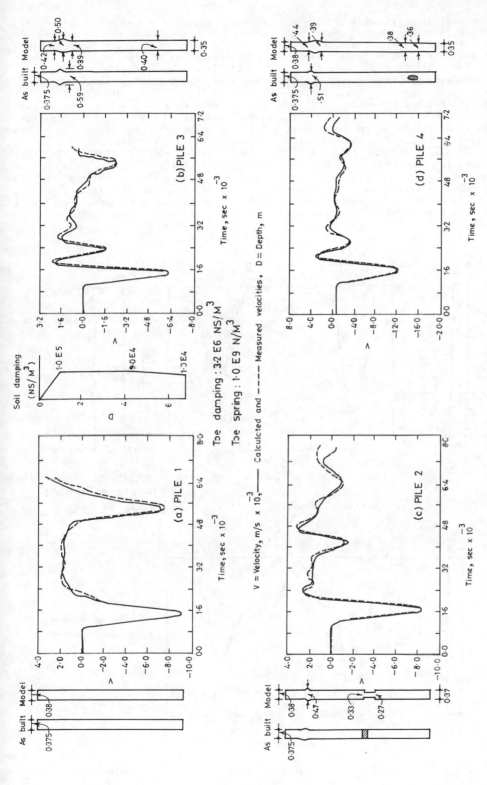

Fig.16 · IT signal matching results of especially cast piles. Vivek Vihar

179

available at this site, the soil model was derived on the basis of sub-soil investigation results and matching the average signal, obtained from the velocity reflectograms of two piles (Fig.10) by neglecting the increased impedance reflected in these signals. The signal matches obtained using this soil model are depicted in Fig. 17. The models show the diameter of pile shafts about 390 to 400 mm which may be considered quite close to the expected dimensions obtained finally in case of compaction piles. Since the increase in shaft diameter, resulting due to driving of cage assembly in freshly laid concrete in these piles depends on various factors and varies from pile to pile, the variation in diameter showed by model piles is obvious. The increase in impedance in both the piles corresponds to an increase in cross-section of model piles. In case of pile (1), the maximum diameter is reflected as 480 mm which is about 36 percent less than the expected built in shape. In the absence of actual measurement of ex-humed pile, it is not possible to explain exact reasons for this difference. The increase in diameter reflected by model is more in case of pile (1) than in pile (2), though the observed signals show higher impedance in case of pile (2) than in pile (1). This observation suggests that the increase in impedance at higher level will be more than the increase at lower level

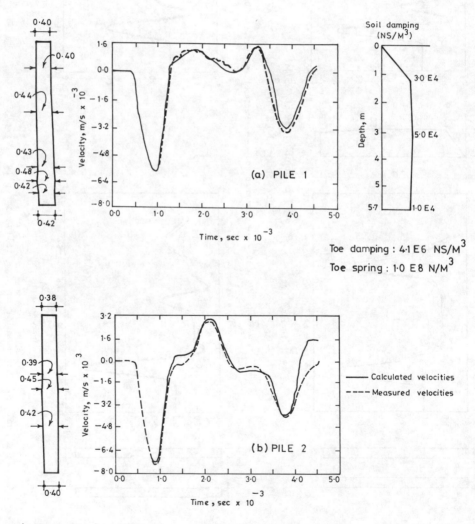

Fig.17 – IT signal matching results of production piles, Paschim Vihar

for the same enlargement in pile cross-section at 2 levels, probably due to dissipation of energy with depth on account of soil resistance and internal damping of pile material.

7 CONCLUSIONS

Low strain integrity testing methodology based on one dimensional stress wave approach as employed by FPDS is simple, quick and requires minimal interference with site activity. The major defects such as cracks, necking, soil inclusions and changes in cross-section, produce their own unique signal in velocity reflectograms and pile can be checked soon after construction. The pile length can also be estimated to a fairly reasonable degree of accuracy if the pile is not too long or skin friction is not too high. However, the minor deficiencies like local loss of cover to pile reinforcement or small inclusions and also type of debris present at pile base are not discovered.

Analysis of velocity reflectograms offer both qualitative and quantitative information. For proper interpretation of integrity test reflectograms, a knowledge of sub-soil investigation results, construction technique and the sequence followed in the pile casting along with the quality of concrete used, is very essential. Qualified personnel who have knowledge of the pile-soil interaction and piling construction techniques are required both for conducting tests in the field and to interprete the test results. The signal matching by the program TNOWAVE which makes use of linear soil model may provide quantitative determination of defects in piles to a reasonable degree of accuracy.

ACKNOWLEDGEMENT

The work reported herein forms normal programme of research of the Central Building Research Institute, Roorkee (UP), India. The authors would like to thank Mr. P. Middendorp, Head Pile Dynamics Section, TNO, IBBC, Netherlands for the guidance provided in conducting integrity tests in the field in initial stages. They are also grateful to various agencies and personnel for making the piles available for testing and providing facilities at the sites during the field work. Thanks are also due to Mr. Surendra Giri and Km. Heera Bhatt for pains taking hardwork in word-processing the paper.

REFERENCES

Bhandari, R.K.; Prakash, C. & Sharma, A.K. (1982), "Failures in Cast-In-Situ Piles", Commemorative Vol. Int. Conf., Mexican S.M. Society, 25th Anniversary of Its Foundation, Mexico.

Bhandari, R.K.; Prakash, C.; Rastogi, P.C. and Sharma, A.K.(1989)," Integrity Testing of Piles by Stress Wave Measurements", IGC-89, Visakhapatnam.

Middendorp, P. and Van Brederode, P.J. (1983), "A Field Monitoring Technique for Integrity Testing of Foundation Piles", Int. Symp. on Field Measurements in Geomechanics, Zurich.

Middendorp, P. and Reiding, F.J. (1988), "Determination of Discontinuities in Piles by TNO Integrity Testing & Signal Matching Techniques", 3rd Int. Conf. on the Application of Stress Wave Theory to Piles, Ottawa.

Mohan, D. (1981), "A Close Look at the Problems of Research and Its Application to Pile Foundations", Indian Geotechnical Journal, Vol. 11, No.1.

Palmer, D.J. and Levy, J.F. (1981), "Concrete for Piling & Structural Integrity of Piles", Piles and Foundations, Thomas Telford Ltd., London.

Prakash, C., Rastogi, P.C. & Sharma, A.K. (1989), "Use of Stress Wave Measurements for Diagnosis and Analysis of Piles", Civil Engineering & Construction Review, Vol.2, No.7, New Delhi.

Rausche, F., Likins, G.E. and Hussein, M. (1988), "Pile Integrity by Low and High Strain Impacts", 3rd Int. Conf. on the Application of Stress Wave Theory to Piles, Ottawa.

Rausche, F. & Seitz, J. (1983), "Integrity of Shafts & Caissons", Symp. on Dynamic Measurements for Capacity and Integrity Evaluation of Piles and Piers, ASCE, Philadelphia.

Reiding, F.J., Middendorp, P. and van Brederode, P.J. (1984), "A Digital Approach to Sonic Pile Testing", Conf. on the Application of Stress Wave Theory on Piles, Stockholm, Sweden.

Seitz, J. (1986), "Low Strain Integrity Testing of Bored Pile", Ground Engineering, November.

Sliwinski, Z.J. & Fleming, W.G.K. (1984), "The Integrity and Performance of Bored Piles", Advances in Piling and Ground Treatment for Foundations Conference, I.C.E., Thomas Telford Ltd., London.

The study of the swelling properties of the altered granite by means of large scale field tests for underground excavations of the largest pumped storage power station in China

Ge Xiurun
Institute of Rock and Soil Mechanics, Academia Sinica, Wuhan, People's Republic of China

ABSTRACT: A brief introduction is given in this paper to the main problem on engineering geology encountered in the underground excavation at Guangzhou Pumped Storage Power Station, i.e. the weak mechanical property and swelling property of the altered granite. The main results of the methods and experiments are given with respect to the swelling properties of the large scale tests carried in-situ of the altered granite. Minerological analysis shows that the content of montmorillonite is closely correlative to the swelling property. The author puts forward a formula to evaluate the swelling pressure of the altered granite and the dynamic expansion mechanism for explaining the damage occurred on the altered granite walls of the underground excavations.

1 GENERAL SITUATION OF GUANGZHOU PUMPED STORAGE POWER STATION*

The Guangzhou Pumped Storage Power Station is the largest one being built in China. The straight distance is only 90 km between Guangzhou and the station, where the topographical and geological conditions are rather ideal. The Station was approved to be built in order to guarantee the safe operation of Guangdong Daya bay nuclear power station and satisfy the need of regulating peak and filling off-peak of Guangdong power system. After its completion, the Station will be connected to the Guangdong electric network by two parallel transmission lines of 500 kv. The economic and social benefits are quite considerable.

The project is mainly composed of dams of upper and lower reservoirs, delivery water system and underground power house. The upper and lower reservoirs are separately situated on the secondary tributaries of Liuxi River, where is a beautiful scenic spot. With plentiful water resources and surrounded by high mountains, the reservoirs are of good natural basins. Both dams of upper and lower reservoirs are reinforced concrete faced rock fill dams.

Both the delivery system and the underground power house are arranged in a thick rock mass of granite. Rock properties consist mainly of coarse biotite granite at the middle stage of Yanshan Mountain. Except for alteration of varying degree occurring in the rock mass around the structural fissure, rock properties are comparatively uniform and integral. The general layout plan and longitudinal section of the tunnel systems and power house are shown in Fig. 1a and 1b.

The installed capacity of the Station at the first stage is 1200 MW, which consists of four pumped storage units of 300 MW each. The maximum net static water head is 535 m. The overall efficiency of the Station is 76 percent.

*This passage is adopted from the 'Brief Introduction of Guangzhou Pumped Storage Power Station' by Guangzhou Pumped Storage Power Station Joint Venture Corporation, November, 1988

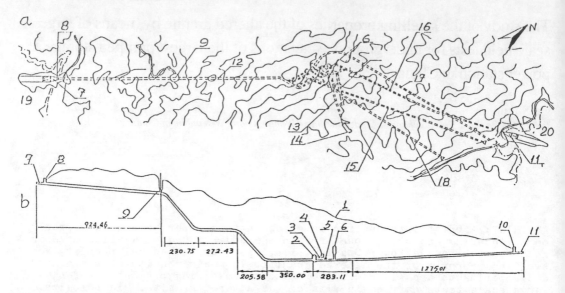

Fig.1 General layout plan (a) and longitudinal section (b) of systems of tunnels and power house

1. groundline, 2. power house, 3. busbar tunnel, 4. transformer gallery, 5. tail gate, 6. surge tank, 7. upper intake, 8. and 10. upper gate chamber, 9. upper surge schaft, 11. lower intake, 12. penstock, 13. cable tunnel, 14. switchyard, 15. emerg. vent. tunnel, 16. tail race tunnel, 17. main access tunnel, 18. exploratory vent tunnel, 19. upper reservoir, 20. lower reservoir

2 ALTERED GRANITE -- MAIN PROBLEM ON ENGINEERING GEOLOGY FOR UNDERGROUND CONSTRUCTIONS

2.1 Engineering outline of the underground constructions

The underground constructions include power house, transformer cavern, surge tank, penstock and tail race tunnel. The underground constructions are very large in size, e.g. the penstock and tail race tunnel are of circular sections, the diameter of which are respectively 8.5m and 9.0m. The width of the main access tunnel reaches as many as 8.0m. The power house and transformer cavern are connected by the busbar tunnel, the section diagram of which is given in Fig. 2. The underground power house is as wide as 21m and as high as 45m.

The caverns are deeply arranged in the granite mass, the overburden is between 330-440m. However, there develop six groups of faults and fissures of NE-NNE and NW-NNW in the granite mass. Altered granite

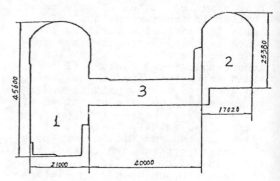

Fig.2 Sectional diagram of the power house (1), transformer cavern (2) and busbar tunnel (3)

with different alterations is found along these faults and fissures.

Due to the large size of the underground constructions, the altered granite belts are unavoidable. While the strength of such a clay bearing altered granite is very poor, the rock will be swollen after absorbing water, most seriously it will be even crumbled into grits. Therefore, the altered granite problem becomes a restrict factor for the long-term stability of the underground constructions, and also the main problem on engineering geology of the project.

2.2 Geological formation of the altered granite

The altered granite is mainly formed after the formation of biotite granite at the third stage of Yanshan Mountain. The underground hydrothermal solution moves along part of the faults and fissures, forming quartzite and calcite veins, at the same time, the hydrothermal alteration and replacement form the altered rock belt including the main vein.

With the process of replacement, the components in the hydrothermal solution vary successively with the gradual decreases of temperature and pressure, forming the altered rock with different mineral compositions at different sites. Hence its mechanical property is very heterogeneous .

Some of the altered granite are in bar distribution, and others in dense parallel belts. The width of the bar distribution generally falls in some ten to tens centimeters, while the dense belts reach some meters to more than ten meters.

2.3 General features of the altered granite

2.3.1 Banding pattern of the altered granite

Regular banding patterns can be found in the altered granite in this region. To sum up, there are five dif-

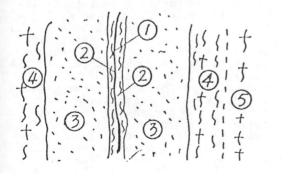

Fig.3 Typical diagram for banding pattern of altered granite

ferent banding as follows:
1. Fine-grained quartzite-calcite,
2. Chlorite and hydromuscovite,
3. Montmorillonite-Kaolinite,
4. Weakly altered granite,
5. Sound granite.
Fig. 3 is the typical diagram for the banding pattern, where the symbols are in correspondence with the text.

2.3.2 Obvious difference between altered and sound graniteS

The alteration leads to the loosening in structure and poorness of elasticity and strength. Therefore, the altered belt is the weak belt in the granite mass.

Since the alteration happened in the granite mass of the Station, the general physico-mechanical properties of the altered granite mass change obviously, for example,

1. the increase of the void ratio: from 0.061(sound granite) to 0.127 (altered granite),
2. the increase of water content: from 2.5% to 4.25%,
3. the decrease of the unit weight of the rock: from $2.6 \times 10^{-2} N/cm^3$ to $2.5 \times 10^{-2} N/cm^3$,
4. the decrease of the longitudinal wave velocity V_p: from 5100--6400M/S to 3540--4060M/S,
5. the decrease of the uniaxial compressive strength: from 90MPa to 17MPa.

2.3.3 Unique feature of swelling and crumbling of the altered granite

The altered granite of the Guangzhou Pumped Storage Power Station contains clayish minerals such as montmorillonite, illite and kaolinite. Particularly, the montmorillonite will be swollen up after encountering of water with the ability of obvious absorption and cation-exchange. After the excavation, the walls of the underground constructions are to be exposed to the air, the montmorillonite will be swollen, causing the structure damage of the altered granite, which would be quite loose, and even crumbled to grits.

In order to guarantee the stability of the underground construction, to make reasonable design for permanent and temporary supports and to adopt the necessary engineering measures, it it by all means important to study the physico-mechanical properties of the altered granite, particularly its swelling property.

3 METHODS TO STUDY THE SWELLING PROPERTY OF ALTERED GRANITE

Since the altered granite of the Station is in bar and belt distribution, the belts are 1-2 meters wide with heterogeneous properties, it is very hard for small sampling experiment in the laboratory to reflect the swelling property of the altered granite. Therefore, the experiment is carried out in-situ with much larger test-pieces.

The tests are arranged in the special adits for the tests, where the geological conditions are quite similar to those of the underground power house. Four types of tests were carried out:

a. free swelling test for the cylindrical test-piece in situ (1 test-piece, No. DP-3),

b. swelling pressure test for the cylindrical test piece in situ (1 test-piece, No. DP-2),

c. free swelling test for the walls of the underground excavations (5 test points, Nos. BP-1, BP-2, BP-5, BP-6 and BP-7),

d. swelling pressure test for the walls of underground excavations (2 test points, Nos. BP-3 and BP-4).

The lithological characters of the above nine test-pieces (points) are of moderately and lightly altered granite. From the test items, it can be divided into two categories, one is the test for free swelling property (a, c), the other is the test for measuring the swelling pressure (b, d). From the methods of the tests and the forms of the test-pieces, it can also be divided into two categories, one is the test for cylindrical test-pieces (a, b), though it is also the large scale test in situ, yet the test conditions are much closer to those in the laboratory, the other is the test for the swelling property of the walls of the underground excavations (c, d). The tests in this category are based on the direct simulation and measurement for the actual swelling of the altered granite around the underground excavations. And the measurement of the swelling pressure provides with the data for the support design and treatment scheme of the underground excavations.

4 STUDY ON THE SWELLING PROPERTY FOR THE CYLINDRICAL TEST-PIECE OF THE ALTERED GRANITE IN-SITU

4.1 Test-piece

Both the diameter and the height of the test-piece are 50cm respectively. At the center of the cylindrical test-piece, there is a hole of 40cm in length with the diameter of 4cm, used for water injection to the central part of the test-piece.

The cylindrical test-piece is manually excavated from the bottom rock mass of the test adits. The lithologic character of DP-2 is of slightly-lightly altered granite and that of DP-3 lightly altered. After shaping the side faces of the test-piece are wrapped with 6 layers of tape, greased in between the layers as antifriction layer. A steel tube of $\phi 60 \times 60 cm$ is capped onto the test-piece, the wall thickness is 15mm. And the inner wall of the tube is also greased. Finally, the cementing sand paste is filled in between the test-piece and the steel tube with 8 water injection holes of $\phi 3 \times 50 cm$ left. The friction resistance of the side wall of the cylindrical test-piece is very small when swelling, and also the lateral deformation is restricted. Since the water injection holes are left and the central water injection hole is drilled, the absorbing condition of the test-piece is rather ideal in the test.

4.2 Free swelling test for cylindrical test-piece of the altered granite

Fig.4 gives the equipment installation diagram of the free swelling test for the cylindrical test-piece. From the diagram the situation of the cylindrical test-piece can also be seen. On the upper surface of the test-piece, two amesdials are arranged, used for the determination of the swelling deformation Bs(mm) of the upper surface of the test-piece. In order to eliminate the swelling deformation influence from the bottom of the test-piece, and the lateral friction influence caused by the large height-diameter ratio, a measuring rod is arranged at the depth of 20cm to upper sur-

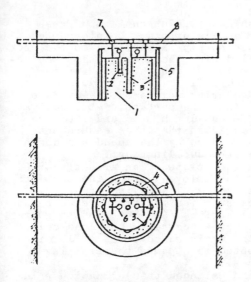

Fig.4 Free swelling test of cylindrical test-piece: 1. test-piece, 2. measurement point in depth, 3. water injection hole, 4. cement mortar layer, 5. steel tube, 6.amesdial, 7. magnetic dial holder, 8. frame

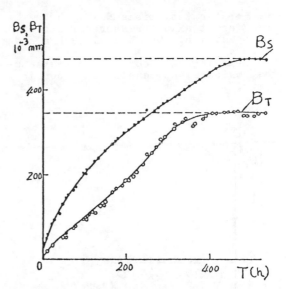

Fig.5 Bs, B_T - T curves of test-piece DP-3

face, and the swelling deformation B_T(mm) at that point is determined by the amesdial. The swelling ratio ω(‰) of the altered granite can be calculated by the following formula

$$\omega = \frac{Bs - B_T}{200} \times 1000 (‰).$$

Fig.5 shows the curve of actual measurement of Bs and B_T of the test-piece DP-3 versus time T.

Since the DP-3 is lightly altered granite, the swelling property is very poor, the swelling ratio is 0.64(‰).

It can be seen from Fig.5 that the swelling of the test-piece tends to be stable after 400hr water absorption.

4.3 Swelling pressure measurement of the cylindrical test-piece of altered granite

The scheme of the equipment installation is shown in Fig.6. The method of measurement is based on the determination of the axial deformation which varies with time under different axial pressure P and well absorption of water, and on the determination of deformation ratio velocity $\dot{\omega}$ (mm/h) of the uniform deformation

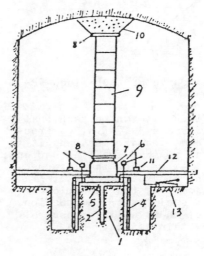

Fig.6 Swelling pressure test of the cylindrical test-piece
1. test-piece, 2. water injection hole, 3. cement mortar layer, 4. steel tube, 5. load plate, 6. amesdial, 7. jack, 8. loading pad, 9. load transfer cylinder, 10. concrete block, 11. magnetic dial holder, 12. frame, 13. pump

stage under that axial pressure, thus forming relational curve P-$\dot{\omega}$. The intersection point of the curve with the axial P are regarded as the lower limit of the swelling pressure of the rock.

For the test-piece DP-2, three different axial pressures P are employed, i.e. 39.7 KPa, 14.9 KPa and 0.0 KPa.

P-ω diagram according to the actual measurement results is shown in Fig.7.

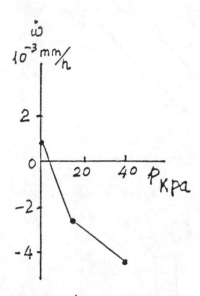

Fig.7 P-$\dot{\omega}$ curve of test-piece DP-2

From Fig.7, it can be seen that when P=14.9 and 39.7 KPa, compressive deformation happens, and it is only when P=0 that swelling happens. By curve P-$\dot{\omega}$ we know the intersection point at axial P are 4.5 KPa. When P=14.9 KPa, the test-piece may turn from swelling to compression, therefore, the swelling pressure must be within 4.5-14.9 KPa.

5 STUDY ON SWELLING PROPERTY TEST FOR WALL SURFACE OF ALTERED GRANITE IN UNDERGROUND EXCAVATIONS

5.1 Principle of test method

As stated above, the test for the wall swelling of excavations is based on the direct simulation and determination of the swelling property of the altered granite around the underground constructions.

Before the swelling test around the cavern walls, loosening surface of the rock wall is demanded to be cleared in an area of 60x60 cm² for every test point until the fresh surface is exposed. Un-interrupted water supply is provided to the rock wall during the test so that the absorption of enough water for the tested wall is maintained.

Measuring points on the surface and into the surface of the wall at different depth of 5 and 10cm are arranged so that the measurement can be done by the amesdials during the free swelling test.

A load plate of an area of 30x30 cm² is employed during the test for the rock wall swelling pressure. Jack supplies the pressure onto the wall surface. And in order to guarantee the accuracy, the load cell with strain gages of 5T or 10T is adopted to measure the swelling force.

The balanced pressurization principle is used to determine the swelling force. The steps are as follows, the initial value is read after installation of the equipments, then spray of water begins, afterwards the swelling deformation values are recorded every other 12hr. If deformation increases, then loading is necessary so as to bring the deformation value at the test point back to the initial value. By repeated measurements and regulations, the deformation value at the test point is kept in correspondance with the initial value from the beginning to the end. Hence, the curve can be obtained in respect to the swelling pressure with the increase of time. And according to its final stable value, the force and pressure of the swelling of the rock wall can be calculated.

5.2 Test equipments

Equipments used for the test of free swelling of the altered granite wall surface around the underground excavation are shown in Fig. 8. Equipment used for swelling pressure of the altered granite wall surface around the underground excavation are shown in Fig.9.

5.3 Results of free swelling test of the altered granite wall

Both the test points of BP-5 and BP-7 belong to the lightly altered granite, although they are exposed

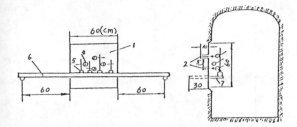

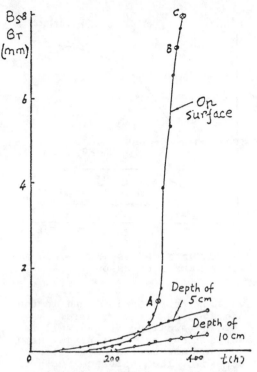

Fig.8 Free swelling test of the altered granite wall surface of the
underground excavation

1. test surface, 2. measurement
point in depth, 3.measurement point
on surface, 4. amesdial, 5. magnetic
dial holder, 6 and 7. frame

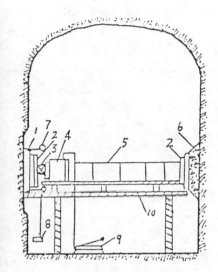

Fig.9 Swelling pressure test of the
altered granite wall surface of the
underground excavation

1 and 2. load plate, 3. load cell,
4. jack, 5. load transfer cylinder,
6. concrete block, 7. amesdial, 8.
strain gage indicator, 9. pump, 10.
frame

Fig.10 Curves of swelling deformation versus T of test point BP-1

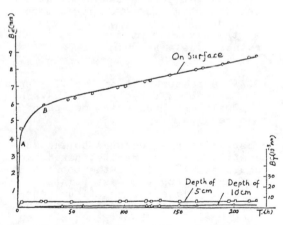

Fig.11 Curves of swelling deformation
versus T of test point BP-2

to the water for a long period, no
swelling deformation is observed.

Test points of BP-1, BP-2 and BP-6
belong to the moderately altered
granite, under the exposure to the
water, obvious deformation is observed, their relational curves of
swelling deformation and time are
respectively given in Figs. 10, 11,
and 12.

These results lead to

a. The swelling of the wall surface around the underground excavation is correlative to the degree
of granite alteration. Generally,
there is no swelling in the lightly
altered granite, whereas obvious

189

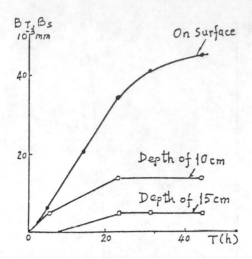

Fig.12 Curves of swelling deformation versus T of BP-6

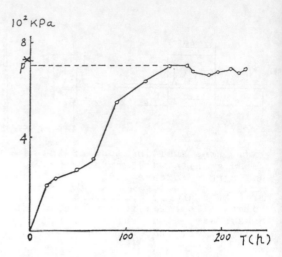

Fig.13 P*-T curve of test point BP-3

swelling property is found in the moderately altered granite.

b. The swelling of the wall is most serious on the surface, with the increase of the depth it sharply decreases. The influence depth of the swelling of the wall surface is about 15cm deep.

c. From the test results of BP-2 and BP-6 it can be seen that at the initial stage the swelling deformation develops sharply, two days later the development decreases to a rather stable stage.

5.4 The measurement of swelling pressure of altered granite cavern wall and processing method of results

Test point BP-3 belongs to the moderately altered granite, naked eye could see that it is the poorest in property among the test points.

Fig.13 gives the relational curve of time T and the nominal swelling pressure P* of BP-3 (P* equals the actual swelling force divided by the area of the load plate). The value of P* reaches as high as 0.69 MPa.

Point BP-4 belongs to lightly altered granite, its actual P* is much smaller, about 0.1MPa. Fig.14 gives the P*-T curve of BP-4.

Since the site of the swelling of the altered granite is not limited within the area under the load plate, therefore, the nominal swelling

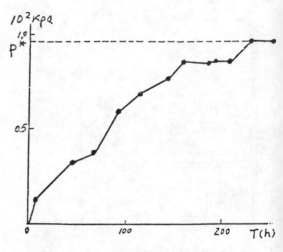

Fig.14 P*-T curve of test point BP-4

pressure P* could not represent the swelling pressure of the altered granite. Further modification should be carried out on P*.

Modification method is based on the basic principles of geomechanics. The model shown in Fig.15 is adopted as the foundation to evaluate the deep area of the rock wall, and α in the Fig. equals $45° - \phi/2$, where ϕ is the angle of internal friction of the altered granite. In accordance with the test results, it is known that the influence depth h can be taken as 10cm. By the volume ABCD in Fig.15, the cross section S(cm²) of its equivalent volume can be deduced. Therefore, the modified

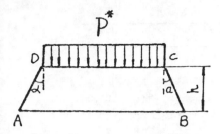

Fig.15 Scheme for modification of swelling pressure

swelling pressure

$$P = P^* \frac{30 \times 30}{S},$$

for BP-3, $\phi=30°$, then P=0.482MPa,
for BP-4, $\phi=40°$, then P=0.072MPa.
The value ϕ of BP-3 and BP-4 can be obtained with respect to the triaxial shear test of the altered granite in situ.

From curve of P*-T, it can be seen that the swelling pressure increases sharply at the initial stage, 10 days later it is kept in a constant value at large.

6 MINERAL COMPOSITION ANALYSIS AND WATER CONTENT DETERMINATION OF ALTERED GRANITE

6.1 Necessity of mineral composition analysis

The above test results tell us that the swelling property is correlative to the alteration degree of the granite. In order to make quantitative analysis for the swelling property, however, it is necessary to make the quantitative analysis for the mineral composition of the altered granite, especially the content analysis of the clayish minerals, which are the main factors causing the swelling of the altered granite, in particular, the montmorillonite.

6.2 Analysis method for mineral composition

Rock samples from these 9 test-pieces or points are studied with X-ray diffraction quantitative analysis. To verify the results of X-ray diffraction analysis, several samples are also contrasted with it by differential thermal analysis with the analysis of the thermal weight. Practice shows that the results obtained by different methods are similar.

X-ray diffractometer used is the product by Phillips Ltd., Netherland. Copper target is employed, voltage of the X-tube: 3415V, current: 12MA, high voltage of the counting tube: 1850V, ratio range: 4×10^2, time constant: 8, sample scanning velocity: 1°/min.

As for the test point BP-3, since the alteration degree is more serious than that of other test-pieces or points, and the swelling pressure actually measured is very high, so, using prudence, other parallel tests for the samples of this test point are carried out in the laboratories of different institutions. And our institute carried out quite a lot of repeated tests for the sample. All of the results of the mineral analysis for the sample of BP-3 are in considerable coincidence with one another.

6.3 Method of water content determination

Both before and after the test, the water content is determined for every test-piece or point. Combustion method is adopted in the determination of water content.

6.4 Results of mineral composition analysis and water content determination

Mineral analysis results of altered granite in every test-piece or point are listed in Table 1.
Results of water content before and after the test for the altered granite of each test-piece or point are listed in Table 2.

7 COMPREHENSIVE ANALYSIS OF SWELLING PROPERTY OF ALTERED GRANITE

7.1 Comprehensive results of swelling property tests for altered granite in-situ

Table 3 gives the comprehensive results of swelling deformation, swell-

Table 1. Mineral analysis results of altered granite (%)

No.	Alteration degree	Montmoril- lonite	Mica Illite	Kaoli- nite	Quar- tzite	Feld- spar	Cal- cite	Dolo- mite
BP-1	Moderately	3.0	11.7	1.3	65	17	few	few
BP-2	Moderately	15.6	16.1	3.1	42.3	20.1	2.8	few
BP-3	Moderately	15.0-22.9	10.0	5.3	45	14.7	2.1	few
BP-4	Lightly	4.3	13.4	2.6	55.5	23.1	few	few
BP-5	lightly	0	11.1	1.9	65	17	few	few
BP-6	Moderately	7.5	11.3	3.9	53	19	few	few
BP-7	Slightly- lightly	0	15.2	2.9	57	17	few	few
DP-2	Slightly- lightly	0	15.1	3	64	13	few	few
DP-3	Lightly	0	12.7	0	72.6	12.2	2.5	2.5

Table 2. Water content before and after the test

Water Cont. (%)	BP-1	BP-2	BP-3	BP-4	BP-5	BP-6	BP-7	DP-2	DP-3
Before	1.33	2.30	2.41	2.80	1.25	1.15	2.85	0.78	2.43
After	4.38	4.87	4.43	3.18	4.92	4.63	4.08	3.69	5.92
Diff.	3.05	2.57	2.02	0.38	3.71	3.48	1.23	2.91	3.49

ing pressure, swelling time, clayish mineral compositions and water contents.

7.2 Analysis of swelling property of altered granite

1. It is evident from Table 3 that the montmorillonite plays the main role in the swelling of the altered granite of the Guangzhou Pumped Storage Power Station. The swelling deformation is correlative to the content of montmorillonite M(%). For instance, M=15.6% at BP-2, the swelling deformation of the rock wall surface of the altered granite reaches 8.9mm in 238h. M=7.5% at BP-6, evident swelling is found in a short period of time, either. And the contents of montmorillonite at BP-5 and BP-7 are nearly zero, although they are exposed to water for a long period of time, no swelling is found in these points.
2. The test results of DP-2 and DP-3 show that swelling is also possible when the content of illite is high.
3. Free swelling test for the wall

around the underground excavations reveals that swelling is most serious at the wall surface, and it decreases sharply with the depth. When depth h reaches 15cm, the swelling is quite faint. Fig.16

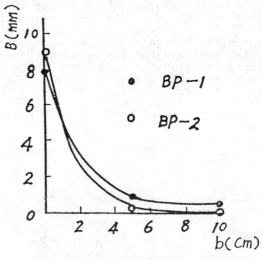

Fig.16 The relationship between the swelling deformation and depth

Table 3. Comprehensive results of swelling properties of the altered granite in situ

No.	Alteration Degree	Water Cont. (%) Before/After	Miner. Cont. (%)			Swel. Pres. P (KPa)	Swel. Deform.		Meas. Time (h)
			Mont.	Kaol.	Mica+ Illi.		Depth (cm)	Value (mm)	
BP-1	Moderately	1.33 / 4.38	3.0	1.3	11.7		0 / 5 / 10	7.991 / 0.967 / 0.414	441
BP-2	Moderately	2.30 / 4.87	15.6	3.1	16.1		0 / 5 / 10	8.883 / 0.006 / 0.002	238
BP-3	Moderately	2.41 / 4.43	15.0- 22.9	5.3	10.0	482	0	2.049*	223
Bp-4	Lightly	2.80 / 3.18	4.3	2.6	13.4	72	0	0.445*	255
BP-5	Lightly	1.25 / 4.92	0	1.9	11.1		0 / 5 / 10	0.0 / 0.0 / 0.0	144
BP-6	Moderately	1.15 / 4.63	7.5	3.9	11.3		0 / 10 / 15	0.045 / 0.014 / 0.005	44
BP-7	Lightly	2.85 / 4.08	0	2.9	15.3		0 / 5 / 10	0.0 / 0.0 / 0.0	298
DP-2	Slightly- lightly	0.78 / 3.69	0	3	15.1	4.5- 15	0	0.234	255
DP-3	Lightly	2.43 / 5.92	0	0	12.7	10-30	0 / 20	0.472 / 0.344	527

* The sum of the quantity of the swelling deformation pressed back at different pressures.

gives the relation between quantities of swelling deformation Bs(mm) and B_T(mm) and depth h(cm).

4. The reason why the swelling of the altered granite excavation wall does not show a deeper influence is due to the poorness of the water permeability of the altered granite. After the absorption, the wall surface turns to block the penetration of water deeper. The wall surface is the free surface, and other surfaces are all restricted, thus it restricts the lateral deformation of the rock, and in turn, this is also unfavorable to form new permeable paths. When more free surfaces exist, depth influence of swelling will increase much more. Therefore, it could be deduced that the swelling of the altered granite at the cross-connecting parts of the underground excavations will bring about more serious consequence.

5. After the swelling, the mechanical property of the altered granite of the underground excavation wall worsens a lot, and it is possible for crumbling, peeling and exfoliation to occur forming new wall surface, thus expanding the influence area of the swelling to the deeper parts. Such dynamic expansion mechanism could result in the damage to the large area and deeper range of the altered granite. The above dynamic expansion mechanism should be considered for the underground excavation's support and treatment when the altered granite is encountered. Necessary measures should also be taken to prevent the damage area by the altered granite to expand to the deeper parts.

6. A significant stage exists for the swelling deformation ratio when the altered granite absorbs water. However, this stage is rather short in time. After that, the swelling deformation ratio slows down gradually.

As for the surface with higher content of montmorillonite, it will eventually turn to peeling and exfoliation.

7. Test results show that the swelling of the altered granite tends to be stable within about 400-500 hours.

8. Fig.17 gives the relation between the swelling pressure P(MPa) and montmorillonite content M(%) of the altered granite in Guangzhou Pumped Storage Power Station.

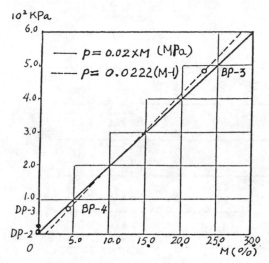

Fig.17 The relationship between the swelling pressure and the content of montmorillonite for altered granite of the Guangzhou Station

Hence a preliminary formula could be obtained to evaluate the swelling pressure of the altered granite in the Station.

$$P = 0.02M$$
or
$$P = 0.0222(M-1)$$

However, since the total number of test points are rather limited, this needs to be further supplemented and verified.

9. Since the montmorillonite content is closely related to the mechanical property of the altered granite, it is suggested that mineral analysis be carried out for all of the altered granite encountered in the underground excavation. And doing so will provide important evidence for the construction support

and treatment measures of the underground excavations of the Station.

10. In addition to engineering geological classification of the altered granite (slightly, lightly, moderately and so on), it is suggested that the classification be added according to the montmorillonite content.

8 MECHANICAL PROPERTIES OF ALTERED GRANITE

Quite a lot of tests and studies have been carried out on the mechanical properties of the altered granite both in situ and in the laboratory. It is worth while to point out the following large scale mechanical tests are carried out in situ:

a. Large scale triaxial compression test in situ (true triaxial test)
Dimension of the test-piece: 50x 50x100cm^3,
Number: 10.
b. Determination of deformation feature of the rock mass in situ
Area of the load plate: 2000cm^2,
Number: 20.
c. Determination of deformation feature of the bore hole wall
Bore hole diameter: 76mm±,
Determined holes: 3,
Number of determination segments: 16.

In regard to the laboratory tests, uniaxial compression tests for 38 rock samples and triaxial compression tests for 30 test-pieces and X-ray diffraction analysis for 14 rock samples as well as lots of physical indices such as unit weight and others are carried out. In order to enhance the accuracy of the laboratory tests, the stiff test machine with electrohydraulic servo-control of Type 815.03 by MTS Ltd., USA is employed. Therefore, the total process curve of the load-displacement for the tested piece is obtained.

Since the mechanical properties are not the focus of this paper, and for the sake of the length, we only give the suggested values of the strength and the deformation features of the altered granite in respect to the comprehension of the large scale tests in situ and laboratory tests (Table 4).

From Table 4 it is also evident that the mechanical properties of

Table 4.

Lithologic Classification		Deformation Feature			Shear Strength		Unit Weight 10^{-2} N/cm³
		Elast. Mod. 10^3 MPa	Deform. Mod. 10^3 MPa	Poisson Ratio	C MPa	ϕ degree	
Moderately Altered Granite	Seriously Altered	2.5	0.5	0.3	0.25	25	2.56
	General	10.0	2.5	0.25	0.4	30	2.56
Lightly Altered		20.0	5.0	0.20	0.7	36	2.56
Slightly-lightly Altered Granite		30.0	10.0	0.20	1.3	40	2.61
Sound Granite		35.0	12.0	0.20	1.5	45	2.61

the altered granite seriously worsen with the alteration degree, and the mechanical indices of the moderately altered granite decrease sharply in comparison with that of sound granite.

9 A BRIEF CONCLUSION

The altered granite, especially the moderately altered granite of Guangzhou Pumped Storage Power Station, is characterized by the poorness in strength, swelling after water absorption and crumbling. Therefore, the weak belts composed of the altered granite is responsible to the stability of the underground excavations.

Considering the altered granite is in bar and belt distribution with much he ogeneous mechanical properties, it is practical for us to adopt the large scale tests in situ, because they have good representativity from the view point of geomechanics.

Since the swelling property tests carried out are closely related to the underground excavations of the project, the results obtained from the tests have already been adopted by designing and construction institutions. The results also provide important evidence for the designing of construction supports and engineering treatment measures.

Tests show that the montmorilonite content of the altered granite is the contributing factor for the properties of mechanics and swelling of the altered granite. It is also the first time to discover the relational expression between the

swelling pressure and the montmorillonite content of the altered granite at the Station.

Tests also show that the swelling of the wall surface decreases sharply with the depth, the depth at which the swelling occurs does not exceed 15cm. However, there exists an area of many free surfaces, thus the problem is made more serious. If the montmorillonite content is higher in the wall surface of the altered granite, peeling and exfoliation would happen after the swelling, forming new wall surface and the swelling area expanding deeper. This dynamic expansion mechanism would result in larger and deeper area of the damage to the altered granite. Therefore, corresponding engineering measures should be taken in respect to the evidence and regularity obtained in the tests so as to guarantee the long-term stability of the underground excavations of Guangzhou Pumped Storage Power Station.

ACKNOWLEDGEMENT

Although I have been in charge of the work, I have to say it is really a collective product. I am deeply indebted to the contributions by Xu Dongjun, Chen Congxin and Yu Heping in the completion of the work. In particular, I would extend my thanks to Guangzhou Electrohydraulic Designing Institution, without the help and support of this Institution, it would be rather difficult to fulfil the work.

The author gratefully acknowledges the support of K. C. Wong Education Foundation, Hong Kong.

2. Embankments, excavations, slopes, and earth retaining structures

Developments in Geotechnical Aspects of Embankments, Excavations and Buried Structures, Balasubramaniam et al. (eds)
© 1991 Balkema, Rotterdam. ISBN 90 5410 019 2

Excavation of submarine ground in the world's largest pneumatic caisson

Shunta Shiraishi
Shiraishi Corporation, Tokyo, Japan

ABSTRACT : 109 000 m3 of excavation was made in the world's largest pneumatic caisson in Tokyo Harbor into the submarine ground of thick soft clay and hardpan. Ground improvements with sand compaction piles and replaced sand were made for preventing uneven settlement of the caisson. Extensive use of unmanned excavation systems with caisson shovels, twin headers and jib-cutters is made in the later stage of excavation at the great depth up to 46.5 m below the mean sea level.

1. INTRODUCTION

This topic concerns the foundation of the off-shore anchorage for 798-m long Tokyo Cross Harbor Suspension Bridge (Fig. 1).

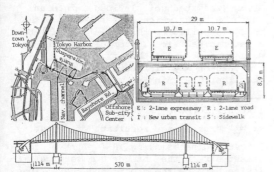

Fig. 1 Tokyo Cross Harbor Bridge

The foundation of the off-shore anchorage is formed by a pneumatic caisson of 70 m in length, 45 m in width and 51 m in height. This caisson was sunk to the depth of 46.5 m below the mean sea level (Fig.2).

The gross volume of the caisson is 160 466 m3 and it is world's largest of this kind. 109 000 m3 of submarine ground below the 12-m deep sea water was excavated in beneath the caisson until it sunk down to the final depth of 34.5 m below the sea bed.

The world's largest before the subject caisson is built was the multi-cell open caisson supporting the Brooklyn-side main-tower of Verrazano Narrows Bridge in

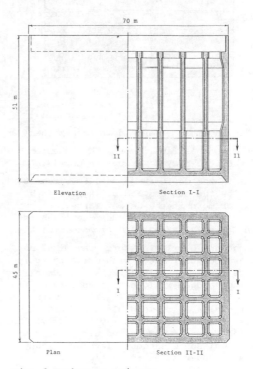

Fig. 2 Anchorage caisson

United States of roughly 135 100 m3 in volume (Gray et al, 1966).

2. SUBMARINE GROUND

The submarine ground at the caisson site

199

is formed by, in turn from the top down-ward, a soft alluvial clay layer A_c of 24 m in average thickness, double layers of sand and gravel A_s and A_g 2 m in average combined thickness and very stiff diluvial clay (hardpan) layer K_c sandwitching a 1.4-m thick very dense sand layer K_s (Fig. 3).

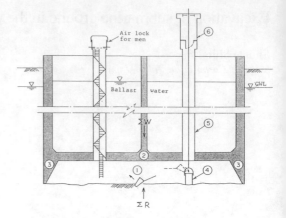

Fig. 4 Pneumatic caisson

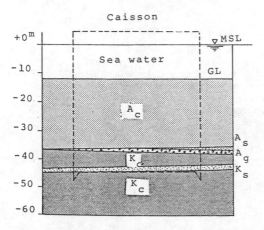

Fig. 3 Soil profile

The layer K_c extends down deeper than the caisson's final depth. The cohesion C of the soft clay A_c is 0.4 kg/cm2 (3.9 kPa) in the depth range shallower than 8 m below the sea bed and 0.5 kg/cm2 (4.9 kPa) or more in the deeper range.

The cohesion C of the deep-laying hard-pan K_c is 10 kg/cm2 (98 kPa) or more. Al-though the overlaying soft clay might be excavated under water with dredging equip-ments like cramshell buckets, the deep-laying hardpan would have been very diffi-cult to excavate under water even with best available submergible equipments. This is the main reason why the excavation procedure under compressed air in a pneu-matic caisson was chosen in place of any type of under-water excavation.

3. PNEUMATIC CAISSON (Dunham 1950)

For the readers who are not familiar with pneumatic caissons, the essential points of pneumatic caissons will be given brie-fly on a simplified example (Fig. 4).

A working chamber ① which is air-tight covered by a ceiling slab ② and coffer-red by cutting edges ③ forms the bottom part of the caisson. Compressed air is filled inside the chamber for holding the ground water from flowing into that when the bottom end of the caisson is sunk down below the ground water table. The excava-tion of the ground is made in the working chamber under compressed air and in dry condition.

The excavated soil is loaded into an earth bucket ④ which, after loaded with soil, is pulled up through a shaft ⑤ and a material air lock ⑥ for discharging the soil outside above the air lock.

A caisson may sink down into the ground when the total sinking weight ΣW (own wei-ght of caisson W_c + weight of outfittings like air locks W_o + weight of ballast wat-er W_w) exceeds the resultant of resisting forces against sinking ΣR (uplift of comp-ressed air in the chamber U + vertical fr-iction between the ground and the periphe-ral wall surface F + total bearing resis-tance of the ground beneath the cutting edges Q).

4. SEATING GROUND FOR CAISSON

A caisson is designed so that the position of the center of gravity of its total sinking weight ΣW coincides with geome-trical center of its base area or the centrum of the uplift U of the compressed air in order to let the caisson sink down in a vertical direction without tilting.

However, a caisson tends to tilt due to uneven settlement caused by non-uniformity of ground.

Once a caisson tilts, although it may be

set back upright by such remedial measure as unbalanced excavation of removing more ground on the high side than on the opposite low side, the caisson is likely to dislocate sideway every time such a remedial measure is taken (Shiraishi 1970).

If a caisson tilts excessively on a soft seating ground and remedial measures are repeated to set it back during the initial course of its sinking, the final dislocation may exceed a tolerable limit. Therefore, the seating ground for a caisson should be uniformly firm enough.

For the subject caisson to be placed in sea water on the soft submarine ground, in particular, the ground should be improved to be much firmer than the ground on land, because the lateral forces of waves and tidal currents surmount the adverse effect to tilt the caisson.

Therefore, extensive improvements had been made to the soft submarine ground before the subject caisson was placed on the sea bed.

5. PREPARATORY WORKS

The following preparatory works had been performed since December, 1986 until January, 1988 before the caisson was placed on the sea bed.

An 8-m wide, 916-m long temporary access treastle leading over water to the caisson site was built on steel pipe-pile bents while the ground improvement works were performed at the caisson site.

Then, the construction platforms surrounding the space where the caisson is to be placed were built on jacketed steel pipe-pile clusters while an 18.5-m high steel shell enclosing the bottom portion of the caisson was built up in the ship-building dock for 200 000-t class tankers located about 40 km south of the caisson site (Fig. 5).

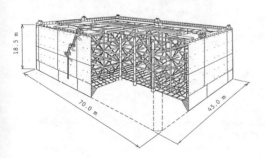

Fig. 5 Steel shell

6. IMPROVEMENTS OF SUBMARINE GROUND

Before the construction platforms were installed, the following improvement works had been made to the submarine ground.

Firstly, 1 890-m2 framing zone of the seating ground for the caisson in the depth-range of 3 m to 15 m below the sea bed was improved by 458 sand compaction piles of 1.5 m in each diameter spaced 2.2 m in triangular layout driven into the ground (Fig. 6, a).

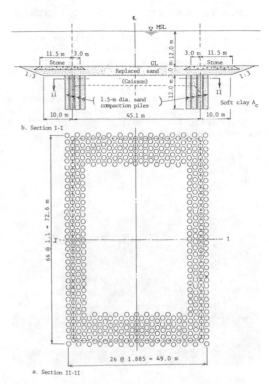

Fig. 6 Submarine ground improvement

By the above improvement, 43% of the in-situ volume or 9 712 m3 of the soft clay A_c was forcibly replaced with the compacted sand and the bearing capacity of the clay was presumed to be increased to 25.3 t/m2 (24.8 kPa) from its natural value of less than 20 t/m2 (19.6 kPa).

The reinforced soil wall thus formed around the void space, which would afterward be brought into existence by excavation in the caisson, was effective for retaining possible inward plastic flow of the soft clay which might have caused heaving of the excavated ground in the caisson.

Secondly, in forming a stable seating mat for the caisson, the top 3-m deep very soft mud had been dredged out over the wide area enveloping the spaces for the caisson and the platforms, and the mud was replaced with sand and partly with crushed stones (Fig. 6, b). The replaced sand and stones were dumped down from above the water and the stone part in the framing zone of the caisson space was compacted with vibrohammers.

7. TOWING, CONCRETING AND PLACING CAISSON

When the steel shell of the caisson was completely built up in the ship-building dock and its gross weight became 4 450 t together with 2 000-t preassembled reinforcement bars for the concrete to be placed afterward in the lower parts of the shell, it was floated up by inundating the dock, pulled out from the dock, towed 45 km over the water of Tokyo Bay, pulled into the caisson space surrounded by the platforms and moored there on January 25, 1988 (Fig. 7 & 8).

Fig. 7 Towing steel shell

Fig. 8 Steel shell, platforms and access treastle

The total amount 27 355 m3 of concrete to be placed inside the steel shell was placed stepwise in 8 lifts during while the steel shell was afloat in the platforms. Then, the steel shell sank down into the sea water gradually as the weight of concrete increased.

When the concrete of the 3rd lift was placed, the draught of the steel shell increased to 11.9 m and the lowest tip of its cutting edge touched down the sea bed. At this stage, 5 900 t of ballast water was poured into the void cells in the concrete of 2nd and 3rd lifts. Then, the cutting edges were forced to penetrate down into the sea bed by 0.5 m by the extra weight of ballast water on April 30, 1988. It made the caisson stable against lateral forces of wind and waves.

When the last 8th lift of concrete was placed on October 7, 1988, the total weight of fully concreted caisson's steel shell including 68 338 t of reinforced concrete, 2 914 t of outfitted steel shell and 5 900 t of ballast water reached the maximum of 77 152 t. Then, the cutting edges were forced to penetrate fully down into the sand mat and the ceiling slab of the working chamber was wholly pressed onto the sand mat at presumed effective pressure of 12.8 t/m2 (12.6 kPa) by the huge weight in excess of the bouyancy of caisson (Fig. 9).

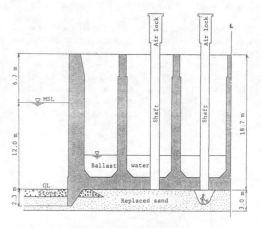

Fig. 9 Caisson pressed onto sand mat

The working chamber at this stage was fully packed with sand.

Air compressors, air pipes and other pneumatic appliances as well as cranes and earth hoppers were installed on the platform during while the 8th lift of reinforced concrete was built up. Air locks and soil discharging equipments were outfitted on and the air pipes connected to the caisson after the 8th lift of concrete was placed.

8. IN-CAISSON EXCAVATION STARTED

When every outfitting was ready for the caisson work, men went down through the central man-air locks and shafts into the sand-packed working chamber under the air pressure of 1.4 kg/cm2 (13.7 kPa) on November 10, 1988 and started to dig open the sand in there with portable tools.

When the dug-open space in the chamber became wide enough for accomodating two in-caisson excavators of backhoe type (Fig. 10), they were brought down into the chamber for further excavation.

Fig. 10 Backhoe-type excavator

The caisson settled inching down as the dug-open space was thus enlarged symmetrically about the geometrical center of the base area. After the 9th lift of reinforced concrete was built up over the top level of steel shell and its weight of 7 123 t was added to make the total sinking weight $\Sigma W = 84\ 275$ t, the bottom end of caisson reached the depth of 3 m below the sea bed at the boundary of the replaced sand and the underlaying soft clay A_c. When the area of dug-open space was enlarged, at the above stage, to 1 736m2 or 55% of the whole base area of 3 156.8 m2 under the air pressure of 1.53 kg/cm2 (15 kPa), the caisson started to sink substantially into the soft clay on December 5, 1988. Since the effective bearing area of the improved soft clay then was 1 420 m2 and the effective sinking force = $\Sigma W - U$ = 84 275 t - 3 156.8 m2 x 15.3 t/m2 = 35 976

t where U is the uplift force of compressed air, the effective bearing pressure corresponding to the unit bearing capacity of the improved clay was 35 976 t ÷ 1 420 m2 = 25.3 t/m2 (24.8 kPa).

If the soft clay A_c had not been improved, unremediable excessive tilting of the caisson due to uneven settlement might have taken place at the above stage. However, the improved clay reacted to the sinking force uniformly and the tilting of the caisson was neglegibly slight. Thence, the lateral dislocation at the top level of the caisson was within 5 cm.

9. CONTROLLING SINKING OPERATION

For assuring stable and even sinking of the caisson, every possible precaution was taken. More than 300 pieces of measuring instruments such as pressure cells, inclinometers, strain gages had been attached previously at key position of the caisson body. Every measured data were transmitted automatically to the central electronic processing units at the interval of every ten seconds to be displayed in graphic and /or digital forms on the TV screens and other indicator panels for monitoring and controlling the caisson-sinking operation. Those displaying units were installed together with controlling devices in the monitoring room which had been built previously on the platform.

The air pressure in the working chamber at the above stage was controlled to keep balance with the hydrostatic pressure of the sea water which fluctuated twice daily due to change in tidal levels. The air pressure then was controlled with automatic pressure regulators within the error of 0.02 kg/cm2 (0.196 kPa). The air pressure was intentionally kept slightly lower than the hydrostatic pressure. Otherwise, there had been a danger of browing out of air through under the cutting edge to induce scoring of sand beneath the tip of cutting edge, which might be the cause of uneven settlement and tilting of the caisson.

The elevation of the four base corners of the caisson body were indicated on the TV display with the accuracy of 1 mm. By monitoring this display, any minute inclination of the caisson body was detected and set back upright by adjusting the distribution of ballast water in six compartments.

The caisson was sunk by around 60 cm once in several days by adding hundreds of tons of ballast water taking hours of time for increasing the sinking force to over-

come the resisting force against sinking
instead of by lowering the air pressure in
the working chamber for reducing the resisting force against sinking, in a matter
of minutes. The latter is a faster and
easier way than the former but it would
have induced disturbance of soft ground
beneath the caisson and consequent tilting
of the caisson body.

When the dug-open area was enlarged by
further excavation to 70% of the whole
base area, the bottom end of the caisson
sank down to the depth of 9 m below the
sea bed on February 7, 1989. 10 caisson
shovels to be operated in the dug-open
area were brought down into the working
chamber one after another until then.

10. CAISSON SHOVEL

A caisson shovel is a compact 40-HP
power shovel with a 150-1 dipper attached
to its telescoping boom which rotates 360°
horizontally, swings 19° upward and 32°
downward and covers the working radius of
4 m (Fig. 11). It hangs on and travels
along a pair of running rails fixed up to
the ceiling slab of working chamber like a
suspended monorail vehicle.

Fig. 11 Caisson shovel

Maximum hourly production rate of a caisson shovel in digging, hauling and discharging soil may be as high as 15 m3 when
digging is easy and hauling distance is
minimum. Average hourly production rate in
the subject huge caisson where average
hauling distance is long, however, may be
5 m3 or less.

A caisson shovel may be disassembled
into 5 pieces each piece being designed to
clear through the narrow passage in a material lock when it is brought down to or
up from the working chamber.

11. LAYOUT OF EQUIPMENTS

7 pairs of longitudinal running rails
for 14 caisson shovels and 2 pairs of lateral rails for 4 caisson shovels along
short side cutting edges had been previously attached and bolted to the bottom
steel cover plate of the ceiling slab
during while the steel shell was built up
in the ship-building dock (Fig. 12).

12 material locks had also been outfitted to the tops of shafts standing up
from the ceiling slab before compressed
air was filled into the working chamber.

A 1-m3 earth bucket clears through each
of the material locks each time it is
pulled up or down.

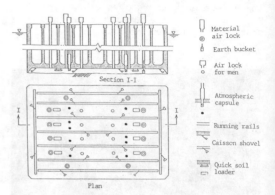

Fig. 12 Layout of equipments

8 material locks in the middle area were
equipped with quick soil loaders, each one
at below a lock. A quick soil loader is
designed to store an earth-bucket full of
soil repeatedly supplied by a caisson shovel in its rectangular open-top container
with belt-conveyer bottom and, when the

Fig. 13 Quick soil loader

204

container is filled up with soil at its low position, it is pulled up above the top of an empty earth bucket and the stored soil is discharged in a stroke of moving the bottom belt conveyer into the bucket in a matter of seconds (Fig. 13).

A quick soil loader saves several minutes of idle time for filling up an earth bucket by repeated supply of soil directly from a caisson shovel every time an earth-bucket full of soil is hauled up out. Therefore, it is remarkably useful for up-raising the soil-hauling-out efficiency.

8 air locks for men and 10 atmospheric capsules for unmanned excavation systems had also been outfitted before the compressed air was filled into the working chamber.

12. UNMANNED EXCAVATION SYSTEM

When the in-caisson excavation with 10 caisson shovels progressed to the extent that the dug-open area ratio increased to 75% and the depth of the bottom end of the caisson increased to 15 m below the sea bed, those 10 caisson shovels were switched to unmanned excavation systems on May 6, 1989. The air pressure in the working chamber then was 2.7 kg/cm2 (26.5 kPa).

Under the high air pressure around 3 kg/cm2 (29.4 kPa), the operators of caisson shovels are allowed to work only one-fourth of their full work-shift hours. It is strictly regulated by the rules for preventing horrible caisson disease. Therefore, their production rate in excavation under such high air pressure decreased to about one-fourth of the standard rate under open air.

In an unmanned excavation system, a caisson shovel is remotely maneuvered by an operator sitting in an atmospheric capsule with a wireless control set like the

manner a radio-controlled toy car is operated.

The atmospheric capsule is a cylindrical air-tight steel cabin hanging down from the piston in an air cylinder which is anchored to and standing up on the ceiling slab (Fig. 14).

The operator sitting in the capsule maneuvers his caisson shovel by watching through its window of reinforced glass with his eyes following the movement of the caisson shovel travelling back and forth by rotating the capsule around to face the window at the shovel.

Fig. 15 Atmospheric capsule, quick soil loader and caisson shovel

Fig. 15 shows a caisson shovel feeding soil to a quick soil loader. Note the operator staring at his caisson shovel through the windows.

Because the air pressure inside the capsule is atmospheric, the operator in it is able to work most of his full work-shift hours and he can excavate 3 times more than the operators exposed to the high air pressure of 3 kg/cm2. In other words, the productivity of an unmanned excavation system is about three times higher than a manned caisson shovel under the high air pressure of 3 kg/cm2.

If there is any danger of the capsule being pushed up forcibly by the ground and crushed due to sudden accidental fall of the caisson, the relief valve attached to its air cylinder opens to release air and the capsule rises up to its highest safe position in the casing.

Until the bottom end of the caisson reached the lowest level of the soft clay layer A_c at the depth of 24 m below the sea bed, in-caisson work with the 10 unmanned excavation systems was continued at the dug-open area ratio less than 100%.

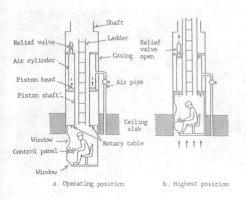

a. Operating position   b. Highest position

Fig. 14 Atmospheric capsule

As long as the excavation was made in the soft clay layer A_c, overall progress of the entire caisson work was not bottle-necked by in-caisson excavation, because the building-up work of the large amount of reinforced concrete body of the caisson on top of its finished part required much more time than the excavation work did.

For instance, while for building up 2.5-m high, 2 482-m3 15th lift of reinforced concrete from 39 m to 41.5 m above the bottom end of the caisson 31 days of time is required, only 14 days of net time was required for excavating 9 470 m3 of the ground in the corresponding depth of 3 m from 17.5 m to 20.5 m below the sea bed. Therefore, there was ample float time in the in-caisson excavation work.

13. UNDERGROUND DEWATERING WITH DEEP WELLS

Until the in-caisson excavation reached the top boundary of the sand layer A_s at the depth of 24 m below the sea bed and the dug-open area-ratio was increased to 95% on Sept. 16, the remaining 8 caisson shovels were brought down into the working chamber one after another. Then they started to excavate the peripherial narrow strips of ground along inside the cutting edges.

Because the operators of those 8 caisson shovels had to be exposed to high air pressure exceeding 3 kg/cm2 (29.4 kPa), their allowable length of working time should be strictly limitted and any substantial work might not be performed by them under the air pressure exceeding 4 kg/cm2 (39.2 kPa).

In order to alleviate this adverse situation, the maximum air pressure was kept at 3.5 kg/cm2 (34.3 kPa) by means of underground dewatering. The underground dewatering was effected by pumping out the ground water in the deep-laying water-bearing layers A_s, A_g and K_s through 6 deep wells which had previously been installed around the caisson.

14. EXCAVATION OF HARDPAN

On September 30, 1989, the in-caisson excavation reached the upper boundary of hardpan layer K_c at the depth of 26 m below the sea bed. It had been presumed very difficult to pick out the hardpan from under the tip end of a cutting edge with caisson shovels, because its resistance against cutting is very heavy. On the other hand, it might be impossible to

let the caisson sink down gravitationally breaking through the very high bearing resistance of hardpan possibly exceeding 50 kg/cm2 (490 kPa) unless the hardpan is cut out from under the cutting edge.

Therefore, the hardpan beneath the cutting edges was sawed out with 4 jib cutters instead of caisson shovels. Each jib cutter was installed on a side of 4 sides of cutting edges framing the caisson. A jib cutter is a newly innovated giant chainsaw thrusted out sawing the hardpan diagonally downward below the cutting edge from the sliding frame mounted on a rack-and-pinion travelling gear (Fig. 16). A jib cutter moves sidewise along the rack rails fixed to the inside overhung face of the cutting edge at the same rate as saw-cutting of the hardpan proceeds.

Fig. 16 Jib cutter at work

It had also been presumed that the progressing rate of digging the very tough hardpan in the middle area of the ground with caisson shovels might be lowered

Fig. 17 Twin header

remarkably. Therefore, the dippers attached to 6 unmanned caisson shovels were
replaced with newly designed twin headers
(Fig. 17). A twin header consists of a
pair of wheel cutters with many cutting
bits planted around the wheels. The wheel
cutters rotate around the axis perpendicular to the axle line of the shovel boom.

The twin-header attached caisson shovels
were much more powerful in digging hardpan
than the dipper-attached ones. Therefore,
the former contributed to supplement
considerably the lowered efficiency of the
latters.

Remaining 12 dipper-attached caisson
shovels, 4 of them being unmanned, scooped
out the debris of hardpan cut out with
twin headers and jib cutters.

With the above equipments, 8.5-m thick
hardpan layer of 26 833 m3 in volume had
been scheduled to be dug out by 4-shift-a-
day work under compressed air and 2-shift-
a-day work with unmanned systems in 75
days of net time or 90 days of gross
period. However, since the excavation of
hardpan was easier and faster than it had
been anticipated, it progressed 21 days
ahead of schedule. Therefore, the in-
caisson excavation totalling 109 000 m3
including the hardpan was completed on
December 9, 1989. Then, the entire period
of this in-caisson excavation of tremendous amount was condensed to 13 months
including 10.7 months for excavating the
soft clay layer A_c and the double layers
of sand A_s and gravel A_g.

Because the extensive ground improvements had been made and the high-precision
control systems were effective for preventing uneven settlement of the caisson and
every precaution was carefully taken for
letting the caisson sink without tilting,
the final dislocation of the caisson body
in horizontal direction was only within
6 cm.

Thus, the in-caisson excavation of unpr-
ecedented volume of submarine ground was
finally accomplished with a great success.

15. REMOVAL OF EQUIPMENTS, LOAD TESTS AND BOTTOM-FILL CONCRETE

Caisson shovels and other equipments
installed in the working chamber including
running rails were dismantled, disassembled and removed out of there. Equipments
and plants outside above the caisson
except pneumatic systems were dismantled
and recovered.

In-situ load tests were performed with
30-cm dia. circular rigid plates placed on

the foundation ground of hardpan. The
average value of ultimate bearing capacities proved by 4 load tests was 656 t/m2.
This test result assures the excellent
stability of the anchorage caisson with
ample factor of safety.

The bottom-fill concrete of 6 670 m3
which corresponds to the seal concrete of
an open caisson was placed fully in the
working chamber on January 20, 1990 in a
long continuous operation without recess.

Finally, the supply of compressed air
was cut off on the following day and air
compressors, air locks and other elements
of pneumatic systems were dismantled and
recovered.

ACKNOWLEDGEMENTS

The writer is very grateful to the estee-
med personnel of the owner of the const-
ruction contract of the subject caisson,
Metropolitan Expressway Public Corporation
(MEPC) and the joint-venture contractors,
Mitsui Kensetsu Co., Ltd., Goyo Kensetsu
Co., Ltd. and Shiraishi Corp. for their
courtesy and assistance offered to him.

Particularily, MEPC played a leading
role in planning, design and construction
management of the huge caisson. MEPC had
also developed the jib cutter and the twin
header in cooperation with the Japan Construction Mechanization Reseach Institute.

The unmanned in-caisson excavation
system and the quick soil loader which had
been designed and developed by Shiraishi
Corp. were employed by MEPC under the
conditions specified.

REFERENCES

Gray, N. et al, 1966. Verrazano Narrows
  Bridge, Construction of tower foundations
  and anchorages. Jour. Constr. Div., No.2,
  Amer. Soc. Civil Engrs. p. 95-117
Dunham, C. W. 1950. Foundations of structures, McGraw-Hill Civil engineering
  series, New York, p. 470-480
Shiraishi, S. 1970. Some particular deep
  excavation in soft ground in Japan. Soils
  and Foundations, Jap. Soc. Soil Mech. &
  Found. Engg. Vol. 10, No.1, p. 1-14

Developments in Geotechnical Aspects of Embankments, Excavations and Buried Structures, Balasubramaniam et al. (eds)
© 1991 Balkema, Rotterdam. ISBN 90 5410 019 2

Full-scale slope failure at Selborne, UK

D.J.Petley
University of Warwick, Coventry, UK

E.N.Bromhead
Kingston Polytechnic, UK

M.R.Cooper & D.I.Grant
University of Southampton, UK

ABSTRACT: A full scale cutting slope failure experiment has been carried out at Selborne, U.K. An extensive instrumentation system comprising inclinometer tubes, string inclinometers, piezometers and wire extensometers was installed, with many of the instruments connected to an electronic data gathering system incorporating alarm level outputs. Failure of the slope was induced by pore pressure recharge. An intensive laboratory programme was performed concentrating on the shear strength properties of the Gault clay. Failure of the slope occurred in July 1989, with total displacements of the sliding mass of about 4 m. Following failure, a series of trenches were excavated through the sliding mass, enabling the precise position of the failure surface to be located; samples were also taken to determine the shear strength on the failure surface.

1 INTRODUCTION

The Selborne Slope Study is a collaborative investigation drawing together expertise from the University of Southampton, the University of Warwick and Kingston Polytechnic. One face of a brickwork's clay pit was steepened to an overall slope of 1:2. An intensive instrumentation system was installed to monitor the behaviour of the 9 metre high slope as it was brought to failure using pore pressure recharge.

This paper describes the general nature of the project and includes an account of the instrument systems adopted. Construction details of individual instruments have not been included as they are not significantly different from readily available standard systems. The paper emphasises the measures adopted to provide multiple options for instrument reading methods; and to establish automatic alarm systems to warn of approaching failure.

It should be noted that the instrumentation in this study was designed to record a single failure event and was not intended for routine performance monitoring. It was therefore essential that, whatever conditions prevailed at failure, the required readings could still be obtained. This requirement influenced several aspects of the system's design.

A large number of laboratory tests have been performed to measure the peak and residual shear strength parameters of the Gault clay. Following the failure in 1989, block samples containing the failure surface were extracted from trench excavations through the slipped mass, and the shear strength along these surfaces was measured in the laboratory.

Analysis of the data obtained is currently being carried out, and it is anticipated that the study will provide valuable information on the mechanisms of failure of cut slopes in overconsolidated clays, and the influence of progressive failure.

Figure 1: Site location

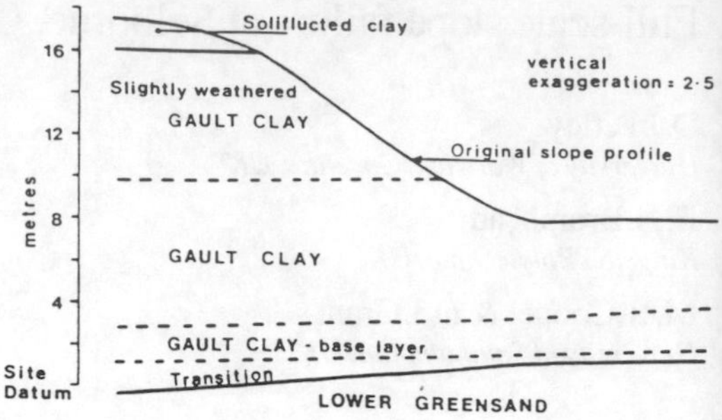

Figure 2: Geological section

(In Figure 2)
Soliflucted clay
16
Slightly weathered
GAULT CLAY
vertical
exaggeration : 2·5
12
Original slope profile
metres
8
GAULT CLAY
4
GAULT CLAY - base layer
Site
Datum
Transition
LOWER GREENSAND

2 THE STUDY SITE AND GEOLOGY

The Selborne Slope Study site is situated 4 kilometres east of Selborne in Hampshire, U.K., as shown on Fig. 1. The study area occupies about one quarter of the west face of the clay pit at the Honey Lane Brickworks of the Selborne Brick and Tile Company. Bed rock is the Gault clay of Cretaceous age.

A feasibility study carried out in 1984/5 had shown the ground profile in the slope to comprise about 2 metres of soliflucted slightly gravelly clay over approximately 6 metres of weathered Gault clay. Unweathered Gault clay forms the lowest 1-2 metres of the slope, and extends at least another 5 metres below pit base level, where a 1-2 metre thick basal layer separates it from a thin transition layer to the lower Greensand. The regional dip is approximately 1.5-2° to the west.

3 GROUNDWATER CONDITIONS

During the feasibility study, nine boreholes were made in the investigation area. One remote borehole was put down in an attempt to confirm the regional dip of the Gault/Greensand interface, but was abandoned above the required depth. Standpipe piezometers were installed in all the boreholes at depths selected to give a good overall impression of the groundwater regime. Water levels in the standpipes were read at two-weekly intervals, when possible,

between January and June 1985. The recorded variations in water levels are shown in Figure 3. The standpipes in boreholes 3 and 4 were periodically flooded by site operations and their results are not known. An interpretation of the standpipe records is given in Figure 4. Figure 4a shows the standpipe tip positions and average water levels plotted on the profile used for the geological section. In Fig. 4b, the elevations have been projected and plotted against the total heads relative to Site Datum. The resulting plot gives a clear indication of a groundwater regime with an essentially downward flow, with a hydraulic gradient of about 0.75, to the underlying Greensand.

4 GEOTECHNICAL DATA

As part of the present research programme, further rotary-cored boreholes were made at the crest of the slope, and continuous undisturbed samples using HWF size core barrel were taken. From examination of the samples, four zones were established:

(a) a soliflucted layer approximately 3m thick;
(b) an upper weathered zone of Gault clay, approximately 2-3m thick;
(c) a lower weathered zone of Gault clay, approximately 6m thick;
(d) unweathered Gault clay.

Water content tests have been made on the samples from the boreholes and the profiles presented in Figures 5a and 5b were obtained. In addition, liquid and plastic limit

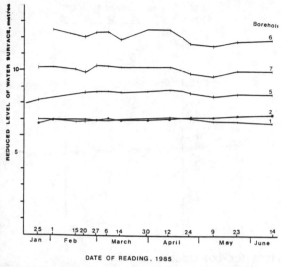

DATE OF READING, 1985

Figure 3: Water level readings in standpipes

determinations were made on selected samples; these results are also included in Figures 5a and 5b.

The laboratory testing programme concentrated on the determination of the peak and residual shear strength parameters in terms of effective stresses, using the samples obtained from the boreholes. Both triaxial (75 mm diameter) and shear box (60 mm square) were used.

From these tests, the following results were

obtained:

A. Soliflucted layer

$$c^1 = 5 \text{ kN/m}^2 \qquad \phi^1 = 21°$$

$$c_r^1 = 0 \qquad \phi_r^1 = 13°$$

B. Upper weathered zone

$$c^1 = 10 \text{ kN/m}^2 \qquad \phi^1 = 22° - 24°$$
$$c_r^1 = 0 \qquad \phi_r^1 = 13°$$

C. Lower weathered zone

$$c^1 = 15 \text{ kN/m}^2 \qquad \phi^1 = 22 - 25°$$
$$c_r^1 = 0 \qquad \phi_r^1 = 14°$$

D. Unweathered Gault clay

$$c^1 = 20 - 25 \text{ kN/m}^2 \qquad \phi^1 = 23 - 26°$$
$$c_r^1 = 0 \qquad \phi_r^1 = 15°$$

Typical results for the upper weathered zone are given in Fig. 6.

Ring shear tests were also performed, and gave results for ϕ_r^1 in agreement with the values reported above. Some tests were carried out to investigate the influence of rate of shearing on ϕ_r^1: these tests indicated that a change in shearing rate from .0001 mm/min to .01 mm/min was accompanied by an increase in ϕ_r^1 of about 2%.

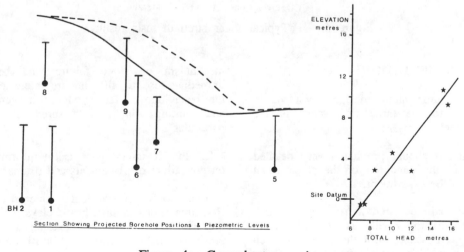

Section Showing Projected Borehole Positions & Piezometric Levels

Figure 4: Groundwater regime

211

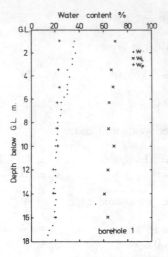
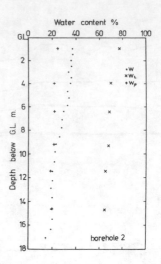

Figure 5: Index properties for boreholes 1 and 2

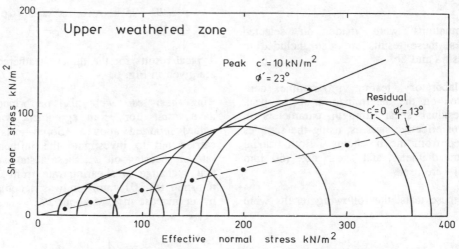

Figure 6: Typical shear strength tests results

5 INSTRUMENTATION

The instrumentation design evolved in response to the three main aims of the experiment:

1. that it should provide a very detailed record of the condition of the slope at all stages of the experiment.

- The main declared aim of the research project was to provide as detailed a case record as possible, with the hope that such detail would "enhance knowledge of the mechanism of slope failure, and be of immediate use to all who are interested in the behaviour of failing slopes in general, and cuttings in stiff fissured clay in particular".

2. that it should give useful information on pre-failure displacements and strains.

- One topic of particular interest with first-time slides in stiff, brittle clays (such as the Gault clay) is the development of a progressive failure mechanism. Thus it was very important that all displacement measuring

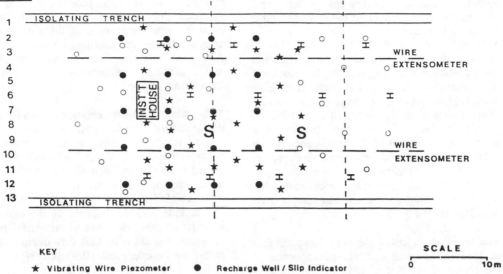

GRID COEFFS	A	B	C	D	E	F	G	H	J	K	L	M	N	P	R	S

KEY

★ Vibrating Wire Piezometer   ● Recharge Well / Slip Indicator

○ Pneumatic Piezometer   **S** String Inclinometer

I Inclinometer Access Tube

SCALE
0 ——————— 10 m

Figure 7: Instrument layout plan

instruments should be in place prior to the commencement of pore-pressure recharge.

3. that the system should produce results up to, and if possible beyond, the final failure event.

- Sufficient redundancy and duplication had to be built into the system to ensure that the "one-off" end point of the experiment would be adequately recorded, even with the highest conceivable level of component failure.

The instrument layout shown in Fig. 5 was developed in response to these requirements. This figure shows the layout in plan view only. Separate cross-sections are presented for each instrument later in the paper.

5.1 General description

The study area was 25 m wide and the cut slope is 9 m high. The instrumented area extended for 18 m back from the crest of the slope and for 6 m beyond the toe. The resulting 1050 m² of plan area contained:

30	Pneumatic piezometers;
30	Vibrating wire piezometers;
12	Inclinometer access tubes;
2	In-place inclinometer strings (total 20 monitoring elements);
10	Wire extensometers (on 2 cradles);
and 20	Recharge wells.

On average therefore the system provided one monitoring instrument to each 8.6 m² of plan area.

This level of instrumentation is not common. The size of most natural slides subjected to detailed study generally leads to sparse coverage. Devin et al, (1988), present a table of inclinometer usage at 13 landslide sites. The highest intensity of coverage noted was equivalent to eight access tubes per hectare, or about one thirtieth of the intensity at Selborne (without considering the string inclinometers). Previous investigations have made use of special study sites to undertake prolonged monitoring, but have rarely been able to instrument so intensively. The excellent study at Saxon Pit (Burland et al, 1977), used five piezometers, three magnet

213

extensometers, five manually read extensometers and three deep levelling points. Surface movement points and photogrammetry completed the coverage on a 200 m long study face.

Perhaps the most intensive instrumentation use in geotechnical engineering is associated with earthfill dams. The instrumentation used to monitor displacement of the Kielder Dam is described by Millmore and McNicol, (1983). The 1140 m long, 52 m high embankment contained 11 inclinometers, 29 extensometers and 62 settlement devices. A total of 185 piezometers of various types were used. On a surface area basis this represents around one hundredth of the Selborne intensity.

The decision to install the very large number of piezometers was a direct consequence of the proposed method of inducing failure. As the slope was to be failed by raising pore-pressures, the existing groundwater regime inferred from previous investigations would not be applicable at failure, nor would modifications based on a small number of piezometer readings be acceptable. It was necessary to provide sufficient coverage to define much more precisely the complete pore-pressure regime within the slope. The effect of each stage of recharge also needed to be closely monitored, so that further increases could be planned and controlled.

The inclinometer coverage was designed to give detailed displacement readings throughout the body of the slope, and through the whole of the pre-failure period. A standard manually read inclinometer probe was used to monitor displacements in 12 access tubes. This gave good spatial coverage while the string inclinometers were intended to give continuously logged movement/depth profiles at two key locations.

Rates of movement and strains within the slip mass were also of interest and it was decided that useful information could be obtained using continuously logged surface mounted wire extensometers.
The vibrating wire piezometers, string inclinometers and wire extensometers were all continuously logged. Great care was taken to design a system having considerable flexibility in reading procedures. It was essential that readings could still be taken even if some component of the data logging system should fail during a critical period. Alarm systems were required to warn on-site and off-site researchers of possibly impending failure.

The whole instrumentation system was designed to a strict cost constraint. The overall study was known to have a maximum possible funding limit of £200,000 and after all other costs had been determined, the sum available for the instrumentation was less than £80,000. Both the total number of instruments and the balance of the complete installation were determined primarily by the stringent requirement that this overall figure could not be exceeded. [1987 prices]

5.2 Piezometers

Two types of piezometers were used in order to give the maximum coverage of pore pressure measurement and recharge control within the cost limits of the research programme. A cross-section showing the arrangement of the piezometer installation is given in Fig. 8.

The 30 vibrating wire piezometers were concentrated in the zone below the slope face and the front part of the crest. This distribution was designed to cover the range of expected positions of the eventual failure surface. Vibrating wire piezometers can be continuously monitored electronically, and in this case they were intended to provide frequent pore-pressure readings in the critical zone up to and during the failure event. It was also hoped that those piezometers sited above the eventual failure surface would continue to monitor post-failure pore-pressures.

The 30 pneumatic piezometers were installed in the areas at the toe, further back from the crest and at depth, areas which it was thought would be less critical. Their main purpose was to extend the area over which pore pressure information would be available

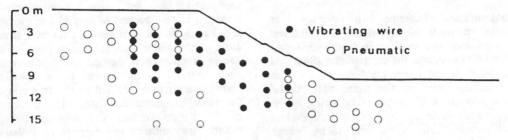

Figure 8:   Piezometer positions

so as to set the detail from the vibrating wire instruments in a wider framework. Although this type of piezometer cannot be read electronically, it does have a considerable cost advantage over the vibrating wire type (approximately one-fifth the cost in this case). By employing this much cheaper system in areas thought to be less important, a considerable cost saving was produced which could be diverted to ensure that critical areas were well served with continuously logged instruments.

The two systems were overlapped, at the toe and the crest, to enable their outputs to be cross-checked and if necessary cross-calibrated. The filter dimensions and grouting up arrangements were the same for both systems. In all cases a 200 mm filter length was formed at the base of the installation borehole. The filter was sealed above with bentonite pellets and the hole filled to ground level with bentonite/cement grout. The short filter length was chosen to give a precise location for the measured pore-pressure.

The pneumatic piezometer layout in the toe area was also intended to model a typical low cost monitoring arrangement. It had been suggested in pre-planning discussions that the onset of instability might be preceded by noticeable changes in pore-pressure in this passive zone. Pneumatic piezometers would represent a low cost means of monitoring for such changes.

5.3 Inclinometers

As with the piezometers two different types of instrument were chosen to give a balance between detail and cost. The positions of the inclinometer installations are shown in cross-section in Fig. 9. Twelve standard plastic inclinometer access tubes were installed to give broad, manually obtained coverage of lateral displacements. The main functions of this part of the system were to give an indication of the way in which the early stages of the failure developed and, by virtue of their greater sensitivity, to give an

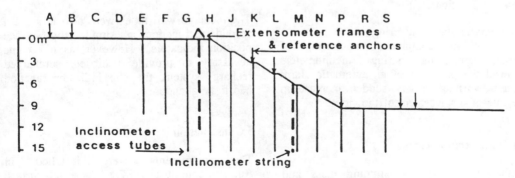

Figure 9:   Inclinometer positions

215

early indication of possible failure.

Continuous electronic monitoring of displacements within the slope was achieved by means of two in-place inclinometer string installations, one at the crest of the slope and the other at the lower third position. Both installations were on the same near central cross-section of the slope. The string at the crest was made up of twelve independent inclinometer sensors on one metre gauge lengths between 1 and 13 m depth, with each sensor separately and continuously monitored. The lower string comprised eight sensors in a similar arrangement between ground level and 8 m depth. The string inclinometers represented the most expensive single items in the system but fulfilled three important functions:

- continuous monitoring of displacements within the slope.
The shortest possible reading cycle for the manually read inclinometers could yield only two or three complete sets of readings per week (if other systems were not to be neglected); this would be insufficient to yield useful information on actual variations in rates of movement.

- taking of continued readings through the failure event.
Manually read inclinometers require access at ground level. It was not intended to allow personnel onto the slope if rapid failure was indicated. Conventional access tubes also cease to allow free passage of the reading torpedo at quite modest distortions. The string inclinometers overcome both these shortcomings by being remotely monitored and by continuing to function at much greater distortions.

- automatic alarm triggering.
By providing continuous monitoring of displacements the string inclinometers allowed the provision of an automatic alarm system warning of any sudden increase in the rate of displacement of the slope.

5.4 Wire extensometers

Surface strains on the slipping mass and larger relative movements between the slip mass and the adjacent ground were monitored by ten wire extensometers, arranged in two groups of five. Each group was based on a datum frame, anchored at the crest of the slope with four mini-piles. The datum frames each carried five gearboxes, each gearbox being driven by a sprocket carrying a chain connected to the free end of the invar extensometer wire. The positions of the datum frames and wire anchorage points are shown in Fig. 9. Relative movements of the anchor points and the datum frame are thus converted into rotations which are detected and measured within the gearboxes, which also contain the signal conditioning circuitry. By selecting the gearing at the datum frame it was possible to fix the required range and sensitivity of each extensometer. In the event all the extensometers, irrespective of range, were found to give stable and reliable output to a precision of 1 mm, which was thought to be the maximum achievable stability of the cradle/wire/anchor system.

The wire extensometers were continuously electronically monitored and were incorporated into the automatic alarm system. It was intended that they would provide a continuous record of amounts and rates of slip mass movement throughout the failure event, hopefully yielding peak velocity components of the slip mass.

5.5 Recharge wells

The recharge wells comprise simple standpipes in long filter zones. The extent of the zones for each row of wells is shown on Fig. 10.

The recharge well tubes were also to serve a secondary function as simple slip surface position indicators. However, as it became necessary to provide a closed, surcharged recharge system, the slip indicator function had to be abandoned.

5.6 Installation

The instruments were installed in August-September 1987. Each sub-surface instrument was placed in its own 150 mm

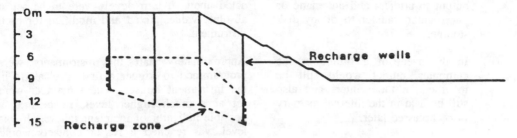

Figure 10:   Positions of recharge wells and recharge zone

diameter hole, drilled by means of a continuous flight auger technique which gave precise control over drilling depth. Instruments under the slope face were installed from temporary benches cut as part of the formation of the eventual slope. Drilling was therefore synchronous with the major earthworks.

5.7 Reading system

The reading system used at Selborne was of particular interest in two respects. Firstly a comparatively high proportion of instruments was connected to the continuous monitoring system and the inclusion of inclinometer and wire extensometer output in the data logging system is unusual. Secondly the fail safes incorporated into the system are considered to be unusually thorough. It must be appreciated that in this application the function of the instrumentation was not just to warn of possible forthcoming failure, nor simply to monitor compliance of an as-built structure with design assumptions; rather the system had to record all details of a major one-off event, which could not be predicted in time or position to an accuracy which would permit reliable pre-programming.

The electronic reading system comprised two parallel sub-systems, each of which carried half of the vibrating wire output and half of the analogue output (from the string inclinometers and wire extensometers). Each sub-system was made up as follows:

- All the instruments were connected to a switch card/multiplexor which handled all

the electronic switching during scanning and also incorporated EEPROMs to refine input signals so as to give fully reduced and corrected output directly in engineering units.

- The switch cards/multiplexors were controlled by programmable data loggers, each containing approximately 170 kbytes of internal memory (enough for about 86,000 readings). In normal operation the data loggers could either output continuously to printers and to a personal computer, or alternatively the readings could be extracted and transferred using independent interrogators. The latter was the preferred mode of operation on site.

The data loggers could be programmed to a variety of logging strategies. Scan rates and intervals for instance, could be varied if necessary according to the rate of change being recorded in any reading. More routinely a "threshold" mode could be used where readings were taken at a scan rate of one set per minute, but only recorded if different from the previously stored value by a pre-set threshold amount. The program also allowed the use of alarm levels as described in the next section.

Considerable attention was given to designing-in alternative reading paths to cover any component failure. The first safeguard was the cross-division between the two data-loggers, so that failure of one data logger would still leave half of the instruments of each type on line. The following reading hierarchy was established:

(i) Normal operation with automatic output to printers and automatic or interrogator transfer to floppy disk storage.

(ii) In the event of a failure of the computer, output would still be recorded on the printers and also still be held in the internal memory to be accessed later.

(iii) In the event of failure of the data logger the switch cards/multi-plexors could still be accessed directly by manual readout units, each with internal memory sufficient to store 500 readings. This system required a plug-in connection to be made in the instrument house. As circum-stances might have led to this house being positioned within the failure zone, special long fly leads were ordered to enable the manual readout units to be operated at a safe distance. They did not prove to be necessary.

(iv) In the extremely unlikely event of a switch card/multiplexor failing at a critical moment the manual readout units could be connected directly to a signal cable ends, but this would depend on the rate and position of failure and the degree of risk involved. Fortunately this option did not have to be attempted.

The entire data logging system, with the sole exception of the personal computer, operated from two continuously recharged 12 volt batteries. These batteries were of the fully sealed type and would have continued to function even if inverted should the instrument house have been affected by the failure.

5.8 Alarm systems

Special facilities incorporated within the programmable data loggers allowed different levels of alarm event to be recognised and acted upon. Alarm levels were to be set at absolute values, fixed and modified for each instrument.

Approximately half the instruments were programmed to trigger a first level alarm in the instrument house. The remainder were set at a much higher level, selected as a possible indicator of incipient failure. If this level was reached a set of contacts would close to activate an auto dialler which in turn would send a pre-recorded message to key telephone numbers. On site audible alarms and floodlighting were also to be triggered at this higher level.

6 RECHARGE SYSTEM

The slope was brought to failure by increasing the pore water pressure in it. This recharge was achieved by means of twenty recharge wells in four rows across the top and upper part of the slope (see Fig. 10). The depths of the wells were chosen to give substantial pore-pressure increases within the body of the slope without giving rise to hydraulic fracture or shallow surface sliding.

Essentially, a simple recharge schedule was followed. Firstly the water levels in the rearmost line of wells was raised to ground level. The response of adjacent piezometers was monitored, and when equilibration had been achieved, the water levels in the next row of wells were raised to ground level. This sequence was repeated until the water levels in all wells had been raised to ground level.

It was proposed that, if failure had not been produced by the recharge of all wells to ground level, then the water levels in all wells were to be raised in 1 metre increments.

A final contingency plan allowed for the drilling of additional recharge wells if necessary. Recharge started in February 1989, and it was found necessary to raise the water levels in the recharge wells to 3 m above ground level before failure of the slope occurred in July 1989.

7 PERFORMANCE OF THE INSTRUMENTATION

The performance of the complete instrumentation system was most satisfactory in that the failure event was recorded in the full level of detail hoped for and expected. In some respects, such as durability and displacement compliance the performance was generally well above expectations. Many points of interest concerning the performance are still coming to light as the data produced is analysed. Among the observations to date we would highlight the following.

(i) Of the 89 independent instruments employed, 86 were still operative at the onset of failure, a reliability for the two year on site period of 98.9% per annum. One of the three lost instruments was damaged by site operations. The other two were either transducer or connection failures. Although still operative the pneumatic piezometers recording negative pressures had become very difficult to keep on line and were disrupted very early in the development of the failure.

(ii) Three piezometers were recovered

Table 1

Instrument Type, Number	Applied Pressure psi	Readout psi
Pneumatic pp1.3	0.0	0.0
	0.3	0.3
	1.0	1.0
	15.1	15.2
	35.0	35.3
	50.0	50.0
Vibrating Wire, S/N 034	0.0	0.0
	10.0	10.0
	20.0	20.02
	30.0	30.10
	40.0	40.26
	50.0	50.34
Vibrating Wire, S/N 010	0.0	0.0
	10.0	10.01
	20.0	20.06
	30.0	30.14
	40.0	40.19
	50.0	50.25

during post-failure excavations and were returned to the manufacturers (Geotechnical Instruments Ltd.) for re-calibration. The results are presented in Table 1 and they show remarkable stability over a twenty-two month period of burial.

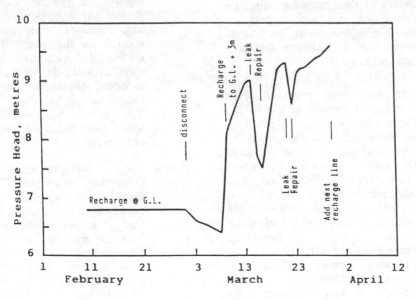

Figure 11:   Records from piezometer P24

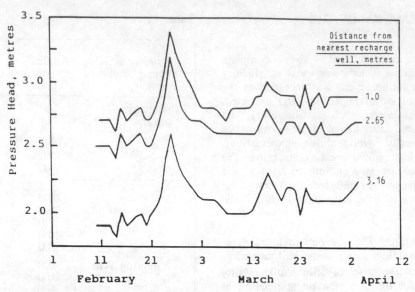

Figure 12:   Comparison of piezometers P3, P5 and P8

(iii) Piezometer readings responded very positively and quickly to applied recharge. A typical response record for a pneumatic piezometer within the recharge zone is shown on Fig. 11.

(iv) Groups of piezometers were consistent within themselves in the pattern of their response. Fig. 12 shows the response records of three piezometers at the same level, but at different distances from recharge wells.

(v) Negative pore pressures were recorded at the toe of the slope. These were probably the result of the stress reduction caused by the substantial excavations in this area. This effect is described by Bishop and Bjerrum (1960) and has subsequently been confirmed by a number of field studies, for example Vaughan and Walbancke (1973). The pneumatic piezometers are not well suited to measuring negative pore pressures. In high permeability soils this problem is easily overcome by taking readings starting from a

high suction, but in low permeability soils the application of suctions to the pneumatic system merely expands the air sack in the tip as the open membrane is drawn closed. (The air sack is provided to allow compressibility so that the membrane which is normally closed can open to register the reading). A reading system had to be devised in which high suctions were permanently applied to the pneumatic leads, closing the membrane and, with time, allowing the air sack volume to come into equilibrium with the membrane in the closed position. This system was successful and will be described in detail elsewhere.

(vi) The inclinometer torpedo calibration remained constant, though some maintenance of wheel bearings did become necessary. Had a longer monitoring period been required then replacement of the running wheels would have been essential. By the end of the project the torpedo had been used for readings equivalent to 17 km of travel.

(vii) Three inclinometer access tubes were very carefully exposed during post-failure excavations. In each case the deformed profile of the tube, was aligned very closely along the exposed failure surface, the tube having been pulled down out of its grout surround within the moving mass and laid flat along the failure surface. Figure 13 was obtained by tracing a photograph and shows the relative positions of the distorted tube, slip plane and empty grout column for an access tube which had undergone around 4 m of relative displacement.

(viii) The string inclinometers and wire extensometers performed their special functions exceedingly well.

Both types of instrument provided continuous displacement/time plots of a type that could not have been obtained with manually read instruments, and both continued to function at very large displacements. Wire extensometers continuously recorded relative movements of up to 150 mm at one minute intervals over one period of 3 hours, and the string inclinometers were still functioning with angular distortions between successive 1 m gauge lengths of up to 42°. A 0.5 m torpedo passed through a maximum angular distortion of 35°, but did so at a great risk to the instrument. At greater distortion the lower part of the access tube became inaccessible whereas the lower parts of the inclinometer strings remained operative even after large distortions had developed higher up.

(ix) During the slope failure the behaviour of the connections to the pneumatic piezometers was very different from that of the vibrating wire connections. The very extensible tubing of the pneumatic instruments deformed and stretched easily to accommodate some very

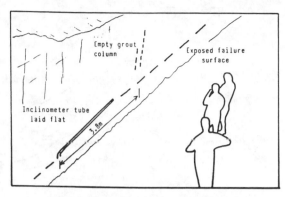

Figure 13: Exposed inclinometer access tube along failure surface

large movements. The vibrating wire signal leads were much less compliant and many snapped very early on in the failure event.

(x) The full hierarchy of reading options was not required, although it was proved in practice routines.

(xi) The reading strategy adopted was the "threshold change" approach with scans at one minute intervals and values only recording if one of the readings was significantly altered.

Instrument flutter frequently exceeded the very tight threshold limits set and about 60% of scans resulted in recorded readings. By excluding the less stable instruments, or setting wider threshold tolerances, this percentage could have been greatly reduced.

(xii) The first alarm level triggered very quickly as the failure developed, and the majority of failure data was obtained from the simple one-minute interval reading strategy on all channels. The enormous quantity of data generated has caused some problems in subsequent data processing. This is acceptable in a research application where detail and duplication are paramount in the design philosophy, but cause problems in the more usual

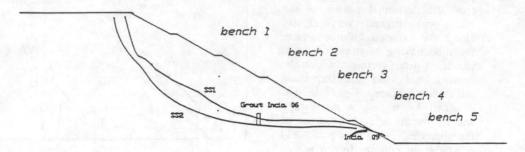

Figure 14:   Shape of failure surface

8 POST-FAILURE INVESTIGATIONS

Following the failure of the slope in July 1989, a number of trenches were excavated through the failed mass, enabling the precise position of the failure surface to be accurately defined. The failure surface was found to be markedly non-circular, as shown in Fig. 14.

A number of large block samples containing the slip surface were taken at this time, and have been used in the laboratory to measure the shear strength on the slip surface.

In addition, block samples up to 150 mm diameter were taken from outside the slipped mass, and these samples have been used for large diameter triaxial tests.

9 CLOSING REMARKS

The failure of the Selborne slope was achieved by pore pressure recharge. Following the failure, a total movement of the slipped mass of approximately 4 m occurred.
    The large quantity of data obtained from the extensive instrumentation system is currently being analysed, and will be presented in later papers.

monitoring applications where delay in accessing data cannot be countenanced.

10 CONCLUSION

The instrumentation systems used for the Selborne cutting stability experiment performed well and have provided a detailed record of a unique experiment. The few shortcomings identified during the data processing are associated more with site pressures and manpower shortages (giving less than planned for reading frequencies on some instruments at some times), and financial constraints (a third string inclinometer would have been extremely useful), than with inadequate instrument performance. There can be little doubt that instrument systems exist to cover all site monitoring requirements for an exercise such as the Selborne experiment. The main difficulties to be overcome are in planning a successful layout for those instruments, in planning an efficient and workable reading schedule where manual reading is employed, and in planning a data access system allowing rapid, user friendly access to electronically logged data.

ACKNOWLEDGEMENTS

The collaborative study is sponsored by the Science and Engineering Research Council by means of three separate grants to the participating institutions. The collaborators (Mr. M.R. Cooper of Southampton University, Dr. D.J. Petley of Warwick University and Professor E.N. Bromhead of Kingston Polytechnic) wish to record their appreciation of the generosity of Colonel J.A. de Benham Crosswell and the Selborne Brick and Tile Company in making such a large part of their active clay pit available for such a long period.

The authors would also like to acknowledge the extensive and invaluable assistance in the form of time, advice, expertise and material support given by Mr. R.C. Weeks and Mr. R.G. Gillard of Geotechnical Instruments Ltd.

The installation of the instrumentation described in this paper was carried out under the direction of Professor E.N. Bromhead, with the assistance of Mr. I.R. Sandman and Mr. D.I. Grant of Southampton University.

REFERENCES

Bishop, A.W., and Bjerrum, L., 1960. The relevance of the triaxial test to the solution of stability problems. Proceedings of the American Society of Civil Engineers Research Conference on Shear Strength of Cohesive Soils, Boulder, Colorado, 437-501.

Burland, J.B., Longworth, T.I., and Moore, J.F.A., 1977. A study of ground movement and progressive failure caused by a deep excavation in Oxford Clay. Geotechnique, vol. 27, 557-591.

Devin, P.E., Pezzetti, G., Ricciardi, C., and Tommasi, P., 1988. Assessing the precision of inclinometric measures by means of an experimental apparatus. Proceedings of the Vth International Conference on Landslides, Lausanne, vol. 1, 393-398.

Millmore, J.P., and McNicol, R., 1983. Geotechnical aspects of the Kielder Dam. Proceedings of the Institution of Civil Engineers, Part 1, 805-836.

Vaughan, P.R., and Walbancke, H.J., 1973. Pore pressure changes and the delayed failure of cutting slopes in over-consolidated clay. Geotechnique vol., 23, 531-539.

Developments in Geotechnical Aspects of Embankments, Excavations and Buried Structures, Balasubramaniam et al. (eds)
© 1991 Balkema, Rotterdam. ISBN 90 5410 019 2

Study on the mechanisms of slope failure due to heavy rainfall by using laboratory and numerical models

Naftali S. Mshana, Atsumi Suzuki & Yoshito Kitazono
Department of Civil and Environmental Engineering, Kumamoto University, Japan

ABSTRACT: Many slope failures occur during or after heavy rain. One of the main causes is considered to be the increase of pore pressure due to seepage of rainwater. Inorder to explicate the mechanism of these slope failures, a $1g$ slope model tests, in which pore pressure distributions in the slope were measured and failure patterns observed, were carried out. In addition, both seepage analysis of rain water and stability of saturated slope by the finite element method was developed. As a result, numerical simulations were consistent with model tests and it was confirmed that the two-dimensional FEM analysis which considered the discontinuity between the base layer and soil layer, and reduction of the cohesion at elements directly affected by piping, was an effective analysis method for prediction of the failure mechanism of slopes.

1 INTRODUCTION

In Japan, many slope failures occur during or after the rainy season from June to end of August. In Kumamoto Prefecture located in central Kyushu, the suburbs of Kumamoto city which are abundant in weathered soil of unwelded tuff (hereinafter called "Haido soil"), catastrophic slope failure has been noted as the main typical disaster.

This paper presents some results of the preliminary stages of a research program carried out to explicate the mechanism of the catastrophic failure of Haido slopes due to heavy rainfall, at laboratory scale. A Laboratory model and a Numerical model are used to facilitate the simulation procedure. Rainfall is simulated by "Sprinkling" while boiling from fractured rocks is simulated by "Piping". Numerical prediction is conducted by using a two-dimensional FEM analysis combining saturated-unsaturated flow analysis and stability analysis. The most interesting feature of the model is its capability to consider the reduction of soil strength due to increase of degree of saturation, the increase of the self-weight of the slope due to seepage and slippage between materials. The present analysis considers the piping effect by reduction of the saturated cohesion of elements with negative mean stress, as recommended by Murata (1989).

2 TYPICAL PROPERTIES OF HAIDO SOIL

2.1 *Origin of Haido soil*

The so called "Haido" is a weathered non-welded tuff originated from the pyroclastic flow deposit of the fourth eruption cycle of the Aso volcano about 30,000 years ago. Haido soil is abundantly distributed near Ueki area in the western flank of Aso caldera, as shown in Figure 1 (Kuroki et al (1983)). The range of its geotechnical properties as given by Kuno (1974) are summarised in Table 1.

Table 1 Range of Geotechnical properties of Haido soil (Kuno, 1974)

PROPERTY	RANGE
Specific gravity, Gs	2.60 – 2.75
Nat. water content Wn (%)	40 – 75
Field density, ρ_t (g/cm^3)	1.35 – 1.80
−74μm sieve (%)	50 – 80
W_L (%)	40 – 70
I_P (%)	10 – 35
Cone index, q_c (kgf/cm^2)	6 – 18
	(0 – 9)
Unconfined compression strength, q_u (kgf/cm^2)	0.1 – 2.2
	(0 – 0.2)
Frict. angle, ϕ_{cu} (°)	15 – 30
Cohesion, c_{cu} (kgf/cm^2)	0.15 – 0.30

NOTE: ( ) −Range for remolded samples

2.2 *Investigated sample*

The physical properties of soil used in the present analysis are shown in Table 2. Figure 2 shows X-ray diffraction pattern on a Haido specimen. From the figure there is proven the existence of Holloysite by $2\theta = 12°$, $20°$, $36°$ and $62°$, and

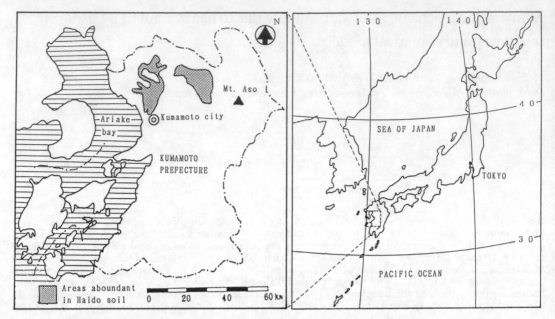

Fig. 1 Areas aboundant with Haido soil in Kumamoto Prefecture (Kuroki et al (1983))

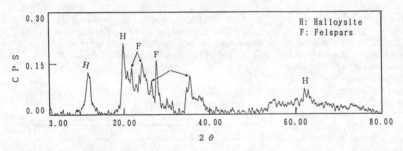

Fig. 2 X-ray analysis pattern

Table 2. Physical properties of the investigated sample[*]

Natural water content,Wn (%)		41.5
Specific Gravity,	Gs	2.497
Consistency		N.P
Gravel	(%)	0.3
Sand	(%)	14.3
Silt	(%)	69.4
Clay	(%)	16.0
Type of soil		V H₁

(*)- Classification by Japanese Standards
(JSF M 111-1990)

trace amount of Felspar between $2\theta = 21°$ and 26°. Halloysite has small specific area and in remolded

state its double T structure can easily be broken, changing most of its void water to free water (Kita et al (1965)). This is believed to be the main cause of the catastrophic failure of Haido slopes during heavy rainfall.

3 RESEARCH METHOD

Test on reduced scale models provides a good insight into the type of phenomena which occur quantitatively, however, they are less informative due to problems of similitude. A simulation procedure is summarised in Figure 3.

Laboratory tests on saturated permeability, water retention and strength parameters gave the basic input parameters to numerical simulation models. Numerical simulations were conducted by external coupling of saturated-unsaturated flow analysis and plane strain stability analysis as in

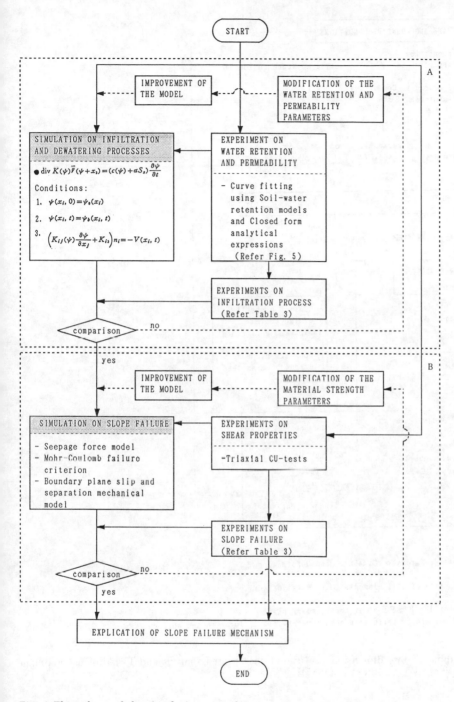

START

IMPROVEMENT OF
THE MODEL

MODIFICATION OF THE
WATER RETENTION AND
PERMEABILITY
PARAMETERS

SIMULATION ON INFILTRATION
AND DEWATERING PROCESSES

● div $K(\psi)\vec{V}(\psi+x_i)=(c(\psi)+\alpha S_s)\frac{\partial\psi}{\partial t}$

Conditions:

1. $\psi(x_i, 0)=\psi_s(x_i)$

2. $\psi(x_i, t)=\psi_b(x_i, t)$

3. $\left(K_{ij}(\psi)\frac{\partial\psi}{\partial x_j}+K_{i3}\right)n_i=-V(x_i, t)$

EXPERIMENT ON
WATER RETENTION
AND PERMEABILITY

- Curve fitting
using Soil-water
retention models
and Closed form
analytical
expressions
(Refer Fig. 5)

EXPERIMENTS ON
INFILTRATION PROCESS
(Refer Table 3)

comparison — no

yes

B

IMPROVEMENT OF
THE MODEL

MODIFICATION OF THE
MATERIAL STRENGTH
PARAMETERS

SIMULATION ON SLOPE FAILURE

- Seepage force model
- Mohr-Coulomb failure
criterion
- Boundary plane slip and
separation mechanical
model

EXPERIMENTS ON
SHEAR PROPERTIES

-Triaxial CU-tests

EXPERIMENTS ON
SLOPE FAILURE
(Refer Table 3)

comparison — no

yes

EXPLICATION OF SLOPE FAILURE MECHANISM

END

Fig. 3 Flow chart of the simulation procedure

Figure 4 (Iseda et al (1985)). Infiltration behaviors are obtained from the solutions of a two-dimensional quasilinear partial differential equation for saturated-unsaturated flow through a porous media; while the failure pattern is derived from the Mohr-Coulomb failure criteria.

It is emphasized here that the computer applications by external coupling introduced in this anal-

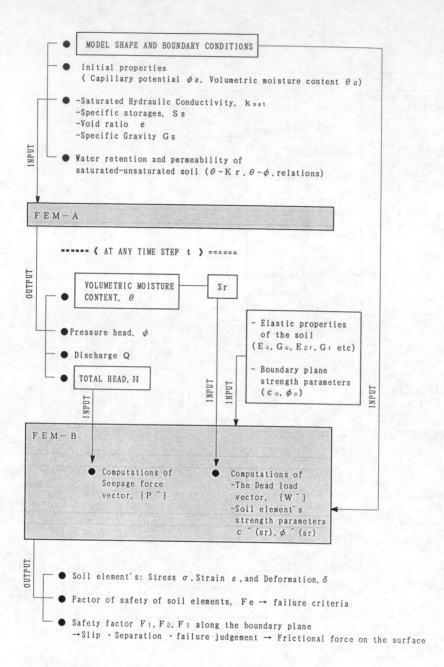

Fig. 4 External coupling of Two-dimensional saturated-unsaturated analysis and Two-dimensional plane strain FEM Stability analysis (Iseda et al (1985))

ysis means that coordinates are taken as fixed. To be perfect the equation should be developed in deforming coordinates, the Darcy velocity being considered as a relative velocity of water with respect to the grains. However, this is a standard approximation that apparently leads to little error (Freeze, 1971).

As shown in Figure 3, sometimes modifications of material input parameters in the range of Laboratory test data is necessary to improve numerical predictions. This is stressed as the model test works in three dimensions.

4 LABORATORY TESTS

Laboratory tests covered: the permeability test, the water retention test, and triaxial compression test.

4.1 Permeability test

Permeability test was carried on remolded samples prepared at the density and degree of saturation in the range covering the ones of the soil layer in the model test.

4.2 Water retention and permeability of unsaturated and saturated soil

Water retention curves and coefficients of permeability of remolded samples were obtained on infiltration by the Column suction method. Since only a limited portion of the soil-water retention curve is measured, the remaining part was predicted by Van Genuchten's curve fitting technics (Figure 5). Van Genuchten proposed a soil-water retention model (Equation (1)) to match the water retention experimental data (Van Genuchten (1980)).

$$\Theta = \left[\frac{1}{1+(\alpha h)^n} \right]^m \tag{1}$$

where $\Theta = (\theta - \theta_r)/(\theta_s - \theta_r)$, θ_s is saturated volumetric water content, θ_r is residual volumetric water content; α, n and m are constant independent parameters ($m = 1 - 1/n$)

θ_r has been considered as the water content at some large negative value of the pressure head, for example, at the permanent wilting point ($\psi = -15,000$ cm or approximately pF 4.2). A value

$\theta_r = 0.185$ was estimated by extrapolating soil-water retention data measured by the centrifugal process.

In substituting equation (1) in predictive conductivity models of Mualem a closed form analytical expression for the relative hydraulic conductivity $K_r(\Theta)$ is obtained (Equation (2)).

$$K_r(\Theta) = \Theta^{1/2}[1 - (1 - \Theta^{1/m})^m] \tag{2}$$

4.3 Shear properties of unsaturated and saturated soil

Triaxial compression test was carried on undisturbed specimens to obtain failure criterion and modulus of deformation. Remolded specimens were prepared at the density and degree of saturation in the range covering the ones of the soil layer in model test. Consolidated undrained (CU) tests were carried on both unsaturated and saturated samples.

The results were adopted in the Numerical simulation model and used as a base of discussions of the catastrophic slope failure mechanism. It is considered important that saturated specimens of Haido soils are characterised by small shear strength at low confining stress.

5 MODEL TEST METHOD

5.1 Compaction and energy test

Variation of density at different compaction efforts was studied in detail before laboratory model tests. A 10 cm soil layer was compacted in 4 layers in a wooden box at 5 compaction energies. Standard compaction carried on the sample gave a maximum dry density $\rho_{d(max)}$ of 1.163 g/cm^3 at optimum water content W_{opt} of 36.0 %. From this test

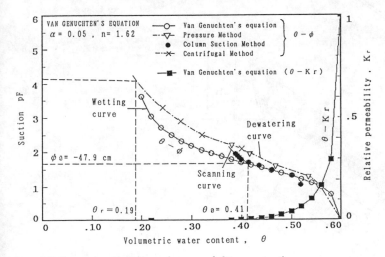

Fig. 5 Soil-water retention and permeability properties

a field density ρ_t of 1.390 g/cm³ (dry density ρ_d = 0.989 g/cm³) corresponding to an approximate compaction energy of 0.822 (Kgf.cm/cm³) was chosen. As shown in Table 1 Haido soils have wet field densities ρ_t ranging from 1.35 to 1.80 g/cm³.

5.2 *Model test apparatus and test procedure*

5.2.1 *Test apparatus*

A trapezoidal steel mold of 88 cm to 152.4 cm length and 60 cm width was used for simulation purposes. The hinged mold can be inclined at various angles (max. 40°), and water supplied from a 7 meters high tank make both sprinkling and piping simulations possible. An artificial rain at any decided precipitation can be applied through the nozzles by adjusting the water supply at the mini-float valve. Piping at a water head up to 1.5 meters can be applied by an adjustable overflow tank. To insure uniform distribution of the water head, an indirect piping through a 2.5 cm thick and 10 cm wide sand filter layer was used

Photo 1 Model test system

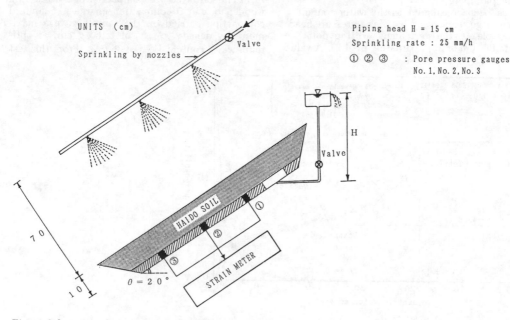

Fig. 6 Schematic diagram of the Model test system

230

(refer Figure 7 for details). The variation of the pore water pressures can be monitored from three points at the base of the molds. A 10 cm soil layer was compacted in 4 layers over a 2.5 cm thick fiber soil-cement layer. To ensure uniform permeability between layers scratching was done between every successive layers.

5.2.2 *Test procedure*

5.2.2.1 *Infiltration test*

An artificial rainfall at a rate of 25 mm/h, was applied on the soil surface; interest has been focused on derivation of Infiltration curves as proposed by Horton. This curve was derived by computing the net inflow from the measured outflow at a specified time interval. Horton, proposed an algebraic equation representing the variation of infiltration, I with time t as follows (Phillip (1956)):

$$I(t) = I_f + (I_i - I_f) \exp(-\beta t) \qquad (3)$$

where I_i and I_f are the presumed "initial" and "final" values of I, and β is a constant parameter.

5.2.2.2 *Failure test*

After the infiltration test the soil layer was brought to failure by piping by a constant water head.

Table 3 Observed phenomena

	OBSERVATION	MD-1	MD-2
INFILTRATION	Visual record (VTR, Photos)	●	●
	Water content at the initial state	●	●
	Pore water pressure	●	●
	Surface deformation at the final state	●	×
	Cone index at the final state	●	×
	Water content at the final state	●	×
FAILURE	Visual record (VTR, Photos)	–	●
	Pore water pressure	–	●
	Surface deformation at the final state	–	●
	Cone index at the final state	–	●
	Water content at the final state	–	●

NOTE: MD-1: Infiltration test(θ =20°)
MD-2: Failure test (θ =20°)
●-Indicates observed phenomenon

Table 3 gives the observed phenomena in the course and end of the experiment. The model test system is as shown in Photo 1 and Figure 6.

6 NUMERICAL SIMULATION METHODS

6.1 *Two-dimensional saturated-unsaturated flow analysis*

From the equation of continuity and Darcy's law it is possible to derive a quasilinear partial differential equation for two-dimensional flow of water in an unsaturated porous medium, equation (4) (Akai et al (1979); Neuman (1973)).

$$div \, K(\psi)\vec{\nabla}(\psi + x_3) = (C(\psi) + \alpha S_s)\frac{\partial \psi}{\partial t} \quad (4)$$

Where

$$\alpha = \begin{cases} 0 & : \text{unsaturated zone} \\ 1 & : \text{saturated zone} \end{cases}$$

$K(\psi)$ is a permeability tensor of second order, $C(\psi)$ is the specific moisture capacity defined as $d\theta/d\psi$, ψ is the capillary potential, x_3 is the gravitational potential , S_s is the specific storage.

The assumptions underlying derivation of the quasilinear partial differential equation of the above two-dimensional saturated-unsaturated flow are that (Singh, 1967):
1. The porous medium is homogeneous.
2. Diffusivity is a unique function of θ.
This is usually not true in nature since a hysteresis effect exist during wetting and drying cycles.
3. Darcy's law is assumed valid.
4. The pore spaces are assumed to be free from air in the sense that air offers no resistance to the flow.
The developed equation considers only the flow of water. The fundamental assumption is that the air phase is continuous and is at atmospheric pressure. This assumption precludes the entrapment of pockets of compressible air in the flow system (Freeze, 1971).
5. Soil properties are assumed to remain constant with time.
Equation (4) must be supplemented by appropriate initial and boundary conditions, which are defined as follows:
1. Initial conditions:
Since hysteresis is not considered in this analysis, ψ is a single valued function of θ and therefore the initial conditions that are required for any given problem are simply,

$$\psi(x_i, \, 0) = \psi_0(x_i) \qquad (5)$$

2. Boundary conditions:
a) Prescribed head boundary

231

$$\psi(x_i, t) = \psi_b(x_i, t) \qquad (6)$$

b) Prescibed flux boundary

$$\left(K_{ij}(\psi)\frac{\partial \psi}{\partial x_j} + K_{i3}\right) n_i = -V(x_i, t) \qquad (7)$$

Where ψ_0, ψ_b and V are prescribed functions, n_i is a unit outer normal vector on the boundary Γ

The relationships $K(\theta)$ and $\psi(\theta)$ can be obtained experimentally as in Figure 5, assuming a single-valued function. Equation (4) with equations (5), (6) and (7) can be solved by the finite element method (FEM). The finite element formulation of this equation may be done most easily with the Galerkin method (Akai et al (1979); Neuman (1973)).

6.2 Two-dimensional stability analysis by FEM

6.2.1 Analysis procedure

The stability of the model test is simulated by a finite element model (see Figure 7), which has a friction boundary plane between the base material and the compacted Haido soil layer.

The analysis considers body forces due to both effective dead load and infiltration, friction forces on the discontinuous boundaries, and reduction of rigidity and elasticity of the fractured zone. In addition, the change of the effective cohesion of each element with the degree of saturation was modelled by hyperbolic equations proposed by Kudara (1983). The dead load of every element is computed from the volumetric water content of nodes while the corresponding seepage force is computed from total water head at the nodes. In the analysis, the body force of the soil material and the seepage force are given as load vectors by converting the selfweight of each triangular element and seepage forces into equal nodal forces.

Iterative calculations were carried by giving certain conditions of convergence to the nodal forces at the friction boundary plane and element stresses in the haido soil layer.

The material properties used in the present analysis are summarised in Table 4.

6.2.2 Material failure criteria

The analysis starts by assuming the soil material an isotropic elastic body, at initial stresses. At failure the elements of local failure (including shear failure and tensile failure) are replaced with cross anisotropic material where anisotropic has developed in the direction of planes of slide or failure and by reducing the rigidity in that direction. The failure criterion adopted in present research is based on Mohr-Coulomb law of sresses and the details are given by Iseda and Tanahashi (1973).

6.2.3 Mechanical model for the boundary plane

Due to the difference in rigidity between the soil and the base material, a mechanical model representing discontinuity was introduced in the analysis (Fukuoka et al, 1977). The mechanical model considers only the controlling conditions of displacement and transmission of stress on the plane of discontinuity. The FEM analysis treated direction of local failure by considering base friction against sliding along the boundary planes between soil and the base.

The discontinuity between the base layer and soil layer is evaluated using the frictional resistance (i.e adhesion and frictional force) between base layer and the soil layer. The discontinuity at the boundary plane, as shown in Figure 8b, is expressed by two nodes which are located on the same coordinates and belonging to different elements. Node $i(U_i, V_i)$ belongs to the elements of soil layer, and node $j(U_j, V_j)$ belongs to the base material. The continuity is evaluated using the dimension comparison between the nodal force (X_i) and the sliding resistance (F_X). The flow chart of the calculation of the restrained forces for a horizontal boundary plane is given in Figure 8c. During iterations, depending on the corresponding conditions, the constrains in nodes i, j are relieved and at the same time base friction F_{xi} in the right direction and nodal force F_{xj} in the left direction (as the frictional force imposed by the soil on the base material) are introduced. Details of this mechanical model have been given by Fukuoka et al (1977) and Iseda et al (1985).

7 RESULTS AND DISCUSSION

7.1 Model test

An artificial rainfall was applied by sprinkling on the surface of the unsaturated slope in the first 130 minutes (Photo 2). The measured infiltration curve against the predicted one is shown in Figure 9. Figure 10 shows the relationship between the time and the pore water pressure due to seepage of rain water. This diagram demonstrates that after 15 minutes the measured pore water pressures increases suddenly. This phenomena is the one which appears in practice, just after the wetting front arriving at the ground water surface. After about 40 minutes, the measured pore water pressures at the bottom of the soil layer reach the steady state. The time lag in the predicted base pore pressure might be due to anisotropy due to compaction procedures. Difficulty was experienced in compacting the ends of a trapezoidal mold.

The slope was piped after 130 minutes of sprinkling. The pore pressure at point ① showed a rather large pore pressure build-up, which indicates that there is a vertically upward component of flow around the piping region. It is considered that the rather large porewater pressure around

Table 4 Material parameters for Numerical simulation

PROPERTY	HAIDO SOIL	S/CEMENT MIX		
Wet density, ρ_t (g/cm^3)	1.390	1.640		
Sat. density, ρ_{sat} (g/cm^3)	1.557	1.650		
Dry density, ρ_d (g/cm^3)	0.982	1.059		
Void ratio, e	1.543	1.623		
Degree of sat., S_{r0} (%)	67.2	93.8		
Coeff. of Perm., k_{sat} (cm/s)	2.62×10^{-4}	1.00×10^{-10}		
Cohesion c^- (kgf/cm^2)	0.15 (.025)	6.0 (6.0)		
Friction angle ϕ^- (°)	35.0 (35.0)	45.0 (45.0)		
Poisson's ratio υ	0.35	0.20		
Young Modulus E_i (kgf/cm^2)	30.0	3600		
Shear modulus G_i (kgf/cm^2)	11.11	1500		
Boundary plane	Cohesion c_0^- (kgf/cm^2)		0.100 (0.0167)	
	Friction angle ϕ_0^- (°)		25.023 (25.023)	

NOTE: ( ) —Values at saturated state

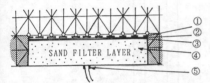

DETAIL—A : PIPING AREA
    Size: 50 cm × 10 cm
① - 2 mm thick Aluminium plate
    (holes ϕ 5 mm at 1 cm center to center)
② - Wire gauze (1 mm square)
③ - Filter paper
④ - Sand filter layer
    (Particle size 0.0074 mm ≦ D < 2 mm)
⑤ - Water supply pipe ϕ 8 mm

NUMBER OF NODES   = 644
NUMBER OF ELEMENTS = 1008
NUMBER OF NODES ON THE DISCONTINUOUS BOUNDARY = 68
(Soil-cement base and HAIDO soil interface)

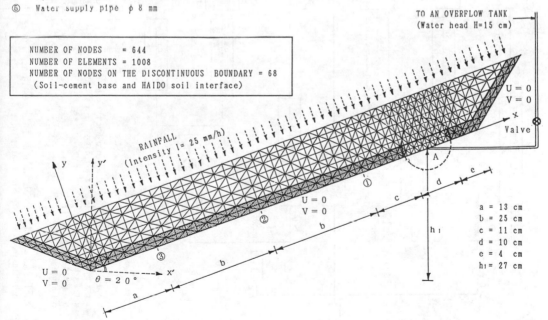

Fig. 7 Plane strain FEM model

a = 13 cm
b = 25 cm
c = 11 cm
d = 10 cm
e = 4 cm
h₁ = 27 cm

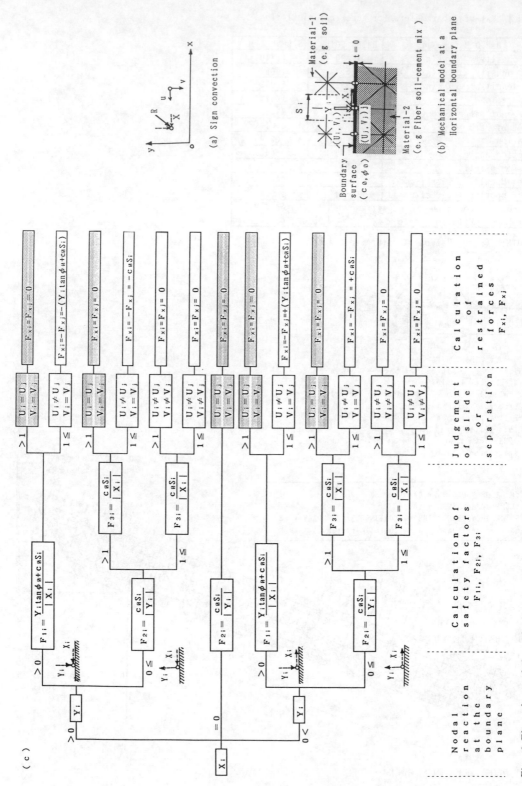

Fig. 8 Flow chart of the calculation of the restrained forces for a horizontal boundary plane (Hatched areas indicates: No slip or No separation)

234

the boiling region after rainfall is one of the main causes of slope failure. The steady state of pore pressures of the wet haido soil is reached in a shorter time than for the dry haido soil. The pore water pressure distributions in the steady state are shown in Figure 11. The first crack appeared in 181 minutes at a location 15 cm from the crest. This was followed by heaving and formation of radial cracks at time 190 minutes. From time 206 minutes, there was gradual flow of the heaved area

and extension of cracks, to a total failure at time 210 minutes (Photos 3, 4 and 5). The sketch of the final state is given in Figure 12.

7.2 Numerical prediction

Predicted failure mechanisms are shown in Figures 13, 14, 15 and 16. Surface concentration of the fracture zone at full saturation (Figure 15)

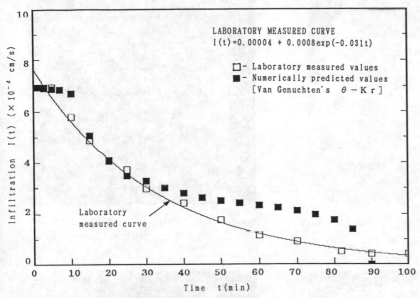

Fig. 9 Measured and predicted infiltration curve during sprinkling

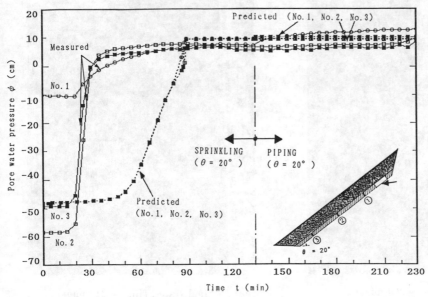

Fig. 10 Developed pore pressure with time

235

Photo 2 End of sprinkling (Time = 130 mins)

Photo 4 Piping (Time = 205 mins)

Photo 3 Piping (Time = 200 mins)

Photo 5 Final state (Time = 210 mins)

can be interpreted as region of concentration of tension cracks. As can be seen in Figures 15 and 16, the failure zone is concentrated around the piping area, which agrees with experimental results.

These results were obtained by reducing the saturated cohesion at elements with negative mean stress. There were no predicted failure zone at normal saturated cohesion as given in Table 4 above.

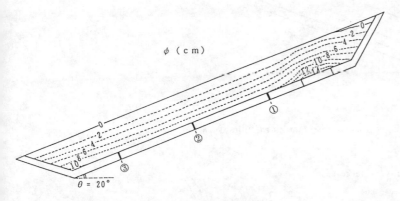

Fig. 11 Computed steady state pore pressure

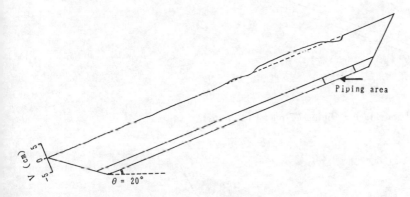

Fig. 12 Cross-section of the final state (end of piping)

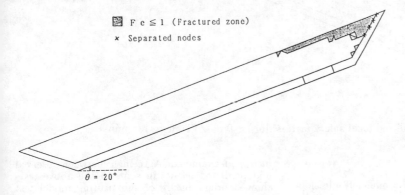

Fig. 13 Predicted fractured zone at full saturation

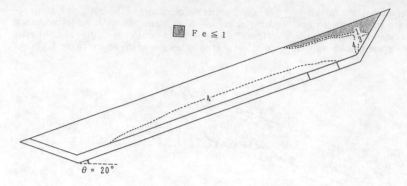

Fig. 14 Contours of the predicted local safety factors at full saturation

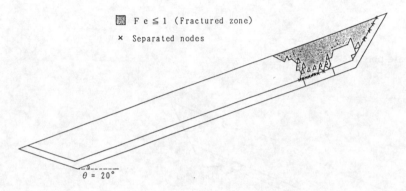

Fig. 15 Predicted fractured zone during Piping (Time = 131 mins)

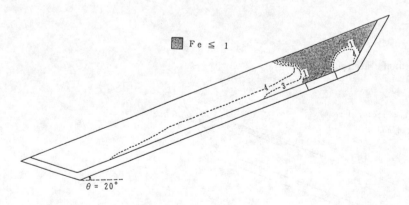

Fig. 16 Contours of the predicted local safety factors during Piping (Time = 131 mins)

8 CONCLUSIONS

A finite element analysis for a model Haido slope has been presented, and the failure mechanism being based on the concept of local safety factors have been examined. Agreement between observed and predicted results in some of the above tests show a bright future of substituting model test with numerical models.

From the results of example problems the following conclusions can be made:

1. The accuracy of this simulation process depends to a high extent on the used water-retention and relative permeability curve. The assumption that the curves are unique causes some deviations as the soil structure keep on changing depending on the level of the seepage forces. On the other-hand, any infiltration by sprinkling cause air entrapment in the pore spaces in soil layer, which offer resistance to flow. These two reasons might be the cause of time lag observed above.

2. The infiltration behaviors of an artificial slope are highly influenced by the size, shape of the model, and anisotropic due to compaction procedures.

3. Numerical simulation is faced by a problem of similarity in physical and mechanical properties.

4. That the failure pattern due to piping can be predicted by reducing the cohesion at saturation with the decrease of the mean stress.

Although the above laboratory infiltration and failure model differ to some extent from the actual field behaviors, agreement between observed and predicted mechanisms at laboratory scale can assure the use of computer simulation at field level, especially to complicated problems involving soil stratification and variations in hydraulic conductivity. Such types of simple and cheap Laboratory simulation models should be encouraged in places where big and sophisticated centrifugal models can not be obtained.

REFERENCES

Akai, K., et al. 1979. Finite element analysis of three-dimensional flows in saturated-unsaturated soils. *The 3rd International Conference on Numerical Methods in Geomechanics*, Aachen: 227-239.

Freeze, R.A. 1971. Three-dimensional, transient, saturated-unsaturated flow in a groundwater basin. *Journal of Water Resource Research, Vol. 7, No.2*: 347-366.

Fukuoka, M., et al. 1977. Earth pressure measurement on retaining Walls. *Proc. of the 9th ICSMFE, Vol. Case History*: 267-270.

Iseda, T. and Tanahashi, Y. 1973. The finite element analysis for the successive failures in natural slopes containing a weak plane. *Journal of Faculty of Engineering, Nagasaki University, Vol. 4*: 81-89 (in Japanese).

Iseda, T., et al. 1985. A study on the gentle slope failure at the Nagasaki heavy rainfall. *Journal of JSSMFE, Vol. 25, No. 2*: 173-184 (in Japanese).

Kita, D., et al. 1969. Geotechnical study on ashy soil in Kyushu district and its lime stabilization. *Conference Papers on Aso Volcanic Ash Soil by the Ministry of Construction, Kyushu Branch*: 69-81 (in Japanese).

Kudara, K. 1983. Study on failure of embankment structures and its counter-measures. *A Doctor thesis submitted to Kyushu University for the Degree of Doctor of Science in Engineering* (in Japanese).

Kuno, G. 1974. Problematic soils in Japan. *Volcanic Ash Soils Committee Series, Chapter 2: Volcanic Ash soils*: 50-51 (in Japanese).

Kuroki, Y. et al. 1983. *Problematic Soils in Kyushu Island. Chapter 11: Haido*: 135-136 (in Japanese).

Murata, S. 1989. Fundamental study on decrease of strength of soils caused by seepage and slope failure due to rainfall. *A Doctor thesis submitted to Kyushu University for the Degree of Doctor of Science in Engineering*: 120-134 (in Japanese).

Neuman, S.P. 1973. Saturated-unsaturated seepage by finite elements. *ASCE, HY12*: 2233-2250.

Phillip, J. 1956. Theory of infiltration: 4. Sorptivity and algebraic infiltration equations. *Soil Science, Nov. 1956*: 257-265.

Singh, R. 1967. Solution of a diffusion equation. *ASCE HY5, Paper 5422*: 43-50.

Van Genuchten, M. 1980. A closed-form equation for predicting the hydraulic conductivity of unsaturated soils. *Journal of Soil Science Society of America, Vol. 44*: 892-898.

Developments in Geotechnical Aspects of Embankments, Excavations and Buried Structures, Balasubramaniam et al. (eds)
© *1991 Balkema, Rotterdam. ISBN 90 5410 019 2*

Time-dependent deformation of anchored excavation in soft clay

S.L. Lee, U. Parnploy, K.Y. Yong & F.H. Lee
National University of Singapore, Singapore

ABSTRACT: Excessive movement of excavation support system may result in construction delays and damages to underground services, roads and near-by buildings. It was observed in the field that after excavation, the support system continue to deform even though there was no further excavation. In this paper, the time-dependent deformation of anchored excavation in soft clay was analysed by using a fully coupled consolidation analysis with elasto-plastic soil model. Computed deformation of the excavation support system was found to be in close agreement with field performance of an anchored sheet pile wall which was carefully monitored during excavation.

1 INTRODUCTION

In Singapore, most excavations in deep deposits of soft clay are supported by sheet piles braced with internal struts or ground anchors. Mana (1978) and Clough & Schmidt (1981) highlighted the various aspects which must be considered in the analysis of deep excavations. These include simulation of excavation sequence, excavation support system and sequence of installation of struts. Field data from such excavations indicate that ground movements and strut loads tend to change with time (Lee et al, 1986). This phenomenon may be attributed to the expansion of the retained soil and the soil below excavation level arising from the dissipation of excess pore pressures.

In this paper an example is presented to describe the analysis of time-dependent deformation of an anchored excavation in soft clay. The effect of excess negative pore pressures generated during soil removal is taken into consideration by using a fully couple Biot consolidation analysis while the soil is assumed to be an elastic-perfectly plastic material obeying the Mohr-Coulomb yield criterion.

2 SITE AND GROUND CONDITIONS

A typical section of the anchored excavation support system is shown in Fig. 1. The excavation was carried out to a depth of 15 m below the river bed.

The subsoil beneath an existing river consists basically of a thick layer of marine clay followed by medium stiff fluvial clay and highly weathered rock of sedimentary origin; the degree of weathering generally decreasing with increasing depth.

Sheet piles (TradeArbed HZ 775A with section modulus of 7960 cm^3/m) were driven to refusal into the weathered sedimentary rock (S4). To enhance stability of the sheet pile against kick-out, toe pins were installed at 2.065 m intervals. The toe pin consists of the steel H-pile inserted into a prebored hole through the double H pile section and then grouted with concrete to integrate with the TradeArbed sheet pile arrangement. Five levels of anchors were installed with lateral spacing of 2.035 m.

After water was pumped out from the cofferdam over a period of one week, equipment was mobilized and the first level of anchors was installed in about 88 days after commencement of pumping operation. The number of days shown in Fig. 1 indicates the time from commencement of pumping operation.

3 ANALYSIS OF ANCHORED EXCAVATION WITH TIME-DEPENDENT BEHAVIOR

3.1 Governing equations

An elastic-perfectly plastic model using the isotopic Mohr-Coulomb yield criteria

241

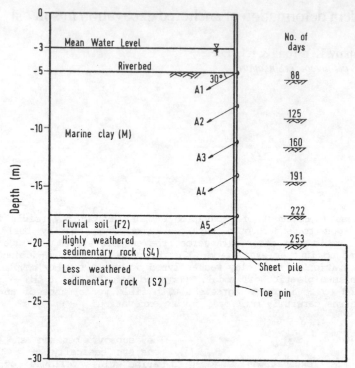

Fig. 1 Soil Profile and Anchoring System

with associated flow rule (Nayak & Zienkiewicz, 1972) is used to describe soil behaviour in the analysis. With this model, the governing equation for consolidation analysis was formulated based on the usual assumptions of saturated soil, principle of effective stress, incompressibility of soil solids and pore water, validity of Darcy's law, and small deformations. Finite element formulation for consolidation in soils had been developed by Sandhu & Wilson (1969) and Britto & Gunn (1987), amongst others. In this study, Britto & Gunn's procedure, which utilises a backward difference time marching scheme, was adopted. With this procedure, the unknown displacement and pore pressure increments at each time step was computed using the following relationship:

$$\begin{bmatrix} K & L \\ L^T & -H\Delta t \end{bmatrix} \begin{bmatrix} \Delta u \\ \Delta p \end{bmatrix} = \begin{bmatrix} \Delta f \\ H \Delta t \ p_o \end{bmatrix} \quad (1)$$

where K = the elasto-plastic stiffness matrix; L = coupling matrix; H = flow matrix; Δu, Δp = unknown displacement and pore pressure increments at the current time step; Δt = size of time step; p_o = excess pore pressure at the end of the previous time step and; Δf = nodal load increments in the current time step.

The initial stiffness algorithm (Nayak & Zienkiewicz, 1972), was used to solve for the unknown displacement and pore pressure increments. With this algorithm, K is replaced by the elastic stiffness matrix and the effects of material nonlinearity added to the load increment vector Δf as an out-of-balance force vector. The solution process was then repeated with the updated load increment vector until the out-of-balance forces are less than a prescribed magnitude.

3.2 Geometric and boundary conditions

The finite element mesh and boundary conditions used in the analysis are shown in Fig. 2. The soil and sheet pile are represented by 8-noded quadrilateral elements and the soil properties obtained from soil investigation are shown in Table 1. The equivalent Young's modulus for the sheet pile wall was obtained by matching the flexural stiffness of the sheet pile

242

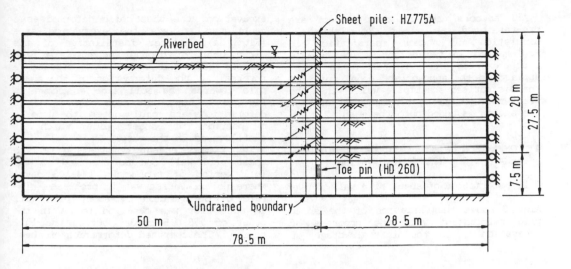

Fig. 2 Finite Element Mesh and Boundary Conditions

Table 1 Soil properties used in analysis

Soil Type	γ (kN/m³)	ϕ'	c' (kN/m²)	Ko	ν'	E' (kN/m²)	k_v (m/sec)	k_h (m/sec)
M	16	22°	0	0.625	0.3	4,333	2×10^{-8}	6×10^{-8}
F2	18	30°	0	0.55	0.25	5,833	1×10^{-7}	1×10^{-7}
S4	20	30°	30	0.80	0.25	90,000	2×10^{-8}	2×10^{-8}
S2	20	40°	100	0.80	0.25	150,000	1×10^{-8}	1×10^{-8}

element to that of the actual sheet pile. In the analysis, the sheet pile wall is considered to be impervious and this is simulated by assigning a much lower coefficient of permeability to the sheet pile elements than the soil elements.

Ground anchors are represented by springs with equivalent stiffnesses, inclined at 30 degrees to the horizontal and fixed at one end. The preload and stiffness of ground anchors used in the analysis are shown in Table 2.

3.3 Construction aspects

Fig. 1 shows the excavation and anchoring sequence of the excavation works. Excavation is simulated by removal of the excavated soil elements from the mesh and application of nodal forces arising from the release of overburden and lateral stresses. The nodal forces are evaluated using the following expression given by Chow (1984)

$$\Delta f = \sum_{i=1}^{m} \left[-\int B^T \sigma \, d(vol) + \int N^T \Delta W \, d(vol) \right]_i$$

(2)

in which Δf = the vector of equivalent nodal forces; B = displacement-strain matrix; σ = the total stress vector; N = element shape function matrix; ΔW = body force vector; and m = the number of excavated elements in the excavation stage.

243

All anchors used in this project are pre-tensioned as shown in Table 2. Simulation of anchor installation is implemented in two stages. The tensile force exerted by the anchors is first applied to the appropriate nodal point of the discretised sheet pile wall. Then the pre-tensioned spring element representing the anchor is added onto the mesh.

4 COMPARISON OF MONITORED AND COMPUTED BEHAVIOR OF ANCHORED EXCAVATION

4.1 Deformation of sheet pile wall

Fig. 3 shows the variation of the sheet pile deformation with the progress of excavation. In the initial stage of excavation, the sheet pile deformed as a cantilever towards the excavation site (Fig. 3a). After installation of the first level of ground anchors, the high preload had the effect of causing a reversal in the deformation of the sheet pile towards the soil side as shown in Fig. 3b. The high magnitude of these deformations are mainly due to the upper 5 m portion of sheet pile which is cantilevered in the water.

The time-dependent deformation of the sheet pile wall can be observed in the full period of two weeks (94-108 days) where no excavation was in progress. The rate of deformation of sheet pile wall in soft clay was observed to be in the order of 0.3-0.5 mm/day during the lull period. The computed deformation of sheet

Table 2 Ground anchor properties

Anchor	WL (kN)	PL (kN)	Stiffness* (kN-m)	Remarks
A1	600	500	6000	WL = Working load
A2	1250	900	8000	PL = Preload
A3	1250	1200	10000	*Anchor stiffness estimated from proof load tests on similar soil in the vicinity
A4	1250	1200	12000	
A5	1250	1200	14000	

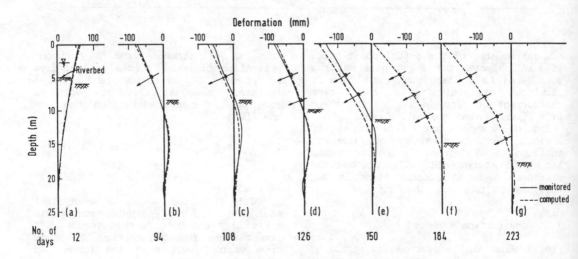

Fig. 3 Time-dependent Deformation of Anchored Sheet Pile Wall

pile wall based on consolidation analysis was found to be in close agreement with the monitored deformation, indicating that the excess negative pore pressures due to excavation process can account for much of the variation of sheet pile wall with time. From Fig. 3e-f-g (there is no monitored values at 184 days and 223 days), it is apparent that the change in deformation of the sheet pile wall is not significant after the installation of the third level of anchors at elevation of -11m. This indicates the importance of preloading in reducing the deformation of the retaining sheet pile wall at deeper stages of excavation.

Fig. 4 shows a comparison of undrained analysis and consolidation analysis at each stage of excavation. It can be seen that the undrained analysis significantly underestimates the deformation of the sheet pile at every stage of excavation.

4.2 Forces in ground anchors

Table 3 shows the computed anchor forces in each stage of excavation by the above consolidation analysis. The forces in the first and the second level of anchors significantly decrease after preload with

Table 3 Computed forces in anchors

No. of days Anchor	Anchor forces (kN/m)					Max. force (kN/m)
	125	160	191	222	253	
A1	313	129	4	0	0	313
A2		436	288	192	150	436
A3			580	465	444	580
A4				561	557	561
A5					598	598

absence of lateral pressure acting on the portion of the sheet pile wall above the river bed. The forces in ground anchors at 3th, 4th and 5th level behave differently from the upper two levels. As the excavation progressed, the deformation associated with the preloading in the lower three levels are small and there is only a slight decrease in the anchor forces after preload.

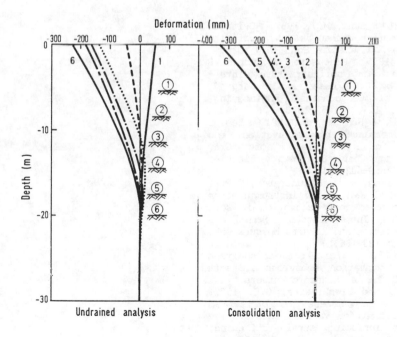

Fig. 4 Comparison of Undrained and Consolidation Analyses

245

5 CONCLUSION

In the present analysis of anchored excavation in soft clay, the close agreement between monitored deformation and the computed results based on consolidation analysis indicates that the excess negative pore pressure due to excavation process can account for much of the variation of sheet pile deformation with time. The observed deformation of the excavation support system in soft clay increases as excavation progresses. Even when there was no excavation, sheet pile deformation of the order of 0.3 to 0.5 mm/day was observed. The small change in the anchor forces coupled with the small deformation of the sheet pile wall during excavation indicates the significance of preloading of ground anchors.

ACKNOWLEDGEMENTS

The research upon which this paper is based is funded in part by the Science Council of Singapore under RDAS Grant No. ST/86/05. The award of a National University of Singapore Scholarship is gratefully acknowledged by the second author.

REFERENCE

Britto, A.M. & Gunn, M.J. 1987. Critical state soil mechanics via finite elements, John Wiley & Sons, New York.

Chow, Y.K. 1985. Discussion on Hybrid FE procedure for soil-structure interaction, by C.S. Desai and S.Sargand, Journal of Geotechnical Engineering, ASCE, Vol. 111, 8:1057-1060.

Clough, G.W. & Schmidt, B. 1981. Design and performance of excavation and tunnels in soft clay. Soft Clay Engineering, Chap. 8, Americ. Elsevier Pub. Co.: 569-602.

Lee, S.L., Yong, K.Y., Karunaratne, G.P. & Chua, L.H. 1986. Field instrumentation for a strutted deep excavation in soft clay, 4th Int. Geotech. Sem.-Field Instrumentation and In-situ Measurement, Singapore: 183-186.

Mana, A.I. 1978. Finite element analyses of deep excavation behaviour in soft clay, PhD Thesis, Stanford University.

Nayak, G.C. and Zienkiewicz, O.C. 1972. Elasto-plastic stress analysis. A generalization for various constitutive relations including strain softening, Int. J. Num. Methods in Eng. 5: 113-135.

Sandhu, R.S. and Wilson, E.L. 1969. Finite element analysis of seepage in elastic media. J.Eng.Mech.Div., ASCE, 95(EM3): 641-652.

Developments in Geotechnical Aspects of Embankments, Excavations and Buried Structures, Balasubramaniam et al. (eds)
© *1991 Balkema, Rotterdam. ISBN 90 5410 019 2*

Evaluation of earth pressure on retaining walls

P.J. Moore
University of Melbourne, Vic., Australia

ABSTRACT: A number of factors which affect the evaluation of lateral earth pressure and retaining wall design was reviewed. The factors include the stiffness of the underlying foundation soil, the use of soft clay backfills, cyclic vertical wall movements, repeated surcharge loading on the backfill surface. The influences of wall geometry and choice of angle of wall friction on wall safety with respect to overturning and sliding have also been examined.

1 INTRODUCTION

When rigid walls are called upon to retain soft clay backfill, the use of the conventional active pressure expressions for cohesive soils are sometimes found to be deficient because of creep effects in the backfill material. A number of experimental studies have been carried out in relation to this problem. The results suggest that while the short term active pressures are amenable to calculation by conventional procedures, the long term earth pressures are not.

The development of active or passive pressure in terms of the amount of lateral wall movement has been explored by a number of workers but there has been relatively little attention paid to the effects of repeated or cyclic wall movements. Practical situations where cyclic movements may occur include navigation locks and dry docks where dominantly upward or downward movement may take place following emptying or filling of the structure. Several field observations of earth pressure acting on such structures have been reported, notably those from sites in the Soviet Union. Design for at-rest earth pressures on these structures is a widely accepted procedure but some of the field observations and an experimental simulation of the wall movements have demonstrated that this is not always correct.

Cyclic loading effects of a different type may occur when surcharge loads are repeatedly applied and removed. These effects have been explored theoretically and experimentally. Some of the con-tributions in this area relate to a single load application only of the surcharge. Further earth pressure increases may occur when cyclic loading is applied and their evaluation by adapting techniques originally proposed for compaction induced earth pressure appears promising.

Also included in this review are the results of some calculations relating to the assessment of safety of rigid retaining walls to overturning and sliding. The effect of a number of factors including wall geometry, wall friction angle and shearing resistance of the backfill, on the calculated factor of safety have been examined.

2 ACTIVE PRESSURE WITH COHESIONLESS BACKFILL

For a dry cohesionless backfill the resultant active force (P_A) acting on the back of a retaining wall may be given by the familiar expression

$$P_A = \tfrac{1}{2}\,\gamma H^2 K_A \qquad (1)$$

where H is the wall height and $\gamma (= \rho g)$ in the unit weight of the backfill soil.

The Coulomb expression is widely used for the active coefficient (K_A) and for a horizontal backfill it simplifies to

$$K_A = (\frac{\sin(\beta-\phi)\sin\beta}{(\sin(\beta+\delta))^{\frac{1}{2}} + (\sin(\phi+\delta)\sin\phi/\sin\beta)^{\frac{1}{2}}})^2 \qquad (2)$$

the symbols being given in the top left corner sketch of Fig. 1.

An alternative technique (Case A) that is used for the determination of the force

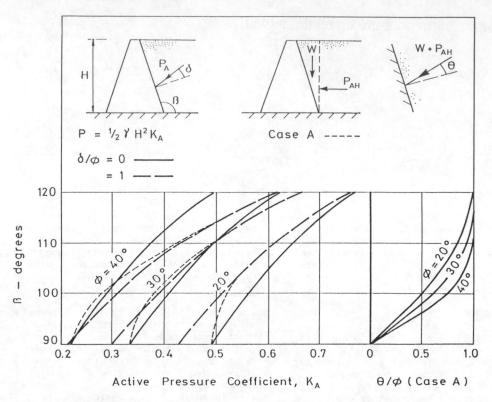

Fig. 1 Factors affecting the Active Pressure Coefficient

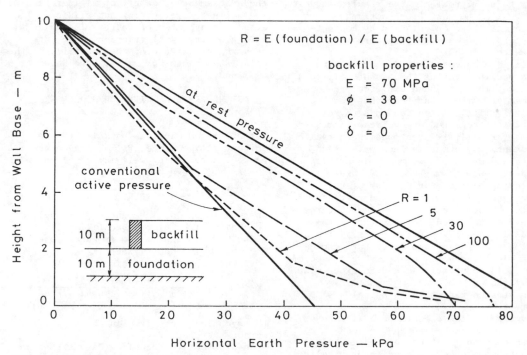

Fig. 2 Effect of Foundation Stiffness on Earth Pressure – Cohesionless Backfill

acting on the back of the wall is illustrated in the sketch in the top right corner of Fig.1. The resultant of the triangular soil weight (W) and the active force (P_{AH}) acting horizontally on the vertical plane through the heel of the wall is evaluated and is expressed in the form of equation (1). This facilitates the calculation of an equivalent K_A value. As shown in Fig. 1 these K_A values are not the same as those values calculated from the Coulomb expression but for large β values they tend to equality with the Coulomb value for $\delta/\phi = 1$. No wall friction angle is included in the Case A calculation but the resultant of W and P_{AH} acts at an angle (θ) to the normal to the back of the wall. As shown in Fig.1 this θ angle does not correspond with the angle of wall friction (δ) in the Coulomb calculation although θ/ϕ does tend towards unity at large values of the angle β.

According to the Coulomb approach, the foundation soil or rock lying beneath the retaining wall and backfill has no influence on the calculated active earth pressures. By means of finite element analyses, Goh (1984) has demonstrated that this is not correct. He examined the pressures developed behind a 10m high wall which was located on a 10m thick layer of foundation soil. The backfill soil properties were the same as those obtained by test on a real soil. As shown in Fig. 2 the pressures developed behind a smooth wall varied between the at-rest pressure and the conventional active pressure depending on the relative stiffness (R) of the foundation soil to the backfill soil. Only when the stiffness of the foundation and backfill soils were approximately equal did the pressure distribution approach that of the conventional active pressure. For stiffer foundation material the earth pressures increase and for a very stiff foundation (R=100) they almost equal the at-rest values. This means that as the foundation material becomes stiffer the earth pressure coefficient (K_A) becomes progressively larger than that given by eqn. (2). For the case of a rough wall similar conclusions regarding the effect of foundation soil stiffness on lateral earth pressure apply.

3 SOFT CLAY BACKFILL

The preceding discussion related to granular backfill only, and while this is preferred for the backfill material circumstances arise where clay soils have to be retained by the wall. For the design of rigid walls retaining cohesive backfills

the Civil Engineering Code of Practice (1951) expressions have been widely used. These expressions were developed for vertical walls with horizontal backfill surfaces. For a saturated clay for which the undrained friction angle ϕ_u is zero, the normal component of the active pressure at a depth of z below the backfill surface has been shown by Packshaw (1946) to be

$$p_a = \sigma_v - 2c_u(1+(c_w/c_u))^{\frac{1}{2}} \qquad (3)$$

in which σ_v = total vertical stress at depth z due to weight of backfill and surcharge; c_u = undrained cohesion; and c_w = wall adhesion.

Equation (3) is used for determination of the short term active pressure distribution. For calculation of the normal component of the long term total active pressure the drained strength parameters c_d and ϕ_d are used in conjunction with the following expression:

$$p_a = \sigma_v' K_a - c_d K_{ac} + u \qquad (4)$$

in which σ_v' = effective vertical stress at depth z; K_a, K_{ac} = coefficients which depend upon the wall friction, wall adhesion and ϕ_d; and u = hydrostatic water pressure at depth z. Equations (3) and (4) are used in situations in which the wall is permitted to yield laterally so that the strength of the backfill soil may be fully mobilized. Where yielding of the wall is prevented many workers have recommended that the retaining wall should be designed to resist the at-rest lateral earth pressure p_o, as given by the following expression

$$p_o = K_o' \sigma_v' + u \qquad (5)$$

in which K_o' = ratio of horizontal to vertical effective stresses. In support of this concept Taylor (1948) considered that the shearing stresses within a clay backfill undergo a slow relaxation so that as the mobilized shear stress decreases the lateral pressure must gradually increase to maintain equilibrium. This lateral pressure increases ultimately to the at-rest value. Taylor, however, did not propose any theoretical or experimental techniques for determination of the coefficient or earth pressure at rest. Terzaghi and Peck (1967) have suggested a value of unity for the coefficient. For many soft clays this value is almost certainly overconservative whereas for over-consolidated soils it may be unconservative. Tschebotarioff (1949) and

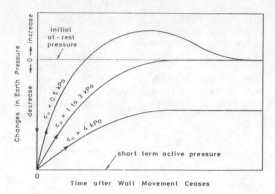

Fig. 3 Effect of Clay Strength on Earth
Pressure Variations

retained soils were saturated kaolin clays
with water contents varying from 80% to
150% and undrained strengths varying from
2.7kPa to 0.4kPa. For backfills with
water contents greater than about 140% decreases in lateral pressures following
wall movements away from the backfill were
not observed. For backfill water content
around 100% (c_u less than about 0.5kPa)
both the lateral pressure and pore pressure
decreased the instant the wall was moved
outward (Fig.3). The magnitude of these
pressure variations was found to depend
upon the magnitude and type of wall displacement. When wall movement was stopped
the lateral pressure and pore pressure
rapidly increased and exceeded the initial
premovement values. Ultimately the pore
pressure and lateral pressure decreased
to the same magnitudes that existed
before the wall was moved.

Vidmar(1963) have conducted model retaining wall tests which have yielded
results that generally confirm Taylor's
shear stress relaxation concept.
Tschebotarioff found at-rest earth
pressure coefficients significantly less
than unity for plastic clays. Unfortunately, Vidmar's experiments were not sufficiently exhaustive to enable a calculation
procedure for lateral pressures from clay
backfills to be recommended. Moore and
Spencer (1972) have described a series of
rigid retaining wall tests in which the

For lower backfill water contents (c_u
values about 1 to 3kPa), the pore pressure and lateral pressure decreased as the
wall was moved in much the same way as in
the previous case ($c_u < 0.5$kPa). After
the wall movement was stopped the pore
pressure and lateral pressure increased
gradually and finally levelled off at the
magnitudes that existed before the wall
was moved (Fig.3).

As illustrated in Fig.4 the conventional active pressure distribution (eqn.3)
provides a reasonable estimate of the minimum lateral pressure reached when the wall

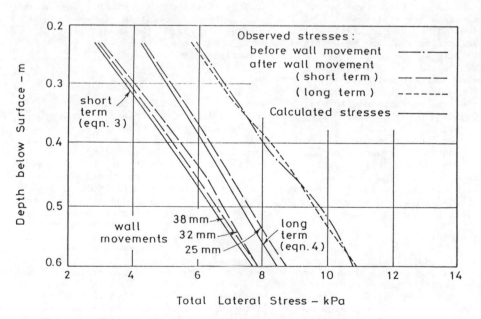

Fig. 4 Typical Earth Pressures in Soft Clay

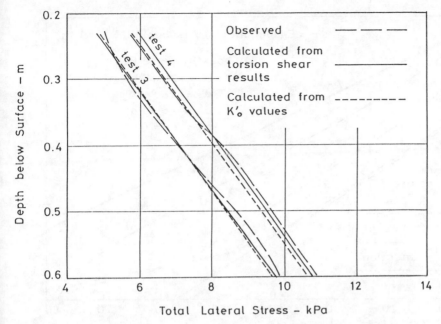

Fig. 5 Typical Long Term Earth Pressures

is moved outwards by a distance approximately equal to 5% of the wall height.
Following the cessation of wall movement
the lateral pressures increase back to the
values existing before wall movement was
initiated. These long term pressures cannot be predicted by means of the conventional equation (4). As shown in Fig.5 the
long term earth pressures can be predicted
by means of equation (5) or from strength
data acquired from special torsion shear
tests as described by Moore and Spencer
(1972).

Retaining wall tests with lower water
content and higher strength (> 4kPa) saturated kaolin clay backfills have been
performed by Moore and Cole (1977). In
some of these tests the outward movement was controlled at a slow (1.7mm/hr.)
rate prior to cessation of wall movement.
Typically as wall movement commenced the
lateral earth pressures decreased from the
at-rest values but did not reach the low
(active) values achieved in tests with
very rapid wall movement. The short term
earth pressures immediately following
rapid wall movement were reasonably predicted by equation (3). However, after
cessation of the wall movement (either
rapid or slow) the earth pressures (see
Fig.3) rose gradually (rapid wall movement
tests)or remained almost unchanged (slow
wall movement case at values less than the
at-rest values. Attempts to predict these
long term earth pressures were not

particularly successful.

When a cohesive backfill is underlain
by a foundation soil, the conventionally
calculated active pressure may significantly underestimate the earth pressure
acting on the wall when foundation soil
stiffness is taken into account. As shown
in Fig.6 for a smooth wall, Goh and Donald
(1984) have demonstrated from finite element analyses, that the earth pressure
approaches the conventionally calculated
distribution only when the stiffnesses of
the backfill and foundation soil are equal.
For stiffer foundation soils the earth
pressure increases and the depth of tension
crack significantly decreases.

4   EFFECT OF CYCLIC VERTICAL WALL
    MOVEMENT

The walls of navigation locks and dry dock
structures undergo cyclic vertical movement as the structure is filled and
emptied. Many of these structures are
built with walls that vary in thickness,
the thin section being at the top. This
means that vertical movement is accompanied by a small horizontal movement of the
wall relative to the backfill soil. There
have been many reported observations of
cyclically varying earth pressures that
develop behind walls of navigation locks
and dry dock structures. For example,
Sinyavskaya and Pavlova (1971) observed
two types of cyclical earth pressure

251

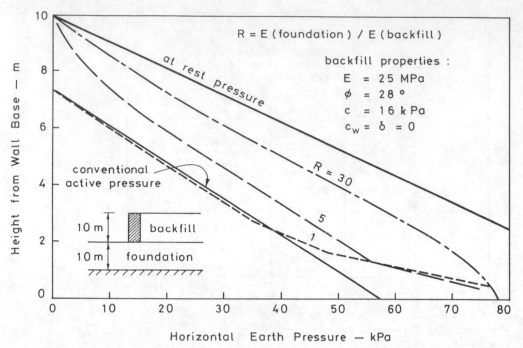

Fig. 6   Effect of Foundation Stiffness on Earth Pressure - Cohesive Backfill

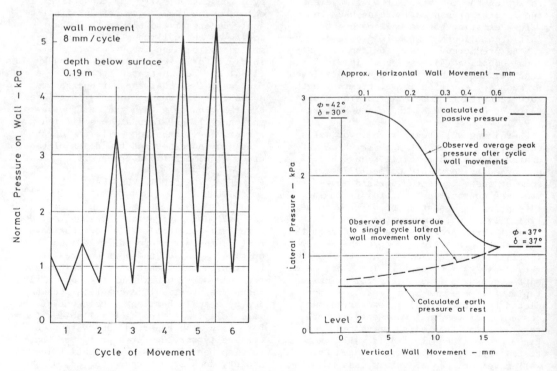

Fig. 7   Typical Earth Pressure Response to Cyclic Wall Movement

Fig. 8   Effect of Magnitude of Cyclic Wall Movement on Earth Pressure

changes against the wall of the Volgograd lock over a ten year observation period. One type was caused by seasonal temperature changes with the maximum earth pressure developing during the summer and caused by wall movements towards the backfill of the order of 2mm. The other type was caused by emptying and filling of the lock, with the maximum earth pressure caused by filling, which produced wall movement toward the backfill also of the order of 2mm. The observed peak earth pressures exceeded the design pressures (earth pressure at-rest) by a factor of more than 3.

From a study of locks on the Volga and Moscow canals Mikhailov and Avdeeva (1973) found that the lateral earth pressure increased in winter up to almost twice the design pressure when the lock was emptied, in contrast to observations elsewhere and in spite of the wall deflecting away from the backfill. This phenomenon of a pressure increase in winter when the lock was emptied was also noted by Moshkov (1974) on the Votkinsk lock. He also observed that emptying the lock caused the structure to rise (movements 4-7mm) and he suspected that this vertical movement was the main reason for the rise in earth pressure. Moshkov found that the observed winter pressures were of the same order as the passive pressures calculated for an upward moving wall by means of the Muller-Breslau equation (which is the same as the Coulomb equation) which gives the passive pressure coefficient (K_p) for horizontal backfill and vertical wall as follows

$$K_p = [\frac{\cos \phi}{(\cos \delta)^{\frac{1}{2}} - (\sin(\phi-\delta)\sin\phi)^{\frac{1}{2}}}]^2 \quad (6)$$

These wall movements were simulated by Moore and Choong (1986) in a laboratory study in which the effects of vertical wall movement with a small horizontal component were observed. They found that the peak earth pressure increased rapidly during the first few cycles of movement as shown in Fig.7, but then exhibited much less variation for successive cycles of movement. The average peak pressure (after about ten cycles of movement) was found to decrease as the amplitude of wall movement increased from 3mm to 16mm (Fig.8). The observed pressures were roughly in agreement with calculated passive pressures based on the Coulomb equation provided the angle of shearing resistance (ϕ) and angle of wall friction (δ) mobilized at small and large displacements respectively were used in the calculations.

5 REPEATED SURCHARGE LOADING

For the evaluation of the horizontal stress distribution on the vertical back of a retaining structure when surcharge loads are applied to the horizontal backfill surface, several procedures are in current use. These include the theory of elasticity, the empirical equations proposed by Terzaghi (1954) and the empirical equations proposed by Rowe (1950). The Rowe procedure which may not be as well known as the others, relates to pressures generated by loaded strip or square footings placed immediately adjacent to the top of a stiff wall. The equations he proposed are

$$K = K_A((5-(z/B))/4)^3 \quad (7)$$

and

$$K = K_A((3.5-(z/B))/3)^2 \quad (8)$$

for strip and square footings respectively, where

z = depth below the top of the wall
B = footing width
K_A = active pressure coefficient

and the horizontal stress (σ_h) is calculated from the vertical stress beneath the footing (σ_v) by the expression

$$\sigma_h = K \sigma_v \quad (9)$$

The effects on lateral earth pressure of repeated vertical loading on a rigid square footing placed at various locations on the backfill surface were explored experimentally by Moore and Chin (1985). The lateral earth pressures were found to increase to a maximum value with increasing number of load cycles. The more highly restrained walls exhibited the greater earth pressures. The magnitudes of the loaded (for footing load in place) and residual (for footing unloaded) pressures increased
(i)   as the distance between the surcharge load and the wall decreased, and
(ii) as the flexibility of wall decreased.

For the footing placed immediately adjacent to the wall, elastic theory was found to significantly underestimate the observed lateral pressures on the wall. As shown in Fig.9(a), the empirical methods of Terzaghi and Rowe gave pressures closer to those observed but the level of agreement was not good. The Rowe distribution provided a convenient envelope to the observed net loaded pressures when the K_o value was used in the calculation instead of K_A.

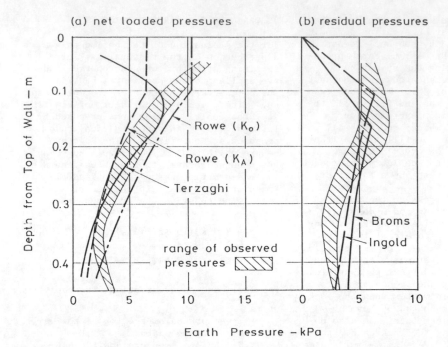

(a) net loaded pressures   (b) residual pressures

Depth from Top of Wall — m

Rowe (K_o)

Rowe (K_A)

Terzaghi

range of observed pressures

Broms

Ingold

Earth Pressure — kPa

Fig. 9   Typical Observed and Calculated Pressures after Repeated Surcharge Loading

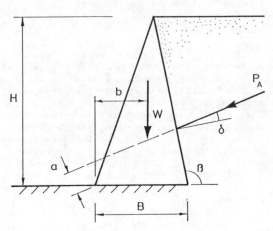

Fig. 10   Definition Sketch

The Broms (1971) and Ingold (1979) techniques, which were developed for the calculation of pressures generated by backfill compaction, were used here for the determination of residual pressures on the wall following removal of the footing load.   The comparison in Fig.9(b) shows that both the Broms and Ingold predictions tend to underestimate the observ-ed pressures at shallow depths and over-estimate the observed pressures at greater depths.

6   FACTORS AFFECTING DESIGN CALCULATIONS

6.1 Overturning

One of the common assessments carried out in retaining wall design is that relating to overturning.   The usual calculation for the overturning factor of safety (F_O) involves taking moments about the toe of the wall.   Referring to the definition sketch (Fig.10), the F_O is given by

$$F_O = W\, b / P_A\, a \qquad (10)$$

For representative values of backfill friction angle (ϕ) and wall geometry (B/H), Fig.11 illustrates the effects of angle of wall friction (δ) and slope of the back of the wall (β) on the overturning factor of safety.   Minimum acceptable values of F_O are normally 1.5 or 2.0 and Fig.11 shows that these values are achieved in most practical cases. F_O values lower than 1.5 can occur however when the base width (B) is small in relation to wall height (H) and very low values of wall friction (δ) apply (see Fig.12).

254

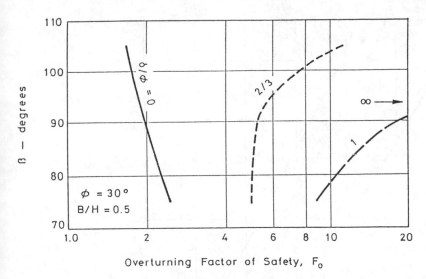

Fig. 11 Effect of Back Slope on Overturning

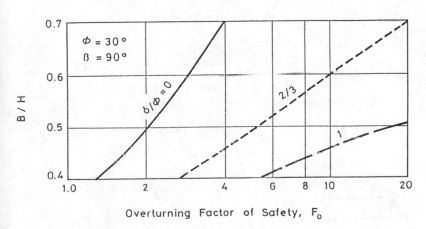

Fig. 12 Effect of the Wall Geometry on Overturning

Some of the disadvantages of using this approach for the evaluation of overturning are that negative factors of safety sometimes appear and different calculated values for F_0 can be found if formulations other than equation (10) are used for the definition of the overturning factor of safety. Because of this, many designers prefer to provide assurance against overturning by requiring the resultant (R) of W and P_A (see Fig.13) to fall within the middle third of the base.

Fig.13 illustrates the effect of inclination of the back of the wall (β) on the eccentricity of the resultant. Figs. 13 and 14 show that problems with the resultant falling outside the middle third

may arise with large values of the angle (β), small values of the angle of wall friction and small values of the B/H ratio. Fig.15 indicates that the wall shape, as measured by (W/H^2) (W is the wall weight per unit length), does not have a large effect on the eccentricity of the resultant. As shown in Fig.16, the resultant may fall outside the middle third if both the backfill friction angle (ϕ) and the angle of wall friction (δ) are low.

In summary, the safety of a rigid retaining wall with respect to overturning is more easily satisfied if

(i)   the backface of the wall leans towards the backfill rather than away from it,

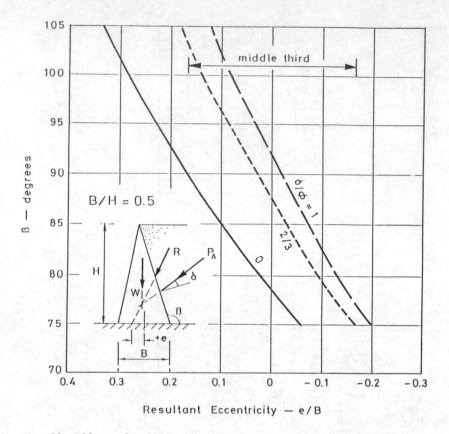

Fig. 13 Effect of Wall Inclination on Eccentricity of Resultant

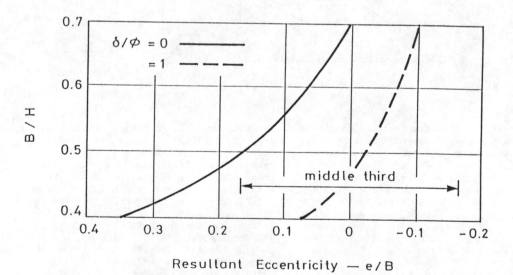

Fig. 14 Effect on Wall Geometry on Eccentricity of Resultant

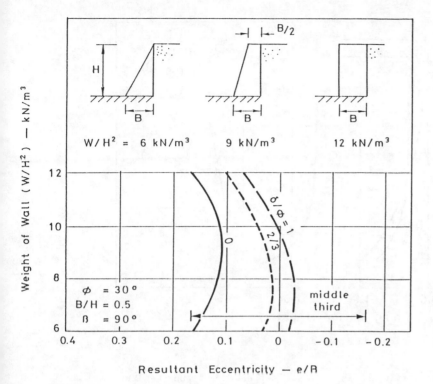

Fig. 15 Effect on Wall Shape on Eccentricity of Resultant.

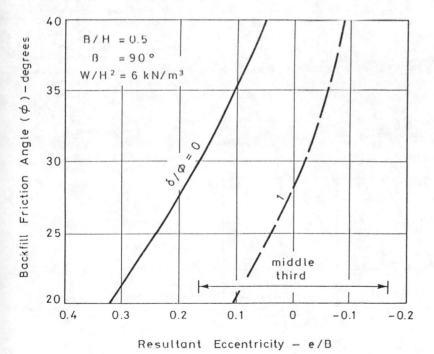

Fig. 16 Effect of Friction Angle on Eccentricity of Resultant.

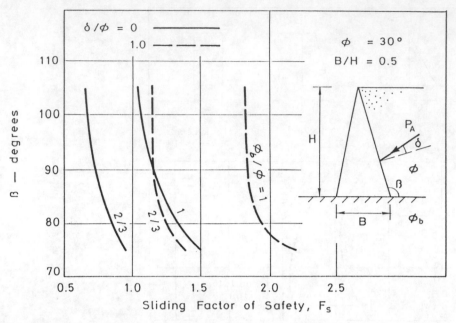

Fig. 17 Effect of Back Slope on Sliding

(ii) the angle of wall friction is relatively large,
(iii) the angle of shearing resistance of
the backfill material is large, or
(iv) the wall base width is large in relation to the wall height.

6.2 Horizontal Sliding
  For the case of a retaining wall with a
planar base founded on a cohesionless
material the factor of safety for sliding
(F_S) may be expressed as

$$F_S = R_V \tan \phi_b / R_H \qquad (11)$$

where R_V, R_H = vertical and horizontal

components of the resultant R, ϕ_b =
friction angle for the foundation soil
  Minimum acceptable values of F_S of
either 1.5 or 2.0 are widely used in design. Fig.17 shows that less than acceptable values of F_S may be obtained with
large values of the angle β, low values of
the friction angle ϕ_b and low angles of
wall friction (δ), Figs.18 and 19 support
some of these findings and also show that
the F_S values fall as (B/H) or the wall
weight (W/H^2) decreases. The sliding
factor of safety is significantly dependent on the backfill friction angle, Fig.
20 showing that F_S decreases as ϕ
decreases.

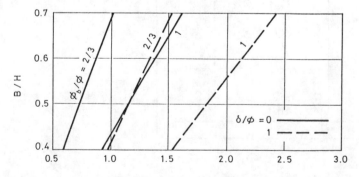

Fig. 18 Effect of Wall Geometry on Sliding

258

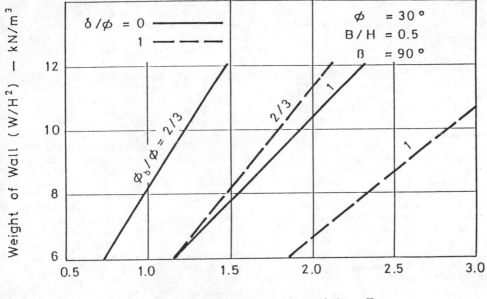

Fig. 19 Effect of Wall Shape on Sliding

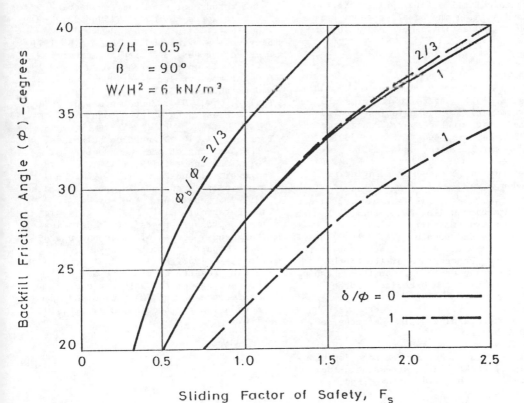

Fig. 20 Effect of Friction Angle on Sliding

7 CONCLUDING COMMENTS

A number of factors which may affect earth pressure and retaining wall design have been reviewed. The findings from this review may be summarised as follows:

a) With either cohesionless or cohesive backfill,the presence of an underlying foundation soil may significantly affect the lateral pressures acting on a retaining wall.

b) With soft clay backfills, the conventional expressions for earth pressure are acceptable for evaluation of short term earth pressure but are not acceptable for long term earth pressure.

c) When retaining walls are subjected to cyclic vertical and lateral (towards the backfill) movement relative to the backfill, the earth pressure may be evaluated by conventional passive pressure considerations provided due allowance is made for the mobilized wall friction.

d)The use of Broms and Ingold approaches for evaluating lateral earth pressure from repeated surcharge loading may lead to overestimation or underestimation of the earth pressure.

e)Retaining wall safety against overturning and sliding may be affected by wall geometry and estimates of wall friction angle.

REFERENCES

Broms, B. 1971. Lateral earth pressure due to compaction of cohesionless soils. Proceedings of the Fourth Conference on Soil Mechanics, Budapest: 373-384.

Earth retaining structures. 1951. Civil Engineering Code of Practice No.2, The Institution of Structural Engineers, London.

Goh, A.T.C. 1984. Finite element analyses of retaining walls, Ph.D. thesis, Monash University, Australia.

Goh, A.T.C. and Donald, I.B. 1984. The retaining wall tension crack problem reanalysed, Proceedings of the Ninth Australasian Conference on the Mechanics of Structures and Materials, Sydney : 39-43.

Ingold, T.S. 1979. The effects of compaction on retaining walls, Geotechnique, 29: 265-282.

Mikhailov, A.V. and Avdeeva, V.I. 1973. Effect of a change in the reaction pressure of backfills on stresses in lock chambers with solid bottoms, Hydro-technical Construction, 1 : 20-26.

Moore, P.J. and Chin, C.H. 1985. Effects of repeated loading on earth pressure, Proceedings of the Eleventh International Conference on Soil Mechanics and Foundation Engineering, San Francisco, 4: 2117-2120.

Moore, P.J. and Choong, Y.T. 1986. Effects of vertical wall movements on lateral earth pressure, Civil Engineering Transactions, Institution of Engineers, Australia, vol. CE28, 3: 250-255.

Moore, P.J. and Cole, B.R. 1977. Active pressure observations on a model retaining wall, Proceedings of Fifth Southeast Asian Conference on Soil Engineering, Bangkok, 235-244.

Moore, P.J. and Spencer, G.K. 1972. Lateral pressures from soft clay, Journal of Soil Mechanics Division, ASCE, vol. 98, SM11 : 1225-1244.

Moshkov, A.B. 1974. Discussion of construction norms : earth pressure on lock chamber walls, Hydrotechnical Construction, 7: 642-648.

Packshaw, S. 1946. Earth pressure and earth resistance, Journal of the Institution of Civil Engineers, London, 233-256.

Rowe, P.W. 1950. The distribution of lateral earth pressure on a stiff wall due to surcharge, Civil Engineering & Public Works Review, 45: 590-592, 654-657.

Sinyavskaya, V.M. and Pavlova, A.E. 1971. Effect of periodic displacements of a lock wall on the earth pressure and reinforcement stresses, Hydrotechnical Construction, 3: 247-254.

Taylor, D.W. 1948. Fundamentals of soil mechanics, John Wiley & Sons, Inc., New York.

Terzaghi, K. 1954. Anchored bulkheads, Transactions of American Society of Civil Engineers, 119: 1245-1280.

Terzaghi, K. and Peck, R.B. 1967. Soil mechanics in engineering practice, John Wiley & Sons, Inc., New York.

Tschebotarioff, G.P. 1949. Large scale earth pressure tests with model flexible bulkheads, Bureau of Yards and Docks, U.S. Department of the Navy.

Vidmar, S. 1963. Relaxation effects on the earth pressure of cohesive soils, Proceedings of the International Conference on Soil Mechanics and Foundation Engineering, Budapest, 103-118.

Developments in Geotechnical Aspects of Embankments, Excavations and Buried Structures, Balasubramaniam et al. (eds)
© 1991 Balkema, Rotterdam. ISBN 90 5410 019 2

Groundwater recharge to prevent subsidence of adjacent structure in a subway excavation

Yusake Honjo
Asian Institute of Technology, Bangkok, Thailand

Akira Morishima
Takenaka Civil Engineering Co., Ltd, Tokyo, Japan

ABSTRACT: In this paper, a construction case record of deep subway excavation using deep well method in which groundwater recharge method was adopted in order to prevent the subsidence of adjacent precision machinery factory is reported. It is also an aim of this report to discuss the points to be considered in adopting groundwater recharge method in prevention of subsidence.

1 INTRODUCTION

The groundwater recharge method is used when excavations are carried out supported by pumping up of groundwater in order to lower the groundwater level (e.g. deep well method). The purposes of the recharge can be:

1. preservation of surrounding groundwater environment from the effect of construction,
2. prevention of subsidence of surrounding ground, and
3. avoidance of high sewage fee by releasing the pumped water into the sewage system.

There is a recognition among construction engineers that the recharge method is a difficult method to employ because of much uncertainty involved in predicting and/or evaluating the performance of this method. The sources of uncertainty can be as follows:

1. difficulty in evaluating transmissivity of concerning aquifer,
2. difficulty in evaluating recharge well efficiency (the definition will be given later), and
3. mechanism of well clogging is not fully understood.

In the subway excavation construction reported in this paper, it was the key issue how to encounter these uncertainties involved in employing the recharge method. Especially, evaluation of the transmissivity of the aquifer and of the recharge well efficiency were the most critical problems. In this construction, full efforts were made to reduce the uncer-

tainties involved in them through the continuous process of "pre-examination" → "test construction" → "actual construction and monitoring". Therefore, it is focused throughout this report that how these uncertainties were reduced as the construction proceeded.

2. DESCRIPTION OF PROJECT

Figs. 1 and 2 show plan and section of this construction project respectively. An underground structure of 11 m in width, 110 m in length and 11.5 m in height was constructed at about 21 m below the ground surface by open cut method. The site was located in downtown Tokyo where the land had been reclaimed on the soft clay. The stratifications of the site from the top were reclaimed fill material, alluvial strata consisting of multiple series of clay layers and sand layers (Yuraku-cho-strata), and diluvial strata (Tokyo-strata). The retaining wall was installed up to 25 m under the ground surface. According to the site investigation done before the construction, every sand layer had confined aquifer. Deep well method had been planned to prevent heaving of $Y\ell$ – C layer and boiling of $Y\ell$ – S_2 layer to draw down the groundwater head in $Y\ell$ – S_2 layer and T_0 – S_2 layer.

However, as shown in Fig. 1, there was a precision machinery factory adjacent to the construction site in which some machines were supported by floating slab foundations. These foundations, although it was possible to adjust the level for

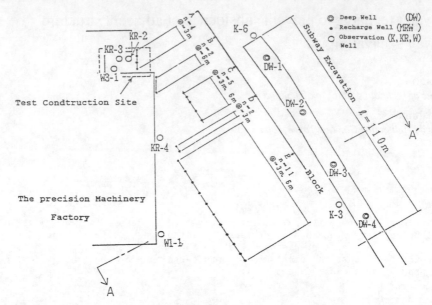

Fig. 1 General plan

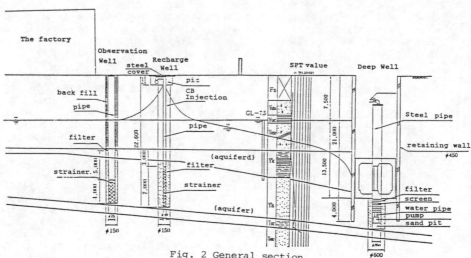

Fig. 2 General section

minor differential settlement, needed to be prevented from differential settlement caused by the construction. The settlement of concern at this time was that caused by consolidation of $Y\ell$ - C layer due to decrease of ground water head in $Y\ell$ - S_2 layer. Several countermeasures were examined at this time including construction of a cutoff wall in front of the factory etc. Finally, it was decided to employ groundwater recharge method because of the expected effectiveness and the economy compared to other alternatives.

3. PRE-EXAMINATION

The result of the pre examination based on the information obtained before carrying out the test construction is summarized in this section. It is instructive to see this result to understand the uncertainties involved in application of the groundwater recharge method to actual construction.

The result is presented in Table 1. In this examination, necessary amount of water to be recharged from the recharge line indicated in Fig. 1 into the aquifer

Table 1 Result of pre-examination

Required total recharge: Q_r
(based on FEM analysis)

$Q_r = C.k.b = 414$ (ℓ/min)

Possible total recharge: Q_p

$Q_p = 2\pi\alpha kbN \dfrac{\Delta h}{\ell n(R/r_w)} = 25\ N$ (ℓ/min)

where,

k = permeability 3.4×10^{-3} cm/sec
($2.1 \times 10^{-3} \sim 6.0 \times 10^{-3}$)
b = thickness of aquifer 7m (4~10 m)
α = well efficiency 0.3 (0.05~0.5 m)
R = influence radius 50m (10~100 m)
r_w = well radius 0.075m
N = number of recharge wells
Δh = injection pressure about 6m by
natural head
C = a constant based on FEM calculation
(17.4×10^3)

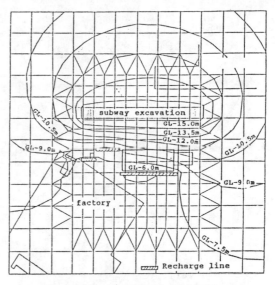

Fig. 3 Result of FEM analysis

in order to maintain the head under the factory more than G.L. - 7.5 m was obtained based on a finite element calculation (F.E.M.) in which the plan of the area was modeled in two dimensions and the head just behind the retaining wall was fixed at G.L. - 17.0 m. On the other hand, the maximum possible recharge for each well was estimated from the stationary solution of the well equation. Based on these two kinds of information, the necessary number of recharge well was determined.

Fig. 3 shows the distribution of head around the factory which was predicted by this FEM claculation. In FEM calculation, the constant head boundary condition, besides the ones stated above, was set to G.L. - 5 m at the edges of the calculated area which is 1 km by 1 km square area of which center is the construction site. In this whole area, the thickness of the aquifer was assumed to be 7 m and the permeability 3.4×10^{-3} cm/sec. It goes without saying that since not all the area was covered by the investigation, there was much uncertainty in parameters employed in the calculation.

According to the calculated result, it was considered sufficient to set 20 recharge wells on the recharge line. However, because of the aforementioned uncertainties involved in the calculation, it was difficult to determine the recharge method as the ultimate solution to prevent subsidence of the factory solely based on this pre-examination.

Among the uncertain factors, the well efficiency (α) and the permeability (k) were the most critical ones. The well efficiency defined here is,

(Well efficiency) =

$$\dfrac{\left(\begin{array}{c}\text{Total head applied} \\ \text{at recharge well}\end{array}\right) - \left(\begin{array}{c}\text{head lost in strainer} \\ \text{filter and well wall}\end{array}\right)}{\left(\begin{array}{c}\text{Total head applied} \\ \text{at recharge well}\end{array}\right)}$$

The well efficiency is considered to be within the range between 0.05 to 0.5; 0.3 is adopted for the pre examination calculation.

It is not unusual in problems concerning groundwater to encounter such high uncertainty. Therefore, it was inevitable to carry out a test construction to examine real feasibility of this method to the problem.

4. TEST CONSTRUCTION

The test construction was carried out by using a part of the recharge line. As shown in Fig. 4, four recharge wells 3 m apart and four observation wells were prepared. Initially, it was attempted to stabilize boreholes during excavation by clear water in order to avoid formation of smear thin zone on well wall. However, it was not possible to prevent collapse in

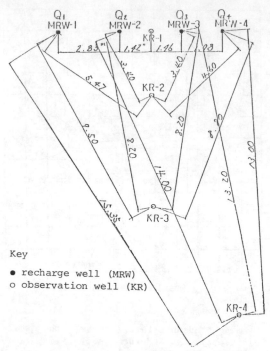

Key

● recharge well (MRW)
○ observation well (KR)

Fig. 4 Distribution of recharge and
observation wells for the
test construction.

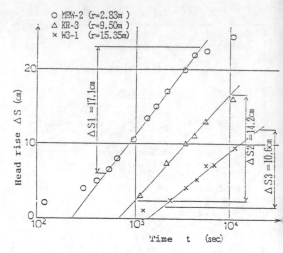

Fig. 5 Time-Head rise relationship

the boreholes only by clear water. As a result, there was no choice but to use bentonite slurry during excavation. The diameter of the boreholes was 15 cm and installed to G.L. - 22 m.

The structure of a recharge well is presented in Fig. 2. The diameter of chloroethylene pipe was 40 mm of which bottom 7 m portion of the pipe was strainer where water injection holes of diameter 6 mm were drilled with the opening ratio of about 3%. The pipe was covered by well screen and then by gravel filter material of appropriate size. Cement bentonite was filled up for the sealing for the upper part.

A water supply tank was used to supply water to 4 recharge wells. Also, a pump was installed in the tank in order to make it possible to recharge water either by the natural head induced by level difference or by injection pressure induced by the pump.

It is common practice to inject water that has been pumped up at the site. However in this construction in order to avoid risk of well clogging by fines contained in pumped up water, clean industrial water was used for the recharge.

Table 2 presents contents of this test construction. The single well recharge test was carried out aiming at evaluation of hydraulic constants at the site and at conformation of effect of pressure injection. In the test to evaluate the hydraulic constants obtained at the site, a constant amount of water was recharged continuously for 3 hours. The methods which are applied to ordinary pumping test were used to evaluate the hydraulic constants. Fig. 5 shows head rise at observation wells with time. Jacob method was employed to estimate the transmissivity ($T = k \cdot b$). The permeability which can be estimated by assuming the thickness of the aquifer to be 7 m was obtained as $1.3 \times 10^{-3} \sim 1.8 \times 10^{-3}$ cm/sec, which is about half of the value used in the pre-examination (see Table 1). The permeability was set to 1.8×10^{-3} cm/sec in analyses done henceforth.

Furthermore, the well efficiency (α) which was estimated based on the theoretical solution shown in Table 1 was about 0.15, which was about half of the value employed in the pre examination. One of the reasons which could have been responsible for this small efficiency was smear zone formed on the well wall by bentonite slurry which had been used for stabilizing bore hole in excavation. It is one of the remaining critical issues in the recharge method to develop well excavation method which can excavate well with sufficiently high well efficiency.

It is economical for the recharge method if one can increase the amount of injection water by pumping. This was tested for double well recharge (2 wells 6

264

Table 2 Contents and results of test construction

Test Item	Purpose	Result	Figure
Single well recharge	1. Estimation of hydraulic constants: Transmissivity (T = hb), storativity (S), permeability (h).	T = 1.3 (cm²/sec) S = 2 x 10⁻³ · k = 1.8 x 10⁻³ (cm/sec)	Fig. 5
	2. Injection pressure ∼ recharge quantity relationship.	possible to increase recharge quantity by increasing injection pressure	Fig. 6
Group well recharge	1. Injection pressure ∼ recharge quantity relationship	possible to increase recharge quantity by increasing injection pressure.	Fig. 6
. 2 wells 6 m apart . 4 wells 3 m apart	2. Confirmation of aquifer head rise by recharge.	The calculated head rise based on the hydraulic constants obtained coincides with the observation.	Fig. 7
	3. Well interference.	Due to the well interference total recharge by 2 wells and 4 wells were almost the same.	Fig. 6

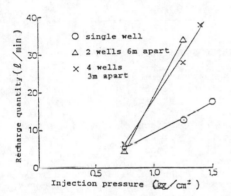

Fig. 6 Injection pressure-recharge
quantity relationship

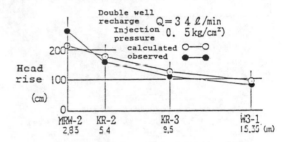

Observation well distance

Fig. 7 Head rise by the double well
recharge

m apart) and for group well recharge (4 wells 3 m apart. Fig. 6 indicates the relationship between the amount of injection and the injection pressure for these two cases. The injection pressure here implies the total of natural head from the ground surface to the aquifer head (0.75 Kg/cm²) and the injection pressure by pumps. In both cases, it was possible to increase the amount of injection in proportion to the applied pressure. It should also be noticed that there was not much difference between the amount of injection by the double wells and by the group wells, which was speculated to be due to the interference of wells. This fact indicates that if there is not sufficient spacing between

recharge wells, the amount of injection for each well reduces considerably due to the well interference.

Fig. 7 presents head rise of the aquifer after 24 hours of recharge by the double wells. In addition, the theoretical solution of head rise based on the group well formula is shown. The group well formula is given as,

$$h_p - H = \frac{1}{2\pi kb} \sum_i Q_i \ln \frac{R_i}{r_i}$$

where

h_p : head at recharge well of concern
H : stationary head outside influence radius
Q_i : injection from i-th recharge well
R_i : influence radius of i-th recharge well

265

Table 3 Examination based on the test construction

Required total recharge (based on FEM calculation):

$$Q_r = C \cdot k \cdot b$$

$$= 17.4 \times 10^3 \times 1.8 \times 10^{-3} \times 7.0$$

$$= 219 \ (\ell/min)$$

Possible total recharge : $Q_p = N \cdot q_p$

Set $q_p = 12 \ \ell/min$ by $0.5 \ Kb/cm^2$ injection pressure from single well based on the test construction result

$$N = \frac{Q_r}{q_p} \times F_s = \frac{219}{12} \times 1.2 = 22$$

24 wells were set including 4 wells used in the test construction. (for notations, see Table 1).

r_i : radius of i-th recharge well
k : permeability
b : thickness of aquifer

The calculated and the observed heads coincided reasonably well, where calculation was made based on the above formula assuming $k = 1.8 \times 10^{-3}$ cm/sec and $R = 50$ m. That is to say the well interference can be evaluated rather well by the above formula.

Table 3 presents the examined result based on the information obtained by this test construction. Following the revised calculation using permeability evaluated by the test construction, the necessary amount of injection was 219 ℓ/min for the whole recharge line. Based on the performance of the recharge wells, it was planned to recharge 12 ℓ/min of water from each well by applying 0.5 Kg/cm² injection pressure. Thus, the necessary number of recharge wells was computed to be 22. The location of these wells are presented in Fig. 2.

As can be understood from Table 3, the most important factors in determining the number of recharge wells are the well efficiency and the maximum possible injection pressure. On the other hand, the permeability governs the absolute amount of water discharged from and recharged to the aquifer. Therefore, it is enevitable to construct wells with high efficiency in establishing on effective groundwater recharge method.

5. RESULTS OF THE CONSTRUCTION

The necessary period of time to lower the ground water level by the deep well method was three month in which no evidence of subsidence was observed in the factory. Therefore, the recharge method was considered to be effective. Fig. 8 indicates head fluctuations of layer during the period the deep well was under operation. The total amount of groundwater discharged and recharged in this period were also presented. From 1st of August to 7th of October, all 4 deep wells were operated; after this period, the wells were gradually stopped, and all of them completely stopped on 27th October. Furthermore, due to unexpectedly high ground water level encountered in the western half of the excavation area, the well point method was applied from 24th of August to 7th of October. One can see the head drop of observation wells behind the retaining wall, namely K-3 and K-6, that the head dropped from G.L.-13 m to G.L. - 19 m and from G.L. - 10 m to G.L. - 15 m respectively in accordance with the increase of the amount of the pumped up water.

The recharge method was controlled based on the observations obtained at three wells that were located in front of the factory, namely W1-1, KR-4 and W3-1 : the amount of injecting water was controlled by injection pressure induced by four pumps. When the deep wells were at full

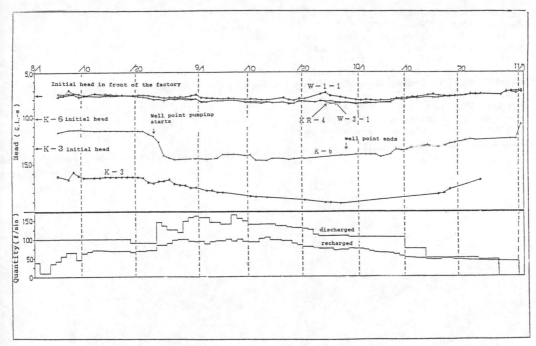

Fig. 8 Head, discharge and recharge fluctuation of $Y\ell-S_2$ layer.

operation, injection pressure of 0.5 ~ 0.4 Kg/cm² was applied to all the recharge wells to increase the amount of recharge to its full capacity, i.e. about 100 ℓ/min. During this time, although it was just for a short period, the observation wells exhibited their maximum drop of about GL - 8.0 m. According to the FEM calculation, it had been predicted that the head beneath the factory would have dropped to - 12 m, if the recharge had not been carried out, which also indicated the effectiveness of the recharge method.

It resulted that the actual amount of water recharged was about half of that had been predicted in Table 3. A reason for this could be, besides the fact that we had taken relatively large permeability for the calculation, the test construction site had had larger transmissivity (or thicker aquifer) compared to the other part. Since there was no decrease in recharge quantity during these three months of construction period, the clogging of wells which had been of concern before the construction seemed not to take place. This was considered due to the use of clean industrial water as the injection water.

6. CONCLUSION

The points to be considered carefully in carrying out recharge method as a countermeasure to prevent subsidence based on the experience gained during this construction are as follows:

1. Every possible effort should be put to make recharge wells as efficient as possible. The efficiency of well is greatly affected by borehole excavation methods.

2. Recharge wells should be sealed carefully so that one can apply high injection pressure.

3. The recharge method at present is considered to be unreliable method because of many uncertain factors involved in estimating the performance of this method to a particular site. However, these uncertainties can be reduced by carrying out a carefully planned test construction.

REFERENCES

Nakazima, S. and K. Seno (1988): "Performance of groundwater recharge method in a subway construction", Kisokoh Vol.16, No. 6, pp. 56-61 (in Japanese).

Morishima, A., Y. Honjo, T. Seshimo and T. Hatano (1988): "Groundwater recharge for a purpose of preventing ground subsidence: Pre-examination, test construction and performance", 43rd annual meeting of JSCE, 3rd Division (in Japanese).

Nakasaki, H., Y. Honjo, T. Ishise and A. Morishima (1986): "A case record of ground water recharge method applied to a subway excavation construction", TAKENAKA Tech. Res. Lab. annual research exhibition (in Japanese).

Developments in Geotechnical Aspects of Embankments, Excavations and Buried Structures, Balasubramaniam et al. (eds)
© *1991 Balkema, Rotterdam. ISBN 90 5410 019 2*

Model tests of slope failure due to seepage water

S. Murata & H. Shibuya
Kumamoto Institute of Technology, Japan

ABSTRACT: Model tests of slope failure were performed on compacted volcanic cohesive soil. The results of the investigation show that: (1) Seepage water is one of the main causes of the slope failure; and depends on both the amount of precipitation and precipitation intensity. (2) The piezometer heads at the places of the slope failure have extremely higher levels than the soil layer surface. (3) The strength of the soil decreases about a half value at places influenced by the seepage water. Therefore, the major cause of the slope failure concern to the groundwater is the decrease of the strength of the ground due to the seepage water.

1 INTRODUCTION

The Japanese Islands, located on the Circum-Pacific seismic belt, adopt a very complicated geological and natural features; and poor ground conditions. With about 75 % of the land covered by the mountains, the country receives very high precipitations in a year. Kyushu Island alone receives an average amount of 2500 mm in a year. The slope failures are frequently occurred by the rainfall, especially heavy rainfall in rainy or typhoon seasons. Since, many people lives near the foots of the slopes where have a possibility of the slope failures, we have serious damages concern to the people, the houses, the property and so on. The defence from these disasters is important as the social problem in Japan.

The mechanism of the slope failures is not exactly clear, because there are many factors concerning the slope failures, for example geological features, natural features, vegetation, groundwater and so on, and these are closely connected each other. And the ground structure and the movement of the groundwater are very complicated, we can't estimate easily. In some failures, we can estimate that the slope failures are caused by the seepage water. The investigations of this slope failure are very few, we are at the beginning of the investigation.

In this paper, we investigated the slope failures due to the seepage water. And the mechanism of the slope failures is also cleared by the model tests.

2 SLOPE FAILURES DUE TO RAINFALL AND INFLUENCE OF GROUNDWATER

On the slope failures due to the rainfall, it is important to know about the characteristics of rainfall, for example precipitation intensity, amount of precipitation, time of precipitation and so on. Typical rainfall concern to the slope failure is as follow: Light rainfall continues several days, and then heavy rainfall, precipitation intensity 40-50 mm/h or more, continues during 2-3 hours or more. The many slope failures occur during the heavy rainfall.

A serious slope failure occurred at Shigeto Tosayamada Kouch Prefecture Japan, On July 5, 1972. The characteristic of the rainfall is shown in Fig. 1. On the previous day of the failure, the amount of precipitation was about 350 mm, the ground had been almost saturated and weakened by the rainfall. In this condition, the amount of precipitation was about 450 mm/day and the precipitation intensity, 50 mm/h or more, had continued 4 hours. The first slope failure occurred during the heavy rainfall, the second serious slope failure occurred after the stop of rainfall. After that, a lot of groundwater flowed out of the several places of the slope (Fig. 2). The slope failures of this type were confirmed at the other places and most of these

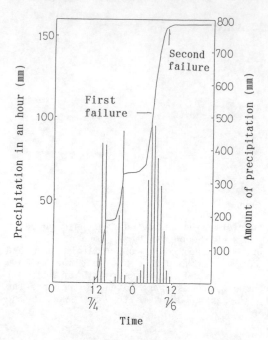

Fig. 1 Relationship between precipitation intensity and amount of precipitation (at Shigeto, Tosayamada, Japan, 1972)

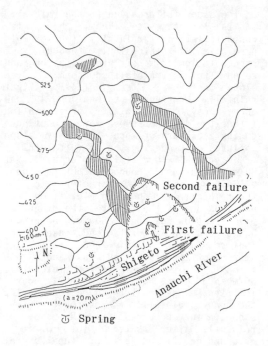

Fig. 2 Slope failure at Shigeto Tosayamada

failures were middle or large-scale, and the slope gradient was comparatively gentle, 30 degrees or so.

Fig. 3 shows the relationship between the scale and the time of the slope failures. The frequency distribution of the small-scale failures has a good correlation with the precipitation intensity. However, as the scale of the slope failures becomes larger, the correlation with the precipitation intensity gradually weakens. This means that the slope failures also occur after the rainfall stopped.

From the following reasons, we have an idea that the seepage water is to be one of the causes of the slope failure.

1. We can see the springs or the holes of piping on the slope surface after the failure.

2. The slope failures occur after the rainfall stopped, and these cases are observed in the large-scale failures.

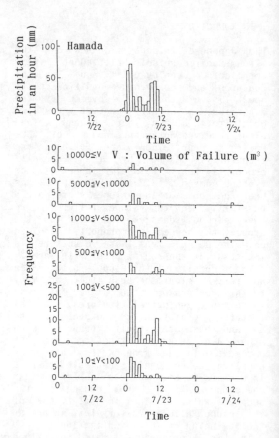

Fig. 3 Relationship between frequency and time of slope failure

3. The frequency distribution of the small-scale failures has a good correlation with the precipitation intensity, However, the correlation with one becomes weak in the large-scale failures.

3 MODEL TESTS OF SLOPE FAILURE DUE TO SEEPAGE WATER

3.1 Test apparatus

Fig. 4 shows the situation of a typical slope concern to the failure due to the seepage water. We simulated this situation in the model tests. Based on this idea, we made tow apparatus shown in Figs. 5 (a) and (b). The former one was used to investigate the general phenomenon of the slope failure and the head of groundwater due to the seepage water. The apparatus was made of the acrylic plate and stuck sands on the bottom to increase the friction. The seepage water was supplied from the small holes on the bottom of the apparatus. All tanks for the seepage water were set up at the same water head. The tanks were able to be moved to any suitable level depending on the thickness of the specimen. Groundwater heads were measured by the piezometers which were set up the bottom of the apparatus.

The latter apparatus was used to find the decrease of the strength of the ground due to the seepage water. The functions of the apparatus were almost same as the former one.

3.2 Sample

The soil samples used in this study were picked up in the campus of our university. The campus was located on the hill consisted of welded tuff from the Aso volcano in Kumamoto Prefecture. The soils are called volcanic cohesive soil and made by the weathering of the welded tuff. The physical properties of the soils were as follows: specific gravity G_s =2.69, uniformity coefficient U_c =52.0, liquid limit w_L =52.7%, plasticity index I_p =11, optimum moisture content w_{opt} =32.7% and maximum dry density ρ_{dmax} =1.351g/cm³ .

3.3 Preparation of specimen and test procedure

The air dried soils were sieved through a 4.76 mm sieve. The soils were then mixed to the optimum moisture content, and compacted to 5 layers in the test apparatus. Then the specimen was saturated by the water from the side of specimen. When the seepage water soaked out to the surface, we considered that the specimen was saturated. After saturation, one side of the apparatus was gradually lifted up to the fixed slope angle and the seepage water was supplied through pipes connected to the holes at the bottom of the apparatus. The piezometer heads were measured at interval of 5 minutes and the situations of the surface were observed by a camera and a video camera at each change.

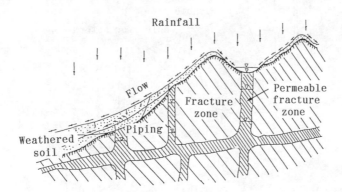

Fig. 4 Situation of a slope under the seepage effect

271

Plan

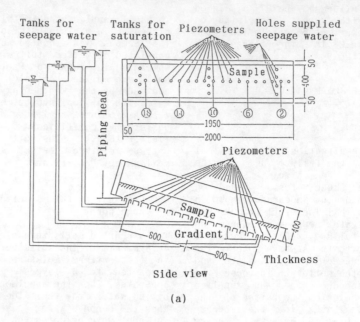

(a)

Plan

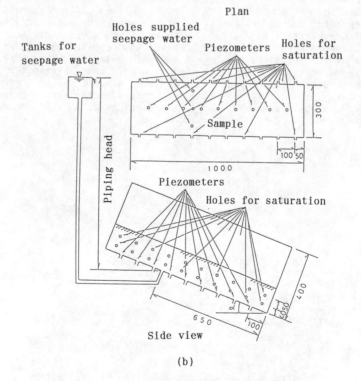

(b)

Fig. 5 Apparatus of model test (unit mm)

4 TEST RESULTS

4.1 Situation of slope failure

4.1.1 Case of slope gradient of 30 degrees

The general slope failure processes were as follow: At first, a small failure occurred at the toe of slope (Photo 1). And the same failures occurred at the middle and top of the slope (Photo 2). The failure of the toe of the slope gradually enlarged (Photo 3). The failures of the middle and top of the slope enlarged with the same pattern (Photo 4). Finally, the failure mud flowed down on the surface of the slope (Photo 5).

Photo. 1

Photo. 2

Photo. 3

Photo. 4

Photo. 5

Photo. 1   Small failure at the toe of slope
Photo. 2   Small failure at the middle of slope
Photo. 3   Enlargement of failure at the toe of slope
Photo. 4   Enlargement of failure at the middle of slope
Photo. 5   Mud flow after failure at the middle of slope

4.1.2 Case of slope gradient of 20 degrees

The general slope failure process was as follow: After piping, a small failure occurred at the toe of slope. The same failure occurred at the middle and top of the slope. These failures did not enlarge after then and the erosion only progressed by the water flowed from the failure places.

4.2 Piezometer heads

The results of piezometer heads are shown in Figs. 6 (a) and (b), at the piping heads of 60 cm and 75 cm respectively. A slope failure did not occur in the former test. However, slope failures occurred in the latter one. In the former test, the piezometer heads near the holes where the seepage water was supplied had almost same level to the soil layer surface. In the latter test, the ones had very higher levels than the surface. This point was extremely different in tests. In all tests which the piping head was 75 cm or more, the slope failures also occurred and the piezometer heads had very higher levels than the soil layer surface.

From these results, the seepage water becomes a cause of the slope failure and

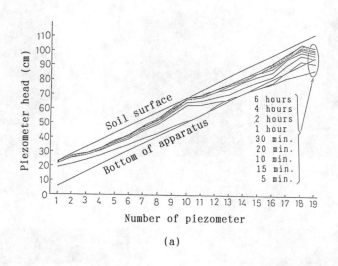

(a)

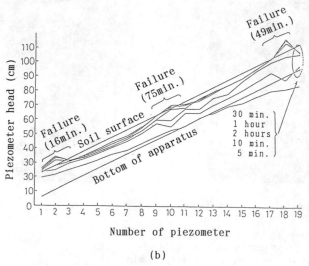

(b)

Fig. 6   Piezometer heads:(a) Piping head = 60 cm and (b) Piping head = 70 cm

the measurement of the piezometer heads can be used to predict slope failures.

4.3 Decrease of strength of soil

The slope failures always occurred near the holes where the seepage water was supplied to the specimen. From this experimental results, it is supposed that a cause of the slope failures lies in the decrease of the strength of the soil due to the seepage water. Therefore, it is necessary to find this decrease of the strength of the soil.

The strength of the specimen was measured by using the apparatus shown in Fig 5 (b) and a cone penetrometer. The tests were performed as follows: Seepage water was supplied from the holes at the bottom of specimen, and the supply was cut off just before the failures occur. Then, the cone penetrometer was inserted at the points of specimen as shown in Fig. 7.

The test results are shown in Figs. 8 (a), (b) and (c), the slope gradient of 0, 20, and 30 degrees, respectively. The cone index at the places where the seepage water was supplied decreased about a half value compared to the one at the places where seepage had less influence. Same results were observed in the other piping heads.

5 CONCLUSIONS

From this investigation, it is found that the slope failure concern to the groundwater was happened by the decrease of the strength of the ground due to the seepage water. The results of this investigation are summarized as the following :

1. The seepage water becomes one of the causes of the slope failure.

2. In the case which the slope failure occurred, the heads of piezometer near the holes for supplying the seepage water had extremely higher levels than the ground. However, in the case which the slope failure did not occurred, it had almost the same level to the ground.

3. The cone index at the places where the seepage water was supplied decreased about a half value compared to the one at the place where seepage had less influence.

4. The slope failures concern to the groundwater were happened by the decrease of the strength of the ground due to the seepage water.

Further progress in this investigation is to find the decrease of the strength of the soil by elemental tests.

REFERENCES

Murata, s., et al. 1988. Influence of seepage water on slope failure caused by rainfall. Tuchi to kiso, Vol.36, No.4: 45-50 (in Japanese).
Shima, M., et al. 1973. Slope failure at Shigeto area. Report of national disaster science: 167-173 (in Japanese).
Shibata, T., et al. 1984. Sediment disasters in Hamada and characteristics of soil at Nakaba area. Report of national disaster science: 38-49 (in Japanese).

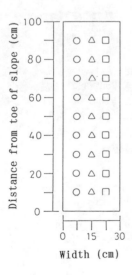

Fig. 7 Points of cone penetrometer measurements

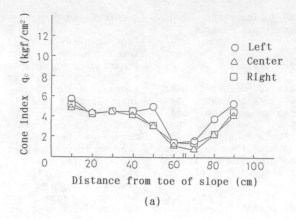

(a)

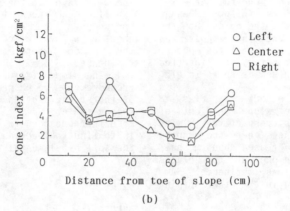

(b)

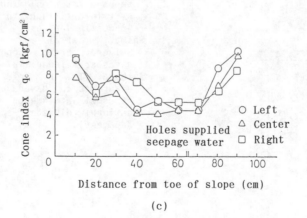

(c)

Fig. 8 Cone index of specimen in model test:
(a) Slope gradient = 0 degree, piping head = 238 cm;
(b) Slope gradient = 20 degrees, piping head = 148 cm and
(c) Slope gradient = 30 degrees, piping head = 170 cm

Model test on sheet-pile countermeasures for clay foundation under embankment

H.Ochiai, S.Hayashi, T.Umezaki & J.Otani
Kyushu University, Japan

ABSTRACT: Sheet-pile countermeasure methods for embankment on soft ground is usually used as the end bearing type method. However, there should be different directions for use of sheet-pile in order to control the stress in ground, effectively. A new test apparatus is developed for the study of sheet-pile countermeasures for clay foundation under embankment. Using this apparatus, a series of model tests is conducted for different types of sheet-pile countermeasures, and the effectiveness of the countermeasures are discussed from the test results.

1 INTRODUCTION

Deformation of clay foundation due to construction of embankment is a serious problem for the surrounding structures. There are different types of countermeasure methods for the problem. The methods with sheet-pile are often used in construction sites, in which there are various ways of applications of the sheet-pile, but the evaluations for the effectiveness of sheet-pile are not always consistent.

The purpose of the paper is to evaluate the effectiveness of the sheet-pile countermeasure methods on the basis of the consideration of both stability and deformation of clay foundations. The laboratory model tests are conducted, in which the effects of sheet-pile length, the inclined angle and the connecting manner with tie rod for sheet-pile countermeasures are discussed (Umezaki et al.(1990) and Ochiai et al.(1990)).

The effects of countermeasure methods are also evaluated by new introduced parameters.

2 TEST APPARATUS AND PROCEDURES

In order to investigate the effect of the sheet-pile countermeasures, a series of the model tests are conducted for different types of sheet-pile in length (Case S-1 and Case S-2), inclined angle (Case S-θ) and connecting manner with tie rod (Case S-T and Case S-T2). It is noted that the case without any countermeasure methods(Case 2) is also conducted for the comparison with those cases. Those are listed in Table 1. The test apparatus consists of the soil bin and the embankment type of loading system as shown in Fig.1 and Photo.1. This apparatus is capable of simulating the behavior in plane strain condition and embankment type of loadings.

Table 1 Test cases

Test case	Without counter-measures	Sheet-pile countermeasures				
	Case-2	CaseS-1	CaseS-2	CaseS-T	CaseS-T2	CaseS-θ
Length of sheet-pile D_s(cm)	none	40.0	20.3	20.3	20.3	20.3
Inclined angle θ (°)	none	0	0	0	0	30
Tie rod method	none	none	none	top	top and bottom	none
Thickness of clay foundation D(cm)	52.9	57.9	54.5	55.1	54.4	55.2
Width of embankment load B(cm)	27.0 (B/D=0.51)	27.0 (B/D=0.47)	27.0 (B/D=0.50)	27.0 (B/D=0.49)	27.0 (B/D=0.50)	27.0 (B/D=0.49)
Installed depth of sheet-pile D_f (cm)	none	40.0 (D_f/D=0.69)	20.3 (D_f/D=0.37)	20.3 (D_f/D=0.37)	20.3 (D_f/D=0.37)	17.6 (D_f/D=0.32)

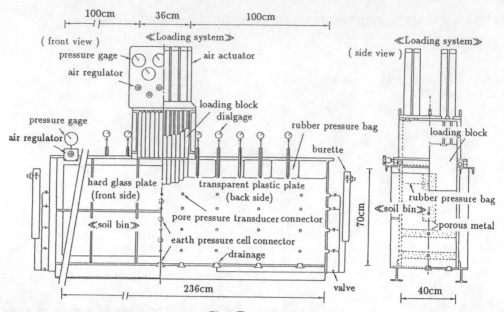

Fig.1 Test apparatus

Photo.1 Test apparatus

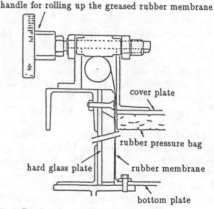

Fig.2 Device for decreasing friction of side wall

2.1 Soil bin

The soil bin used here is scaled in 236cm wide, 40cm long and 70cm high as shown in Fig.1. Consolidation pressure and overburden presssure are applied by using 24 pieces of rubber bags. The rubber bags can cause large deformation on soft clay foundation during both stages of consolidation and loadings. In order to satisfy one-dimensional consolidation and plane strain condition, aluminum pipes of 1 cm diameter are placed in a row under the rubber bags.

The friction of side wall is decreased by using the new designed device with a handle for rolling up the greased rubber membrane as shown in Fig.2. Water level and drainage can be also controlled in the apparatus.

2.2 Loading system

A loading system consists of 12 loading blocks with air actuators, and the load is applied independently by each air actuator, as shown in Fig.1. Following features are highlighted for this system:

1. Embankment type of loading under the overburden pressure is simulated by using both the loading system and the rubber bag system.

2. Loading processes of embankment construction in site are simulated by controlling the air regulator.

3. Several types of loading distributions are simulated by applying the different amount of load in each loading block.

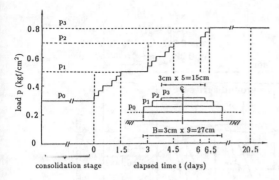

Fig.3 Loading pattern

2.3 Test procedures

Ariake clay, which is one of the typical sensitive clay in Japan, is used as foundation soils in the test. The physical properties of the clay used here are as follows; specific gravity: $G_s=2.66$, natural water content: $w_n=96.0$ %, liquid limit: $w_L=68.8$%, and plasticity index: $I_p=29.3$. Ariake clay in slurry condition is filled in the soil bin until initial thickness of 68cm. Then, overburden pressure of $p_0=0.3$kgf/cm² is applied during 2.5 months under drained condition at the surface and bottom of the ground. The model steel sheet-piles of 1mm thickness are installed at the positions of both toes of the embankment, and embankment load of the width $B=27$cm is applied step-by-step($p_1+p_2+p_3=0.2+0.2+0.1$kgf/cm²) in 6.5 days as shown in Fig.3.

2.4 Measurements

Vertical and lateral displacements at the ground surface and in the ground, bending moment of the sheet-pile, horizontal stress in plane strain direction, and pore water pressure are measured in the tests. The vertical displacements at the ground surface are measured by dialgages, while the rest of displacements are plotted on the transparences attached to the side walls. The bending moment of the sheet-pile is obtained by strain gages. The horizontal stress and the pore water pressure are also measured by pressure transducers.

3 BEHAVIOR OF CLAY FOUNDATIONS DUE TO EMBANKMENT LOADS

Deformation properties of the ground without any countermeasures are investigated to obtain the basic data for evaluating the effectiveness of the countermeasure methods. Here in the paper, the vertical and lateral displacements of the ground and influence extent of ground deformations are discussed.

In order to know the strength property in the ground before applying the embankment load, the cone penetrometer tests were conducted at a distance of 80 cm from the embankment center. Test results show that the cone index was not constant with the ground depth. This means that the dissipation of the pore water in the consolidation stage has not been complete in all test cases, so that the remains of the settlement due to the consolidation pressure are considered in the test results.

The test results of the vertical displacement at the ground surface and the lateral displacement in the ground are shown in Fig.4. The vertical displacement of the ground occurs until a distance from the embankment toe, and the upward displacement dose not appear because the embankment load is applied step-by-step in 6.5 days.

It is realized from the same figure that the lateral displacements occure until the depth $D \fallingdotseq 40$cm ($D/B=1.5$) and the maximum values appear at the depth $D=20\text{-}25$cm($D/B=0.75\text{-}1.0$) from the ground surface. It is also observed that the lateral displacement at the distance of 60cm from the embankment center after 20.5 days is smaller than that after 6 days. This means that the settlement due to consolidation continues even after the end of loading, whereas the lateral displacement dose not increase.

The directions of the displacement vector at the distance of 70 cm from the center are almost vertical downward as shown in Fig.5. The influence extent in horizontal direction of the deformations is about 70cm, because it is about 2.5 times of the embankment width($B=27$cm), and 1.2 times of the ground depth($D=52.6$cm), respectively.

4 SHEET-PILE COUNTERMEASURE METHODS

A series of model tests with different types of sheet-pile countermeasures are conducted for the soft clay ground showing the deformation properties described above.

4.1 Tested countermeasure methods

The main effects of the sheet-pile countermeasures are to change the stress conditions in the ground and to decrease both vertical and lateral displacements of the ground. In order to discuss the effect of the sheet-pile length, two cases of the length $D_s=20$ and 40cm are examined. As described above, the lateral displacement without any countermeasure occures until the depth $D \fallingdotseq 40$cm and the maximum values appear at the depth $D=20\text{-}25$cm from the ground surface. These lengths of sheet-pile $D_s=20$ and 40cm correspond to those depths. The sheet-pile methods with tie rod are also tested as the method for controlling the deformation of the ground. In these cases, the length of

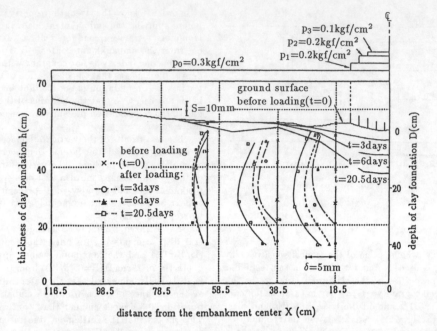

Fig.4 Lateral displacements in clay foundation
(without any countermeasures)

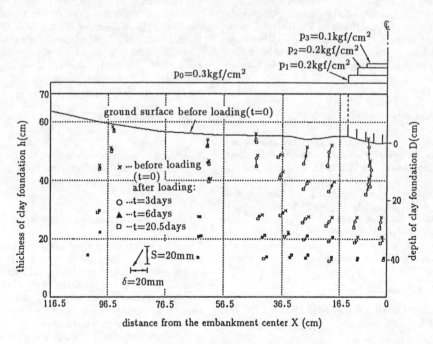

Fig.5 Displacement vectors in clay foundation
(without any countermeasures)

the sheet-pile is constant of $D_s = 20$cm. One is the case(Case S-T) of the sheet-pile in connection with tie rod at the pile top and the other is the one(Case S-T2) with tie rod at both top and bottom of the sheet-pile. Photo.2 shows the assembly of the model sheet-pile for Case S-T2. Piano-wire is used for the tie rod, and the connection of the cross tie rod at the pile top is set after installing the sheet-pile. And also, as a method of controlling the stress in the ground, the test with batter sheet-pile (Case S-θ), which is inclined towards the embankment center by 30°, is conducted as shown in Fig.6.

4.2 Behavior of clay foundations with sheet-pile countermeasures

(1) Parameters for the ground behavior
Here in the paper, following values shown in Fig.7 are introduced to evaluate the behavior of clay foundations:

Vertical displacement at the ground surface

$$S_x = \alpha \cdot S_0 = \alpha' \cdot S_{n0} \qquad (1)$$

Lateral displacement at the ground surface

$$\delta_x = \beta \cdot S_0 = \beta' \cdot S_{n0} \qquad (2)$$

Influence extent of deformation at the ground surface

$$L = \gamma \cdot D = \gamma' \cdot B \qquad (3)$$

where S_x, δ_x, L, S_0 and S_{n0} change with time. A subscript x means the horizontal distance from the embankment center. S_0 is the settlement of the ground surface at the embankment center, and S_{n0} is the one for the case without any countermeasure methods. It is also noted that D is the thickness of the clay foundation, B is the embankment width, and α, β, γ, α' β' and γ' are the coefficients.

The basic idea of selecting the parameters are based on the consideration of evaluating the ground deformation in terms of vertical and lateral components. It is considered from the test results that the amount of vertical and lateral displacements are close related to that of the ground settlement at the embankment center, so that the settlement at the embankment center S_0 may be used as one of the influence factors for the deformation properties. For the cases of the sheet-pile countermeasures, those displacements are normalized by the settlement S_{n0} for evaluating the effectiveness of countermeasures, quantitatively.

(2) Behavior of clay foundations with sheet-pile countermeasures
For comparisons of the coefficients, $\alpha' = S_x/S_{n0}$ and $\beta' = \delta_x/S_{n0}$ are plotted against the coefficient $\gamma = L/D$ in Fig.8 for tested cases of Table 1. The behavior

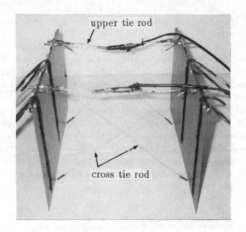

Photo.2 Assembly of model sheet-pile (Case S-T2)

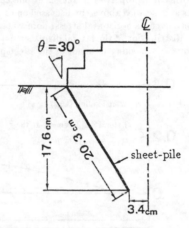

Fig.6 Batter sheet-pile countermeasure (Case S-θ)

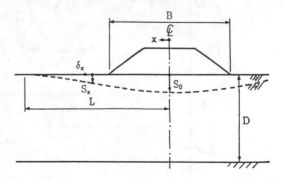

Fig.7 Parameters for the evaluation of the ground behavior

of the sheet-pile countermeasures are summarized as follows:

The vertical displacement of clay foundations is discontinuous at the position of the sheet-pile, and the distributions of the deformation show the different tendency in each case. The maximum value of lateral displacement at the ground surface occures at the toe of the embankment for all cases. In comparison with the case without sheet-pile, the settlements under the embankment for the cases of the sheet-pile countermeasures except for Case S-2 and Case S-θ are large, whereas the settlements around the embankment and lateral displacements in the ground except for Case S-T2 are small.

4.3 Mechanism of the ground deformation in countermeasures

The deformation properties in the sheet-pile countermeasures as described above are discussed on the stress conditions, such as horizontal stress, pore water pressure, distribution of the lateral displacement in the ground and bending moment of the sheet-pile.

(1) Stress in the ground under the embankment

In order to investigate the differences of the stress propagation by countermeasure methods, the stress increment in the ground due to the embankment load are discussed in terms of the horizontal stress, $\Delta \sigma_2$, in plane strain direction at the embankment center. Fig.9 shows $\Delta \sigma_2$ at the distance H=18cm from the ground base after loading (t=6.5 days) for each countermeasures. As shown in this figure, the stress concenteration is caused by using tie rod in the area between two sheet-piles, and it is also realized for the case of long sheet-pile. For Case S-θ, which the length of sheet-pile is the same as the other cases except for Case S-1, the stress is less induced in comparison with the other cases. It may be considered herein that the stress concentration causes the large amount of the settlement at the embankment center.

(2) Excess pore water pressure

Fig.10 shows the excess pore water pressure in the ground. The measured values of pore water pressure at H=28cm for Case S-T2 are larger than that of the other cases. It may be considered, therefore, that the large displacements in Case S-T2 are resulted from the small effective confining stress.

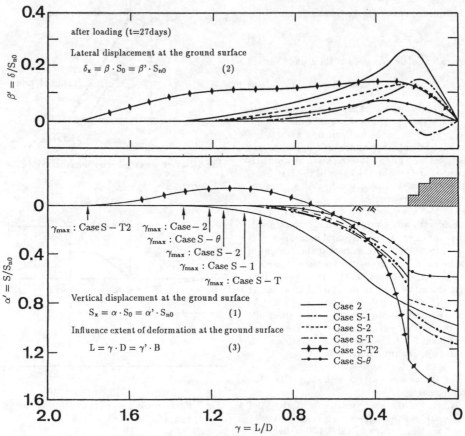

Fig.8 Behavior of foundations under embankment load

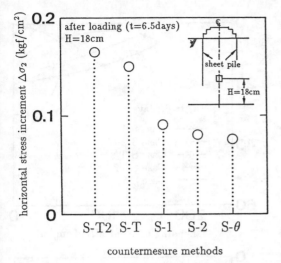

Fig.9 Horizontal stress increment under embankment
center

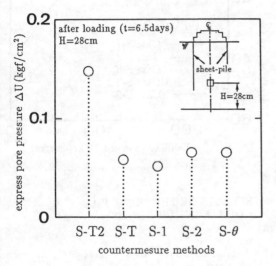

Fig.10 Excess pore water pressure under embankment
center

(3) Lateral displacement at the toe of embankment
    and the bending moment of sheet-pile

The distributions of the lateral displacement at the
toe of the embankment are shown in Fig.11 at 6.5 and
27 days from the beginning of the loading.

In the case without any countermeasures, the lateral displacement in the ground is still increasing even
at 27 days but that at the ground surface is almost

constant at the end of loading(t=6.5days). While the
lateral displacement at the ground surface is decreased
by the sheet-pile countermeasures, but that in the
ground depends on the directions of use of the sheet-
piles.

The relation between the lateral displacements δ_1
and δ_z at ground surface and in the ground is important for discussing the effectiveness of sheet-pile
countermeasure. Fig.12 shows the relation between
δ_1 and δ_{max}, in which δ_{max} is the maximum value of
lateral displacement δ_z in the ground and is chosen
as the representative value of δ_z. The relation is different for each countermeasure method, so it may be
considered that an evaluating parameter of the lateral
dispalcement in the ground for the sheet-pile countermeasure methods is not δ_1, but δ_{max}.

Fig.13 also shows the distributions of the bending
moment of sheet-pile. The large amount of bending
moment is appeared for the cases with tie rod (CaseS-
T and CaseS-T2). The same tendency is measured in
the horizontal stress increment as shown in Fig.9. In
comparison of CaseS-θ with CaseS-T, which are different in inclined angle and connecting manners, the
maximum values of bending moment are almost equal
for the two cases, whereas the different stress concentrations are caused as shown in Fig.9.

Based on the discussions described above, the mechanism of ground deformation with each sheet-pile countermeasure is summarized as follows:

For Case S-1, where the length of sheet-pile is about
two-thirds of the ground thickness, the large value of
horizontal stress acts on the sheet-pile. A rotation of
sheet-piles is caused by the horizontal stress with respect to the top, because both sheet-piles are not connected to each other. Thus, the large lateral dispace-
ment occurs in the ground.

Case S-2, where the length of the sheet-pile is about
one-third of the ground thickness, is effective for decreasing lateral displacements, both at the surface and
in the ground. Therefore, the settlement at the toe of
embankment is smaller than that of the case without
any countermeasures.

For Case S-T, where sheet-piles are connected to
each other at the top of piles, the stress concentration
is induced in the ground between two sheet-piles, and
the large amount of bending moment is developed on
the sheet-piles. As a result, the lateral displacement
in the ground is small as much as that for Case S-2.

For Case S-T2, where both top and bottom of sheet-
piles are connected to each other, the lateral dispalcement in the ground is large because of the remarkable stress concentration between two sheet-piles. Besides the large amount of lateral displacement is induced under the bottom of the sheet-piles.

For CaseS-θ, the batter sheet-pile is capable of reducing the vertical stress due to embankment load.
Therefore, not only the lateral displacement and settlement at the toe of embankment but also the settlement at embankment center are reduced remarkably.

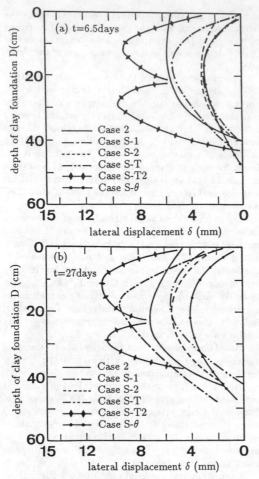

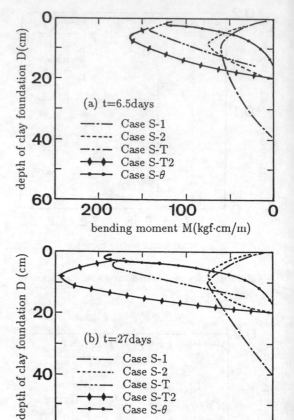

Fig.11 Lateral displacements in clay foundations
at the toe of embankment

Fig.13 Bending moment of sheet-piles

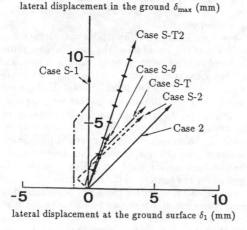

lateral displacement in the ground δ_{max} (mm)

Case S-T2

Case S-1

Case S-θ

Case S-T

Case S-2

Case 2

lateral displacement at the ground surface δ_1 (mm)

Fig.12 Relationships between δ_1 and δ_{max}

5. EFFECTS OF THE SHEET PILE COUNTERMEASURES

5.1 Parameters for the effects of sheet-pile countermeasures

As described above, the amount of lateral displacement is one of the indispensable factors for evaluating the effects of countermeasures. Here in the discussion, the maximum lateral displacement δ_{max} in the ground is at the toe of embankment, which is the value at the depth D=15cm in the tests, is used as a representative value of lateral displacement in the ground.

The following parameters are also focused on from the discussions on the deformation properties of the ground as shown in Fig.14. They are the settlements at the center and toe of the embankment, S_0 and S_1, and the influence extent from the embankment center, L. The nondimensional values, S_0/S_{n0}, S_1/S_{n0},

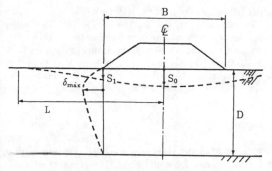

Fig.14 Parameters for the effects of sheet-pile countermeasures

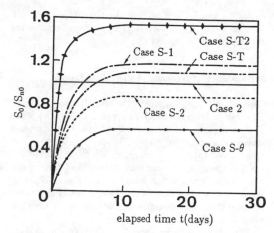

Fig.17 Effect for settlement at the embankment center

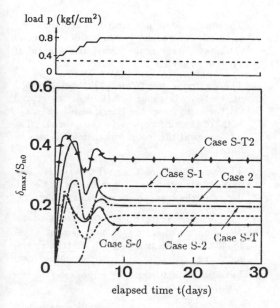

Fig.15 Effect for lateral displacement in clay foundations

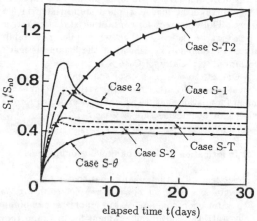

Fig.18 Effect for settlement at the toe of embankment

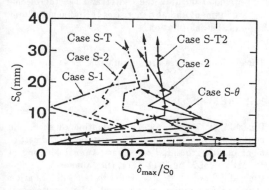

Fig.16 The construction controlling diagram with S_0 and δ_{max}/S_0

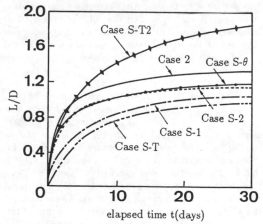

Fig.19 Effect for influence extent of deformation

δ_{max}/S_{n0} and L/D, which are the time dependent values, are used for the evaluation of the effects of the sheet-pile countermeasures.

5.2 Stability of the embankment foundations

The values, δ_{max}/S_{n0}, are plotted for all cases in Fig.15. The values increase quickly except for Case S-1 when the embankment load is applied, and they decrease during the constant value of load. Finally, they become constant after loading, in which the value S_{n0} itself converges to a constant value. The lateral displacements for Case S-2, Case S-T and Case S-θ are always smaller than that of the case without any countermeasures. The effectiveness is significant for Case S-θ.

The construction controlling diagram by Matsuo et al. (1977) is shown in Fig.16 as a relationship between S_0 and δ_{max}/S_0. The degrees of stability of embankment foundations are large for all cases because the embankment load is applied step-by-step in 6.5 days. These degrees for countermeasures are larger than that of the case without sheet-pile. The stability of the foundation right after the loading is much improved, especially for Case S-1.

One of the major effect of sheet-pile countermeasures is generally to increase the stability of embankment foundation, but the effect is not expected for the sheet-pile countermeasure of Case S-T2.

5.3 Deformation of the embankment foundations

(1) Settlement at the embankment center
The values, S_0/S_{n0}, for all cases are plotted in Fig.17. The values for countermeasures increase monotonically until the end of embankment loading, and they become constant afterwards. As realized from the figure, the maximum value of S_0/S_{n0} in each case for countermeasures except for Case S-θ is 0.9-1.5. Thus, the settlements at the embankment center for countermeasures are larger than that of the case without countermeasures, so that the effects of decreasing settlement at the embankment center are not expected for the sheet-pile countermeasures except for the batter type (Case S-θ).

(2) Settlement at the toe of the embankment
It is considered that the settlement at the toe of embankment, S_1, is a value for evaluating the settlement around the ground. The values, S_1/S_{n0}, are plotted in Fig.18. The value after loading is about 0.7 for the case without sheet-pile, while these values are 0.4-0.6 for all the other cases except for CaseS-T2. Thus, the sheet-pile countermeasures reduce the settlement at the toe of embankment, effectively. The settlement for Case S-T2 is large due to the large amount of lateral displacement caused by the strss concentration between two sheet-piles.

(3) Influence extent of the deformation
The values, L/D, are plotted in Fig.19. The value for the case wthout any countermeasures is about 1.2, whereas the values for countermeasures except for Case S-T2 are 0.8-1.0. It may be considered that the influence extent is also reduce by sheet-pile countermeasures except for Case S-T2.

6 CONCLUSIONS

A series of model tests on sheet-pile countermeasures for clay foundation under embankment were conducted, where the thickness of the clay foundation was about twice of the embankment. Main conclusions are as follows:

1. The countermeasure (Case S-1), where the length of sheet-pile is about two-thirds of the ground thickness, decreases the lateral displacement at the ground surface.

2. The countermeasure (Case S-2), where the length of sheet-pile is about one-third of the ground thickness, decreases the lateral displacements both at the surface and in the ground. The settlement at the toe of embankment is also reduced.

3. The countermeasure (Case S-T), where the tops of sheet-piles are connected each other by tie rod, decreases the influence extent remarkably.

4. The countermeasure (Case S-θ), where the sheet-pile is inclined towards the embankment center by 30°, decreases not only the lateral displacement and settlement at the toe of embankment but also the settlement at embankment center remarkably.

5. Whereas the countermeasure (Case S-T2), where both top and bottom of two sheet-piles are connected each other by tie rods, increases the settlement both at the center and the toe of embankment and also increases the influence extent of the deformation.

Those effectivenesses are summarized in Table 2.

REFERENCES

Umezaki, T. et al. 1990. Evaluation of sheet-pile methods as a countermeasure for stability and settlement of soft ground, Proceedings, Annual Meeting of JSSMFE, Vol.2, pp.1211-1214, (in Japanese).
Umezaki, T. et al. 1990. Model tests on deformation of soft ground under embankment, Western Regional Division Report of Japan, Group for the Study of Natural Disaster No.9, pp.50-57, (in Japanese).

Table 2 Effectiveness of the sheet-pile countermeasure methods

Test case	CaseS-1	CaseS-2	CaseS-T	CaseS-T2	CaseS-θ
Sheet-pile length D_s (cm)	40.0	20.3	20.3	20.3	20.3
Inclined angle θ (°)	0	0	0	0	30
Tie rod method	none	none	top	top and bottom	none
Stability of the embankment foundation	◎	○	○	×	◎
Lateral displacement at the ground surface	◎	○	○	×	◎
Settlement at the embankment center	×	○	×	×	◎
Settlement of clay foundations	○	◎	◎	×	◎
Influence extent of deformation	○	○	◎	×	○

◎ : significant effective
○ : effective
× : not effective

Ochiai, H. et al. 1990. Sheet-pile countermeasre for stability and settlement of soft ground, Proceedings of the Symposium on the Application of New Materials and New Methods for Ground Disaster Prevention, Nishinippon Institute of Technology, pp.77-86, (in Japanese).
Matsuo, M. et al. 1977. Diagram for construction control of embankment on soft ground, Soils and Foundations, Vol.17, No.3, pp.37-52.

3. Tunnels and buried structures

Developments in Geotechnical Aspects of Embankments, Excavations and Buried Structures, Balasubramaniam et al. (eds)
© *1991 Balkema, Rotterdam. ISBN 90 5410 019 2*

Behavior and simulation of sandy ground tunnels

Toshihisa Adachi
Kyoto University, Japan

Atsushi Yashima
Gifu University, Japan

Keisuke Kojima
Tokushima University, Japan

ABSTRACT: In the present study, to understand the actual behavior of a shallow, sandy ground tunnel and to establish a new design method for a thin, flexible shotcrete lining, laboratory model tests and numerical case studies using different types of analytical models were performed.

1 INTRODUCTION

A basic principle in tunneling is that a tunnel should be supported by the surrounding ground as much as possible. Namely, a tunnel must be sustained by the shear strength of the ground, and steel supports and/or linings should only play a role of assistant members in maintaining the strength of the ground. Shotcrete linings have been used based on the above principle, and their effectiveness is well recognized. However, the design method of shotcrete linings has not been established and the supporting mechanism, especially for shallow, sandy ground tunnels, is not well understood mainly because mechanical behavior of the surrounding ground including interaction between the ground and linings has not been clarified and no proper analytical methods have been developed so far. It is, therefore, very important to understand the actual behavior of the surrounding ground during the excavation phase and establish an analytical model to describe this behavior as accurately as possible.

To investigate the mechanical behavior of a shallow, sandy tunnel, laboratory model tests were performed. Chapter 2 concerns a series of model tests conducted in ground composed of dry sand by simulating the linings with thin tracing paper. In Chapter 3, circular tunnel excavation simulated by using a diameter reducible device preliminarily installed in the ground prepared by piling up aluminum rods is explained. Since it was found from experimental results that tunnel construction in cohesionless ground caused a particular pattern of discontinuous displacements, the

use of joint elements in the analysis to accomodate these discontinuities was proposed (Adachi et al., 1985). Also in Chapter 3, ability of the proposed numerical method to simulate the deformation behavior of model tests is demonstrated.

When a tunnel is excavated below ground water level, a drainage well is sometimes dug to lower the water level before excavation work. In Chapter 4, the effect of a ground water level lowering method for sandy ground tunnels is investigated and simulations of tunnel excavation with different ground water levels that were carried out are presented.

Chapter 5 concerns simulations of actual subway construction work in sandy ground by the proposed numerical method. Analytical results are discussed in relation to installation time of the linings.

In recognition of the difficulties and uncertainties associated with the determination of sandy ground tunnel design parameters from laboratory experiments and site explorations, the present study proposes a numerical procedure for the determination of non-linear constitutive parameters by back-analysis based on measurements made in the ground surrounding the tunnel in Chapter 6. Consequently, the present procedure of back-analysis based on site measurements in the early stages of sandy ground tunnel construction will permit fine tuning of the design parameters, thereby improving the design of subsequent stages of construction.

Finally, results of these chapters are summarized in Chapter 7.

2 EXPERIMENTAL STUDY ON A THIN AND FLEXIBLE TUNNEL SUPPORT SYSTEM

2.1 Introduction

The mechanical efficiency of a thin, flexible tunnel support system such as shotcrete lining and rock bolts was investigated on the basis of experimental work. Model tests of tunnel excavation in which shotcrete lining and rock bolts were simulated by pieces of thin paper were carried out in dry sand ground. From the results, it was found that, 1) even very thin and flexible tracing paper closed the ring lining and had a remarkable effect on tunnel stability, and, 2) the effect of rock bolts appears only when they are placed so as to penetrate into the outside of a plastic zone developed in the surrounding ground.

2.2 Basic principles of tunneling

We all have memories from childhood of amusing ourselves by tunneling in the sand. We learned by this experience that it was a task of extreme difficulty to tunnel through completely dry or completely saturated sand. This difficulty discovered in childhood is essentially nothing different from that encountered with actual tunnel construction. Even for dry sandy ground, however, the basic principle of tunneling, i.e., a tunnel should be supported by the surrounding ground as much as possible, must never be changed.

In recent years, NATM (New Austrian Tunneling Method) (Rabcewicz,1969) has been widely used, especially in Japan. The method may be characterized by the following three points.

1. It is based on the basic principle of tunneling, i.e., A tunnel should be supported by the surrounding ground as much as possible.

2. For realizing the basic principle of tunneling and preserving the strength of surrounding ground, flexible support systems such as shotcrete and/or rock bolts are efficiently applied based on the second principle of tunneling, i.e., Permit elastic deformation in ground, but exclude loosening it.

3. Field measurements (viz., observational method) are efficiently applied in order to confirm whether the flexible support system is satisfactorily functioning and to indicate the proper timing for operations such as final tunnel lining.

Both basic principles of tunneling stated above are not the original ideas only of NATM, but have been cultivated from experience in tunnel construction throughout the world, extending over many years. It can be said that the significance of NATM is in the realization of these principles by making the best use of shotcrete, rock bolt and ductile steel support.

In any case, it is important to clarify precisely and understand the mechanical behaviors of the surrounding ground in connection with tunneling. Therefore, first of all, we would like to explain the second principle.

Generally speaking, as schematically illustrated in Figure 2.1(a), the stress-strain curves of ground materials show a strain-hardening-softening type when the confining pressure is less than the transitional stress of the material. The volume change is compressive in nature at the initial stage of shear deformation before the peak strength is reached and shows remarkable dilatancy, viz., resulting in a loosening of the strength. Also, the stress-strain relations of ground materials are affected by the strain rate. Thus, immediately after excavating a tunnel, deformation of the surrounding ground is governed by the stress-strain relation indicated by the strain rate (high) in Figure 2.1(b), whereas the stress relaxation from point P to point Q will take place later when the ground deformation is restrained by placing tunnel lining.

Now, let's consider the case when the surrounding ground with the aforementioned mechanical properties and lining are working together against the shear stress $(\sigma_\theta - \sigma_r)$ developed by tunnel excavation as shown in Figure 2.1(c). It is easily understood that the total bearing strength of both ground and lining is required to be larger than the acting stress, in order to stabilize the tunnel. Since bearing stress of the ground varies with the deformation as given by the curve OBC in Figure 2.1(d), the portion of stress which should be taken over by the lining also changes with the tunnel wall displacement. The relationship between the bearing stress by lining and the tunnel wall displacement is given in Figure 2.1(e), this is the characteristic line known as Fenner-Pacher curve in NATM. This figure shows that the best way to place lining is to aim at point B where the ground exhibits the maximum strength. Con-

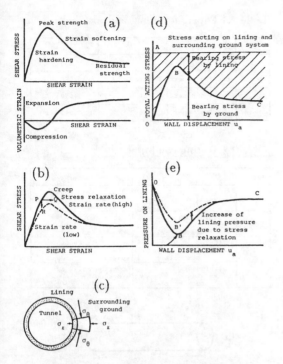

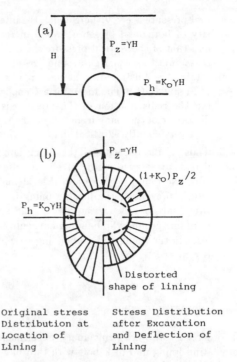

Figure 2.1 Schematical representation of the basic principle of tunneling, (a) Stress-strain relationship of ground material, (b) effect of strain rate on stress-strain relation, (c) lining and ground system against acting stress condition, (d) changes in lining and ground bearing stresses with tunnel wall displacement, and (e) change of lining pressure with tunnel wall displacement.

Figure 2.2 Peck's imaginary experiment, (a) anisotropic latent stress condition in ground, and (b) pressures and deflections of hypothetical lining (after Peck, 1969).

versely speaking, the acting pressure on the lining becomes the minimum value. This is the explanation for the second principle of tunneling, Permit elastic deformation in ground, but exclude loosening it.

It is also shown in Figure 2.1(e) that the lining pressure increases with time, because stress relaxation occurs in the ground later when the ground deformation is restrained by placing lining. This is one reason why the tunnel lining pressure increases with time especially for rate sensitive ground materials.

As previously mentioned, the specific feature of NATM lies in the best use of a flexible support system such as shotcrete, rock bolt or ductile steel support. It has been recognized that the flexible, thin shotcrete lining placed immediately after excavation is very effective not only for avoiding detrimental movement of the ground but

also for preventing the ground from deteriorating. Peck (1969) developed an ingenious concept for the mechanical effect of thin flexible lining by considering a series of imaginary experiments as follows:

1. Consider a mass of soil with a horizontal ground surface and assume the latent stress state in the soil is anisotropic (i.e., $K_0 < 1$, where K_0 is the coefficient of earth pressure at rest) as shown in Figure 2.2(a).

2. Push a circle lining, which is perfectly flexible but capable of supporting appreciable ring stress in compression, into the soil without disturbance.

3. Since no disturbance is involved, the state of stress in the soil is the same as the latent stress. The distribution of radial pressure on the lining is shown in the left hand side of Figure 2.2(b).

4. Now suddenly remove the soil from inside the tunnel. Since the circular flexible lining can be in equilibrium only when the ra-

293

dial pressure is everywhere equal, the intensity of horizontal pressure should increase and that of the vertical pressure should decrease until an equal all-around pressure is attained as illustrated in the right hand side of Figure 2.2(b). The mean for accomplishing the redistribution of the pressure is a distortion of the lining from a circular shape to a very slightly elliptical shape.

5. Thus, if the capacity of the lining is sufficient to carry the ring stress associated with the radial pressure and if local buckling is prevented, the flexible lining is entirely satisfactory for support of the surrounding ground. No bending moments can exist in the lining and no bending strength is required.

2.3 Model test procedure

In order to test Peck's imaginary experiments, simulating lining and rock bolt with thin paper, model tests were carried out in dry sand. For the tests, a rectangular box container, 80cm in height, 90cm in width and 30cm in depth, and a brass pipe with a diameter of 8cm were prepared, and dry Keisa sand was used as the ground material, assuming the ground conditions against tunnel stability to be the worst.

In Table 2.1, the type, weight and thickness of paper used as lining and rock bolt are given, however Kent Paper No.1 was only used as rock bolt. Four types of tests, A, B, C, and D were carried out as given in Table 2.2. Namely, test A was to examine how the effect of lining on the tunnel stability changes with the thickness of lining. Test B was to look into the influence of loosening the surrounding ground on stability of the tunnel. Test C was to evaluate how tunnel stability is improved by placing rock bolt in addition to lining. Test D was to investigate the influence of the variation of arrangement of rock bolts on tunnel stability (Table 2.3). In test D, the length of rock bolt was fixed at 8cm. The tests were carried out by the following procedure.

1. Put the brass pipe in the container as shown in Photo.2.1(a).

2. In test A, a paper lining, as in Figure 2.3(a), or in test B, a system with paper lining and paper rock bolt, as in Figure 2.3(b), is rolled

Table 2.1 Papers for lining and rock bolt.

	Paper type	weight (g/m^2)	thickness (mm)
No.1	Kent paper	154.3	0.180
No.2	tracing paper	60.0	0.058
No.3	tracing paper	50.0	0.050
No.4	tracing paper	40.0	0.045

Table 2.2 Testing condition.

Test	Supporting Condition	Parameter
A	lining only	lining thickness, overburden
B	lining only + looseness (δ)	looseness, overburden
C	lining + rock bolt	rock bolt length, paste, overburden
D	lining + rock bolt	configuration of rock bolt, overburden

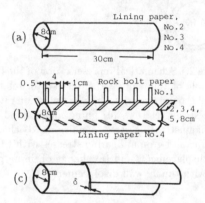

Figure 2.3 Size and arrangement of lining and rock bolt, (a) for test A, (b) for test C, and (c) for test B.

around the brass pipe. Photo.2.1(b) shows this stage in the test.

3. Sand is put into the container and tamped every 10cm until the prescribed overburden is reached. Overburden here means the depth from ground surface to the springline elevation, and was varied at 10, 15, 20, 25, 30, 40, and 50cm in this study.

4. Finally, to simulate progress of the tunnel face, only the brass pipe is pulled out 2cm by 2cm, leaving the paper lining in the ground. The pulled out length of the pipe at the time when the tunnel collapses is recorded as the face advancement. In test B, the pa-

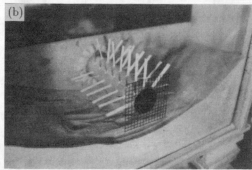

Photo.2.1 Testing procedure, (a) inserting the brass pipe, (b) rolling the lining pipe, and (c) preparation completed.

per with a thickness of δ-mm is pulled out to loosen the surrounding ground artificially before the simulation of tunnel excavation.

2.4 Test results

Figure 2.4 shows the results of test A for the cases when tracing paper No.2, No.3 and No.4 were

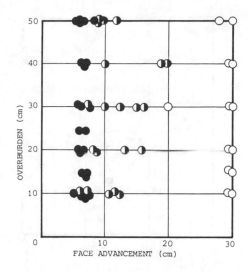

Figure 2.4 Relation of tunnel advancement at tunnel collapse vs overburden for test A.
○No.2, ○No.3, ○No.4

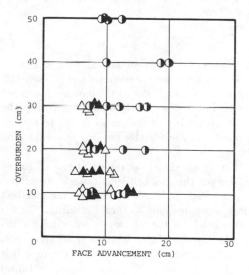

Figure 2.5 Relation of tunnel advancement at tunnel collapse vs overburden for test B.
△δ=3mm, △δ=1mm, ○δ=0mm.

used as lining paper. For the No.4 lining, the tunnel collapses at the face advancement of around 5cm, irrespective of the dimensions of overburden. It is clearly seen that the face advancement increases when the lining becomes thicker. From this figure, thin flexible lining is very effective in

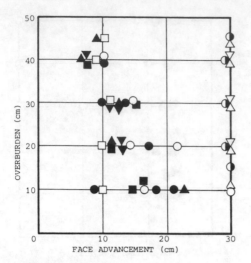

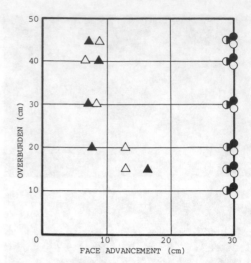

Figure 2.6 Relation of tunnel advancement at tunnel collapse vs overburden for test C.
(without paste) ○L=2cm, ⊔L=3cm, △L=4cm, ▽L=5cm, ○L=8cm
(with paste) ●L=2cm, ⊔L=3cm, ▲L=4cm, ▼L=5cm.

Figure 2.7 Relation of tunnel advancement at tunnel collapse vs overburden for test D.
(without paste) ○Basic type, ○Type 1, △Type 2, ▲Type 3, ○Type 4.

stabilizing the sandy ground tunnel.

Figure 2.5 shows the results of test B. From this figure, it is found that as the loosening area becomes larger, the tunnel becomes more unstable. This unstability of the tunnel is due to the appearance of considerable bending moment in the lining.

Figure 2.6 shows the results of test C. In the case when the rock bolt length is 2~5cm without a paste margin, it can be said that the rock bolt does not contribute to tunnel stability. On the other hand, for the case with a paste margin, the stabilizing effect by rock bolt is easily recognized from the figure. In the case when the rock bolt length is 8cm without a paste margin, the tunnel does not collapse up to the face advancement of 30cm. It is concluded that the effect of rock bolts appears only when they are placed to penetrate

Table 2.3 Rock bolt arrangement.

Basic Type	Type 1	Type 2	Type 3	Type 4

into the outside of a plastic zone developed in the surrounding ground.

In test C, we investigated the influence of variation of rock bolt length on tunnel stability with a basic type rock bolt arrangement shown in Table 2.3. In test D, influence of the variation of rock bolt arrangement on the tunnel stability was investigated. Figure 2.7 shows the results of test D. From this figure, rock bolts placed at an inclination of 45° from the splingline are found to be the most effective. On the other hand, rock bolts placed at the tunnel crown and splingline do not contribute to tunnel stability. In this test case, the length of rock bolt was fixed of 8cm. Therefore, it is supposed that rock bolt placed at the tunnel crown stays inside of the plastic zone.

The experimental result with No.4 paper in test A (in Figure 2.4) agrees well with the results of Murayama's lowering panel experiment (Murayama and Matsuoka, 1974) in which he used aluminum rods of 5cm in length having various diameters to simulate the sandy ground. What he found from his experiments are as in the following. As shown in Figure 2.8, aluminum rods within Region-1 descend in the same amount as that of the lowering panel. When Region-1 descends, Region-2 loosens in the mass and flows toward Region-1, but the descending speed of Region-2 is not remarkable as that of Region-1. How-

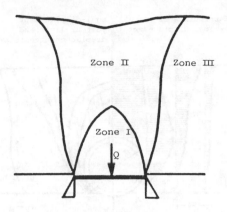

Fig.2.8 Various regions developed in sandy ground (Murayama and Matsuoka, 1974)

ever, Region-3, the outer part of Region-2, does not show any flowing movement. Futhermore, the vertical load Q acting on the lowering panel significantly changes depending on the width B of the lowering panel. However, it does not change due to a difference in overburden and is nearly equal to the weight of the aluminum mass in Region-1. Comparing the results of test D with Murayama's experimental result, it can again be said that the effect of rock bolt becomes remarkable only when it is placed so as to penetrate into the outside of Region-1. From this experiment, we can learn that in order to guarantee tunnel stability in sandy ground, controlling the lining thickness is practically easier and a better method than placing longer rock bolts.

3 BEHAVIOR AND SIMULATION OF SANDY GROUND TUNNEL

3.1 Introduction

To obtain basic data for field measurements and the design methods of sandy ground tunnels, a series of laboratory model tests with a diameter reducible device preliminarily installed in the ground and two different types of finite element analyses were performed. In the case of shallow overburden, a large displacement took place even at the ground surface, while the larger displacement zone reduced in size when the overburden became deeper. The use of joint elements in the analyses was found to be more effective than that of usual elasto-plastic elements to describe

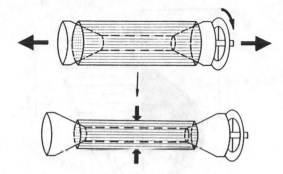

Figure 3.1 Diameter reducible device to simulate tunnel excavation.

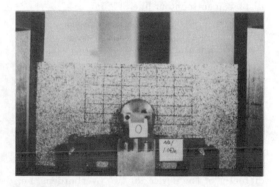

Figure 3.2 Apparatus and specimen of aluminum rod mass used for experimental studies.

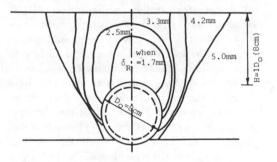

Figure 3.3 Enlargement of equi-displacement contour lines of 2mm during the reduction of the tunnel diameter.

the discontinuous movements occurring in sandy ground due to tunneling.

297

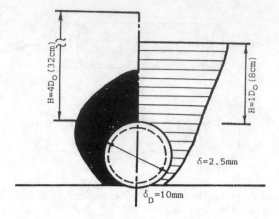

Figure 3.4 Experimental results of equi-displacement contour line of 2.5mm.

3.2 Experimental studies

A circular tunnel excavation was simulated by using a diameter reducible device (Figure 3.1) preliminarily installed in the ground prepared by piling up aluminum rods of 5cm in length with diameters of 1.6 and 3.2mm, as shown in Figure 3.2. The tests were carried out to see how the surrounding ground behaves when reducing the diameter of the device for two overburdens, i.e., H=1D and 4D (where D is the tunnel diameter, 8cm).

Figure 3.3 shows how the equi-displacement contour lines of 2mm enlarge while reducing the tunnel diameter for a shallow overburden (H=1D). It was found that a large displacement zone enlarges from the tunnel crown toward the ground surface very rapidly and reaches the ground surface at the reduction of the tunnel diameter of 3.3mm.

The equi-displacement contour lines of 2.5mm obtained when reducing the tunnel diameter by 10mm are shown in Figure 3.4 for two different overburdens, i.e., H=1D and 4D. The vertical displacement distributions in the positions, 0, D/2, D and 3D/2 from the center line of the tunnel axis are given in the left half of Figure 3.5, while the equi-displacement contour lines are in the right half. One can see that larger displacement zones developed when the overburden became smaller.

3.3 Analytical studies

Elasto-plastic finite element and joint element anal-

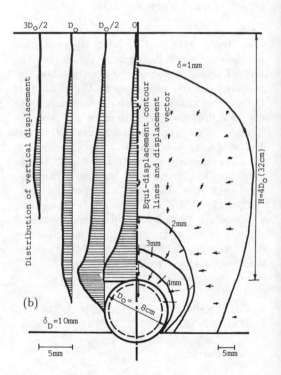

Figure 3.5 Experimental results of vertical displacement distributions and equi-displacement contour lines, (a) H=1D, and (b) H=4D.

yses were performed. In both, a finite element mesh as shown in Figure 3.6 was used.

3.3.1 Elasto-plastic element analysis

The initial stress state in the ground was assumed to be a K_0-condition (K_0=0.5). The prescribed

298

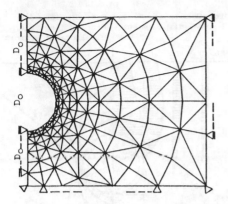

Figure 3.6 Finite element mesh (H=1D).

Table 3.1 Material parameters used in analyses.

	Aluminum
Unit Weight γ (gf/cm^3)	2.18
Young's Modulus E (gf/cm^2)	$500 + 10^2\sigma_m$
Poisson's Ratio ν	1/3
Cohesive Strength c (gf/cm^2)	0
Internal Friction Angle ϕ (°)	30
Coefficient of Earth Pressure at Rest K_o	0.5

displacements were given for the nodal points on the tunnel wall in the case of an aluminum rod ground. Drucker-Prager's yield criterion (Drucker and Prager,1952) was applied and material parameters are summarized in Table 3.1.

3.3.2 Joint element analysis

The discontinuous ground movements in sandy ground generally take place due to tunneling. To describe this behavior, Goodman's joint elements (Goodman and St.John, 1977) were applied in the analyses.

Yield shear stress, τ_y, is determined as a function of the normal stress according to the Mohr-Coulomb failure law:

$$\tau_y = c + \sigma_n\tan\phi \quad \sigma_n \geq 0: \text{ compression}$$
$$\tau_y = 0 \quad \sigma_n \leq 0: \text{ tension}$$

$$(3.1)$$

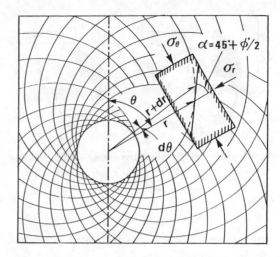

Figure 3.7 Trajectories of failure lines around the tunnel.

where c is the cohesive stress and ϕ is the friction angle of the joint element.

The finite element mesh was the same as the elasto-plastic analysis, except joint elements were distributed on all of the boundaries of triangular finite elements. The joint elements on the tunnel wall were arranged so as to be at an angle of $(45° - \phi/2)$ to the maximum principal stress, σ_θ, and parabolically extended into the surrounding ground. If the stress state is hydrostatic and the Mohr-Coulomb failure criterion is admissible, the failure surface appears at the angle of $(45° + \phi/2)$ to the maximum principal stress plane. Considering a small sample in a polar coordinate system, as illustrated in Figure 3.7, the trajectory of the failure line is expressed as

$$\frac{r \cdot d\theta}{dr} = \tan\alpha.$$

$$(3.2)$$

Considering the relation of $\tan\alpha = \cos\phi'/(1 - \sin\phi')$ and integrating Eq.(3.2) with the boundary condition of $r = R$, we get

$$r = R \cdot exp(\frac{\cos\phi'}{1 - \sin\phi'})$$

$$(3.3)$$

where R is the radius of the tunnel. Figure 3.7 shows trajectories of failure lines obtained by Eq.(3.3). Although the stress state in real ground is complicated and usually not hydrostatic, joint elements in this study are arranged according to

299

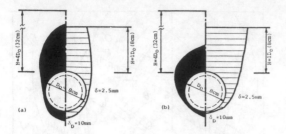

Figure 3.8 Analytical results of equi-displacement contour lines of 2.5mm, (a) elasto-plastic analysis, and (b) joint element analysis.

trajectories of failure lines illustrated in Figure 3.7.

The material parameters used in the analyses are given in Table 3.1, and the elastic-rigidity of joint elements in the normal direction was selected as 100 times larger than that of the triangular element.

3.4 Analytical results and discussion

The analytical results are given in Figures 3.8, 3.9, and 3.10. The equi-displacement contour lines of 2.5mm for H=1D and 4D are shown in Figure 3.8. Comparing Figure 3.8 with Figure 3.4, the joint element analysis was found to explain the test results better than the elasto-plastic analysis. In Figure 3.9, the joint element analysis was also recognized to better simulate the test result for H=1D, not only in the vertical displacements but in the equi-displacement contour lines as well. The advantage of the joint element analysis is clearly seen for H=4D in Figure 3.10.

To establish a design method of flexible shotcrete linings, simulation of the model test discussed in Chapter 2 was carried out. The analyses for dry sand ground with an overburden of H=50cm were performed for two different lining rigidities of $EA = 10$ and $10^4 gf$. Other material parameters are shown in Table 3.1.

Plastic zones and plastic joint elements developed in the surrounding ground are shown in Figure 3.11 for the case of $EA = 10gf$. The actual phenomenon, i.e., the loosening zone developing upwards in the case of a shallow, sandy ground tunnel, can be well seen as simulated by the joint element analysis.

The displacement distributions of the thin lin-

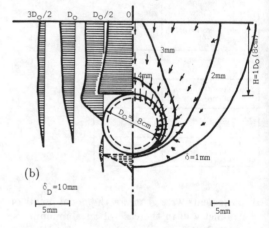

Figure 3.9 Analytical results of vertical displacement distributions and equi-displacement contour lines for H=1D, (a) elasto-plastic analysis, and (b) joint element analysis.

ing are shown in Figure 3.12. The actual displacement pattern of the lining in a K_0 initial stress state is elliptical as shown in Figure 3.13. Namely, larger convergences take place at the crown and invert portions than at the springline. The joint element analysis better simulates this fact, too.

The distributions of axial force acting in the linings are given in Figure 3.13. In the case of a very thin flexible lining ($EA = 10gf$), the results by the joint element analysis show a very small

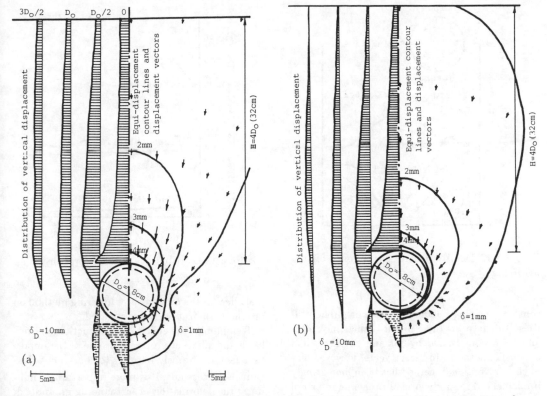

Figure 3.10 Analytical results of vertical displacement distributions and equi-displacement contour lines for H=4D, (a) elasto-plastic analysis, and (b) joint element analysis.

force in the lining and very uniform distribution (Peck, 1969).

From the results, the joint element analysis better simulated tunneling in sandy ground than the elasto-plastic analysis. This means that a good device is necessary in the analysis of ground behavior where discontinuous movements may take place.

4 CONSIDERATION OF THE INFLUENCE OF GROUNDWATER LEVEL

4.1 Introduction

As a result of expansion of the major population regions in Japan, it has become necessary to construct new railway systems. It is becoming more common that underground railways are planned and constructed in consideration of the environment preservation especially in commercial and residential areas.

In underground railway construction, it is bet-

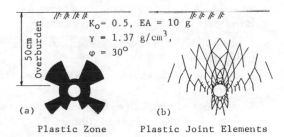

Figure 3.11 Plastic zones and plastic joint elements developed in the ground, (a) elasto-plastic analysis, and (b) joint element analysis.

ter to leave the earth covering as thin as possible from the point of view of construction cost and maintenance cost for drainage, ventilation and traffic safety as well as the convenience of passengers. Since the major cities of Japan are located in wide alluvial plains where soft ground with a high groundwater table prevails, tunnel should be built safely without damaging adja-

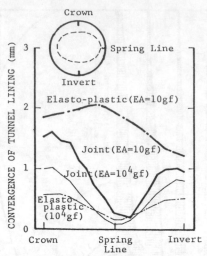

Figure 3.12 Distributions of convergence of tunnel lining.

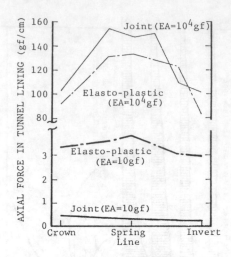

Figure 3.13 Distributions of axial force in lining.

cent or overlying buildings, streets or public utilities. In the past, these urban tunnels have usually been constructed by cut and cover method or shield method. In recent years, however, use of the conventional method has been increasingly implemented in sandy ground tunneling in lieu of the shield method and this has resulted in reduction of the tunneling costs to a great extent.

When tunneling through ground with a thin cover and high groundwater table, it is necessary to investigate the following problems carefully:

1. the stability of the tunnel face and surrounding ground,
2. problems related to underground water conditions, and
3. the effect of tunneling on adjacent or overlying structures and public utilities.

To work out these problems, it is most important to understand the actual behavior of the surrounding ground during tunnel excavation and to establish an analytical method which can simulate the behavior as accurately as possible.

Tunnel construction causes discontinuity especially in cohesionless sandy ground. In Chapter 3, a way to simulate the discontinuous movement in sandy ground was proposed by using joint elements in analysis and its efficiency was shown.

In this chapter, first, the analytical method is extended to include due consideration of the mechanical interaction between groundwater and the surrounding ground. As an application exam-

ple, effect of the groundwater lowering method on ground stabilization is investigated.

To minimize the effects of construction on overlying and adjacent properties is one of the important factors. Namely in order to construct urban tunnels successfully, it is more important to minimize the deformation of surrounding ground(e.g. surface settlement, etc.) than to minimize the earth pressure on tunnel supports or tunnel linings. Accomplishment of this by using the so-called characteristic line method (NATM) is often discussed.

4.2 Analytical method

The framework of the analytical method adopted in this chapter is the same as that developed by Akai and Tamura (1978) and used in a previous tunnel problem (Adachi et al.,1979), except for introducing a joint element in the analysis (Adachi et al., 1985) and Adachi (1985). Expressed in other words, this method introduces a consolidation theory to the method mentioned in Chapter 3, to simulate the influence of groundwater. The analytical method is thoroughly derived on the basis of generalized Biot's equations for consolidation (Biot,1941).

As described in Chapter 3, Goodman's joint elements were used and arranged on some borders between triangular or rectangular elements, to simulate the discontinuity developed in the surrounding ground during tunnel excavation. For

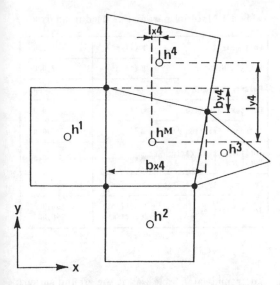

Figure 4.1 Geometrical configuration appeared in equation of continuity.

normal continuous elements (triangular and rectangular elements) , Akai and Tamura's method (1978) can be adopted with slight modification to deal with leaning triangular elements. Based on Biot's equations and the variational principle, they formulated the extended element stiffness equation for multidimensional consolidation analysis as follows

$$[K_e]\{\Delta u\}|_{t+dt} + \{B_v\}\gamma_w h|_{t+dt} =$$
$$\{\Delta f\}|_{t+dt} + \{B_v\}\gamma_w h|_t \qquad (4.1)$$

$$\{B_v\}^T\{\Delta u\}|_{t+dt} + \alpha\gamma_w h|_{t+dt} - \Sigma\alpha_i\gamma_w h_i|_{t+dt} = 0 \qquad (4.2)$$

where $[K_e]$ is the elemental stiffness matrix, $\{\Delta u\}$ is the increment of nodal displacement vector, h is the total water head of the element, h_i is the total water head of the neighboring element, $\{B_v\}$ is the vector to calculate the element volume change from nodal displacements, $\{\Delta f\}$ is the increment of the nodal force, γ_w is the unit weight of groundwater, and t and $t + dt$ denote discretized time. Eq.(4.1) is the equation of the equilibrium to express that the nodal forces by the effective stress and the pore water pressure increment have to be equal to the nodal load increments. On the other hand, Eq.(4.2) is the equation of continuity to represent that the volume change of an element must be equal to the total flow rate from neighboring elements into the element. Assuming

anisotropic permeability and referring to Figure 4.1, α and α_i in Eq.(4.2) are expressed as

$$\alpha = dt(\Sigma k_x b_{yi}/l_{xi} + \Sigma k_y b_{xi}/l_{yi}) \qquad (4.3)$$

$$\alpha_i = dt(k_x b_{yi}/l_{xi} + k_y b_{xi}/l_{yi}) \qquad (4.4)$$

where k_x and k_y are coefficients of permeability of x and y directions, and dt is the increment of time step.

However, extended element stiffness equations, Eqs.(4.1) and (4.2) are not able to be applied to joint elements, because joint elements have a different feature from continuous ones. Consequently, a few assumptions must be introduced as follows:

1. joint elements have no pore water pressure, except when calculating the initial stress of the element, and

2. from the standpoint of movement of pore water, a joint element is ignored, if that joint element is located between two continuous elements. Namely, pore water can flow only between two continuous elements.

These assumptions are based on the fact that joint elements take up no volume. From the above assumptions, the extended stiffness equation for a joint element can be written as

$$[K_{ej}]\{\Delta u\}|_{t+dt} = \{\Delta f\}|_{t+dt} \qquad (4.5)$$

in which $[K_{ej}]$ is the elemental stiffness matrix for the joint. It is worth noting that joint elements do not have a continuity equation, because they don't have any pore water.

For continuous elements, the elasto-plastic or elasto-viscoplastic constitutive law can be used in this analytical method, however, the elastic constitutive law was adopted for the finite elements in this study.

On the other hand, the elasto-(perfectly) plastic constitutive model, the same as mentioned in Chapter 3, is adopted for joint elements. It is based upon Mohr-Coulomb's criterion, but its details are omitted here.

4.3 Numerical example

The plane strain condition was assumed in all examples. Figure 4.2(a) shows the initial and

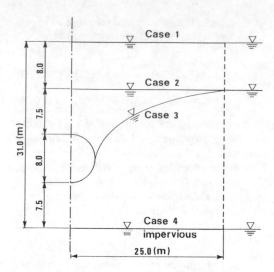

Figure 4.2(a) Initial and boundary conditions of groundwater.

Table 4.1 Material parameters used in analyses.

Unit Weight	γ (tf/m³)	2.0
Young's Modulus	E (tf/m²)	2,000
Poisson's Ratio	ν	0.333
Cohesive Strength	c (tf/m²)	0.0
Internal Friction Angle	ϕ' (°)	30.0
Coefficient of Earth Pressure at Rest	K_θ	0.5
Permeability Coefficient	k (cm/sec)	10^{-2} , 10^{-4}
Joint Stiffness	k_n (tf/m²) k_s (tf/m²)	150,000 150,000

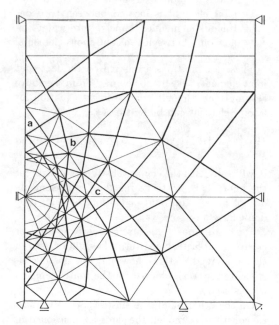

Figure 4.2(b) Finite element mesh used in analysis.

boundary conditions for the problems discussed in this chapter. A tunnel with an 8m diameter was excavated with 15.5m of earth covering under four different initial groundwater conditions, indicated as Case-1 to 4, that is, Case-1, the ini-

tial groundwater table was at the ground surface; Case-2, it was at 7.5m above the tunnel crown; Case-3, it changed parabolically from the tunnel springline into the surrounding ground as corresponding to a case of partial dewatering; and Case-4, it was at the base rock. The soil layer through which a tunnel passes was assumed to be sand ($k = 0.01 cm/sec$) or silt ($k = 0.0001 cm/sec$) in this study. The finite element mesh used in the problems is as shown in Figure 4.2(b). The thick solid lines in the mesh correspond to the joint elements. The material parameters used in the analysis are summarized in Table 4.1, in which k_n and k_s are the joint stiffness parameters for the normal and tangential directions. Tunnel excavation in the analysis was simulated by releasing the initial nodal stress from the nodal points on the tunnel periphery by 4 stress release steps. The complete stress release(100%) was assumed to have taken place for one day. In order to investigate qualitatively the effect of dewatering, the problems were analyzed for the case of excavation without any support.

4.4 Effect of groundwater level

Effect of the underground water table was investigated for tunneling in sandy layer with a coefficient of permeability of $0.01 cm/sec$. Figure 4.3 shows the distributions of the initial mean effective stress and the initial pore water pressure p_w in the adjacent surrounding ground along the tunnel periphery. Naturally, in Case-1 with a high groundwater table, the mean effective stress value

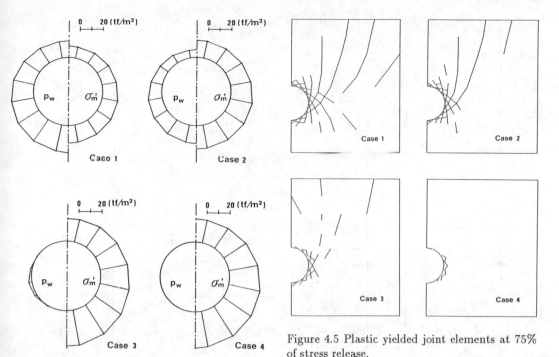

Figure 4.3 Initial mean effective stress and pore water pressure around tunnel periphery.

Figure 4.5 Plastic yielded joint elements at 75% of stress release.

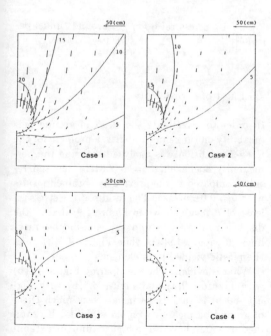

Figure 4.4 Displacement vectors and equi-displacement contour lines in surrounding ground at 75% of stress release.

is small and the pore water pressure is relatively high, while those values are reversed in Case-3 and Case-4 with a lower groundwater table. Considering the strength dependency on the effective confining stress, it can be easily estimated that the stable ground condition will be preserved under the lower groundwater table condition.

In Figure 4.4, the equi-displacement lines and the displacement vectors calculated at 75% of the stress released are given for the four cases. Although the computed results at 100% of the stress released is not given here, in Case-1 the soil, especially above the tunnel, flakes off and fall down into the tunnel opening due to liquefaction. From Figure 4.4, effect of underground water head on ground movements is more remarkable above than beneath the tunnel. It is obvious also that the larger displacements can be seen under a high groundwater table condition.

The plastic yielded joint elements developed in the ground at 75% of the stress released are shown in Figure 4.5. In the cases with a high groundwater table, i.e., Case-1 and 2, the yield joint elements become interconnected from the tunnel wall to the ground surface, while in Case-4 the yielded joint elements appear only around the tunnel springline.

305

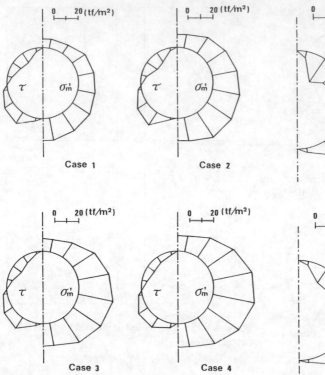

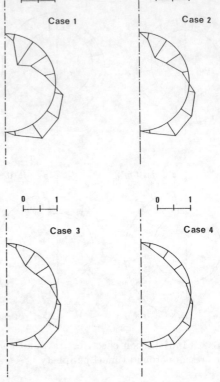

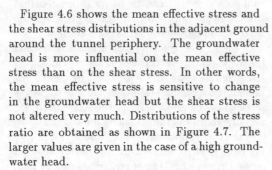

Figure 4.6 Mean effective stress and shear stress around tunnel periphery.

Figure 4.7 Stress ratio (τ/σ'_m) around tunnel periphery.

Figure 4.6 shows the mean effective stress and the shear stress distributions in the adjacent ground around the tunnel periphery. The groundwater head is more influential on the mean effective stress than on the shear stress. In other words, the mean effective stress is sensitive to change in the groundwater head but the shear stress is not altered very much. Distributions of the stress ratio are obtained as shown in Figure 4.7. The larger values are given in the case of a high groundwater head.

Figure 4.8 shows the change of the mean effective stress distribution in the ground between the ground surface and the tunnel crown with a progress of the stress release. In the cases with a high groundwater head, the magnitude of the mean effective stress is smaller even before the excavation started (0% of the stress release), and it is worth noting that in Case-1 the mean effective stress becomes nearly zero at the complete stress release state (100%). This satisfied the condition for liquefaction of the ground above the tunnel.

4.5 Effect of permeability

In order to investigate the effect of ground permeability on tunneling stability, a problem for the Case-2 condition was analyzed by using the permeability coefficient $k = 0.0001 cm/sec$ and results compared to the previously obtained results with $k = 0.01 cm/sec$. The results at the stress release of 75% are shown in Figure 4.9, that is, the displacement vectors and the equi-displacement lines in Figure 4.9(a), while Figure 4.9(b) gives the plastic yielded joint elements.

What is evident when comparing Figure 4.4(b) with Figure 4.9(a) and Figure 4.5(b) with Figure 4.9(b) is that there in not very much effect of ground permeability on the stability. If examined in detail, however, a higher tunneling stability can be preserved when the permeability coefficient takes a smaller value.

Figure 4.10 shows the pore water pressure change

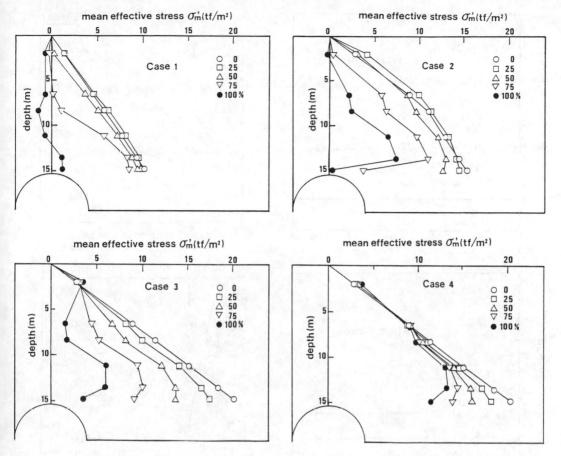

Figure 4.8 Change of mean effective stress distribution in the ground above tunnel crown with a progress of stress release.

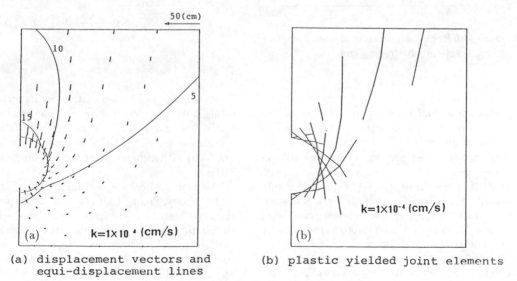

(a) displacement vectors and   (b) plastic yielded joint elements
    equi-displacement lines

Figure 4.9 Analytical results for k=10⁻⁴cm/sec.

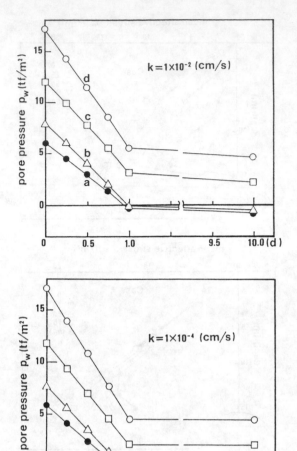

Figure 4.10 Pore water pressure change with time at four points in surrounding ground.

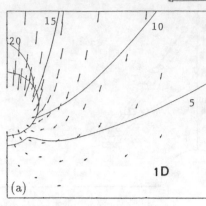

(a) displacement vectors and equi-displacement lines

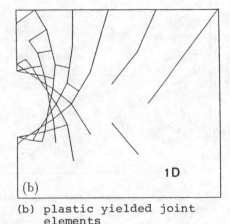

(b) plastic yielded joint elements

Figure 4.11 Analytical results for H=1D.

with time at four points, a, b, c and d located in the ground 2m away from the tunnel wall as shown in Figure 4.2. In the case of $k = 0.0001$ cm/sec, the excess pore water pressure still remains one day after the excavation started, i.e., at 100% stress release, but the magnitude is very small. When tunneling with a usual daily advance rate (100% of the stress release completed for one day in the analysis) through sandy or silty layers whose permeability coefficient is in the region of $0.01 - 0.0001 cm/sec$, it is concluded that effect on the ground stability is very slight.

4.6 Effect of earth covering

To investigate the effect of earth covering on tunneling stability by comparing with the results of Case-2, we analyzed a problem under the following conditions. The depth of cover was about 1D(7.5m) with a groundwater table at the ground surface.

The computed results are illustrated in Figure 4.11. Namely, Figure 4.11(a) shows the displacement vectors and equi-displacement lines, while the plastic yielded joint elements are given in Figure 4.11(b); it is obvious that when the cover becomes thinner, the displacement becomes larger, and the number of plastic yielded joint elements increase.

5 APPLICATIONS TO ACTUAL CASE STUDIES

5.1 Introduction

As a result of expansion of the major population regions and the destructive rise of land prices in Japan, it has become necessary to develop underground, the so called "GEO-FRONT" including very deep regions.

Areas around the north-eastern part of Chibaprefecture have a rapidly increasing population as suburbs of Tokyo (and Chiba), resulting in train congestion during morning and evening rush hours which has reached the saturation level. Accordingly, several new railway systems were planned in these areas as shown in Figure 5.1. The Toyo Rapid Transit Line is a new access line, which was planned to ease the congestion and is scheduled to be directly connected with the Tozai Line and Soubu Line of Japan Railway (J.R.). In addition, Hokusou Development Railway Line is an extended line 11.94km in length, starting from Takasago station of the Keisei Electric Railway and terminating at the Kita-Hatsutomi station of Hokusou Development Railway which was planned to reach New Tokyo International Airport.

Both railway lines were planned to go underground for certain lengths because they pass through commercial or residential regions. These sections, therefore, consist of tunnels that were decided to be constructed by conventional method (NATM) from the standpoints of environmental preservation and reduction of construction cost. Since these tunnels pass at a relatively shallow depth(less than 10m) beneath busy streets in commercial or residential sections, the construction should not excessively affect adjacent or overlying properties.

In this chapter, in order to find a suitable excavation method to minimize ground movement, the analytical results will be compared with the measured results and observations obtained from trial tunneling and the analyses will be discussed. In section 5.2, analyses are carried out to simulate the behavior of Narashinodai Tunnel of the Toyo Rapid Transit Line which was excavated on trial using two different divisions of heading. On the other hand, in section 5.3, simulational calculations are executed for excavation of Kuriyama Tunnel of Hokusou Development Railway Line to provide suitable tunnel lining techniques.

Figure 5.1 Map of two new railway lines with tunnel locations.

5.2 Effects of divisions of heading on tunnel stability

5.2.1 Narashinodai Tunnel

As shown in Figure 5.1, the Toyo Rapid Transit Line which connects Nishi-Funabashi and Katsutadai in Chiba prefecture, covering 16.2km, 10.8km above ground and 5.4km underground was constructed. The middle part of the section was planned to go underground because the line passes through a commercial and residential section. The construction work is divided into cut-and-cover work and tunnel work. For the former, the soil continuous underground wall method is adopted to minimize the ground level change near the excavation site, considering the underground water level. For the tunnel named Narashinodai Tunnel, conventional method (NATM) is applied from the standpoint of environmental preservation and reduction of construction cost.

Since this tunnel passes at a relatively shallow depth beneath busy streets in commercial and residential regions, the construction should not excessively affect adjacent or overlying properties. In order to find a suitable excavation method to minimize the ground movement, a section was excavated on trial by using two different divisions of heading.

5.2.2 Geological conditions

The soil conditions and topography along the tunnel alignment are shown in Figure 5.2. The cover

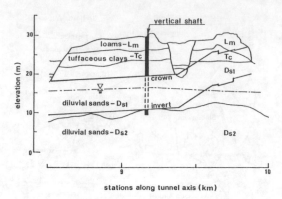

Figure 5.2 Soil conditions and topography along the tunnel alignment

Table 5.1 Material parameters for sand (DS$_1$) used in analyses.

Unit Weight	γ (tf/m^3)	1.8
Young's Modulus	E (tf/m^2)	200+200σ'_m
Poisson's Ratio	ν	0.35
Cohesive Strength	c (tf/m^2)	2.0
Internal Friction Angle	ϕ ($°$)	32
Joint Stiffness	k_n (tf/m^2) k_s (tf/m^2)	20,000 20,000

over the tunnel ranges from zero to 10m. The soils vary from tills, loams, clays and sands with a groundwater table near the tunnel springline. The tunnel passes through an upper sandy layer, D_{s1} with N-value (SPT blow number) less than 30. The D_{s1} layer is unconsolidated hydrous sand layer formed in the fourth diluvial epoch, which is generally called the Narita Sand Layer. The material parameters of D_{s1} are summarized in Table 5.1, which were obtained from soil explorations and soil tests. The D_{s1} layer is homogeneous, showing a uniformity coefficient of about 2 and little fine particle inclusion rate. The D_{sh} layer is located near the tunnel foundation where a lot of shell is included. The D_{sh} layer is water permeable, showing typical features of an aretesian conditioned layer. Judging from these characteristics of geology and water logic, the excavation has to be executed under considerably adverse conditions.

5.2.3 Tunneling procedures

The tunnel in the standard section is horseshoe shaped, 10m wide and 8.5m high. Cross section area after completion is $52m^2$ and the excavated one about $72m^2$. Trial tunneling was carried out from a vertical shaft (Kita-Narashinodai Shaft locates at 9.185km, indicated in Figure 5.2) for about the first 40m by a division of heading, the so-named CD-method (with Center Diaphragm) and for the next 25m by the CDS-method (with Center Diaphragm and Struts) as illustrated in Figure 5.3.

In the CD-method, tunneling was carried out by taking out section [1] and providing primary support which consists of II-shape steel support (150mm) and shotcrete of 20cm thickness as well as a center diaphragm made of H-shape steel support (125mm) and shotcrete of 12cm thickness as shown in Figure 5.3. Similar procedures were continued in tunneling sections [2], [3], [4], [5], and [6]. Excavation of the right half section, that is, sections [4], [5] and [6], was performed 20m behind the working face of the left half section.

On the other hand, what distinguishes the CDS-method from the CD-method is the heading sequence and installation of struts between the primary support and center diaphragm at each excavation stage. It should be noted also that cycle length to reach full tunnel section was 15m (taking about 20 days) in the case of CDS-method, while it was 30m (about 30 days) in the CD-method. It can be said that tunneling by the CDS-method was performed smoothly.

Two kinds of supplemental construction methods have been adopted before the cutting face arrives. One is a chemical fluid injection method to consolidate the ground, and the other is a dewatering method to deal with underground water.

5.2.4 Comparison of computed and measured results

The analytical method developed in the previous chapter was applied to simulate the behavior observed during the above mentioned tunneling. The material parameters adopted in the analyses are given in Table 5.1. Poisson's ratio is assumed to be 0.35 and joint stiffness parameters are giving as $k_n=k_s=2.0\times10^4$tf/m^3 by taking Young's modulus for the sands into account.

The primary support (H-shaped steel support with shotcrete), the center diaphragm (H-shaped

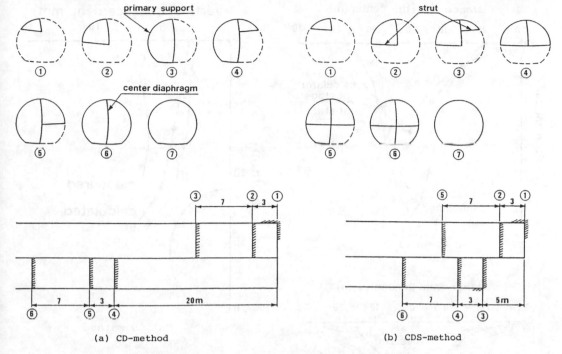

Figure 5.3 Division of heading of excavation sequence.

Table 5.2 Parameters for primary support, center diaphram and struts used in analyses.

	Primary support	Center Diaphragm	Struts
EA(tf)	235,200	188,370	63,630
EI(tf·m²)	620	338	178

steel support with shotcrete) and the struts (H-shape steel beam) are modeled as beam elements in the analyses and their parameters, that is EA (A:cross-sectional area) and EI (I:the geometrical moment of inertia) are set as indicated in Table 5.2.

The initial stress state in the ground was assumed to be K_0- condition ($K_0=0.54$). As previously mentioned, the tunnel excavation was simulated by releasing the initial nodal stresses, however, it is very difficult to judge at which numerical step the beam elements by way of primary support, center diaphragm and struts are provided. As a result of this consideration, beam elements were assumed to be introduced at the stress released rate of 67%.

The lateral distribution of surface settlement is shown in Figure 5.4 to compare the measured and calculated results at the excavation stage [6].

Relatively close agreement between measured and calculated values was obtained by the CDS-method, while in the CD-method the analysis underestimated surface settlement. The measured and calculated distributions of vertical displacement in the ground between ground surface and tunnel crown are given in Figure 5.5. It is seen in the figure that analysis for the CD-method considerably underestimates the displacement but that by the CDS-method slightly overestimates them.

Figure 5.6 shows axial stresses developed at some portions in each support element and the measured values are also indicated for the case by the CDS-method. The measured values agree well with the calculated results in the tunnel crown arch section, but good agreement can not be seen in other portions.

Although the ground behavior depended on the difference in division of heading, a large difference can not be predicted by the analysis. A reason why a larger movement took place in the case by the CD-method can be considered due to its slower tunneling advance rate. In order to take account of this effect, an analysis was carried out by taking the time to provide the supports at 100% of stress released condition. In other words,

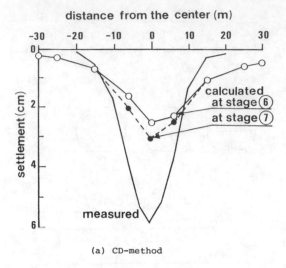

(a) CD-method

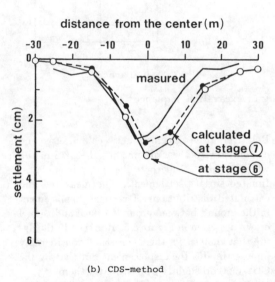

(b) CDS-method

Figure 5.4 Lateral distribution of surface settlement.

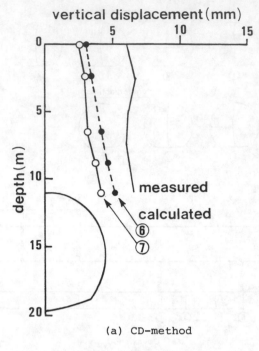

(a) CD-method

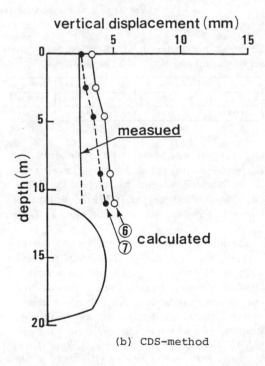

(b) CDS-method

Figure 5.5 Measured and calculated distributions of vertical displacement in the ground above tunnel crown.

the supports are provided immediately before excavation of the next section. The calculated results together with the measured values are shown in Figure 5.7. The results give a better estimation than that by the previous analysis. Namely, this tells us that it is very important to form the inner structural shell around the tunnel periphery as quickly as possible. Also it is necessary in the analyses to take account of degrees of skill in tunneling.

Next, the influences caused by removal of the

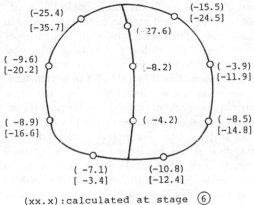

(-25.4) (-15.5)
[-35.7] [-24.5]
         (-27.6)

(-9.6)   (-8.2)   (-3.9)
[-20.2]   [-11.9]

(-8.9) (-4.2) (-8.5)
[-16.6]   [-14.8]

    (-7.1)   (-10.8)
    [-3.4]   [-12.4]

(xx.x):calculated at stage ⑥
[xx.x]:calculated at stage ⑦
   "-":denotes compressive stress(tf)

(a) CD-method

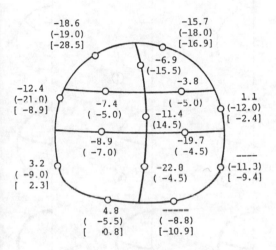

-18.6   -15.7
(-19.0)   (-18.0)
[-28.5]   [-16.9]
           -6.9
          (-15.5)
            -3.8
-12.4
(-21.0)   -7.4   (-5.0)   1.1
[-8.9] (-5.0) -11.4   (-12.0)
                 (14.5)   [-2.4]
          -8.9   -19.7
         (-7.0) (-4.5)
3.2   ----
(-9.0) -22.0   (-11.3)
[2.3] (-4.5)   [-9.4]

    4.8   -----
  (-5.5)   (-8.8)
  [0.8]   [-10.9]

xx.x :measured at stage ⑥
(xx.x):calculated at stage ⑥
[xx.x]:calculated at stage ⑦
  "-":denotes compressive stress(tf)

(b) CDS-method

Figure 5.6 Axial stress developed in support elements.

center diaphragm should be estimated. The calculated results corresponding to this tunneling stage [7] have been indicated in the figures already given. It is expected from the figures that the ground above the tunnel crown will raise about 5mm in the case by the CDS-method, but further surface settlement will occur in the CD-method,

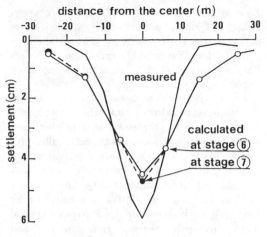

(a) lateral distributions of surface settlement

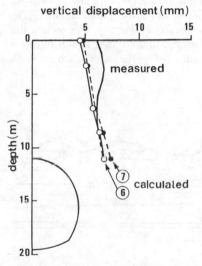

(b) Vertical displacement above tunnel crown

Figure 5.7 Analytical results at 100% of stress release and measured ones.

and the axial stresses in the primary support will increase when the center diaphragm is removed.

5.3 Effects of installation timing and stiffness of primary supports on tunnel stability

5.3.1 Kuriyama Tunnel

As shown in Figure 5.1, Hokusou Development

313

Railway Line was planned to connect Takasago station in Chiba-prefecture and New Tokyo International Airport (Narita Airport) by installing an extended line that directly connects to the Keisei Electric Railway which accesses the metropolitan region in Tokyo.

The Kuriyama Tunnel was planned to run in the northeast of Takasago station, covering 1,810m under the Kuriyama plateau at 30m above sea level. The plateau around the tunnel is called Shimofusa Plateau which is a diluvial plateau with a relatively flat top. At the beginning, this section was planned to be constructed by cut-and-cover method, however, the construction was done by tunneling work except for the Kuriyama station house, considering the geological condition and the location where the tunnel passes beneath a residential area at relatively shallow depth. For the tunneling work, NATM with short bench cut and side pilot methods was partly applied. The construction section in question is a section under the residential area located at 3,860m from Takasago station, where the short bench cut method is applied.

soil of the Kanto Loam whose origin is volcanic ash is widely distributed to cover the plateau in large scale. The tunnel passes through an upper sandy layer, D_{s1} with a groundwater table near the tunnel crown. The material parameters of D_{s1} are summarized in Table 5.2 in section 5.2.2.

Since the characteristics of geology and water logic are poor, chemical fluid injection and the dewatering methods are scheduled to be used. Pipe roof method was also adopted before the cutting face arrives for the supplemental construction method. The tunnel in the standard section is horseshoe shaped, 10.2m wide and 8.7m high and the area of excavated cross section is about 72m^2. In the short bench cut method, tunneling was carried out symmetrically by taking out each

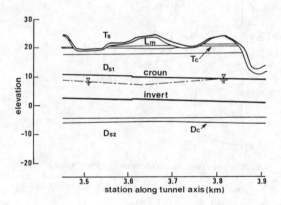

Figure 5.8 Soil conditions and topography along the tunnel alignment.

5.3.2 Geological conditions and tunneling procedure

The soil conditions and topography along the tunnel alignment are shown in Figure 5.8. The cover over the tunnel ranges from zero to 15m. The geological condition is almost similar to those of the Narashinodai Tunnel. Namely, it is mainly composed of sand stratum above which the viscous

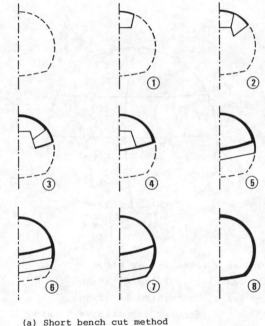

(a) Short bench cut method

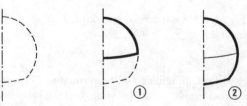

(b) Top heading method (Hypothetical)

Figure 5.9 Division of heading of excavation sequence.

section and providing primary support which consisted of H-shape steel support (150mm) and shotcrete of 20cm thick as shown in Figure 5.9 (a).

On the other hand, Figure 5.9 (b) shows the hypothetical top heading excavation method for the contrast tunneling method. In this case, tunneling was carried out by taking out the top half section at one time and providing primary support and shotcrete. Subsequently, the bottom half section is excavated similarly.

5.3.3 Comparison of computed and measured results

In order to construct urban tunnels successfully, it is more important to minimize the deformation of surrounding ground than to minimize the earth pressure on tunnel supports or lining. As mentioned in section 5.2, it becomes clear that the main reason for the difference of the observed behavior in actual tunneling is caused by the difference of tunnel advance rate rather than the difference of division of heading. While the choices relating the tunneling method are decided mainly based on the experience and judgment of the engineers, there is not enough quantitative information to support the decision.

In this section, various factors regarding the design of tunneling and details on the simulation results of actual and hypothetical tunneling are provided.

Effect of Installation Timing of the Primary Supports

First, effect of the installation timing of the primary supports were investigated for tunneling with the short bench cut method in a sandy layer. To investigate that effect on tunneling stability, a problem from Kuriyama Tunnel was analyzed using the following four kinds of installation timing, that is, the beam elements were assumed to be introduced at the stress released rate of 0%, 33%, 67% and 100%.

Figure 5.10 shows the equi-displacement contour line calculated at full excavation stage [8] for four cases. The plastic yielded joint elements developed in the ground at final excavation stage are shown in Figure 5.11. In both figures, they illustrate the results at the excavation stage [6] for the case of a stress released rate of 100%, because the soils above the tunnel flakes off and falls down

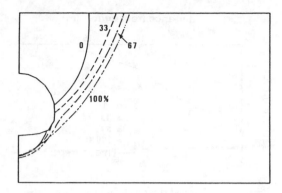

Figure 5.10 Equi-displacement contour lines of 2cm calculated at full excavation stage.

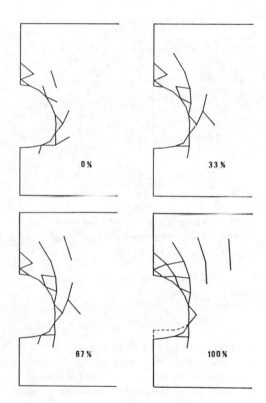

Figure 5.11 Plastic yielded joint elements developed in the ground at full excavation stage.

at the excavation stage [7]. It is seen in both figures that when the stress released rate becomes larger, the displacement becomes larger, and the number of plastic yielded joint elements increase.

The lateral distributions of surface settlement

315

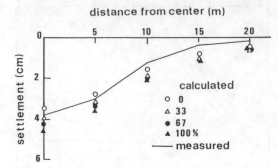

Figure 5.12 Lateral distribution of surface settlement.

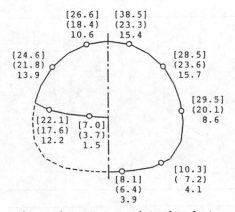

beam elements were introduced at
the stress released rate of
[xx.x]:0% , (xx.x):33% , xx.x : 67%

Figure 5.13 Axial stress developed in support elements.

are shown in Figure 5.12 to compare the measured and calculated results at final excavation stage. Relatively close agreement between measured and calculated displacement was obtained at the neighboring tunnel center line, on the other hand the analysis overestimated the surface settlements at a far distance from the tunnel center. The influences that resulted due to the four stress released rates of introducing beam elements to surface settlements were found not to be large in the case of the short bench cut method. The axial stresses developed at some portions in each support element at excavation stage [4] for three stress released rates are given in the left half of

Figure 5.13, while those ones at excavation stage [8] are shown in the right half. It is seen in the figure that the axial stresses decrease linearly with a delay in the installation timing.

Secondly, influence of the installation time of the primary supports were investigated for the hypothetical tunneling as shown in Figure 5.9 (b). The hypothetical tunneling method has a different character from the actual short bench cut method, that is, the tunneling was carried out by taking out large sections at one time. The installation time of the primary support was set up as the following three kinds of stress released rates, i.e., 0%, 33% and 67%.

Figures 5.14~5.17 are plotted in the same manner as Figures 5.10~5.13 for the short bench cut method. It is seen in those figures that when the stress released rate becomes larger, the displacement becomes larger, and the number of plastic yielded joint elements increase. The influences that resulted due to the three stress released rates of introducing beam elements to the tunneling stability for this case were found to be larger than in the case of the top heading method. Figure 5.17 shows the axial stresses developed at some portions in each support element at top excavation stage [4] and full excavation stage [8]. The axial stresses of this case are larger than those of the case of short bench cut method. Comparing Figures 5.10~5.13 with Figures 5.14~5.17, it is evident that the primary supports affect tunneling stability remarkably in this case, because they make a closed structure just after installation. On the contrary, the primary supports have not made a closed structure until excavation stage [4] for the short bench cut method.

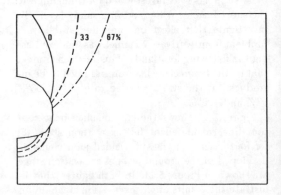

Figure 5.14 Equi-displacement contour lines of 2cm calculated at full excavation stage.

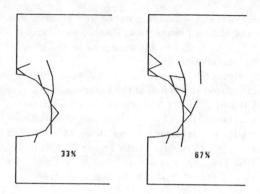

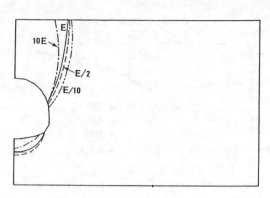

Figure 5.15 Plastic yielded joint elements developed in the ground at full excavation stage.

Figure 5.18 Equi-displacement contour lines of 2cm calculated at full excavation stage.

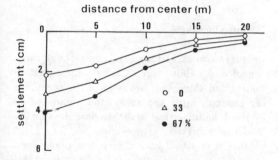

Figure 5.16 Lateral distribution of surface settlement.

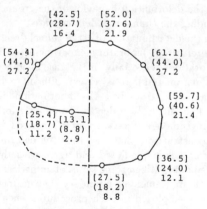

beam elementswere introduced at
the stress released rate of
[xx.x]:0% , (xx.x):33% , xx.x:67%

Figure 5.17 Axial stress developed in support elements.

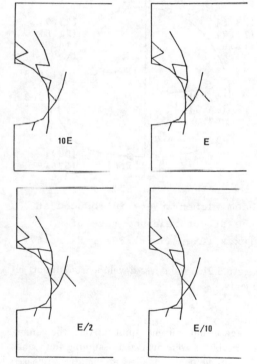

Figure 5.19 Plastic yielded joint elements developed in the ground at excavation stage [8].

Effect of Stiffness of the Primary Supports

Effect of the stiffness of primary supports were investigated for tunneling by the short bench cut method as shown in Figure 5.9 (a). In order to in-

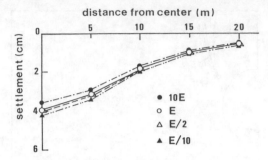

Figure 5.20 Lateral distribution of surface settlement.

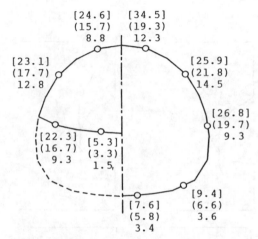

beam elements were introduced at
the stress released rate of
[xx.x]:0% , (xx.x):33% , xx.x :67%

Figure 5.21 Axial stress developed in support elements.

vestigate that influence qualitatively, the tunneling problems were analyzed assuming four kinds of beam elements installed at the stress released rate of 33%. Stiffness of the beam elements were set up as one tenth, one half, the same and ten times normal stiffness.

Figure 5.18 shows the equi-displacement contour line calculated at final excavation stage [8], by use of the four kinds of stiffness of the beam elements . The plastic yielded joint elements developed in the ground at stage [8] are shown in Figure 5.19. The lateral distributions of surface settlement are shown in Figure 5.20. The influ-

ences that resulted due to the four kinds of stiffness of beam elements to displacement and stability are found to be smaller than the effect of the installation time.

Figure 5.21 shows distribution of the axial stresses calculated at one half of the normal stiffness. The left half of Figure 5.21 shows the axial stresses at excavation stage [4] for three installation times, while the right half shows those at excavation stage [8]. Comparing with Figure 5.13, it is seen that the axial stresses decrease only about 10% than those calculated with regular stiffness.

6 ESTIMATION OF DESIGN PARAMETERS FOR EARTH TUNNEL BASED ON MONITORED DISPLACEMENT DURING EXCAVATION

6.1 Introduction

In order to properly design and construct an earth tunnel under thin cover, it is very important to understand the actual behavior of the surrounding ground during the excavation phase and to establish an analytical model to describe this behavior as accurately as possible.

Since it is a well known fact that tunnel construction in a cohesionless ground causes a particular pattern of discontinuous displacements, it has been proposed to use joint elements in the analysis to accommodate these discontinuities. In Chapters 3 and 4, ability of the method to simulate the deformation behavior of the ground surrounding the tunnel was demonstrated.

In the determination of earth tunnel design parameters from laboratory experiments and site explorations (especially in the case of joint stiffness parameters), it is a very difficult and uncertain task to analyze the deformation behavior of the field. In order to compensate for this, the revision of soil parameters is practiced during the construction phase, so that the deformation may be better predicted and agree with the observed data. When some types of numerical analysis (finite element or boundary element method, etc.) is employed in such procedures, such revision of material constants requires too much effort in heuristic trials, especially in the case where the surrounding ground consists of multiple layers.

In order to overcome such shortcomings, some methods have already been proposed. Gioda (1985) proposed a procedure for inverse analysis which works well if the number of soil parameters is not large. Sakurai et. al. (1985) presented a method to evaluate the plastic zone occurring around underground openings. Gioda and Sakurai (1985) presented recent developments in numerical techniques for back-analysis in the field of geomechanics including tunneling.

However, these studies are not able to treat the non-linear constitutive model nor the discontinuous behavior occurring in sandy or cohesionless ground.

In this chapter, we present a numerical procedure for the determination of non-linear constitutive parameters by back-analysis based on measurements made in the ground surrounding the tunnel. Consequently, the present procedure of back-analysis based on site measurements in the early stages of an earth tunnel construction will permit fine tuning of the design parameters, thereby improving the design of subsequent stages of construction.

6.2 Description of the procedure

6.2.1 Constitutive model for joint element

It is important to establish a numerical procedure to simulate the discontinuous displacement behavior observed in the ground during tunnel construction in cohesionless soil. As mentioned in Chapter 3, it becomes evident that the joint element analysis is better for simulating the discontinuous behavior in sandy ground than the usual elasto-plastic analysis. In that joint element analysis, the elastic constitutive law was adopted in the continuous element and the elasto-perfectly-plastic constitutive model (bi-linear model) was adopted in the joint elements.

In the joint element model proposed by Goodman and St. John (1977) material parameters can be introduced directly. Although every constitutive expression for joint elements has been used in conjunction with an elastic-perfectly-plastic model based on the Mohr-Coulomb criterion, this approach has not been taken in this study since it is impossible to back-analyze non-linear constitutive parameters as long as the deformation modulus is only a function of stress.

Because the stress is almost independent of the deformation modulus in a problem of homogeneous elastic media, it is impossible to back-analyze the deformation moduli from the monitored stresses.

Also in non-linear constitutive material, the deformation modulus is not a dominant factor in determination of the stress value. As a consequence, the hyperbolic stress-strain relation is adopted in a tangent direction to the joint. The most general representation of a hyperbolic stress-strain equation may be the constitutive model proposed by Duncan and Chang (1970). Their model makes the deformation modulus decrease with an increase in the principal stress difference. And the final strength which corresponds to the asymptote of the hyperbolic equation, is governed by the Mohr-Coulomb failure criterion specified by stress. Hence a modification of Duncan and Chang's model is required. The modified point is that the deformation modulus decreases with the increase in the joint tangential displacement rather than the increase in stress difference. As can be seen in Figure 6.1 the deformation modulus decreases by increasing tangential joint displacement. The tangent modulus corresponding to any point on the stress-tangential joint displacement curve, such as shown in Figure 6.1, is expressed as

$$k_s^t = \frac{k_s^i s^2}{(s + k_s^i |u_j|)^2} \sigma_n' \qquad (6.1)$$

$$\tan\phi' = R_f \cdot s \qquad (6.2)$$

where k_s^t is tangent modulus of rigidity, k_s^i is initial tangent modulus of rigidity, s is shear strength ratio (the asymptotic value of shear stress), ϕ' is internal friction angle, R_f is failure ratio, σ_n' is effective normal stress of joint elements, and u_j

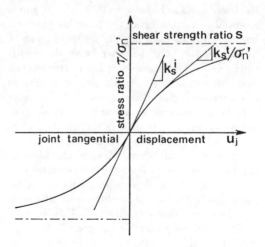

Figure 6.1 Stress-displacement relation for joint element.

319

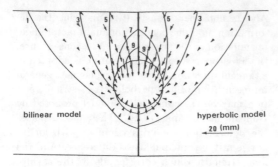

Figure 6.2 Displacement vectors and equi-displacement contour lines in surrounding ground at 75% stress release.

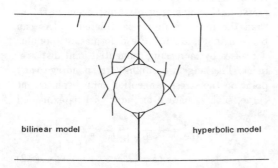

Figure 6.3 Plastic yielded joint elements at 75% stress release.

is tangential joint displacement. Eq.(6.2) is only valid for the cohesionless joints. The value of R_f has been found to be between 0.75 and 1.0, and is essentially independent of the confining pressure, however, it is assumed that R_f always takes unity in this chapter. On the other hand, the constitutive model for normal direction is defined such that the zero tensile strength and large stiffness in compression conditions are met.

Because it is necessary to verify the above mentioned hyperbolic constituive model, its performance in hypothetical case studies are shown here. Figure 6.2 compares analytical results of equi-displacement contour lines and displacement vectors by use of the hyperbolic model and a bilinear one in the case of overburden depth equaling the tunnel diameter (H=1D). This example is the same one applied in Chapter 3. It is difficult to clarify the difference of the two results. The plastic yielded joint elements developed in the ground at 75% stress release are shown in

Figure 6.3. For the hyperbolic model, the yielded joint elements are decided by stress ratio τ/σ_n of exceeding 75% of the shear strength ratio. Although it is not easy to compare both results, this hyperbolic model has enough ability to simulate the tunneling in sandy ground.

6.2.2 Formulation of the problem

Consider a geotechnical problem in which material parameters are unknown ones to be estimated based on a limited number of field measured displacements. Assuming a hypothetical set of material parameters, one can obtain nodal displacements $\{u\}^n$ at the time step n by means of the finite element method with incremental formulation.

$$\{u\}^n = \{u\}^{n-1} + [K]^{-1}\{\Delta f\}^n \qquad (6.3)$$

where $[K]$: global stiffness matrix, $\{\Delta f\}$: nodal force increment vector and superscript (-1) indicates inversion.

The finite element method which can treat continuous elements, beam elements, truss elements and joint elements was used to model the soil-structure system. It was assumed that the material parameters for beam and truss elements are known, since they represent artificial materials. Hence unknown material parameters to be estimated are initial tangent modulus k_s^i and the shear strength ratio s for joint elements and the Young's modulus E and Poisson's ratio ν for continuous elements.

A set of material parameters can reasonably be determined so that the sum of squares of differences between calculated and measured displacements is minimized, if it can be assumed that these measured displacements can be equally weighted regardless of the quality or reliability of their measurement and the locations of measured points. Thus, the back-analysis problem is formulated as an optimization problem with objective function and constraints as

$$\text{minimize} \qquad J = \sum_n^{N_t} \sum_i^{N_d} (u_i^n - U_i^n)^2 \qquad (6.4)$$

$$E > 0, \quad 0 < \nu < 0.5, \quad k_s^i > 0, \quad s > 0 \qquad (6.5)$$

where J : objective function, N_t : number of time steps or construction steps, N_d : number of measured values of displacement at a certain time step, U_i^n : field measured displacement at the node i at the time step n, and u_i^n : calculated displacement corresponding to U_i^n. By solving an optimization problem consisting of Eqs.(6.4) and (6.5), a set of material parameters are obtained which can represent the behavior of the overall ground during tunnel excavation.

6.2.3 Numerical analysis

It is not easy to solve the formulated optimization problem analytically, because of its high nonlinearity in the objective function. There is a numerical procedure called a conjugate gradient technique proposed by Fretcher and Reeves (1964) which can be effectively used in this optimization problem. The procedure is as follows :

1. Set the initial values of decision variables (material parameters)

$$\{x\}_0 = \{E, \nu, k_s^i, s\}_0. \qquad (6.6)$$

2. Calculate the gradients of objective function J on the material parameters $\{x\}$

$$\{g\}_m = \{\partial J/\partial x\}_m \qquad (6.7)$$

where m implies the iteration number.

3. Calculate the conjugate gradients $\{d\}_m$ as

$$\{d\}_m = -\{g\}_m + \frac{\{g\}_m^T\{g\}_m}{\{g\}_{m-1}^T\{g\}_{m-1}}\{d\}_{m-1} \qquad (6.8)$$

in which superscript T indicates transpose.

4. Modify the value of the material parameters as

$$\{x\}_{m+1} = \{x\}_m + \alpha_m\{d\}_m \qquad (6.9)$$

where α_m has to be determined so that $\{x\}_{m+1}$ minimizes the objective function locally. When $\{x\}$ violates the constraints, they take the boundary values.

5. Repeat (2) to (4) until a chosen approximation becomes satisfactory.

The gradient of the objective function J on E is calculated as

$$\partial J/\partial E = 2\sum_{n}^{N_t}\sum_{i}^{N_d}(u_i^n - U_i^n)\partial u_i^n/\partial E \qquad (6.10)$$

$$\partial u_i^n/\partial E = [\partial[K]^{-1}/\partial E\{\Delta f\}^n + [K]^{-1}\partial\{\Delta f\}^n/\partial E]_i \qquad (6.11)$$

where $()_i$ denotes the i-th element of vector and

$$\partial[K]^{-1}/\partial E = -[[K]^{-1}\partial[K]/\partial E[K]^{-1}]^T \qquad (6.12)$$

Since the global stiffness matrix $[K]$ is formed by superposing the element stiffness matrix $[K_e]$, the term $\partial[K]/\partial E$ in Eq.(6.12) is also obtained by linear superposition of $\partial[K_e]/\partial E$. And material parameters are only involved in the constitutive equation, then

$$\partial[K_e]/\partial E = \int_{vol}[B]^T\partial[D]/\partial E[B]dv \qquad (6.13)$$

where $[B]$: matrix to calculate strain from nodal displacements, $[D]$: matrix to specify stress-strain relation, and vol : volume of the element. Namely, it is only necessary to calculate the gradient of stress-strain matrix $[D]$ on material parameters. When assuming plane strain condition, $\partial[D]/\partial E$ for the continuous elements are given as

$$\partial[D]/\partial E = \frac{(1-\nu)}{(1+\nu)(1-2\nu)} \times$$

$$\begin{bmatrix} 1 & \nu/(1-\nu) & 0 \\ \nu(1-\nu) & 1 & 0 \\ 0 & 0 & (1-\nu)/2/(1-\nu) \end{bmatrix} \qquad (6.14)$$

$\partial[D]/\partial\nu$ can be also calculated in a similar manner. Of course, the gradient of stress-strain matrix for the continuous element on k_s^i or s are zero. On the other hand, it needs few modifications to calculate the gradient of the stress-strain matrix for joint elements $[D_j]$, because joint elements have a different type of stress-tangential joint displacement (shear displacement along a joint) matrix. The stress-tangential joint displacement matrix is defined as

$$[D_j] = \begin{bmatrix} k_s^t & 0 & 0 \\ 0 & k_n & 0 \\ 0 & 0 & l^3 k_n/4 \end{bmatrix} \qquad (6.15)$$

where k_n : joint stiffness of normal direction, l : length of the joint. Then, its gradient on the initial stiffness k_s^i can be written as

$$\partial[D_j]/\partial k_s^i = \begin{bmatrix} 1 & 0 & 0 \\ 0 & 0 & 0 \\ 0 & 0 & 0 \end{bmatrix}\frac{\partial k_s^t}{\partial k_s^i} \qquad (6.16)$$

Assuming the gradients of normal joint stress on k_s^i, $\partial\sigma_n'/\partial k_s^i$ are zero, the gradient $\partial k_s^t/\partial k_s^i$ can be calculated from Eqs(6.1) and (6.2). The gradient

321

on s is also calculated in a similar manner. It is easy to extend the present procedure to a case where the soil deposit consists of multiple layers.

6.3 Application to hypothetical case studies

To verify that the proposed procedure is valid, its performance in hypothetical case studies is demonstrated below. The condition of plane strain is assumed in all examples. First, assuming a hypothetical set of soil parameters, E, ν, k_s^i and s, one can calculate nodal displacements by means of the finite element analysis. Subsequently, relevant displacements are used as if they were field measurement data and back-analysis is performed considering a geotechnical problem in which the soil parameters, E, ν, k_s^i and s are unknown quantities to be found. However, since Poisson's ratio neither significantly fluctuates in a wide range nor largely affects the displacements, it is assumed to be a known parameter throughout the following case studies.

Example-1

At first, it should be confirmed that the back-analysis procedure estimates the material parameters correctly, if the hyperbolic constitutive model has some stress dependency. Figure 6.4 shows the model and location of measurement points. Element 3 is a joint element. Although the joint elements have no thickness, it is illustrated as if they have a certain thickness to clarify their concept. The white circle indicates the location of a measurement point of vertical displacement, and the black circle shows the one for horizontal displacement. The loads illustrated in Figure 6.4 were assumed to act in 5 steps by one over five. The material parameters used in this model are summarized in Table 6.1, which are the values to be estimated (correct material parameters). In this table Model-a has no stress dependency (σ_n=constant=1.0), and Model-c indicates the hyperbolic model described in section 6.2, having stress dependency (σ_n=normal joint stress).

Figure 6.5 illustrates the iteration behavior by the present procedure of back-analysis using Model-a and c. As shown in Figure 6.5, the correct material parameters are obtained in both models. Although the hyperbolic model (Model-c) is dependant on normal stress, the tangential joint dis-

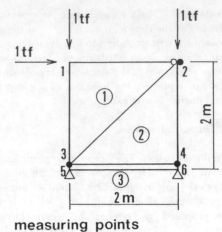

measuring points

● lateral movement
○ settlement

Figure 6.4 Finite element model for hypothetical example-1.

Table 6.1 Material parameters for example-1.

	Model-a	Model-c
Young's modulus E(tf/m^2)	10000.0	
Poisson's ratio ν	0.2	
initial tangent modulus of rigidity k_s^i(1/m)	100.0	100.0
shear strength ratio s	0.7	0.7 ($\phi'\fallingdotseq$35°)
initial stress σ_N^0(tf/m^2)		1.0

Table 6.2 Material parameters for example-2.

Young's modulus E(tf/m^2)	260.0
Poisson's ratio ν	0.3
initial tangent modulus of rigidity k_s^i(1/m)	50.0
shear strength ratio s	0.5($\phi\fallingdotseq$27°)

placement works dominantly in the present procedure and is able to estimate the correct material parameters.

322

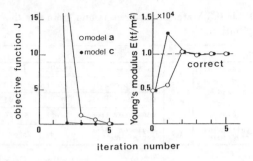

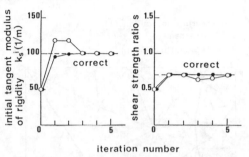

Figure 6.5 Iteration behavior for example-1.

Example-2

This example considers a hypothetical tunnel excavation in homogeneous sandy ground. Figure 6.6 illustrates the finite element mesh and location of measuring points. Table 6.2 shows the material parameters to be estimated (correct values). The bold lines show the joint elements. The simulation of tunnel excavation is done by unloading the initial stress at the top face of tunnel in 3 steps and at the bottom face in 2 steps.

Case-1 : In this case, the present procedure is applied using the data collected after full excavation of the cross section. Figure 6.7 shows the iteration behavior by the present procedure of back-analysis. In Figure 6.7, the symbol (I) shows the iteration process using all observational displacements shown in Figure 6.7. On the other hand, the symbol (II) shows the iteration behavior using the data measured at nodes 17,11,23 and 121 where measurement is easier than other locations in Figure 6.7, because those nodes are located on the ground surface or cutting face. In this case the present procedure furnishes nearly correct values of soil parameters after only several iterations.

measuring points
● lateral movement
○ settlement

Figure 6.6 Finite element model for hypothetical example-2.

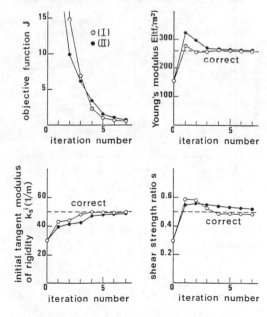

Figure 6.7 Iteration behavior for example-2 (case-1).

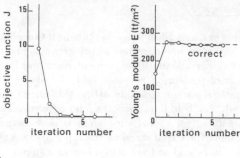

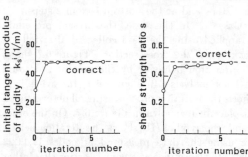

Figure 6.8 Iteration behavior for example-2 (case-2).

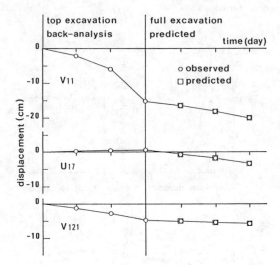

Figure 6.9 Predicted and measured displacement for example-2 (case-2).

Table 6.3 Material parameters and grain size distribution.

Narita Group (sand)				
gradation(%)		unit weight (tf/m³)		1.779
gravel	0.2	Young's modulus (tf/m²)		2300.0
sand	90.8	Poisson's ratio		0.33
silt	6.0	cohesive strength(tf/m²)		2.10
clay	3.0			
uniformity coefficient	2.2	angle of shear resistance (°)		35°54'

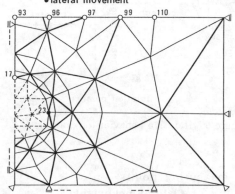

Figure 6.10 Finite element model for actual example.

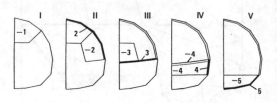

Figure 6.11 Simulation of tunnel excavation.

Case-2 : Figure 6.8 shows the iteration behavior using data measured during excavation of the top portion of the tunnel. The correct material parameters are obtained in this case, too. Figure 6.9 shows the projected displacement based on the material parameters identified by this procedure. The back-analyzed material parameters can predict the behavior of the ground after top excavation. These results indicate that the present procedure can successfully estimate the material parameters and can accurately predict future behavior by back-analysis based on field monitoring displacements.

324

6.4 Application to actual case studies

6.4.1 Applied tunneling (Kuriyama Tunnel)

The actual example given here is a cross-section of Kuriyama Tunnel on the J.R. Hokusou line at Narashino, Chiba-Prefecture. The section is located 3,860m from Takasago Station. The tunnel is excavated in a diluvial sandy deposit belonging to the so-called Narita group. The material parameters and grain size distribution of that layer determined by laboratory tests are listed in Table 6.3. Figure 6.10 illustrates the finite element model and the location of measuring points. Joint elements are shown by the bold lines. Since this section was excavated by the top heading method, simulation of the tunneling is modeled as shown in Figure 6.11.

Although the present procedure requires that every measuring point has to have an observational displacement at every time step, there were no measurement data prior to the arrival of the cutting face without surface settlements. Consequently, it was necessary to introduce an extrapolation device to estimate missing data. Figure 6.12 shows the surface settlement rate with progress of the cutting face. This relation is made by use of the average value of observational settlement that is measured at many points near this cross section. The measured displacements inside the tunnel cross section (convergence of the cross section) are extrapolated based on this relation to account for the release of initial stress in the soil mass prior to the arrival of the cutting face.

In this actual example it is assumed that Young's modulus is given by $E = E_0 \times \sigma'_m$, where σ'_m is the mean effective stress of the element. The primary supports (H-shape steel support with shotcrete) are modeled as beam elements in the analysis and their material parameters are set as the values determined by JIS (Japan Industrial Standards). The initial value of each material parameter is set based on Table 6.3 and Poisson's ratio is fixed at 0.33.

6.4.2 Application results

Case-1 : At first, the present procedure is applied using the data collected after full excavation of the cross-section. Figure 6.13 shows the iteration behavior of the procedure. Figure 6.13 also contains the soil parameters adjacent to tunnel springline which have been obtained from labora-

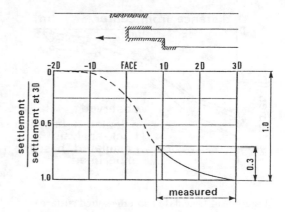

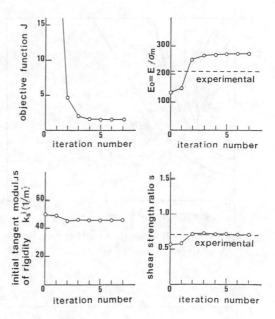

Figure 6.12 Surface settlement rate with progress of cutting face.

Figure 6.13 Iteration behavior for actual example (case-1).

tory tests. As seen in Figure 6.13, Young's modulus obtained from laboratory tests is substantially smaller than the back-analyzed one. On the other hand, the experimented internal friction angle is in fairly good agreement with the back-analyzed value. However, the comparison has been done assuming that the failure ratio, R_f, is unity for this actual case. It is considered that the effect of the sampling disturbance appeared to Young's modulus to be larger than the shear strength.

325

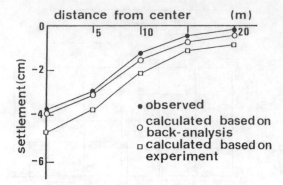

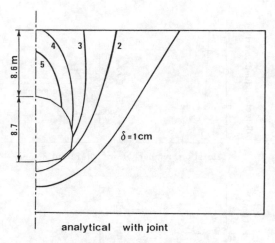

Figure 6.14 Calculated and measured surface settlement for actual example (case-1).

Figure 6.15 Equi-displacement contour lines based on back-analyzed material parameters.

The calculated surface settlements based on soil parameters estimated from back-analysis are shown in Figure 6.14 together with those based on soil parameters obtained from laboratory tests and measured in the field. The back-analyzed parameters simulate the surface settlement with good accuracy, but the experimental ones give large displacements. Figure 6.15 shows the equi-displacement contour lines at full excavation calculated by the estimated soil parameters, while Figure 6.16 illustrates the ones based on the model test mentioned in Chapter 3 for the same overburden condition (H=1D). As can be seen in those figures, the back-analyzed material parameters are able to simulate behavior of the surrounding ground during the excavation. The left half of Figure 6.17 illustrates the distribution of joint elements that the stress ratio exceeded to one half of the shear strength ratio during excavation of the top portion of the cross section, while the right half shows those at full excavation. However, there is no joint element where the stress ratio exceeded 75% of the shear strength ratio. The excavation was done restraining looseness.

Case-2 : The procedure of back-analysis is applied using the data collected during the excavation of the top portion of the cross section. Figure 6.18 illustrates the performance of this procedure. Each material parameter is estimated as almost the same value as in Case-1. The predicted surface settlements at full excavation based on the back-analyzed soil parameters are shown in Figure 6.19 together with those measured in the field. The projected displacements based on the

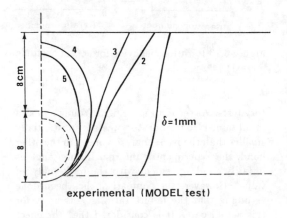

Figure 6.16 Equi-displacement contour lines based on model test.

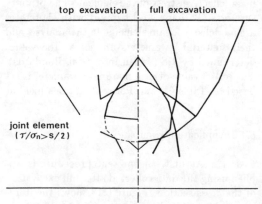

Figure 6.17 Distribution of joint elements in which stress ratios exceed 75% of shear strength.

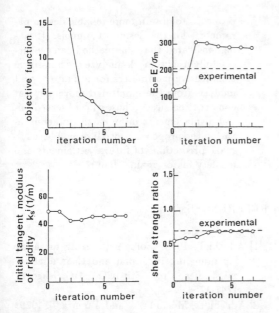

Figure 6.18 Iteration behavior for actual example (case-2).

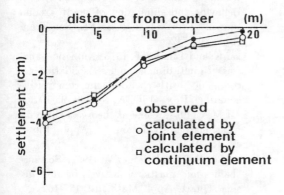

Figure 6.19 Calculated and measured surface settlement for actual example (case-2).

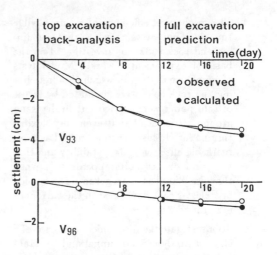

Figure 6.20 Predicted and measured displacement for actual example (case-2).

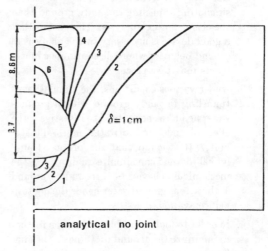

Figure 6.21 Equi-displacement contour lines based on back-analyzed material parameters (continuous elements only).

back-analyzed material parameters are shown in Figure 6.20 together with those measured in the field. The back analyzed material parameters can predict the behavior of the ground after top excavation stage.

Figure 6.19 also contains the calculated surface settlement by using a continuous element only (no joint element is used in the back-analysis procedure). Figure 6.21 shows the equi-displacement contour calculated without joint elements. Comparing Figure 6.16 with Figure 6.21, it is consid-

ered necessary to introduce the joint elements to simulate behavior of the surrounding ground during tunnel excavation.

7 CONCLUSIONS

From this study, the following conclusions are derived.

327

1. The mechanical efficiency of a thin flexible tunnel support system such as shotcrete lining and rock bolts was investigated on the basis of experimental work. Model tests of tunnel excavation in which shotcrete lining and rock bolts were simulated by pieces of thin paper were carried out in dry sand ground. It is found that even very flexible thin paper closing the ring lining has a remarkable effect on tunnel stability and that the effect of rock bolts appears only when they are placed so as to get into the outside of a plastic zone developed in the surrounding ground.

2. To investigate the mechanical behavior of a shallow sandy ground tunnel and to establish the design method for a thin, flexible shotcrete lining, laboratory model tests and two types of finite element analyses were performed. From the results, it was found tha the joint element analysis is better in simulating tunneling in sandy ground than the elasto-plastic analysis. This means that a good device is necessary in the analysis of ground behavior where discontinuous movements may take place.

3. Under various conditions, the simulation of tunneling in sandy ground with a groundwater head was performed to investigate the effect of underground water on ground stability. It is found that the proposed analytical method can simulte qualitatively the mechanical behavior of surrounding ground with various groundwater heads during tunnel excavation.

4. In order to find a suitable excavation method to minimize the ground movement, the analytical results were compared with the measured results from the trial tunneling. From the results, the installation timing of primary support is found to have a great influence on ground stability when a large section is taken out at the same time, while its effect dcreases if the small division of heading is adopted. To stabilize a tunnel, it is necessary that the primary supports and lining make a closed structural shell around the tunnel as quick as possible and the surrounding ground mobilizes its own strength as much as possible.

5. A numerical procedure by back-analysis was proposed to obtain more reliable design parameters for the design of tunnels. It was demonstrated that the method can be successfully used to back-analyze the non-linear constitutive parameters for a joint element analysis from the monitored movements at construction sites. Thus, with observed initial displacements in an earth tunnel excavation the proposed procedure enables us to make predictions of future settlements and a safety factor against failure.

REFERENCES

[1] Adachi,T. 1985. Some supporting methods for tunneling in Japan and their analytical studies. Proc. 5th ICONMIG, Nagoya, 1747-1754.

[2] Adachi,T., Mochida,Y. and Tamura,T. 1979. Tunneling in fully-saturated soft sedimentary rocks. Proc. 3rd ICONMIG, Aachen, 599-610.

[3] Adachi,T., Tamura,T. and Yashima,A. 1985. Behavior and simulation of sandy ground tunnel. Proc. 11th ICSMFE, San Francisco, 709-712.

[4] Akai,K. and Tamura,T. 1978. Numerical analysis of multi-dimensional consolidation accompanied with elasto-plastic constitutive equation. Proc. JSCEA, No.269, 95-104.

[5] Biot,M.A. 1941. General theory of three dimensional consolidation. J. Appl. Phys., 12, 155-164.

[6] Drucker,D.C. and Prager,W. 1952. Soil mechanics and plasticity analysis in limit design. Quart. Appl. Math., 10, 157-162.

[7] Duncan,J.M. and Chang,C.Y. 1970. Nonlinear analysis of stress and strain in soils. Proc., ASCE, No.SM5, 1629-1653.

[8] Fretcher,R. and Reeves,C.M. 1964. Function minimization by conjugate gradient. Computer Journal, 7, 149-154.

[9] Gioda,G. 1985. Some remarks on back analysis and characterization problems. Proc., 5th ICONMIG., Nagoya, 1, 47-61.

[10] Gioda,G. and Sakurai,S. 1987. Back analysis procedures for the interpretation of field measurements in geomechanics. Int. J. for Numerical and Analytical Methods in Geomechanics, 11, 555-583.

[11] Goodman,R.E. and St.John,C. 1977. Finite

element analysis for discontinuous rocks. Proc. Numerical Methods in Geotechnical engg., McGraw-Hill, 148-175.

[12] Murayama,S. and Matsuoka,H. 1974. Earth pressure on tunnels in sandy ground. Journal. JSCE, No.187, 95-108.

[13] Peck,R.B. 1969. Deep excavation and tunneling in soft ground. Proc. 7th ICSMFE., State of the arts volume, Mexico, 225-290.

[14] Rabcewicz,L.Y. 1969. Stability of tunnels under rock load. Water Power, June, July and August.

[15] Sakurai,S., Shimizu,N. and Matsumuro,K. 1985. Evaluation of plastic zone around underground opening by means of displacement measurements. Proc., 5th ICONMIG., Nagoya, 111-118.

Developments in Geotechnical Aspects of Embankments, Excavations and Buried Structures, Balasubramaniam et al. (eds)
© 1991 Balkema, Rotterdam. ISBN 90 5410 019 2

Earth pressure acting on buried pipes

J.Tohda
Osaka City University, Japan

ABSTRACT: This paper introduces the author's ten years study on earth pressure acting on buried pipes. Earth pressure concentration on the buried pipe, particularly on its top and bottom is extensively discussed by means of a full-scale field test, centrifuge model tests, and an elastic analysis. The full-scale field test revealed that high earth pressure concentrations on the pipe top and on the pipe occurred owing to the sheet-pile extraction and they caused the cracks in the concrete pipes. The centrifuge model tests using a rigid model pipe clarified the effects of the following six factors on the earth pressure concentrations: type of pipe installation, roughness of pipe surface, cover height, thickness of sand bedding, ditch width, and density of sand ground. The elastic analysis explained successfully the mechanism of the earth pressure concentrations and provided a new concept for considering the earth pressure on buried pipes.

1 INTRODUCTION

Since earth pressure acting on buried pipes is a difficult subject to study owing to both the complexity of the soil-pipe inter action problem and combination of various factors, its actual state has not been made clear yet. In this situation, failures of buried pipes have occurred continuously, though many earth pressure theories such as Marston-Spangler theory (Spangler 1948) were proposed hitherto.

In the latter-half of 1970's, sewer concrete pipes, buried by the open excavation method using sheet-piling and set on concrete cradles, often cracked when sheet-piles were extracted after backfilling. The author investigated the cause of pipe failures through a full-scale field test (Tohda et al. 1981 and 1985a) and found that the sheet-pile extraction induced significant earth pressure concentration on the pipe top together with an increase in the total vertical load carried by the pipe (earth pressure concentration on the pipe), which caused the pipe failures.

Thus, the earth pressure concentrations were investigated by means of a series of centrifuge model tests (Tohda et al. 1985 b, 1986a and 1988), and the following re-sults were obtained:

1. High earth pressure concentrations similar to the field test data were observed during the sheet-pile extraction.

2. Pressure concentration on the pipe bottom also occurs when pipes are installed on soil beddings.

3. The earth pressure concentration on the pipe top and bottom always occur in higher or lower intensity in any types of pipe installations.

4. The theories of earth pressure and design loads proposed hitherto do not correspond to the real phenomenon, because they neglect the earth pressure concentration on the pipe top and bottom.

These test results made it necessary to innovate the traditional theory of earth pressure acting on buried pipes. Therefore, the earth pressure on buried pipes was analyzed using a two-dimensional elastic model (Tohda et al. 1986b). The results clearly showed the mechanism of the earth pressure concentrations, giving a new concept of the earth pressure on buried pipes.

In this paper, these three studies will be introduced and the design of buried pipes will be discussed on the basis of these studies.

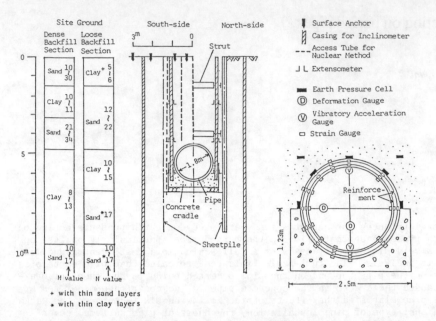

Fig.2-1   Construction Profile and Instrumentation at the Test Sections

Table 2-1   Backfill Soil

Primary Properties	G_s	U_c	Max. grain size (mm)	Percentage of Particles				W_{opt} (%)	γ_{dmax} (tf/m³)
				Gravel	Sand	Silt	Clay		
	2.69	30	19	30	54	12	4	13.4	1.88

Secondary Properties		w (%)	γ_t (tf/m³)	e	ϕ_d	c_d (tf/m²)
	Dense Backfill	12	1.70	0.77	34°15'	0
	Loose Backfill	8	1.58	0.84	33°00'	0

2 A FULL-SCALE FIELD TEST

2.1 Outline of the test

Fig.2-1 shows the construction profile and instrumentation at the test section. Concrete pipes having an inner diameter of 1.8 m and wall thickness of 0.127 m were buried in a ditch 7 m deep and 3 m wide, the cover height being 4.5 m. The pipes were set on an in-situ concrete cradle of 180° bedding angle.

The site ground was diluvial terrace, and the ground water table was about 7.5 m deep below the ditch bottom throughout the test. The backfill soil was decomposed granite, whose properties are shown in Table 2-1. Two test areas with dense and loose backfill were prepared. In the dense backfill area, the soil was placed in every 20cm layer and compacted heavily with a rammer. In the loose backfill area, the soil was placed in every 1 m layer and compacted lightly.

At the sheet-pile extraction with a vibro-hammer (40 KW), a large quantity of soil was brought up in the hollows of sheet-piles, leaving a considerable amount of vacant spaces in the ground. There was trouble in extracting several sheet-piles near the measuring sections, and these piles were re-extracted 4 days later (re-extraction of sheet-piles).

2.2 Observed results

Fig.2-2 shows the change in earth pressure measured in the two test sections. In both sections, high earth pressure concentration

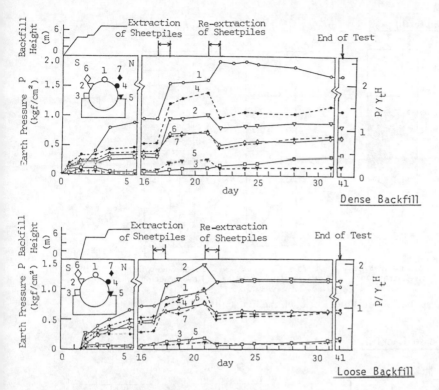

Fig.2-2   Measured Earth Pressure

on the pipe top occurred owing to both the extraction and the re-extraction of sheet-piles; the maximum earth pressure at the pipe top was recorded 2.5 times as great as the overburden pressure $\gamma_t H$ in the dense backfill section and, 1.6 times, in the loose backfill section.

As a result of this high earth pressure concentration, the strain and deflection of the pipe increased markedly and longitudinal cracks were found at the inner top surface of almost all the pipes in both test areas just after the sheet-pile extraction. The re-extraction of sheet-piles, furthermore, yielded to new cracks at the inner bottom surface of the pipes.

The sheet-pile extraction caused the movement of the soil filling up the vacant spaces which the sheet-piles left behind in the ground, resulting in considerable settlement of the backfill.

Fig.2-3 illustrates the distribution of void ratio measured by a nuclear method (RI) in the two backfill sections before the extraction and just after the re-ex-

traction of sheet-piles. In the dense backfill section, the dense backfill soil mass over the pipe maintained its high density throughout the test, while the backfill soil at the pipe side loosened owing to the sheet-pile extraction.

2.3 FE analysis of the test

The test result was analyzed by using finite element (FE) method based on linear elastic theory. Fig.2-4 shows FE models for the two test sections in the two test stages. The conditions for the analysis were determined as follows:
   1. The boundary conditions both between the soil and the concrete structures (pipe and cradle) and between the pipe and the cradle were assumed to be smooth, while the boundary condition between the soil and the sheet-piles was assumed to be fixed.
   2. The extent and values of elastic moduli in the backfills and natural ground were determined as shown in Fig.2-4 and Table 2-2 on the basis of the test results.

The calculated results conformed well to

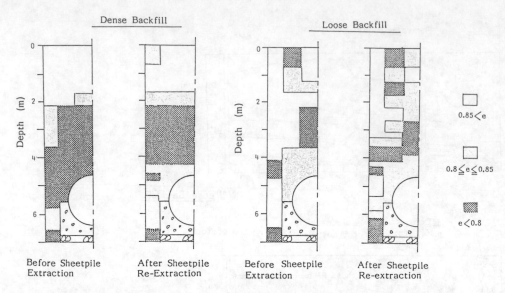

Fig.2-3 Measured Distribution of Void Ratio in the Backfill

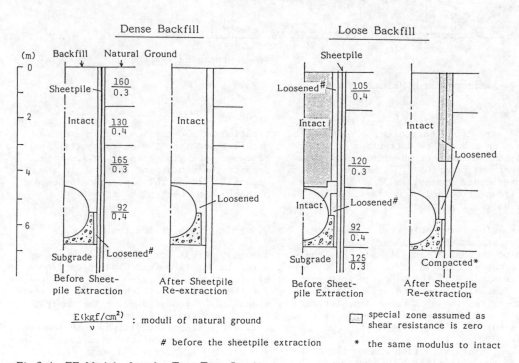

$\dfrac{E(kgf/cm^2)}{\nu}$: moduli of natural ground

☐ special zone assumed as shear resistance is zero

before the sheetpile extraction

* the same modulus to intact

Fig.2-4 FE Models for the Two Test Sections

the measured ones as shown in Fig.2-5, suggesting that the assumptions applied in the analysis were reasonable. The calculated vertical and horizontal loads are also illustrated in the right-hand of each figure.

Just after the re-extraction of sheetpiles, the calculated vertical loads on the pipe show the unique mountain-shaped pattern as a result of the pressure concentration on the pipe top, which is utter-

334

Table 2-2 Elastic Moduli of Soils

E : kgf/cm²

Soils			Dense Backfill	Loose Backfill	Comments
Intact Backfill		E	110	30	determined considering both results of triaxial compression tests and observed settlement of backfills at the sheetpile extraction
		ν	0.3	0.2	
Loosened Backfill		E	30	5	determined from measured distributions of void ratio
		ν	0.3	0.2	
Subgrade		E	92	125	determined from results of plate loading tests
		ν	0.4	0.3	
Natural Ground		E	92-165	92-125	determined from N values
		ν	0.3-0.4	0.3-0.4	
Narrow zone holding sheet-pile initially	Before sheet-pile extraction	E	5	5	assumed by trial from observed behavior of the ground*
		ν	0.3	0.2	
	After sheet-pile extraction	E	2.2	2.2	
		ν	0.3	0.2	

* Initially, this zone transmitted shear force (through friction of the sheetpile) from the backfill to the natural ground. After the sheetpile extraction, this zone was vacated transmitting no stress. It was filled again with soils at the end of test.

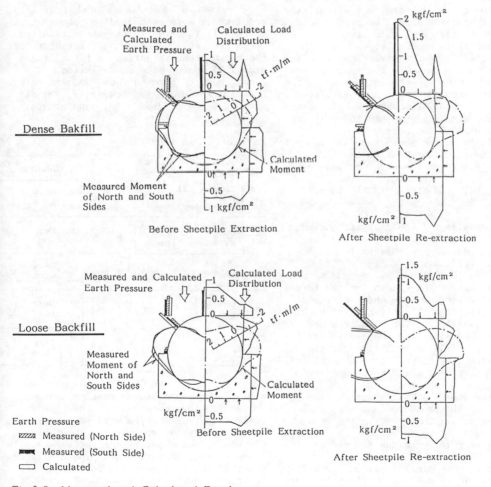

Dense Bakfill

Measured and Calculated Earth Pressure

Calculated Load Distribution

Measured Moment of North and South Sides

Calculated Moment

Before Sheetpile Extraction

After Sheetpile Re-extraction

Loose Backfill

Measured and Calculated Earth Pressure

Calculated Load Distribution

Measured Moment of North and South Sides

Calculated Moment

Before Sheetpile Extraction

After Sheetpile Re-extraction

Earth Pressure

▨ Measured (North Side)

▬ Measured (South Side)

▭ Calculated

Fig.2-5 Measured and Calculated Results

ly different from the uniform load ordinarily assumed in the current design of buried pipes.

2.4 Discussion on the earth pressure concentration

It should be noted that the density of the backfill affected the degree of the earth pressure concentration due to the sheet-pile extraction both on the pipe top and on the pipe (the latter concentration means an increase of the total vertical earth pressure carried by the pipe). The effect of these two pressure concentrations on the maximum bending moment, Mmax, in the pipe section was estimated by using the calculated loads for "just after the re-extraction of sheet-piles", as follows:

1. Under the mountain-shaped vertical loads, Mmax became 1.45 times the value for the case under the uniform load without the pressure concentration on the pipe top in the dense backfill and, 1.26 times, in the loose backfill respectively.

2. The calculated total vertical load was 1.27 times the value for the case of γHD without the pressure concentration on the pipe in the dense backfill and, 1.10 times, in the loose backfill respectively. Here, γH is the overburden pressure and D is the outer diameter of the pipe.

3. Multiplying the above two coefficients, the resultant Mmax became 1.84 times the value for the case without the two pressure concentrations in the dense backfill and, 1.39 times, in the loose backfill.

The test results and the analysis showed the mechanism of the earth pressure concentrations due to the sheet-pile extraction as it follows:

1. In the dense backfill section, the sheet-pile extraction eliminated the forces acting before between the sheet-pile and the soil (the backfill soil and the natural ground), resulting in the high pressure concentrations after the sheet-pile extraction; the backfill in both sides of the pipe loosened, owing to the re-extraction of the sheet-piles, which increased the pressure concentrations.

2. In the loose backfill section, the magnitude of the pressure concentrations was rather lesser than that in the dense backfill section, which was due to the arch action between the pipe and the ground.

Furthermore, the preliminary FE calculations with an assumption of fixed boundary condition between the pipe and the soil generated no earth pressure concentration on the pipe top in any cases. Therefore, the smooth boundary on the pipe surface was concluded to be one of the conditions to produce the high earth pressure concentration on the pipe top.

None of the theories or design formulae for the total vertical earth pressure exerted on buried pipes explain the phenomenon of the earth pressure concentrations. Moreover, non-uniformity of load distribution was neglected in the current design procedure, which gives definitely unsafe design criteria.

3 CENTRIFUGE MODEL TESTS

3.1 Outline of centrifuge model tests

Five series of two-dimensional centrifuge model tests were conducted, as shown in Table 3-1, to investigate parametrically the effect of the following six factors on the earth pressure acting on buried rigid pipes: 1)type of pipe installation, 2)roughness of pipe surface, 3)cover height H, 4)thickness of sand bedding H_b, 5) ditch width B_d, and 6) density of sand ground.

Table 3-1 Test Conditions

Series	Pipe Installation	Pipe Surface	H	H_b (cm)	B_d	Ground	Number of Test
A	Ditch -S Ditch -0 Embankment	Smooth	9	4	13 —	Dense Loose	6
B	Ditch-S Embankment	Smooth Medium Rough	9	4	13 —	Dense Loose	12
C	Ditch-S Ditch-0 Embankment	Smooth	4.5 9 12	4	13 —	Dense Loose	18
D	Ditch-S Embankment	Smooth	9	1 2 4	13 —	Dense Loose	12
E	Ditch-S Ditch-0	Smooth	9	4	13 17 21	Dense Loose	12

Fig.3-1 shows models and testing systems for the following three types of pipe installations: ditch type with sheet-piling (Ditch-S), ditch type without sheet-piling (Ditch-0), and embankment type (Embk.). In these models, a rigid model pipe having an outer diameter of 9 cm was buried in both dense and loose model grounds of dry sand as shown in Table 3-2 and Table 3-3.

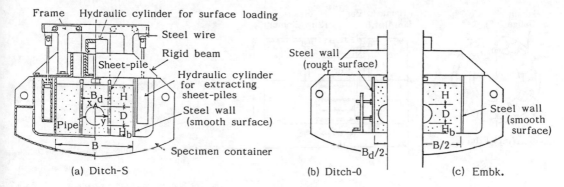

Fig.3-1 Models for Three Types of Pipe Installations

Table 3-2 Properties of Silica Sand

G_s	Grain Size	U_c	ρ_{dmax}	ρ_{dmin}
2.65	0.24-1.4mm	1.75	1.58t/m³	1.32t/m³

Table 3-3 Density and Strength of Sand and Friction Angle ϕ_p against the Pipe Surface

Ground	ρ_d	c_d	ϕ_d	ϕ_p		
				Smooth	Medium	Rough
Dense	1.55 t/m³	0	47°	17°	42°	45.5°
Loose	1.43	0	36°	16°	40°	44.5°

Table 3-4 Dimensions of Standard Model (unit : cm)

	D	H	H_b	B_d	B
Model	9	9	4	13	34
Prototype	270	270	120	390	1020

Table 3-4 shows the dimensions of the standard model together with those of the corresponding prototype. The models were built in a scale of 1/30 of the prototype, and put into the centrifugal acceleration field of 30 g using the centrifuge shown in Fig.3-2 (Mikasa 1984). In the Ditch-S model, a pair of model sheet-piles 5 mm thick was extracted simultaneously to produce the high earth pressure concentrations.

Fig.3-3 shows the rigid model pipe, which was designed to measure both normal and tangential earth pressures acting on the pipe surface. The pipe surface was finished very smoothly. Both a rough surface pipe and a medium rough surface pipe were prepared by pasting two kinds of sand paper on the smooth surface pipe.

3.2 Earth pressure change during sheet-pile extraction

The change in normal earth pressure σ measured in the Ditch-S model is shown in Fig. 3-4. This was the case of the standard model using the smooth surface pipe and the dense ground condition. In the course of the sheet-pile extraction, the pressure σ at the pipe top reached its peak value, which was over two times greater than the overburden pressure γH; the degree of this pressure concentration was very similar to the data observed in the dense backfill in the former field test. Furthermore, the high pressure concentration on the pipe bottom also occurred during the sheet-pile extraction and maintained its high degree after the sheet-pile extraction.

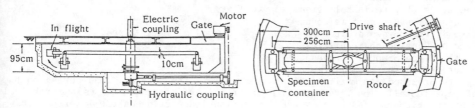

Fig.3-2 Centrifuge

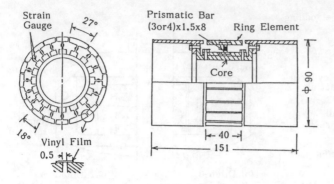

Fig.3-3 Rigid model pipe (unit : mm)

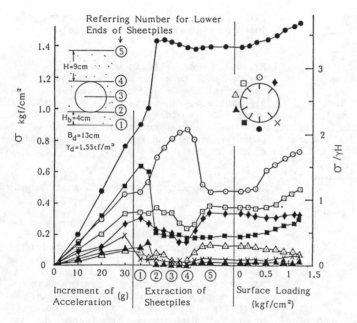

Fig.3-4 Change in Normal Earth Pressure Measured in the Ditch-S Model

3.3 Effect of the type of pipe installation

Fig.3-5 illustrates the distributions of the measured normal and tangential earth pressures, σ and τ, in polar coordinates for the three types of pipe installations, where the tangential pressure τ is counted as positive in the downward direction. The data noted as "during sheet-pile extraction" in the Ditch-S model correspond to those measured when the normal pressure σ on the pipe top reached its maximum. In all types of pipe installations, the normal pressures σ at the pipe top and bottom were considerably greater than those at the pipe

sides; the tangential pressure τ was very small at every points.

The measured pressures σ and τ gave the vertical and horizontal loads as it follows:

$$p_v \text{ and } p_r = \sigma + \tau \cdot \tan\theta,$$

$$p_h = \sigma - \tau \cdot \tan\theta$$

Here, p_v is the vertical load acting on the upper-half of the pipe, p_r is the vertical reaction load acting on the lower-half of the pipe, and p_h is the horizontal load.

338

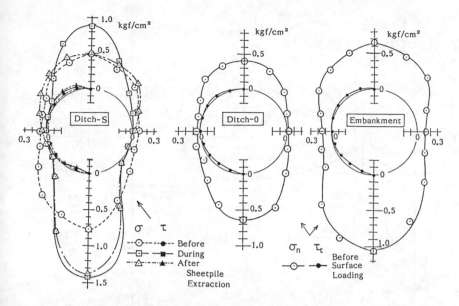

Fig.3-5 Distribution of Measured Normal and Tangential Earth Pressures for Three Types of Pipe Installations

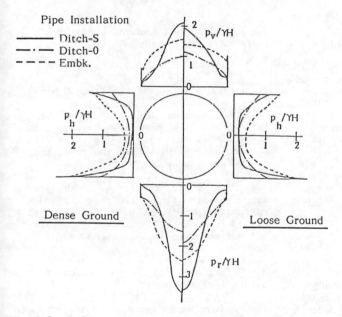

Fig.3-6 A Series: Measured Loads for Different Types of Pipe Installations

The angle θ was measured from the pipe top. Fig.3-6 shows the distribution of these loads normalized by γH for the three types of pipe installations. The right and left sides of the figure are the data for the dense and loose sand ground, respectively.

The earth pressure concentration on the pipe top and bottom was the highest in the Ditch-S model and the lowest in the Ditch-0 model. The effect of the ground density on the load distribution was very small.

339

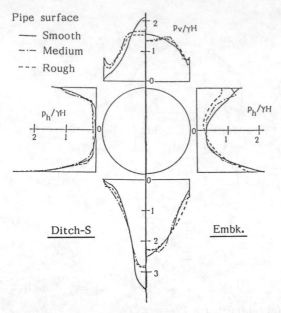

Fig.3-7 B Series: Measured Loads for Different Roughness of Pipe Surface

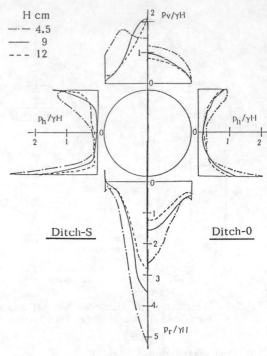

Fig.3-8 C Series: Measured Loads for Different Cover Height

3.4 Effect of roughness of pipe surface

Fig.3-7 shows the measured loads in the Ditch-S and Embk. models for the different roughness of pipe surface under the dense ground condition. The measured vertical loads, p_v and p_r, on the smooth surface pipe were the mountain-shaped pattern as a result of the high pressure concentration on the pipe top and bottom. On the other hand, the measured vertical loads, p_v and p_r, on the rough and medium rough surface pipes were trapezoidal shapes with lesser pressure concentration on the pipe top and bottom.

3.5 Effect of the three geometries in pipe installation

Figs.3-8, 3-9 and 3-10 show the change of the measured loads due to the cover height H, thickness of sand bedding H_b, and ditch width B_d for the dense ground condition, as it follows:

1. Cover height: In the Ditch-S model, the vertical load p_v acting on the upper-half of the pipe when H=4.5 cm was the shape of the letter M quite different from other cover heights. This characteristic feature was also observed in the Embk. model (the figure is omitted here), while

it was not observed in the Ditch-O model.

2. Thickness of sand bedding: In the Ditch-S and Embk. models, the load p_r at the pipe bottom increased its value with the decrease of H_b, while the loads p_v and p_h were independent of H_b.

3. Ditch width: The degree of the pressure concentration on the pipe top and bottom in the Ditch-S model decreased with the increase of B_d, while such a tendency was not observed in the Ditch-O model.

3.6 Effect of earth pressure concentrations on the bending moment

The maximum bending moment, Mmax, produced in the pipe section was calculated using the principle of least work under the following two external load conditions: (a) both the measured earth pressure and the pipe weight were applied, and (b) only the measured earth pressure was applied. In the latter case, the intensity of the vertical reaction pressure p_r was determined by multiplying the measured p_r by the ratio of $(1-W/P_r)$, where W is the pipe weight and

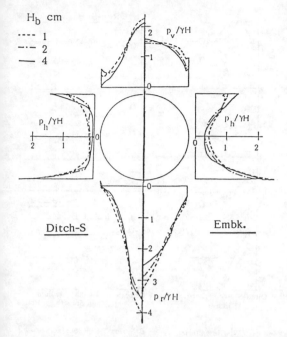

Fig.3-9   D Series: Measured Loads for Different Thickness of Sand Bedding

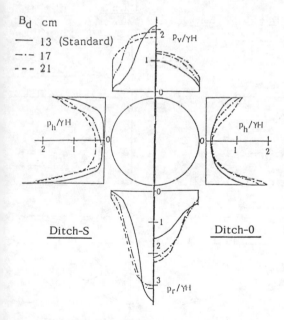

Fig.3-10   E Series: Measured Loads for Different Ditch Width

P_r is the total vertical reaction load obtained by integrating the measured p_r.

In order to estimate the effect of the two earth pressure concentrations on Mmax, the following two coefficients, K' and K, were defined as it follows:

$$K' = Mmax / \{ (Q / D) \cdot (D / 2)^2 \}$$

$$K = Mmax / \{ (\gamma H) \cdot (D / 2)^2 \}$$

where Q is the total vertical load, which was given as $Q = P_v + W$ in the case (a), and $Q = P_v$ in the case (b); P_v is the total vertical load obtained by integrating the measured p_v; D is the outer diameter of the pipe; γH is the overburden pressure.

The coefficient K' represents only the effect of the load distribution on Mmax, because it is normalized by the total vertical load Q. Furthermore, the above two equations derive the following correlation:

$$K = K' \cdot (Q / \gamma HD)$$

Thus, the coefficient K represents the effects of both the load distribution and total vertical load, namely the two pressure concentrations, on Mmax.

The degree of the two pressure concentrations and the coefficients K' and K for all tests are summarized in Fig.3-11 and Fig.3-12. In these figure, the dash-dotted lines show the values (the mean values for the dense and loose ground) calculated according to the Japanese current design methods for rigid pipes (JSWAS 1970 and MAFF 1988), where the total vertical loads P_v are specified as follows: 1) $P_v = \gamma H B_d$ for the Ditch-S type, 2)Marston's projection formula for the Embk. type, and 3)Marston's ditch formula for the Ditch-0 type. The pipe weight is usually neglected.

With regard to the load distribution, all these design methods assume the uniform load distribution, though the bedding angle is variable. Here, the JSWAS standard was applied, which assumed a uniform vertical load over the pipe and a uniform vertical reaction load over a range of 120° for the bedding angle without the horizontal load. This load distribution gives the value of the coefficient K' as 0.275.

Comparison of the design values with the test results shows that the design values are in consistent with the most test cases; particularly in the Ditch-S model, the current design tends to give smaller and less

341

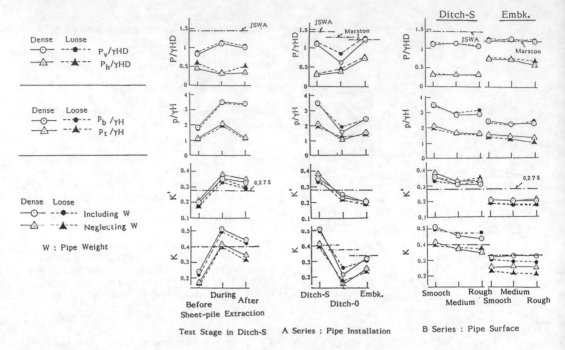

Fig.3-11 Change in Earth Pressure and Maximum Bending Moment under Various Conditions (Ditch-S Standard Test, A Series and B Series)

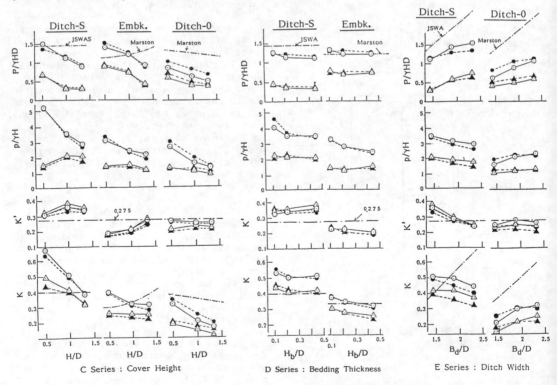

Fig.3-12 Change in Earth Pressure and Maximum Bending Moment under Various Conditions (C, D and E Series)

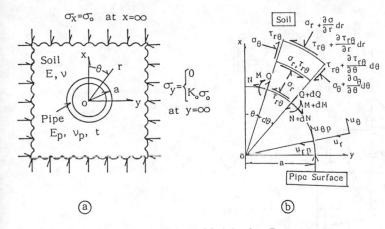

Fig.4-1 Two-dimensional Elastic Model of a Buried Pipe

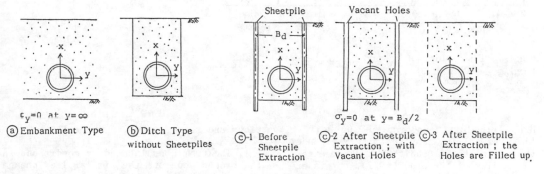

(a) Embankment Type $\varepsilon_y=0$ at $y=\infty$

(b) Ditch Type without Sheetpiles

(c)-1 Before Sheetpile Extraction

(c)-2 After Sheetpile Extraction ; with Vacant Holes

(c)-3 After Sheetpile Extraction ; the Holes are Filled up.

$\sigma_y=0$ at $y=B_d/2$

(c) Ditch Type with Sheetpiles

Fig.4-2 Installation Conditions of a Buried Pipe

safe values of K for the pipe than the test results, because it neglects the earth pressure concentration on the pipe top and bottom.

4 ELASTIC ANALYSIS ON A TWO-DIMENSIONAL MODEL

4.1 Model and conditions for the analysis

The earth pressure on buried pipes was analyzed on the basis of the elastic theory using a two-dimensional symmetrical model shown in Fig.4-1. Where, the x and y axes are taken as the vertical and horizontal directions, respectively. The analysis was carried out under the plane strain condition, and the following three factors were applied:

1. The interface boundary condition between the pipe and the soil was given as both smooth and fixed boundaries. The separation between the pipe and the soil was not considered.

2. The boundary condition on the lateral boundary plane at $y=\infty$ was given for the two types of pipe installations as follows: $\sigma_y=0$ for the ditch type with sheet-piling, and $\varepsilon_y=0$ ($\sigma_y=K_o\sigma_x=\nu/(1-\nu)\cdot\sigma_x$) for the embankment type. These two lateral boundary conditions correspond to those of (c)-2 and (a) in Fig.4-2, respectively.

3. The deformation property of the pipe and soil was given in terms of both Poisson's ratio ν of the soil and the flexibility index K of buried pipes. The index K was defined as it follows:

$$K = E / (S_f / a^3)$$

343

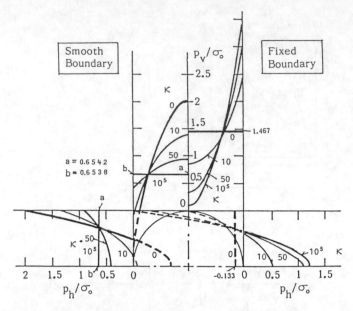

Fig.4-3 Distribution of p_v and p_h for Several κ Values
($\sigma_{y(y=\pm\infty)}$ =0, ν =1/3)

Here, E: elastic modulus of the soil, S_f: flexural stiffness of the pipe wall under the plane strain condition, and a: external radius of the pipe (=D/2). S_f is expressed as follows:

$$S_f = E_p t^3 / \{ 12 (1 - \nu_p^2) \}$$

in which E_p: elastic modulus of the pipe, ν_p: Poisson's ratio of the pipe, and t: wall thickness of the pipe. The value of the index κ was changed from zero to the infinite; κ=0 corresponds to the case when the pipe is a rigid body, and k=∞ corresponds to the case when a hole exists in the ground without pipe.

4.2 Calculated results

The distributions of the calculated vertical and horizontal loads, p_v and p_h, are illustrated non-dimensionally for several κ values in Fig.4-3 when σ_y=0 at y=∞ and ν=1/3, and in Fig.4-4 when ε_y=0 at y=∞ and ν=1/3. The left and right side of the each figure shows the case under the smooth and fixed interface boundary condition between the pipe and the soil. These figures show that the earth pressure depends markedly on the three factors applied in the analysis (the lateral and interface boundary conditions and the κ value), as it follows:

1. The two lateral boundary conditions in the ground generate the similar load change pattern with the change of κ. However, the range of the load change is considerably greater in Fig.4-3 than in Fig.4-4.

2. The smooth interface boundary generates high earth pressure concentration on the pipe top (and bottom) in the case of κ= 0 when the pipe is perfectly rigid, particularly in Fig.4-3. With the increase of κ, the loads p_v and p_h become uniform and the total vertical load P_v decreases.

3. The fixed interface boundary generates uniform loads when κ=0. With the increase of κ, the load p_v shows the concave shape, while the load p_h shows the convex shape, and the value of P_v decreases.

4.3 Comparison of the measured and calculated loads

The measured loads p_v and p_h in the centrifuge model tests are shown in Fig.4-5 for the Ditch-S model and in Fig.4-6 for the Embk. model, which are the data for the standard model and dense ground condition; the left and right side of each figure

344

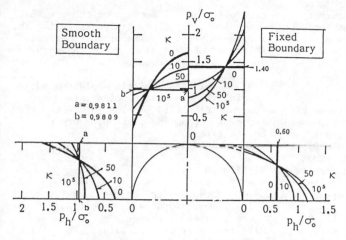

Fig.4-4 Distribution of p_v and p_h for Several κ Values ($\varepsilon_{y(y=\pm\infty)}$ =0, ν =1/3)

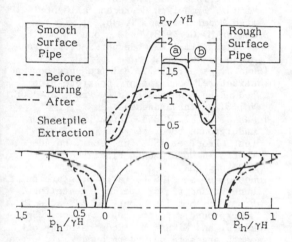

Fig.4-5 Measured Loads for the Ditch-S model in Centrifuge Model Test

shows the data for the smooth surface pipe and the medium rough surface pipe, respectively.

Here, these measured loads were compared with the calculated loads when κ=0 in Fig. 4-3 and Fig.4-4 under the condition that the overburden pressure γH in the tests was given as the boundary stress σ_o at x=∞ in the calculation, as it follows:

1. For the Ditch-S model: The measured loads on the smooth surface pipe (the solid curve in the left side of Fig.4-5) conformed well to the calculated loads with the

smooth interface boundary (the curve noted as κ=0 in the left side of Fig.4-3). On the other hand, the measured loads on the medium rough surface pipe (the solid curve in the right side of Fig.4-5) did not conform to the calculated loads under the two interface boundary conditions (the two curves noted as κ=0 in Fig.4-3). However, the flat central part of the measured p_v (ⓐ in Fig.4-5) was close to the calculated p_v with the fixed interface boundary, while the sloping side part (ⓑ in Fig.4-5) was close to the calculated p_v with the smooth interface boundary. This suggests that the sand slid on the side surface of the medium rough surface pipe.

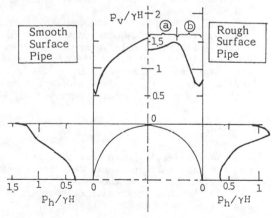

Fig.4-6 Measured Loads for the Embk. Model in Centrifuge Model Test

2. For the Embk. model: The measured loads (Fig.4-6) and the calculated loads (κ =0 in Fig.4-4) for the Embk. model showed a similar correlation to that for the Ditch-S model described above.

Thus, the calculation results gave a reasonable explanation for the change of the measured loads due to both the type of pipe installation and roughness of the pipe surface. Furthermore, the measured loads during the sheet-pile extraction in the Ditch-S model were close to those in the field test. Therefore, the calculation results were concluded to represent exactly the actual earth pressure on buried pipes.

4.4 Consideration on the earth pressure concentrations

The elastic analysis revealed that higher earth pressure concentrations are generated when κ is smaller, the interface boundary on the pipe smoother, and σ_y on the lateral boundary in the ground smaller. This shows clearly that the critical earth pressure concentrations due to the sheet-pile extraction, observed in the field test and centrifuge model tests, were caused by the following three conditions: 1)the high rigidity of the pipe, 2)the smooth interface boundary on the pipe, and 3)the small horizontal earth pressure at the ditch wall acting after or at the sheet-pile extraction.

Furthermore, the elastic analysis gave us a new interpretation for the drastic change of earth pressure due to the type of pipe installation, observed in the centrifuge model tests, that it is caused by the difference in the lateral boundary condition in the ground.

Traditional earth pressure theories or conventional design formula ignore or consider inadequately the effect of the three conditions employed in the analysis, and as a result, they can not explain the mechanism of the earth pressure concentrations.

5 CONCLUSIONS

1. The full-scale field test revealed that the high earth pressure concentrations on the pipe top and on the pipe occurred owing to the sheet-pile extraction and they caused the cracks in the concrete pipes.

2. The centrifuge model tests using the rigid model pipe clarified the effects of the following six factors on the pressure concentrations: type of pipe installation, roughness of pipe surface, cover height, thickness of sand bedding, ditch width, and density of sand ground.

3. The elastic analysis explained successfully the mechanism of the pressure concentrations and provided a new concept for considering the earth pressure on buried pipes.

REFERENCES

JSWAS (Japan Sewage Works Association) 1970. Design standard of reinforced concrete pipe for sewage.
MAFF (Ministry of Agriculture, Forestry and Fisheries) 1988. Design standard of Pipeline.
Mikasa M. 1984. Two decades of centrifugal testing in Osaka City University. Proc. of Int. Sympo. on Geotechnical Centrifuge Model Testing. Tokyo.
Spangler M. G. 1948. Underground conduits —an appraisal of modern research. Trans. of ASCE. Vol.113.
Tohda J. et al. 1981. Soil pressure on, and structural behavior of, concrete pipe bedded on concrete cradle. Proc. of JSCE. Vol.310. (in Japanese)
Tohda J. et al. 1985a. Earth pressure on underground concrete pipe in a field test. Proc. of ASCE Int. Conf. on Advances in Underground Pipeline Eng. Madison. USA.
Tohda J. et al. 1985b. Earth pressure on underground rigid pipe in a centrifuge. Proc. of ASCE Int. Conf. on Advances in Underground Pipeline Eng. Madison. USA.
Tohda J. and Mikasa M. 1986a. A study of earth pressure on underground rigid pipes by centrifuged models. Proc. of JSCE. Vol.376/3-6.
Tohda J. and Mikasa M. 1986b. A study of earth pressure on underground pipes based on theory of elasticity. Proc. of JSCE. Vol.376/3-6.
Tohda J. et al. 1988. Earth pressure on underground rigid pipes: Centrifuge model tests and FEM analysis. Proc. of Int. Conf. on Geotechnical Centrifuge Modelling. Paris.

Developments in Geotechnical Aspects of Embankments, Excavations and Buried Structures, Balasubramaniam et al. (eds)
© 1991 Balkema, Rotterdam. ISBN 90 5410 019 2

Large pipes buried in soft soil – Prefabricated pipes in river Nile sediments

H.-U. Werner & F. Eichstädt
Dorsch Consult, Consulting Engineers, Munich, Germany

ABSTRACT: The case history of a 10 km long sewer line in Helwan, Egypt, is presented. It is especially referred to the geotechnical problems, occuring during the construction of the sewer, which consists of prefabricated, reinforced concrete pipes with an inner diameter of between 3.05 m and 3.40 m.

In areas, where soft soil was encountered, large settlements of the pipes were observed. The findings were analysed and it was found, that not only the consolidation of the subsoil was responsible for the deviation of the sewer, but also a number of other factors.

This paper deals with the problems of and the appropriate solutions for large diameter pipes, buried in soft soil.

1 INTRODUCTION

One of Egypt's major environmental protection measures presently under construction, is the Helwan Waste Water Project.

Helwan, a town south of Cairo, formerly was a well known health resort. After a rapid industrial development the city is now a centre of heavy industries. (see fig. 1)

Despite continued efforts to provide adequate waste water drainage and treatment, many parts of the area were deficient in or without essential facilities.

To protect Egypt's major water source, the river Nile, a waste water masterplan was developed for the greater Helwan area, about 12 years ago.

This paper is dealing in general with problems, occuring when large diameter pipe sewers have to be constructed in irregular alternating stratification and in particular with difficulties encountered, when a part of the pipeline of the Helwan main sewer had to be laid in extremely soft soil.

As the consulting engineer for the design is also the site supervisor and has additionally been entrusted with the project management, the problems that arose could be solved in close cooperation with the contractor.

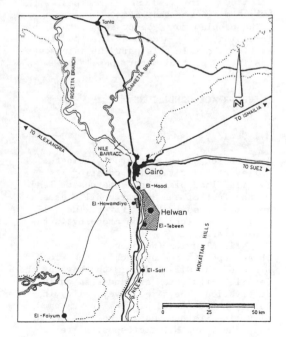

Fig.1 Location map

2 THE PROJECT

The general layout of the main system of
the Helwan Waste Water Project is shown on
figure 2.

The main sewer with inner diameters of
up to 3.40 m has its flow direction from
the north to the south, and this is
against the natural slope of the Nile
valley. This unusual solution was governed
by other overruling criteria of the project objectives.

From the main pumping station (No. 1)
the sewage will be transported through a
double pressure pipeline to the treatment
plant, which is dimensioned for a capacity
of 350.000 m^3 in the first stage and
555.000 m^3 in the second stage. The puri-
fied water will be pumped to a compensation reservoir, from where it is distributed to an irrigation area.

The material selected for the branch
sewers are:
- glazed stoneware pipes for a diameter
  of up to 1200 mm
- reinforced concrete pipes for a diameter of up to 2000 mm

The main sewer was designed to have a
cast-in-situ horseshoe profile. But in the
first two construction sections it was
constructed to an alternative of the
successful bidder and was made from prefabricated, reinforced circular pipes with
a 1.45 m wide contact area.

The problem zone, discussed in the
following, is located near the pumping
station No 1. In this part the main sewer
has its largest diameter, namely 3.4 m.

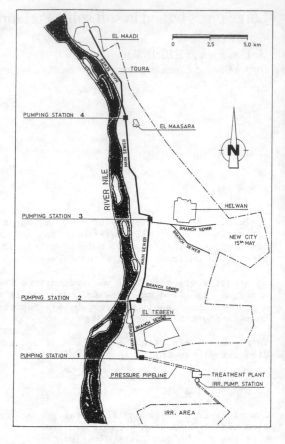

Fig.2: Helwan sewage scheme

3 SUBSURFACE CONDITIONS

3.1 Geology

The project area is located within the
valley of the river Nile covering its
fertile flood plain and some desert land
at the foot of the Mokattam Hills escarp-
ment to the east of the river. The following formations are outcropping in the
project area:
  - Nile Alluvium (Holocene)
These deposits occupy much of the flood
plain of the Nile and form the base of the
fertile soils which are under intensive
cultivation. They are clayey, silty in
nature and composed of fine quartz sand,
clay minerals and heavy minerals. On the
top, the Young Nilotic Deposits are
partially coverd by a thin mantle of
aeolian sand.

  - Waadi Alluvium (Holocene)
They are found in the project area along
the courses of waadis. Such deposits are
composed of coarse sand and sandy loam
mixed with cobble, gravels and rock fragments.
  - Older Nile Terraces (Pleistocene)
These deposits are essentially composed of
quartz sand mixed with gravels and are
intercaleted with clay lenses.
  - Pliocene Deposits
This Pliocene Section, 20 m thick, consists of sand and marly limestone.
  - Outcropping Bedrocks
They are dominant in the EL - Mokattam
Hills and are composed of sandy marly
limestone.

3.2 Encountered stratification

To get a clear idea of the subsurface,
numerous borings and laboratory tests were

carried out during the period of design and construction. The maximum exploration depth for the sewer was around 20 m.

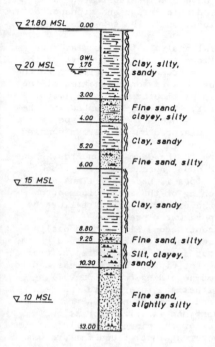

Soil Profile
B1A

Fig.3: Soil profile

According to the results of the borings and following the general geology, sand and gravel layers alternate with strata of clay and silt, which occur in irregular sequence and variable thickness. The thickness of the cohesive strata varies between a few decimeters and some meters. In some cases they were found directly below the surface, in other cases in a depth of more than ten meters.

The density of the non cohesive soils varies mostly between medium dense and very dense.

The consistency of the cohesive layers mainly lies between very soft and semi solid, whereby in a single drillhole all degrees of consistency might occur and a great fluctuation must be expected even within a section of a few meters. From the geological point of view, these layers can be characterised as normal consolidated soils with hardly any geological preloa-

ding. In previous parts of the old Nile river beds, clay strata may occur, which are not yet completely consolidated.

3.3 Ground water

The ground water level, influenced by the river Nile, the so—called quartenary ground water, is found at a depth of 0.2 m to 5.0 m below the surface. Due to the existence of permeable layers, dewatering is generally needed for deep excavations.

Several chemical analysis were carried out on the ground water. The degree of aggressiveness against concrete was classified according to the German standard, DIN 4030, and the results vary through the whole range:
- slightly aggressive
- heavily aggressive
- very heavily aggressive.

4 GENERAL GEOTECHNICAL PROBLEMS

4.1 General aspects

In the case of extremely and unpredictably varying stratification, generally a great flexibility concerning the methods of excavation, shoring, foundation and dewatering is required.

This paper is refering to the construction of the sewers and in particular to the sections which are in areas of soft soil.

4.2 Bearing capacity and settlements

The minimal geological preloading is a very important factor, which has to be taken into consideration for evaluating the bearing capacity of the cohesive layers. On the one hand such layers are of high compressibility, and high loads will cause large settlements; on the other hand the shear strength is low in comparison to an over consolidated soil.

Normally settlements should be a subordinated problem for sewer constructions, because the weight of the excavated material is greater than the weight of the pipes and their contents.

But whilst lowering the ground water table, the hydraulic uplift is taken away. Consequently settlements due to the load of the pipes and the backfill will occur during the time of ground water lowering, especially in soft soils and in the case of long during ground water lowering periods. When the ground water rises again

and the soil and the pipe become submerged, the load decreases and the settlements will be finished.

The amount of settlements also depends
on the handling of the excavated bottom.
If it remains free of load even for more
than a few hours only, it could heave and
loose its natural preloading. Consequently
the already small bearing capacity would
decrease and increase settlements. Therefore measures to prevent the heaving are
to be taken into consideration.

In some cases it is worth while to
determine the sensitivity of the soil
against dynamic loadings (liquefaction
effects due to earthquakes or due to
vibrations occuring during construction).

4.3 Ground water lowering

As mentioned in the last chapter, the
period of time during the ground water is
lowered, must be minimized in order to
reduce settlements caused by the omission
of submerge.

The steady-state ground water level,
which was mostly encountered near the
surface, was to be lowered to at least
0.5 m below excavation bottom. This meant
in general more than 6 m for deep sewers
with large diameters. Gravity wells were
necessary for the sand and gravel layers
and/or wellpoints with vacuum operation
for the silty and clayey sand layers.
Since the latter can only be used down to
a suction depth of approx. 6 m, they had
to be stacked for greater depths of application.

Due to the alternating stratification,
care had to be taken when lowering the
ground water, so as to prevent a break-up
of the excavation bottom, caused by the
hydraulic pressure below the interlaid and
nearly impermeable clay and silt layers.
The upward water pressure at the bottom of
the clay layers had to be reduced to a
value, less than the weight of the remaining soil between the excavation bottom
and the underside of the impermeable
layer, to prevent the excavation bottom
from "piping".

Observation wells or piezometers are a
prerequisite in order to control the
effectiveness of the dewatering system and
the safety against piping.

An excellent operating dewatering system
is one of the main factors for the stability of the excavation bottom. Especially
for the here encountered soil conditions.
If the water table is not to be lowered
down to at least 0.5 m below the foundation level, and this is coupled with dynamic
energy, introduced into the ground, the

soil might lose its bearing capacity by
becoming soft.

5. EXECUTION OF THE WORK

5.1 Constructional details

The pipes have been prefabricated in an
on-site production plant and transported
to their location in the sewer line on the
road.

Structural analysis for circular sewer
cross sections pressume a support of the
pipe by its surrounding backfill up to the
impost. For this reason compaction of the
backfill in particular up to the impost
was necessary. This was guaranteed by the
operation of a deep-vibrator. (type
"Mono", manufacured by the German company
Keller)

According to the German standards, especially DIN 4033, a sand- or gravel bed
below the pipes is necessary. Thus avoiding softening of the excavation bottom
and also the danger of ground failure by
reducing pore water pressure.

In the contractors alternative design
the sand bed had a thickness of 5.0 cm. As
this was not enough for a suffient drainage below the pipes, the thickness had to
be modified according to the prevailing
subsoil situation, in order to ensure a
suitable and stable foundation.

To prevent non-elastic deformations of
the soil below the pipes, a sufficient
safety factor against ground failure was
required:

- In non-cohesive soils, this was
achieved, when the natural soil or the
refilled soil had an angle of internal
friction of ≥ 30 (medium dense). To
prove the density of the sand, dymanic
penetration tests were conducted on the
excavation bottom in advance of
the pipe laying. A driving rod with a
pile shoe area of 5 cm^2 was used. If the
number of blows per 10 cm penetration,
was greater than 12, the density of the
sand was considered to be sufficient.

- If cohesive layers were encountered directly below foundation level,
then their undrained shear strength was
determined by vane tests. An undrained
shear strength of $c_u \geq 25$ kN/m^2 was
fixed as criterion. This had to be
achieved down to a depth of at least
0.8 m below the underside of the pipe.
$C_u \geq 25$ kN/m^2 was regarded as an average
result of at least 3 values but no single
value smaller than $c_u = 20$ kN/m^2 was
permissible.

- If the above mentioned criteria

were not reached, load distribution
measures became necessary, which enlarged
the   foot width of 1,45 m to at least
3.0 m. This was achieved either by
concrete slabs with a width of at least
3.0 m or by adequate soil exchange to a
depth of 1 m below the underside of the
pipes.

5.2 Shield method

The successful contractor proposed among
other protection measures, to use a shield
for the support of the trench walls. It
was the largest ever constructed U-shaped
shield of that type and should allow a
continuous overall work process with an
average speed of 13 m sewer construction
per day. This promised the advantage of a
short period of disturbance of the soil,
but turned out to be rather inflexible
relating to changes in the subsoil conditions. A longer than expected starting
period was needed, due to some steering
problems. A successfull use of the shield
could only be observed in areas :

- with no need of special preparatory
  works for the foundation of the
  pipes,
- with hardly any crossing service
  lines,
- where a simple gravity well system
  was sufficient for the required
  ground water lowering.

To use the shield method, it was necessary to lower the ground water 50 - 60 m
ahead of the shield. For this reason the
ground water lowering started at least 3
to 4 days before pipe laying.
  The standard system was to locate gravity wells at 15 m intervals on both sides
of the trench. Borings with a diameter of
700 mm and a depth of 19 m were carried
out. The filter-pipes had a diameter of
380 mm and a length of 2.0 m. The average
water discharge was between 4 and 5 l/s.
  In soft soil areas an additional depth
of only 0.5 m could be excavated by the
shield at the beginning. If soft soil was
encountered below the pipes and a soil
exchange for example of 1.0 m or 1.5 m
underneath the pipe footing was necessary,
the additional excavation must be done
manually. To avoid expensive and time
consuming manual work, another method to
reduce settlements in areas of soft soil
was proposed. Instead of soil exchange
reinforced concrete slabs were used below
the pipe footing. They had lengths up to
4.0 m, a width of 1.25 m and thicknesses
up to 0.3 m.

6 PARTICULAR PROBLEMS

6.1 Situation

Near to the pumping station, which will
deliver the waste water with a lift of
22 m, through a twin-pipe pressure pipeline to the treatment plant, the encountered soil conditions were extremely bad.
  Soft and pulpy clay and silt, intercala-
ted by layers of silty fine sand and in
some cases also coarse sand, were to be
expected above and up to several meters
below foundation level.
  After unsuccessful trials to excavate
the trench without shoring (collapse of
one sloped side of the trench due to the
insufficient dewatering system) and taking
into account the experience gained by the
use of the shield, the contractor finally
used sheet piles for supporting the trench
walls.

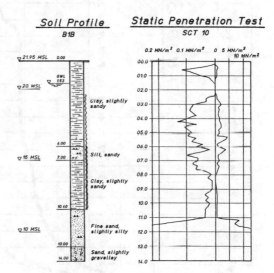

Fig.4: Situation at the main problem zone

12.0 m long sheet piles were driven into
the ground by the use of a vibrating pile
hammer.
  As a result of the bad soil conditions
below the 3.40 m diameter pipes, load
distribution measures became necessary. In
the beginning concrete slabs with a width
of 3.0 m were used below the pipe footing
with an additional soil exchange of 0.6 m
beneath the slabs. Then the method was
changed and the soil was excavated down to
1.3 m below the underside of the pipes and

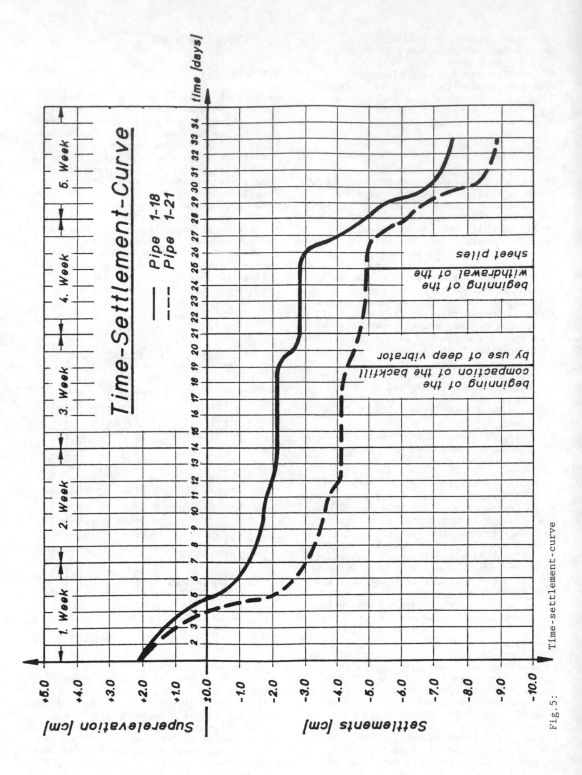

Fig.5: Time-settlement-curve

replaced by a special granular material
with an angle of internal friction of
$\geq 30°$. The granular material was compacted
by static rollers in the lower 0.5 m of
the soil exchange.

6.2 Settlements during construction

All pipes had been laid with an initial
superelevation of 2.0 cm, equivalent to
50 % of the expected settlements.

The invert levels of the pipes were
measured daily. From the measured settlements it was noticed, that already after
some days the pipes had got settlements
of 3.5 cm to 5.5 cm. When withdrawing the
sheet piles the settlements sharply increased to figures between 7.5 cm and
8.8 cm (see Fig. 5).

6.3 Preliminary studies

In order to avoid unacceptable settlement,
the origin of the settlements were analysed. It was supposed, that the total
settlement consists of four parts:
- The first part of the settlements
resulted from the high compressibility of
the soft strata. (consolidation)
- The second part stems from the
fact that during the withdrawal of
the sheet piles a thick layer of soft
material which had adhered to the
sheet piles when withdrawing, was removed
together with the sheet piles. According
to this observation, it was supposed,
that the soil below the pipes, still
under pressure from the weight of the
pipes and the one metre backfill above
the pipes, partly encroached into the
space, vacated by the sheet pile and the
adhered soil (balance of soil volume).
- The third part of settlements could
be found in the inadequate dewatering
system. During the additional excavation
for the soil exchange piping occurred at
the excavation bottom. Particles of
fine sand were sluiced out of the deeper
layers, and it was supposed, that this
was one of the main reasons, that the
settlements could not be decreased as
calculated by adopting soil exchange
measures (subrosion).
- The fourth possible part of settlements could have been the influence of
vibration while withdrawing the sheet
piles (possible liquefaction effect).

6.4 Immediate measures to reduce settle-
ments

Issued from on-site decisions, the following work procedures were put into
practice:
- Performance of additional soil investigations.
- The dewatering system was improved in
order to avoid piping.
- The superelevation was partly increased, so that it would result in an
adequate gradient of the sewer.
- Pipes with too much settlement were
relaid.
- During withdrawing of the sheet piles
gravel (40/100 mm) had to be inserted
down to the foot of the sheet piles with
the help of the Keller vibrator, in order
to prevent settlements stemming from
volume deficit.
- A test was carried out on a stretch
of approximately 10 m as the above
mentioned second part of contribution to
settlements was deemed to prevail the
fourth one.
- Intensification of invert level ni-
vellements.

After having finished the trial,
additional settlements up to nearly 2.5
cm were observed. As this figures were
partly due to further consolidation settlements, withdrawing of the sheet piles
and simultaneously inserting gravel by
assistance of the Keller vibrator obviously did not clearly lead to a reduction of the settlements.

6.5 Detailed soil investigations

Four additional borings and numerous
static penetration tests were carried out.
Disturbed and undisturbed samples were
taken from the excavation bottom and a
laboratory test program was run on samples
obtained from the borings and from the
excavation bottom.

As an example for the stratification,
encountered in the drill holes, reference
is made to Figure 4.

Static penetration tests were carried
out every 50 m's along the route and
beside the settling pipes. From the interpretation of the latter named tests, a
relation was set up between the already
measured settlements and the results of
the tests.

In addition, static penetration tests
were performed in the direct surroundings
of the bore holes. They aimed at calibra-
ting the friction and the cone resistance
to the encountered stratification.

A representative diagram of the static

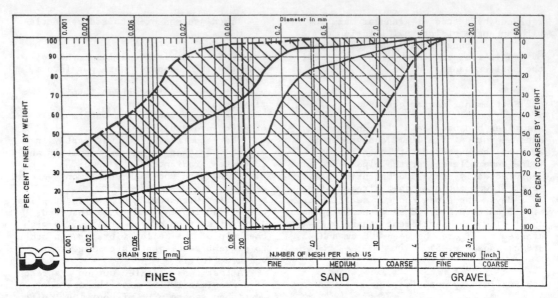

Fig.6 : Size range of the soil

soundings is given on figure 4. It was carried out beside the boring No. B1B, which is also shown on figure 4.

The soil below the formation level consisted of grey to black silty fine sand and grey to black clay. Clay and sand were following in an irregular sequence.

In the laboratory the following properties of the silty fine sand and the silty clay were determined:

```
- silty fine sand
Uniformity Coefficient        3.5 - 4.5
Wet Density                   17.8 - 18.0 t/m³
Natural Moisture Content      35.7 - 39.1 %
Liquid Limit                       -
Plastic Limit                      -
Consistency Index                  -
Loss - On - Ignition          5.2 - 5.3 %
- silty clay
Uniformity Coefficient             -
Wet Density                        -
Natural Moisture Content      31.1 %
Liquid Limit                  43.9 %
Plastic Limit                 23.9 %
Consistency Index             0.64
Loss - On - Ignition          8.1 %
```

The silty fine sand was characterised as poorly graded. The determination of the "Atterberg Limits" was not possible because of the small amount of cohesive particles.

It was compressible and of a low bearing capacity. According to its grain skeleton this type of soil was sensitive to lique-

faction, meaning that it was endangered in becoming liquid when combined with water and dynamic energy.

The cohesive samples could be classified as clay (DIN 18122: TA-TL, USCS: CL-CH) mostly of high, in few cases of low plasticity. The consistency was mainly between soft and pulpy.

According to the low consistency the bearing capacity of the clay was very small. In some cases the clay was sensitve to water.

6.6 Interpretation of the test results
    and consequences

Looking at the test results, it can be stated, that in some cases, where the silty finesand was encountered in a depth equal to the depth of the sheet piles, a collapse of the grain skeleton might have occurred. This was caused by dynamic energy, introduced into the soil by the vibrator and the vibrating pile hammer. Accordingly vibrations should be avoided or at least minimized.

The ramming of the sheet piles by the use of a vibrating hammer could be continued as this would lead to a pre-settlement before placing the pipes.

However, it was instructed to clean the sheet piles and to treat them in a manner as to reduce to the greatest possible extent the previously observed adherence of the soil.

For the withdrawing of the sheet piles
the best solution would have been to use a
hydraulic equipment, for example a "Pile-
master". However, this type of machine
could not be made available in a short
time.

It was decided that the compaction of
the backfill had to be made by light
equipment layer by layer while avoiding
the use of the Keller vibrator.

From the results of the static penetration tests it could be seen that the depth
of the soft soil decreased northwards i.e.
in the direction of the pipe laying.

Due to the fact that there was a long
delivery time as well as for a vibration-
less sheet piling equipment and for a
complete sheet pile space grouting equipment some further proposals have been
discussed:

- gravel piles
- drilled concrete piles
- closed sheeting

- gravel piles
It was proposed by the contractor to
install 2, 3 or 5 parallel rows of piles,
according to the depth of the bearing
stratification. The foundation level would
consist of a gravel layer of 1.0 m thickness, this being necessary for load distribution reasons.

However, the consultant had technical
objections, because mortared gravel piles
need an undrained shear strength of the
surrounding soil of at least $c_u \geq$
15 kN/m^2. This could not be guaranteed by
the contractor.

- drilled concrete piles
According to the structural analysis, the
concrete piles should have a diameter of
1.20 m, an embedded length of 4.0 m and a
reinforced pile head and pile cap. The
distance of the piles had to be 2.5 m.

The extremely high costs of this solution made it uneconomical.

- closed sheeting
A shored trench would be excavated below
an approximately 2.5 m deep sloped trench.
The shoring would consist of 6.0 m long
sheet piles which would remain in the
ground. At the bottom a 35 cm reinforced
concrete slab would be installed and
connected to the sheet piles. The sheet
piles would be struted temporarily during
construction and the calculated settlements would be about 4 cm.

This method was connected with an unacceptable time of stand still due to the
delivery time of the sheet piles.

6.7 Second on site trial: bentonite-grouting

Due to the costs and to the equipment that
was available on site, a bentonite-injec-
tion-method was chosen.

At first a trial should be carried out,
whereby a stretch of about 10 m of the
trench would be prepared for withdrawal
with the assistance of injecting cement-
bentonite-suspension into the ground at
the foot of the sheet piles, in order to
fill the cavities remaining in the soil,
when the sheet piles were removed. The
spacing of the injection tubes would be
about 0.5 m.

In the initial stages, some problems were
encountered in bringing the injection
pipes into the correct position. Some
tests had to be made on the sides of the
trench, so as to develop the right working
procedures.

Then the injection tubes were driven
into the ground by the use of water down
to the foot of the sheet piles.

During the time of injection the pressure of the suspension had always to be
kept well above 3 bars, otherwise the
effect would be in doubt.

The mixture for 1 m^3 suspension was
defined as:

150 - 200 kg cement
 20 - 30 kg bentonite
650 - 800 kg filler (stone powder)
650 - 700 kg water

During the test the following properties were measured and monitored:

- the pressure of the suspension,
measured on the ground surface
- the suspension flow (always to
be moving during the withdrawing of
the sheet piles)
- the quantity of the suspension in
relation to the number of the withdrawn
sheet piles
- settlement measurements of the pipes
(to be continued in the same way as
during the other type of withdrawing
operation)

To minimize the bad influence of dynamic energy on the soil, the withdrawal
was made with the vibration energy as low
as possible.

After evaluating the results, it was
decided to continue the above described
method. It should be carried out as long
as the thickness of the clay layers and
the presents of the silty sands, tending
to liquefaction, necessitate it.

It was stated, that a soil exchange of
1,3 m had to be carried out.

The settlements, including the influence of the construction, were predicted between 5.0 cm and 10.0 cm and the pipes should be laid with a superelevation of up to 10 cm.

Continuous tests and measurements should be carried out with the view to possible amendments of the decision already made.

7 CONCLUSIONS

Numerous factors have to be taken into consideration while designing and constructing large pipes which are to be buried in soft soils:

a. Although the pipes were laid at great depth, the relationship of depth to width of the excavated trench varies between figures of 1.0 and 1.5. This being caused by the large diameter of the pipes. For this reason stress and pressure calculation of the soil cannot be made according to normal geotechnical characteristics, but have to take into consideration the special conditions.

b. As main sewers are usually located in areas with a ground water table near to the surface, high standards are necessary for shoring and ground water lowering systems. The importance of the above mentioned increases as soil conditions become difficult. Especially when irregular alternating stratification of permeable sand and gravel layers, quick sand, water sensitive silts and nearly impermeable clays of soft consistency are encountered.

c. The settlements of the pipes, causing a deviation of the sewer, are not only a result of the consolidation, but also the following reasons have to be taken into account:
- encroaching of the soil below the pipes into the space vacated by the sheet piles and adhered soil
- subrosion effects, caused by the insufficient dewatering ("piping")
- liquefaction effects and because of this a collapse of the grain skeleton, caused by dynamic energy, which is introduced into the ground by technical equipment
- long duration of construction periods can influence the elevation of the excavated trench bottom
- heaving due to enlarged horizontal and decreased vertical stress below the excavation bottom
- wide spread ground settlements due to ground water lowering
- workmanship that does not take due care, in regard to general production procedures

d. Consolidation settlements can be predicted (this is dependent on the availability of exact soil data) and generally equalized by superelevation. The other factor compiled above, must be taken into account by adequate measures to minimize their influence to the lowest possible extent.

e. As no criteria for the deviation of the pipes exist in the current international standards, the determination of the acceptable deviations has to be made with due regard being paid to the hydraulic conditions and to the conditions of the pipe construction.

f. Large diameter pipes are to be designed according to structural analysis. Different construction stages have to be taken into account.

g. Insufficient compaction of backfill, forming an integral part of large pipe constructions, increase the strain of the structure twofold:
- increase of the pressure on the top of the pipe, due to disappearence of the arching effect of the backfill above the pipe.
- reduction of horizontal counterpressure
both with the potential to deform the pipe

h. Special attention must be paid during the design of the pipe. The loads of the backfill and the "uplift effect" (caused at the end of the ground water lowering procedure) must be taken into account, especially refering to the empty pipe.

Analyses of the longitudinal bending strain of pipelines subjected to vehicle loads

Nobuo Takagi
Research & Development Institute, Tokyo Gas Co., Ltd, Japan

Abstract: The failure of small diameter pipes may be caused by longitudinal bending strains induced by heavy vehicles. However, it is difficult to evaluate the effect because of the presence of joints. The paper deals in detail with a matrices analysis which allows easy interpretation of the problem. The analysis includes two features; the stress distribution of wheel loads on the pipe is precisely calculated on the basis of Fröhlich soil stress theory, and a rotational stiffness of pipe joints is introduced. The paper also presents a simpler method for practical use so that the maximum bending strain of pipes beneath the wheel load can be easily predicted. Both the methods are compared with detailed field measurements.

1 INTRODUCTION

The main purpose of this paper is to present two analytical methods which can estimate the longitudinal strain in shallow pipelines subjected to heavy vehicle loads.

The gas mains as well as other utility pipelines in Japan are regulated to be buried at depths of 1.2m or more by the Japanese Road Act. As pipe materials have much been improved in the last two decades, a great concern is paid to reducing the current pipe depth under minor less traffic roads in order to reduce construction costs. Annual investment for the mains of Tokyo Gas Company is about several hundred million US dollars, approximately 80 per cent of which is used for the construction works such as digging trenches, backfilling with sand, repairing the pavement and so on. If a shallower installation were to be allowed, say, at a depth of about 0.8m like the practice in Western countries, it is expected that costs could be reduced by about 20 per cent.

Although the failures of small diameter gas mains, up to 200 or 300mm, are mainly caused by the excessive longitudinal bending moments, no design method for the longitudinal deformations of the main exists in either legislation or recommendation except in the Petroleum Pipeline Business Law(PPBL). The paper describes a numerical method which can evaluate the longitudinal bending strain of pipes having mechanical joints, and presents another simple prediction method. Both the methods are compared with measurements obtained during detailed field experiments. The experiments described were carried out to investigate the longitudinal strains developed in three different types of pipes due to static and rolling wheel loads. Other factors relative to the depth of pipe and the weight of load were also examined.

2 A NUMERICAL METHOD

2.1 Load distribution and its approximation by Fourier expansion

According to Takagi(1989), the vertical stress in a sandy subgrade associated with vehicle loadings is well evaluated by the Fröhlich theory. The basic equation is given by Eq.(1), where the stress concentration factor is assumed to be 5. Sand is generally used in Japan for a backfilling material of trenches.

$$\sigma_v = \frac{\nu Q}{2\pi} \frac{z^\nu}{(x^2 + y^2 + z^2)^{(\nu+2)/2}} \quad (1),$$

where, x, y and z : coordinate system,
z vertical downwards,
x direction along
pipe axis,
σ_v : vertical stress in
subgrade,

ν : stress concentration
factor(=5),
Q : wheel load.

The load(q) acting on the pipe, per unit length, can be calculated by integrating Eq.(1) along y-axis over its diameter(D), where the radius(R)=D/2.

$$q(x) = 2 \int_0^{D/2} \sigma_v \, dy = \frac{5Q}{\pi} \frac{z^5}{(x^2+z^2)^3}$$

$$\{ \frac{R}{(x^2+R^2+z^2)^{1/2}} - \frac{2}{3} \frac{R^3}{(x^2+R^2+z^2)^{3/2}}$$

$$+ \frac{1}{5} \frac{R^5}{(x^2+R^2+z^2)^{5/2}} \} \qquad (2),$$

Since the wheel load has a contact area of several hundreds square centi meters, it is really not a point load as Q but a distributed one. Both, however, present little difference in the soil stress at the depths of 60cm or more. The load q associated with more than one wheel can be easily calculated by the superposition of Eq.(2), considering their individual positions. When the load q is substituted into the beam on elastic foundations theory which will be introduced later, a complete solution of the pipe deformation will be obtained. It is, however, difficult to solve the equation in the direct form. The load q may be approximated by the form of the Fourier expansion(Eq.(3)) in order to handle easily,

$$q(x) = \frac{a_0}{2} + \sum_{i=1}^{n} \{ a_i \cos(\frac{i\pi x}{S})$$

$$+ b_i \sin(\frac{i\pi x}{S}) \} \qquad (3),$$

where a_i and b_i are both the Fourier coefficients, and [-S,S] is the span of expansion, and n is the maximum wave number to be taken into account. Because of the difficulty in calculating the coefficients in this form, they were obtained by assuming that the Eq.(2) is linear within every small divided span(Δx).

2.2 Stiffness equation of the beam on elastic foundations theory

It is broadly known that the following beam on elastic foundations theory provides a good tool to analyse the longitudinal deformation of buried pipelines. The theory consists of a simple beam and a linear elastic body, having a one-dimensional spring of a reaction coefficient(k), which

are considered as the pipeline and the subgrade, respectively.

$$EI \frac{d^4 z_p}{dx^4} + kD z_p = q(x) \qquad (4),$$

where, E : Young's modulus of pipe material,
I : moment of inertia about cross section of pipe,
k : subgrade reaction coefficient,
z_p : pipe deflection,

and other notations as previously described. When substituting Eq.(3) into the right hand term of Eq.(4), a general solution of the differential equation will be expressed in the following form.

$$z_p = \{G\}\{x\}^T + f(x) \qquad (5),$$

where, $\{G\} = (G_1, G_2, G_3, G_4)$
$\{x\} = (\cos\lambda x \cosh\lambda x, \cos\lambda x \sinh\lambda x,$
$\sin\lambda x \cosh\lambda x, \sin\lambda x \sinh\lambda x)$,
$\lambda = \sqrt[4]{(kB/4EI)}$,
$f(x) = \frac{1}{kB} [\frac{a_0}{2} +$
$\sum_{i=1}^{n} \{ \alpha_i a_i \cos(\frac{i\pi x}{S}) +$
$\alpha_i b_i \sin(\frac{i\pi x}{S}) \}]$,
$\alpha_i = \frac{1}{1 + \frac{1}{4}(\frac{i\pi}{\lambda L})^4}$.

Among these parameters, four elements of the vector{G} are determined by the boundary conditions given to the elastic beam.

Fig.1 shows a member of length(l) which is imaginarily separated from the beam. Giving deflections(δ), rotations(θ), moments(m) and shear forces(P) to the both nodes i and j, the vector{G} is eliminated. An element stiffness equation is then provided as given in Eq.(6).

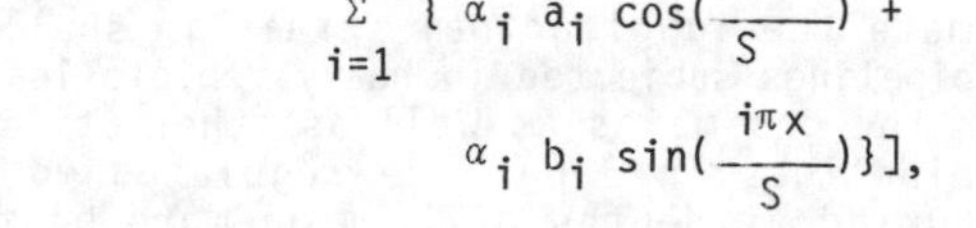

x=0   x=1
δ_i, θ_i δ_j, θ_j
i   member length l   j

Fig.1 Boundary conditions at both ends of a member

$$\{P\}^T + \{r\}^T = 2EI\lambda [K] \{\delta\}^T \qquad (6),$$

where, $\{r\}^T = 2EI\lambda ([K] \{f_1\}^T - \{f_2\}^T)$,

$$\{P\} = (P_i/\lambda, m_i, m_j, P_j/\lambda),$$
$$\{\delta\} = (\lambda\delta_i, \theta_i, \theta_j, \lambda\delta_j),$$
$$\{f_1\} = (\lambda f(0), f'(0), f'(1), \lambda f(1)),$$
$$\{f_2\} = (f'''(0)/2\lambda^2, -f''(0)/2\lambda, f''(1)/2\lambda, -f'''(1)/2\lambda^2).$$

Crofts et al.(1977) have shown the element stiffness matrix[K] was given in Eq.(7) as a function of λl. Eq.(6) means that the deflection vector$\{\delta\}$ is a linear function of the load vector$\{P\}$, accompanied with the element stiffness matrix[K] and the equivalent nodal force vector$\{r\}$.

$$[K] = \begin{vmatrix} \dfrac{Z_2}{Z_3} + \dfrac{Z_1}{Z_4} & \dfrac{1}{2}(\dfrac{Z_4}{Z_3} + \dfrac{Z_3}{Z_4}) \\[2mm] & \dfrac{1}{2}(\dfrac{Z_1}{Z_3} + \dfrac{Z_2}{Z_4}) \\[4mm] \text{symmetric} \\[6mm] \dfrac{1}{2}(\dfrac{Z_4}{Z_3} - \dfrac{Z_3}{Z_4}) & \dfrac{Z_2}{Z_3} - \dfrac{Z_1}{Z_4} \\[2mm] -\dfrac{1}{2}(\dfrac{Z_1}{Z_3} - \dfrac{Z_2}{Z_4}) & \dfrac{1}{2}(\dfrac{Z_4}{Z_3} - \dfrac{Z_3}{Z_4}) \\[2mm] \dfrac{1}{2}(\dfrac{Z_1}{Z_3} + \dfrac{Z_2}{Z_4}) & -\dfrac{1}{2}(\dfrac{Z_4}{Z_3} + \dfrac{Z_3}{Z_4}) \\[2mm] & \dfrac{Z_2}{Z_3} + \dfrac{Z_1}{Z_4} \end{vmatrix} \qquad (7),$$

where, $Z_1 = \cosh(\lambda l) + \cos(\lambda l)$,
$Z_2 = \cosh(\lambda l) - \cos(\lambda l)$,
$Z_3 = \sinh(\lambda l) + \sin(\lambda l)$,
$Z_4 = \sinh(\lambda l) - \sin(\lambda l)$.

3 CONSIDERATION OF PIPE JOINT ROTATION

3.1 Rotational stiffness of joint

Fig.2 shows two members of the pipeline connected with a certain joint to each other. If the joint is of a pin type, then rotation will freely occur and the moment at the joint will be zero. In practice, however, some moment of resistance(m) acts

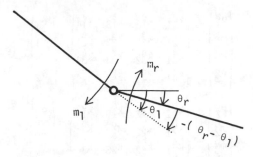

Fig.2 Rotation at a mechanical joint

at the joint. According to many laboratory tests, the moment is almost proportional to the rotational difference at the joint shown(Attewell et al. 1986) in Eq.(8).

$$m_l = M(\theta_r - \theta_l)$$
$$m_r = -M(\theta_r - \theta_l) \qquad (8).$$

The proportional constant(M) is called 'rotational stiffness', which is obtained by a bending test and varies with the type of joint and its diameter.

3.2 Incorporation of rotational stiffness in the stiffness equation

The previous section clarified the relation between the displacement and load vectors in a single member of a pipeline. Since the real pipeline consists of several members and joints, not only has Eq.(6) to be constructed along the whole members, but the rotational stiffness also has to be incorporated in the equations. Fig.3 shows a pipe system consisting of two members with a joint. For simplicity it is assumed that $\{r\}=0$, $2EI\lambda =1$ and $\lambda =1$ in Eq.(6), and the following equations will then be given for both the members a and b.

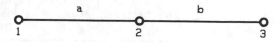

Fig.3 A pipe system with two members

359

$$
\begin{Bmatrix} P_1^a \\ m_1^a \\ m_2^a \\ P_2^a \end{Bmatrix} =
\begin{vmatrix} k_{11}^a & k_{12}^a & k_{13}^a & k_{14}^a \\ k_{21}^a & k_{22}^a & k_{23}^a & k_{24}^a \\ k_{31}^a & k_{32}^a & k_{33}^a & k_{34}^a \\ k_{41}^a & k_{42}^a & k_{43}^a & k_{44}^a \end{vmatrix}
\begin{Bmatrix} \delta_1 \\ \theta_1 \\ \theta_2^a \\ \delta_2 \end{Bmatrix} \quad (9a),
$$

$$
\begin{Bmatrix} P_2^b \\ m_2^b \\ m_3^b \\ P_3^b \end{Bmatrix} =
\begin{vmatrix} k_{11}^b & k_{12}^b & k_{13}^b & k_{14}^b \\ k_{21}^b & k_{22}^b & k_{23}^b & k_{24}^b \\ k_{31}^b & k_{32}^b & k_{33}^b & k_{34}^b \\ k_{41}^b & k_{42}^b & k_{43}^b & k_{44}^b \end{vmatrix}
\begin{Bmatrix} \delta_2 \\ \theta_2^b \\ \theta_3 \\ \delta_3 \end{Bmatrix} \quad (9b).
$$

Secondly, superposition of the above two equations produces a total stiffness equation(Eq.(10)). Since the rotations and the moments at the node 2 are both discontinuous, Eq.(10) turns out to be a simultaneous equation with seven unknowns.

$$
\begin{Bmatrix} P_1^a \\ m_1^a \\ 0 \\ 0 \\ 0 \\ m_3^b \\ P_3^b \end{Bmatrix} =
\begin{vmatrix}
k_{11}^a & k_{12}^a & k_{13}^a & k_{14}^a & \cdot & \cdot & \cdot \\
k_{21}^a & k_{22}^a & k_{23}^a & k_{24}^a & \cdot & \cdot & \cdot \\
k_{31}^a & k_{32}^a & k_{33}^a+M & k_{34}^a & -M & \cdot & \cdot \\
k_{41}^a & k_{42}^a & k_{43}^a & k_{44}^a+k_{11}^b & k_{12}^b & k_{13}^b & k_{14}^b \\
\cdot & \cdot & -M & k_{21}^b & k_{22}^b+M & k_{23}^b & k_{24}^b \\
\cdot & \cdot & \cdot & k_{31}^b & k_{32}^b & k_{33}^b & k_{34}^b \\
\cdot & \cdot & \cdot & k_{41}^b & k_{42}^b & k_{43}^b & k_{44}^b
\end{vmatrix}
\begin{Bmatrix} \delta_1 \\ \theta_1 \\ \theta_2^a \\ \delta_2 \\ \theta_2^b \\ \theta_3 \\ \delta_3 \end{Bmatrix} \quad (10).
$$

As Eq.(10) is simplified for the explana-tion, the following two remarks have to be considered in the analysis. The vector{r} must be added to the left hand term of Eq.(10), and the rotational stiffness M should be divided by $2EI\lambda$ to be dimensionless.

Unknown elements of the vectors { δ } and {P} can be obtained when solving the total stiffness equation related to a given problem, in consideration of the remarks above. The solutions obtained are all about the nodes in this stage. When the solved vector { δ } is substituted into Eq.(5) or its derivative, the vector{G} for any member will then be calculated. The deformations at any value for x can, therefore, be analysed by Eq.(5).

4 COMPARISON TO MEASUREMENTS

Detailed field experiments were carried out by the author(Takagi 1989) to investigate the longitudinal strains developed in shallow buried pipelines due to static and rolling wheel loads. The analytical method presented in the previous chapter is evaluated here by applying it to some of the experimental results.

4.1 Outline of the experiment

Two types of pipeline were installed at depths of 30, 60, 90 and 120cm; ductile iron ones of 100mm diameter and steel ones of 25mm diameter, both with mechanical joints. Each pipeline comprised five 4m lengths. The middle part of each pipeline had 12 pairs of strain gauges, which were glued to both crown and invert surfaces in the direction of the pipeline. The three axle truck used in the experiment had double tyres on both rear axles. Two types of tests were carried out; i.e. stationary mode and running mode. Here we refer only to the pipe strains only at 60cm depth obtained in the stationary test. Figs.4 to 7 show the

Table 1. Dimensions and mechanical properties of pipes used.

pipe material	outside diameter (cm)	inside diameter (cm)	pipe length (cm)	Young's modulus (MN/m^2)	rotational stiffness (kN.m/rad)
ductile iron	12.14	10.44	400x5	$1.47 \ast 10^5$	$1.54 \ast 10^5$
steel	3.40	2.76	400x5	$2.06 \ast 10^5$	$5.85 \ast 10^3$
steel	40.64	39.06	500x1	$2.06 \ast 10^5$	---

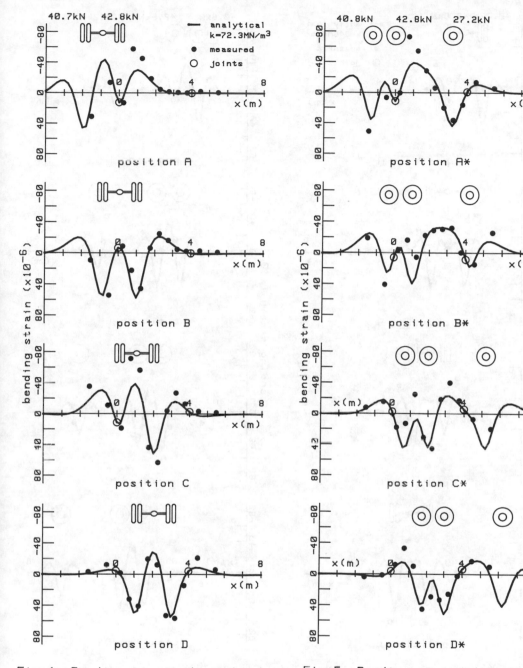

Fig.4 Bending strain distribution of the 100mm diameter
ductile iron pipe(A to D series)

Fig.5 Bending strain distribution of the 100mm diameter ductile iron pipe(A* to D* series)

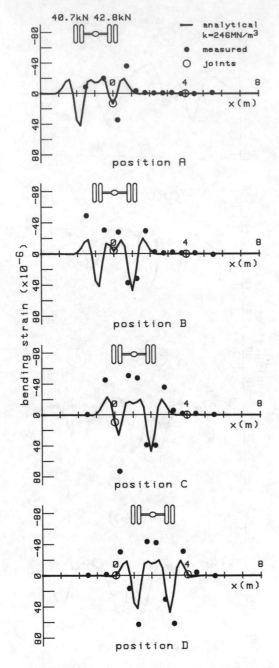

Fig.6 Bending strain distribution of the 25mm diameter steel
pipe(A to D series)

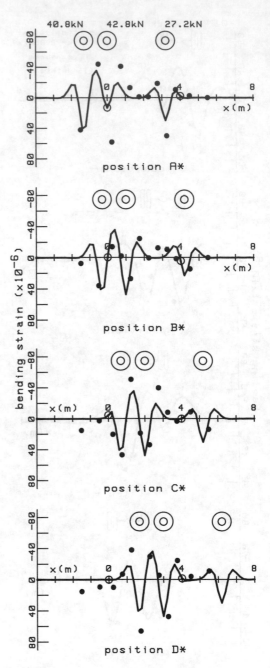

Fig.7 Bending strain distribution of the 25mm diameter steel
pipe(A* to D* series)

bending strain distribution measured in the two types of pipeline. The wheel placements above the pipeline were divided into two directions; one(* mark) was the three right wheels loading and the other(non-mark) was the left and right driving wheels loading. In each direction, four different positions (A, B, C and D) of 1m intervals were tested above the pipeline in order to examine the effect of the joint. The reason why the positions were varied was that the pipeline behaviour was considered to depend on a relative distance between the loaded position and joint locations.

4.2 Conditions of analysis

Table 1 shows the main dimensions and material properties of the pipelines. According to the circular plate loading test at the three different depths in the backfilled sand, the average subgrade reaction coefficient($\bar{k}$) was approximately 47MN/m^3 for the first 2.5mm settlement. Applying this $\bar{k}$ to the Vesic method(Attewell et al. 1986 and Vesic) described later, the effective subgrade reaction coefficients(k_{eff}) of 72 MN/m^3 and 223MN/m^3 are evaluated for the ductile iron and the steel pipes, respectively.

The Fourier coefficients a_i and b_i of Eq.(3) were calculated by assuming the span is [-800,800]cm, the small interval $\Delta x=$ 10cm and the maximum wave number n=30. The origin in the x direction was set at the midpoint between the both wheels of the driving axle during the A to D series of tests, while it was set at the midpoint between the front and the driving axles during the A* to D* series of tests. The other wheels were not taken into account because they have little effect on soil stress onto the pipeline. The double tyres were assumed to act as a single load. The weight of each wheel load is shown in Figs.4 to 7. Calculation was carried out by a 16bit computer.

4.3 Discussion

Figs.4 to 7 show the numerical calculations as well as the strain measurements of the pipelines at a depth of 60cm in the stationary mode test. The calculations simulate well the bending strain profile in any test condition, although there is a little difference between the calculations and the measurements. The profile does not shift with its shape remaining unchanged, associated with the different loading positions. The profile changes little by little

due to the existence of the joints.

The analysis generally gives a better agreement in the case of the ductile iron pipe than in case of the steel pipe. The calculation is the best at the positions B and D, where neither wheel locates just above any joint. The analysis, however, underestimates in other positions where either wheel acts just above the joint. In the latter case a bigger force acts at the joint and increases its rotation as well. A better approximation would have been obtained, if a larger rotational stiffness, M, had been used in the analysis.

Although the analysis for the steel pipe is less accurate than that for the ductile iron pipe, it still gives a good simulation for the strain profiles. One major reason for the lower estimation is thought to be associated with the reliability of the value for k. A large error may occur when evaluating k for the small diameter pipe from the plate loading test having ten times larger diameter than that of the pipe.

5 A SIMPLE METHOD FOR CALCULATING LONGITUDINAL BENDING STRAIN

From a practical point of view, the analytical method described above is rather complicated to apply to the problem. The practical concern is normally focused on the maximum bending strain beneath the wheel. Another simpler method will, therefore, be required in order to predict the maximum strain. The previous experimental results further suggest that the following assumptions may be made for the pipelines considered here;

(a) It is adequate, for the small diameter pipelines, to take account of only a single tyre load for the evaluation of the longitudinal bending strain,
(b) The mechanical pipe joints may be neglected, because the joints had little effect on the bending strain,
(c) The surface vehicle load is regarded as a concentrated load.

The above assumptions greatly simplify the calculation of bending strain in the buried pipes. If the vertical pressure due to a surface vehicle load is considered to act along the pipe as a triangular load as shown in Fig.8, then the maximum bending moment(M_{max}) will be given by the Hetenyi formula(Hetenyi 1946). Its basic principle is also based on Eq.(4). Since the maximum bending strain(ε_{max}) equals $M_{max}D/2EI$, then,

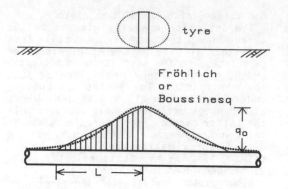

Fig.8 An assumed triangular load on pipes

$$\varepsilon_{max} = \frac{M_{max}\ D}{2EI} = \frac{q_0}{2L^2k}\ \lambda L$$

$$\{\ 1 - \exp(-\lambda L)(\cos\lambda L + \sin\lambda L)\ \}\quad (11).$$

In Eq.(11), q_0 and L represent the maximum load per unit pipe length and a half width of the triangular load respectively, as shown in Fig.8. These parameters can be

easily derived by Eq.(2) and its integration along x-axis (see Takagi 1989). For example, in case of the Fröhlich theory, these parameters are presented as follows;

$$q_0 = \frac{5Q}{\pi\ z}\ (\ R/t - 2R^3/3t^3 + R^5/5t^5\ )\quad (12),$$

$$L = \frac{2}{5}\ z\ \frac{Rz/t^2 + \arcsin(R/t) + 2Rz^3/3t^4}{R/t - 2R^3/3t^3 + R^5/5t^5}$$
$$(13),$$

where, $t^2=R^2+z^2$.

The value of subgrade reaction coefficient(k) is very important, because the maximum strain is inversely proportional to this parameter(see Eq.(11).) The value for k may be evaluated by Vesic(Attewell et al. 1986 and Vesic 1961). At infinite depth, and for subgrade Poisson's ratio of 0.5, the effective modulus (k_{eff}) is

$$k_{eff}D = 1.03\ (E_gD^4/EI\ )^{1/12}\ \bar{K}B\quad (14).$$

where E_g is an elastic property of the foundation. On the other hand this may be obtained from $\bar{K}$ for a circular plate(diameter B) loading test provided that Poisson's ratio is assumed.

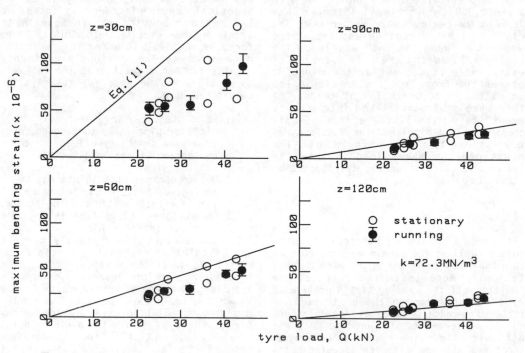

Fig.9 Maximum bending strain of 100mm diameter ductile iron pipes and tyre load

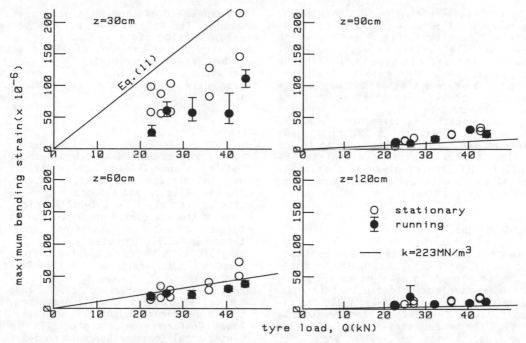

Fig.10 Maximum bending strain of 25mm diameter steel
pipes and tyre load

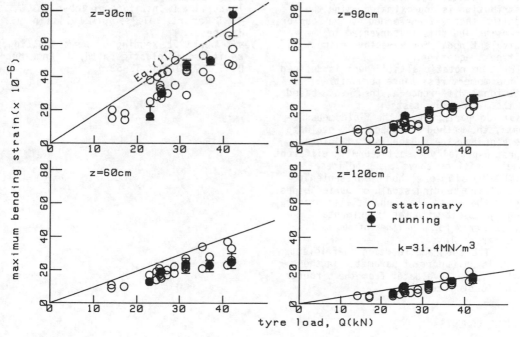

Fig.11 Maximum bending strain of 400mm diameter steel
pipes and tyre load

Using this effective foundation modulus, the simple method was applied to the experimental results in Figs.9 and 10, which give the relationship between the wheel load and the maximum pipe strain beneath it. Fig.11 shows the result of another experiment where a similar test was carried out using steel pipes of 5m lengths and 400mm diameter(see Takagi et al. 1989). The pipe specification is given in Table 1: and an average $\bar{K}$ was about 62.8MN/m^3. The straight lines in the figures indicate the maximum bending strains calculated by Eq.(11). The strains given by the simple method agree well those actually measured particularly at depths of 60cm or more.

6 CONCLUSIONS

Two methods were presented to estimate the longitudinal deformation of the small diameter pipelines with mechanical joints which are subjected to the traffic load. Being applied to the detailed field experiments, both the methods can reasonably predict the bending strain behaviour of the pipelines. The main conclusions are:-

(1) The accurate method is developed by incorporating the rotational stiffness of the joint and the Fröhlich soil stress theory into the stiffness equation for the beam on elastic foundations. The stress distribution is approximated using the Fröhlich theory transformed by the Fourier expansion and this is converted to the equivalent nodal force vector in the stiffness equation.

(2) The rotational stiffness is defined as the moment resistance per unit rotational difference at the joint, based on the laboratory tests.

(3) Comparing it to the field measurements, the method can adequately evaluate the bending strain behaviour of jointed pipes, especially associated with different wheel positions above the pipelines.

(4) The simple method is presented to predict the bending strain of buried pipes due to the wheel load, and it also gives a good agreement with the experimental measurements for pipelines of 400mm diameter or less.

(5) The subgrade reaction coefficient, which is an important parameter in the analyses, was evaluated from the circular plate loading test.

ACKNOWLEDGMENTS

The author is grateful to Dr B.M.New of the Transport and Road Research Laboratory who read the draft and made many useful comments. The help of the staff in Senju Pipeline Office of Tokyo Gas is greatly acknowledged for the provision of a site suitable for the experiment.

Any views expressed in this paper are not necessarily those of the Tokyo Gas Co., Ltd.

REFERENCES

Attewell P.B. et al. 1986. Soil movements induced by tunnelling and their effects on pipelines and structures, p.129. Blackie, Glasgow and London.
Crofts J.E. et al. 1977. Lateral displacement of shallow buried pipelines due to adjacent deep trench excavations, Geotechnique Vol.27, pp.161-179.
Hetenyi M. 1946. Beams on elastic foundation, p.17, University of Michigan Press, Ann Arbor.
Takagi N. 1989. An experiment on the longitudinal deformation of small diameter buried pipelines subjected to vehicle loads, Conference on Instrumentation in Geotechnical Engineering, held on 3-5 April 1989, at University of Nottingham, U.K., pp.449-463.
Takagi N. et al. 1989. An experiment on the circumferential deformation of buried steel pipelines subjected to vehicle loads, R & D Institute of Tokyo Gas, R & D Report, Vol.33, pp.121-141(in Japanese).
Vesic A.B. 1961. Bending of beams resting on isotropic elastic solid, J. Eng. Mech. Div., ASCE87, pp.35-53.

Developments in Geotechnical Aspects of Embankments, Excavations and Buried Structures, Balasubramaniam et al. (eds)
© 1991 Balkema, Rotterdam. ISBN 90 5410 019 2

TBM tunneling and supports in ravelling and squeezing rock

Noppadol Phien-wej
Asian Institute of Technology, Bangkok, Thailand

Edward J.Cording
University of Illinois at Urbana-Champaign, Ill., USA

ABSTRACT: Three TBM's were used to excavate jointed medium hard shale of Stillwater Tunnel, Utah of the U.S. Bureau of Reclamation. The first telescoping full shield TBM with incorporated concrete segments suffered severe problems in a fault zone due to ravelling and squeezing ground and had to be abandoned and the contract was terminated after conducting an instrumentation program to re-evaluate the ground conditions. Two shrinkable shield TBM's were selected to complete the tunnel in a new contract and proved to be successful. A short shield TBM with steel rib support was satisfactorily capable of advancing in all types of ground condition at the rate as high as 70 meters per day. An innovative blade-gripper TBM allowed the installation of concrete segments and was capable of advancing freely in ravelling and squeezing ground. Light expended circular steel ribs with minor reinforcements were capable of providing initial support to squeezing ground. Complete backfilling of concrete segmented rings was difficult to achieve early due to raveling of fractured shale in the crown area. Delayed completion of backfilling left soft spot above the segment crown thus created high load eccentricity in the arch areas of the segmented rings resulting in bending cracks of arch segments at small rock load. However, with prompt grouting these cracks did not jeopardize long-term performance of the segments.

1. INTRODUCTION

The 13-km-long Stillwater Tunnel is located at the upstream end of the Strawberry Aqueduct, Central Utah Project of the U.S. Bureau of Reclamation (USBR). The aqueduct, which consists of 7 tunnels, 5 pipelines and 6 diversion dams is a transmountain water-collection system designed to convey water from reservoirs located along the aqueduct line to the existing Strawberry Reservoir, from which the water is to be distributed to the Wasatch Front area of the State of Utah. Stillwater Tunnel is lined with concrete lining, 2.25-m inside diameter transitioned to 2.48 m. Before construction, the tunnel was intended to be a pioneer project in the U.S.A. for rapid excavation in weak rock using a TBM incorporated with concrete segmented liner owing to its potential cost and time savings. The contract for construction of the tunnel was awarded to Harrison Western-Copper Joint Venture in 1976. The construction began in 1977 using drill-and-blast excavation for a short reach

from the inlet end as a precaution in excavation through the South Flank fault, a major fault zone of the region, and in 1978 using a TBM with concrete segments from the outlet end. However, the contract had to be terminated by and for the convenience of the U.S. Government in 1979 due to severe tunneling difficulties in a faulted shale area in the forms of repeated stalling of the TBM and severe damage to the concrete segments. At that time, only 4.16 km of the tunnel had been excavated: 1.92 km from the inlet portal and 2.24 km from the outlet portal.

The contract for completion of the tunnel was awarded to Traylor Brothers and Fruin-Colnon Joint Venture in 1982. Excavation of the remaining 8.6 km tunnel was successfully accomplished between October 1982 and September 1983, using two TBM's which were particularly designed for advancing in heavy ground. Placement of final concrete liner and grouting were continued through the fall of 1984. Although the construction cost of the Completion Contract was within the engineer estimate, the overall cost of the tunnel became one of the most expensive in the U.S. tunnel history.

2. GEOLOGIC SETTING

Stillwater Tunnel is on the southwestern side of the Uinta Mountain of the Rocky Mountain System. The tunnel begins at the Rock Creek, the site of the Upper Stillwater Reservoir, passes through a 780 m high ridge and ends at the valley wall of the North Fork Duchesne River.

Stillwater Tunnel penetrated through Precambrian rock formations. The tunnel alignment, S67°W, is subparallel to the beddings of the rock strata which generally dips $10°$ to $20°$ to the south. From the inlet portal the tunnel penetrated rocks of the Uinta Mountain Group which consisted of sandstone, argillite and lenses of shale, siltstones and quartzite. The tunnel remained in these rocks for approximately 600 m before encountering a 70-m-wide disturbed zone of the steeply dipping South Flank fault, the major geologic discontinuity along the tunnel alignment. Upon leaving the fault the tunnel entered younger rocks of the Red Pine Shale Formation and remained in this rock formation throughout the remaining length, but successively crossed younger beds at an apparent dip of approximately $4°$. The Red Pine Shale consisted primarily of a dark gray to black, medium to hard, thinly bedded to laminated, well indurated illitic shale with numerous thin (few centimeters to meters) interbeds of sandstone and siltstone. The intact shale, water content = 2-3 %, had unconfined compressive strength between 50 to 88 MPa (63 MPa on avg.), Schmidt Hardness 22 and Total Hardness 18. The sandstone interbeds had unconfined compressive strength above 90 MPa, Schmidt Hardness 40 and Total Hardness 80.

Besides the South Flank Fault, the geologic structures along the tunnel alignment on the ground surface consisted of numerous smaller high-angle faults. At tunnel level as many as 12 fault or large shear zones intersected the tunnel. Most of these zones traversed the tunnel near perpendicularly.

A classification shown below was developed as a means for grouping the shale masses and aiding the determination of their potential ground responses.

Classification of Rock Mass of Stillwater Tunnel

Class	Description	Length Encountered
I	Siltstone to sandstone, silty to sandy shale, widely jointed (1-3 m spacing).	26%
II	Shale to siltstone, moderetely jointed (0.3 to 1 m spacing), some sheared, slickensided surfaces.	38%
III	Shale, closely jointed (0.05 to 0.3 m spacing), some thin shears.	26%
IV	Shale, closely jointed and sheared shear zones have some with (several centimeters) and contain crushed or soft materials. Wide shear zones with large amount of clay gouge are classed as IVb	10%

The shale exhibited a variety of tunnel ground responses, namely, mild stress slabbing, loosening and raveling upon excavation, light to moderate squeezing in shear zones, minor swelling in wet shear zones, and minor slaking of wet fractured shale. The siltstone and sandstone were generally massive and experienced only minor stress slabbing. Raveling and squeezing of the shale were the two significant ground responses that caused some difficulties.

3. TBM'S AND SUPPORT SYSTEMS

3.1 First Contract

The TBM used in the First Contract was Robbins Model 92-192 (Figure 1) which incorporated a 7.2 m long, articulated, telescoping, full-barreled shield with a diameter of 2.85 m. The shield consisted of three separate sections: front, central, and telescoping rear sections. The forward thrust of up to 4.45 MN was provided by two wall grippers extending out from the telescoping shield section. The telescoping shield section also provided a shelter under which concrete segments were erected. The variations in diameter along the shield body occurred when the rear section was telescoping during shoving the cutterhead forward. The TBM also contained a full circle auxiliary thrust ring provided to assist in seating concrete segments into position.

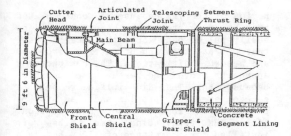

Fig. 1 Telescoping shield TBM, first contract

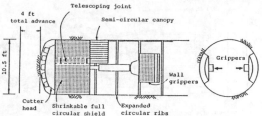

Fig. 3 Shrinkable short shield TBM, completion contract

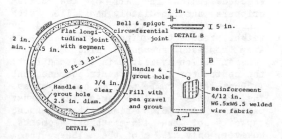

Fig. 2 Precast concrete segmented lining

The TBM-excavated tunnel was supported with a precast concrete segmented liner (Figure 2). Each segmented ring, 0.9 m wide, consisted of 4 pieces of unbolted 125 mm thick segments. The segments were reinforced principally for handling and erecting bending stresses.

For a load factor of 1.7, the concrete segmented rings were designed to be capable of sustaining uniform ground pressure up to 1.14 MPa (Marushack and Tilp, 1980). The segments were not designed for longitudinal thrust to aid the advance of the TBM.

The 75 mm wide annulus between segments and rock surface was to be immediately backfilled with pea gravel and subsequently grouted with cement to ensure uniform loading of the segmented rings.

3.2 Completion Contract

Due to time constraint the contractor selected to excavate the tunnel from both ends using two TBM's that were particularly designed to be capable of advancing in heavy ground. The TBM's were developed under a consultation between Traylor Brothers Inc. and Tyman Fikse Mechanical Consultants of Seattle, Washington. The shields of both TBM's were shrinkable so that they could shrink diametrically to accommodate inward ground movement to

prevent building up of high ground pressure. Both TBM's permitted the erection of support close to the tunnel face under the shelter of the shields. The TBM's also provided access for spiling tunnel roof around and ahead of the TBM, if needed. However, face and roof instability never occurred throughout the excavation of both TBM's.

3.3 Inlet TBM

The TBM used in the inlet excavation (Figure 3) had a 1.8-m-long full-barreled shield with longitudinal joints at both springlines. Servo-hydraulic rams were fitted at the joints to allow the shield to shrink vertically up to 75 mm in diameter in order to maintain rock pressure on the shield below 0.62 MPa. The cutterhead was thrusted forward by two grippers pressing against the tunnel walls (approximately 5.4 m from the face). The grippers were designed to provide a normal forward thrust of 4.27 MN.

The TBM allowed the installation of rock bolts and steel ribs as initial supports. In the first 510 m of the excavation, a pattern of 1.2 m long spilt set bolts was installed behind the shield while the TBM was advancing through siltstone and sandstone (Class I-II). As the TBM advanced further into jointed and brittle shale (class III), the ground required a greater amount of rock bolting. The shale tended to break into small pieces and revel down, making it difficult to drill holes for rock bolts. Subsequently, the support was changed to 3-piece expanded light circular steel ribs (M4x13) spaced 1.2 m on center with timber lagging inside the flanges. The steel ribs were assembled under the 1.2-m-long canopy tail of the shield and expanded against the rock surface in front of the grippers, approximately 3.9 m from the face, as soon as the canopy shield moved clear of the ribs. The steel ribs were used throughout

369

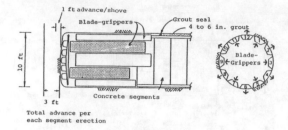

Fig. 4 Blade-gripper TBM, completion
contract

the rest of the excavation even in other
short sections of siltstone and sandstone
that actually did not necessitate the
steel support. The high rate of face
advance of the TBM and the difficulty to
accurately identify the changing ground
condition as it was developing in the
heading within a hundred meter made
changing supports to optimally suit the
ground condition unfeasible. The use of
the steel ribs throughout the remaining
excavation was selected in order maintain
the high rate of face advance by using
only one support system that had proved to
be capable of providing initial support to
all classes of ground.

3.4 Outlet TBM

To incorporate the use of the $2 million
worth of concrete segments inherited from
the First Contract the contractor innova-
tively designed and built the "blade-
gripper" TBM, (Figure 4), from parts of
the telescoping shield TBM that remained
lodged in the tunnel. The modification
consisted of reducing the shield area and
increasing the thrust capacity of the
machine, using an entirely new gripper
and shield design. Twelve 3.6 m long,
0.70 m wide and 37.5 mm thick blade-
grippers were placed around the perimeter
of the machine to provide reaction for
thrusting the cutterhead forward and to
also serve as a full shield to support
rock surrounding the machine. The blade-
grippers were controlled individually by
hydraulic rams that could exert contact
pressures up to 0.69 MPa against the
tunnel wall. The grippers provided total
thrust for the cutterhead up to 7.56 MN.
Each gripper could be moved forward and
backward 0.3 m and radially up to 0.15 m.
Though the TBM was designed primarily
for installation of concrete segments, the
contractor chose to install initial steel
support in the early trial excavation

period of the TBM because he contemplated
that by using steel ribs, the TBM could be
easily pulled back from the face in case
that it had steering or cutting problems
and needed modifications. The contractor's
approach proved to be a wise decision
because the TBM did initially have a
severe steering problem and had to be
pulled back for a major modification.
After TBM readvancing, steel ribs were
installed for a distance of 330 m before a
proper steering control of the TBM was obtained and the support was changed to
concrete segments. The segments were
erected inside the tail of the blade-
grippers and immediately backfilled with
sand-cement-flyash grout mix (6.3:1:4 by
weight) before the grippers moved forward
away from each segmented ring.

As an option available in the contract,
the contractor designed a stronger segment
(150-mm-thick) to be used in heavy areas.
The segments had 1.55 times the thrust
capacity and 3.75 times the bending
capacity of the 125-mm-thick segments.
However, only 22.5 m of the 150-mm-thick
segments were erected because of the
difficulty in assembling within the tail
end of the blade-grippers as well as the
fact that ground pressure was not high
enough to necessitate the use of the
segments as indicated by the instrumentation result (Phien-wej, 1987).

The contractor also prepared a compressible grout mix to be used in squeezing
areas to accommodate rock closure and
reduce load on the segments. The grout
mix, which was obtained by adding
polystyrene foam beads to the regular mix,
was expected to compress by as much as 75
mm radially if the ground pressure
exceeded 2.07 MPa (collapse strength of
the cement matrix surrounding the foam
beads). However, the compressible grout
mix was used only once, on a trial basis,
in a 105 m wide faulted area (Class IV)
that exhibited moderate squeezing. Even
in this area, which was one of the
heaviest sections of the tunnel, instrumentation results indicated that rock
pressure was not high enough to
necessitate the use of the compressible
grout mix.

4 PERFORMANCE OF TBM'S AND SUPPORTS

4.1 First Contract TBM

The TBM began to excavate the tunnel from
the outlet portal in July 1978 and was
abandoned in May 1979. The TBM stalled at
least 14 times throughout its excavation.

Stalling of the TBM usually occurred following an interruption of face advance, such as weekend shutdown and equipment or power failure (Marushack and Tilp, 1980). To overcome stalling, the advance of the TBM was assisted by thrusting off against the segments with the auxiliary thrust ring. However, in a more difficult ground, this measure practice was not sufficient to free the TBM, but instead leading to segment damage. In such conditions, hand excavation around the upper perimeter of the shield was employed.

In February 1979 after 2100 m of face advance the TBM began to advance into a disturbed zone of shale at a depth of 660 m. The ground was described as fractured, blocky and brittle shale which consisted of several shear zones. In this area the TBM stalled several times with increasing difficulties to get free as the ground conditions were progressively worse. Hand excavation to free the TBM aggravated ground disturbance, which led to deeper zones of loosened rock and higher rock load. Ultimately, the TBM was severely stalled when the front of the shield embedded in a 5.4 m thick cross-cutting seam of fault materials. The TBM advanced only 80 m during the last month of excavation.

A number of concrete segments installed in this reach were severely damaged. Longitudinal bending crack of the arch segments was the most predominant mode of damage, although the last twenty-two segmented rings were badly cracked by excessive bearing pressure created by the auxiliary thrust ring and had to be replaced.

Complete pea gravel backfilling of the segmented rings in this fractured and sheared shale was difficult and usually not obtained, especially in the crown area, due to large overbreak and fallen pieces of shale trapped behind the segments. Pea gravel was not capable of filling voids between trapped shale pieces and thus soft spots were present in the crown area behind the segments. The use of somewhat angular pea gravel resulted in friction between pea gravel grains making it difficult to move pea gravel into all void spaces. Construction reports indicated that placement of pea gravels was also delayed in some of the area. The placement of cement grout, the supplemental backfilling, had not been carried out at all in the area.

The deficient segment backfilling, the offset (up to 50 mm) along the longitudinal joints and the low bending capacity of the segments led to bending

cracks running across the segments. Without obtaining subsequent timely full backfilling, bending damage of the segments continued to propagate due to the build-up of rock pressure from ground squeezing, and became severely extensive and the segments had to be replaced.

With the mentioned difficulties, the USBR then decided to stop further TBM advance and embarked a drill-and-blast exploratory and instrumentation reach in front of the TBM. After the re-evaluation of the ground condition and based on the instrumentation results the USBR decided that the TBM would not be capable of successfully excavating the remaining length of the tunnel and the First Contract was terminated. It was also claimed by the contractor that the shale exhibited unforeseen squeezing which made the TBM and segments incompatible with the ground encountered. However, based on the investigation

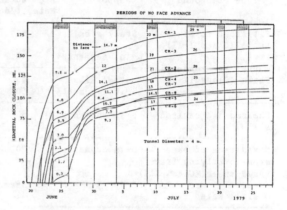

Fig. 5a History of horizontal rock closure

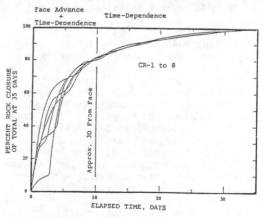

Fig. 5b Percent rock closure of total at 35 days

371

Table 1 Summary of advance rate of short shield TBM

Type of Ground	Support	Avg. Advance Rate, ft/day	Max. Advance Rate, ft/day	Penetration Rate, ft/hour
Siltstone to sand- stone (ClassI-II)	Rock Bolts	120	200	15-16
	Steel Ribs	138	220	
Shale (Class I-III)	Steel Ribs	148	232	20-30
Sheared and faulted Shale (Class III-IVb)	Steel Ribs	147	228	20-25
Overall Rate		135	232	
Utilization (percent)	37			

1 foot = 0.3048 meter

Table 2 Summary of advance rate of bladegripper TBM

Type of Ground	Support	Avg. Advance Rate, ft/day	Max. Advance Rate, ft/day	Penetration Rate, ft/hour
Siltstone to sandy shale (Class I-II)	Steel Ribs	53	75	10
	6-in. Segments+	20	45	
Shale (Class II-III)	Steel Ribs*	43	68	7-10
	5-in. Segments	50	70	10-15
Sheared and faulted shale	5-in Segments	50	65	10-15
Mixed-face of sand- stone overlying frac- tured shale (Class-III)	5-in Segments	20	22	1.5-5
Overall Rate		41	75	
Utilization (percent)	20			

* Rates in the first 475 ft of excavation.
+ Early in segment installation.

of ground condition, construction records, and the instrumentations results the authors believe that the inability of the TBM to advance in the sheared shale was mostly initiated by rock pieces and wedges falling down and jamming onto the shield not the squeezing pressures. The long, stiff and variable-diameter shield of the telescoping shield TBM created a high potential for rock jamming and then the build-up of high skin friction on the shield in blocky and squeezing shale. The first stalling occurred approximately 180 m from the outlet portal in a blocky shale mass during an attempt to realign the TBM back on grade. At that location, the ground cover was only 150 m, which suggested that the stalling of the TBM was not entirely caused by ground squeezing because the ground stresses at the location would not be high enough to produce serious squeezing condition.

In authors' opinion, the convergence measurement in the instrumentation reach (Figure 5) indicated clearly that the high rock load on steel ribs which was recorded to increase rapidly with time was primarily due to the effect of face advance not the creep of the ground. The tunnel face was advanced slowly but the support was erected immediately behind the face. This could lead to high rock load on support even in the elastic rock. The instrumentation results in the Completion Contract revealed that even in the worst ground section squeezing was considered to be only moderate.

4.2 Completion contract

The performance of the two TBM's in the Completion Contract are summarized in Tables 1 and 2.

4.3 Inlet side excavation by short shield TBM

The TBM successfully excavated 7.52 km (86 percent of the total tunnel length in the Completion Contract) at an average rate of 34.5 m per 24-hour working day, with the maximum rate of 69.6 m per day. The TBM never experienced steering or stalling problems throughout its excavation. The TBM practically advanced through all classes of shale (Class II-IVb) at the same rate. The degree of jointing and shearing of the shale did not have an influence on the advance rate of the TBM. The average advance rate in the shale was 44.4 m/day. The rates were slightly lower in sandstone to siltstone (Class I-II) : 41.4 and 36.0 m/day with steel support and rock bolt support, respectively. The utilization of the TBM was 32 percent.

In faulted shale, high ground loads built up rapidly on the shield and shrank the shield by as much as 50 mm diametrically. The ability of the TBM to allow early support installation and fast rate of advance proved to be very crucial in the stability control of the heavy and squeezing shale using the relatively light steel support. Early support installation close to the face helped prevent excessive ground disturbance and preserved the insitu properties of the shale mass. The fast rate of face advance allowed quick access for repairing and reinforcing the overstressed and distorted steel ribs in order to prevent failure. Because of small tunnel size, full access for working on the support was not available until the complete passage of the trailing gear of the TBM, which was approximately 60 m long. The fast rate of face advance of the TBM allowed the access within 48 hours.

The light expanded circular steel ribs (M4x13, 1.2 m o.c.) proved to be an adequate initial support system in most parts of the tunnel. The ribs, when expanded tight against the smooth machine-cut rock wall, had capacity near their pure thrust capacity because of low bending moments. In light squeezing areas (Class III), the steel ribs were capable of stabilizing the tunnel although yielded early, and experienced some distortions.

In sheared and faulted shale (Class IV), high initial rock pressure combined with overbreak resulted in early severe buckling and distortion of the ribs. Although the contractor tried to block overbreak as soon as possible, the operation sometimes could not keep up with the fast rate of face advance. Thus the ribs were left under unfavorable loading conditions. Rib distortion was progressive due to ground squeezing which led to large-scaled failure of the ribs in a 150-m-wide faulted area. However, the tunnel did not collapse. The area had to be remined and steel jump sets (W6x20) were installed. In another similar ground section, timely and continuous repair and reinforcement was able to curb the distortion of the light ribs from growing into failure.

The reinforcement consisted of placing longitudinal bracing (100-mm steel channels), replacing timber lagging, resetting steel ribs and welding steel straps onto the inner surface of the ribs as a circumferential reinforcement. Longitudinal bracing was the primary means of reinforcement, which proved to be crucial for the performance of the high-stressed ribs. Most of the steel ribs as installed behind the TBM had very minimum longitudinal bracing because timber lagging was minimal on the side walls as the ground was generally capable of standing up by itself even in faulted areas.

4.4 Outlet excavation with blade-gripper TBM

The TBM began to cut into the original face left from the First Contract in late January 1983. However, only after 10.5 m of advance, the TBM had to be pulled back behind the original face for a 2-month major modification to the blade-gripper system and cutterhead because the TBM tended to dive and its direction could not be controlled. The blade-grippers were shortened from 4.5 m to 3.6 m and a guide ring cap was installed behind the cutterhead to house the front end of the grippers so that their direction could be better controlled. Steering of the TBM was further assisted by an on-board computerized laser guidance system, manufactured by Zed Instruments, Inc., England. The system continuously monitored the orientation and position of the TBM relative to a laser beam to millimeters and projected the TBM's position 3 m ahead of its current position. Consequently, the operator was able to make early small corrections in the TBM misalignment. The use of the guidance unit resulted in the ability to control the TBM alignment within 25 mm.

The TBM proved to be capable to advancing freely in all types of ground classes and allowing the installation of concrete segments without any serious problems, although the TBM required frequent maintenance due to its complicated gripper hydraulic system. Due to the delay in assembling and modification the TBM only advanced for a total distance of 1170 m during its 4 months of operation. To save time, the contractor decided to have the Inlet TBM excavate the tunnel 1500 m more than originally planned.

The maximum advance rate of the TBM was 21 m/day with the average of 10.8 m/day. The advance rates of the TBM were practically the same (15 m/day on the average) in all types of shale (Class II-IV) when installing the 125-mm-thick segments. A much lower advance rate (6 m/day) was obtained when installing the 150-mm-thick segments in the early period of segment installation, partly due to the difficulty in erecting the segments.

Even though the TBM proved to be capable of advancing in all types of ground, the overall performance of the TBM was not so satisfactory due to the delay in assembling, modification and frequent mechanical problems. The utilization of the TBM was only 13 percent (not including 3 months delay in assembling).

One drawback of the TBM was the tendency of the blade-grippers to disturb the fractured and sheared shale due to the high flexibility of the shield formed by the blade-grippers and frequent movement which, in turn, worsened loosening and raveling conditions in the crown. Small broken pieces of shale tended to fall out through the gaps between the gripper blades and caused overbreak in the crown. Steel straps were welded along the edges of the blade-grippers in the roof area to seal off the initial 100-mm-wide gaps in order to reduce rock pieces from falling off and producing large overbreak. This led to more loosened rock pieces trapped behind the shield.

Overbreak and loosened pieces of shale trapped behind the blade-grippers reduced tractions for the grippers to shove the cutterhead forward and caused difficulties

in obtaining early completion of good backfilling of the segmented rings.

Low traction resulting from raveling of the shale caused a very difficult situation in advancing the TBM through a 15 m mixed-face section of highly fractured shale overlying a sandstone layer. In a short distance of this mixed-face section, the traction provided by the grippers was not sufficient to shove the cutterhead forward, therefore the two invert blade-grippers were anchored down to the sandstone with rock bolts to provide additional reaction for shoving the cutterhead.

Ravelling of the shale posed significant difficulty in segment backfilling. Backfilling of the segments was begun immediately after every ring installed before the gripper-shield moved forward clear from the ring. Raveling produced overbreak in the crown that made it difficult to seal the annulus for grout backfilling. The volume of grout brought to the heading for each segmented ring was not always sufficient to fill the additional volume created by overbreak. Loose rock pieces produced by raveling remained in the crown and left voids between rock pieces which were difficult to fill with the stiff grout mix. Even though the cement grout mix could penetrated and fill voids much better than the pea gravel used in the First Contract, it still could not penetrate small voids. The loose rock left ungrouted formed a soft zone in the crown area of the segment ring. Grouting in the crown area of some segmented rings installed in fractured or sheared shale (Class III-V) had to be delayed by as much as 5 rings behind the TBM.

In heavy ground areas (Class IV), the delayed completion of full backfilling of the segments caused large ring movements (horizontal contraction and vertical expansion), longitudinal joint offset (up to 45 mm) and bending cracks (Figure 6). The segments tended to deflect upward into the soft zone in the tunnel roof resulting in high load eccentricity in the arch area of the segmented ring (Figure 7).

Forty-six percent (414 rings) of the segmented rings installed in the Completion Contract experienced longitudinal cracks across the width of the arch segments due to high initial bending stresses and low bending capacity. The cracks opened up soon after the rings emerged from the tail of the TBM (within 1 to 3 rings behind the TBM) and continued to propagate and widen with further face advance. The instrumentation results showed that these bending

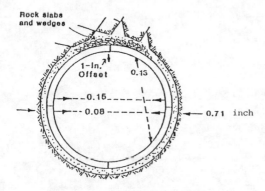

OUTLET TEST SECTION 3, INCOMPLETE GROUTING IN CROWN

3 DAYS AFTER INSTALLATION

Fig. 6 Recorded movement of an instrumented segmented ring

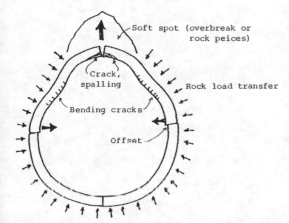

Fig. 7 Schematic depiction of segment distortion due to soft zone above crown

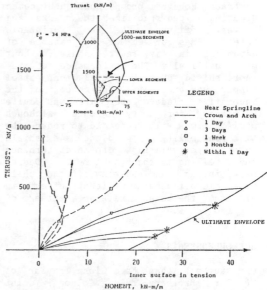

Fig. 8 Load level and eccentricity of segments

moment. The instrumentation results indicated that the long-term rock pressure in heavy ground was even below 35 percent of the design capacity of the segments.

CONCLUSIONS

The failure of the TBM Tunneling in the First Contract was due to the incompatibility of the TBM and the segment installation procedures with the ground. Stalling of the TBM was attributed to both rock pieces jamming on the long stiff shield body and the build-up of ground pressure due to ground disturbance as well as squeezing. The prime ground responses that caused the problem were loosening and ravelling rather than squeezing. In fractured shale where significant loosening and raveling occurred in the crown area, good backfilling of the concrete segmented ring with pea gravel was extremely difficult to achieve which led to bending cracks. With high ground pressure build up with time on these poorly backfilled rings, the cracks propagated resulting in severe damage of segments. Even with the use of immediate sand-cement-flyash backfill, complete backfilling of the segments was still difficult to achieve resulting in bending cracks developing on many of the arch segments installed in the Completion

cracks formed at very low rock pressure but due to very high eccentricity due to poorly backfilled zones above the segmented rings (Figure 8).

Up to 8 cracks developed in a segment and the width ranged from hairline to 2.5 mm (mostly thinner than 0.1 mm). The bending cracks typically became stable within 30 m from the tunnel face following the completion of full grout backfilling and hardening of the grout. The longitudinal bending cracks that initially developed on the segments did not jeopardize subsequent ability of the segments to support the ground since a more uniform loading of the segments was later obtained by complete backfilling. In fact the cracks would act as additional ring joints and help reduce bending

375

Contract. However, prompt subsequent cement
grouting helped stabilized the cracks and
prevent failure. The two shrinkable
shield TBM's proved to be capable of
advancing in heavy faulted shale without
much difficulty. The performance of the
short shield TBM with initial light steel
support was excellent. The concept of the
innovative blade-gripper TBM was quite
promising for heavy ground although
modifications will be needed to reduce the
high flexibility of the blade-made shield
in order to avoid high ground disturbance
and ravelling in the tunnel roof. Degree
of jointing and shearing of the shale
mass did not show significant effect in
altering the advance rates of either of
the two TBM's.

ACKNOWLEDGMENTS

The authors thank Mr. Glen Traylor of
Traylor Brothers-Fruin Colnon Joint
Venture to provide the opportunity and
financial support for them to conduct the
study. Also thanks are extended to Mr.
John McDonald, Stanley Prescott and
Jackson Graham, the Traylor Brothers
personnel on the project site.

REFERENCES

Cording, E.J., Mahar, J.W. and Fernandez,
   G. (1982). Evaluation of ground condi-
   tions and their effect on excavation and
   support, Stillwater Tunnel Completion
   Project, Report for Traylor Brothers-
   Fruin-Colon Joint Venture.
Cording, E. J. and Shannon and Wilson,
   Inc. (1984), Summary of tunneling con-
   ditions, observations and instrumenta-
   tion results in Stillwater Tunnel, Com-
   pletion Contract, Final Report Vols. 1,
   2 and 3 for Traylor Brothers-Fruin Colon
   Joint Venture.
Marushack, J.M. and Tilp, P.J. (1980),
   Stillwater Tunnel- A progress report,
   U.S. National Committee on Tunnel Tech-
   nology, Newletter No. 31, September.
Phien-wej, N. (1987), Ground response and
   support performance in a sheared shale,
   Stillwater Tunnel, Utah, PhD. Thesis,
   University of Illinois at Urbana-
   Champaign.
U.S. Bureau of Reclamation (1981), Con-
   struction and foundation materials test
   data and Stillwater Tunnel instrumenta-
   tion data for Stillwater Tunnel Comple-
   tion, Bonneville Unit, Utah, Central
   Utah Project, Specifications 40-C2035.

Developments in Geotechnical Aspects of Embankments, Excavations and Buried Structures, Balasubramaniam et al. (eds)
© *1991 Balkema, Rotterdam. ISBN 90 5410 019 2*

Behavior of a buried flexible steel pipe under groundwater level

Masami Fukuoka & Yoshinori Imamura
Science University of Tokyo, Noda City, Chiba, Japan

Takaki Omori
Kumagai Gumi Co., Ltd, Shinjuku, Tsukudo, Tokyo, Japan

ABSTRACT: A steel pipe 2m in diameter and 20 mm in thickness surrounded by "panel type earth pressure gauges", which can simultaneously measure both normal and tangential components of earth pressure of a backfill soil against a pipe was embedded in soil. After measuring earth pressure of soil in the dry state, groundwater level was raised by pouring water into the bottom of the backfill. The tangential components of earth pressure changed remarkably, but the normal components of earth pressure did not change much, contrary to our prediction. The earth pressure from the side direction decreased with the rise of groundwater. The pipe was elongated horizontally, and the bending moments of the pipe increased accordingly. The apparent earth pressure distribution used for designing flexible pipes gives a triangular shape, but the measured one was similar to the trapezoidal shape just like the earth pressure on the back of retaining walls. The earth pressure was changed by a strong earthquake which happened while the test was being conducted.

1 INTRODUCTION

Studies of buried rigid or flexible pipes have been done since the beginning of the century (Spangler, in 1933 and 1941). Many researchers have studied earth pressure distribution around buried pipes. Most pipes have been designed using the experimental formula developed by Spangler in 1973. Yet, despite many improvements in their design and construction, buried pipes are still sometimes damaged by earth pressure. The causes of this damage were not satisfactorily analysed, because the earth pressure distribution around buried pipes could not be measured accurately. There have been many experiments measuring earth pressure around buried pipes. In 1977 and 1978, James and Bagge did experiments using an earth pressure gauge developed at Cambrige University, which also measured both types of stress. In 1988, Shmulevich and Galili measured both types of stress using plane-stress transducers which were similar to

the Cambrige soil transducer used by Bransby. Though these load cells are suitable for a small scale in-door experiment, their sizes are too small to measure earth pressure on pipes accurately, if a stress concentration appears.

A manual for plastic sewage pipes was published earth pressure applied to the assumed planes is shown in Fig.1. Figure 2 shows earth pressure on the assumed planes and the pipe. The reduced earth pressure at a point P on the surface of the pipe is given by the following equations.

$$n = q \cdot \sin^2\theta + p \cdot \cos^2\theta \qquad (1)$$
$$s = (q-p)\sin\theta \cdot \cos\theta \qquad (2)$$
$$p = q(1-\sin\theta) \qquad (3)$$

Where,
q: intensity of earth pressure
p: intensity of earth pressure
n: normal component of earth pressure
s: tangential component of earth pressure

The apparent earth pressures, p and q, are obtained by the following equations.

$$p = n - s \cdot \cot \theta \qquad (4)$$

$$q = n + s \cdot \cot \theta \qquad (5)$$

Where, $0 < \theta < \pi/2$

These equations can be used when the surface of a pipe is evaluated the apparent earth pressure diagram for design.

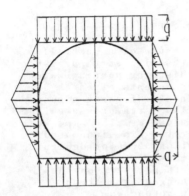

Note: q:earth pressure

Fig.1 The apparent design earth pressure applied to the assumed planes(Manual for plastic sewage pipes, National Land Development Technology and Research Center, Japan, in Japanese)

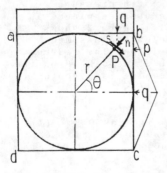

Fig.2 The apparent design earth pressure and the reduced earth pressure and on the pipe

Let us try to change the apparent earth pressure distribution into the reduced earth pressure distribution on the surface of the pipe. The earth pressure distribution for each quarter quadrant is the same. Therefore, the distribution for the first quadrant is given in Fig.3 and 4. Figure 3 gives the relationship between the apparent earth pressure for design on the assumed planes and the reduced earth pressure on the pipe. The horizontal axis is angle θ in degrees and the vertical axis is the ratios n/q and s/q. Figure 4 gives the earth pressure distribution on the surface of the pipe computed by the use of equations (1),(2) and (3). The earth pressure distribution is illustrated with arrows together with curves, which represent the vertical and tangential components. It is not clear whether the design earth pressure diagram takes into account groundwater or not. It is impossible to add the effect of groundwater in Figs.3 and 4, assuming that the design earth pressure diagram is drawn for the case of nonsaturated soil. As it was very difficult to measure the earth pressure on pipes accurately, the design apparent earth pressure was given by many experiments measuring deformation of pipes. The apparent earth pressure diagram on the vertical wall(bc) is triangular, because the horizontal axis of the pipe is extended horizontally. The authors conducted a field measurement of earth pressure using "panel type earth pressure gauge". The pipe used was a flexible plastic pipe. Steel sheet piles were driven as the excavation was being conducted. After placing the pipe at the bottom of the excavation, the pipe was backfilled with sand. Figure 5(a) illustrates the earth pressure distribution in the form of apparent earth pressure by using equations (4) and (5). The earth pressure increased remarkably when the sheet piles were pulled up. Figure 5(b) shows the earth pressure distribution after the sheet piles were pulled. The earth pressure measured after the sheet piles were pulled was very different from the earth pressure used for design. The groundwater was pumped out during construction, but it was restored afterwards. The earth pressure changed accordingly. It was necessary to investigate how the recovery of groundwater effects increase of earth pressure. The purpose of this experiment is to clarify this point.

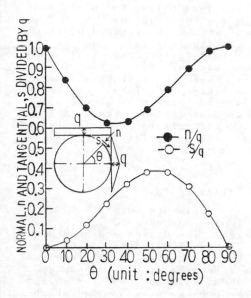

Fig.3 The relationship between the apparent earth pressure for design on the assumed planes and the reduced earth pressure on the pipe

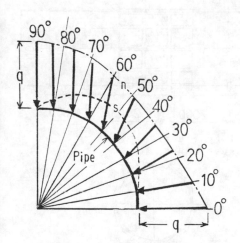

Fig.4 The reduced earth pressure on the pipe. The arrows show resultant earth pressure and n and s are normal and tangential component of earth pressure

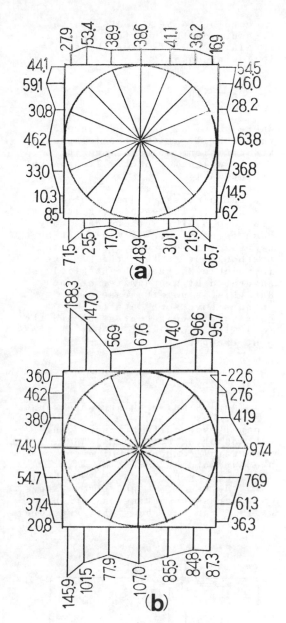

Note: Pipe diameter:2.4m, the depth from the ground surface:2.6m, backfill soil: crusher-run γ=19.3kN/m^3

Fig.5 Earth pressure distribution of a plastic sewage pipe(Unit:kPa)
(a) Before extracting sheet piles
(b) After extracting sheet piles

According to text books, earth pressure on the backfill of a retaining wall is given by the following equation.

$$P = \frac{1}{2} k \gamma H^2 \qquad (6)$$

Where, P:total earth pressure on the backfill of the retaining wall. k:coefficient of earth pressure. γ:unit weight of backfill. H:height of the wall.

The earth pressure distribution is triangular. The value of k is about 0.3 to 0.6. The authors have many case records, but it would not be necessary to discuss it here. The textbooks state that the k value is invariable whether the backfill is saturated by groundwater or not. But this is not the case. The authors have some data to corroborate this contention and the following experiment is one of the best examples.

2 EXPERIMENTAL PHASE

2.1 Panel type earth pressure gauge

A panel type earth pressure gauge for measuring both normal and tangential components of earth pressure at the same time was developed by Professor M.Fukuoka in 1971. He used this gauge to measure earth pressure acting on retaining walls. Figure 6 shows the structure

of the panel type earth pressure gauge. It consists of two panels and six load cells. The lower panel is fixed on the pipe and four bolts connect the lower panel to the upper panel. The six load cells, situated between the upper panel and lower panel, consist of two bearing load cells, two plate load cells and two ring load cells. Normal stress is measured by the two bearing load cells and the two plate load cells. Tangential stress is measured by the two ring load cells. When the upper panel moves on soil-pipe interface, both normal stress and tangential stress are measured at the same time. As Fig.7 shows, the authors installed ten of these panels completely around the steel pipe.

2.2 Experimental outline

As an experimental pipe, a flexible steel pipe was used, which is shown in Fig.7. The diameter of the pipe was 2000mm. The length was 6000mm, and the pipe material thickness was 20mm. The area of the embankment was 208m^2 (13m×28m) and the height of the embankment was 4.9m, which

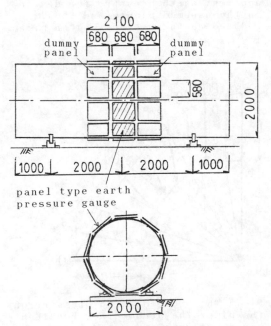

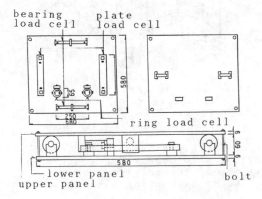

Fig.6 Structure of the panel type earth pressure gauge (Unit:mm)

Fig.7 Arrangement of the panel type earth pressure gauges(Unit:mm)

is shown in Fig.8. Figure 9 shows the front view and side view of the pipe and the embankment. To prevent leakage from the embankment when water was added, a geotextile PPF#300 and a geomembrane were laid down. All of the panels were covered with geomembranes so that sand and water could not get into the panels. Similarly, both ends of the pipe were covered with geomembranes and balks. In addition, the authors used sand as a backfill soil around the pipe. The soil was compacted in three separate passovers using a tamping roller which weighed 0.8kN. PVC pipes were installed to pour water into the embankment and to measure groundwater level around the pipe in the embankment. Table 1 shows the soil properties of the backfill.

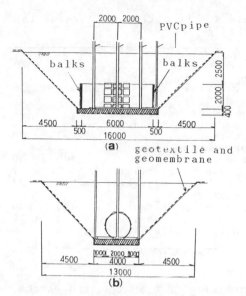

Fig.9 Structure of the experimental embankment (Unit:mm)
(a) Side view (b) Front view

Fig.8 Experimental embankment on the grounds of the Science University of Tokyo

Table 1. The soil properties of the backfill

Water content	(%)	12.70
Wet density	(kN/m^3)	18.40
Dry density	(kN/m^3)	16.30
Void ratio		0.57
Degree of saturation	(%)	57.30
Apparent cohesion intersept	(kPa)	30.00
Apparent angle of internal friction	(°)	38.00
Uniformity coefficient		2.67
Coefficient of curvature		1.04
Specific gravity		2.57
Optimum water content	(%)	16.00
Maximum dry density	(kN/m^3)	1.70
Procter's relative density	(%)	96.0

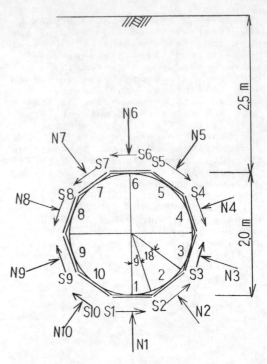

Fig.10 Normal and tangential components of earth pressure

3 BEHAVIOR OF PIPE AND DEFORMATION DUE TO RISING GROUNDWATER LEVEL

3.1 The earth pressure distribution on the surface of the pipe

Figure 10 shows the front view of the pipe which illustrates normal and tangential components of earth pressure. The authors installed ten of these panels around the pipe and put panel number 1 at the bottom of the pipe. The panels were numbered 1-10 in a counterclockwise sequence. The authors measured forces (N1-N10 and S1-S10) acting on each panel using these load cells. The dimension of each load cell is 0.58m×0.68m and the area is 0.39m^2. Stresses acting on these panels are obtained from the total forces, N and S, divided by the area of the panel(0.39m^2). These load cells were sealed by geomembranes so that the measurements would be constant in the groundwater conditions. The authors carried out the calibration of these load cells when they were installed on the pipe. The measured error was less than 10%. Figure 11 shows the result gathered from the panels which measured normal forces. The horizontal axis shows normal forces and the vertical axis shows groundwater depth. When groundwater level rose from -4.5m to -0.54m below ground surface, normal forces changed gradually. But the amount of their change was not always large. Figure 12 shows the result gathered from the panels which

measured tangential forces. The amount of change of tangential forces is variable. The authors expected that S1 and S6 would not change, but, in fact, a small change was measured. S2 and S10 increased until groundwater level rose to the top of the pipe, and decreased afterwards. S3 and S9 were variable at first but did not change afterwards. S4 and S8 were invariable. S5 and S7 decreased gradually. Judging from the change in these tangential forces, change of volume, stress and physical properties in the backfill soil causes internal deformation of the pipe with the rising of groundwater level. In addition, these effects caused the change of tangential forces on the surface of the pipe. The change of pore water pressure against stress is smaller than the change of tangential forces.

The panel type earth pressure gauges can measure earth pressure more correctly than the standard type earth pressure gauges which cause a stress concentration. It

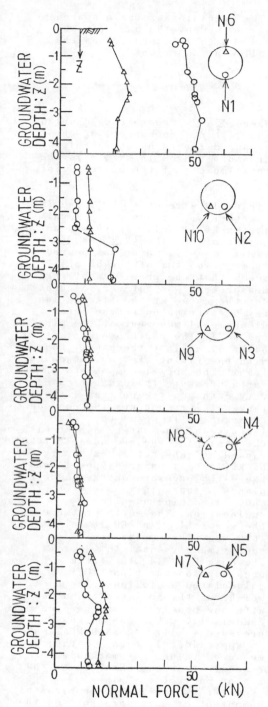

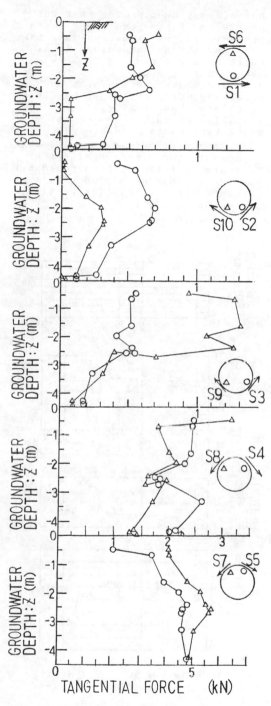

Fig.11 Change of normal forces
acting on the panel type earth
pressure gauges
Note: N1-N10: Measured forces,
n1-n10: Components of earth
pressure, N/A=n,A=0.58m×0.68m

Fig.12 Change of tangential forces
acting on the panel type earth
pressure gauges
Note: S1-S10: Measured forces,
s1-s10: Components of earth
pressure, S/A=s,A=0.58m×0.68m

has already been proven that the precision of the panel type earth pressure gauges is much higher than the conventional one dimensional earth pressure gauges. This is shown in the results of earth pressure measurements for a retaining wall. The dummy panels which are shown in Fig.7 were installed at both sides of the panel type earth pressure gauges. This was necessary so as to be able to measure two dimensional earth pressure. But the authors could not measure two dimensional earth pressure perfectly as the results are shown in Fig.13.

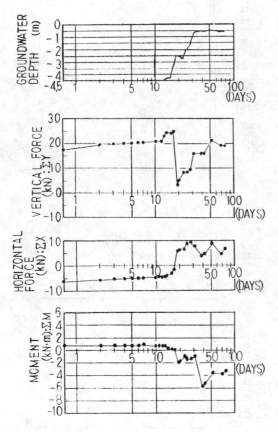

Fig.13 Change of ΣX, ΣY, ΣM with the rising of groundwater level,
Note:
ΣX:Total horizontal component of measured forces on panels
ΣY:Total vertical component of measured forces on panels
ΣM:Total moment of measured forces on panels

The precision of the measurement is checked by the use of Newton's law, which is expressed as the following equations.

$$\Sigma X=0 \qquad (7)$$
$$\Sigma Y=0 \qquad (8)$$
$$\Sigma M=0 \qquad (9)$$

Where, ΣX, ΣY denote total horizontal and vertical components of the measured forces on the panels of the pressure gauges in kN. ΣM is the total moment of the measured forces in kN·m.

Figure 13 gives the ΣX, ΣY and ΣM together with groundwater level in the backfill soil. There was no change of the the ΣX, ΣY and ΣM variables until water was poured from the bottom of the backfill. It seems that the precision of the measurement was not very accurate. But the cause of the fluctuation may be attributed to the three dimensional movement of the pipe and soil. The earth pressure of the backfill against the panel changed with the rising of the groundwater level. The normal components of the earth pressure (N) did not change very much. On the contrary, the tangential components (S) changed remarkably. It seems that the increase of the water content with the rising of the groundwater level caused an internal movement of the backfill soil mass. Figure 14 shows the earth pressure on the assumed planes. Figure 15 is a sketch of the buried pipe used for explanation. The earth pressure on the vertical plane EF is 18.4×2.5= 46.0 kPa. This value is approximately equal to the value shown in Fig. 14. The earth presssure on the plane BC is caluculated as follows. The total weight of the soil and the pipe with the panels inside the area ABCD is 117.8 kN. The earth pressure is 117.8/2=58.9 kPa, which is approximately equal to the measured value. A remarkable stress concentration was observed at the center part of the pipe.
Now let us examine the horizontal component of earth pressure on the pipe. Assuming the vertical planes AB and CD in Figs. 14 and 15 is an assumed plane of a retaining wall, earth pressure acting at points E and F is expressed by the equation:

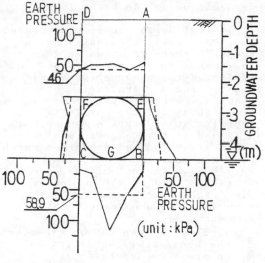

Fig.14 The earth pressure on the
assumed planes

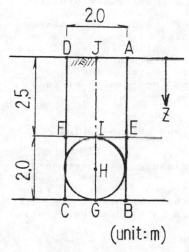

Fig.15 A sketch of the buried pipe
for explanation

$$p = k \gamma z \qquad (10)$$

Where, γ :unit weight of the soil,
k :coefficient of earth pressure,
z :the depth from the ground surface.

Taking $\gamma = 18.4 \text{kN/m}^3$, z=2.5m, k is
0.3, 0.4 0.5, the value of p is
obtained as 13.8, 18.4 and 23.0kPa,
respectively. The measured
horizontal earth pressure at the
point E and F in Fig.14 are 8.75
and 27.5 kPa, respectively, and the

average is 18.0 kPa. The measured
value is approximately equal to the
computed value of k=0.40.
 Next, let us examine the effect of
raising the groundwater level. The
submerged unit weight of soil is
obtained as $\gamma \text{sat}=20 \text{kN/m}^3$ by Table 1.
Earth pressure on the planes EF and
BC in Fig.16 is caluculated as 50.0
and 62.6 kPa, respectively. The
increments of earth pressure when
submerging is 8.7 and 6.3 percent.
The measured value was smaller than
the computed one. The horizontal
diameter of the pipe was elongated
by the reduction of horizontal
earth pressure, as described later,
and the vertical diameter was
contracted at the same time. The
soil above the pipe did not settle
following the change of the
diameters. Friction on the assumed
vertical walls AF and DF increased
gradually. This phenomena is well
demonstrated by Fig.17. The earth
pressure at both sides of the plane
EF decreased first and the tendency
of reduction graduated to the
center part of the pipe.
 As to the horizontal earth
pressure on both sides of the pipe,
Fig.18 shows the earth pressure
acting at the face EB. Especially
at the lower part of the pipe, the
earth pressure decreases while
groundwater level rises at the
center of the pipe. It seems that
the backfill soil under the pipe

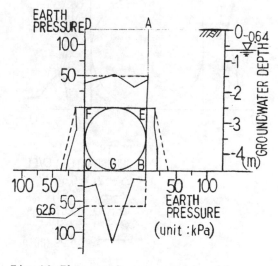

Fig.16 The earth pressure on the
assumed planes when groundwater
level rises

moved in the direction of the arrow and was caused by the seepage of water. The authors expected that this would be brought about by both the shrinkage and the decrease of the modulus of deformation of the backfill soil caused by absorbing water. On the other hand, the earth pressure at the upper part of the pipe increased temporarily. But finally the earth pressure decreased on the whole.

Lets return to the subject of earth pressure on retaining walls When water rose to -4.5m below ground surface, the modulus of deformation in the backfill soil decreased. As the result, horizontal earth pressure decreased, too. Conversely horizontal earth pressure above groundwater level increased. This phenomena was graduated to the upper parts with rising groundwater level. For example, when groundwater level was the highest, if the horizontal coefficient of earth pressure:k is 0.45 and if k is able to apply the apparent horizontal submerged earth pressure, earth pressure is 20.9 kPa at the top of the pipe and it is 37.7 kPa at the bottom of the pipe. These earth pressures are very small when compared to the theoretical ones.

Such phenomena can not be explained, if the horizontal coefficient of earth pressure decreases in groundwater conditions. The theoretical water pressure at the bottom of the pipe is 61.3 kN but the measured total earth pressure was smaller than 25.0 kPa. Influence of groundwater against the horizontal earth pressure can not only be applied to buried pipes but also to retaining walls. The authors have measured the same phenomena as the results of this experiment.

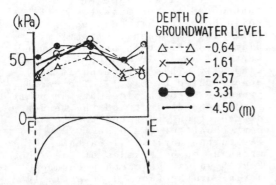

Fig.17 Earth pressure on EF plane

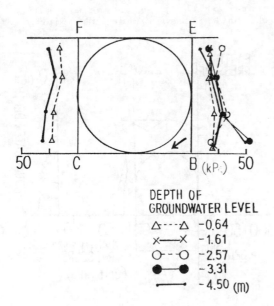

Fig.18 Earth pressure on EB and FC plane

3.2 Stress and strain acting on the pipe

Wire strain gauges were attached on the four points of the center of the pipe as shown in Fig.19. Figure 20 and 21 show the strain distribution in the circumferential

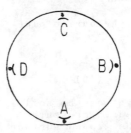

Note: A,B,C,D:Positions of wire strain gauges.
),Circumferential direction
•,Axial direction

Fig.19 Positions of wire strain gauges

386

and axial direction respectively. The horizontal axis represents strain and the vertical axis represents groundwater depth from the groundsurface of the backfill. Figure 20(a) shows the behavior of strain at the top and the bottom of the pipe in the circumferential direction. When water is not poured, both strain become tensile. But when groundwater rises -2.5m below ground surface, both strain become compressive temporarily. The maximum of tensile strain is the rise of the groundwater level is 300×10^{-6} with rising groundwater level. Figure 20(b) shows the behavior of strain at the right and the left side of the pipe in the circumferential direction. Both strain become compressive and it seems that compressive stress in the axial direction is larger than bending stress. The initial strain at the left side becomes tention. This is an inverse phenomenon and it seems to bring out under the influence of compacting backfill soil around the pipe. This strain became compressive with rising groundwater level. To the contrary, strain at the right side decreased when groundwater level rose 2.5m

below ground surface. Figure 21(a) shows the relationship between groundwater depth and strain in the axial direction at the top and the bottom of the pipe. Figure 21(b) shows the relationship between groundwater depth and strain in the axial direction at the right and the left of the pipe. These results show that the pipe moves variable as a three dimensional object. The bending stress of the pipe in the right and left direction may be more contributing to change of the pipe than the bending stress of the pipe in the top and bottom of the pipe.

4 RELATIONSHIP BETWEEN EARTH PRESSURE AND BENDING MOMENT OF PIPE

Castigliano's theorem can be applied to obtain the relationship between earth pressure and bending moment of a pipe. First of all, the three equations, $\sum X=0$, $\sum Y=0$, $\sum M=0$, should be hold to keep stability. As the condition of stability was kept three dimensionally in this

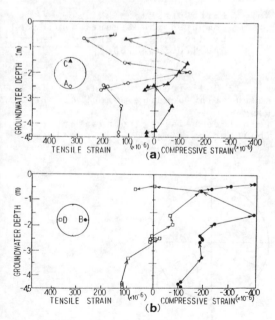

Fig.20 Change of wire strain gauges in the circumferential direction following the rise of groundwater level

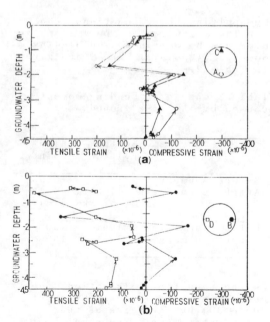

Fig.21 Change of wire strain gauges in the axial direction following the rise of groundwater level

experiment, the measured earth
pressure did not satisfy the
conditions of stability. The
measured part of the pipe could
keep stability by the force
transmitted from the neighboring
parts of the pipe besides the earth
pressure from the backfill. It was
impossible to measure the force
transmitted from the neighboring
pipe. The magnitude of the force
seemed to be relatively small
compared with the earth pressure.
The modified earth pressure used
for the caluculation is given in Fig.
22. The tangential forces are
included in the normal forces. The
following equations were derived
for analysis. A concentrated load,
P , is applied from top and bottom
as shown in Fig.23. The bending
moments at points B, D and C are
expressed as follows:

$$M_B = \frac{1}{\pi} Pr = 0.318Pr \qquad (11)$$

$$M_D = (\frac{1}{\pi} - \frac{1}{2}) \sin \varphi Pr \qquad (12)$$

$$M_C = (\frac{1}{\pi} - \frac{1}{2})Pr = -0.182Pr \qquad (13)$$

By applying these equations, the
bending moment at any points on the
pipe can be obtained. The bending
moments at points B and C are given
in Table.2.

Table 2 Change of bending moment
caused by the rise groundwater
level

	Bottom Point B	Side Point C
Before submerge	10.60	-6.90
After submerge	11.85	-9.18
	(Unit:kN m/m)	
Before submerge	100%	100%
After submerge	118%	133%

As the result of calculation, the
bending moments increase by about
20 to 30% by the raise of the
groundwater level. The results of
calculation are different from the
measured strain. These may be due to
the three dimensional motion of the
pipe.

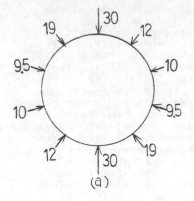

(a)

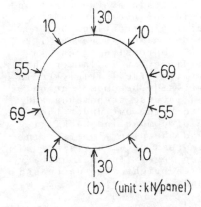

(b) (unit:kN/panel)

Note: Earth pressure (kN) are
computed the figures indicated
by dividing the area of the
panel(0.58m×0.68m)

Fig.22 Modified earth pressure for
calculating bending moment of the
pipe
(a)Before raising groundwater level
(b)After raising groundwater level

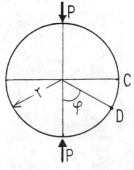

Fig.23 Sketch for caluculating
bending moment of a pipe by the use
of a concentrated load

5 SOIL PROPERTIES DURING THE TEST

Figures 24 and 25 illustrate results of cone penetration tests performed during the backfilling operation. The elevations of ground surface at the times of the test were 0.4, 1.3, 2.9, 3.2 and 4.5 m. Average value of cone resistance was 100 kPa, and maximum and minimum values were 50 and 200 kPa, respectively. The coefficient of deformation, Es, which is obtained by the use of the equation Es=7.1+1.2qc. In order to investigate change of mechanical properties of soil by saturation, the cone penetration test was performed at the ground surface near the measured portion of the

pipe. The submerged part of the soil gave higher water content and lower values of cone resistance, as shown in Fig.26. Water was supplied from the bottom of the backfill carefully, but it was impossible to guarantee that the backfill soil was fully saturated everywhere.

6 BEHAVIOR OF PIPES WHEN AN EARTHQUAKE OCCURS

It is clear that the behavior of pipes is complicated when an earthquake occurs. Pipelines are a very important part of our everyday lives. On February 6, 1987, a large earthquake occurred in the Tohoku region of Japan. We were able to accumulate valuable data about the

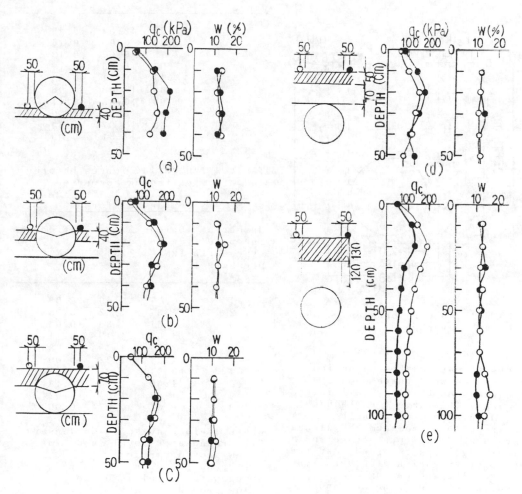

Fig.24(a)-(e) Cone resistance and water content at each layer

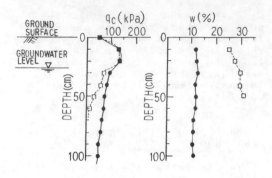

Fig.25 Change of cone resistance
and water content by raising of
groundwater level

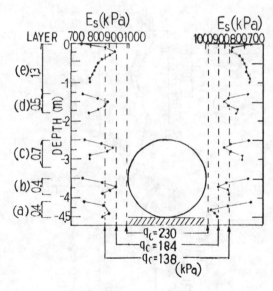

Fig.26 Relationship between Es and
qc at each layer

effects of earthquakes on buried
pipes. There were three separate
earthquakes on that day, all in the
same area. The largest earthquake
had a magnitude of 6.9 on the
Richter scale. A maximum 220gal was
recorded at Miyagi prefecture and a
maximum 28gal was recorded near the
grounds of the Science University
of Tokyo. The increasing or
decreasing change of normal and
tangential forces could be measured

by ten of the panel type earth
pressure gauges. Table 3 and 4
shows the amount of change in
normal and tangential forces when
the earthquake occurred. These
changes are complicated on both
sides of the pipe. Tangential
forces acting on both the upper and
the lower part of the pipe decreased.
On the other hand, they increased
on both sides of the pipe.

Table 3 Normal forces before and
after the earthquake occurred

	Normal forces		
Number of the panel type earth pressure gauge	Before quake (kN)	After quake (kN)	Difference (kN)
1	42.77	42.32	-0.45
2	6.69	6.93	+0.24
3	7.75	7.62	-0.13
4	7.02	6.83	-0.19
5	8.17	10.51	+0.23
6	18.86	17.88	-0.98
7	10.84	10.95	+0.11
8	6.87	6.37	-0.50
9	10.35	10.97	+0.62
10	11.54	11.51	-0.03

Table 4 Tangential forces before
and after earthquake occurred

	Tangential forces		
Number of the panel type earth pressure gauge	Before quake (kN)	After quake (kN)	Difference (kN)
1	0.77	0.69	-0.08
2	1.37	1.32	-0.05
3	1.81	1.89	+0.08
4	4.08	4.19	+0.11
5	2.37	2.29	-0.08
6	1.31	1.22	-0.09
7	3.91	3.75	-0.16
8	1.60	1.74	+0.14
9	1.02	1.04	+0.02
10	0.01	0.07	+0.06

where 'Before quake'is on January
31. Groundwater level is-0.57m
below ground surface.
'After quake'is on February 7.
Groundwater level is-0.52m.
below ground surface.

CONCLUSIONS

The steel pipe with a diameter of
2 m and a thickness of 20mm
surrounded by " panel type earth
pressure gauges" which measure the
normal and tangential components of
earth pressure was buried under
the ground with a backfill having
2.5m in depth. Earth pressure
changed when groundwater level rose.
More tests of this kind should be
repeated before getting final and
general conclusions. The following
conclusions were obtained
temporarily as the results of this
test.
1. The measured earth pressure
distribution was quite different
from the conventional apparent
earth pressure diagram. Especially,
a stress concentration was observed
at the bottom of the pipe,
tangential forces were acting at
the surface of the pipe, and so on.
2. The mechanical properties and
the volume of the backfill soil
were changed with the rise of the
groundwater. As a result, an
internal deformation took place,
which seems to be a cause of the
increase of tangential components
of earth pressure.
3. The vertical components of
earth pressure did not change
appreciably, contrary to our
predictions. The earth pressure in
the horizontal direction decreased,
because the backfill soil was well
compacted during the filling
operation. The horizontal diameter
of the pipe was elongated. The
bending moment of the top and
bottom increased by about 20 % and
that of the sides decreased by
about 30 %.
4. The aim of the experiment was
to measure two dimensioanl earth
pressure, but a three dimensional
earth pressure acted on the pipe.
The pipe surrounded by the panels
not only received the earth
pressure but also the force
transferred from the neighboring
portion of the pipe. It seems to be
necessary to take into this account,
and increase the safety factor
without depending upon the
conventional two dimensional design
computation.
5. A relatively large earthquake
occured at the test site during the
test period. The record of earth
pressure measurement after the
earthquakes is described.
6. The panel type earth pressure
gauges proved useful to measure
earth pressure on pipes, because
stress concentration does not
appear.

ACKNOWLEDGEMENTS

The test was supported by Dr.Akira
Saito of Nippon Kokan Co.Ltd. who
gave us financial support. Mr.Masao
Hashimoto of Tokyo Shiki Kougyo
company manufactured the panel type
pressure gauges. The students of
the Science University of Tokyo
participated in this test for many
years. The authors would like to
extend their deep appreciation to
those who helped the test.

REFERENCES

Spangler,M.G., & Handy 1973 Loads
    on underground conduits, Soil
    Engineering Third Edition
M.Fukuoka & Y.Yosida 1977.Reserches
    on earth pressure on cantilever
    retaining wall with cohesive soil
    as backfill-surcharge, earthquake.
    Journal of the Faculty of
    Engineering, the University of
    Tokyo(B), Vol.33, No.2, p15-48.
I.Shmulevich & N.Galili 1986.
    Deflections and bendings moments
    in buried pipes. Journal of
    Transportation Engineering Vol.112,
    No.4, July.
M.Fukuoka, et al 1981. Earth
    pressure measurements on
    retaining walls. Case History
    Volume, Proceedings of the 9th
    International Conference on Soil
    Mechanics and Foundation
    Engineering, p237-296.
M.Fukuoka & Y.Imamura 1984. Earth
    pressure on retaining walls and
    buried pipes. Int. Conf. Case
    Histories in Geotechnical
    Engineering, University of
    Missouri-Rolla, p.355-362.
Mehdi,S & Zarghamee,M 1986. Buried
    flexible pipe with nonuniform
    soil support. Journal of
    Transportation Engineering
    Vol.112,No.4,July.
Watkins,R.K., & Spanger,M.G., 1958
    Some characteristics of the
    modulus of passive resistance of
    soil: A study of similitude.

Highway Reserch Board Proceedings, Vol.37, p.576-583.

I.Shmulevich & N.Galili 1986. Soil stress distribution around buried pipes. Journal of Transportation Engineering, Vol.112, No.5, September.

J.K.Jeyapalan & B.A.Boldon 1986. Performance and selection of rigid and flexible pipes. Journal of Transportation Engineering, Vol.112, No.5, September.

A.K.Howard 1977. Modulus of soil reaction values for buried flexible pipe. Jounal of Geothechnical Div. Proc. of ASCE.

Y.Esashi & K.Nishi 1979. Behaviour of large size intake pipes buried in sea bed. The Japanese Society of Soil Mechanics and Foundation Engineering, Vol 30, No.3 (in Japanese) p.63-69.

H.Nanbu & A.Sakano 1982. Field load tests on buried steel pipe(2000mm). The Japanese Society of Soil Mechanics and Foundation Engineering, Vol.30, No.4 (in Japanese) p.5-11.

James,R.G & Lansen,H 1977. Centrifuga model tests of buried rigid pipes. International Conference on Soil Mechanics and Foundation Engineering Vol.1, Tokyo, Japan p.567-750.

Bagge,G., & Funghange,L., 1978. Measured loads on a buried circular pipe with base. The 9th International Congress on Concrete, Vienna, Austria.

M.Fukuoka 1988. Earth pressure and effective stress. The Japanese Society of Soil mechanics and Foundation Engineering, Vol.36 No.6, June,(in Japanese) p.37-42.

Developments in Geotechnical Aspects of Embankments, Excavations and Buried Structures, Balasubramaniam et al. (eds)
© 1991 Balkema, Rotterdam. ISBN 90 5410 019 2

Shotcrete mix design, process of shotcreting for tunnel support (Narmada Power House Tunnel)

A.V.Shroff
University of Baroda, India

M.J.Pandya
Narmada Hydropower, India

ABSTRACT: The support system consisting of tensioned expansion rock bolts and shot-creting layer with welded wire mesh is best suited for power house tunnel. The design of shotcrete mix comprising of cement:sand: gravel with chemical admixture based on ideal frame work provide best compatibility with strength and setting time showing low rebound potential.

1. INTRODUCTION

The Sardar Sarovar Project envisages construction of concrete gravity dam, 155m high above the deepest foundation, across river Narmada (21 degree 50 minutes N 73 degree 45 minutes E) in Bharuch district of Gujarat State of India. Project contemplates the constru-ction of an underground power house underneath the right bank of the river Narmada. The underground construction of power house was necessiated due to the increase in length of the spillway dam, non-availability of suitable site for surface power house on the left bank and other techno-economic reasons.

The power house complex comprises of components like intake dam, steel lined pressure shaft (Penstrock) machine hall in underground power house cavern, control room, cable spreading areas and bus galleries, tail race tunnel, draft tube tunnels, collection pool, exit tunnels, open cannel, pressure shaft for penstock, acces tunnel to power house, penstock, transformer yard and switchyard etc. The layout and cross section of machine hall is shown in Figure 1. Other components include Bus Shaft, Cable Tunnel, Drainage and Grouting Gallery, Lift Well, Control Room, Ventilation Shaft, etc. As the cavern will be surrounded by water, a drainage gallery is proposed which would serve as an access for curtain grouting on the river side and collection pool side.

2. REGIONAL GEOLOGY

The geological investigations reveal fairly competent but highly jointed porphyritic basalt, which has been intruded by two dolerite dykes and one narrow vertical trap (basalt) dyke. The contacts between this dolerite dyke and the adjoining basalt rock are not cemented and clearly show shear zones on each side of the dyke, being more prominent, weathered and open along the north side of the dyke, overhanging the location of the power house. The two dolerite dykes are massive and unweathered, with widely spaced joints. The area of the power house between the two dolerite dyke is formed of various basalt flows which dip locally in this area. The basalt is highly jointed with two major and two minor joint sets. This will likely to result in formation of wedges dipping into the power house cavity. The bottom of the machine hall will rest on a dolerite dyke which apparently is a competent rock. (M.M. Shah et al, 1984). Thus the rock is competent for an unlined underground excavation of huge size is used for the machine hall, although the closely ely jointed nature of the rock and the fact that some of the joints are open require special precautions during the excavation of the cavern, to avoid loosening of the rock and sliding of wedges into the cavity. For that the support system provided is shown in Figure 2.

393

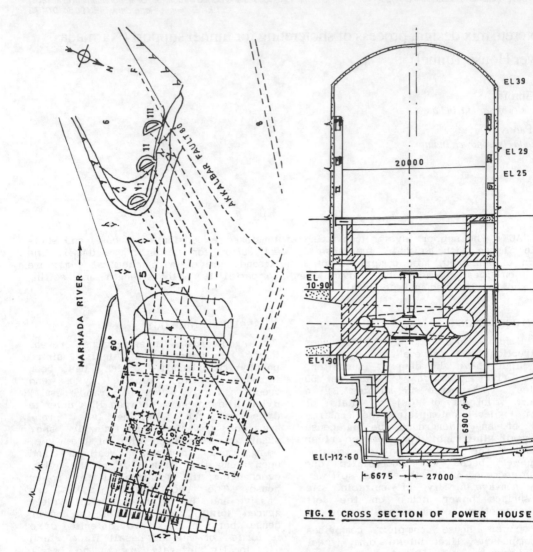

FIG. 2 CROSS SECTION OF POWER HOUSE

1 PENSTROCK
2 MACHINE HALL
3 DRAFT TUBE TUNNEL
4 COLLECTION POOL
5 EXIT TUNNEL
6 TAIL CHANNEL
7 DRAINAGE & GROUTING GALLERY
8 ACCESS TUNNEL
9 CABLE TUNNEL

FIG.1 GEOLOGY AND LAYOUT OF UNDERGROUND
POWER HOUSE

3. STAGE OF SUPPORT SYSTEM

3.1 Design Principle :

The support system consisting of tensioned
expansion rock bolts and shotcrete
with welded wiremesh is used on the
principles of New Austrian Tunnelling
Method to utilize the surrounding rock
mass as main load bearing component.
The rock mass quality can be evaluated
by Baroton's Engineering rock mass
classification (NGI) as well as by
Bieniawski (CSIR) Rock Mass Rating
(RMR) method for determining the support
system for roof arch and walls. The
shortcrete is the primary stabilizing
means which interacts with the ground
and prevent loosening of the rock
especially between the bolts. The welded
wiremesh fabric is a stiff mesh which
acts as the structural reinforcement
with shotcrete. The shotcrete and wiremesh together forms a continuous support
to the rock in between bolts and provide
an impervious barrier for any seepage.

3.2 Stages of Support System :

For power house cavern, pressure shaft, Draft tube tunnel, Access tunnel and Exit tunnels the details of shotcreting provided is enumerated in the table no.1. Sequence cycle of work adopted is profile making, drilling holes, charging, blasting, defuming, mucking, Geomapping, shotcreting, Rockbolting etc.

To make support system effective immediately after making a blast which will expose the roof of the walls of the cavern, and within a maximum time of four hours after the blasting the surface is scaled and 25mm thick (or 50 mm) shotcrete layer is applied over the entire exposed surface of the rock (first stage rock support). This operation is performed prior to or simultaneously with the beginning of the mucking operation.

Table 1 : Magnitude and Thickness of shotcreting

Particulars	Pressure Shaft	Machine hall	Draft Tube Tunnel	Access Tunnel
Shotcrete thickness	38 mm thick	76 mm 50 mm & 38 mm	38 mm thick	100 mm 438 mm thick
Magnitude of shotcrete in sq. m (approx).	19,000	30,000	35,000	7,000

The groutable rock bolts is applied following the application of the first shotcrete layer or simultaneously with the same as second stage of support. The rock bolts are completely installed and tightened within a maximum lapse of 12 hours after the blast and covered all the exposed rock surface within a maximum distance of 1 to 1.5 m from the excavation front or heading. The third stage of rock support consist of retightening and grouting the rock bolts, installing linkwire of welded wire mesh reinforcement supported by the bolt plates and then applying the two additional layers of shortcrete 25 mm each Figure 3. This tertiary treatment is applied as soon as practicable after blasting.

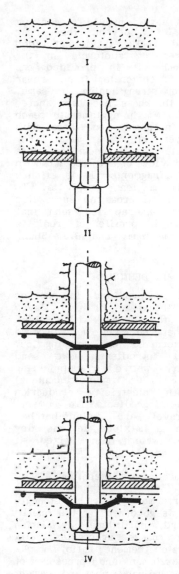

FIG.3 SEQUENCE OF ROCK SUPPORT

3.3 Support System Cavern :

In machine hall roof of power house cavern, 25 mm dia, 6 m long high yield expansion shell type groutable rockbolts are provded with a spacing of 1.75 m centre in staggered manner which are tensioned to 14 tons. The crown support pressure accordingly works out to be 0.93 kg/sq.cm. Two layers of shotcrete each of 38 mm thick with welded wire mesh are applied. The foot wall side of the shear zone is stiched by longer rock bolts of 7 to

10.5 m so as to provide minimum anchorage of 3 m beyond the shear zone in sound rock. It is provided one more layer of shotcrete in the shear zone rock. It is provided one more layer of shotcrete inthe shear zone portion for strengthening the same. The shotcrete thickness has been increased 85 mm 123 mm in agglomerate reach and 152 mm in shear zone area with additional layer of wire mesh. For the agglomerate band occuring close to the roof in view of its contact with weathered rock fragments, it is stiched by using 7-11 m long rockbolts. To fill up the rockfall area the successive shotcrete layers are applied such that the ultimate arch profile of roof is achieved (N. Ramswamy and B.J. Shah, 1987).

4. SHOTCRETE MIX DESIGN

The shotcrete is provided in one or more layers, thickness of each layer being 25 mm to 38 mm with welded wiremesh in between. Dry mix process is usually exercised or carrying out shotcrete. All ingredients like sand and gravel are wetted by sprinkling water (about 3 to 6 % moisture content) to avoid static electricity production at nozzle. The ingredients i.e cement, sand and gravel are weight-batched and mixed in a mixture in the dry condition. The water being mixed at the nozzle.

4.1 Materials used for shotcreting

Shotcrete performance i.e. strength and durability is governed by the quality of cement and admixture, gradation of sand and coarse aggregates, mix proportion, water cement ratio, workmanship of nozzleman and pressure of air and water. Ordinary portland cement conforming to IS 269-1976 is used for shotcrete. Fine aggregates i.e. sand conform to IS 383-1976 is used for shotcrete. Fine aggregates i.e sand conform to IS 383-1973 and graded evenly from fine to coarse as per zone II and III grading is used. The F.M. of sand shall be between 2.4 to 2.8 only. The coarse aggregate shall generally confirm to the grading given in IS 9012-1976 for maximum size of gravel upto 10 mm.

4.2 Framework approach

Framework approach for shotcrete mix design is developed by author to facili-

tate the site engineer for developing appropriate shotcrete mix. In this method, the first step is to determine the desired properties of the resultant cement admixture combinations alongwith appropriate fine and coarse aggregate proportion in presence of proper water: cement ratio. Next, the framework is developed to balance set time and compressive strength with minimum specified rebound when shotcrete mix is shooted to the surface of application. This is in idealisation of shotcrete mix formed in a given system, and is the most representative element for the ideal system that can be visualized. This allows the relative percent of each ingradient to be determined. Finally appropriate ingredients species are mixed in the system based on the ideal properties of the resultant mix.

The ideal shotcrete mix must have the following framework:

1. The initial set in the Vicat Needle Test should occur within 3 minutes.
2. The final set in the Vicat Needle Test should occur within 12 minutes.
3. The shotcrete mix should be considered acceptable, if the average strength from three consecutive tests is equal to atleast 85% of the design strength (strength specified in Tender) and no single test is less than 75% of the design strength which shall confirm to the following minimum design strength:

   i) Strength of the order of 105 kg/sq.cm should be attained at the age of one day.
   ii) The ultimate strength (28 days) should not be lower than 280 kg/sq.cm.
   iii) The coefficient of variation of strength test result shall be less than 15% and no single value of strength shall be less than 75% of design strength.
4. The ultimate 28 days strength should not be lower than 30% of the strength of an unaccelerated specimen. This restricts the accelerator dosage within 6%.
5. Shotcrete specimens shall be dense, impermeable with less shrinkage, uniform and non-liminated.
6. The above strength should be balanced with the minimum percentage of rebound (15 to 25% for vertical applications).

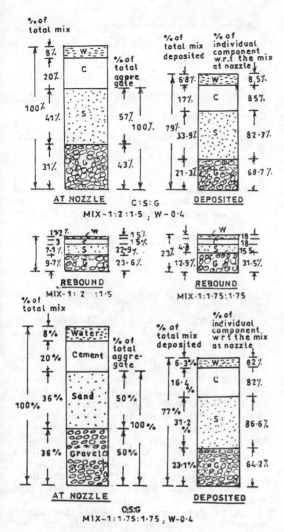

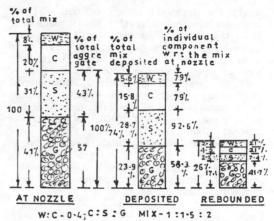

FIG.5 BAR CHART OF REBOUND AND IN PLACE
MATERIALS OF SHOTCRETE DRY MIX

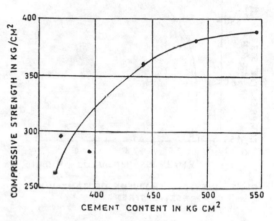

FIG.6 CEMENT CONTENT V/S COMPRESSIVE
STRENGTH

FIG.4 BAR CHART OF REBOUND AND
IN PLACE MATERIALS OF SHOCRETE
DRY MIX

7. The water cement ratio variation should be within the range of 0.35 to 0.55. The percentage of water with relation to sand is of the order of 16 to 17 per cent.

4.3 Variation of gravel or sand in total aggregate

Refer Figure 4, it indicates that with the increase in gravel percentage or decrease of sand in total aggregate of 1:3.5, the compressive strength increased. The compressive strength of 276 kg/sq.cm at 29% gravel increased to 379 kg/sq.cm at 43% gravel. Though percentage increase of gravel or decrease of sand causes better compaction, dense shotcrete, higher bond and good flexureal and compressive strength alongwith low shrinkage and permeability, it abnormally affects percentage deposition of shotcrete on the surface. The rebound potential of 20% at gravel 43% shoots upto 26% at 57% gravel. Referring to the bar graphs–figure 5 and 6 it clearly indicates that the maximum gravel rebound potential is observed at 50 and 55% of input gravel.

4.4 Variation of cement level in shotcrete mix

Keeping fine aggregate (sand) : coarse

aggregate (gravel) ratio constant (57:43) and water cement ratio (0.4) constant, the increase of cement level increases compressive strength (Figure 7). The rate of increase is faster upto the cement level increase from 0 (366 kg/sq. cm) to 19% (436 kg/sq.cm). Afterwards, the rate of increase is gradual tending to a constant value of compressive strength equal to about 380 kg/sq.cm at cement level of 32% (485 kg/cub.cm). Increase of cement level keeping same proportion of sand and gravel having constant fineness modulus and gradation may not change the rebound of cement upto some cement level.

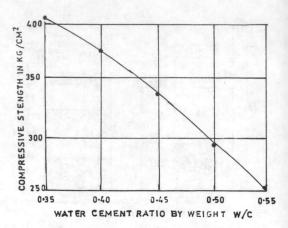

FIG.8 WATER CEMENT RATIO V/S COMPRESSIVE STRENGTH

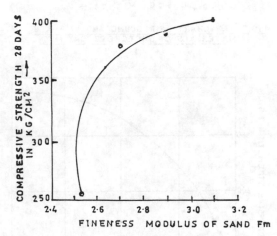

FIG.7 FINENESS MODULUS OF SAND V/S COMPRESSIVE STRENGTH

4.5 Variation of fineness modulus of fine aggregate

It indicates that increase of fineness modulus of sand increases compressive strength (Figure 8). The compressive strength of 250 kg/sq.cm at F.M. of 2.54, increases to 382 kg/sq.cm at fineness modulus of 2.91 which tends to remain same for any increase in fineness modulus of sand thereafter. The reduction in fineness modulus can be considered to be a measure of the amount of rebound. For a shotcrete mix having cement: total aggregate ratio 1:3.5, water cement ratio 0.4 and sand: coarse aggregate in the proportion of 57 and 43 percent, it is observed that the fineness modulus of sand ranging between 2.7 to 2.9 seems to be optimum with respect to consistent unconfined strength and low rebound potential of 15% to 20%.

4.6 Variation of water in shotcrete mix

For maximum density sufficient water must be added to overcome bulking. At the point, maximum density is reached, the mass will become plastic and will begin to flow or sag. It is essential that mix should be gunned at wettest stable consitancy or "point of incipient sag". If the water content is decreased below this point, bulking occurs or air is trapped, causing a less dense product. The present investigation believes that serious failure of shotcrete may be due to this cause. If the shotcrete is placed too dry, aggregates will not embed itself into the previously applied shotcrete and tends to "ricochett off". As the water content is increased, the mixture in-place becomes more plastic and will accept the incoming aggregates. As water content is increased, rebound is decreased. Figure 9 shows water: cement ratio against compressive strength. In concurrence with the above analysis, as the water:cement ratio is increased, the compressive strength falls to balance the compressive strength and rebound. It seems that the optimum water:cement ratio works out to be 0.4 to 0.5 which gives compressive strength within the limit of IS code i.e. between minimum 280 kg/sq.cm and rebound percentage between 15 to 25%.

4.7 Variation of admixtures (sodium aluminate + sodium carbonate + calcium carbonate)

Keeping cement sand and gravel in

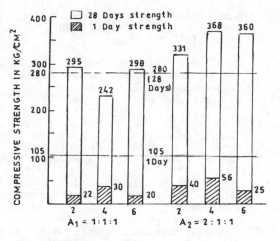

FIG.9 BAR CHART OF COMPRESSIVE STRENGTH
IN KG/SQ CM V/S ADMIXTURE PERCENTAGE

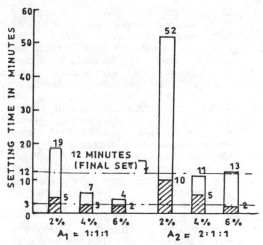

ADMIXTURE USED IN PROPORTION
$Na_2Al_2O_3 : Na_2CO_3 : CaCO_3$

FIG.10 BAR CHART OF SETTING TIMES IN
MINUTES V/S ADMIXTURE

the proporion of 1:2:1:5 with W:C ratio 0.4, the influence of variation of admixture is reported. In the above groups of admixtures, in general, the strength time data indicate that though the strength is increasing with time, but the loss of short term and long term strength with admixture is observed. This loss is severe in 1:1:1 as compared to 2:1:1 and 1:2:1. In 2:1:1 the loss of strength is moderate. It is observed in general that initial and final setting doorcases with Increase of admixture percentage in all the groups. In group 1:1:2 particularly 5% and 6% admixture satisfies strength as well as setting time compatibility indicating low rebound potential. Above results also illustrate and analysed graphically in bar graphs Figure 10.

4.8 Cube strength, core strength of test panel, split tensile strength and Flexural strength.

Data indicates linear relationship between core-equivalent cube strength and split tensile strength of the core. Split tensile strength lie in the range of 80 to 95 kg/sq.cm at 28 days. The average flexural strength sawed from shotcreted panel observed as 69.54 kg/sq.cm.

5. PROCESS OF SHOTCRETING

Dry mix process is usually exercised for caraying out shotcrete. All ingredients like sand and gravel are wetted by sprinkling water (about 3 to 6% moisture content) to avoid static electricity produced at nozzle. The ingredients are weigh-batched and mixed in a mixture in the day condition. The water being mixed at the nozzle.

Process of shotcreting is usually carried out by Aliva 304 Hydraulic spraying machine, U.K. (Figure 12). It is having robot STABILATOR for shotcreting of unstable zones from a distance. The machine arm is designed according to the modular system and consists of 6 main groups like spraying arm, vehicle mounting platform, cab with seat to be mounted on platform, value kit with control box hydraulic unit with electric motor and head light unit. It is able to perform a full rotation i.e. full circle and hence provide the transverse advance of the spraying nozzle.

6. CONCLUSIONS

1. The support system has been suggested based on the recommendations by Bartonand Bieniawski. The system can be evolved on the basis of emperical design methods by US Army Corps of Engineer's manual on "Rock Reinforcements".

2. The support system consisting of tensioned expansion rock bolts and shotcrete with welded wire mesh is used on the principles of New Australian Tunnelling method to utilize the surrounding rock mass as main load bearing component.

3. Though percentage increase of gravel causes better compaction dense shotcrete, higher bond and good compressive and flexural strength along with low shrinkage and permeability, it abnormally affects percentage deposition of shotcrete on the surface i.e. it increases rebound potential.

4. Compressive strength increases with increase of fineness modulus of sand in total aggregate but it leads to higher rebound.

5. The variation of water in shotcrete should be such that the shotcrete mix should be maximum dense plastic and mix shall be gunned at "wettest stable consistency" or "point of incipient sag".

6. The chemical admixture consisting of sodium aluminate, sodium carbonate and calcium carbonate in the proportion of 1:1:2 with 5% dosage have indicate best compatibility with cement in terms of setting time and strength with low rebound potential within ideal framework.

7. From this investigation it seems that cement:sand:gravel in the proportion of 1:2:1.5 with cement ratio of 0.4 having above chemical admixture in the dosage of 5% of cement weight can be considered to be the best as it satisfies ideal framework.

REFERENCES

1. Pandya M.J. (1989): "Shotcrete mixdesign for Tunnel support, ME-Thesis submitted to M.S. University of Baroda, May 1989.

2. Litwin A and J.J. Shideler (1966): "Laboratory study of shotcrete". ACI-SP-14, Shotcreting.

3. Kapoor and Patodia (1984): "Studies on the application of shotcrete in shafts of power house II of Mahi Hydel power project" - 51st Annual R & D session, CBIP, New Delhi, India.

4. Desai N.B. (1982): "Rock fill Dam and support system at Sardar Sarovar Project"., IGC-82 conference held at Surat, India.

5. Ramswamy N. and Shah B.J. (1987): "Geological investigation for Sardar Sarovar (Narmada) Project Hydro Power".

6. Shah M.M. (1982): "Geological investigations at Sardar Sarovar Narmada Dam Project".

7. Sudhindra and Shri (1984): "Shotcrete mix design for Chukha Hydel Project, Bhutan". 51st Annual R & D session, CBIP, New Delhi, India.

Developments in Geotechnical Aspects of Embankments, Excavations and Buried Structures, Balasubramaniam et al. (eds)
© 1991 Balkema, Rotterdam. ISBN 90 5410 019 2

Geotechnical experiences from construction of large diameter underground water pipe in soft Bangkok clay

Noppadol Phien-wej & A. S. Balasubramaniam
Asian Institute of Technology, Bangkok, Thailand

Wichai Suthipongkiat
Bangkok Metropolitan Waterworks Authority, Thailand

ABSTRACT: Large ground movements developed from cut-and-cover excavation using braced sheet pile wall for laying large buried steel pipes in Bangkok clay. The ground movements which were in the order of 2.5-6% of the excavation depth were largely attributed to the characteristics of bracing and problems of void backfilling during sheet pile removal. The flexible steel pipes were designed according to the Iowa formula. However, the actual deflection response of the pipes which were installed in relatively narrow trench was totally opposite to the assumption in the formula. The pipe deflection was largely affected by ground movements associated with bracing removal and sheet pile pulling as well as backfill compaction above the pipe rather than stiffness of backfill on the sides of the pipes.

1. INTRODUCTION

Between 1976 and 1983, Metropolitan Waterworks Authority (MWA) constructed a huge tunnel project to provide a main water transmission system for Bangkok. The project which consisted of 34-km-long concrete-lined tunnel placed in stiff clay layer was one of the longest length soft ground tunnel project in the world. Considerable experiences of tunneling in Bangkok subsoil were gained from this project. In 1986, MWA embarked an extension of the main water transmission system, the Water Transmission Conduit project, in which the methods of cut-and-cover excavation and pipe jacking for installation of a large underground conduit in soft Bangkok clay were adopted for the first time. The initial phase of the project which consisted of 32 km of pipe laying and pipe jacking was completed in 1988. The selection of the design parameters and construction method and procedures without prior experience in these construction areas had led to numerous problems and difficulties during construction. Nevertheless, the experience from this pioneer project provides useful information which can be used to improve the design concepts and construction procedures for both the buried pipe and the temporary retaining structures for deep excavation in Bangkok clay. This paper is intended to present some of the experience gained from the Water Transmission Conduit that the first author obtained during his involvement in the later stage of the project.

2. WATER TRANSMISSION CONDUIT PROJECT

The project involved the installation of large diameter steel pipes in the soft Bangkok clay. This conduit project is to complete the loop of main water transmission system for the Bangkok area on the east bank of the Chao Phraya River (Fig. 1). The project was let out in 10 separate contracts.

Cut-and-cover was used as the main method of pipe installation because most of the conduit alignment could be placed in the relatively open area (80% of total length) and along the light traffic roads (20 % of total length). Pipe jacking was used for conduit sections crossing beneath main roads and canals. With the opportunity to utilize these construction methods instead of tunneling MWA was able to save the project cost by more than half. In the cut-and-cover sections the pipes (invert elevation) were generally placed at 3.6 to 5.5 m. below the ground surface for the 2 m-to 3.2-m-diameter pipes, respectively (Table 1). The inverts of the pipes installed by pipe jacking were 5.9 to 10.3 m. deep. In some areas where there was

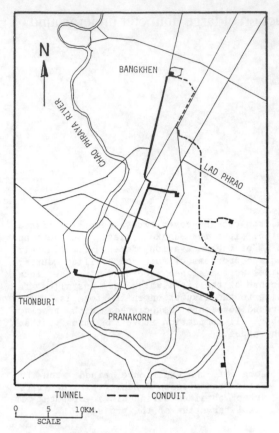

Contract No.	Total Length (m)	Pipe Dia. (m)	Pipe Wall Thick (mm)	Depth of Trench (m)	Depth of Soft Clay (m)
G-MC-1A-R	2551.6	3200	27	5.50 (8.50)*	14.5
G-MC-1B-R	4808.0	3200	27	3.80-5.50 (10.70)*	14.5
G-MC-1C-R	2459.0	3200	27	5.50 (6.95)*	14.50-16.0
G-MC-2A-R	2230.4	3000	26	5.20 (8.70)*	14.0
G-MC-2B-R	3051.9	3000	26	5.20 (8.70)*	15.0
G-MC-3A	3685.2	2300	16	3.70 (9.20)*	16.0-18.0
G-MC-3B	3376.6	2300	19	4.40 (10.30)*	14.0-19.0
G-MC-2A	3177.6	2000	19	4.15 (8.00)*	14.5
G-MC-2B	2976.2	2000	19	3.60 (8.65)*	15.0
G-BC-1	3025.0	2000	13	3.70 (8.00)*	14.5

Table 1 Pipe sizes and excavation depths

* Maximum depth of trench for concrete encased pipes or of pipe jacking pits.

TUNNEL ——— CONDUIT

0 5 10 KM.
SCALE

Fig. 1 Bangkok water main transmission
project.

necessity imposed by the existing surrounding conditions, the pipes had to be
placed at deeper depth and they were
encased in concrete blocks in order to
prevent excessive pipe deflection and pipe
uplift. In such cases the pipe inverts
were placed at 4.9 to 10.3 meter below the
ground surface. Braced sheet pile wall
was used as the temporary support for the
trench excavation for pipe laying and for
driving and arriving pits for pipe jacking.
Ground movements induced by the trench
excavation were quite significant and
caused some damages to adjacent road
pavements and buildings resting on shallow
foundation.

The aspect of the ground movement associated with excavation in the conduit
project is presented herein the following.
In addition the performance of the large-
diameter pipe observed during placement,
trench backfilling and sheet pile pulling
is reported and compared with the design
concept, which will be shown later that

they are very different. The detail of the
pipe jacking method with blind shield
which has proven to be very successful in
pipe installation in the Bangkok soft clay
is not included in this paper.

3. SUBSOIL ALONG THE CONDUIT ALIGNMENT

Soil borings of at least 15 meters deep
were drilled at 200 m interval throughout
the whole length of the conduit alignment.
The typical subsoil profile consists of 1-
2 m. thick of fill materials or weathered
clay followed by soft clay which extends
9-13 meters from the ground surface.
Underlying the soft clay layer is a thin
layer of medium clay which is subsequently
underlied by a stiff clay layer which
exists at 14-19 meter below the ground
surface. The unconfined compressive
strength of the very soft to soft clay is
in the range of 2 to 4 t/m². The water
content is 50 to 80 %, and plasticity
index 40 to 60.

4. BRACED SHEET PILE WALL AND GROUND MOVE-
MENTS

Braced sheet pile walls were specified by
the consulting engineers as a means of
soil retaining structures for the trenches
and pits excavation for pipe laying.
Trench braced wall was used in the pipe
laying section which required the minimum
trench width of 2 times the pipe diameter.
The detail of the typical trench and backfill of the pipe are shown in Fig. 2. In

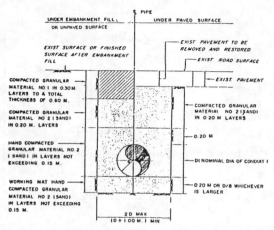

Fig. 2 Requirements for conduit trench
and backfill.

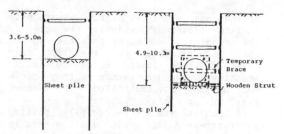

a) Typical section b) Concrete encased section

Fig. 3 Sheet pile braced-wall for pipe
laying trenches.

Table 2 Types and sizes of sheet pile wall

Types of Work	Excavation Depth (m)	Sheet Pile Type	Length (m)
Braced excavation for trench of typical pipe laying section	3.60-5.50	FSP II, III, IV	9.60-14.00
Braced excavation for trench of pipes with concrete encasement	4.90-10.30	FSP II, IV	12.00-16.00
Braced excavation for driving pit and arriving pit of pipe jacking	5.90-10.30	FSP,III, IV	14.00-20.50

* Property of sheet piles/meter of wall

	Section Modulus cm³/m	Thickness mm	X-area cm²
FSP Type II	874.00	10.50	153.00
FSP Type III	1340.00	13.00	191.00
FSP Type IV	2270.00	15.50	242.00

the typical section of pipe laying, the
construction constraints imposed by the
depth of the trench (3.6 - 5.50 m) and
the size of the pipe (dia. = 2.0 - 3.2 m)
led to the selection of one level of
bracing which was placed above the pipe
(Fig. 3a). This led to a large free span
of the sheet pile wall between the brace
and trench bottom which posed a possibility of large ground movement. For soft
Bangkok clay the maximum free span of wall
should not be greater than about 2.0 to
2.5 meters according to the recommendation
given by Peck (1969) to avoid large ground
movements. The sheet pile wall for the 5.5
meter deep trench had a free span of 4.5
meters which was very excessive. With
this constraints of bracing, the stability
of sheet pile wall largely dependent on the
passive soil resistant acting on the
penetration length of the wall below the
bottom of excavation, and the wall could
then be designed as the anchored bulkhead
wall. As a consequence, the required penetration depth of the wall below the base
of trench was about 1.5 times the trench
depth due to the presence of thick soft
clay below the base. The sizes of the
sheet pile and the required lengths used
in the project are summarized in Table 2.

For the deeper trenches used in the
concrete-encased pipe sections, close-
spaced braces were used for the whole
depth, among which the lowest brace being
temporary (Fig. 3b) due to fear of base
instability. This temporary brace was to
be removed to allow room for pipe
installation after casting concrete base
slab to provide a bearing pad for rakers.
The bracing of the sheet pile wall of the
driving and arriving pits for pipe jacking
was similar to the encased pipe trench.

In general, the installation of the
bracing system in the project did not
follow the suitable procedure in order to
limit ground movements. The struts were
not properly prestressed, instead they
were only tightened to the wale (longitudinal steel I-beam used to transfer soil
load on wall between strut points to the
struts) by means of screw-tightening or
timber wedging. Numerous wide gaps between
sheet piles and wales were left open
without wedging. This bracing installation
procedure would trigger additional ground
movements.

It is well known that removal of sheet
pile will leave void in the ground, at
least equal to the volume of sheet pile
being removed. In addition, the use of
long penetration length of the sheet pile
below the trench bottom in Bangkok clay
resulted in a high tendency of clay to
adhere to the surfaces of the U-shaped
sheet pile. This led to thick slabs of
clay being clogged in the sheet pile and

pulled out along with the sheet pile causing additional voids in the ground. The condition was worst for the longer sheet pile penetrated into the stiff clay layer. These voids tended to cause additional ground movements if they could not be properly backfilled. For sheet pile in soft clay, backfilling of the voids during and after sheetpile pulling is extremely difficult and generally unsuccessful because the voids tend to cave in immediately below the sheet pile tip. In this project void back-filling was attempted by water spraying sand piled on the ground surface against the sheet pile during sheet pile pulling. This method did not seem effective because the sand could not effectively flow down between sheet pile and soil to fill the voids created below the pile. It was generally observed that most amounts of sand were washed away on the ground surface instead.

With the above mentioned difficulties for bracing and construction methods, large ground movements were observed to occur for most of the trench excavation for the typical pipe laying sections where the sheet pile wall acted as anchored bulkhead wall rather than the "typical braced wall". Only minor ground movements occurred for the braced-wall systems used for deep trenches sections for concrete encased pipes and driving and arriving pits for pipe jacking (Arayasiri, 1987), brought about by the use of proper bracing throughout the trench depth.

The effect of large ground movements associated with the method of trench excavation and support for the typical pipe sections became evident when the excavation was conducted along road pavements and/or adjacent to buildings with shallow foundation. The excavation caused considerable damages, in term of large settlement, to the road pavements and to the buildings. According to the settlement records along two roads (Tonglau and Tiam Roommitr Roads), obtained by surveying in response to the instruction by the authors, large ground movements in the term of maximum ground settlements after construction were generally in the order of 2.5 to 5 percent of the excavation depth. In some areas where the sheet pile walls were not so stable, the maximum ground settlements reached up to 9 percent of the excavation depth.

Fig. 4 shows the typical final settlement profile and magnitudes across the pipe trench along Tonglau Road (Contract # G-MC-3B) induced by the 4.4 meter-deep excavation using one-level-braced wall (12 m long) to lay 2.3 m-diameter pipe. The

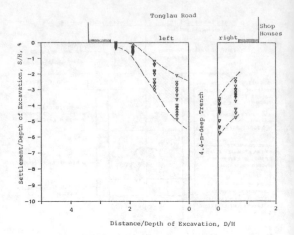

Fig. 4 Ground settlements along Tong Lau road total.

final ground elevation surface was recorded after sheet pipe pulling. It can be seen from the figure that the maximum ground settlements occurring right at the trench wall were mostly in the order of 10-25 cm (2.5-6.0% of excavation depth), and ground settlement zone extended more than 10.5 meters from the trench (2.5 x depth of excavation). It can also be noted that the levels of settlement were quite similar for both traffic (left) and non-traffic (right) sides of the trench. The large ground movements had severely damaged the road pavements, mostly in the form of tilt. The concrete pavement along the whole road (about 2 km) needed to be jacked up or totally replaced. It is also interesting to note that this large magnitude of ground movement did not cause any evident damage to the 2-3 storey shop house buildings located along the trench, about 8 meters from the trench wall. These buildings are supported by 21-m-long concrete piles.

Slightly smaller level of ground settlement than the above case was recorded along Tiam Roommitr Road (G-BC-1) where a shallower trench (3.7 m deep) was excavated and supported with 9-m-long sheet pile to place 2-m-diameter pipe. Fig. 5b shows the range of the final settlement profile of the road pavement recorded after sheet pile pulling at which time the maximum ground settlements next to the trench wall were typically in the order of 2 to 5% of the excavation depth (i.e. 7.5 to 18 cm of settlement). In unstable trench section due to the need to use a shorter sheet pile, the maximum settlements were up to 7 to 9 percent of

404

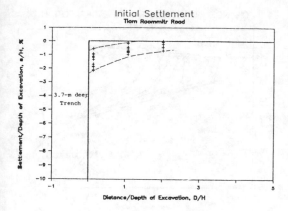

Initial Settlement
Tiam Roommitr Road

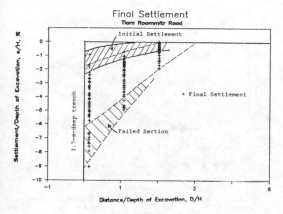

Final Settlement
Tiam Roommitr Road

Fig. 5 Ground settlements along Tiam Roommitr road.

the excavation depth and concrete slab
sidewalks were severely damaged. The
earlier measurements which were taken just
after excavation to the bottom of the
trench showed that the maximum settlements
at that time were only in the range of 0.7
to 2.2 % of the excavation depth (Fig.
5a). This indicate that the increases in
the amount of settlement occurring during
pipe laying, trench backfilling strut
removal and sheet pile pulling (the
process took 2 to 3 days) were mostly in
the range of 1.0 to 2.0 times the initial
settlement after excavation except for the
unstable trench section where the increase
could be much more. These large increases
in settlement were partly due to creep of
the clay, high flexibility of the sheet
pile wall and voids created by sheet pile
pulling. Unfortunately, there were no
record of the ground movement just prior
to strut removal and sheet pile pulling to
really compare these effects. However, the
effects of latter two seem to be more
significant.

The removal of the strut above the pipe
during backfilling would make the sheet
pile to act as a cantilever wall. With
the narrow backfill on the sides of the
flexible pipe, the effective soil reaction
modulus of the backfill would be small so
the wall deflection would be large. The
level of ground movements induced in the
trench excavation in the project was much
more than the typical range expected for
braced-wall in very soft to soft clay with
no basal heave, which would be only about
1 to 2 percent of excavation depth (Peck
1969). According to the method of ground
movement prediction for braced wall in
soft clay with tight bracing, good workmanship and without construction difficulties suggested by Mana and Clough (1981)
which is based upon wall flexibility and
factor against basal heave (FS), the
maximum ground settlements induced by the
excavation in the two cases which had FS
about 1.4 and 1.7, respectively when considering 1 t/m surcharge load, would be
about 1.0 and 0.7 % of the excavation
depths, respectively. A much larger settlements recorded in the actual conditions
were due to construction difficulties and
constraints mentioned earlier.

5. PIPE PERFORMANCE DURING CONSTRUCTION

The large diameter steel pipes used in
this conduit project were considered flexible type. As common for flexible pipes,
the stresses in the pipe wall brought
about by pipe deflection induced by
backfill load and surcharge above the
pipe are more critical than the stresses
during the service period. Therefore, the
pipe were designed according to the well
known Iowa formula which is to limit the
amount of bending stresses in the buried
pipe induced by horizontal diametral outward deflection of the pipe caused by
vertical load from overburden and surcharge.
Under such a load the pipe will be squashed
vertically and the soil around the pipe on
both sides will provide passive resistance
to prevent large pipe deflection. Therefore, the stiffness of the backfill
material on the sides of the pipe is very
important for good performance of the
pipe, and is included in the Iowa formula
as modulus of soil reaction. The Iowa
formula is as,

$$\Delta x = \frac{D.K \ W_t r^3}{EI+0.061E'r^3}$$

where

Δx = Increase in horizontal diameter
W_t = Total vertical load
EI = Pipe wall stiffness
E' = Modulus of soil reaction
r = Pipe radius
K = Bedding factor
D = Time lag factor (only for dead lead)

With the formula, pipe deflection is used as the control parameter in the design and construction. The pipes in the conduit project were designed to limit the diametral pipe deflection below 3%.

Since there was no previous experience in buried flexible pipe in Bangkok clay, the designer assumed the pipe would behave according to the formula and selected the E' value for the backfill sand based upon the suggested values given by the U.S. Bureau of Reclamation (Howard, 1977) which is also adopted by the ASTM. In general practice, the compacted zone of backfill material is required to extend about 1 to 2 pipe diameters on each side of the pipe to assure an adequate zone of good soil stiffness. However, in this project the width of the trench was required at 2 pipe diameters at the minimum, probably to reduce excavated soil volume. The specified backfill material was well-graded medium to coarse sand compacted to 93% Modified Proctor density. According to Howard this material would have the values of E' in the range of 140 to 200 ksc. In view of the potential of loosening of the backfill caused by voids created by sheet pile pulling the designer adopted the reduced E' value of 24 ksc for the design in the project; the value was only slightly larger than the value of soft Bangkok clay. However, the heavy compaction requirement of the backfill and the selection of the reduced E' value for conservatism reason in the design of the pipe using the Iowa formula were meaningless as will be shown in the following discussion.

The changes in diameters of two pipes, i.e. 2-m-diameter and 3-m-diameter pipes in Contract # G-BC-1 and G-MC-B-R during backfilling and sheet pile pulling are shown in Figs. 6 and 7, respectively, and summarized in Table 3. Pipe diameters at various stages of backfilling were measured with a rod extensometer. The measurements show that the horizontal and vertical diametral changes of each pipe were similar in magnitude but opposite in direction.

It can be seen that both pipes behaved similarly in response to the backfilling above the pipes and sheetpile pulling. The process of backfilling above the pipe

CHANGE IN HORIZONTAL PIPE DEFLECTION

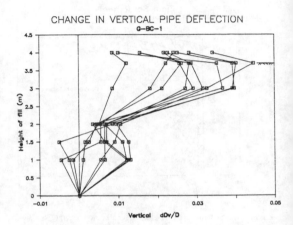

CHANGE IN VERTICAL PIPE DEFLECTION

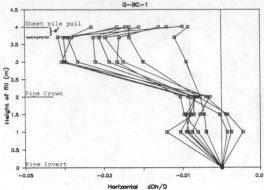

Fig. 6 Horizontal and vertical diametral pipe deflection, 2-m-diameter pipe.

Table 3 Measurement of changes in horizontal pipe diameters

Cause of Deflection	Horizontal Diameter Change, $\Delta D/D$ %	
	2-m-Diameter Pipe G-BC-1	3-m-Diameter Pipe G-MC-2B-R
1) Own weight Deflection	0.6	0.5
2) Backfill to crown	-0.4 to -1.1	+0.4 to -0.4
3) Backfill from crown to surface	-1.6 to -3.2	-0.4 to -1.5
4) Sheet pile pulling	+0.5 to +1.0 (Mostly)	+1.4 to +2.0 (Mostly)
5) 3) + 4)		
6) Ring Flexure Stiffness EI/D^3, kPa	4.8	11.5

to the surface level resulted in significant horizontal squashing of the pipes while sheet pile pulling led to some recovery by outward deflection of the horizontal pipe diameter. The mode of pipe deflection during the trench backfilling above the pipe contradicted to the typical response of flexible pipes which is assumed in the Iowa formula.

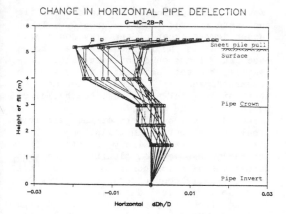

CHANGE IN HORIZONTAL PIPE DEFLECTION
G—MC—2B—R

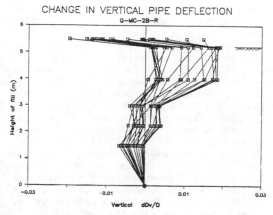

CHANGE IN VERTICAL PIPE DEFLECTION
G—MC—2B—R

Fig. 7 Horizontal and vertical diametral
pipe deflection, 3-m-diameter pipe.

The shortening of the horizontal pipe
diameters during backfilling from the pipe
crown to the surface ranged from 1.6-3.2%
of pipe diameter (equivalent 3.2-6.4 cm)
for the 2-m-diameter pipe and 0.4-1.5%
(1.2 to 4.5 cm) for the 3-m-diameter pipe.
The smaller change of the large pipe was
possibly attributable to its large
flexural ring stiffness (EI/D^3 = 0.115
kg/cm for the 3-m-diameter, 0.048 kg/cm
for 2-m-diameter). This observed mode of
pipe deflection can be explained by two
reasons, i.e. inward lateral movements of
the sheet pile wall following strut removal, and heavy backfill compaction above
the pipes. There was only one level of
strut placed 1 meter above the pipes.
Removal of the strut with the flexible
pipe encased in narrow backfill already in
place was likely to cause large inward
lateral movement of the sheet pile wall
and ground and thus the pipe. The lateral
ground movement would approximately amount
to one half of the pipe diameter change.

Heavy backfill compaction requirement
above the pipe level in the relatively
narrow trench in the project was likely to
cause the backfill on the sides of the
piles to further density and bulge toward
the pipe as high compaction was exerted on
the backfill above. It seems that the
effect of strut removal was more predominant than the heavy compaction even though
the record of pipe diameters shown in
Figs. 6 and 7 do not have sufficient
record points to clearly indicate so.

Sheet pile pulling after the completion
of trench backfill created voids below the
sheet pile and resulted in the tendency of
the soils on both sides to move in and
fill the voids. With narrow sand backfill
on the sides of the existing vertical egg-
shaped pipes, there was high tendency of
the sand to move outward into the voids
resulting in horizontal extension of the
pipe diameter. Sheet pile pulling caused
the increase in horizontal pipe diameter
by generally 0.5-1.0 % of the diameter
(1-2 cm) for the 2-m-diameter pipe and
1.4-2.0% (4.2-6 cm) for the 3-m-diameter.
The larger movements for the latter
clearly showed the effect of the use of
thicker and deeper sheet pile which would
create larger voids.

The measurements showed that the changes
in the horizontal pipe diameters induced
by backfilling from the pipe top to the
ground surface and then sheet pile pulling
ranged from 0.5 to 3.0% in diametral
contraction for the 2-m-diameter pipe and
0.5% in contraction to 1.8% in extension
for the 3-m-diameter pipe.

The horizontal pipe deflections induced
by backfill placement on the sides of the
pipe from the invert to the crown ranged
from 0.4 to 1.1 % diametral contraction
for the 2-m-diameter pipe and from 0.4%
diametral extension to 0.4% contraction
for the 3-m-diameter pipe. The observed
increase in the horizontal diameter of
some 3-m-diameter pipes was probable the
result of flooding of the pipe trench due
to heavy rain during backfilling. Notice
that the pipe experienced upto 0.7%
diametral extension when the pipe was
flooded and the backfill was upto the
springline.

With the self weight deflection around
0.5-0.6% pipe diameter for both pipes, the
final shape of the pipes were in vertical
oval shape by upto 2.7% diameter deflection
for the 2-m-diametral pipe and in round to
horizontal oval shape by upto 2.5%
diameter deflection for the 3-m-diameter.

The measurements of the diameters of all
the pipes installed in the soft clay in
the project show that the final shapes of

the pipes after construction were of both vertical and horizontal ovals which were brought about by the effects of backfilling above the pipe and sheet pile pulling, the amounts of which depended on significantly on the construction methods and procedures. Because the deflection response of the flexible pipe installed in the soft clay in this manner is totally different from that assumed in the Iowa formula, which is based on the response observed for typical flexible pipes installation, it would be senseless if the pipes are to be designed according to the formula. The deflection of the pipe depends very little on the stiffness of the backfill on the side of the pipe. Instead, the pipes should be designed to withstand the deflections induced by soil movement caused by strut removal and sheet pile pulling. Heavy compaction just above the pipe on the sides is unnecessary and may be harmful, thus should be avoided. The effect of voids created by sheet pile pulling, which could cause severe outward deflection of the pipe, especially when large and long sheet pile is used, should be minimized by using a good method of void refilling during sheet pile pulling.

CONCLUSIONS

The experience from the installation of the large diameter buried steel pipes in the Water Transmission Conduit Project in Bangkok clay using trench excavation with braced sheet pile walls has shown two important aspects regarding the behavior of ground movements and the pipe deflection response during construction. The selection of the braced sheet pile walls which had a large wall span between bracing and trench depth and the wall acted like anchored bulkhead type resulted in large ground movements. The initial settlements after trench excavation were in the order of 1-2 % of the trench depth. The ground settlements occurring during backfilling and sheet pile removal were very significant, amounting to 1 to 2 times the initial settlement. The placement of narrow backfill on the sides of the pipe and the flexibilities of the pipe and the sheet pile wall in the soft clay resulted in additional ground settlement when the bracing were removed during trench backfill. In addition, large voids created below the sheet pile during the pile removal led to further ground settlement. This could be very significant for long and large sized sheet pile penetrating in stiff clay layer, in which case large amounts of clay clogged to the sheet pile and were removed from the ground. To a lesser degree creep of the clay could also contribute to the increase in the ground movement. The ground movement caused some damages to the adjacent concrete road pavements and small buildings founded on short piles, but no damages to buildings on long concrete piles.

The deflection pattern of the flexible steel pipes during installation in the narrow trench was entirely different from generally observed for flexible pipes with normal installation conditions. The pipes were laterally compressed and underwent an increase in vertical dimension upon backfilling from the pipe top to the ground surface. This observed pipe deflection was probably attributed to the removal of bracing and the heavy backfill compaction above the pipe. The magnitude of the deflection depended on the flexural ring stiffness of the pipe, the width and stiffness of the backfill on the sides of the pipe and the flexural stiffness of the sheet pile wall. The Iowa formula is not valid and can not be used to design the flexible pipes installed and deformed in such conditions. The removal of the sheet pile wall resulted in lateral expansion of the pipe as the backfill moved outward to fill the voids created by the sheet pile. The amount of the diametral pipe expansion depended largely on the size and length of the sheet pile. The longer and larger the sheet pile, the larger the deflection.

REFERENCES

Arayasiri, M. (1987). "Sheet Pile for Water Conduit Project: A Case Study" Proceedings of Seminar on Sheet Pile, Engineering Institute of Thailand (in Thai).

Howard, A.K. (1977). "Modulus of Soil Reaction Values for Buried Flexible Pipe", ASCE Journal, Vol. 103, No. GT 1, January, pp. 33-43.

Mana, A.I. and Clough, G.W., (1981). "Prediction of Movements for Braced Cuts in Clay", ASCE Journal, Vol. 107, No. GT 6, June, pp. 759-777.

Peck, R.B., (1969). "Deep Excavations and Tunneling in Soft Ground", Proceedings of 7th International Conference on Soil Mechanics and Foundation Engineering, State-of-the-Art Volume, pp. 225-290.

4. Soil and rock improvements

Developments in Geotechnical Aspects of Embankments, Excavations and Buried Structures, Balasubramaniam et al. (eds)
© 1991 Balkema, Rotterdam. ISBN 90 5410 019 2

Consolidation of clayey soils using combined dewatering method

H. Kotera, Y. Wakame & T. Sakemi
Tasei Corporation, Tokyo, Japan

T. Matsui
Osaka University, Japan

ABSTRACT: This paper deals with the vertical drain method which is one of the soft soil improvement methods. The authors try to ascertain the differences in the consolidation process between various installation methods of vertical drains through   field and laboratory test data.

Many papers have described effects of remoulding or disturbance of clay by vertical drain installation on the consolidation process. As a result, it becomes clear that the difference in consolidation process between various drain installation methods is caused by smear or disturbance of clay just around the drains. Therefore, in the present paper the authors investigate strength of clays in between vertical drains after soil improvement work and try to estimate a zone of smear or disturbance of the clay.

Another effect on the consolidation process is caused by the drain well-resistance. Few papers are available on its field observations, while some papers on the analytical solution. In this paper, the influence of lowering the groundwater level with vertical drains and surcharge embankment on the consolidation process is discussed through field and laboratory test results.

Next, a theoretical solution of the consolidation process, including the effects of smear, well-resistance and dewatering, is developed, being based on Kjellman and Hansbo's method (1981). Then, the availability of the new consolidation equation is examined through some field and laboratory test data. As a result, it is confirmed that the forced dewatering method in combination with the vertical drain method and preloading method is significantly effective for reducing well-resistance and for promoting consolidation.

1 INTRODUCTION

Soft clay consolidations by vertical drain methods with preloading have been established practically in theories, design and construction methods. However, the effects of drain installation methods and drain materials on the consolidation behavior have not sufficiently been examined yet.

Displacement drains, which are driven by vibratory or percussion action with closed mandrel, disturb a soft clay just around vertical drains, while non-displacement drains which are driven by an auger machine or a water jet do not disturb the soft clay. Therefore, non-displacement methods are better than displacement methods from view point of reducing clay disturbance.

Recently, the demand for deeper vertical-type of drains, more than 30 m, are increasing. Consequently, driving machines are becoming heavier and taller. Non-displacement methods have a depth of 20 m of marginal capacity for economical reasons, which can not be used for the installation of these deeper type drains.

Soil improvement work by vertical drains in Japan is almost all carried out by using a displacement method with closed mandrel. As vertical drains become deeper, the effect of well-resistance can not be ignored, and also the clay disturbance effect of vertical drain installation has to be taken into serious consideration in the design.

Simons (1965) reported that the coefficient of consolidation (c_v), for simulated displacement drains was 20 % lower than that for non-displacement drains about a soft gray silty marine clay with a sensitivity of 6. Orrje and Broms (1967) reported that the undrained shear strength was decreased about 20 % by the clay disturbance due to vertical drain installation, which occurred in the region of about 1.5 times vertical drain diameter. Consequently, vertical drain installation has an influence on c_v, m_v (coefficient of compressibility) and k_c (permeability of the soil). It is difficult, however, to determine these parameters of the soil in the disturbed (smeared) zone.

The theoretical analysis of vertical drains was solved by Barron (1948) who established his solution on free strain including effects of smear and well-resistance. Yoshikuni and Nakanodo (1974) also solved an equal strain case which was based on Barron's method. However, both solutions are very complicated equations including Bessel function. Onoue (1988) proposed a numerical analysis in multilayered anisotropic soils considering well-resistance.

In this paper, the authors try to propose a new analysis of consolidation process, including the effects of smear, well-resistance and dewatering. Field observations and laboratory experiments demonstrate that the proposed consolidation equation is a practically useful expression of consolidation process with effects of smear, well-resistance and dewatering.

2 ANALYTICAL SOLUTION OF CONSOLIDATION CONSIDERING EFFECTS OF SMEAR, WELL-RESISTANCE AND DEWATERING

Hansbo (1981) have further developed Kjellman's solution (1948) which was based on the equal strain theory, and have proposed a solution including smear and well-resistance. The authors try to further develop Hansbo's solution, and to propose a solution considering effects of smear, well-resistance and dewatering.

Figure 1 shows a schematic illustration of drain well and the water flow, in which the radius of drain well is r_w, the radius of influenced zone by each drain well is R, the disturbed zone by driving mandrel is r_b. To derive an analytical solution, following basic assumptions are made;

1) Deformation occurs only in the vertical direction,
2) Soil is saturated,
3) Water flows horizontally to the drain well,
4) Permeability of soil is homogeneous even in the disturbed zone, and
5) Horizontal inflow of water to the drain well equals to vertical outflow of water from the drain well.

According to Darcy's law, the rate of radial water flow toward the drain well is expressed as follows:

$$v_r = k_c \times i_r$$

where
v_r: rate of radial pore water flow toward the drain well
k_c: coefficient of permeability of the soil
i_r: hydraulic gradient in the soil at a distance r
r: radius from the drain center

In the disturbed zone $(r_w \leqq r \leqq r_b)$, the rate of radial water flow is represented by

$$v_{r1} = \frac{k_{c1}}{\gamma_w} \times \frac{\partial U_1}{\partial r} \qquad (1a)$$

and in the non-disturbed zone $(r_b \leqq r \leqq R)$ by

$$v_{r2} = \frac{k_{c2}}{\gamma_w} \times \frac{\partial U_2}{\partial r} \qquad (1b)$$

where
γ_w: unit weight of the water
U : excess pore water pressure at a distance r

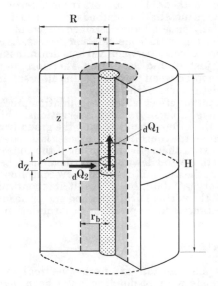

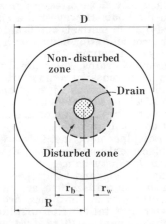

Fig 1. Schematic illustration of drain wells and water flow

Suffixes 1 and 2 correspond to the disturbed zone and the non-disturbed zone, respectively.

Because both equations of $(1a)$ and $(1b)$ are continuous at $r=r_b$,

$$\frac{k_{c1}}{\gamma_w}\times\frac{\partial U_1}{\partial r}=\frac{k_{c2}}{\gamma_w}\times\frac{\partial U_2}{\partial r} \qquad (1c)$$

An assumption that the volume change in the saturated soil equals to the dissipated pore water pressure leads to the following equation in the disturbed zone $(r_w\leqq r\leqq r_b)$, that is

$$2\pi r\times v_{r1}=\pi\times(R^2-r^2)\times\frac{\partial\varepsilon}{\partial t} \qquad (2a)$$

and in the non-disturbed zone $(r_b\leqq r\leqq R)$

$$2\pi r\times v_{r2}=\pi\times(R^2-r^2)\times\frac{\partial\varepsilon}{\partial t} \qquad (2b)$$

From Eqs. $(1a)$, $(1b)$ and Eqs. $(2a)$, $(2b)$, the following equations can be derived.

$$\frac{\partial U_1}{\partial r}=\frac{\gamma_w}{2k_{c1}}\times(\frac{R^2}{r}-r)\times\frac{\partial\varepsilon}{\partial t} \qquad (3a)$$

$$\frac{\partial U_2}{\partial r}=\frac{\gamma_w}{2k_{c2}}\times(\frac{R^2}{r}-r)\times\frac{\partial\varepsilon}{\partial t} \qquad (3b)$$

When the hydraulic gradient of excess pore water in the vertical drain at the depth,z, is assumed as $\partial U_1/\partial z$, the water volume of outflow from a small element dz of drain well, dQ_1, equals to that of inflow to the small element, dQ_2 (see Fig.1).

The water volume of outflow from the drain well is influenced by its hydraulic gradient,i_w. Therefore,

$$dQ_1=-\frac{q_w\times i_w}{\gamma_w}\times\frac{\partial^2 U_1}{\partial z^2}dz\times dt$$

$$=-\frac{\pi r_w^2\times k_w\times i_w}{\gamma_w}\times\frac{\partial^2 U_1}{\partial z^2}\times dz\times dt \qquad (4a)$$

where H: drain length
q_w: drain capacity
$\quad=A_w\times k_w=\pi r_w^2\times k_w$
A_w: cross-sectional area of the drain
k_w: permeability of the drain well
When drains are not circular such as a plastic board, the reduced permeability of the plastic board drain well, k_w' is given by the following equation, instead of k_w.

$$k_w'=\pi\times a\times b\times k_w/(a+b)^2$$

where
$\quad a$: width of the plastic board drain well
$\quad b$: thickness of the plastic board drain well
$\quad i_w$: hydraulic gradient in the drain well $=H/(H-h_0)$
$\quad\quad H=H/2$ is perfectly penetrated drain case
$\quad\quad H=H$ is imperfectly penetrated drain case
$\quad h_0$: lowered water level in the drain well

The water volume of inflow from the surrounding soil to the drain well at the time, t, is given as follows:

$$dQ_2=\frac{2\pi r_w\times k_{c1}}{\gamma_w}\times\frac{\partial U_1}{\partial r}dz\times dt \qquad (4b)$$

As $dQ_1=dQ_2$ at $r=r_w$,

$$\frac{r_w\times k_w\times i_w}{2k_{c1}}\times\frac{\partial^2 U_1}{\partial z^2}+\frac{\partial U_1}{\partial r}=0 \qquad (4c)$$

Integration of Eq. $(3a)$ or $(3b)$ gives

$$U=\frac{\gamma_w}{2k_c}-[\{R^2\times\ln r-\frac{r^2}{2}\}+A]\frac{\partial\varepsilon}{\partial t} \qquad (5)$$

From Eq. $(3a)$ and Eq. $(4c)$, a new boundary condition at $r=r_w$ is given as follows:

$$\frac{\partial^2 U_1}{\partial z^2}=-\frac{\gamma_w}{r_w\times k_w\times i_w}\times(\frac{R^2}{r_w}-r_w)\times\frac{\partial\varepsilon}{\partial t} \qquad (6a)$$

$$\frac{\partial^2 U_1}{\partial z^2}=-\frac{\gamma_w}{k_w\times i_w}\times(n^2-1)\times\frac{\partial\varepsilon}{\partial t} \qquad (6b)$$

where n is R/r_w.

To develop a consolidation equation considering the effect of dewatering, being based on Eq. $(3a)$, the following boundary conditions are considered.
1) Perfectly penetrated vertical drains to an underlain sand layer.

As shown in Fig.2a, the soft clay layer has a thickness of h_1, and a waterhead of h_0 is lowered in the underlain sand layer. In this case, the boundary conditions on the surface of the drain well are as follows:

$$0\leqq z\leqq h_0 \qquad U(r,0,z)=-\gamma_w\times z \qquad (7a)$$

$$h_0\leqq z\leqq h_1 \qquad U(r,0,z)=-\gamma_w\times h_0 \qquad (7b)$$

2) Imperfectly penetrated vertical drains to an underlain sand layer.

As shown in Fig.2b, vertical drains of length h_3, are stopped at the midway of the soft clay layer of thickness h_2, and an waterhead of h_0 is lowered in the upper sand layer. In this case, the boundary condition is

$$0 \leqq z \leqq h_3 \qquad U(r,0,z) = -\gamma_w \times h_0 \qquad (7c)$$

$$h_3 \leqq z \leqq h_2 \qquad U(r,0,z) = -\gamma_w \times h_0 \times \frac{h_2 - z}{h_2 - h_3} \quad (7d)$$

3) Perfectly penetrated vertical drains to an artesian pervious sand layer.

As shown Fig.2c, the vertical drains perfectly penetrate to the soft clay layer of thickness h_4, connecting with the artesian pervious sand layer. The waterhead of h_0 is lowered in the vertical drain. In this case, the boundary conditions are

$$0 \leqq z \leqq h_0 \qquad U(r,0,z) = -\gamma_w \times z \qquad (7e)$$

$$h_0 \leqq z \leqq h_4 \qquad U(r,0,z) = -\gamma_w \times h_0 \times \frac{h_4 - z}{h_4 - h_0} \quad (7f)$$

All above-mentioned boundary conditions are generally expressed as follows:

$$U(r,0,z) = az + b \qquad (7g)$$

where,   at $z = 0 : U_1 = b$ $(7h)$
at $z = H : U_1 = aH + b$ $(7i)$

Then Eq. (6b) is solved under these boundary conditions, as follows:

$$U_1 = -\frac{\gamma_w}{k_w \times i_w} \times (n^2 - 1) \times \frac{\partial \varepsilon}{\partial t} \times \frac{(z^2 - H \times z)}{2} + az + b \quad (8a)$$

Substitution of Eq. (8a) into Eq. (5) gives

$$U_1 = -\frac{\gamma_w}{2k_{c1}} \frac{\partial \varepsilon}{\partial t} [\frac{2k_{c1}}{k_w i_w} \times (n^2 - 1) \times \frac{(z^2 - Hz)}{2}$$

$$+ \{R^2 ln \frac{r}{r_w} - \frac{(r^2 - r_w^2)}{2}\}] \qquad (8b)$$

As for U_2, the following equation can be derived under the condition that U_2 is equal to U_1 at $r = r_b$.

$$U_2 = -\frac{\gamma_w}{2k_{c2}} \frac{\partial \varepsilon}{\partial t} [\frac{2k_{c2}}{k_w i_w} \times (n^2 - 1) \times \frac{z^2 - Hz}{2} + \{R^2 ln \frac{r}{r_b}$$

$$- \frac{(r^2 - r_b^2)}{2}\} + \frac{k_{c2}}{k_{c1}} \{R^2 ln \frac{r_b}{r_w} - \frac{(r_b^2 - r_w^2)}{2}\}] \quad (9)$$

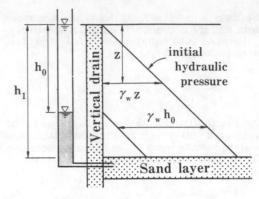

Fig 2a. Perfectly penetrated verticaldrains to an underlain sand layer

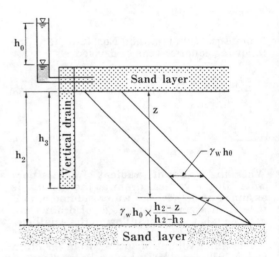

Fig 2b. Imperfectly penetratedvertical drains

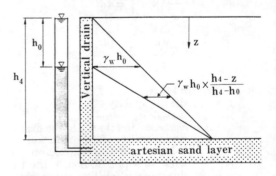

Fig. 2c. Perfectly penetrated to artesian pervious sand layer

The mean excess pore water pressure Ua of the soil at any depth z can be obtained by the following integration

$$\pi Ua(R^2 - r_w^{\,2}) = \int_{r_w}^{r_b} 2\pi U_1 \times r \times dr + \int_{r_b}^{R} 2\pi U_2 \times r \times dr \quad (10)$$

where Ua: the mean excess pore water pressure at any depth z

Eq. (10) becomes

$$Ua = \frac{\gamma_w}{8k_{c2}} \times \frac{\partial \varepsilon}{\partial t} \times D^2 \times W + az + b \quad (11)$$

where

$$W = -\frac{8k_{c2}}{k_w \times i_w} \times \frac{1}{D^2} \times (n^2 - 1) \times \frac{(z^2 - H \times z)}{2} + \frac{k_{c2}}{k_{c1}}$$

$$\times \{\frac{n^2}{(n^2-1)} \ln s + \frac{1}{(n^2-1)} \times (-s^2 + 1 + \frac{s^4}{4n^2} - \frac{1}{4n^2})\}$$

$$+ \frac{n^2}{(n^2-1)} \times (\ln \frac{n}{s} - \frac{3}{4} + \frac{s^2}{n^2} - \frac{s^4}{4n^4}) \quad (12)$$

s : ratio of radius of disturbed zone to radius of drain well $= r_b/r_w$
$H = H/2$: perfectly penetrated drain case
$H = H$: imperfectly penetrated drain case

where Ua_0: the initial mean excess pore water pressure at any depth, z
      T_h: time factor in radial consolidation $(= c_h \times t/D^2)$
      c_h: coefficient of consolidation in the horizontal direction $(= k_c /(\gamma_w m_v))$
      m_v: coefficient of compressibility
      D : diameter of the influenced soil

From Eq. (11), the solution obtained on excess pore water pressure including the effects of smear, well-resistance and dewatering is represented as follows:

$$Ua - (az+b) = Ua_0 \times EXP(-\frac{8T_h}{W}) \quad (13)$$

The mean excess pore water pressure of all the soil is given

$$Ua_t - (\frac{aH}{2} + b) = Ua_{0t} \times EXP(-\frac{8T_h}{ESW}) \quad (14a)$$

where Ua_t: the mean excess pore water pressure of all the soil
      Ua_{0t}: the initial mean excess pore water pressure of all the soil
      ESW : a factor of consolidation including smear, well-resistance and dewatering as follows:

$$ESW = \frac{8k_{c2}}{k_w} \times \frac{(H-h_0)}{H} \times \frac{1}{D^2}(n^2-1) \times \frac{H^2}{12} + \frac{k_{c2}}{k_{c1}}$$

$$\times \{\frac{n^2}{(n^2-1)} \ln s + \frac{1}{(n^2-1)} \times (-s^2 + 1 + \frac{s^4}{4n^2} - \frac{1}{4n^2})\}$$

$$+ \frac{n^2}{(n^2-1)} \times (\ln \frac{n}{s} - \frac{3}{4} + \frac{s^2}{n^2} - \frac{s^4}{4n^4}) \quad (14b)$$

$H = H/2$: perfectly penetrated drain case and when h_0 is deeper than $H/2$, h_0 is taken as $H/2$.
$H = H$: imperfectly penetrated drain case and when h_0 is deeper than H, h_0 is taken as H.

The mean degree of consolidation, U_h, for all the soil is

$$U_h = 1 - EXP(-\frac{8T_h}{ESW}) \quad (14c)$$

Equation $(14a)$ suggests that consolidation forces acting on the soft clay are both the preloading embankment and the lowered groundwater head, and also that the magnitude of the consolidation force on the soft clay by the dewatering is equivalent to the decreased water head in the drain wells.

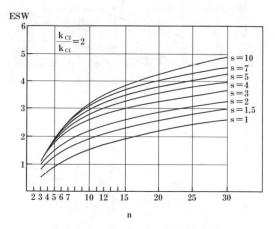

Fig 3a. Relationship between ESW and n in the case where $k_{c2}/k_{c1} = 2$

The numerical computations were performed for the case where $k_{c2}/k_{c1} = 2,3,4,5,7$ the disturbed zone $s = 1$ to 10 and $n = 3$ to 30. Figures 3a,3b,3c,3d and 3e show the relationship between ESW and n changing k_{c2}/k_{c1} and s as parameters.

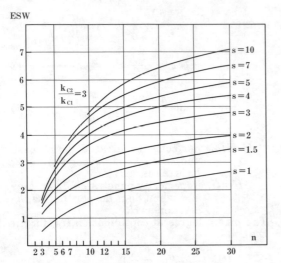

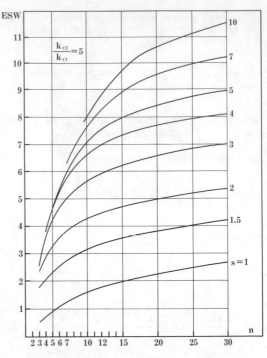

Fig 3b. Relationship between ESW and n in the case where $k_{c2}/k_{c1} = 3$

Fig 3d. Relationship between ESW and n in the case where $k_{c2}/k_{c1} = 5$

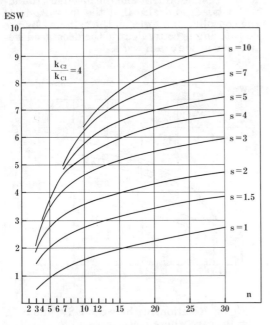

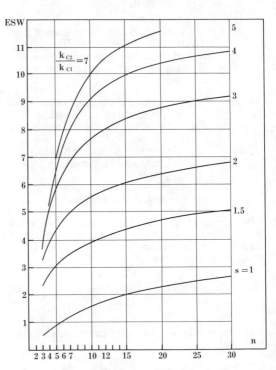

Fig 3c. Relationship between ESW and n in the case where $k_{c2}/k_{c1} = 4$

Fig 3e. Relationship between ESW and n in the case where $k_{c2}/k_{c1} = 7$

3 DISTURBANCE OF SOFT CLAY BY VERTICAL DRAIN INSTALLATION

The authors have tried to confirm the disturbance of soft clay by two vertical drain installation methods, using field tests at Goi in Chiba Prefecture. The test site has a top sandy soil of 1 m deep, and below this layer a very soft clayey soil of 4.2 m deep exists, and an alluvial sand layer (N-value in SPT is more than 20) is underlain.

Tests were carried out using two installation methods. One is the installation of sand drains by continuous auger driving (Non-displacement drains) and surcharge embankment. The other is the installation of sand drains by vibratory driving (Displacement drains) and surcharge embankment. Sand drains in both cases are 43 cm in diameter and 8 m in depth, which are installed 2.5 m in equilateral triangle spacing. Embankment surcharge is 4 m high.

After the end of the consolidation work, block samplings of clays between sand drains were carried out up to G.L.-4 m by open excavation work for both vertical drain installation methods. Differences of the unconfined strength in both cases are shown in Fig. 4. The disturbed zone is not cleared from this figure. However, the differences are comparatively large in the zone from the surface of the sand drain to 50 cm from the center of the drains.

This suggests that the consolidation in the non-displacement method is faster than the displacement method. Aboshi et al (1979) reported on the relationship between the drain spacing and coefficient of consolidation by displacement sand drains, and concluded that as n (drain spacing ratio R/r_w) is greater than 7, the change in c_v was not significant.

When soil improvement with a vertical drain and surcharge embankment is designed, the c_v and c_h values have to be decided. However, this decision is not easy. The value of c_h is sometimes taken more than twice of the c_v value by the oedometer test. A back-analysis is indispensable for the soil improvement work by the preloading and vertical drain method.

In this paper, the authors try to take the same value for c_v and c_h because of simplicity, and to examine the availability of Eqs. (14a,b,c) which include the effects of smear, well-resistance, and dewatering through data in field tests and laboratory tests.

4 FIELD CASE HISTORY 1 (Goi in Chiba Prefecture)

The relationship between settlements and elapsed time of the field test at Goi in Chiba Prefecture is shown in Figs.5a,5b and 5c.

The coefficient of consolidation, c_v, is obtained to be 250 cm²/d by back-analysis in the case of embankment surcharge without sand drains (Fig.5a). The permeability of drains (k_w) are 3.8×10^{-3} cm/s which is obtained from Creager's 20 % grain size table, and the permeability of non-disturbed clay (k_{c_2}) is 6.6×10^{-7} cm/s which is obtained from the oedometer test under around the actual effective stress. Outer radius of the influenced soil (R) is 131.25 cm. Drain spacing ratio, n, is 6.16.

As for non-displacement vertical drains (continuous auger), the smear effect is not considered but the well-resistance effect is considered. ESW value in this case is calculated to be 1.134 in Eq. (14b). Figure 5b shows the measured and calculated settlements, in which c_v (=c_h) is 150, 250 and 350 cm²/d with a ESW value of 1.134. The calculated settlement in case where c_v (=c_h) is 250 cm²/d and ESW is 1.134 agrees well with the measured one. Consequently, the availability of Eqs. (14a,b,c) to represent the consolidation process is confirmed.

As for displacement vertical drains (vibratory driving) , not only smear effect but also well-resistance effect have to be considered. In this case, disturbed zone ratio (s) and ratio of disturbance ($K'=k_{c2}/k_{c1}$) have to be decided. Several values of s and K' are chosen and the calculated settlements are plotted in Fig. 5c with the measured settlements.

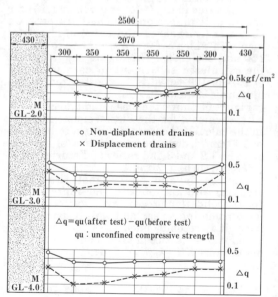

Fig 4. Variations of unconfined compressive strength in both cases

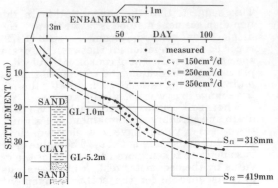

Fig 5a. Settlements and elapsed time relationship at place Goi without sand drains

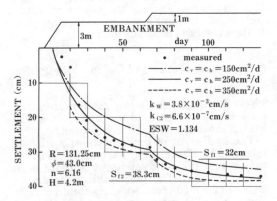

Fig 5b. Settlements and elapsed time relationship at place Goi with sand drains by non-displacement method

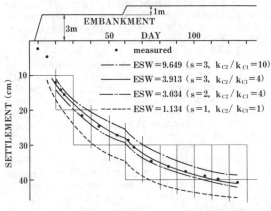

Fig 5c. Settlements and elapsed time relationship at place Goi with sand drains by displacement method

As a result, the best fitted curve can be obtained in case where s is 3 and $K'(=k_{c2}/k_{c1})$ is 4 ($ESW=3.913$).

That is, the disturbed zone has 1.5 times casing diameter from the center of a vertical drain, and also the permeability in the disturbed zone is one-fourth of that in the undisturbed zone.

This back-analysis also suggests that Eqs. ($14a,b,c$) are effective for representing the consolidation process including the smear and well-resistance effects.

5 FIELD CASE HISTORY 2 (Kuki in Saitama Prefecture)

This project is a road construction on soft soil. The very soft clay soil, in which the N-value in SPT is nearly zero, deposits from the ground surface to G.L.-13.8 m. A pervious sand layer exists below the soft clay layer. In this case, sand drains were installed by both non-displacement (continuous auger) and displacement (closed mandrel) methods with a bore hole diameter of 43 cm.

The non-displacement method was carried out on a 2.2 m spacing (effective radius,R, is 124.3 cm, n is 5.78). The displacement method was carried out on a spacing of 1.4 m square (R is 79.1 cm, n is 3.68). The permeability of sand drain, k_w, is $3.8×10^{-3}$ cm/s and the permeability of the undisturbed clay (k_{c2}) is $2×10^{-7}$ cm/s. Both were obtained in the same way as in Field case history 1.

ESW value in Eq. ($14b$) is given as 1.101 for the non-displacement method. Figure 6a shows the measured and calculated settlements, in which $c_v (= c_h)$ are taken to be 100, 150 and 200 cm²/d. This figure suggests that the calculated settlement in case where c_v ($=c_h$) is 100 cm²/d is   best fitted to the measured settlement.

Figure 6b shows the measured and calculated settlements for the displacement method. In the calculated curves, c_v ($=c_h$) is chosen to be 100 cm²/d, disturbed zone ratios (s) are taken to be 1,2 and 3 and ratios of disturbance K' ($=k_{c2}/k_{c1}$) to be 1, 2 and 4.

Figures 6a and 6b suggest that Eqs. ($14a,b,c$) are also effective for representing the consolidation process with vertical drains for considering smear and well-resistance effects. It is recommended that the disturbed zone ratio (s) is chosen to be 3 and the ratio of disturbance $K'(=k_{c2}/k_{c1})$ is 4.

6 FIELD CASE HISTORY 3 (Noda in Chiba Prefecture)

In this field test, two kinds of displacement methods with a closed mandrel by vibratory

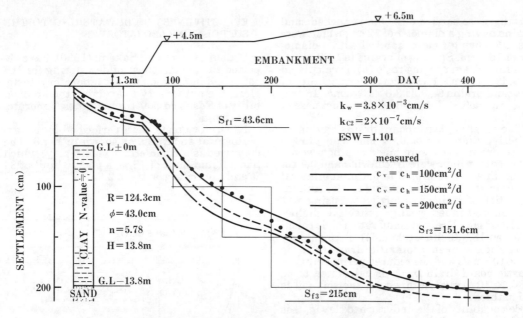

Fig 6a. Settlements and elapsed time relationship at place Kuki
with sand drains by non-displacement method

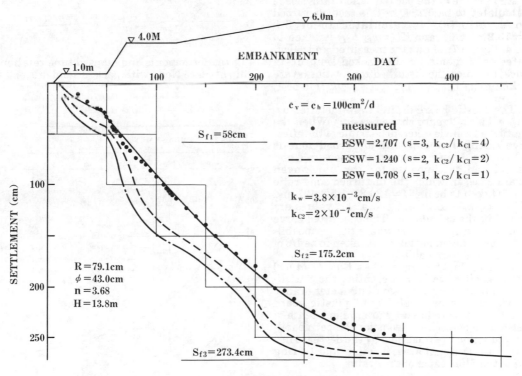

Fig 6b. Settlements and elapsed time relationship at place Kuki
with sand drains by displacement method

419

driving were carried out. One is packed sand drains having a diameter of 12 cm, which were installed using a close mandrel with a diameter of 15.2 cm. Spacing of drains is 1m square, and then drain spacing ratio (n) is calculated as 9.42 $(1.0 \text{ m} \times 1.13/12\text{cm})$. Permeability of the sand drains (k_w) is 3.8×10^{-3} cm/s. In this case, the soft soil deposits had a thickness of 6.5 m.

The other is plastic board drains having an effective diameter of 6.84 cm. The plastic board drains used were made of a non-woven geotextile, which were 10 cm wide and 0.7 cm thick. Spacing of drains is 0.9 m equilateral triangle one (effective spacing is 94.5 cm). Installation of drains was carried out using a 11 cm square-shaped casing (effective diameter:12.43 cm). Drain spacing ratio (n) is calculated as 13.8 $(0.9\text{m} \times 1.05/6.84)$. Permeability of the non-woven geotextile drain (k_w) is 2.0×10^{-2} cm/s and the reduced permeability of plastic board drain (kw') is calculated to be 3.84×10^{-3} cm/s. In this case, the soft soil is 9.5 m thick.

Permeability of the undisturbed clay (k_{c2}) is 2×10^{-7} cm/s and the coefficients of consolidation, $c_v (=c_h)$, is taken to be 60 cm²/d that are determined from the oedometer test under the actual effective stress.

ESW value of the packed sand drain case is calculated to be 5.234. In this case, disturbed zone ratio (s) is calculated to be 3.8 $(=15.2 \times 1.5/6)$ and $K'(=k_{c2}/k_{c1})$ is taken to be 4. Figure 7a shows the measured and calculated settlements for the packed sand drain case, in which the disturbed zone ratios (s) are taken to be 1, 2, 3.8 and $K'(=k_{c2}/k_{c1})$ to be 1, 2.5, 4.

The plastic board drains were installed using a 11 cm square-shaped casing. When the disturbed zone is taken to be triple of the effective casing diameter, the disturbed zone is 186.5 mm. Figure 7b shows the measured and calculated settlements for the non-woven drain case, in which the disturbed zone ratio (s) is taken to be 5.45 $(=124.3 \times 1.5/3.42)$ and $K'(=k_{c2}/k_{c1})$ to be 4. Several k_w values are taken in the calculation. The best fitted curve can be obtained in case where the permeability of non-woven geotextile is taken to be 1/6 of the reduced permeability (k_w').

Figures 7a,7b suggest that Eqs. (*14a,b,c*) are also effective for representing the consolidation process even for small diameter drains. However, the permeability of plastic board drain must be reduced by around one tenth of estimated one. The reasons for the permeability reduction are imagined by the decrease in the cross-sectional area, bending and tearing off of the filter paper of the drains.

7 EFFECTIVENESS OF DEWATERING TO THE SOFT CLAY IN LABORATORY TEST

Matsui, Kotera and Sakemi (1990) have reported the effectiveness of dewatering for the soft clay improvement by laboratory tests. Here, the authors try to examine the applicability of Eqs. (*14a,b,c*) through those laboratory test data.

Laboratory tests were performed by an experimental apparatus as shown in Fig.8. The clay used is a commercial kaolin clay, which has a liquid limit of 75 %, a plastic limit of 34 % and a specific gravity of 2.65.

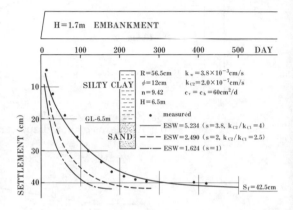

Fig 7a. Settlements and elapsed time relationship at place Noda with packed sand drains

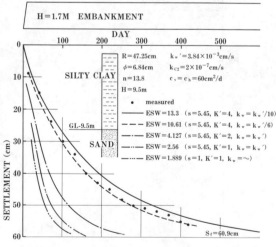

Fig 7b. Settlements and elapsed time relationship at place Noda with plastic board drains

The drain of non-woven geotextile is a tube-shaped having a diameter of 2 cm. The lower end of the drain is fixed in the drainage layer of sand at the bottom of the test bin. Drain spacing ratio (n) of this laboratory test is 24.35. In this laboratory test, a disturbed zone of clay around the vertical drain does not exist, because the drain is installed before making the model ground.

Such five tests as 1) preloading, 2) preloading with the dewatering, 3) preloading with vertical drains, 4) the dewatering in combination with preloading and vertical drains, and 5) the dewatering with vertical drains, were carried out. The initial water content of clay is about 150 %. The surcharge is 9.8 kPa in cases 1,2,3,4 and 2.16 kPa in case 5. The water head of the sand layer is lowered to 20 cm above the sand layer in cases 2, 4 and 5.

Figure 9a shows the relationship between the settlements and elapsed time in cases 1 and 2. The solution using Terzaghi's one-dimensional consolidation theory fits the test results very well, if $c_v = 80$ cm²/d is used. The permeability of the non-woven geotextiles(k_w), which is the same material as in above-mentioned Field case history 3, is 0.02 cm/s. The permeability of non-disturbed soil (k_{c2}) is 5×10^{-7} cm/s.

ESW value in Eq. (14b) of case 3 is calculated to be 2.477 with consideration of well-resistance effect. When the permeability of the plastic board drain is taken as 1/30, ESW is 3.382. In case 4, ESW is given as 2.446 if the permeability of the plastic board drain is taken en to be infinite due to the effect of dewatering. Figure 9b shows the measured and calculated settlements of cases 3 and 4. Both settlements agree well each other in case where the drain permeability is reduced.

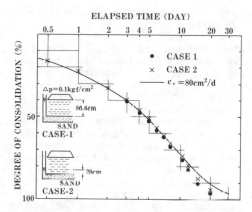

Fig 9a. Degree of consolidation and elapsed time relationship in cases 1 and 2

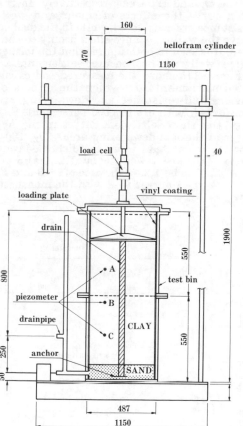

Fig 8. General view of the laboratory experimental apparatus

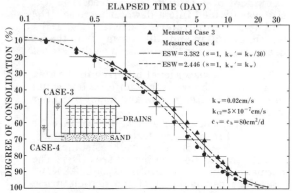

Fig 9b. Degree of consolidation and elapsed time relationship in cases 3 and 4

Discussion on data of case 5 is eliminated, because in case 5, the settlement behaves as if the vertical drain is not effective. This may come from in sufficient vacuum pressure by the dewatering.

The combined dewatering is an effective means for reducing well-resistance, for promoting consolidation and for strengthening soft clay by the lowered waterhead loading. Equations (14a,b,c) are effective for representing the consolidation process of the combined dewatering method. The permeability of the plastic board drains has to be reduced. However, to decide the amount of permeability reduction is not easy and more field test data should be collected. For the practical design method at the present time the cross-sectional area of the plastic board drains is sometimes reduced to around a half of the estimated value.

8 FIELD CASE HISTORY 4 (Urayasu in Chiba Prefecture)

Kotera, Sakemi and Matsui (1989a, 1989b) have reported a field test of the combined dewatering method. The applicability of Eqs. (14a,b,c) including the effects of smear, well-resistance and dewatering to the field data will be confirmed below.

The test site, which has an area of 240 ha, has a very soft cohesive and sandy soil of over 50 m thick. The soil improvement work started with spreading a 1 m thick sand mat on the dredged very soft clay (Fc layer) of 6.1 m thick. Below the Fc layer, a coastal loose and pervious sand of 9.1 m thick (As), a marine deposit of soft clay of 37 m thick (Ac) and a diluvial sand deposit (Ds) are underlain. Sand drains reach 19.8 m deep of Ac layer (Ac$_1$). Sand drains, which are 50 cm in diameter, 3.5 m square in spacing ($n = 7.91$) and 36 m in length, were installed and a preloading embankment of 3 m high was placed. The test site was divided into a dewatering area (D$_1$) and two non-dewatering areas (D$_2$, D$_3$), which are 150 m and 130 m away from the D$_1$ area, respectively. Twelve deep wells were set up to dewater the D$_1$ area.

The groundwater level lowering in the area surrounding the wells was uniform and equal to about 6 m after 3 days and about 8 m after ten days.

The effects of smear and well-resistance are considered in non-dewatering areas (D$_2$, D$_3$) and the effects of smear, well-resistance and dewatering in the dewatering area (D$_1$). The permeability of sand drains (k_w) is 3.8×10^{-3} cm/s. The permeability of the non-disturbed clay (k_{c_2}) is 8.7×10^{-7} cm/s and c_v ($= c_h$) is 600 cm^2/d in Fc layer. The value of k_{c2} is 3.9×10^{-7} cm/s and c_v ($= c_h$) is 320 cm^2/d in Ac$_1$ layer.

Disturbed zone ratio (s) are taken to be 3 and K' ($= k_{c2}/k_{c1}$) to be 4. The lowered waterhead (h_0) is 8 m at the D$_1$ area. ESW is 4.353 in the Fc layer and 4.752 in the Ac$_1$ layer of the D$_2$, D$_3$ areas where the smear and well-resistance are considered. ESW is 4.33 in the Fc layer and 4.582 in the Ac$_1$ layer of the D$_1$ area where the smear, well-resistance and dewatering are considered.

Figure 10a shows the measured and calculated settlements of the Fc layer at the D$_1$ area. In the solid curve the effects of smear, well-resistance and dewatering are considered, while in the dashed curve the well-resistance effect is only considered. It is noticed that the former agrees well with the measured settlement before the dewatering work, but the latter agrees well with that during the dewatering. It is difficult to understand about this behavior through only this data, but it is at least confirmed that the dewatering promotes the consolidation significantly. Figures 10b,c show settlement curves at the D$_2$ and D$_3$ areas, respectively. In the solid curve the effects of smear and well-resistance are considered, while in the dashed curve the well-resistance effect is only considered. It is seen in both figures that the former agrees well with the measured settlements.

Figure 11a shows the measured and calculated settlements, in which the effects of smear, well-resistance and dewatering are considered. Both are in good agreement. Figures 11b,c show settlement curves in the Ac$_1$ layer of the non-dewatering area (D$_2$, D$_3$). For the calculated curves, the disturbed zone ratio (s) are taken to be 1,2,3 and K' ($= k_{c2}/k_{c1}$) to be 1,2,4. The case of s = 3 and K' ($= k_{c2}/k_{c1}$) = 4 agrees well with the measured one.

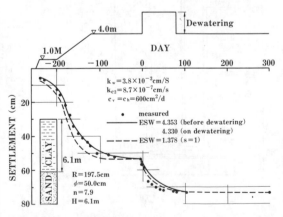

Fig 10a. Settlements and elapsed time relationship of Fc layer by combined dewatering method at D$_1$ area in place Urayasu

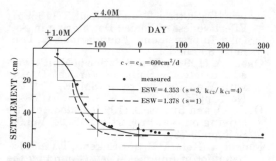

Fig 10b. Settlements and elapsed time relationship of Fc layer by sand drains at D_2 area in place Urayasu

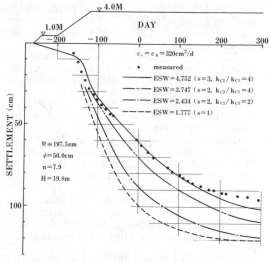

Fig 11b. Settlements and elapsed time relationship of A_{c1} layer by sand drains at D_2 area in place Urayasu

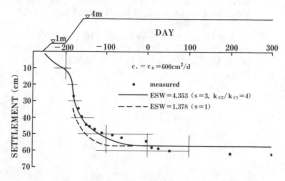

Fig 10c. Settlements and elapsed time relationship of Fc layer by sand drains at D_3 area in place Urayasu

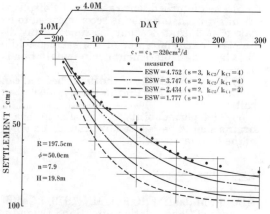

Fig 11c. Settlements and elapsed time relationship of A_{c1} layer by sand rains at D_3 area in place Urayasu

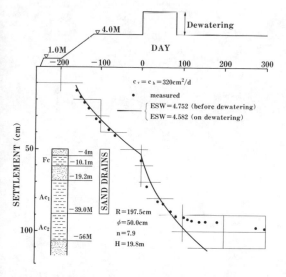

Fig 11a. Settlements and elapsed time relationship of A_{c1} layer by combined dewatering method at D_1 area in place Urayasu

In this case history, Eqs. (*14a,b,c*) are effective for representing the consolidation process with the effects of smear, well-resistance and dewatering.

CONCLUSIONS

The authors have developed a consolidation theory considering the effects of smear, well-resistance and dewatering. Field case histories and laboratory tests were performed to confirm the applicability of the derived consolidation equation. The main conclusions are summarized as follows:

1) It was confirmed that the proposed Eqs. (14a,b,c) are effective for representing the consolidation process with the effects of smear, well-resistance and dewatering.

2) Equations (14a,b,c) have good applicability to the cases from small diameter of vertical drain to larger ones.

3) Equations (14a,b,c) suggest that consolidation forces acting on the soft clay are both the preloading embankment and the lowered groundwater head, that the magnitude of the consolidation force by the dewatering is equivalent to the decreased water head in the drain wells, and that dewatering is effective to reduce the well-resistance, being followed by their confirmation through field and laboratory data.

4) The disturbance of clay around vertical drains in the non-displacement method like continuous auger driving is less than the displacement method like the vibratory driving, and consequently, the consolidation in the non-displacement method is faster than the displacement method.

5) The disturbed zone ratio (s) by displacement drain installation is recommended to be taken as 3, that is about 1.5 times casing diameter from the center of the casing. While that by non-displacement drain installation is 1.

6) The permeability of the disturbed soil can be taken to be 1/4 of that in the non-disturbed soil.

7) The permeability of the plastic board drains should be reduced by around one tenth of the estimated one.

REFERENCES

Aboshi, M.,Ishii, I.,Hanai, M. (1979) : Coefficient of Consolidation after Sand Drain Installation,14th Annual Conference of JSSMFE, pp301-304. (in Japanese)

Barron,R.A. (1948) : Consolidation of Fine-Grained Soils by Drain Wells, Trans. ASCE, Vol.113, pp718-742.

Hansbo, S. (1981) : Consolidation of Fine-Grained Soils by Prefabricated Drains, Proc. 10th ICSMFE, Stockholm, Vol.3, pp677-682.

Kjellman, H. (1948) : Consolidation of Fine-Grained Soils by Drain Wells, Trans. ASCE, Vol.113, Contribution to the Discussion.

Kotera, H.,Sakemi, T. and Matsui, T. (1989a) : A Precompression Method of Soft Ground Improvement Using Dewatering, Quaternary Engineering Geology, 25th Annual Conference,Edinburgh.

Kotera, H.,Sakemi, T. and Matsui, T. (1989b) : Effectiveness of Forced Dewatering in Soft Clay Improvement, Foundations & Tunnels - 89,London.

Onoue,A. (1988) : Consolidation of Multilayered Anisotropic Soils by Vertical Drains with Well Resistance, JSSMFE, Vol. 28, No. 3, pp75-90

Orrje, O. and Broms, B. (1967) : Effects of Pile Driving on Soil Properties, ASCE, Vol.93, SM5,pp59-73.

Matsui, T.,Kotera, H. and Sakemi, T. (1990) : Behavior of Improved Clay Ground by the Combined Dewatering Method,Proc.10th Southeast Asian Geotechnical Conference, Vol.1,pp93-100 .

Simons, N.E. (1965) : Consolidation Investigation on Undisturbed Fornebu Clay, Norwegian Geotechnical Institute, Oslo Publication No.62.

Yoshikuni, H. and Nakanodo, H. (1974) : Consolidation of Soils by Vertical Drain Wells with Finite Permeability, JSSMFE, Vol.14, No.2.

Developments in Geotechnical Aspects of Embankments, Excavations and Buried Structures, Balasubramaniam et al. (eds)
© 1991 Balkema, Rotterdam. ISBN 90 5410 019 2

Some examples of field tests for soil improvement methods in Japan

M. Aoyama, M. Nakamura, M. Kuwabara & M. Nozu
Fudo Construction Co., Ltd, Tokyo, Japan

ABSTRACT: The soil improvement methods commonly applied in Japan are sand drain method, sand compaction pile method, and deep mixing method. The summary of these methods, and significance and the planning method of field tests for soil improvement methods are described. Then some examples of field tests in Japan using these methods are reported. Field tests are carried out prior to the main construction for the purpose of clearing the uncertainty and making the optimum decision on the design procedure and construction for the main construction.

1. INTRODUCTION

In geotechnical engineering, it is often said that "Nothing can be determined without measured data". Geotechnical engineering can be said as the study which has made progress by concentrating on how to clear the uncertainty in materials, soil tests, and design procedures concerned.

In Japan, the soil improvement technique has made a remarkable progress because of a great number of soft ground area and its complicated soil profile. Above all, in the large scale projects recent years in Japan at the Tokyo bay and Osaka bay areas, the application of the soil improvement method is not avoidable. Owing to the progress of advanced measurement technique and data process technique, a practical method to effectively utilize prior information on the construction of the structure on a soft ground has also been developed. Now it is general that the main construction is commenced after an optimum decision making has been done based on the data of field test in order to clear uncertain factors, such as original of the soil and ground properties, their required functions, and application of soil improvement design and construction.

However, there still are several difficulties in the design procedure, the evaluation method of the soil improvement effect and the execution management system at the main construction. In this paper, the significance and the outline of the planning of field tests on some typical soil improvement methods in Japan are introduced. Furthermore, some examples of the prototype tests on a soft clayey ground are reported.

2. TYPICAL SOIL IMPROVEMENT METHODS IN JAPAN

The soil improvement methods which are most commonly used in Japan are the sand drain method, sand compaction pile method, and deep mixing method. Characteristics, history, design procedures and current situations of these construction methods are explained in brief below.

2.1 Sand drain method

In the sand drain method, sand piles which are formed in the soft clayey soil accelerate consolidation in the ground by its drain effect and the surcharge, and increase the strength of the ground. Since the sand drain method was introduced to Japan in 1952, it increased the number of execution with the development of coastal industrial zones which started in the late 1950's (Ishi et al.1953). Although there still are some subjects left to be discussed about the effectiveness of this method, it has been the principal soil improvement method in Japan because its application method and management method have been clarified to some extent(Aboshi 1983). Today, the applications have been tried to the deep soft ground which exceeds 30 meters in thickness or to high water content ground, of which natural moisture content exceeds the liquid limit(Matsuo et al.1984).

2.2 Sand compaction pile method

In the sand compaction pile method (called SCP method hereafter), well compacted sand piles with relatively large diameter are formed in the soft ground. This method can be applied to both sandy grounds and clayey grounds. For the loose sandy ground, the SCP method can increase bearing capacity, prevent settlement and soil liquefaction, and increase horizontal subgrade reaction. On the other hand, for the clayey ground, it can increase the shear strength, bearing capacity, stiffness of ground and also can reduce the settlement by forming a composite ground consisting of sand piles and clayey ground.

This method was started in Japan in 1956. Initially, it was only applied to improve soft grounds on land. After many developments for the better method it came to be applied to the offshore since 1966(Ichimoto et al.1980). Through more theoretical and experimental studies, it has become widely applied to the large construction projects for road and harbor construction.

In sandy grounds, the SCP method increases the density and the N-value of sandy ground by vibration effect and filling effect of pile materials. The SCP method for sandy ground is designed aiming at filling effect of materials into sandy ground. As the fineness of sandy ground are not originally taken into consideration, a new design method which takes the content of fine particles into consideration has been proposed recently by Mizuno et al.(1987).

In the design procedure of SCP method for clayey ground, improved ground is generally considerd as a composite ground mentioned above. The design procedures of the shear strength and settlement of the composite ground are shown in Table 1(Sogabe 1981).

Table 1   Shear Strength and Settlement in Composite Ground

Items	Shear strength	Settlement
in composite ground	$\tau = (1-a_s) \cdot c + a_s \cdot (\mu_s \cdot \sigma_s + \gamma_s \cdot z) \cdot \tan \phi_s \cdot \cos^2 \theta$ $c = c_o + \Delta c$ $= c_o + \mu_c \cdot \sigma_c \cdot U \cdot c/p$	$S = \beta \cdot S_o$ $= 1/(1 + (m-1)) \cdot S_o$
in high replacement ratio		for $a_s \geqq 60\%$, $S = \beta \cdot S_o$ $= (1-a_s) \cdot S_o$

Where

τ ; shearing stress of composite ground

a_s ; $a_s = A_s /(A_s + A_c)$, called "replacement ratio" as shown below

c_o ; cohesion of original ground

c ; cohesion of clay

μ_s ; stress concentration coefficient of sand ($= m/(1+(m-1))$)

μ_c ; stress reduction coefficient of clay ($= 1/(1+(m-1))$)

σ_s ; vertical stress on the sand piles as shown below

σ_c ; vertical stress on the clayey ground

σ ; vertical loading intensity

γ_s ; unit weight of sand pile

ϕ_s ; internal friction angle of sand pile

m ; stress distribution ratio

U ; degree of consolidation

c/p ; rate of strength of increase

β ; settlement reduction coefficient

S_o ; settlement of original ground

S ; settlement of composite ground

θ ; angle as shown below

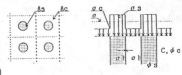

Concept of composite ground

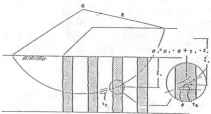

Stability calculation of   soft ground

Recently, due to development of the larger sized construction machinery, the improvement can be conducted in much deeper ground than ever. On the land construction, it is applicable as deep as 35 m, while in offshore application as deep as 65 m. On the other hand, the smaller construction machinery has been developed to reduce the influence of vibration. Moreover, in order to accomplish an adequate improvement against the variation of the soil properties, an automatically controlled SCP method has been developed, and even no-human-operator method (robotization) are now developing (Kawakami et al. 1988).

2.3 Deep mixing method

In deep mixing method (called DM method hereafter), the ground is improved by making use of the chemically bonded action with mandatory mechanically in-situ stirring and mixing soft clayey soil with the stabilizer, such as cement milk or mortor.

The study for development of this method was started in around 1965, and it was put into practical use in about 1975(Terashi 1982). In this method, although the increase of the shear strength of the clayey soil at inter-pile can not be expected, an unconfined compressive strength of the improved ground increases to several kgf/cm² in a short period, and decreases compressibility. This method has an advantage that does not affect the surrounding environment as it does not use vibration.

With regard to the design procedure, the improved ground by DM method is considered to be composite ground in case of pile shaped improvement in which unconfined compressive strength of improved soil is designed to be several kgf/cm²(shown in Fig.1). By the way, when unconfined compressive strength of improved soil is designed to be several tens kgf/cm², such as in a large scale harbor construction requiring a heavy weight structure to be permanently supported, the design procedure is made assuming the improved ground by the DM method as a rigid body. This is because the rigidity between improved soil and original ground are significantly different. The block shaped improvement is also commonly used(in shown in Fig.1). The design procedure of block shaped improvement is shown in Fig.2 (Terashi 1982). Since a special blended cement has been developed in recent years, the

application of DM method has been tried to apply to problem soils such as organic soils to which original DM methods were difficult to apply with the specificaton of the normal portland cement.

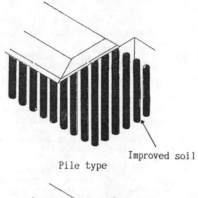

Pile type

Improved soil

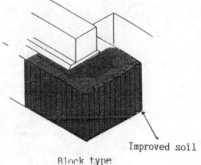

Block type

Improved soil

Fig.1 Soil Improvement Type by DM method

3. SIGNIFICANCE AND PLANNING OF FIELD TEST ACCOMPANYING SOIL IMPROVEMENT

3.1. Significance of field test

Field test means here prototype test conducted using actual ground and structure, in case that there is a peculiarity in the soil, the construction scale is bigger than ever before, and there is no certain measure which can estimate its behavior during and after construction. It is also conducted to study the confirmation and improvement of design procedure.

If the soil improvement has been made, the data, given by the soil test, of soil strength change and settlement speed before and after the soil improvement, are often different from those given by the calculated or predicted data on the design. That is because soil properties vary by location and errors may occur in testing and analysis. Therefore, field tests are necessary before actual main construction. As a result of the field tests, the behavior of stability, deformation and pore water pressure are given. Through the estimation of the degree of consolidation and the amount of residual settlement, and through the reconsideration of various factors of the soil, more practical design of the construction is planned, which may lead to actual main construction.

Although there are some tests which do not rely on the field test, such as a centrifuge loading test and a simulation methods using computer, but these tests have not come to practical use yet. Therefore, the field test has a very important role in removing uncertainty, and in making an appropriate plan for practical design and construction. Further, it is also important to clarify the purpose and technical problems of the field tests before their construction. Table 2 shows the technical points of field test on the soft clayey ground in the loading test when soil improvement is also carried out.

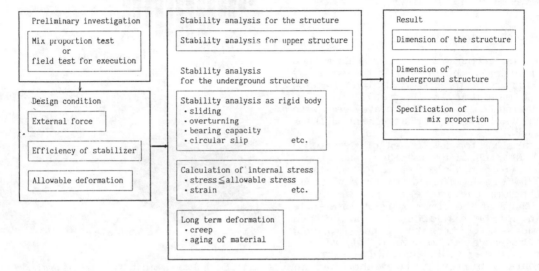

Fig.2 Design Procedure for Soil Improvement as Foundation

Table 2 Technical Points of Field Tests on Soft Clayey Ground

Items	Contents	Sand drain method	SCP-method	DM-method
Investigation	Soil properties of improved soil	Soil properties of improved soil (C_v, M_v, q_u)	Soil properties of inter-pile (C_v, M_v), Pile strength, Replacement ratio	Soil properties of improved soil, Pile strength, Replacement ratio
Design	Sliding failure, Settlement, Increase of soil strength, Deformation, — Estimation of soil improvement and establishment of design procedure	Progress of settlement, Increase of soil strength	Pile strength, Replacement ratio, Stress distribution ratio, Increase of soil strength, Progress of settlement, Soil property of upheaval soil	External force acting on improved soil, Strength of improved soil, Replacement ratio
Execution	Execution management, Stability control, Settlement prediction, Soil investigation, Environmental condition	Permeability of pile material, Continuity of pile	Permeability of pile material, Continuity of pile	Mixing proportion Aging Volume of stabilizer (cement milk or cement mortor)

3.2 Planning of field tests

(1) Decision of field test scale

In consideration of the following main construction, field tests should be executed in an appropriate scale and place. About 10% of the construction cost has been invested in each field test so far, and it seemed standard that 1–3% of construction cost is paid for measurement expense as shown in Fig.3(Shibata ed. 1979).

The scale of the test body is planned to be large enough to have the same stress distribution as may occur in the ground in the main construction. The place which has an average properties of the ground or the worst ground condition in the construction area is generally selected as the test site.

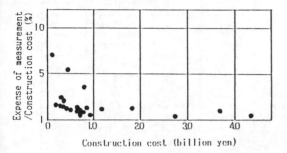

Fig.3 Construction Cost and Expense of Measurement in Observational Method (after Shibata ed. 1979)

(2) Measurement planning

In the loading test of clayey ground, items in Table 3 are to be measured with regard to the technical points. The measuring instruments and the frequency of measurement are shown in Table 4.

The measuring instruments should satisfy the following requisites;

· They should be available in the range up to the maximum deformation that is expected under loading.

· Objects to be measured are mostly transitory, so certainty is surely required.

· They should be operative for long term measurement and monitoring.

Table 3 Technical Points and Measuring Instruments

Problems	Item of field measurement	Management method and prediction method	Measuring item	Measuring instruments and soil investigation
Sliding failure	Variation of soil properties, Error of estimation	Safety factor of structure, Load, Deformation, Crack and upheaval	Circular arc method / Stability control	Measurement of formation / Field observation / Measurement of displacement	Leveling / Displacement peg, / Extensometer, Inclinometer
Settlement	Variation of soil properties, Error of estimation	Final settlement, Degree of consolidation, Residual settlement	Settlement analysis, Prediction of settlement, Dissipation of pore water pressure, Changing of moisture water content etc.	Measurement of settlement / Measurement of stress	Settlement plate, Differential settlement gauge / Piezometer, Earth pressure cell
Increase of soil strength	Variation of soil properties, Error of estimation	Soil strength, Surcharge, Degree of consolidation	Soil investigation, Embankment height, Dissipation of pore water pressure,	Measurement of stress / Soil investigation, Sounding	
Deformation	Variation of soil properties, Error of estimation	Deformation of surrounding area	Control using measured data	Test for physical properties, Shear test, Consolidation test	Boring, Soil sampling Sounding, / Soil test
Environmental assessment (vibration, noise)	Variation of soil properties, Error of estimation	Vibration and noise at aiming point	Control using measured data	Measurement of vibration and noise	Vibration and noise level meter

Table 4 Measuring Instruments and Frequency of Measurement

Measuring Instruments and soil investigation	Frequency of measurement (in case of filling)								Remarks
	during construction					Term of maintenance	Term of measurement		
	during soil improvement	during filling	after filling						
			to 1 month	from 1 to 3 months	after 3 months				
Leveling									
Displacement peg,	1 time/day	1 time/day (2 times/ day)	1 time/day (1 time/ 3 days)				for 1month after filling		();if it is nec-essary
Extensometer,	1 time/day	1 time/day (2 times/ day)	1 time/day (1 time/ 3 days)				for 1month after filling		();if it is nec-essary
Inclinometer	before/after soil improvement	1 time/day (1 time/ week)	1 time/ week				for 1month after filling		
Settlement plate, Differential settlement gauge	1 time/day (2 times/ day)	1 time/day (2 times/ day)	1 time/day (1 time/ 3 days)	1 time/ week	1 time/ month	2 ~4 times/ year or 1 time/year	till removal of surcharge		();if it is nec-essary
Piezometer	1 time/day	1 time/day	1 time/day (1 time/ 3 days)	1 time/ week	1 time/ month	1 time/ 3 months	till removal of surcharge		();if it is nec-essary
Earth pressure cell	1 time/day	2 times/ day	1 time/ week				till completion		
Boring, Soil sampling Sounding, Soil test	before/after soil improvement	right after filling	right after completion of filling						
Vibration and noise level meter									

Table 5 Items in Installation of Measuring Instruments

Items	Contents
1.Arrangement of measuring object	• Technical problem • Quantity of measuring instruments
2.Condition of Execution	• Order of construction • Natural condition(ex.tide and snowfall) • Reasonable cable route
3.Management	• Area to be measured • Arrangement of measuring instruments
4.Maintenance	• Alternative cable route • Equipment for long term measurement

As for the installation point of the measuring instruments, items to be discussed are shown in Table 5. An example of installation of the measuring instruments in an embankment is shown in Fig.4.

3.3 Utilization of the field test results

The following show the items to be paid attention to when the results of the field tests are to be utilized in the main construction:

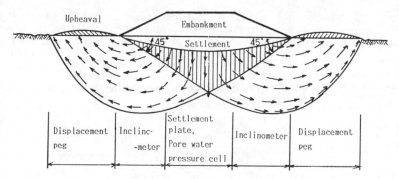

Fig.4 Example of Instllation in an Embankment

• In case when the data given by field test are different from those given based on the prediction, this cause should be searched for and the data given based on prediction should be reviewed from the viewpoint of geotechnical engineering.

• Some prediction method of failure and stability control methods for embankments on the soft clayey grounds and some prediction methods of final consolidation settlement have been proposed. Each prediction method and management method has its reason and background, therefore it is necessary to understand them correctly and set up these methods in compliance with the purpose of field tests at the construction (Kurihara 1986, Kato 1986).

4.EXAMPLES OF PROTOTYPE FIELD TEST ACCOMPANYING SOIL IMPROVEMENT

4.1. Type of field tests

The significance of field test is to aim at a practical designing and construction under an appropriate planning at the field. From these points, field tests are classified into three groups shown in Table 6. In the case when the establishment of design procedure is main object, indoor tests (centrifuge loading test, etc.) or small scale tests are

carried out, whereas field tests are carried out for the confirmation and improvement of the design procedure. In this chapter, at first, two examples of field tests to confirm the design procedure are introduced by quoting documents published in Japan. Secondly, an example where uncertainty was coped with is introduced. In this example, the design procedure was revised based on the information given by the field test in order to get an optimum condition for the main construction.

4.2. Loading test on improved ground by low replacement ratio SCP method

(1) Purpose of the field test

Construction cost, term, and deformation of SCP method whose replacement ratio is as low as 20-40% are shown in Fig.5 for comparison with other methods. When the construction term is relatively long and a certain degree of deformation of the structure can be allowed, SCP method with low replacement ratio is economical. However, the method whose replacement ratio is around 70% has been commonly used. Because, theoretical and experimental research on low replacement ratio SCP method have not been made for

Table 6 Classification of Field Test

Purpose of Field Test	Experiment method
To establish the design procedure	Centrifuge loading test or computer simulation
To confirm the design procedure	Prototype loading test
To upgrade the accuracy of the design procedure	

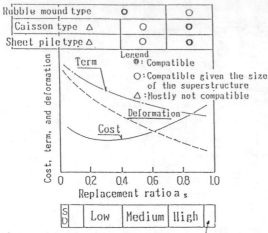

Rubble mound type	O	O
Caisson type △	O	●
Sheet pile type △	O	●

Low : Low replacement ratio
Medium : Medium replacement ratio
High : High replacement ratio

Legend
●: Compatible
O: Compatible given the size of the superstructure
△: Mostly not compatible

.Conventional replacement

Fig.5 Standing of the Low Replacement Ratio SCP Method (after Takahashi et al. 1988)

the offshore application. Therefore, field tests were executed to confirm the design procedure of low replacement ratio SCP method(Takahashi et al. 1988, Okada et al. 1989).

The improved ground by the low replacement ratio SCP method is regarded to be composite ground consisting of sand piles and clayey soil. For the evaluation of improvement effect, it is important to clarify the properties of this composite ground. In this field test, the following two technical matters were to be studied;
· To confirm consolidation characteristics of the composite ground.
· To confirm shearing characteristics of the composite ground at the ultimate condition.

(2) Outline of the field test

a. Soil profile

The soil profile at the test site consists of a bearing stratum of mud stone, sand and gravel layer of 0-2m in thickness, and an alluvial clayey layer of about 10m in thickness above the gravel layer. The alluvial clayey layer is the aim of improvement on this field test. The soil properties at test site is shown in Fig.6.

b. Test method

The test body is shown in Fig.7. After leveling of sand blanket, SCP at the replacement ratio of 25% (sand pile diameter; 1.7m, pile interval; 3.0m) were formed in the ground. Next to this SCP area another SCP was formed at the ratio of 70% (sand pile diameter; 2.0m, pile interval; 2.1m) in order to fix the direction of sliding failure.

The ground was loaded in two steps afterwards. For the first step loading, it was loaded 3tf/m^2 in order to get consolidating characteristics of the

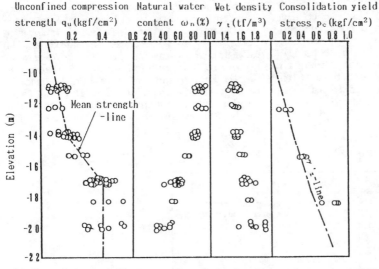

Fig.6 Original Soil Properties (after Takahashi et al. 1988)

432

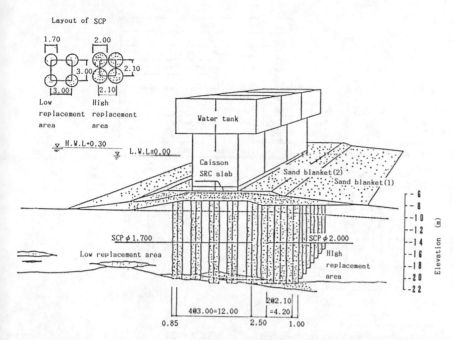

Fig.7 Illustration of the Loading Body(after Takahashi et al. 1988)

improved ground by SCP method. The test
ground was left as it was for 10 months,
and during this period, the behavior of
the ground was observed and monitored.

The second step loading was carried out
to get the shearing characteristics at the
ultimate condition of the improved ground
by SCP. The second step loading being
added above the first one until the
improved ground was failed, and the
behavior during this step was also
observed and monitored.

c.Field observation and soil investigation

In order to get the consolidating
characteristics and the shearing
characteristics at the ultimate condition,
the soil investigation was carried out at
the following steps; at the original
ground, right after the SCP driving, at
the first step loading, during
consolidation, just before and after the
second step loading. These soil
investigations were carried out inside
and outside the range of improvement.

The measuring sites were selected aiming
at the expected large scale deformation
in order to monitor the behavior of the
improved ground from the start of
consolidation till failure. The measuring
points and instruments are shown in
Table 7 and Fig.8, respectively.

Table 7 Measuring Instruments and Items
(after Takahashi et al. 1988)

Name of Instrument	Items to be measured	Object to be measured
Inclinometer	Underground displacement	Shear surface
Inclinometer for caisson	Inclination of loading body	Stability of loading body
Differential settlement gauge	Settlement	Settlement characteristics
Load cell	Pressure of loading body	Pressure of loading body
Earth pressure cell	Vertical earth pressure	Stress distribution ratio
Piezometer	Exess pore water pressure	Dissipation process
Water level recorder	Ground surface water pressure, Water level in caisson, Water level in water tank	Vertical displacement, Loading weight, Loading weight
Tidal level recorder	Tidal level	Tidal level adjustment
Displacement peg	Surface displacement	Sliding range

433

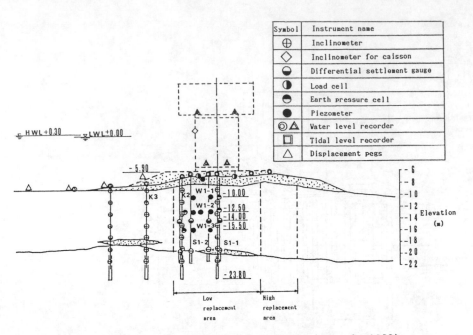

Symbol	Instrument name
⊕	Inclinometer
◇	Inclinometer for caisson
◒	Differential settlement gauge
◑	Load cell
◓	Earth pressure cell
●	Piezometer
◎△	Water level recorder
▢	Tidal level recorder
△	Displacement pegs

Fig.8 Arrangement of Instruments (after Takahashi et al. 1988)

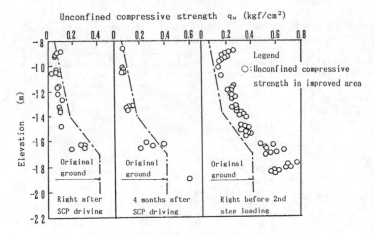

Fig.9 Changes in Clay Strength (after Takahashi et al. 1988)

(3) The result of the field test

a. The time-soil strength relationship

Fig.9 shows the comparison of the unconfined compressive strength measured at each loading step and after loading on the basis of unconfined compressive strength of the original ground. The following can be found from this figure.
• Immediately after the installation of SCP, the unconfined compressive strength of the improved soil showed a 30% decrease due to the disturbance caused by piling.
• 4 months after the installation of SCP, the unconfined compressive strength recovered to 90% of the original strength.
• Just before the step loading, the unconfined compressive strength increased by 80% of the original strength.

b. Failure mode of the improved ground

• Fig.10 shows the loading-settlement

434

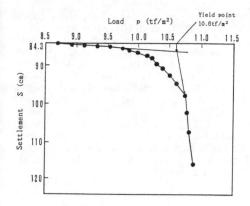

Fig.10 Relation Load and Settlement
(after Okada et al. 1989)

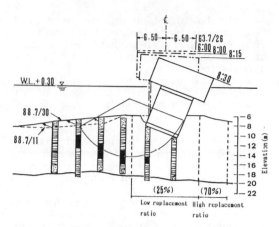

Fig.12 Sliding Surface due to Failure
(after Okada et al. 1989)

relationship at every five minutes after
the calibration of loading. The yield
load was estimated approximately 10.6 tf/
m^2, using this loading-settlement curve.
· Fig.11 shows the movement of
displacement pegs placed at the sea bed.
The maximum displacement was 2.5m. Since
the length of the region where the
displacement was almost same as that of
the loading body, the cylindrical failure
is estimated to have occurred.

circular arc calculation using
shear strength shown in Table 1.
· Fig.13 shows the change of the
underground horizontal displacement with
depth during the second step loading. As
this figure is in accordance with the Fig.
12, the sliding failure is estimated to
have occurred at this site.

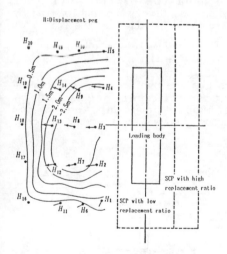

Fig.11 Shape of the Displacement due to
Failure (after Okada et al. 1989)

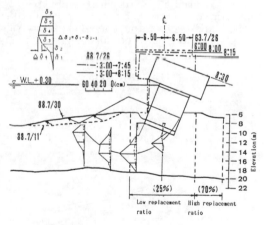

Fig.13 Deformation of Tested Area due to
Failure (after Okada et al. 1989)

· Fig.12 shows the area where the
measured strength was significantly low
immediately after the second step loading,
based on the result of the soil
investigation. The area forms a critical
circle in shape, which is obtained by

(4) Summary of the field test

The test results are summarized as follows.
· The strength of the original ground
decreased by 30% after the SCP
installation and recovered to the
original level after five months.
Moreover, by adding consolidation loading,
strength increment can be obtained
quantitatively.

· The failure mode of improved soil under the ultimate loading is circular in shape which is in accordance with a critical circle obtained by circular arc calculation.

From the above test results, the efficiency of the design procedure of SCP method has been confirmed.

4.3 Loading tests on the improved ground by deep mixing method

(1) Purpose of the field test

At offshore construction by DM method, the entire volume of soft clayey ground is improved in most cases. However, by reducing the volume of the ground to be improved, the construction cost would be reduced and the construction period of the large-scale harbor facilities would be shortened. The field test was carried out in order to observe the behavior of the soil under loading and the failure mode and also to confirm the applicability of design procedure.

The following are the technical problems on conducting the test.

· To get the behavior of improved soil by cost-saving DM method.

· To confirm the structural stability of the improved soil by cost-saving DM method and to examine the efficiency of the present design procedure.

(2) Outline of the field test

a. Soil profile

The original ground is composed of an alluvial clayey layer (8.0m to 18.0m in depth from the ground level), a sand layer (18.0m to 20.0m), a diluvial clayey layer (20.0 to 24.0m) and a sand-and-gravel layer (deeper than 24m). In the alluvial clayey layer the moisture content is over 80% and the cohesion is $C=0.15Z$ (tf/m²), taking $Z=0$ at 8m under the ground level. In the sand layer N-value is $N \fallingdotseq 8$, and in the diluvial clayey layer the cohesion is $C=2.0+Z$(tf/m²), taking $Z=0$ at 20m in depth under the ground level. In the sand-and-gravel layer N-value is $N \geq 20$. The sand-and-gravel layer is considered to be the bearing stratum.

b. Test procedure

The test body is illustrated in Fig.14. The length of the test body in the normal direction is 30m and the width of the surcharge in the direction perpendicular to the normal line is 50m. After each step of loading, the test body was left to stand for two to three days to conduct measuring.

c.Field observation and soil investigation

Earth pressure and water pressure acting on loading caissons and DM improved soil, and the behavior of the soil were measured consecutively during loading by the instruments installed in the soil.

(3) The result of the field test

Fig.15 shows the horizontal displacement and the settlement presented by Hirao(1985). The front toe of the improved soil showed settlement, whereas

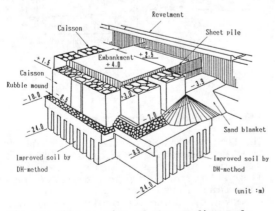

Fig.14 Illustration of the Loading Body (after Hirao 1989)

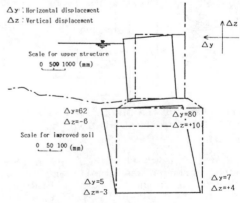

Fig.15 Deformation of Loading Body (after Hirao 1989)

the heel of the improved soil was lifted. The upper part of DM improved soil was slanted forward by 8 to 6cm.

(4) Summary of the field test

From the result of various measurements and the analysis, it has been confirmed that the improved soil by cost-saving DM method behaved as a rigid body until the final loading. The stability analysis method in the present procedure as shown in Fig.2 has been proved to be effective. These results have been applied to further studies of design procedures.

4.4 Loading test of the improved ground by sand drain method

(1) Purpose of the field test

Here, the case of a LNG tank construction on the thick soft reclaimed ground as an example is introduced(Matsuo et al:1984) It was easy to make a prediction of about of 3m settlement, if left as it is, because the construction site was still fresh after reclamation, being in the unconsolidated condition.

Therefore, it was assumed to be difficult to form sand pile in fresh reclaimed clayey ground. Then the special sand drain (called CF drain) method of which the upper part (corresponding to fresh reclaimed layer) packed drain method and lower part was sand drain method, and preloading by embankment was applied in order to complete consolidation settlement prior to the tank base construction. The CF drain method is an improved form of sand drain method.

Since the ground is a soft clayey layer of over 33m in thickness which is undergoing consolidation process, a large scale field test must be carried out to get necessary information for the design and execution planning of the main construction.

The basic conditions concerning the design of this project were as follows:
· The range of ground improvement is limited within the area where main structures will locate as shown in Fig.16.
· No consolidation settlement should occur after the removal of the surcharge.
· The construction term of the surcharge is 3 months excluding the term of ground improvement and the surcharge is left for 4 to 5 months.
· Safety factor of the circular slip surface is $F_s \geq 1.1$ to carry out stability management of the embankment by

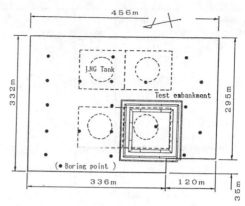

Fig.16 Plot Plan of Test Embankment and Boring Point (after Matsuo et al. 1984)

field observation.

Then the specification of the ground improvement was decided as follows:
· CF drain method ; 40cm in diameter, 40m in length (upper 16m packed), 1.8m square arrangement
· Height of the embankment ; 6m
The main technical requirements to satisfy these conditions were as follows:
· To establish a superior design which makes it capable to construct a 6m high embankment on the very soft clayey layer safely and speedily, and to establish a stability control method by field observation during the embankment construction.
· To clarify the limit of application of the consolidation theory to large-scale settlement of over 5m and decide the soil coefficients in the design calculation for the main construction.
· To establish a superior method for estimating settlement and for judging the time of the surcharge removal.
· To solve various problems on making an execution plan for a large-scale earthwork, especially to confirm the effect and efficiency of the new method, upper part packed sand drain method.

(2) Outline of the field test

a. Soil profile

Soil investigation was carried out at the locations shown in Fig.16 and the result obtained is shown in Fig.17. The soil profile is composed of a thin layer of sand at the surface, a reclaimed clayey layer, an alluvial layer of 25m in thickness and alluvial sand and gravel layer. The diluvial sand and gravel layer

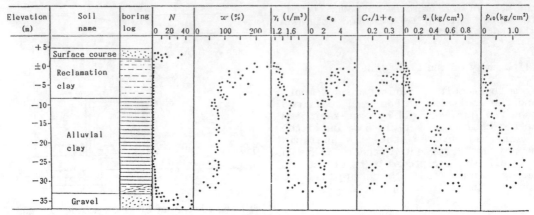

| Elevation (m) | Soil name | boring log | N 0 20 40 | w (%) 0 100 200 | γ_t (t/m³) 1.2 1.6 | e_0 0 2 4 | $C_c/1+e_0$ 0.2 0.3 | q_u (kg/cm²) 0 0.2 0.4 0.6 0.8 | p_{c0} (kg/cm²) 0 1.0 |

N:Standard penetration test p_{c0}:consolidation yield stress ω:Natural water content γ_t; Moisture density

e_0:Void ratio C_c:Compression index q_u:Unconfined compressive strength

Fig.17 Original Soil Properties (after Matsuo et al. 1984)

is considered to be the bearing stratum of the soil. The natural water content of the reclaimed clayey layer is as high as the liquid limit, W_L =100%, so that soil sample of reclaimed clayey layer narrowly stands itself. The depth from the surface to the bearing stratum is 37m, of which the thickness of the soft clayey layer is about 33m.

b. Test procedure

The test embankment was located at the place shown in Fig.16, where the soil has a typical soil profile in the LNG tank construction site. The scale of the test embankment, which has the area to measure settlement, pore water pressure and the soil strength, were set to have the same stress distribution at the bottom of the consolidated layer as may occur in the soil during the main construction. The settlement was measured during and after the construction of the test embankment.

c.Field observation and soil investigation

The field observation was carried out to solve the aforementioned technical problems. The technical problems and the measuring instruments, method and frequency are shown in Fig.18(JSMFE 1990). The arrangement of the measuring instruments is shown in Fig.19.

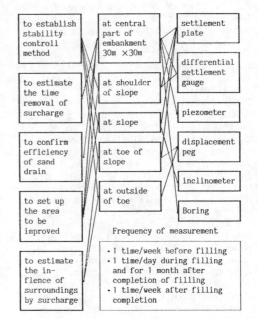

Fig.18 Plan for Field Observation

(3) The result of the field test

a. Stability control during the embankment

The settlement and horizontal displacement were measured to observe the stability during the embankment construction(Tsukada et al.1984). The result is shown in Fig.20. Fig.21 shows an example of the stability control process using the diagram for construction control of embankment proposed by Matsuo and

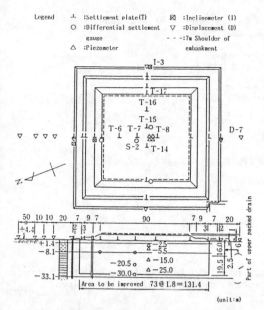

Legend
⊥ :Settlement plate(T)   ⊠ :Inclinometer (I)
○ :Differential settlement ∇ :Displacement (D)
  gauge   - - - :7m Shoulder of
△ :Piezometer   embankment

I-3

Area to be improved 73@1.8=131.4

(unit:m)

Fig.19 Arrangement of Measuring Instrument
(after Matsuo et al. 1984)

Kawamura(1977), taking horizontal
displacement(δ) and settlement(ρ) as
parameters. P/P$_f$ is more than 1.0, the
embankment goes to failure. When P/P$_f$ is
smaller than 1.0, the embankment is
judged to be stable. Here, P is
embankment loading and P$_f$ is the
embankment loading at the time of failure.
  The obtained results are as follows.
  · The construction of a 6 to 7m
embankment can be completed within 3
months.
  · The stratum which controls the
stability of soil is the reclaimed
clayey layer.
  · Settlement plates, differential
settlement gauges and inclinometers are
useful in stability management.
Displacement pegs are also useful as
supplement.
  The values shown in Table 8 are
specified for stability control.

b. Analysis and estimation of settlement

The following informations concerning
settlement are obtained.
  · Among many methods for predicting
settlement, the hyperbolic method is the
best method. Using the data for 1.5
months after the construction of the
embankment, the highly precise estimation
can be obtained as shown in Fig.22.
  · When the amount of settlement at any

time is estimated statistically based on
the result of consolidation test, an
accurate estimation is difficult to
obtain because of the wide variation of
estimated settlement and decrease of
settlement speed, although the measured
final settlement is in the distribution of
estimated final settlement as shown in
Fig.23.
  Therefore, it is concluded to be the
most appropriate construction method that
the main construction is designed using
the coefficient of consolidation(c$_v$)
obtained by inverse analysis of the result

Legend   ○ :Displacement peg (D-7) ○ :Settlement plate(T-15)
          ● :Inclinometer   (I-3) △ :Settlement plate(T-17)

          $\Delta \delta / \Delta t$ (cm/day);Horizontal displacement speed
          $\Delta \rho / \Delta t$ (cm/day);Settlement speed

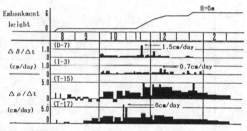

Fig.20 Example of Measured Data
(after Matsuo et al. 1984)

Legend
○ ;Using total settlement as ρ P$_f$; Embankment load
● ;Using settlement of reclaimed   at failure
            layer as ρ P ; Embankment load

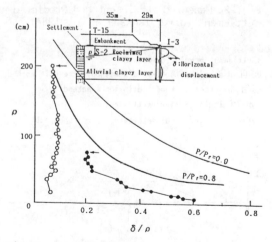

Fig.21 Example diagram for construction
control of embankment
(after Matsuo et al. 1984)

Table 8 Standard for Stability Control at Site (after Matsuo et al. 1984)

Stability control method		Standard for stability control
Fundamental stability control method	Matsuo,Kawamura's $\rho - \delta / \rho$ method	• P/P$_f$ ≦0.9 in case of using total settlement • P/P$_f$ ≦0.85 in case of using settlement of reclaimed layer
Supplementary stability control method	$\triangle \rho / \triangle t$ method	• $\triangle \rho / \triangle t$ ≦5.0cm/day
	$\triangle \delta / \triangle t$ method	• $\triangle D / \triangle t$ ≦1.0cm/day • $\triangle d / \triangle t$ ≦1.0cm/day

where, ρ ;total settlement
  δ ;maximum horizontal displacement measured by inclinometer
  $\triangle \rho$;increment of settlement per day
  $\triangle D$;increment of maximum horizontal displacement
    measured by inclinometer
  $\triangle d$;increment of horizontal displacement measured by displacement peg
  $\triangle t$;day

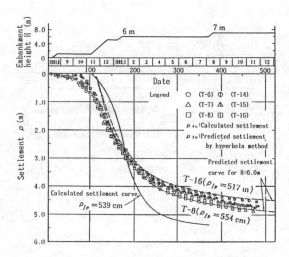

Fig.22 Measured Settlement and Predicted Settlement (after Tsukada et al. 1984)

<Settlement prediction method>
▥ ;Asaoka's method,Honda's method,curve rule method
▤ ;Hyperbola,√t method(Taylor's method)
ρ_{fc}:Calculated final settlement
ρ_{fr}:Predicted final settlement by measured data

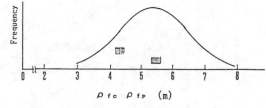

Fig.23 Distribution of Calculated Final Settlement and Predicted Final Settlement (after Matsuo et al. 1984)

of the field test and the construction is proceeded under field observation.

(4) Summary of the field test

The following are the efficiency observed during the main construction which was executed taking the result of the field test into consideration(Kanatani et al.1986).
· The stability control method planned beforehand made it possible to build up the embankment within the scheduled period.
· The required degree of consolidation was obtained by setting the time of the surcharge removal, based on the settlement estimation method planned beforehand.
· Almost no-settlement was observed for the period of 18 months after the surcharge removal so that it is considered that the ground was improved into the required condition.

5. CONCLUSION

The abstract of the soil improvement methods commonly applied in Japan, and significance and the planning method of field tests for soil improvement methods are described. Then some examples of field tests in Japan by newly remodeled sand drain method, SCP method and DM method are also reported in detail. As mentioned in each field test example, field tests are carried out prior to the main construction for the purpose of clearing the uncertainty and making the optimum decision on the design procedure and construction for the main construction. In the future, field test will be more important.

The summary of this paper is as follows:
(1) The soil improvement methods which are commonly used in Japan are sand drain method, SCP method, and DM method.
(2) The significance of field tests is to make the optimum plan for more reasonable design and construction.
(3) Field tests are classified into three categories in terms of the purposes; to establish the design procedure, to confirm the design procedure, and to upgrade the accuracy of the design procedure.
(4) The examples of prototype field tests using SCP method (low replacement ratio), DM method and sand drain method are introduced. The first two field tests aimed at confirming the design procedure and the last one aimed at upgrading the accuracy of the design procedure.

REFERENCES

Aboshi,H (1983)," Technical Problem of Deep Soil Improvement, The sand drain is effective method or not?",Tsuchi to Kiso, Vol.31,No.2,pp.69~74 (in Japanese).

Hirao,H (1985), " Field Test of Soil Improvement Method for Construction of Kansai International Air Port",Technical Report on Construction Technique (in Japanese).

Ichimoto,E and Suematsu,N (1980), " Frontiers of Execution Engineeriq Development,II . Engineeriq Development Examples in Japan",4. COMPOZER Method, JSCE, Vol.65,No.8, pp. 26 ~30 (in Japanese).

Ishi,Y and Yanai,Y (1953)," Sand Drain Method", Tsuchi to Kiso,Vol.1,No.3,pp.16 ~22,(in Japanese).

JSSMFE (1990)," How to Plan for Field Observation",pp.49 ~79(in Japanese).

Kanatani,Y.,Tamaki,K and Suzuki,H (1986), " Design and Execution of Yokkaichi LNG Tank base",Civil Engineering in Electric Power,No.200,(in Japanese).

Kato,F (1986)," Technical Subject on Observational Method",Proceeding of JSSMFE Chuugoku Branch,Vol.4,No.1,pp.63 ~72(in Japanese).

Kawakami,T.,Katsuhara,M and Isoda,T (1988), " NEW Automated Operration and Feed back Control in Soil Improvement Method", Proceedings 5th ISRC.,Vol.2,pp.801~810

Kurihara,N (1986)," Observational Method with Micro-computer",JSSMFE,pp.183~208 (in Japanese).

Matsuo,M.,Tsukada,J.,Kanatani,Y and Shouno, H (1984), " Soil Improvement Test Work for Construction of Yokkaichi LNG Tank Base", JSCE,Vol.69.No.4, pp.9~15 (in Japanese).

Matsuo,M and Kawamura,K (1977), " Diagram for Construction Control of Embankment on soft ground", soils and Foundations, Vol.17,No.13,pp.37 ~52

Mizuno,Y.,Suematsu,N and Okuyama,K (1987), " Design Method of Sand Compaction Pile for Sandy Soils Containing Fines", Tsuchi to Kiso,Vol.35,No.5,pp.21~26, (in Japanese).

Okada,Y.,Yagyu,T and Kouda,Y (1989), " Field Rupture of Soil Improved by Sand Compaction Pile Method with Low Sand- Replacement Ratio",Tuchi to Kiso,Vol.37, ~No.379,pp.57~62(in Japanese).

Shibata,T ed.,(1979), " Observational Method", The Nikkan Kougyou,Ltd.,pp.210 ~pp.215(in Japanese).

Sogabe,T (1981)," Technical Problem on Sand Compaction Pile Method in Design and Execution", Technical Report for a Discussion in 36th Annual Meeting of JSCE (in Japanese).

Takahashi,K and Shiomi,M (1988)," Field Loading Test on Improved Ground by Low Replacement Ratio SCP-method in Maizuru Port",Civil Engineering,Vol.43,No.10,pp. 81~89(in Japanese).

Terashi,M (1982), " Deep Mixing Method, State of the Art Report on Soil Improvement Method",pp.53~68,JSSMFE, (in Japanese)

Tsukada,J.,Tamaki,K and Kanatani,Y (1984), " Soil Improvement Work and Stability Control in Yokkaichi LNG Tank Base", Civil Electric Power,No.190, (in Japanese)

Developments in Geotechnical Aspects of Embankments, Excavations and Buried Structures, Balasubramaniam et al. (eds)
© 1991 Balkema, Rotterdam. ISBN 90 5410 019 2

Soil improvement for tank foundation on sand deposit

Y. Ozawa, S. Sunami & M. Kosaka
Civil Engineering Office, Nikken Sekkei Ltd, Japan

ABSTRACT: A full-scale loading test was conducted prior to the construction of a large petroleum reserve storage base, applying loads equivalent to the actual tank load to decide the compaction method of sand deposits and to predict the degree of settlement. Based on the results, the design method for the tank foundation was proposed and validated by water loading tests of the tanks. Consequently, tank foundations were able to be designed safely and very economically.

1 INTRODUCTION

In order to assure a stable supply of oil in Japan, large scale petroleum storage bases have been constructed around the country. Since large settlement of tanks would cause serious problems to the function of steel floating roof tanks, precise studies of settlements occurring have been conducted at the base already constructed (Ozawa et al, 1985).

The base studied here is located alongside a coastal sand dune belt and is a large scale facility with 30 tanks on 150 hectares of land reclaimed with sand of relatively uniform grain size. Each tank is 82.5 m in diameter, 24.5 m in height and 113,000 kiloliters in volume.

Fig.1 shows the arrangement of tanks on the site. The soil conditions at the site presented the following engineering problems:

Fig.1   View of storage base

1. Since the ground was a loose sand deposit, it needed to be compacted to an N value of 15 or over to prevent liquefaction.
2. The settlement characteristics of the compacted sand had to be determined.

To solve these problems, trial compaction tests were conducted, with various specifications for the vibratory compaction method. Based on the results, several effective specifications were selected and a loading fill equivalent to the design tank load was then applied. The settlement was measured using three types of device, and lateral movement was also measured during loading. Based on the measurements obtained, the appropriate modulus of elasticity and Poisson's ratio were determined for the improved layers.

The settlements measured during subsequent water loading tests of the tanks agreed extremely well with the settlement predicted using the parameters obtained above.

2 SOIL PROFILE

As shown in Fig.2, the soil layer presenting engineering problems for the tank foundation was a reclaimed sand layer which was formed from loose and uniform medium to fine brownish sand, and contained thin, dark gray, lens-shaped silt layers in places. N values of the reclaimed layer were from 5 to 30 above the ground water level and from 6 to 10 below water level. Compared to naturally sedimented soil, this layer was   nearly uniform sand deposit with little variation in its stratigraphic profile.

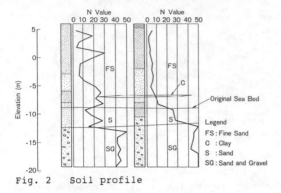

Fig. 2   Soil profile

3 EFFECTIVENESS OF SOIL IMPROVEMENT

Three potential soil improvement methods were tested:the Sand Compaction Pile(SCP) Method, the Rod Compaction(RC) Method and, as an optional auxiliary method for compacting the top layer, the Mammoth Tamper(MT) Method with 2 minutes tamping durations per point was also tested.

For each of the 11 specifications shown in Table 1, a trial compaction test was performed, and the distributions of N value before and after improvement were investigated. Fig.3 shows how the soil with originally average N value of 8 was changed under each SCP specification, i.e. replacement ratio. This figure shows the following results:

1. Taking the scatter of N values into account, a pile pitch of 2.4 m with sand charge of 0.4 m³/m provided N values of 15 or greater.

2. N values became constant at a pile pitch of 2.4 m with charge of 0.4 m³/m.

3. An improvement specification with a pile pitch of 2.4 m and charge of 0.4 m³/m would therefore be most economical.

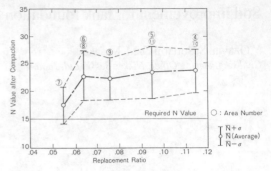

Fig.3   N value after SCP

As shown in Fig.4, the RC method also provided the desired N value of 15 with a pile pitch of 2.4 m, but it was found to be difficult to control the amount of sand and thus the SCP method was judged to be superior.

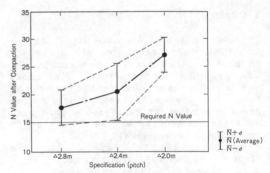

Fig.4   N value after RC

The MT method also proved to be effective for improving the top layer, as shown in Fig.5. Therefore SCP with MT was adopted as the compaction method.

Table 1   Specification for compaction

Area	Compaction Method	Volume of Charge (m³/m)	Pile Pitch (m)	Replacement Ratio
①		-----	△2.0	----
②	R.C	-----	△2.4	----
③		-----	△2.8	----
④		0.5	△2.0	0.115
⑤	S.C.P	0.5	△2.2	0.095
⑥		0.4	△2.4	0.064
⑦		0.4	△2.6	0.055
⑧		0.4	△2.4	0.064
⑨	S.C.P & M.T	0.4	△2.2	0.076
⑩		0.5	△2.0	0.115
⑪		0.5	△2.2	0.095

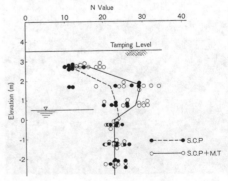

Fig.5. Effect of mammoth tamper

4 FULL LOADING FILL TEST

Based on the results of trial compaction tests, five soil improvement specifications were selected for testing to determine the settlement characteristics. Because the specification with a pile pitch of 2.4 m and sand charge of 0.4 m³/m was considered to be the most likely candidate, this specification was chosen as the central replacement ratio among the five specifications. A loading of fill equivalent to a full tank load was then constructed so as to cover all the improved areas. The five specifications and the fill profile are shown in Fig. 6.

Area	Pile Pitch (m)	Replacement Ratio
A	2.2	0.095
B	2.2	0.076
C	2.4	0.064
D	2.6	0.055
E	2.8	0.047
F	2.4	0.064

Volume of Charge
A : 0.5 m³/m
Others : 0.4 m³/m

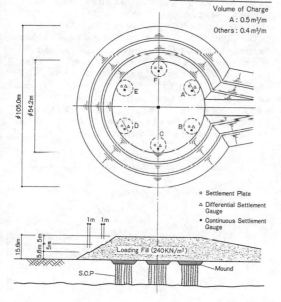

○ Settlement Plate
△ Differential Settlement Gauge
● Continuous Settlement Gauge

Loading Fill (240 KN/m²)

S.C.P.   Mound

Fig. 6   Plan and section of test fill

4.1 Settlement measurement

Fig.7 shows the settlement measurement methods used in each area. Settlement measurement was performed using three types of device, to cross-check and ensure precision. (1) Settlement plates were placed on the top of mound layer, and gave total settlement. (2) The screw anchors of continuous settlement gauges were driven into a layer with N values of greater than 50 below the improved soil layer, thus recording settlement of only the improved soil layer. (3) Measurement elements of differential settlement gauges were located on top and below the mound, below the ground water level, in the middle of the reclaimed layer, and at the bottom end of the SCP.

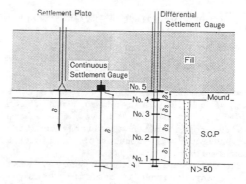

Fig. 7   Settlement measurement

The settlement vs. time curve obtained from the continuous settlement gauges is shown in Fig.8. The amount of settlement for the improved areas was about half that of the original unimproved area. Although the differences in the amount of settlement among the five specifications were small, the amount of settlement decreased with increasing sand replacement ratio, with the single exception of Area B. On the basis of the results of the trial compaction tests and loading test, a soil improvement specification of SCP with a pitch of 2.4 m and sand charge of 0.4 m³/m, plus the MT method, was selected.

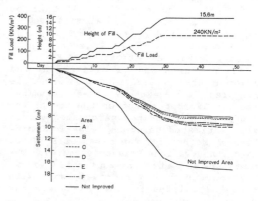

Fig. 8. Settlement vs time curve

445

4.2 Relationship between N value and E

A conventional method applying the theory of elasticity is generally used for prediction of settlement under structures on sand. Many equations for estimating the modulus of elasticity E from N values have been proposed in the past. For example, D'Appolonia (1968) proposed Eq.(1), while Bowles (1986) did Eq.(2). Eq.(3) is commonly used in Japan.

$$E = 1060 \cdot N + 21600 \quad (KN/m^2) \quad (1)$$

$$E = 500 \cdot N + 15 \quad (KN/m^2) \quad (2)$$

$$E = (2500 \sim 3000) \cdot N \quad (KN/m^2) \quad (3)$$

The value of modulus of elasticity E varies greatly depending upon which equation is used. It was therefore desired to establish a reliable equation for predicting the modulus of elasticity E from N values.

In the case described above, most settlement occurred in the improved layer and settlement of the layer below the improved layer was very small. Settlement of the improved layer was considered to be one dimensional, because the width of the fill was much greater than the thickness of the improved layer. Therefore the equation describing elastic settlement S is;

$$S = \frac{\Delta P \cdot H}{\alpha \cdot E} \quad (4)$$

$$\alpha = \frac{(1-V)}{(1+V) \cdot (1-2v)} \quad (5)$$

where ΔP is pressure, H is layer thickness, and v is Poisson's ratio. Poisson's ratio is generally set at a value between 0.2 and 0.4. However, the value of α in equation (5) changes greatly from 1.11 (v=0.2) to 2.14 (v=0.4) depending on Poisson's ratio, and therefore estimation of Poisson's ratio is as important as estimation of E.

The measured and calculated lateral movement are compared in Fig.9, where the measured lateral movements were obtained from inclinometers set on the toe of the loading fill, and the calculated values were estimated using the elastic Finite Element Method for the cases of E=30,000 KN/m², and various Poisson's ratios. The measured pattern and values of deformation agreed extremely well with those calculated in the case where the Poisson's ratio was 0.3.

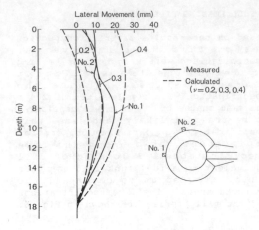

Fig. 9   Measured and calculated lateral movement

The relationship between the amount of settlement measured by the differential settlement gauges and the N values is plotted in Fig.10. Since the layer between Settlement Elements No.1 and No.2 includes thin clay layers, settlement of this layer(marked by "◇" in Fig.10) was larger than would otherwise be expected from the N value of the layer. These values were thus excluded from the analysis. From this figure, E=3,000·N is found to be a suitable equation for estimating E, while the equations proposed by Bowles and D'Appolonia estimate a smaller E value.

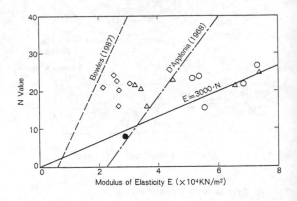

Fig. 10   E-N relationship

4.3 Predicted and measured values

The predicted settlements at the walls of the tanks were calculated. After completion of 30 tanks, water loading tests were performed and the actual settlements at the walls of the tanks were measured. From the N values under each of the tank , the moduli of elasticity were obtained. For all tanks, a uniform settlement of 25 mm was taken as the settlement of the deep layers with N values of greater than 50. The relationship between the measured and predicted values is depicted in Fig.11, which shows that the predicted values agreed comparatively well with the measured ones.

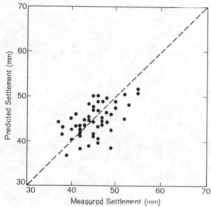

Fig. 11 Measured and predicted settlement for water test

5 CONCLUSIONS

The key results and conclusions of the trial compaction tests and full scale loading test were   summarized below:
1. The relationship between the replacement ratio of SCP and increment in N values due to compaction was able to be clearly determined.
2. The parameters necessary to estimate settlements were able to be obtained and were verified during tank water loading tests.
3. The modulus of elasticity E was found to be predicted precisely by the relationship $E=3,000 \cdot N$ (KN/m^2).
4. A Poisson's ratio of 0.3 showed an extremely good correspondence with measured values of lateral movement in the fine grain sand deposits of the site.
5. A full scale loading test was found to be very effective for designing the most economical tank foundations.

ACKNOWLEDGMENT

The authors wish to thank the Japan National Oil Corporation for giving us the valuable opportunity to perform this study. The authors also express their appreciation to Mr.I.Morita for his continuous consultation and encouragement during the test.

REFERENCES

Bowles, Joseph E. 1986. Elastic foundation settlements on sand deposits, ASCE, GT6 : 846-860.
D'Appolonia, D.J. et al 1968. Settlement of spread footings on sand, ASCE, SM3 : 735-760.
Ozawa et al 1985. Prediction of settlement of tank foundations, ICSMFE, Vol.4 : 2223-2226. San Francisco.

Developments in Geotechnical Aspects of Embankments, Excavations and Buried Structures, Balasubramaniam et al. (eds)
© 1991 Balkema, Rotterdam. ISBN 90 5410 019 2

Behavior of the improved ground by the Deep Mixing Method under embankment loading

Yusuke Honjo, Chung-Ho Chen, Lin Der Guey, D.T. Bergado & A.S. Balasubramaniam
Asian Institute of Technology, Bangkok, Thailand

Ryosuke Okumura
Takenaka Civil Engineering and Construction Co., Ltd, Tokyo, Japan

ABSTRACT: A full-scale test embankment on Deep Mixing improved ground with two different configurations, namely wall type and pile type, were constructed and subjected to the embankment loading. Field observations were made in order to elucidate the behavior of the improved ground with the different configurations. The measured results indicated that : (1) The lateral movement and settlement of wall type improved ground are less than that of pile type improved ground, when subjected to the same loading. (2) The deformation pattern of pile type improved ground is tilting, i.e. simple shear type deformation, where the bending stress is developed in the improved soil, whereas the deformation pattern of wall type is sliding. Both of these results indicated that the wall type improved ground is more effective in resisting against horizontal loading than the pile type improved ground.

1 INTRODUCTION

Cement mixing method has been used to improve the properties of soils for several decades. However, application of this method to improve soft clay for foundation ground only started in late 1970's when the deep mixing method (DMM) was developed in Japan . (Endo, 1976; Terashi et al., 1979; Kawasaki et al., 1981; Suzuki, 1982).

This method (DMM) originally was developed to improve the soft ground for port and harbor structures, DMM is now applied to foundation of structures built on land such as embankments, buildings, and storage tanks.

The typical configuration of improved ground for port and harbor structures are either block, wall, or grid type as shown in Fig. 1. The improved ground for port and harbor structures are configurated in this way in order to resist against large horizontal force induced by back fill.

On the other hand, the stabilized ground applied to the foundations of on-land structures is mainly pile type, because the loads acting on the improved ground on on-land structures are mainly vertical. Also, this type of the stabilized ground is easier and more economical to construct. However in some cases, horizontal forces cannot be neglected; for such cases a rational design method has not yet been established.

The purpose of this study is to investigate the behavior of the improved ground by Deep Mixing Method under horizontal loading condition induced by a structure such as an embankment. Furthermore, to obtain basic data to establish the design method for the improved ground which is subjected to relatively small horizontal forces through carrying out embankment loading test on the improved ground on AIT campus.

2 BACKGROUND

2.1 Material properties of the stabilized soil

The engineering properties of the DMM stabilized soil which need to be taken into account in design can be summarized as follows: The strength of the stabilized soil (measured, say, by unconfined compression strength) is improved remarkably. With the cement contents normally employed in the actual constructions (i.e. 100 Kg to 200 Kg of cement to 1 m^3 of soil), the strength of soil can be easily increased 50 times or more. However, the stabilized

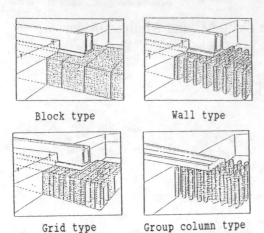

Block type   Wall type

Grid type   Group column type

Fig. 1 Various pattern of treatment.

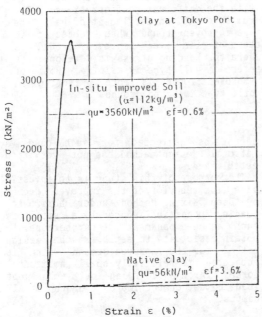

Fig. 2 Typical stress-strain curves of native clay and in-situ improved soil in unaxial compression test

soil becomes much stiffer and brittler compared to the original soil. (Fig. 2). The consolidation yield stress of the stabilized soil is also much higher than that of the original clay. It may be appropriate to state that the stabilized soil behaves much like a heavily overconsolidated clay because of its stiffness and high strength before the yielding and distinguished strain softening behavior after the yielding. This property is considered to be attributed to the bonding effect of the cement agent.

2.2 Problems in the improved ground design

The engineering properties of the stabilized soil described above bring a difficult problem in designing effective and yet economical improved ground. In order to design economical stabilized ground, one needs to reduce the volume of the improvement. However, if one reduces the improvement volume, it is inevitable to design the improved ground as composite ground consists of the stabilized soil part and the soft soil part. In other words, one need to fully take into account the interaction of these two foreign parts in solving this problem. This is very difficult task to solve because these two parts are completely different in the stiffness and in the strength as explained above.

In order to review the design methods proposed for this improved ground, it may be convenient to distinguish following three different loading conditions to which this improved ground is exposed:

1. The improved ground subjected to vertical loading only

This is a case the improved ground is used, for example, foundation of buildings. In this loading condition, pile type improved ground illustrated in Fig. 1(d) is employed. The design procedure for this case is proposed by Broms and Boman (1975, 1979) and also by Terashi and Tanaka, (1981, 1983a). Although the former studied the lime columns by in-situ tests and the latter investigated the cement columns by the laboratory model tests, they came out with rather similar conclusions that the stabilized soil part can be considered as pile foundation made of a low strength material.

2. The improved ground subjected to large horizontal loading

The improved ground employed for foundation of dikes in port and harbor is a typical example of this type.

The improved ground is constructed either block, wall or grid configurations (Fig. 1) so that it has sufficient resistance against large horizontal force which is induced by back fill. The design procedure for this type of ground is established in Japan in late 1970's and has been successfully used in many projects (Endo, 1976; Suzuki, 1982).

The design procedure is presented in Fig. 3. The main portion of this procedure consists of three parts, namely C3, C4, and C5 in Fig. 3. In C3, the stability of the

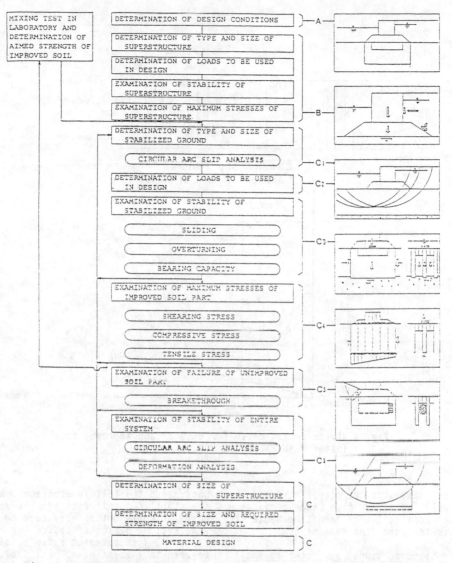

Fig. 3 Outline of simplified design procedure (after TAKENAKA, 1985)

improved ground as a whole is examined by considering it as a sort of gravity type structure. It is assumed that the stabilized soil part and the soft soil part of the improved ground behaves as one body. In C4, the stress induced in the stabilized soil part is calculated to ensure it is lower than the design strength of the stabilized soil; underlying assumption here is that all the external forces acting to the improved ground is concentrated to the stabilized soil part. In C5, stability of the soft soil part is examined by checking the stability of this part against the break through. The size of the improved

ground is mainly determined in C3, whereas the improvement ratio is given in either C4 or C5.

Basic idea of this design method is to absorb all the stress induced by the external forces in the improved ground by the stabilized soil part. This makes design calculation easier because it is not necessary to know the full interaction between the stabilized soil part and the soft soil part. This is reasonable as long as the horizontal force is large enough to induce shear stress in the improved ground far high to be resisted by the soft soil part. However, it could be uneconomical

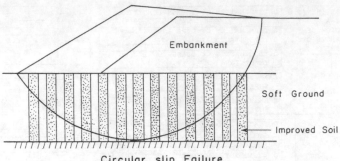

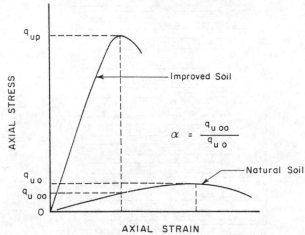

Fig. 4 Stress-strain curve for natural and improved soil
(after TERASHI et al, 1983).

for the case when the stress induced is relatively small so that the resistance from the soft soil part cannot be neglected, which is the case discussed next.

3. The improved ground subjected to small horizontal loading

If the improved ground is employed as foundation of moderately high embankment, it should resist against not only vertical load induced by the weight of the embankment, but also horizontal loads. It is apparent that the design procedure introduced in 2.2 can be applied to determine the configuration of the improved ground for this case. However, the improved ground design by this procedure is tend to be expensive due to high improvement ratio. In order to come out with more economical configuration of the improved ground, one needs to fully take into account the contribution of the soft soil strength to the design procedure. This asks one to

understand the full interaction behavior of the stabilized soil part and the soft soil part, which is not an easy task to solve.

A typical design procedure for what is called "low improvement ratio conditon" can be seen in Terashi et al., (1983b). It is based on the circular slip method (Fig. 4), and the following design strength is recommended to be adopted:

$$C = C_p A_p + \alpha C_O (1-A_p) \qquad (1)$$

where:
C_p : shear strength of the stabilized soil
C_O : shear strength of the in-situ soft soil
α : strength reduction factor (see Fig. 4)
A_p : the improvement ratio

To the authors' understanding, this design method was developed from the design method for the sand compaction pile method (Aboshi and Suemoto, 1985). However, the material properties of sand pile and the DMM stabilized soil are completely different.

Therefore, the effectiveness of the design method has to be examined very carefully.

In these few years, several in-situ measurements on the DMM improved ground have been published. (Amano et al., 1986; Tanaka et al., 1986; Unno et al., 1987; Yoshioka et al., 1988; Kotani et al., 1988). Some of the resutls are summarized in Figs. 5, 6 and 7. All these 3 cases, the dominant mode of the deformation of the stabilized soil piles is simple shear type tilting which is induced by the horizontal force by the embankment. It is natural for the improved ground to deform in this way because the shear stiffness of the pile type improved ground is controlled by the soft soil between the piles. In addition to this tilting type deformation, the stabilized soil piles are exposed to bending which can be seen from the deformation curves of these piles. This is apparently caused by uneven distribution of horizontal force and the surrounding soft ground stiffness.

Kotani et al. (1988) reported a failure case of the pile type improved ground (Fig. 8(a)). The area of improvement had been determined to eliminate all the circular slip surfaces whose safety factors were less than 1.2. However, the failure took place as shown in Fig. 8(b). It is the authors' opinion that in this failure case, large shear deformation occured in the improved ground (i.e. tilting of the piles) which resulted the heave of the toe and the settlement of the embankment. If the improved soil had enough shear stiffness to redistribute the force induced by the embankment and the excavation to the surrounding ground, such failure would not have taken place.

As can be seen from these observations, design procedure of the DMM stabilized ground subjected to small horizontal loading condition is not yet established. The circular slip method using the soil strength of Eq. (1) is not correctly taking account of the typical deformation pattern of this improved ground under this condition.

The test embankment construction presented in the next section was planned to study the behavior of the improved ground under moderately large horizontal loading condition (Chen, 1990).

3. TEST EMBANKMENT CONSTRUCTION

3.1 Soil conditions at the site

The test embankment site is located on the AIT campus 42 Km north of Bangkok. Fig. 9 illustrates the soil profile at the site

which is a typical Bangkok clay profile of the central Chao Phraya Plain. Under 2 m of the weathered crust, there is about 6 m of soft Bangkok clay layer whose natural water content is 70 to 80 %. This soft clay layer is underlain by the stiff layers.

The strength of the soft clay layer measured by the various methods is presented in Fig. 9. In this case, the undrained shear strength obtained by field vane test with Bjerrum's correction, unconfined compression stregnth, and SHANSEP method gave similar values, which is about 2 t/m .

3.2 Construction of DMM improved ground and embankment

The strength of the laboratory mixed cement stabilized soil was obtained for the clay sampled at 2.5 m, 4.5 m and 7.0 m depth for different cement content (a). The results shown in Fig. 9 are the average unconfined compression strength for a = 10 % (i.e. about 100 Kg of Portland cement to 1 m of soft clay) with the curing period of 28 days. For comparison, the unconfined compression strength of in-situ mixed samples is also presented in Fig. 9. The cement content at the construction was 100 Kg per 1 m of the soft clay. In this construction, the in-situ strength was roughly half of the lab. mixed sample strength, which coincides the experience in elsewhere (e.g. Kawasaki et al, 1981).

The plan and the section of the improved ground as well as the test embankment are shown in Fig. 10. The configuration of the improved ground for the north side is the pile type, whereas that of the south side is the wall type; both of them have the same number of the improved columns to sustain the same force induced by the embankment. The height of embankment is 5 m and the embankment toe was excavated 2 m to remove the crust, and to increase the lateral deformation as well as to maintain plane strain condition as much as possible.

The construction of the DMM piles was done in December 1989. The details of the construction condition is tabulated in Table 1. The construction progress of the embankment and its toe excavation is presented in Table 2.

3.3 Instrumentation

The instrumentation layout is presented in Fig. 11. The main objective of this measurement was to trace the deformation of the improved ground; 8 inclinometer pipes and 14 settlement gages were installed for this purpose.

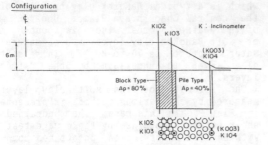

Configuration

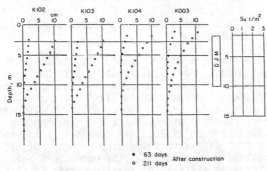

Measured Results

• 63 days
○ 211 days   After construction

Fig. 5 The configuration and measured results of HAYASHIMA interchange (after TANAKA et al, 1986).

Configuration

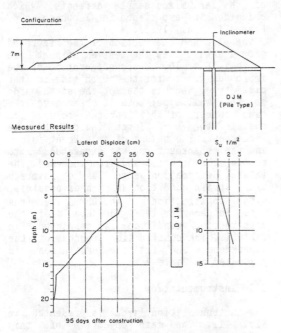

Measured Results

Lateral Displace (cm)

S_u t/m²

95 days after construction

Fig. 6 The configuration and measured results of the test embankment adjacent to HAYASHIMA interchange (after AMANO et al, 1986).

Configuration

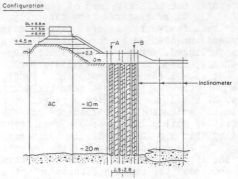

Measured Results

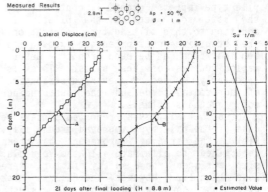

21 days after final loading (H = 8.8 m)

• Estimated Value

Fig. 7 The configuration and measured results of loading test on pile type improved ground (after YOSHIOK et al, 1988).

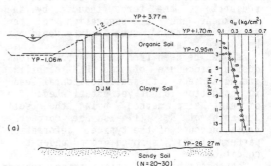

(a)

In - Situ Soil Condition and Ground Improvement
(After KOTANI et al. 1988)

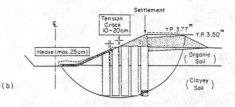

(b)

Fig. 8 Failure condition (after KOTANI et al, 1988).

454

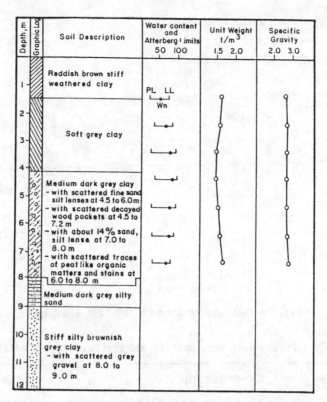

Soil Boring Log

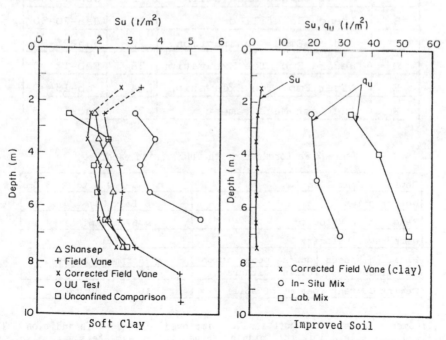

Soft Clay

Improved Soil

Fig. 9 Comparison of unconfined compression strength between lab-proposed and in-situ cement-treated soil.

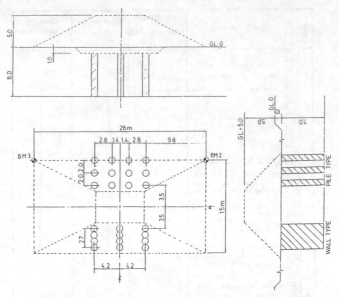

Fig. 10 Plan and section of the test embankment.

Table 1 Progress of construction and instrumentation activity.

Stage	Description	Time (days)	Date
A	After Trench Excavation	2	Jan-7-90
B	Trench Backfilled	15	Jan-20-90
C	End of Emb. Construction	20	Jan-25-90
D	Before Emb. Toe Excavation	38	Feb-12-90
E	After Emb. Toe Excavation	44	Feb-18-90
F	The Last Measurement	71	Mar-17-90

Table 2 Performance of the DMM mixing machine.

Depth of Stabilization		7 m (GL-1.0 m to GL-8.0m)
No. of piles		ϕ 1m x 24
Penetration Velocity		0.67 m/min.
Withdrawal Velocity		1.0 m/min.
Rotation speeds of blades	during penetration	38 rpm
	during withdrawal	38 rpm
Cement Content (per m³ of Soil)		100 or 150 kg/m³

The inclinometers were placed both inside the stabilized soil pile and in the middle of the soft soil part (Fig. 11) so that one could observe the detailed behavior of the improved ground. In addition, 27 stand pipe type piezometers were also installed to evaluate the excess pore water pressure build up and its dissipation.

456

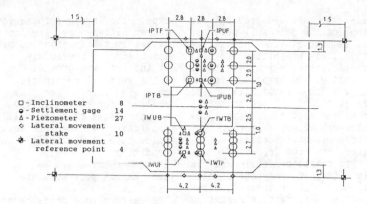

Fig. 11 Instrumentation layout.

Legend:
□ - Inclinometer   8
● - Settlement gage   14
△ - Piezometer   27
◇ - Lateral movement stake   10
⊕ - Lateral movement reference point   4

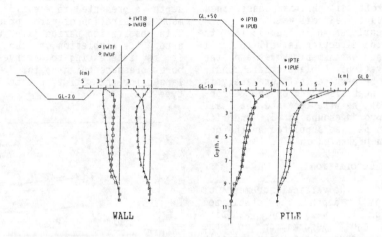

Fig. 12 Comparison of lateral movement on improved and unimproved ground (71 days).

4. RESULTS OF THE MEASUREMENT AND DISCUSSION

4.1 Lateral deformation

The lateral movement of the ground detected by the 8 inclinometers are plotted in Fig. 12. The loci of these inclinometers at 2 m depth are also shown in Fig. 13. The following can be stated:

(1) The deformation of the pile type improved ground is larger than that of the wall type.
(2) The behavior of the wall type stabilized soil part is observed to be a parallel shift, whereas that of the pile type stabilized soil parts is tilting and is developing bending moment inside the columns which can be expected from the curvature of the deformation.

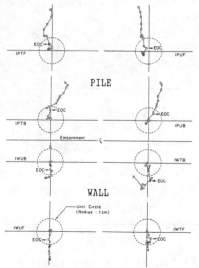

Fig. 13 Lo'ci of lateral movements at depth of 3 m.

(3) In the pile type improved ground, the deformation of the soft soil part is the same as that of the stabilized soil part. It is exhibiting lateral flow type deformation which is typical for the soft ground exposed to an embankment loading. The stabilized soil piles are just moving together with the surrounding soft soil, and does not seem to work effectively to resist against the horizontal loading.

(4) On the other hand, the deformation of the wall type stabilized soil part is completely different from that of the soft soil part. It is apparent that the stabilized soil part is resisting against the horizontal force by its shear stiffness, and the deformation of the soft soil part is constrained by the stabilized soil wall.

(5) It is clearly shown in Fig. 13 that the deformation direction is well controlled by the embankment shape and the toe excavation. A plane strain condition can be assumed in future analysis.

(6) What has been discussed on the deformations of the pile type and the wall type improved ground in (3) and (4) above can be also supported by the observation presented in Fig. 13.

4.2 Vertical deformation

Fig. 14 indicates the vertical movement of the ground at 1 m depth. The stabilized soil part for the wall type exhibits approximately equal deformation at the front and at the back, whereas for the pile type, the larger deformation at the back. These differences are considered to be caused by the deformation pattern differences indicated previously for the two types of the improved ground.

The vertical deformation of the soft soil part is larger for the pile type and smaller for the wall type. Therefore, it is speculated that the wall type gives more constraint to the soft soil part than the pile type under this kind of loading condition.

4.3 Excess pore pressure

The excess pore pressure measured at the centre of the soft soil part of the two types of improved ground at three different depth is presented in Fig. 15. There is not much dissipation of pore pressure within this observation period (i.e. until 51 days after the completion of the embankment). It is interesting to observe the excess pore pressure build up for 7.0 m depth is larger for the wall type than for the pile type.

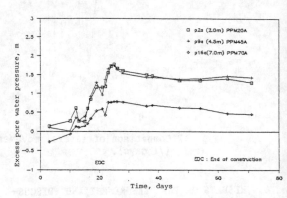

Fig. 15(a) **Excess pore water pressure various with time group 2a(PPMA).**

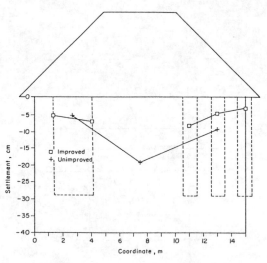

Fig. 14 **Comparison of settlement of improved and unimproved part at 1m depth (71 days).**

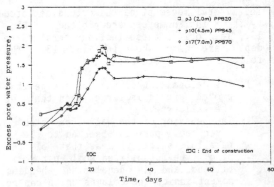

Fig. 15(b) **Excess pore water pressure various with time group 3 (PPB).**

458

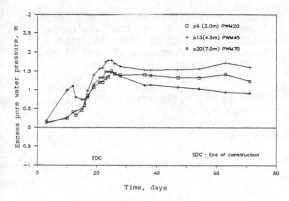

Fig. 15(c) Excess pore water pressure various with time group 6 (PWM).

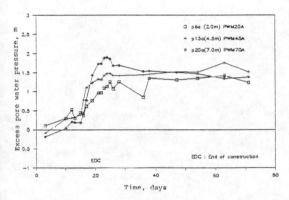

Fig. 15(d) Excess pore water pressure various with time group 6a (PWMA).

In Fig. 16, the peak excess pore pressure at the different locations is plotted across the embankment. It could be said that the peak excess pore pressure is higher for the pile type at the shallower depth. However, it is reversed at the deeper depth.

One of the purposes to install piezometers in the improved ground was to obtain indirect indication of the stress concentration to the stabilized part and the soft soil part. From this stand point, the lower pore pressure build up in the wall type (this is only true for the shallower part) could be considered to indicate higher stress concentration of the external load to the stabilized soil, although this is not valid explanation for the deeper part. The excess pore pressure measurements still need to be looked into to understand the behavior.

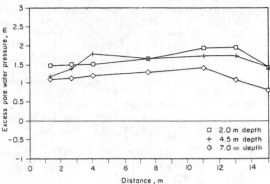

Fig. 16 Peak excess pore pressure along embankment section.

5. CONCLUSIONS

Based on the work completed so far the following conclusions can be drawn:

(1) The deformation pattern of the pile type improved ground is different from that of the wall type improved ground. The deformation pattern of the pile type improved ground is tilting and it moves together with surrounding soft soil when subjected to horizontal force. On the other hand the deformation pattern of the wall type improved ground is sliding, and the movement of the soft soil part is restricted by the stabilized soil wall, i.e. it provides some restrains to the soft soil part.

(2) The tilting pattern deformation of the pile type improved ground can be considered as simple shear deformation. Therefore the shear stiffness and the shear strength of this type of improved ground is controlled by the property of the soft soil. Furthermore, due to the uneven distribution of horizontal movement along the depth, the bending stress is induced in the stabilized pile columns.

6 ACKNOWLEDGEMENT

This is a part of the results of a research at AIT sponsored by TAKENAKA Civil Engineering and Construction Co., Ltd. The authors are grateful for their support.

7 REFERENCES

Aboshi, H. and Suemoto, N. (1985):   The

State of the Art on Sand Compaction Pile Method, Proc. 3rd NTI International Geotechnical Seminar, pp. 1-12, Singapore.

Amano, H., Morita, T. Tsukada Y. and Takahashi, Y. (1986): Design of Deep Mixing Method for High Road Embankment-Hyashima I.C. - , Proc. Annual Domestic Conference of JSSMFE, pp. 1999-2002 (in Japanese).

Broms, B.B. and Boman, P. (1975): Lime-Stabilized Columns, **Proc. 5th Asian Regional Conference on SMFE**, Vol. 12, No. 4, pp. 23-32.

Broms, B.B., and Boman, P. (1979): Stabilization of Soil with Lime Column, **Ground-Engineering**, May, pp. 23-32.

Chen, Chung-Ho, (1990): Behavior of the Improved Ground by Deep Mixing Method under Embankment Loading, AIT Thesis No. GT-89-17, pp. 184.

Endo, M. (1976): Recent Development in Dredged Material Stabilization and Deep Chemical Mixing in Japan, Life-long Learning Seminar Soil and Silt Improvement, Univ. of California, Berkeley.

Kawasaki, T., Niina, A., Saitoh, S. Suzuki, Y. and Honjo, Y. (1981): Deep Mixing Method Using Cement Hardening Agent, **Proc. 10th ICSMFE**, pp. 721-724, Stockholm.

Kotani, E., Stukakoshi, H. and Inoue, H. (1988): A Consideration of Design of Deep Mixing Method, Proc. Annual Domestic Conference of JSSMFE, pp. 2273-2274 (in Japanese).

Suzuki, Y. (1982): Deep Chemical Mixing Method Using Cement as Hardening Agent, Proc. Sym. on Soil & Rock Improvement Techniques Including Geotextiles, Reinforced Earth and Modern Piling Method, AIT, Bangkok, Thailand.

Tanaka, H., Nagao, H., Tsukada, Y., Morita, T., Sato, T. and Kikuchi, Z. (1986): Behavior of Deep Mixing Method Column for High Road Embankment-Hyashima I.C.-, Proc. Annual Domestic Conference of JSSMFE, pp. 141-142 (in Japanese).

Terashi, M., Tanaka, H. and Okumura, T. (1979): Engineering Properties of Lime-Treated Marine Soils and D.M. Method, **Proc. 6th Asia Regional Conf. SMFE**, Vol.1, pp. 191-194.

Terashi, M. and Tanaka, H. (1981): Ground Improved by Deep Mixing Method, **Proc. 10th ICSMFE**, Vol. 3, pp. 777-780, Stockholm.

Terashi, M. and Tanaka, H. (1983a): Settlement Analysis for Deeping Mixing Method, **8th Europ. Conf. on SMFE**, pp. 955-960.

Terashi, M., Fusetani, H. and Noto, S. (1983b): Deep Mixing Method : Practice and Problem, **Tsuchi-To-Kiso**, Vol. 31, No. 6, pp. 57-64, in Japan.

Unno, Y., Tsuruta, K., Kobayashi, S. and Gotoh, M. (1987): Prediction of Behavior of An Retaining Wall on DMM Improved Ground, Proc. Annual Domestic Conference of JSCE, III pp 812-813, In Japan.

Yoshioka, H., Yoshinaga, K. and Okubo, H. (1988): Loading Test on Improved Ground of Pile Type Using Deep Mixing Method, Proc. Annual Domestic Conference of JSSMFE, pp. 2283-2284 (in Japanese).

Developments in Geotechnical Aspects of Embankments, Excavations and Buried Structures, Balasubramaniam et al. (eds)
© 1991 Balkema, Rotterdam. ISBN 90 5410 019 2

Mechanically stabilized earth (MSE) and other ground improvement techniques for infrastructure constructions on soft and subsiding ground

D.T. Bergado, C.L. Sampaco, M.C. Alfaro & R. Shivashankar & A.S. Balasubramaniam
Asian Institute of Technology, Bangkok, Thailand

SUMMARY: The presence of thick deposits of soft clay and the effect of ground subsidence due to the excessive pumping of groundwater cause numerous foundation problems to infrastructure constructions in Chao Phraya Plain, Thailand in the form of large subsoil compressions and the associated differential settlements as well as slope instability. To mitigate such natural hazards, several schemes of mechanically stabilized earth (MSE) are proposed for such constructions as underground pipeline, road culverts, road embankments, dikes along irrigation canals, and approach embankments to overpasses and viaducts. The MSE method can be also combined most appropriately with the subsoil improvement using either lime/cement piles, granular piles, or vertical band drains. The combined improvement schemes can increase shear strength, reduce total and differential settlements, and minimize lateral spreading. Moreover, the aforementioned improvement schemes can be a viable alternative to the existing method of using precast, reinforced concrete pile foundation.

1 INTRODUCTION

Most major cities in the Southeast Asian region are located in the coastal plains with thick deposits of soft clays. These deposits are found in most countries of the region (Fig. 1) such as the Chao Phraya Plain in Thailand, Mekong Delta in Cambodia and Vietnam, Malaysian Coastal Plain, Philippine Central Plain, and Indonesian Coastal Plain. Majority of the infrastructure constructions in the coastal plains are road embankments, flood control dikes, landfills, embankments along irrigation canals, underground pipelines, road culverts, etc.

For the usual constructions on soft and subsiding ground, pile foundations are commonly used. This results in more expensive project costs and large differential settlements between pile supported structures and the ground supported structures. As an alternative, mechanically stabilized earth (MSE) using geogrid reinforcements is proposed. This method consists of reinforcing the soil using steel or polymer (plastic) ma-

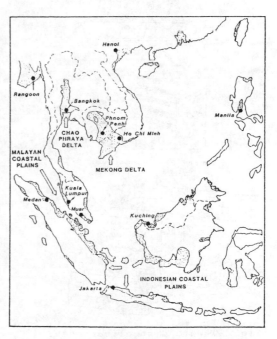

Fig. 1. The distribution of recent clays in Southeast Asia (After Cox, 1970)

461

terials. The reinforcement which is strong in tension combines with the soil which is strong in compression, forming a very strong and semi-rigid composite material. The tension in the reinforcement is mobilized by the interaction between the reinforcement and the soil in the form of friction or adhesion and bearing resistances. The grid reinforcement usually generates pullout resistance up to 6 times higher than the strip reinforcement (Nielsen and Anderson, 1984). Among the grid reinforcement, steel grids are found to be superior than the polymer grid in terms of low extensibility and higher tensile strengths (Fowler et al, 1986). Since the mobilization of the interaction between the soil and reinforcement depends on the relative strains generated in the reinforcement, the steel grid reinforcement is preferable. The scheme for this method is shown in Fig. 2.

To obtain the necessary data for analysis, full scale laboratory and field load tests as well as laboratory pullout tests were performed. The test results are presented together with the modeling and prediction of soil-geogrid interaction including applications of MSE constructions. Part of the data were obtained during the sabbatical leave of the first author at Saga University, Japan. At the later part of this paper, several ground improvement techniques that can be combined with the MSE construction are described and presented.

2 LOAD TESTS ON UNREINFORCED AND REINFORCED GRAVEL

The load test set-up on both unreinforced and reinforced gravel using either double reinforcement or geocell are shown in Fig. 3 (Bergado et al, 1988b). Types 1, 2, and 3 refer to the unreinforced, geocell, and double reinforcements, respectively. The schematic diagrams for Types 4 and 5, which are also shown in Fig. 3, refer to the geocell and double reinforced gravel, respectively, resting on the reconstituted soft clay. The reinforcement consisted of Tensar polymer grids. The gravel specimen has plan dimensions of 150 cm by 150

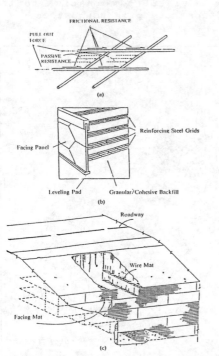

Fig. 2. Welded steel bar mesh, steel grids reinforcement with concrete facing and welded wire wall.

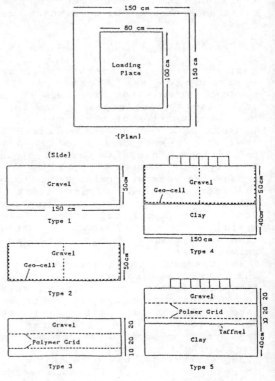

Fig. 3. Load tests on unreinforced and reinforced gravel.

cm and a height of 50 cm. The unit
weight of the gravel was maintained
at 1.8 t/m^3 as much as possible. The
resulting load-settlement relationships from the load tests are plotted
and summarized in Fig. 4. The gravel
with geocell yielded the largest deformation due to the difficulty of
compacting the corner sections. Also,
the cell sizes of the geocell maybe
too large in relation to its whole
dimensions. The double reinforced
gravel performed better with less deformation and was subsequently recommended.

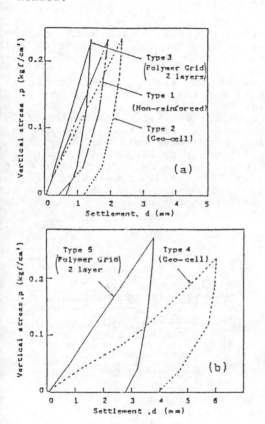

Fig. 4. Load-settlement curves from
the load tests.

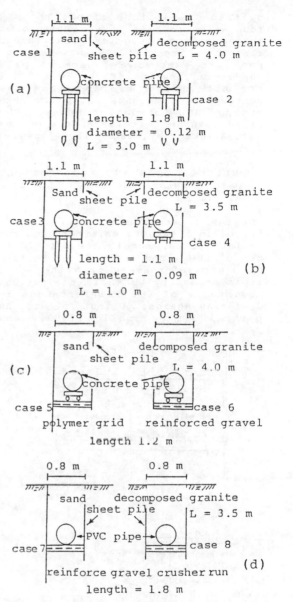

Fig. 5. Sewage pipeline on soft and
subsiding ground with reinforced foundations.

3 FULL SCALE TEST ON UNDERGROUND PIPELINE CONSTRUCTION

The usual foundation of underground
sewage pipeline on soft and subsiding
ground consists of concrete pipe on
pile foundations of either long (Fig.
5a) or short (Fig. 5b) dimensions
with either sand or decomposed granite backfill (Bergado et al, 1988c).

Alternative designs consist of concrete pipes (Fig. 5c) or PVC pipes
(Fig. 5d) resting on mechanically
stabilized earth (MSE). Full scale
field tests were done on cases 1 to 8
as shown in Fig. 5 wherein the movements at the top of the pipeline were
monitored. The measured movements as
plotted in Fig. 6a indicate the applicability of the mechanically sta-

bilized gravel foundation. The MSE foundation in this case consisted of Tensar polymer geogrids with crusher run gravel backfill. The movements of the more expensive scheme of pile supported pipeline (Figs. 6a and 6b) are not much different from the corresponding amounts using the much cheaper MSE foundation scheme. The field loading tests were done on the soft Ariake clay, Saga City, Japan.

4 PULLOUT TESTS AND APPLICATIONS USING TENSAR AND BAMBOO GEOGRIDS AT AIT

For the internal stability of reinforced soil structures, there are two failure modes, namely: tensile failure of the wires and pullout failure of the reinforcement from the soil. It is necessary to know the pullout resistance of the reinforcement in designing against pullout failure. As pointed out by Ingold (1983), the

pullout test is the most realistic model for assessing the performance of low extensibility geogrids.

A type of pullout apparatus was developed at the Asian Institute of Technology (AIT) by Bergado et al (1987) based on the model of Peterson and Anderson (1980). The cell was made of reinforced concrete which was open at the top and front. The cell was approximately 0.8 m wide by 1.0 m long and 0.9 m high with two steel columns fixed on each side as shown in Fig. 7a. The normal pressure was applied by means of hydraulic jack reacting against a steel reaction beam simulating the desired overburden pressure. The reinforcement was clamped by two steel angles and the pullout force was applied by another hydraulic jack reacting against the steel supporting frame which was fixed in front as shown in Fig. 7b.

Both the weathered clay and lateritic soil backfills were compacted inside the pullout cell to 95% of standard proctor density at optimum moisture content with lift thickness of 15 cm. The desired normal pressure was applied and allowed to come to equilibrium for about 10 minutes. The pullout force was applied at a dis-

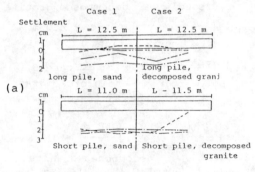

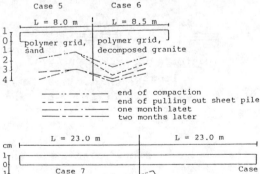

Fig. 6. Settlements of sewage pipeline on reinforced foundations.

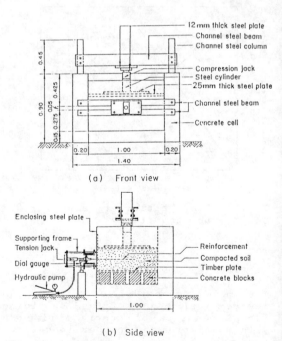

(a) Front view

(b) Side view

Fig. 7. The pullout test cell (Dimension in meters).

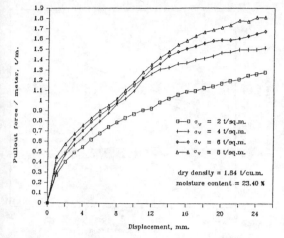

Fig. 8. Force-displacement curves
for pullout test no. 3 of
Tensar SS2 geogrids with
weathered clay backfill.

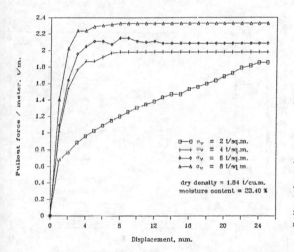

Fig. 9. Force-displacement curves for
pullout test no. 1 of bamboo
grids with weathered clay
backfill.

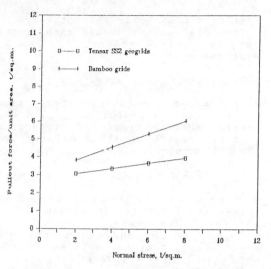

Fig. 10. Comparison of pullout test
results for Tensar SS2 geogrids and bamboo grids with
clayey sand backfill.

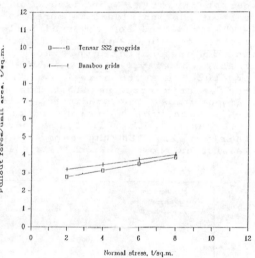

Fig. 11. Comparison of pullout test
results for Tensar SS2 geogrids and bamboo grids with
weathered clay backfill.

placement rate of 1 mm/min. The pullout force and horizontal displacements were recorded at one minute interval until the reinforcement was
pulled to a displacement of 2.5 cm.
The normal pressure ranged from 2 to
10 tsm.

The size of the Tensar geogrids was
0.46 m by 0.46 m. The bamboo grids
were having dimensions of 0.275 m by
0.60 m with cross-sectional dimension
of 0.5 cm by 1.0 cm for each individual member. The mesh geometry was 10
cm by 15 cm.

Using bamboo grids, the force-
displacement curves had more well-
defined peaks than those using Tensar
geogrids as shown in Figs. 8 and 9,
respectively. Moreover, it was found
that the bamboo grids have higher
pullout resistance than Tensar SS2
geogrids provided that each has the
same plan area as shown in Figs. 10
and 11, for lateritic soil and

weathered clay soil, respectively.

The total pullout resistance of the geogrids reinforcement, F_t, can be expressed as:

$$F_t = F_f + F_b \quad \cdots \cdots \quad (1)$$

where: F_f is the adhesion resistance of longitudinal member and F_b is the bearing resistance of the transverse member. Bergado et al (1987) derived the following:

$$F_t = (2\mu A_f + m A_b N_c)\, C_u \quad \cdots \quad (2)$$

where: μ is the adhesion factor between soil and reinforcement, A_f is the total plan area of the geogrid only, C_u is the undrained cohesive shear strength of the soil as defined by Ingold (1983), m is the total number of the transverse members, N_c is the bearing capacity factor for a strip footing embedded in the soil which is equal to 7.5 (Ingold, 1983), and A_b is the cross-sectional area perpendicular to the direction of pull of the individual members.

The predicted and observed pullout resistance are compared in Figs. 12 and 13 using lateritic soil and weathered clay, respectively. The predicted results agreed well with the experimental results. For the bamboo grids, the predicted pullout resistance with adhesion factor of unity using both soil types, agreed fairly well with the experimental results as shown in Figs. 14 and 15.

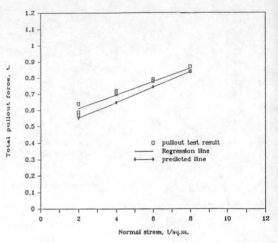

Fig. 13. Comparison of predicted and observed pullout force for Tensar SS2 geogrids with weathered clay backfill.

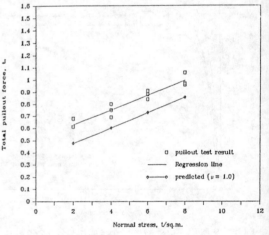

Fig. 14. Comparison of predicted and observed pullout force for bamboo grids tested with lateritic soil ($\mu = 1.0$).

5 REINFORCED CANAL EMBANKMENT, PATHUM THANI, THAILAND

A case study on the application of the aforementioned pullout test results was done in Amphor Nong Sua, Pathumthani Province, Thailand. The infrastructure is a road embankment on the bank of an irrigation canal. The difference of water level in the irrigation canal during the rainy and dry seasons is 2 m. The slip failure occured during low water level as

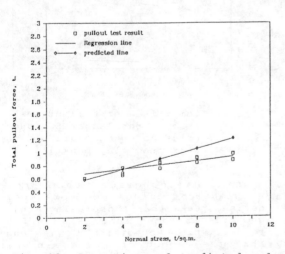

Fig. 12. Comparison of predicted and observed pullout force for Tensar SS2 geogrids with clayey sand backfill.

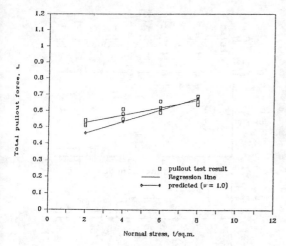

Fig. 15. Comparison of predicted and observed pullout force for bamboo grids tested with weathered clay ($\mu = 1.0$).

shown in Fig. 16. The Public Works Department of Thailand repaired the embankment by using geogrids reinforced soil as shown in Fig. 17. The design of the remedial measures and the reinforcement consisting of Tensar SS2 geogrids were donated by Netlon Ltd. A reanalysis of the mechanical stabilization scheme was done by Bergado et al (1987). The soil profile together with the basic soil properties are shown in Fig. 18.

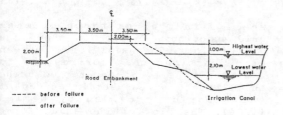

Fig. 16. Geometry of embankment before and after failure.

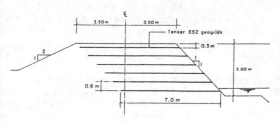

Fig. 17. Tensar SS2 geogrids layout as proposed by Netlon Limited.

Depth, E	Graphic Log	Soil Description	Unit Weight (kN/m³) 19 20	Water Content (%) 50 100	Plastic Limit (%) 20 40	Plasticity Index (%) 20 40	Specific Gravity (G) 2 4	Field Vane Shear Strength, kN/m² 10 20
1		Reddish brown and yellowish brown soil, weathered clay	o	o		o	o	
2								
3			o	o		o	o	o
4		Very soft light gray clay with silt seams,			o	o		o o
5			o	o		o	o	o oo
6		decayed woods and						o oo
7		decomposed roots	o	o		o	o	o o
8								o o
9			o	o		o	o	o o
10								

🗆 lateritic soil   🗆 spot decayed woods and decomposed roots
🗆 yellowish brown soft clay   🗆 shell
🗆 silt seams

Fig. 18. Soil profile at the site of case study.

The field vane shear strength slightly increased with depth. The total stress analysis was employed in conjunction with the Modified Bishop Method of slope stability analysis at various conditions and the results are tabulated in Table 1.

Table 1. Summary of reanalysis of a case study

Analysis		Factor of safety
Unreinforced embankment (consider tension crack)	slope failure side slope 2:1 (H:V) w/o surcharge	0·98
	slope failure side slope 2:1 (H:V) w/ surcharge	0·89
	slope failure side slope 1:1 (H:V) w/ surcharge	0·69
Reinforced embankment	bearing capacity failure	2·02
	slope failure side slope 1:1 (H:V) w/ surcharge	1·44
	tension failure	2·40
	pull-out failure	8·74

The external stability of the embankment was analyzed by checking the factor of safety against bearing capacity failure and deep slope stability, yielding values of factor of safety of 2.00 and 1.44, respectively. The slope failure analysis was carried out using the method suggested by Jones (1985). For the internal stability, the tensile force in the reinforcement and the pullout resistance were evaluated. The factor of safety against tensile failure was 2.40 and the corresponding value for the pullout resistance was 8.74 as shown in Table 1.

6 PULLOUT TEST USING TENSAR GEOGRIDS AT SAGA UNIVERSITY, SAGA CITY, JAPAN

The apparatus for the pullout test is shown in Fig. 19 which was designed by the first author during his sabbatical in Saga University, Saga, Japan (Bergado et al, 1988c). The size of the geogrid reinforcement was 80 cm by 75 cm. A total of 14 strain gages were bonded to the geogrid in two lines. For the gravel backfill, a total of 4 tests were performed, namely: GM1 with 6 and 8 tsm normal pressure, SR2 with 6 tsm, and SS35 with 6 tsm. For decomposed granite, SS35 was used under vertical stresses of 2, 4, and 6 tsm for a total of 3 tests.

Fig. 19. Pullout test set-up.

When the geogrid is pulled out from the backfill soil, the pullout resistance is mobilized in both grid junction and ribs as shown in Fig. 20a. The pullout resistance on both sides of the geogrid are modelled to be concentrated at the grid junctions as shown in Fig. 20b (Ochiai and Sakai, 1987). The pullout force, F_t, exerted on the polymer grid, produces displacement, X_i, at each grid junction.

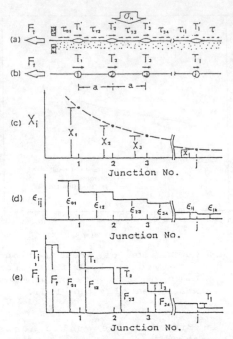

Fig. 20. Analytical procedure of pullout test results.

The displacements of the grid junctions (Fig. 20c) can be measured in the pullout test under constant vertical stress. The strains of the geogrid between junctions can be calculated by:

$$\varepsilon_{ij} = (X_i - X_j)/a \quad \ldots \ldots (3)$$

where a is the distance between grid junctions. Figure 20d shows the strain plots. The axial force between grid junctions that correspond to the strain, ε_{ij}, is determined using the standard stress-strain curves of the geogrids provided by Netlon (1984). The plot of the axial force ($F_i - F_{i+1}$) represents the pullout resistance, T_i, mobilized at each grid junction. The shear stiffness, k_s, can be obtained from the plot of T_i versus displacement, u, by the following equation:

$$2\tau = 2k_s u \quad \ldots \ldots \ldots (4)$$

where τ is the shear resistance.

The pullout force, F_t, for gravel backfill with Tensar GM1 reinforcement was plotted against strain. Figure 21 shows the strain plots under normal stress of 6 tsm at

468

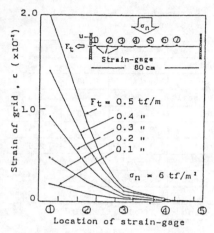

Fig. 21. Typical strain plots from pullout test.

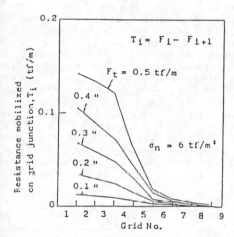

Fig. 22. Typical pullout resistance mobilized on grid junctions.

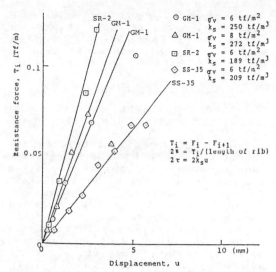

Fig. 23. Shear stiffness for gravel backfill.

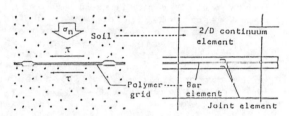

Fig. 24. FEM model of reinforced soil.

different pullout force, F_t. The calculated pullout resistance, T_i, is plotted in Fig. 22 and the shear stiffness, k_s, derived from the pullout resistance is plotted in Fig. 23 versus displacements for GM1, SR2, and SS35. The k_s value obtained for SR2 was the highest followed by GM1 and SS35. The above process demonstrated the calculation of k_s.

Figure 24 shows the finite element model of the reinforced soil (Ochiai et al, 1988). The geogrid is transformed into a joint element on both sides representing the interface with the soil and a truss element in the middle transmitting axial force only. The joint element expresses the transmission of the compressive force and sliding against shear displacement. The relationship between shear displacement $\{u\}_{s,n}$ of the joint element (length=ℓ) and the incremental loading $\{F\}_{s,n}$ is given in the following expression:

$$\{F\}_{s,n} = [K]_{s,n} \{u\}_{s,n} \quad \cdots \quad (5)$$

where the stiffness matrix $[K]_{s,n}$ is expressed by

$$[K]_{s,n} = 1/4\,\ell
\begin{bmatrix}
k_s & 0 & k_s & 0 & -k_s & 0 & -k_s & 0 \\
 & 2k_s & 0 & 0 & 0 & 0 & 0 & -2k_n \\
 & & k_s & 0 & -k_s & 0 & -k_s & 0 \\
 & & & 2k_n & 0 & -2k_n & 0 & 0 \\
 & & & & k_s & 0 & k_s & 0 \\
 & (\text{sym}) & & & & 2k_n & 0 & 0 \\
 & & & & & & k_s & 0 \\
 & & & & & & & 2k_n
\end{bmatrix} \quad (6)$$

The joint element has two unit stiffness, a normal stiffness, k_n, and a shear stiffness, k_s. The shear stiffness can be determined from the results of the laboratory pullout test.

7 REINFORCED GRAVEL FOUNDATION FOR CULVERT CONSTRUCTION, JAPAN

A project of Japan Ministry of Construction consisted of a box culvert construction along the road crossing a canal in Saga Plain, Kyushu Island, Japan. The culvert structure is founded on a 15 m thick, soft and compressible Ariake clay on subsiding ground. Ordinarily, pile foundation is used in this type of construction. However, in this study, it was recommended to use the cheaper reinforced gravel foundation using two layers of Tensar GM1 geogrids (Fig. 25) so that the differential settlements will be minimized. The FEM model described earlier was used for the analysis. The parameters used are tabulated in Table 2. As shown in Fig. 25, the double reinforced gravel foundation has a thickness of 50 cm whose bottom is located 2.75 m below the ground level (line A-A in Fig. 25).

Table 2. Parameters used in the FEM analysis

	$E(t/m^2)$	ν	$\gamma(t/m^3)$
Asphalt	2.0×10^5	0.38	2.3
Base course	1.75×10^4	0.43	1.8
Stabilized soil	1.0×10^4	0.40	1.8
Gravel	1.5×10^4	0.43	0.8
Clay	3.0×10^2	0.45	0.6
Culvert	3.0×10^6	0.17	2.5

	$k_s(t/m^3)$	$k_n(t/m^3)$
Joint element	2.50×10^2	1.0×10^4

	$E(t/m^2)$	$A(m^2)$
Bar element	1.8×10^5	1.3×10^{-3}

The results of the FEM analysis are plotted in Figs. 26 and 27 whose reference is line A-A of Fig. 25. The settlements resulting from the body force plus external force are shown in Fig. 26 for both unreinforced and reinforced cases at a uniform load of 6 tsm. The reinforced case yielded less differential settlements. A different scheme was used in Fig. 27 wherein the culvert was excluded from the external force application. The settlements with reference to line A-A (Fig. 25) resulting from external loads only as shown in Fig. 27 indicate less differential settlements in the mechanically stabilized case.

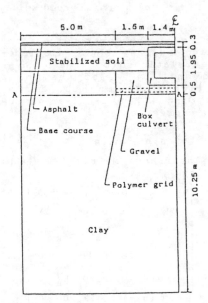

Fig. 25. Box culvert construction with reinforced gravel foundation on soft Araike clay.

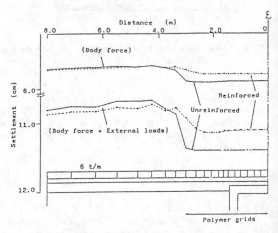

Fig. 26. Settlements at line A-A under uniform loading.

470

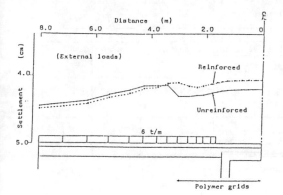

Fig. 27. Settlements at line A-A without load in the box culvert.

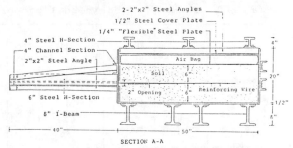

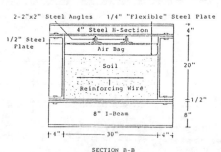

Fig. 28. Different views of the pull out testing cell along the longitudinal (section A-A) and transverse (section B-B) directions.

8 PULLOUT TESTS AND APPLICATIONS OF WELDED WIRE MECHANICALLY STABILIZED EARTH

Welded wire steel grids generate frictional resistance in its longitudinal member and passive resistance in front of the trans-verse member. Chang et al.(1977) was the first to report that the pullout resistance of the welded wire grid was approximately six times greater than for strip reinforcements. In this study, the pullout tests were conducted using a 50"x30"x20" (LxWxH) test cell designed and manufactured at the Asian Institute of Technology (Bergado et al, 1989). The schematic diagram of the pullout apparatus is shown in Fig. 28. The vertical stress was supplied by air bags sandwiched between flexible, 1/4" thick metal plates. The pullout force, measured by means of an electrical load cell, was applied by means of an electrically-controlled hydraulic cylinder with a maximum capacity of 50,000 lbs (22,730 kg). The horizontal displacement of the mat was monitored using a linear variable differential transformer (LVDT) and the pulling rate was 1 mm/min. The data acquisition system consisting of the 21X micrologger recorded both the mat displacement and the axial strains in the longitudinal and transverse members through bonded electrical resistant strain gages. A typical instrumented mat used in the pullout tests is shown in Fig. 29. The reinforcement used were fabricated locally using the readily available mild steel bars. Tensile strength tests

conducted axially on these bars indicate the occurrence of yield stresses at low strains in the order of 0.4 to 0.5%. A modulus of elasticity of 194,737 MPa was calculated corresponding to 0.70 of the yield strength. The backfill materials consist of weathered Bangkok clay and lateritic residual soil whose properties are tabulated in Table 3. The backfill materials were compacted at 95% standard Proctor densities at both dry and wet sides of optimum.

Typical stress-strain relationships for dry and wet side compactions are shown in Figs. 30 and 31, respectively. It can be observed that the pullout strength in the wet side compaction is very much lower compared to the dry side. The plots of strain with distance from the facing are typically shown in Figs. 32 and 33 for the dry and wet side compactions, respectively. The results indicate linear variation of strains along the longitudinal members ranging from 0.01% to 0.2%. The levels of strains reduced when the backfill compaction varies from dry side to wet side of optimum. The dry side exhibited steeper slope of strain variation with distance.

471

Table 3. Summary of basic soil properties

soil Descripttion	Gs	Wp (%)	Wl (%)	Ip (%)	% Passing No.200	$\gamma_{d\ max}$ (t/m³)	Wopt (%)	Direct Shear Test				UU Triaxial Test (95%)	
								Dry Side (95%)		Wet Side (95%)		c (t/m²)	φ (degree)
								c (t/m²)	φ (degree)	c (t/m²)	φ (degree)		
Redish Gray Weathered Clay (CL)	2.67	21	45	24	83	1.60	23.3	8.9	32.2	7.3	16.5	11.80	31.5
Lateritic Residual Soil (GC)	2.61	39	23	16	18	1.93	11.5	3.54* (8.80)#	56.83* (40.20)#	1.43*	47.47*	8.0	32.5

* Low Normal Pressure (0.2 to 1.8 t/m²)   # High Normal Pressure (1 to 13 t/m²)

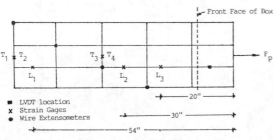

Fig. 29. Typical schematic diagram of an instrumented mat.

- LVDT location
× Strain Gages
● Wire Extensometers

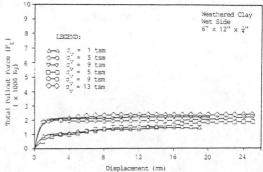

Fig. 31. Typical load-displacement curve for weathered clay compacted at the wet side of optimum.

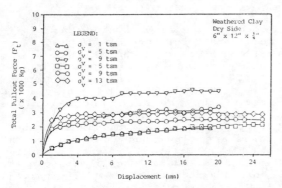

Fig. 30. Typical load-displacement curve for weathered clay compacted at the dry side of optimum.

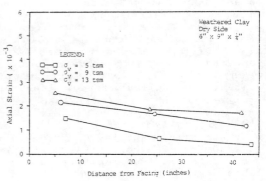

Fig. 32. Variation of axial strain with distance from the facing for weathered clay compacted at the dry side of optimum.

Having stronger soil, the dry side compaction generated higher strains in the reinforcements and lower mat displacements than in the wet side compaction.

Peterson and Anderson (1980) proposed that there is frictional or adhesion resistance at the longitudinal member and bearing resistance in front of the transverse members. The passive resistance generated is similar to that of a strip footing rota-ted to the horizontal direction (Fig. 34a). The frictional or adhesion resistance, F_f, is taken as:

$$F_f = 2\mu\sigma_v A_f \quad \ldots \ldots \quad (7)$$

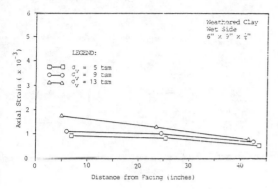

Fig. 33. Variation of axial strain with distance from the facing for weathered clay compacted at the wet side of optimum.

where μ is the adhesion factor between the soil and reinforcement, and A_f is the total surface area of the longitudinal member outside the failure plane. The bearing resistance, F_b, was derived by Peterson and Anderson (1980) as follows:

$$F_b/(Nwd) = cN_c + \sigma_v N_q \quad \cdots \quad (8)$$

where N is the number of transverse wires, w is the width of the reinforcing mats, d is the diameter of the wire and N_c and N_q are Terzaghi's bearing capacity factors.

Jewell et al (1984) suggested to base the failure mechanism for low strains to failure mechanism associated with punching failure mode in the soil as shown in Fig. 34b. Rowe and Davis (1982) have studied the bearing stresses on vertical surfaces loaded horizontally for the case of anchor plates embedded in the soil using elasto-plastic finite element analysis. The bearing stress of the soil on grid members is assumed to be similar to the base pressure on deep foundations. For cohesive-frictional soil, the equation derived by Rowe and Davis (1982) can be expressed as:

$$F_b/(Nwd) = C_u F_c' + hF_\gamma' \quad \cdots \quad (9)$$

where C_u is the soil adhesion which is computed to include the effect of surcharge, F_c' is the factor to account for the effect of cohesion on anchor behaviour, and F_γ' is the factor for the effect of soil weight.

The prediction of passive resistance of the transverse members pro-

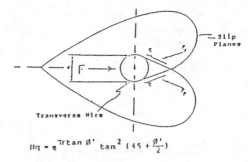

$$Nq = e^{\pi \tan \phi'} \tan^2 (45 + \frac{\phi'}{2})$$

(a) Bearing Capacity Failure Mode
(after Peterson & Anderson 1980)

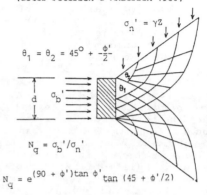

$$\theta_1 = \theta_2 = 45° + \frac{\phi'}{2}$$

$$N_q = \sigma_b'/\sigma_n'$$

$$N_q = e^{(90 + \phi')\tan \phi'} \tan (45 + \phi'/2)$$

(b) Punching Failure Mode
(after Jewell et al 1984)

Fig. 34. Failure mechanisms with respective expressions for N_q.

posed by Peterson and Anderson (1980) seemed to overestimate in the dry side compaction (Fig. 35). The model, however, yielded good agreement with the experimental results for the wet

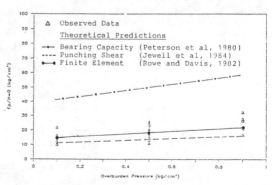

Fig. 35. Experimental vs. theoretical passive resistance of transverse members for weathered clay compacted at the dry side of optimum.

473

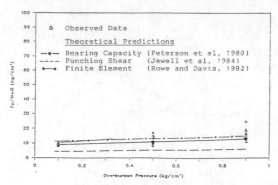

Fig. 36. Experimental vs. theoretical passive resistance of transverse members for weathered clay compacted at the wet side of optimum.

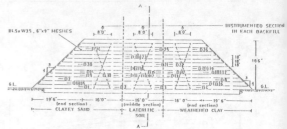

Fig. 37. Front section of the welded wire wall.

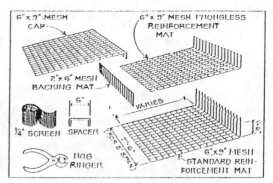

Fig. 38. Accessories used for the construction of welded wire wall (after Hilfiker Co., 1988).

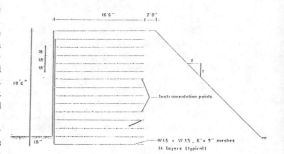

Fig. 39. View of welded wire wall along section A-A showing the instrumented welded wire mat layers.

side compaction (Fig. 36). The closer prediction in the wet side compaction reflected the full mobilization of the passive resistance due to larger mat displacements during the pullout test. It was also observed that full mobilization of passive resistance occurred for smaller diameter bars. The punching failure mode proposed by Jewell et al (1984), which can develop for smaller mat displacements, seems to closely predict the actual behaviour of the soil-grid interaction in the dry side compaction (Fig. 35). The model slightly underestimated the passive resistance in the wet side compaction (Fig. 36). The prediction using finite elements according to Rowe and Davis (1982) seems to yield the best agreement for both dry side and wet side compactions (see Figs. 35 and 36).

A welded wire mechanically stabilized earth (MSE) embankment was constructed inside the campus of the Asian Institute of Technology in conjunction with the USAID Research Project. The experimental embankment was divided into three sections, each about 16 ft long (4.88 m), and utilizing three different backfill materials, namely: clayey sand, lateritic soil, and weathered clay (Fig. 37). The parts of the welded wire wall reinforcements are shown in Fig. 38. It consisted of a vertical welded wire MSE wall facing on one side and a sloping unreinforced wall in the opposite side as shown in Fig. 39. A minimum of about 7 instrumented layers for each backfill soil types were provided. The reinforcement mats consisted of W4.5 x W3.5 wires with 6"x9" grid openings and were instrumented with strain gages at various locations as shown in Fig. 39. Each reinforcement units has dimension of 8' wide and 18'9" long. A schematic layout of the typical instrumentation of the test embankment is shown in Fig. 40.

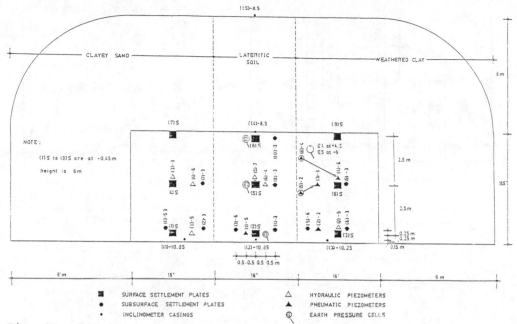

Fig. 40.

Legend (Fig. 40):
- ■ SURFACE SETTLEMENT PLATES
- ● SUBSURFACE SETTLEMENT PLATES
- · INCLINOMETER CASINGS
- △ HYDRAULIC PIEZOMETERS
- ▲ PNEUMATIC PIEZOMETERS
- ⊘ EARTH PRESSURE CELLS

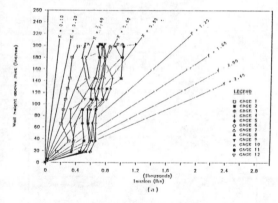

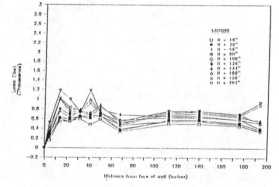

Fig. 41. (a) Plot of tension against wall height above mat for weathered clay (Mat No. 2).

(b) Plot of tension against distance from the face of the wall for weathered clay (Mat No. 2).

From the measured strains, the tension in the wire can be computed, given the modulus of elasticity of steel, and the cross-sectional area of the wire. The computed tensions are plotted for each mat as typically shown in Fig. 41 for the weathered clay backfill in relation to embankment height and distance from the face.

Lateral movements of the subsoil and the embankment were measured using a Digitilt inclinometer. Three of these (I1 to I3) were installed vertically near the face. Two others were installed in the opposite side (Fig. 40). The plots of the lateral movements are given for inclinometer I2 in Fig. 42. The maximum lateral movement in the subsoil occurred at 3.0 m depth, the weakest part. After 29 days since the end of construction, the lateral movements ranged from 110 to 120 mm. The lateral move-

475

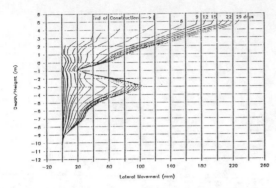

Fig. 42. Plots of depth/height against lateral movement for inclinometer no. 2

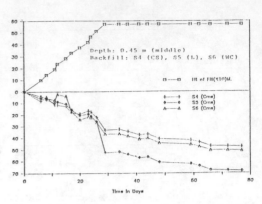

Fig. 44. Observed surface settlements at the center of each section of the wall (S4 to S6).

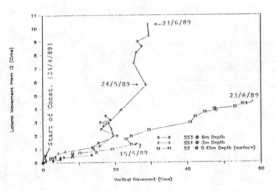

Fig. 43. Lateral movement plotted against the vertical movement of wall (inclinometer no. 2).

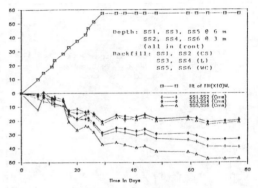

Fig. 45. Observed subsurface settlements of the soft clay foundation (SS1 to SS6).

ment plotted against vertical settlements is shown in Fig. 43. This indicates that the soft clay subsoil is being squeezed out from beneath the embankment, simultaneously with the dissipation of excess pore pressures.

Settlements were measured by levelling survey with reference to a benchmark. The surface settlements at the front near the face have been more or less identical. However, the observed maximum surface settlement occurred at the very center (S5) as plotted in Fig. 44 and the shape of the settlement surface is somewhat like a bowl shape. The subsurface settlements are plotted in Fig. 45.

Pneumatic and hydraulic piezometers were used to measure excess pore pressures. These piezometers were installed at different locations (Fig. 40). The excess pore water pressure decreased at a very slow

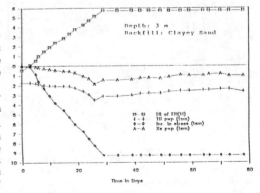

Fig. 46. Measured excess porewater pressure developed in the soft clay foundation (hydraulic piezometer no. 3).

rate immediately after the end of construction (Figs. 46 and 47). The dissipation of excess pore pressure

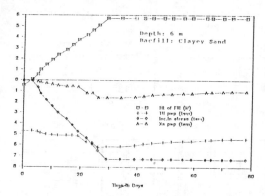

Fig. 47. Measured excess porewater
pressure developed in the
soft clay foundation (hy-
draulic piezometer no. 4).

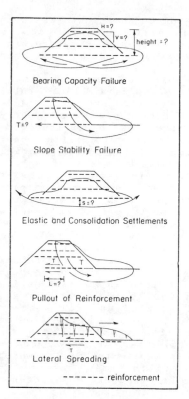

Fig. 48. Design of mechanically stabilized embankment on soft
ground.

indicated the occurrence of the
process of consolidation. Also shown
in these figures were the increase in
vertical stress as calculated by the
method of Poulos and Davis (1974). In
all cases, the excess pore pressures
were found to be far below the surcharge load. The excess pore pressures were also affected by the
seasonal fluctuation of the groundwater level.

To determine the vertical pressure
distribution along the base of the
wall, SINCO pneumatic total earth
pressure cells were installed along
the centerline of the embankments
perpendicular to the wall face
beneath the lateritic soil backfill.
The cell was 9 inch in diameter and
0.434 inch thick, made of stainless
steel. The vertical pressure distribution as observed from the three
cells in the front (E1), middle (E2),
and back (E3) during the embankment
construction varied according to the
changes in the loads during the
different stages of construction.

9 ADVANTAGES OF MSE CONSTRUCTION ON
  SOFT GROUND

As stated previously, the main foundation problem in coastal plain areas
is the presence of thick, soft clay
deposits which is very weak and compressible material. Associated with
the low strength of the subsoil is
the problem of low bearing capacity
and slope instability. To solve these
problems, the conventional approach
is to use very gentle slope of 3H to

1V. Moreover, the embankments cannot
be constructed very high because of
slope instability or sliding problems
(see Fig. 48). Mechanical stabilization by reinforcing the embankment
will allow steeper slope of 1H to 1V
which entails less volume of fill and
savings in material costs as well as
construction of higher embankments.
Due to the compressible clay foundation, earth structures on soft clays
such as road embankments can be subjected to large settlements to as
much as 2 m in 10 years. These settlements are caused by compression of
the soft clay due to the imposed
loads of the embankments as well as
the ground subsidence due to pumping
of groundwater. Since the embankments
cannot be constructed very high
because of stability problems, costly
reconstruction is needed to raise the
embankments above maximum flood level
after 10 years. Utilizing reinforced
earth embankment, the stability will
be improved. Thus, higher embankments
can be constructed extending the

477

design life of earth structures to as
much as 30 years. Another problem
occurring on earth structures resting
on soft clay foundation such as road
embankments is the phenomenon of lateral spreading of the embankment
that will contribute to large total
differential settlements (Fig. 48).
This problem can be prevented by utilizing reinforced earth construction.
Furthermore, reinforced earth construction can tolerate differential
movements common in areas of subsiding ground. Reinforced earthfills
will function as a stiff raft floating over compressible stratum with
the reinforcements providing tensile
strength to the earthfills. This method of construction also redistributes the load to become more uniform, thereby, reducing differential
settlements. Thus, reinforced earthfill can also form as an alternative
foundation for residential houses,
industrial buildings, oil storage
tanks, and other lightly-loaded
structures. The present state-of-
practice utilizes precast, reinforced concrete pile foundation which
can create differential movements
between the pile supported structures
and the surrounding soil due to the
consolidation or compression of the
clay and ground subsidence.

10 IN-SITU GROUND IMPROVEMENT TECH-NIQUES

The MSE constructions can be effectively combined with other in-situ
underground improvement techniques
for maximum efficiency. Several below
ground improvements have been
studied, namely: lime/cement piles,
granular piles, and vertical drains
and are described below.

Lime/cement piles operate on the
principle that the calcium ions in
both lime and cement react with the
clay through the processes of ion exchange and flocculation as well as
pozzolanic reaction. The divalent
calcium ions replace the monovalent
sodium and hydrogen ions in the
double layer surrounding each clay
mineral. Thus, fewer number of divalent calcium ions is needed to neu-
tralize the net negative charge of
each clay minerals reducing the size
of the double layer and increasing
the attraction of the clay particles
leading to flocculated soil structure. Furthermore, the silica and

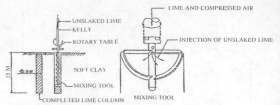

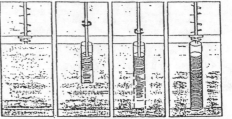

(A) LIME/CEMENT PILE INSTALLATION BY SWEDISH METHOD

(B) LIME/CEMENT PILE INSTALLATION BY JAPANESE METHOD

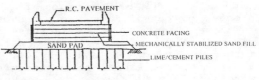

(C) TYPICAL COMBINATIONS

Fig. 49. Installation of lime/cement
piles.

alumina in the clay react with the
lime, forming such cementing agents
such as calcium silicates and calcium
aluminate hydrates in a process
called pozzolanic reaction. The organic content of the soft Bangkok
clay seemed to vary from 2 to 5% with
occasional value of 9 % while the
salt contents can be as much as 0.5
to 2 % (5 to 20 g/ℓ). The results of
the testing program at the Asian
Institute of Technology have found
that adding 5 to 10 % of quicklime is
the optimum mix proportion (Balasubramaniam et al, 1988). The addition
of quicklime increased the unconfined
compressive strength to about 5 times
as much and increased the preconsolidation pressure by 3 times as much.
The coefficient of consolidation also
increased by 10 to 40 times and the
effective strength parameters also
in-creased, especially the angle of
internal friction from 24° to 40°.
Moreover, it was found from actual
experiments that the admixtures in
the clay did not diffuse into the
surrounding soil and thus, preventing
pollution of the ground. Mixing 10%
cement with the soft Bangkok clay

increased the unconfined compressive strength up to 10 times and increased the preconsolidation pressures by 2 to 4 times as much. An increase in the coefficient of consolidation by about 10 to 40 times was also observed (Law, 1989). Usually, cement is more effective than lime when the organic content in the clay exceeds 8%. Lime/cement piles are constructed either by dry jet mixing or wet jet mixing method (Miura et al, 1987). In the former, the cement or quicklime powder is injected into the deep ground through a pipe with the aid of compressed air and then the powder is mixed with the clay mechanically by means of the rotating wings as demonstrated in Fig. 49. In the latter method, the slurry cement is jetted into the clay by a pressure of about 20 MPa from a rotating nozzle.

Granular piles are composed of compacted sand and gravel inserted into the soft clay foundation by displacement method. It has the main advantage over the reinforced concrete pile in that it redistributes the stresses to the surrounding ground but still retaining some form of stress concentration in the more stiffer material of the granular pile. The pile deforms by bulging into the clay and distributes the stresses at the uppper portion of the subsoil strata rather than transfer ring the stresses to the deeper layers, thus causing it to reinforce the soft ground and at the same time causing the soil to support it (Bergado et al, 1988e). Results of the laboratory and field tests conducted at the Asian Institute of Technology (Bergado et al, 1988a), indicated that the stress concentration factor which is the ratio between the stresses in the pile and the clay, varies from about 2 to 5. The strength and bearing capacity increased by 4 to 5 times and the settlement was reduced by 30 to 50%. Typical applications of granular piles on soft ground are illustrated in Fig. 50.

Vertical drains served to shorten the paths of drainage for the less permeable soft clay, accelerating the rate of consolidation such that most of the settlements occur before or during construction stage. Vertical drainage can be made by inserting sand drains or prefabricated band drains into the soft clay. Preloading is required to create hydraulic

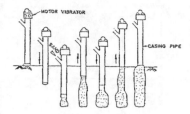

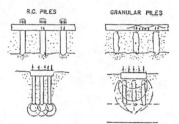

(A) INSTALLATION OF GRANULAR PILES

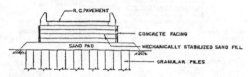

(B) COMPARISON OF CONCRETE AND GRANULAR PILES

(C) TYPICAL COMBINATIONS

Fig. 50. Granular piles (sand or gravel piles).

gradients for the pore water in the clay to follow into the drains and can be applied by means of embankment surcharge or by the use of vacuum pressure (Moh et al, 1987). This method has been successfully tested at the Asian Institute of Technology in both the laboratory and field conditions (Bergado et al, 1988d). In this case, most of the subsoil compression due to the embankment loading is completed in the first 6 to 10 months with only negligible amounts remaining for the rest of the project life. The typical highway applications of prefabricated vertical drains are illustrated in Fig. 51.

11 RESEARCHES ON GROUND IMPROVEMENT TECHNIQUES

Current researches on different techniques of ground improvement is currently going on at the Geotechnical and Transportation Engineering Division of AIT to mitigate such natural hazards such as ground subsidence and compressible foundation subsoil. Among the soil improvement techniques

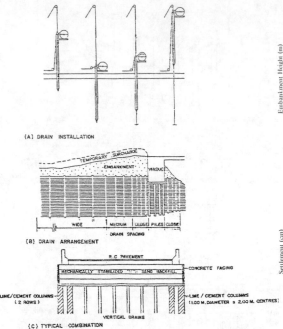

(A) DRAIN INSTALLATION

(B) DRAIN ARRANGEMENT

(C) TYPICAL COMBINATION

Fig. 51. Vertical drains with preloading and lime/cement piles.

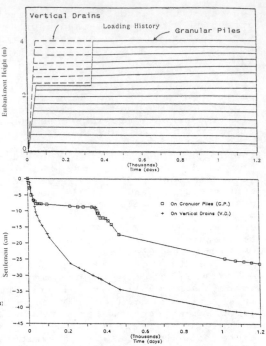

Fig. 52. Settlement behavior of embankments on granular piles
and on vertical drains.

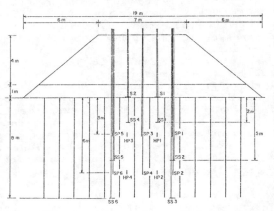

Fig. 53. Section view of embankment
on prefabricated Alidrains.

that can be applicable are reinforced
earth (mechanically stabilized
earth), granular piles, vertical
drains, and lime or cement piles.
Already, test embankments have been
constructed on soft Bangkok clay improved with granular piles (Bergado
et al, 1988a) and vertical Mebra band
drains (Bergado et al, 1988d) inside
the AIT campus. A comparison of the
settlement behaviour of the two embankments on granular piles and on
Mebra drains is shown in Fig. 52. A
third test embankment was recently
completed using a different type of
vertical band drains (Alidrain) which
was installed using both large and
small mandrel sizes. Figures 53 and
54 show the section view and plan
view of the embankment, respectively.
The observed time-settlement is shown
in Fig. 55. Meanwhile, prototype models are being tested in the AIT soil
engineering laboratory concerning reinforced earth, granular piles, vertical drains, and lime/cement stabilized soil.

The U.S. Agency for International
Development (USAID) through USAID
(Thailand) under the Program on
Science and Technology Cooperation
has granted the Asian Institute of
Technology a three-year research pro-

ject with total cost of US$150,000.00
through Dr. Dennes T. Bergado of the
Geotechnical and Transportation Engineering Division as the principal
investigator. The main objective of
this research project is to develop
the design and construction guidelines for using locally available
weathered Bangkok clays including lateritic soils and clayey sands, which
are classified as cohesive soils to

480

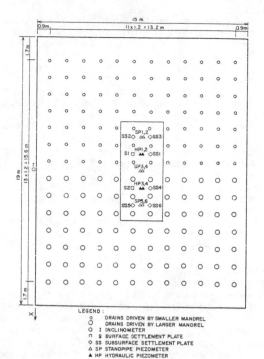

LEGEND :
o   DRAINS DRIVEN BY SMALLER MANDREL
Ö   DRAINS DRIVEN BY LARGER MANDREL
O   I INCLINOMETER
⊓ S SURFACE SETTLEMENT PLATE
◇ SS SUBSURFACE SETTLEMENT PLATE
△ SP STANDPIPE PIEZOMETER
▲ HP HYDRAULIC PIEZOMETER

Fig. 54. Plan view of embankment on prefabricated Alidrains.

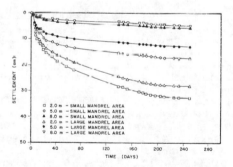

□ 2.0 m - SMALL MANDREL AREA
○ 5.0 m - SMALL MANDREL AREA
▲ 8.0 m - SMALL MANDREL AREA
△ 2.0 m - LARGE MANDREL AREA
● 5.0 m - LARGE MANDREL AREA
▽ 8.0 m - LARGE MANDREL AREA

Fig. 55. Observed time-settlement curve for embankment on prefabricated Alidrains.

be used as backfill materials with welded steel bars as grid reinforcement. The main work includes construction of large scale pullout box in the laboratory for pullout tests to study the interaction between the backfill soil and the reinforcement using advanced electronic data measurements and collection. Preliminary results and analyses of the data on pullout tests have been published (Bergado et al, 1989). Also, a full scale welded wire wall embankment system with a height of 5.5 m was

recently constructed at the campus of AIT. The wall is shown in Fig. 37 and consists of a vertical face with sloping sides. Field pullout tests will be performed on dummy reinforcements embedded at different depths of the wall. The corrosion rate will also be monitored on buried dummy reinforcements. This research work forms a logical extension to the current knowledge on soil reinforcement as understood for granular soils, which develop high frictional resistance between the soil and reinforcement elements.

12 CONCLUSIONS

The existing natural hazards in coastal plain areas such as the presence of soft ground and the effect of ground subsidence due to excessive pumping of groundwater, require different schemes of soil and ground improvements for infrastructure constructions to prevent slope instability and minimize total and differential settlements. It is concluded that the appropriate scheme will be a combination of either lime/cement piles or granular piles for improvement in the subsoil with overlying mechanically stabilized (reinforced earth) embankment fill. It is noted that both lime/cement piles and granular piles are also basically a form of earth reinforcement. The vertical drains combined with preloading are quite effective in precompression and preconsolidation of the soft clay subsoil but this scheme often needs some time to execute. The various soil/ground improvement techniques described in this paper are economical and viable alternatives to the existing method of using precast, reinforced concrete piles to support embankments, especially in the transition units between the pile-supported viaducts (overpasses) and the at-grade road embankments.

REFERENCES

Balasubramaniam, A.S., Bergado, D.T., Buensuceso, B. and Yang, C.W. (1988). Ground Improvement Techniques for Approach Road Design and Rehabilitation of Road Embankment. Proc. Symp. on Roads, Bridges and Highways for 21st Century, Bangkok, 14 pp.

Bergado, D.T., Bukkanasuta, A. and Balasubramaniam, A.S. (1987). Laboratory Pullout Tests Using Bamboo and Polymer Geogrids Including a Case Study. Geotextiles and Geomembranes Journal, Vol. 5, pp. 153-189.

Bergado, D.T., Miura, N., Panichayatum, B. and Sampaco, C.L. (1988a). Reinforcement of Soft Bangkok Clay Using Granular Piles. Proc. Intl. Geotech. Symp. on Theory and Practice of Earth Reinforcement, Fukuoka, Japan, pp. 179-184.

Bergado, D.T., Sampaco, C.L., Miura, N. and Sakai, A. (1988b). Reinforced Gravel Foundations for Box Culvert and Sewage Pipeline Construction on Soft and Subsiding Ground. Proc. 1st Indian Geotextiles Conference, Bombay, India.

Bergado, D.T., Sampaco, C.L., Miura, N. and Onitsuka, K. (1988c). Interaction of Geogrid Reinforcement with Gravel and Decomposed Granite Backfill with Applications. Proc. Intl. Conf. on Geomechanics of Tropical Soils, Singapore, pp. 351-357.

Bergado, D.T., Miura, N., Singh, N. and Panichayatum, B. (1988d). Improvement of Soft Bangkok Clay Using Vertical Band Drains Based on Full Scale Test. Proc. Intl. Conf. on Eng'g. Problems of Regional Soils, Beijing, China, pp. 379-384.

Bergado, D.T., Singh, N., Sim, S.H., Panichayatum, B. and Sampaco, C.L. (1988e). Improvement of Soft Bangkok Clay Using Vertical Drains Compared with Granular Piles. Geotextiles and Geomembranes Journal (in press).

Bergado, D.T., Cisneros, C.B., Shivashankar, R., Alfaro, M.C. and Sampaco, C.L. (1989). Pullout Resistance of Steel Geogrids with Weathered Clay Backfill. Proc. Symp. on the Application of Geosynthetic and Geofibre in SE Asia, Kuala Lumpur, Malaysia.

Chang, J.C., Hannon, J.B. and Forsyth, R.A. (1977). Pull Resistance and Interaction of Earthwork Reinforcement and Soil. Transportation Research Board, 56th Annual Meeting.

Cox, J.B. (1970). The Distribution and Formation of the Recent Sediments in Southeast Asia. Proc. 2nd Southeast Asian Conference on Soil Eng'g., Singapore, pp. 29-47.

Fowler, J., Peters, J. and Franks, L. (1986). Influence of Reinforcement Modulus on Design and Construction of Mohicanville Dike No. 2. Proc. 3rd Intl. Conf. on Geotextiles, Vienna, Austria, pp. 267-271.

Ingold, T.S. (1983). Laboratory Pullout Testing of Grid Reinforcement in Clay. Geotechnical Testing Jour., Vol. 6, No. 3, pp. 112-119.

Jewell, R.A., Milligan, G.W.A., Sarsby, R.W. and Dubois, D. (1984). Interaction Between Soil and Geogrids. Proc. Symp. on Polymer Grid Reinforcement. Thomas Telford Ltd., London.

Jones, C.J.F.P. (1985). Earth Reinforcement and Soil Structures. Butterworths Book Co., London.

Law, K.H. (1989). Strength and Deformation Characteristics of Cement-Treated Clay. M.Eng. Thesis No. GT-88-6, AIT, Bangkok.

Miura, N., Bergado, D.T., Sakai, A. and Nakamura, R. (1987). Improvement of Soft Marine Clay by Special Admixtures Using Dry and Wet Mixing Methods. Proc. 9th Southeast Asian Geotech. Conf., Bangkok, Thailand.

Moh, Z.C. and Woo, S.M. (1987). Preconsolidation of Soft Bangkok Clay by Non-Displacement Sand Drains and Surcharge. Proc. 9th Southeast Asian Geotech. Conf., Bangkok, Thailand.

Netlon Ltd. (1984). Test Methods and Physical Properties of Tensar Geogrids. Technical Guideline of Netlon Ltd., England.

Nielsen, M.R. and Anderson, L.R. (1984). Pullout Resistance of Wire Mats Embedded in Soil. Report Submitted to the Hilfiker Co., Utah State University, Logan, Utah.

Ochiai, H., Hayashi, S., Ogisaku, E. and Sakai, A. (1987a). Analysis of Polymer Grid Reinforced Soil Retaining Wall. Proc. 8th Asian Regional Conference, Tokyo, Japan.

Ochiai, H. and Sakai, A. (1987b). Analytical Method for Geogrid Reinforced Soil Structures. Proc. 8th Asian Regional Conference, Tokyo, Japan, pp. 483-486.

Peterson, L.M. and Anderson, L.R. (1980). Pullout Resistance of Welded Wire Mats Embedded in Soil. Utah State University, Logan, Utah.

Poulos, H.G. and Davis, E.H. (1974). Elastic Solutions for Soil and Rock Mechanics. John Wiley & Sons Inc., New York.

Rowe, R.K. and Davis, E.H. (1982). The Behaviour of Anchor Plates in Sand. Geotechnique, Vol. 32, No. 1, pp. 25-41.

Developments in Geotechnical Aspects of Embankments, Excavations and Buried Structures, Balasubramaniam et al. (eds)
© 1991 Balkema, Rotterdam. ISBN 90 5410 019 2

Interaction of steel geogrids and low-quality, cohesive-frictional backfill and behavior of mechanically stabilized earth (MSE) wall on soft ground

D.T. Bergado, A.S. Balasubramaniam, K.H. Lo, R. Shivashankar, C.L. Sampaco & M.C. Alfaro
GTE Division, Asian Institute of Technology, Bangkok, Thailand

ABSTRACT: This paper deals with the behavior of MSE test wall consisting of steel grids reinforcement embedded in cohesive-frictional backfill soil that was constructed on soft Bangkok clay. The mechanism of interaction between steel grids reinforcement and low-quality, cohesive-frictional backfill is being modelled and analyzed. Laboratory pullout tests were performed using large pullout box designed for this study. Field pullout tests were also conducted on dummy reinforcements embedded in a full scale mechanically stabilized earth wall utilizing three different backfill materials, namely: clayey sand, lateritic residual soil, and weathered clay backfill. The magnitude of mobilized pullout resistance as well as the strains induced in the reinforcing elements were strongly influenced by the variation of the base pressure of the wall which is, in turn, affected by the arching effects of the reinforcements and the pattern of deformation of the soft clay foundation. Laboratory pullout tests generally provided conservative approximation of actual pullout resistance in comparison to the field pullout test results. Finite element analyses were also done to simulate and verify the load-displacement of the reinforcements and the behavior of the MSE wall. The overall agreement between the experimental and theoretical results were quite satisfactory. The measured wall settlements and lateral movements, after more than a year, were quite excessive but were found to have no adverse effects on the overall wall performance. Thus, it can be concluded that MSE wall and embankment systems using steel grids reinforcements embedded in locally-available, cohesive-frictional backfill soils is a viable and economical alternative scheme for infrastructure construction on soft ground.

1 INTRODUCTION

Most major cities in the Southeast Asian region are located in the coastal plains with thick deposits of soft clays. These deposits are found in most countries of the region (Fig. 1) such as the Chao Phraya Plain in Thailand, Mekong Delta in Cambodia and Vietnam, Malaysian Coastal Plain, Philippine Central Plain, and Indonesian Coastal Plain. Most of these coastal plains are characterized by a recent deposit of marine sediments consisting of a topmost layer of compressible clay underlain by alternating layers of stiff clay and dense sand with gravel. Typical of these features is the flat, deltaic-marine deposit of the Chao Phraya Plain in Thailand, wherein Bangkok metropolis is located, which covers a width of about 200 km and a north-south dimension of about 300

Fig. 1 The distribution of recent clays in Southeast Asia.

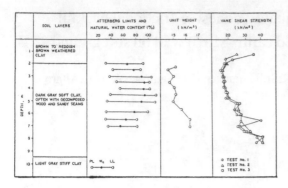

Fig. 2 Typical subsoil profile and geotechnical properties at AIT campus.

km. The typical near surface subsoil profile at the campus of the Asian Institute of Technology (AIT), located about 45 km north of Bangkok is shown in Fig. 2 with the corresponding geotechnical properties. The presence of a thick layer of soft clay can pose considerable problems to infrastructure constructions within the coastal plains because of its high compressibility and low strength. Majority of these infrastructure constructions are road embankments, flood control dikes, landfills, embankments along irrigation canals, etc. In the case of road embankments, such infrastructures are subjected to height restrictions of about 3.4 m with a gentle slope of 3H to 1V to avoid stability failures. But even with this height, highway embankments can undergo excessive settlements of about 2 m in about 10 years time, sinking below their maximum flood level, and thus, requiring costly reconstruction and maintenance works (Bergado et al, 1990a). On the other hand, rapid expansion and high cost of land often compels for steep and high embankments without the normally wide and flat slopes. To alleviate such problems, mechanical stabilization by earth reinforcement (mechanically stabilized earth) can be a viable alternative. The technique offers specific technical, economic, and aesthetic advantages compared to the more conventional methods, making the principle to be presently adopted for construction of retaining walls, embankment slope repairs, retention of excavations and in-situ slope stabilization. The deformation response characteristics of mechanically stabilized earth structures often provide technically attractive solutions on sites with poor foundation soils owing to their extreme tolerance of large deformations, both laterally and vertically, compared to conventional retaining walls. Their flexibility also allows the use of lower factor of safety for bearing capacity design. Moreover, the reinforcement in the soil tend to distribute the stresses in the foundation subsoil, hence, eliminating the use of concrete piles for lightly-loaded structures. Furthermore, mechanical stabilization by steel grids reinforcement will minimize differential movements within the earthfill, reducing the occurrence of cracking in the pavements.

Mechanically stabilized earth (MSE) consists of reinforcing the soil using steel or polymer (plastic) grid materials. The reinforcement which is strong in tension combines with the soil which is strong in compression, forming a very strong and semi-rigid composite material. The tension in the reinforcement is mobilized by the interaction between the reinforcement and the soil in the form of friction or adhesion and bearing resistances. The grid reinforcements usually generates pullout resistance up to 6 times higher than the strip reinforcements (Nielsen and Anderson, 1984). Among the grid reinforcement, steel grids are found to be superior than the polymer grid in terms of low extensibility and higher tensile strengths (Fowler et al, 1985). Since the mobilization of the interaction between the soil and reinforcement depends on the relative strains generated in the reinforcement, steel grids are preferable.

High quality granular backfill materials are suitable for mechanically stabilized earth constructions. However, these backfill materials are not often readily available especially within the coastal plains and thus, are expensive due to high transportation cost. The use of locally-available, poor-quality, cohesive-frictional backfill materials is, therefore, imperative for economic reasons. Granular backfill materials are considered high in quality and suitable for such construction because they are free-draining and consequently the stress transfer between the reinforcement and soil backfill is said to be immediate as lift of the backfill is placed, and shear strength increase will not lag behind vertical loading. They usually behave as elastic materials in the load level of practical interest such that no post construction movements associated with internal yielding or readjustments are anticipated. On the contrary, fine-grained materials are considered as poor in quality for use in mechanically stabilized earth constructions owing to their poor drainage characteristics and often exhibit elastoplastic or plastic behavior which increases the possibility of post-construction movements due to creep. In addition, as soils become more fine-grained, their resistivity, which is an important factor controlling the rate of galvanic corrosion, generally decreases which is indicative for aggressive soils. With the introduction of grid reinforcements and modern methods of corrosion protection, the use of poor quality backfill materials in mechanically stabilized earth construction has gradually attracted much attention in the recent years due to their added economic advantages and feasibility in other places especially those situated within coastal plains.

The basic design criteria for MSE structures involves satisfying: (i) external stability and (ii) internal stability (Lee et al, 1973; McKittrick, 1978; Anderson et al, 1986a,b; Mitchell and Villet, 1987). External stability is evaluated by considering the entire reinforced soil mass as a semi-rigid structure which is checked for with the conventional criteria such as (a) overturning, (b) sliding, (c) bearing capacity, and (d) deep stability (conventional slope stability with a failure surface below the reinforced mass). The internal stability of reinforced soil structures requires an evaluation of: (a) tension in the reinforcing elements and (b) pullout resistance of reinforcing elements. As pointed out by Ingold (1983), the pullout test is the most realistic model to study the soil-reinforcement interaction problems.

This paper is a partial result of the on-going research project at the Asian Institute of Technology, concerning the investigation of the potential use of widely available, but poorer in quality, backfill materials in walls and embankments stabilized mechanically by welded wire mesh (steel geogrids) reinforcements. The research project involves the construction of a full-scale welded wire wall/embankment system on compressible foundation and the study of the soil-reinforcement interaction by laboratory and field pullout tests. Results of the laboratory pullout tests have been published earlier (Bergado et al, 1989a,b; Bergado et al, 1990b,c,d,e).

2 THE MECHANICALLY STABILIZED EARTH (MSE) TEST WALL

2.1 Description, construction and instrumentation Program

The test embankment is divided into 3 sections along its length, consisting of about 14.64 m long at the top, and

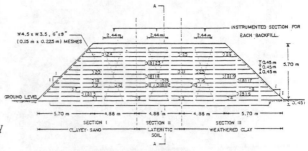

Fig. 3 Front sectionof the mechanically stabilized earth (MSE) wall.

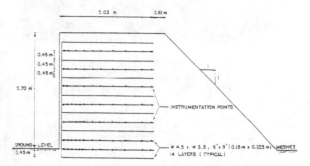

Fig. 4 Typical cross-section of the wall showing instrumentation points in the reinforcing mats.

composed of 3 different backfill materials (Fig. 3). It has a vertical welded-wire reinforced wall with a wire mesh facing unit on one side, and a sloping, unreinforced embankment along the opposite side as shown in the section view in Fig. 4. Table 1 summarizes the relevant geotechnical properties of the backfill materials used in the embankment. The reinforcing mats used were 2.44 m wide by 5.0 m long galvanized, welded wire mesh of W4.5 x W3.5 (diameter of 6.0 mm and 5.40 mm, respectively) wires on a 6 x 9 inches (0.15 x 0.225 m) grid openings. Seven of these mats were instrumented with strain gages for each section as shown in Fig. 4. The embankment construction involved the placement of the reinforcement, with the bent-up portion forming a part of the facing elements. The first layer of the reinforcement mats was laid 0.45 m below the ground level. Backing mats and screens were provided along the vertical side of the wall and backfilled with pea gravel which extended up to 0.45 m towards the embankment to prevent erosion of the cohesive backfill soil. The fill layers between reinforcing mats were placed and

485

Fig. 5 Photograph of the mechanically
stabilized earth (MSE) wall
(a) front view and (b) oblique view

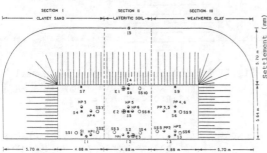

DEPTH OF INSTRUMENTATION IN METERS (BELOW GROUND SURFACE)

LEGEND	No. OF INSTRUMENTATIONS									
	1	2	3	4	5	6	7	8	9	0
▪ SURFACE SETTLEMENT PLATES (S)	0.45	0.45	0.45	0.45	0.45	0.45	0.45	0.45	0.45	–
O SUBSURFACE SETTLEMENT PLATES (SS)	5.30	3.00	6.00	3.00	6.00	3.00	3.00	3.00	6.00	3.00
⊕ INCLINOMETER CASINGS (I)*	10.85	10.85	10.25	8.50	8.50	–	–	–	–	–
⊕ HYDRAULIC PIEZOMETERS (HP)	5.00	5.00	3.00	6.00	7.00	4.00	–	–	–	–
⊕ PNEUMATIC PIEZOMETERS (PP)	5.00	2.00	3.00	6.00	12.00	4.00	–	–	–	–
⊕ EARTH − PRESSURE CELLS (E)	0.45	0.45	0.45	–	–	–	–	–	–	–

* UP TO TOP OF EMBANKMENT   * 5 & 6 IN EMBANKMENT

Fig. 6 Schematic layout of field instrumentation for the MSE wall.

compacted in 3 equal compaction lifts up
to a total thickness of about 0.45 m
corresponding to the vertical spacing of
the reinforcing grids. Each lift was
compacted by a combination of a hand
tamper and a roller to the specified
density of about 95% standard Proctor
compaction. The placement moisture content
was maintained within 1-2% on the dry side
of optimum as verified by the Troxler
nuclear densitometer. Figure 5 shows a
photograph of the completed MSE test
embankment. An instrumentation program,

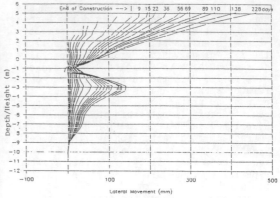

Fig. 7 Typical plots of wall lateral
movement against depth or height
(Inclinometer No. 2).

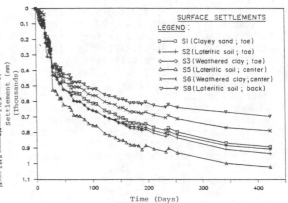

Fig. 8 Plots of surface settlemetns with
time.

primarily consisting of strain measurements
on the seven instrumented blayers in each
section, was developed to evaluate the
performance of the MSE test embankment. In
addition, dummy reinforcing mats as shown
in Fig. 3 were also instrumented and
embedded at different levels of each section for field pullout tests. Additional
dummy reinforcing mats for corrosion observation were embedded at different locations for later retrieval. The schematic
layout of the field instrumentation is
given in Fig. 6.

2.2 Lateral displacements, settlements, and excess pore pressure

The typical plots of the lateral movements
for inclinometer I2 are shown in Fig. 7
which is similar to that of inclinometers
I1 and I3. After 228 days from the end of

construction period, the maximum outward lateral movement measured at the top of the wall face was about 300 mm. The maximum lateral movement in the subsoil at 3 m depth was about 110 mm, which indicated the potential location of shear failure surface corresponding to the weakest soft clay layer. The rate of lateral movement in the subsoil was, however, observed to be decreasing with time. The direction of the subsoil lateral movements in I4 and I5 are opposite and of smaller magnitudes compared to that near the face which seems to indicate that the soil is being squeezed out from the front and from the back, but predominantly from the front corresponding to the heavier load. The surface settlement-time relationships at different sections of the embankment are shown in Fig. 8. The surface settlements at the front along the longitudinal section beneath the interconnected wire mesh facing of the wall (S1,S2,S3) have been almost identical with a magnitude of 0.90 m. Similarly, the subsurface settlements at 6 m depth were also found to be nearly identical at 0.25 m. However, at 3 m depth below the same longitudinal section, settlement plate SS4 at the middle section settled an amount of 0.51 m which is comparably lower than its adjacent settlement plates SS2 and SS6 which settled at 0.67 m and 0.75 m, respectively. The maximum surface settlement occurred at the center plate S5 below the middle section (lateritic residual soil) such that the overall settlement pattern at the surface indicates a dish-like configuration. The vertical (d) and lateral (δ) deformations were plotted in d-(δ/d) coordinates, similar similar to the diagram for construction control of embankments on soft ground (Matsuo and Kawamura, 1977). As shown in

Fig. 9, the plots are way below the critical boundary curves of $p_j/p_f = 0.90$ which was the suggested failure criterion for assessing the safety of embankments. Piezometer readings taken at different locations beneath the wall indicated that the porewater pressures continued to dissipate, though at a very slow rate as typically shown in Fig. 10.

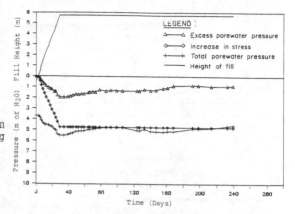

Fig. 10 Typical plots of excess pore pressure with time.

2.3 Earth pressure at the base of the wall

To determine the vertical pressure distribution at the bottom of the wall, SINCO pneumatic total pressure cells were used (Fig. 6). During the construction of the first 4 layers, the base pressure at 0.50 m behind the face (E1) was increasing (from 1 to 29 kN/m²) and higher than E2 and E3 (each recorded a constant pressure of 1 kN/m²), implying that the center of pressure is located near the face. After 8 layers, the center of pressure tends to have been shifted backwards as observed from the base pressure readings, probably due to the increase in embankment weight, and with it, the increase in the surface settlements at the center. By the end of the twelfth layer, the base pressures recorded in all the three cells were nearly the same at 55 kN/m². The surface settlements near these points were also about the same at 26 mm. However, towards the end of construction, E2 recorded a base pressure of 70 kN/m² which is greater than 63 kN/m² at E1, and much greater than the pressure of 50 kN/m² at E3, resulting to the drastic increase of settlement at S5. It has been observed throughout the post-construction phase that any abrupt increase in E2 is followed

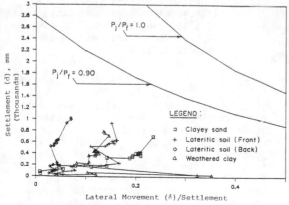

Fig. 9 Safety assessment of wall performance during construction.

by a release of pressure at E1. When E1 starts increasing from its lowest value, there will be at first a slight release in the value of E2, and thereafter, E2 again starts increasing gradually at first for a while and then at some stage, an abrupt increase. This whole process seems to develop a cyclic variation in the base pressures caused by arching effects due to interconnection of reinforcements at the facing units of the wall. This process is expected to continue until consolidation of the subsoil is completed. Any abrupt increase in the value of E2 coupled with large settlement therein at the center, was also reflected by the sharp increase in porewater pressures at the center and the variable mode of strain generation in the reinforcing mats as discussed in the next section. The base pressure distributions are plotted in Fig. 11.

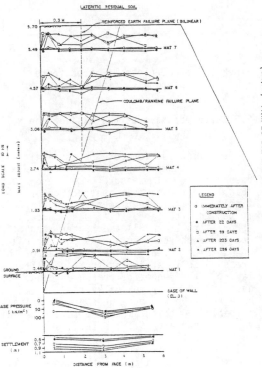

Fig. 11 Variations of reinforcement tension, base pressure, and settlements at different periods for the middle section.

2.4 Tension in the reinforcing wires

The 21X datalogger with a multiplexer and storage module was used to store the data from the temperature compensating electrical-resistant strain gages which were mounted diametrically opposite each other. The reinforcement tensions immediately after construction and four other periods after construction, are depicted in Fig. 13 for the middle section of the test embankment, including the settlement profiles and base pressures. After 22 days from the end of construction, it is seen that some reinforcing mats displayed a sudden release of stresses in all sections corresponding to the abrupt decrease in earth pressure near the face (E1) to almost a zero value, with E2 and E3, retaining almost the same magnitude. After 89 days, the earth pressures at 3 locations were all found to increase drastically at almost the same rate which subsequently increased the strains in the reinforcement for all layers. The maximum lateral pressures immediately after construction are plotted with depth in Fig. 12 for the middle section, and were compared to existing earth pressure theories

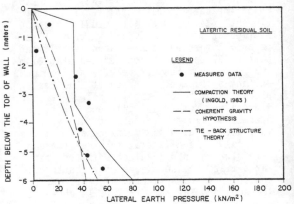

Fig. 12 Comparison between the measured lateral earth pressures immediately after construction, and the various theoretical expressions (middle section).

on reinforced soil structures as discussed by Jones (1985). The measured values immediately after construction were higher than the coherent gravity and tie-back structure hypotheses but seems to be closely predicted by the compaction theory proposed by Ingold (1983). It is interesting to note how arching effect has altered the lateral pressures in the lateritic soil section, which is supposed to be the strongest backfill used in the wall, by yielding lower measured lateral pressures than the other 2 sections. This

effect was verified from field pullout
tests on dummy reinforcing mats located at
the center section which indicated con-
trasting results with theoretical expectations in terms of lower pullout resistances
than both weathered clay and clayey sand,
and strikingly, decreasing pullout resistances with increasing overburden pressures.
On the other hand, the pullout resistances
from laboratory pullout tests show increasing pullout capacity with increasing
normal pressures (Bergado et al, 1989a,b;
1990b,c,d,e).

2.5 Lateral earth pressure coefficient, K

Typical variations of reinforcement tension
during construction as each lift of the
backfill was placed above the mat are
given in Fig. 13 for the middle section.

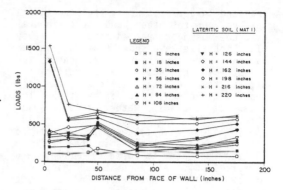

Fig. 14 Variation of reinforcement tension
with distance from the face of wall
at different fill heights for mat
no. 1 in the middle section of the
wall.

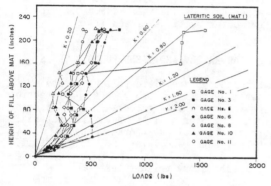

Fig. 13 Variation of wire tensions with
height of fill for mat no. 1 in
the middle section of the wall.

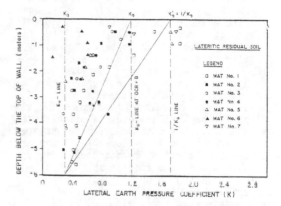

Fig. 15 Measured lateral earth pressure
coefficient (K) during construction of the wall at the middle
section.

The graphs were replotted to show the
variation of tension with distance from
the face of the wall as shown in Fig. 14.
The maximum value of K was obtained
corresponding to the height of the backfill
above each reinforcing mat and plotted as
shown in Fig. 15 for the lateritic residual
soil at the middle section of the wall.
The plots of K indicate an increasing
trend as we approach the top of the wall,
with most of these values way above the
limiting active value (K_a). Similar trend
was obtained for the other 2 sections of
the wall. This variation is significantly
different from those reported for welded
wire walls with high quality backfill on
comparatively hard foundations (Anderson
et al, 1987) as well as for reinforced
earth walls (McKittrick, 1978). These deviations may be attributed to the flexibility of the foundation subsoil and to the
residual pressures induced by compaction.

Foundation compressibility can enhance
lateral displacement of the wall face as
the test embankment is constructed. The
lateral movement necessary to develop the
fully active case (K_a) had been reported
by Terzaghi (1934) to be only a minimal
fraction of the wall height (H/1000). Any
further movement of the face will increase
the lateral pressure (Carder et al,
1977,1980; Terzaghi, 1934). In this study,
the maximum lateral movement measured for
the wall, immediately after construction,
amounted to about 0.15 m which is much
higher than the required displacement of
0.0057 m to develop the fully active case.
For a grid-reinforced soil wall, this
continuous outward wall movement may cause
the full mobilization of the passive
resistance of transverse members, thereby

489

inducing larger strains in the longitudinal members. On the other hand, experimental evidence using compacted sand and silty clay backfills have shown that the maximum lateral earth pressures throughout the height of the wall are significantly higher than were calculated from the at-rest (K_O) values, especially at the top meter of the wall (Carder et al, 1977, 1980; Murray and Boden, 1979; Ingold, 1983). These high values were attributed to the residual stresses induced by compaction as was also found by Murray and Farrar (1990). Although these magnitude of residual stresses depend on the type and size of the compaction plant employed, the maximum stresses at the very top of the wall cannot exceed the lateral pressure calculated using the coefficient of earth pressure for unloading, K_O'(Broms, 1974; Rowe, 1954). Carder et al (1977, 1980) assumed K_O' to have a value of $1/K_O$, where K_O was calculated from the Jaky (1948) equation ($K_O = 1-\sin \phi'$). Figure 15 typically shows the envelopes of K for the middle section which appeared to have a similar trend as the other 2 sections of the wall. The K-envelope varying from K_a to K_O' may be taken as the upperbound. The other envelope which varies from K_a to K_O, where K_O was calculated assuming that the compacted backfill is overconsolidated with an OCR of 8, seems to be more appropriate. These recommended envelopes for K may be applicable to walls constructed up to 6 m height on soft and compressible foundation, and utilizing poor quality backfill materials having both friction and cohesion. This result may also be overconservative for walls which will undergo only very small lateral movements. In addition, the values of K reported herein were calculated from the construction data of the wall and did not take into account the increase in vertical pressures due to overturning moment. If such a factor is considered, lower back-figured values of K will result.

2.6 Maximum tension line

The response generated by the wall due to foundation compressibility creates a unique situation wherein existing theories on earth pressures may not be directly applicable. Current design methods use either classical earth pressure theories such as those of Coulomb and Rankine, or an empirical design method which usually involves the assumption of an equivalent fluid pressure distribution. While this situation is seldom, if not encountered in practice, it can be said that the present design method presupposes the very specific combinations of a rigid wall rotating actively about its toe with a backfill acted on by gravitational forces only. For reinforced soil walls, the maximum tension line was reported to define a failure surface or wedge of a Coulomb/Rankine type failure plane, bilinear failure plane, or the log spiral failure plane. Any of these conditions may not be satisfied if foundation flexibility gets into the picture as in the case of the base pressure distributions cited earlier. It was, however, found that even with this variable mode of earth pressure at the base of the wall, the maximum tension line for this study (Fig. 11) seems to closely conform to the log spiral failure plane (farther from the face than the Coulomb/Rankine failure plane) at the lower half of the wall, and conform to the reinforced earth failure plane (closer to the face than the Coulomb/Rankine failure plane) for the upper half of the wall, especially at the two outer sections (I and III).

3 LABORATORY AND FIELD PULLOUT TESTS

Laboratory and field pullout tests were carried out to study the interaction between the steel geogrids reinforcement and the poor quality backfill materials consisting of weathered Bangkok clay, clayey sand, and lateritic residual soil.

3.1 Laboratory pullout tests

In the laboratory, pullout tests were conducted using a 50"x30"x20" (LxWxH) test cell made up of 1/2" thick steel plates and was designed and fabricated at the Asian Institute of Technology. The schematic diagram of the pullout apparatus is shown in Fig. 16. The vertical stress was supplied by an air bag located between the 1/4" thick, flexible metal plates. The pullout force, measured by means of a load cell, was applied to the test specimen using an electrically-controlled hydraulic cylinder mounted against the supporting frame of the pullout cell. The horizontal displacement of the mat was monitored using the linear variable differential transformer (LVDT) and was pulled out at the rate of 1 mm/min. The data acquisition system (21X micrologger with AM416 multiplexer) recorded both the mat displacement and the axial strains in the longitudinal and transverse members by means of strain gages. The tests were performed on varying wire sizes, mesh geometry and compaction conditions of the backfill materials. The reinforcements used in this study were fabricated locally using

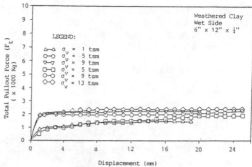

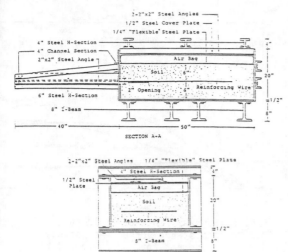

Fig. 16 The pullout testing cell used in laboratory pullout tests.

readily available mild steel bars. Tensile strength conducted on these bars indicated occurrence of yield stresses at low strains in the order of 0.2% to 0.3% and a modulus of elasticity of 30 x 10 psi. The rein-forcements consisted of 1/4", 3/8", and 1/2" diameter steel bars with varying aperture sizes of 6"x9", 6"x12" and 6"x18". The backfill materials were com-pacted at 95% standard Proctor densities at both dry and wet sides of optimum. Typical stress-strain relationships for dry and wet sides of optimum compaction for the weathered clay backfill are shown in Figs. 17 and 18, respectively. It can be observed that the pullout resistance in the wet side compaction is very much lower

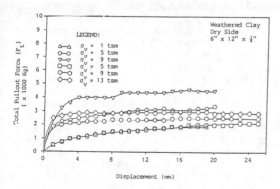

Fig. 17 Typical load-displacement curves for weathered clay compacted at the dry side of optimum.

Fig. 18 Typical load-displacement curves for weathered clay compacted at the wet side of optimum.

compared to the dry side although both conditions was found to increase with increasing confining normal stresses, as observed for good quality granular backfill materials. The same trend was also found for the other types of backfill materials used in this study. However, lateritic residual soil backfill seems to have higher pullout resistance than the clayey sand and weathered clay backfills, in descending order. Reinforcements with their transverse members removed were also tested to determine the contribution of the longitudinal members through adhesion/ friction with the soil. In all the tests conducted, the soil-reinforcement interac-tion indicated the dominant role of passive resistance contributed by the transverse members to the total pullout resistance. The friction resistance of the longitudinal members was found to have a minimal contri-bution to the total pullout resistance. It was also found that the mobilized interac-tion depends on the relative displacements between the reinforcement and the backfill, wherein the smaller diameter bars were effective in enhancing the full mobiliza-tion of passive resistance. The 6"x9" mesh geometry seems to be the most efficient of all the grid sizes that were used. The steel grids pulled out with linearly decreasing axial strains from the point of load application, in the order of 0.01 to 0.02% as expected for inextensible rein-forcements.

3.2 Field pullout tests

Field pullout tests were carried out in conjunction with laboratory tests to verify the soil-reinforcement interaction under actual soil condition using field prototypes. The tests were conducted on dummy reinforcing mats embedded at different locations along the vertical face of the wall as shown in Fig. 3 and were performed in the same manner as in the laboratory, utilizing the same pulling machine and data acquisition system. Typical field pullout test set-up is shown

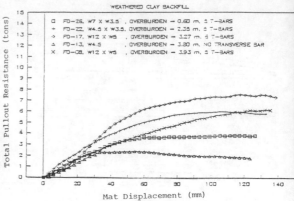

Fig. 20 Summary plots of field pullout resistance versus the mat displacement for the weathered clay backfill.

Fig. 19 Typical set-up of pullout test in the field.

in Fig. 19. The typical stress-strain relationship for the weathered clay backfill is shown in Fig. 20. The pullout resistance is seen to increase with increasing overburden pressure which confirms the laboratory tests and theoretical expectations. The same trend was observed with the clayey sand backfill. On the contrary, the lateritic residual soil backfill showed a decrease in pullout resistance with increasing overburden pressure. This is attributed to the arching effects due to the interconnection of the interface of the reinforcements and higher subsoil deformations at the middle section. This was verified when the pullout resistance of dummy reinforcements embedded at the lower parts of the wall was considerably less compared with the clayey sand and weathered clay backfill sections with the same overburden

pressures as well as the same reinforcement geometry and size. Without arching effects, this should unlikely to happen considering that lateritic residual soil has higher strength which consequently could give higher pullout resistance as observed in laboratory tests.

3.3 Comparison of laboratory and field pullout tests

It was observed that the pullout resistance produced in the field pullout tests were higher than in the laboratory under the same backfill and overburden pressure conditions. One reason could be due to the embedment length of the reinforcement employed in the field pullout test, which is double than that in the laboratory. Longer reinforcements have higher peak pullout resistance for the same overburden pressure (Chang et al, 1977). Other reasons could be the variation in sample compaction, effects of pea gravel near the wall face, boundary conditions and scale effects. As the reinforcement is pulled out from the pullout box, lateral pressure can develop against the rigid front face, leading to arching of the soil over the inclusion which reduces the vertical stress on the reinforcement, and consequently, decrease the pullout resistance (Juran et al, 1988; Palmiera and Milligan, 1989). It was also observed that the laboratory pullout tests yielded a peak pullout resistance at relatively low displacement compared to the field pullout tests. This is attributed to the fact that more elongation in longer reinforcement will result in more displacement to reach the yield load for the same overburden pressure.

4 NUMERICAL MODELLING OF MSE WALL/EMBANKMENT SYSTEM

4.1 Computer program description and modifications

The NONLIN1 computer program was written and developed to analyze numerically the soil-structure interaction problems and permits solution to problems involving relative movements between structural media such as concrete and steel as well as soil media. The program implements the analytical method suggested by Ochiai and Sakai (1987) which combines the joint element expressing the property of the discontinuous plane, with the bar element transmitting axial force only. Material nonlinearity is considered in the program and the "initial stress" iterative technique for solving nonlinear problems as proposed by Zienkiewicz et al (1969) is adopted. The program was specifically designed for modelling the soil-reinforcement interaction in a pullout test and can be used efficiently to predict pullout resistance of the reinforcement of a given geometry. Necessary debugging modifications were also made to use the program in IBM/3080 mainframe computer system. Modifications were also made in this study so that it can be used to analyze the MSE test embankment and to model the pullout tests in the laboratory (Lo, 1990). Consequently, the vertical facing of the welded wire wall subjected to a transverse load was modelled by using a special 2 or 3-node one-dimensional bending element (Hermitian element), having a length L and a section property EI that combines elastic modulus E with the area moment I. Two degrees of freedom that are both translational and rotational were assumed at each node of the element. By considering the behavior as linearly elastic, the constitutive matrix $[D]^e_B$ for the Hermitian element can be expressed as:

$$[D]^e_B = EI \tag{1}$$

For such an element, the element stiffness matrix $[K]^e_B$ is given by Cheung and Yeo (1979) as:

$$[K]^e_B = EI \begin{bmatrix} 12/L^3 & 6/L^2 & -12/L^3 & 6/L^2 \\ 6/L^2 & 4/L & -6/L^2 & 2/L \\ -12/L^3 & -6/L^2 & 12/L^3 & -6/L^2 \\ 6/L^2 & 2/L & -6/L^2 & 4/L \end{bmatrix} \tag{2}$$

A subroutine SMATBN was then developed to form the element stiffness matrix $[K]^e_B$ of the Hermitian element and was added to the program NONLIN1. Subroutines were also developed to calculate the shape functions and their derivatives for 2- and 3-node bending elements.

4.2 Input parameters for the model elements

The following input parameters were employed to define the material properties of all the model elements:

Soil Elements. The three types of backfill material used in the wall were included and were represented by an elasto-plastic material model obeying a Mohr-Coulomb failure criterion. Hyperbolic parameters suggested by Duncan and Chang (1970) were also adopted in the analyses. Stress-strain parameters of the soils from UU triaxial tests were used to account for partial saturation and the short-term undrained loading condition. A constant value of 0.36 was assumed for the Poisson's ratio of the backfill materials. The undrained elastic modulus E_u of the foundation subsoil was assumed to be a function of the undrained field vane shear strength, S_u, i.e. $E_u = \alpha S_u$, where lies between 70 and 250 for Bangkok clay (Balasubramaniam and Brenner, 1981). From correlations on pressuremeter and the vane shear tests, a value of α of 145 was obtained by Bergado et al (1986). The undrained Poisson's ratio was taken as 0.49.

Interface Elements. The interface between the soil and structure was represented by one dimensional joint elements to allow for relative displacement between the soil and structure if the mobilized shear stress at the interface equals or exceeds that obtained from Mohr-Coulomb strength theory. The shear stiffness of the joint element was estimated based on the method recommended by Ochiai and Sakai (1987). **Figure 21** shows the variation of

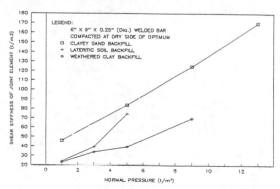

Fig. 21 Variation of shear stiffness of joint element with normal pressures.

the estimated values of the shear
stiffness of joint element for all the
backfills. The normal stiffnesses of the
joint elements were assumed to be a higher
value of 100 times their shear stiffnesses.
The material parameters c and φ of the interface were set to be the same as those
of the surrounding backfill.

Reinforcing Elements. Three parameters
were required to define the bar material
representing the steel bar mat reinforcements. The modulus of elasticity and yield
stress were set equal to the known value
for steel. The area per unit length of the
wall face was calculated based on the
actual cross-sectional area of longitudinal
steel per bar mat and the horizontal
center-to-center spacing between bar mats
to convert the three-dimensional discrete
bar mats into two-dimensional representation.

Beam Elements. The modulus of elasticity, yield stress, and sectional area per
unit width of wall were taken equal to the
value used in reinforcing elements, while
the moment of inertia was assumed to be
3.8×10^{-10} m^4/m for W4.5 bars.

4.3 Finite element mesh and boundary idealization

The finite element mesh was established
based on the geometry of the structure
under consideration, the zones of expected
high stress gradients, the zones of interest for computed stresses and deformations, the practical limitations of
program capacity and the required running
times. Mesh boundary conditions were
selected to appropriately model the expected deformations and were set far enough
from the reinforced soil zone so as to
have negligible influence on the problem.
The typical finite element meshes used for
the pullout tests and in analyzing the
welded wire reinforced wall are shown in
Figs. 22 and 23, respectively.

4.4 Results of numerical modelling

Only the end-of-construction behavior of
the mechanically stabilized wall was
modelled based on total stress finite
element method which takes into account
the nonlinear inelastic behavior of the
soil and slippage between soil and
reinforcement. An elasto-perfectly plastic
model using Mohr-Coulomb yield criterion
was used to describe the soil behavior in
the analysis. The interface between soil
and reinforcement and soil-wall facing
were represented by a joint element of
zero thickness and with slip behavior
being defined by a Mohr-Coulomb failure

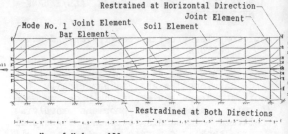

No. of Modes : 152
Nos of Element : a) Soil Element = 194
                 b) Joint Element = 31
                 c) Bar Element = 11

Fig. 22 Typical finite element mesh for
numerical modelling of soil-
reinforcement interaction in the
pullout test.

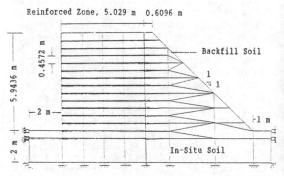

Fig. 23 Typical finite element mesh for
numerical modelling of the MSE
wall/embankment system.

criterion. The reinforcement was considered
as a linear elastic material with axial
stiffness but negligible flexural rigidity.
One dimensional bending element was employed to model the face of the wall. In
addition, the effects of material nonlinearity were incorporated into the
numerical algorithm by adjusting the
initial stress matrix.

Load-Displacement Response. The comparison between the predicted load-displacement
curves and experimental results for 6"x9"x
1/4" diameter steel mesh is typically shown
in Fig. 24. The results show a good agreement between the experimental and analytical values from finite element analysis.
The maximum difference in pullout load
between the experiment and prediction was
about 15%. In order to define completely
the load-displacement response in the
pullout test, five displacement increments
were applied to the nodal point at the
free end of the reinforcement just in
front of the pullout box.

494

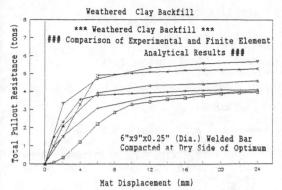

Fig. 24 Comparison of experimental and FEM prediction of load-displacement curves for weathered clay backfill.

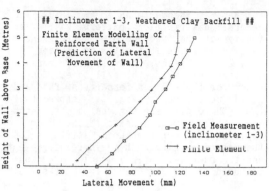

Fig. 25 Comparison of the measured and FEM prediction of wall lateral movements at inclinometer station I3.

Lateral Movement of the Wall Face. The MSE test embankment was constructed such that each section of the facing panel that was added to form the facing element was held in place to prevent lateral movement, while the corresponding level of backfill was added. Appropriate modelling of the facing element is considered as the main factor for the lateral deformations. One-dimensional bending (Hermitian) elements were properly introduced to simulate the typical facing units of the welded wire wall, i.e. bent pronged mats, backing mats, and screens as adopted in the field. The typical profile of lateral movements of the wall face at the end of construction stage from the finite element results is illustrated in Fig. 25 for the middle section. For comparison, the measured lateral movements at inclinometer I2 is also shown. Generally, the shape of the finite element result profiles agreed well with the measured profiles for all sections along the wall. The higher values measured in the field maybe attributed to the partial drainage in the soft clay foundation at the early stage of construction. In other words, a certain amount of consolidation settlement had already taken place prior to the completion of the MSE test embankment. The incremental sequence of reinforced soil wall construction also caused each new soil layer and facing panel to be placed on a previous layer that had already undergone some lateral deformations. In general, the maximum lateral movement predicted by the finite element method occurred at some 4 to 5 m above the base of wall while the actual lateral movement profile gradually in-creases to a maximum value at the top of the wall.

Tension Along the Longitudinal Members. The FEM prediction of the tension force distribution along the longitudinal bar for mat #2 in the weathered clay section during the end of construction is shown in Fig. 26. The plot was produced after plotting the points representing the predicted bar element tension at the center of each bar element, and then connecting the points with a smooth curve. The tensile stress obtained by FEM, which was based on a continuous reinforcing sheet covering the entire area at a given reinforcement level of the wall, was converted to actual tensile stress in the discrete bar-mat by multiplying it with the horizontal center-to-center spacing of the reinforcements, and dividing by the longitudinal cross-sectional steel area

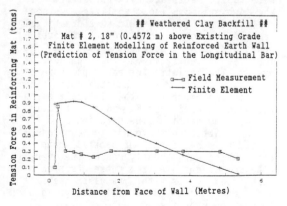

Fig. 26 Comparison of the measured and FEM prediction of reinforcement tension in the weathered clay section (Mat No. 2).

495

per bar mat (Schmertmann et al, 1989). The general distribution pattern was zero tension at the free end of the reinforcement, with steadily increasing tension up to a maximum value, then slightly decreasing tension towards the wall face. Comparison of the tension force from FEM and field measurements (Fig. 26) shows that the FEM results are somewhat higher than the field measurements. The crude approximation of the shear stiffness on the interface element between reinforcement and the backfill material could have contributed to the difference in the results.

5 CONCLUSIONS

The MSE test wall/embankment with its stability never endangered, both during construction and post-construction phases inspite of the excessive settlements and lateral movements, is an ample proof that mechanically stabilized system can be effectively used to reinforce poor quality and marginal quality backfill materials on soft foundations as was also confirmed from laboratory and field pullout tests conducted in this study. Laboratory pullout tests on the three backfills with normal presures up to 130 kN/m^2 proved that even with such poor quality backfill materials, the pullout resistances increased with increasing confining normal pressures as observed for good quality backfill materials, but were much affected by the moisture content and degree of compaction of the soil. The laboratory pullout test generally provided conservative approximation of the actual pullout resistance. The overall agreement between the experimental and theoretical results were quite satisfactory. Furthermore, if the subsoil was also stabilized by some effective method, it would have drastically reduced the total and differential settlements and lateral movements, resulting in improved stability of the embankment system. The variation of earth pressure at the base of the wall, the strains in the reinforcing members, the lateral earth pressure coefficient K, and the location of maximum tension line were found to be strongly affected by foundation compressibility and the effects of backfill compaction, which subsequently caused the overall behavior to deviate from those observed on reinforced walls with granular backfill and constructed on a relatively good foundation subsoils.

6 ACKNOWLEDGMENTS

Grateful acknowledgments are extended to USAID (Thailand) for providing the necessary funds to support our present research activities.

7 REFERENCES

Anderson, L.R., Sharp, K.D. & Harding, O.T. (1987). Performance of a 50-ft. High Welded Wire Wall. Geotech. Spec. Publication No. 12 (ASCE), Ed JP Welsh, pp. 280-308.

Anderson, L.R., Sharp, K.D., Woodward, B.L. & Winward, R.F. (1986a). Performance of the Rainier Avenue Welded Wire Retaining Wall. Dept. of Civil Eng'g., Utah State Univ., Logan, Utah, U.S.A.

Anderson, L.R., Liu, J. & Bay, J.A. (1986b). Experimental Evaluation and Design Recommendations for Syro Anchored Retaining Wall System. Syro Steel Company, Centerville, Utah, U.S.A.

Balasubramaniam, A.S. & Brenner, R.P. (1981). Consolidation and Settlement of Soft Clay. Soft Clay Engineering, Brand, E.W. & Brenner, R.P. (eds.), Elsevier Scientific Publishing Company.

Bergado, D.T., Khaleque, M.A., Neeyapan, R. & Chang, C.C. (1986). Correlations of In-situ Tests in Bangkok Subsoils. Geotechnical Engineering Journal, Vol. 17, pp.1-37.

Bergado, D.T., Cisneros, C.B., Shivashankar, R., Alfaro, M.C. & Sampaco, C.L. (1989a). Pullout Resistance of Steel Geogrids with Weathered Clay Backfill. Proc. Symp. App. Geosynthetic and Geofibre in SE Asia, Petaling Jaya, Selangor Darul Ehsan, Malaysia, pp. 1-26-1-34.

Bergado, D.T., Sampaco, C.L., Alfaro, M.C., Shivashankar, R. & Balasubramaniam, A.S. (1989b). Welded Wire Wall and Embankment System with Poor Quality Backfills on Soft Clay. Third Progress Report Submitted to the U.S. Agency for Intl. Development, Bangkok, Thailand, 109 pp.

Bergado, D.T., Ahmed, S., Sampaco, C.L. & Balasubramaniam, A.S. (1990a). Settlements of Bangna-Bangpakong Highway on Soft Bangkok Clay. J. Geotech. Eng. Div. (ASCE), 116(1), pp. 136-155.

Bergado, D.T., Sampaco, C.L., Alfaro, M.C., Shivashankar, R. & Balasubramaniam, A.S. (1990b). Welded Wire Wall and Embankment System with Poor Quality Backfills on Soft Clay. Fourth Progress Report Submitted to the U.S. Agency for Intl. Development, Bangkok, Thailand, 134 pp.

Bergado, D.T., Sampaco, C.L., Shivashankar, R., Alfaro, M.C., Anderson, L.R. & Balasubramaniam, A.S. (1990c). Interaction of Welded Wire Reinforcement with Poor Quality Backfill. Proc. 10th Southeast Asian Geotech. Conf. (1), Taipei, pp. 29-34.

Bergado, D.T., Cisneros, C.B., Sampaco, C.L., Alfaro, M.C. & Shivashankar, R. (1990d). Effect of Compaction Moisture Contents on Pullout of Steel Geogrids with Weathered Clay Backfill. Proc. 4th Intl. Conf. on Geotextiles and Geomembranes, The Hague, Netherlands.

Bergado, D.T., Sampaco, C.L., Shivashankar, R. & Balasubramaniam, A.S. (1990e). Welded Wire Wall and Embankment System with Poor Quality Backfills on Soft Clay. Fifth Progress Report Submitted to the U.S. Agency for Intl. Development, Bangkok, Thailand.

Broms, B.B. (1974). Lateral Earth Pressures due to Compaction of Cohesionless Soils. Proc. 4th Budapest Conf. on SMFE, pp. 373-384.

Carder, D.R., Pocock, R.G. & Murray, R.T. (1977). Experimental Retaining Wall Facility-Lateral Stress Measurements with Sand Backfill. Rep. No. LR 766, Trans. and Road Res. Lab., Crowthorne, 19 pp.

Carder, D.R., Murray, R.T. & Krawczyk, J.V. (1980). Earth Pressures Against an Experimental Retaining Wall Backfilled with Silty Clay. Rep. No. LR 946, Trans. and Road Res. Lab., Crowthorne, 21 pp.

Chang, J.C., Hannon, J.B. & Forsyth, R.A. (1977). Pullout Resistance and Interaction of Earthwork Reinforcement and Soil. Trans. Res. Record 640, Trans. Res. Board, National Research Council, Washington, D.C., pp. 1-7.

Duncan, J.M. & Chang, C.Y. (1970). Nonlinear Analysis of Stress-Strain in Soils. Journal of Soil Mechanics and Foundation Div., ASCE, Vol. 96, No. SM5, pp. 1629-1653.

Fowler, J., Peters, J. & Franks, L. (1985). Influence of Reinforcement Modulus on Design and Construction of Mohicanville Dike No. 2. ICG, Vienna, Austria, pp. 267-271.

Ingold, T.S. (1983). The Design of Reinforced Soil Walls by Compaction Theory. The Structural Engineer, 61A(7), pp. 205-211.

Jones, C.J.F.P. (1985). Earth Reinforcement and Soil Structures. Butterworks and Co. Ltd., London, U.K.

Juran, I., Knochennus, G., Acar, Y.B. & Arman, A. (1988). Pullout Response of Geotextiles and Geogrids (Synthesis of Available Experimental Data). Geosynthetics for Soil Improvement, Geotechnical Special Publication No. 18, ASCE, pp. 92-111.

Lee, K.L., Adams, B.D. & Vagneron, J.J. (1973). Reinforced Earth Retaining Walls. J. of the Soil Mech. and Found. Div., ASCE, Vol. 99, No. SM10.

Lo, K.H. (1990). Modelling of Laboratory and Field Pullout Test of Steel Geogrid Reinforcements. M.Eng. Thesis, Asian Institute of Technology, Bangkok, Thailand.

Matsuo, M. & Kawamura, K. (1977). Diagram for Construction Control of Embankment on Soft Ground. Soils and Foundations, Vol. 17, No. 3, pp.37-52.

McKittrick, D.P. (1978). Reinforced Earth: Application of Theory and Research to Practice. Proc. Symp. on Soil Reinforcing and Stabilizing Techniques, Sydney, Australia.

Mitchell, J.K. & Villet, W.C.B. (1987). Reinforcement of Earth Slopes and Embankments. National Cooperative Highway Research Program Report 290, Trans. Research Board, National Research Council, Washington, D.C.

Murray, R.T. & Boden, J.B. (1979). Reinforced Earth Wall Constructed with Cohesive Fill. Proc. Intl. Conf. Soil Reinf., Vol. 2, Paris, pp. 569-577.

Murray, R.T. & Farrar, D.M. (1990). Reinforced Earth Wall on the M25 Motorway at Waltham Cross. Proc. Instn. Civil Engrs., Part I, Vol. 88, pp. 261-282.

Nielsen, M.R. & Anderson, L.R. (1984). Pullout Resistance of Wire Mats Embedded in Soil. Report Submitted to the Hilfiker Company, Utah State University, Logan, Utah, U.S.A.

Ochiai, H. & Sakai, A. (1987). Analytical Method for Geogrid-Reinforced Soil Structures. Proc. 8th Asian Reg. Conf., Kyoto, Japan, pp. 483-486.

Palmiera, E.M. & Milligan, G.W.E. (1989). Scale and Other Factors Affecting the Results of Pullout Tests of Grids Buried in Sand. Geotechnique, Vol. 34, No. 3, pp. 515-524.

Rowe, P.W. (1954). A Stress-Strain Theory for Cohesionless Soil with Application to Earth Pressure At-Rest and Moving Walls. Geotechnique, Vol. 4, No. 2, pp. 70-88.

Schmertmann, G.R., Chew, S.H. & Mitchell, J.K (1989). Finite Element Modelling of Reinforced Soil Wall Behavior. Geotechnical Engineering Report No. UCB/GT/89-01, Univ. of California, Berkeley, U.S.A.

Terzaghi, K. (1934). Large Retaining Wall Tests (I): Pressure of Dry Sand, Engineering News Record, Vol. 112,, 136-140.

Zienkiewicz, O.C., Valliappan, S. & King, O.P. (1969). Elasto-plastic Solutions of Engineering Problems: Initial Stress, Finite Element Approach. Intl. J. for Num. Methods in Engineering, Vol. 1, pp. 75-100.

Developments in Geotechnical Aspects of Embankments, Excavations and Buried Structures, Balasubramaniam et al. (eds)
© 1991 Balkema, Rotterdam. ISBN 90 5410 019 2

A computational method for the design of passive bolts

B. Indraratna
Division of Geotechnical and Transportation Engineering, Asian Institute of Technology, Bangkok, Thailand

ABSTRACT: Analytical approach for the design of fully grouted bolts in underground excavations is presented, which involves elasto-plastic concepts and the mechanics of bolt/ground interaction. The theory elucidates the influence of shear stress distribution, bolt pattern and geometry on tunnel convergence. Laboratory simulation and field measurements have provided convincing evidence to support the theory.

1. Introduction

The objective of this paper is to introduce a method for the design of fully grouted rock bolts in deep circular excavations. The analytical theory is based on elasto-plastic concepts and the mechanics of bolt-rock interaction, where the influences of shear stress distribution, bolt pattern and tunnel geometry are considered.

Passive bolts considered in this paper are fully grouted untensioned frictional members, which develop load as the surrounding rock deforms. The axial load is developed as a result of the transmission of shear stresses from the rock to the bolt surface. Generally very small relative displacements are adequate to mobilize maximum bolt tension. Passive bolts are particularly effective in stabilizing fractured or yielding ground because in such conditions active or pretensioned bolts cannot achieve a good mechanical anchorage.

2. Analytical Model

The development of the analytical model is based on the application of elasto-plastic concepts (continuum behaviour) to deep circular excavations in homogeneous, isotropic material subjected to uniform stress fields ($k_o = 1$). In addition, the rock material is assumed to have a brittle mode at the peak stress with a subsequent residual strength. This material behaviour is explained by the "Elastic, Brittle-

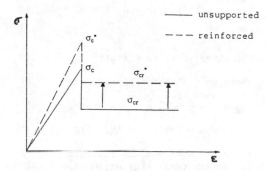

Fig. 1 Elastic, brittle-plastic model

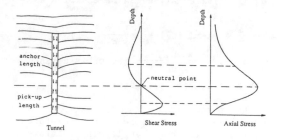

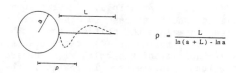

Fig. 2 Shear and axial stress distributions along a fully grouted passive bolt

Plastic Model", as shown in Figure 1. The shear stress distribution along the bolt surface is determined by equilibrium considerations, and an appropriate model is illustrated in Figure 2. Near the tunnel boundary, the upward shear stresses resist the movement of the rock towards the cavity, hence this part of the bolt is named as the pick-up length. The shear stresses on the inner portion of the bolt maintain equilibrium, where the bolt is resisted against motion (anchor length) by the surrounding rock. The maximum tension load occurs at the neutral point, where the relative displacement between the bolt and the ground is effectively zero.

The equilibrium condition for the elastic, brittle-plastic material following a linear Mohr-coulomb failure criterion can be developed by combining the following equations:

$$\frac{d\sigma_r}{dr} + \frac{\sigma_r - \sigma_\theta}{r} = 0$$

for equilibrium (plane strain) and

$$\sigma_\theta = m.\sigma_r + s.\sigma_c$$

for linear failure criterion,

where, $m = (1+\sin \phi)/(1-\sin \phi)$.

Hence, $\dfrac{d\sigma_r}{dr} + (1-m)\dfrac{\sigma_r}{r} = \dfrac{s.\sigma_c}{r}$

Symbols are listed in Appendix 1.

If a circular tunnel is reinforced by a bolt pattern consisting of identical bolts with equal spacing around the circumference and along the tunnel axis, then the equilibrium equation can be modified as:

$$\frac{d\sigma_r}{dr} + [1-m(1+\beta)]\frac{\sigma_r}{r} = \frac{s\sigma_c}{r}(1+\beta)$$

where, $\beta \dfrac{\pi.d.\lambda.a}{S_L.S_T}$ factor is defined as the bolt density parameter.

The parameter λ is similar to the coefficient of friction in reinforced earth or the bond angle in reinforced concrete, and is named as the bolt friction factor. In principle, λ relates the mobilized shear stress to the stress applied normal to the bolt surface. The bolt density parameter (β) reflects the relative density of the bolts with regard to the tunnel perimeter. In reality, the quantity β/λ (dimensionless) varies from 0.10 to 0.40 as given in Table 1. For example, for a tunnel of radius 5.0 m excavated in a fractured rock mass (ϕ =

Table 1 Summary of β/λ ratios determined from several case histories

a (m)	d (mm)	S_L (m)	S_T (m)	β/λ	L (m)	Source
10.4	29	1.5	1.5	0.41	7.6	Bawa and Bumains (1972)
4.5	26	1.5	2.0	0.12	3-6	John (1976)
1.65	25	0.9	0.9	0.16	1.8	Freeman (1978)
5.10	25	1.0	1.0	0.37	6-9	Ito (1983)
2.00	16	0.8	0.8	0.16	1.5	Sun (1984)
2.60	18	0.9	0.9	0.18	1.8	Sun (1984)
3.0	20	1.0	1.0	0.19	2.5	Liu and Huang (1984)

35°) supported by passive bolts (d = 32 mm) at spacings of S_L = 1.5 m and S_T = 1.0 m, the magnitude of β/λ becomes 0.33.

The bolt length (L) is independent of the density parameter (β), because the effectiveness of a bolt depends on its length relative to the extent of the zone of yielded or overstressed rock. Both the neutral point on the bolt surface and the radius of the plastic zone surrounding the cavity are functions of the bolt length. Grouted bolts create a zone of improved reinforced material in the region of the pick-up length (upward shear stresses), where the apparent friction angle and the uni-axial compressive strength of the rock mass are thereby enhanced.

3. Stabilization of the tunnel wall vicinity

The extent of the plastic zone is directly related to the rock mass properties (σ_c and ϕ). Therefore, any improvement of the rock strength should reduce the volume of yielded rock. Consequently, the plastic zone around a reinforced opening must be smaller than that of an unsupported counterpart. Hence, this reduced yield zone is defined as the 'Equivalent Plastic Zone', associated with the reinforced opening. The bolt density parameter (β), the bolt length (L), the radial distance to the neutral point (ρ), the tunnel radius (a) and the in-situ field stress (σ_o) influence the propagation of the equivalent plastic zone.

As illustrated in Figure 3, the radius of the equivalent plastic zone (R*) must be divided into three categories depending on the location of the elasto-plastic boundary with respect to the neutral point:

Category I: R* < ρ < a+L
Category II: ρ < R* < a+L
Category III: R* > a+L

500

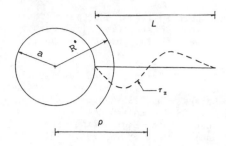

Category (I): $R^* < \rho < (a+L)$ (minimal yielding)

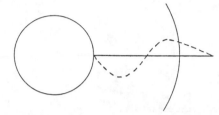

Category (II): $\rho < R^* < (a+L)$ (major yielding)

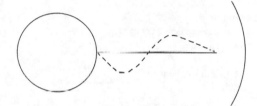

Category (III): $R^* > (a+L)$ (excessive yielding)

Fig. 3 Categorization of the extent of yielding

In category I (minimal or controlled yielding), the extent of the equivalent plastic zone is relatively small, hence stability of the tunnel wall is not affected significantly. However, in category III (excessive yielding), the complete bolt is enclosed within the yielded material and therefore the tunnel wall displacements may become unacceptable. A complete mathematical treatment is given by Indraratna and Kaiser (1989) and a brief summary is presented in Appendix 2.

Tunnel wall displacements can be predicted from the following equation after the equivalent plastic zone radius (R*) has been determined for the appropriate yield category:

$$U_a^*/a = (\frac{1-\nu}{2G})s\sigma_C \{1 + M[(R^*/a)^{m+\alpha} - 1]\}$$

$$(\frac{1-\nu}{2G})\ \sigma_C(1-s)\,[R^*/a]^{1+\alpha}$$

where: $M = (m+1)/(m+\alpha)$.

Table 2 Influence of bolt density on tunnel wall convergence (a/L=1.3)

S_T/L	S_L/L	β	% Reduction of Convergence
Unsupported		0.000	0.1
0.50	1.00	0.073	10.9
0.50	0.50	0.145	19.8
0.50	0.33	0.220	32.3
0.50	0.25	0.291	63.5

Table 2 summarizes typical percentage reductions of tunnel convergence with respect to the bolt density, for a tunnel excavated in a yielding material with the properties: $\phi = 32^\circ$, $\nu = 0.25$, $E/\sigma_C = 420$, $S = 0.90$, $\alpha = 2.0$, $L/a = 0.77$, $\lambda = 0.36$ (smooth rebars) and $\sigma_O = 14$ MPa. It shows that when short bolts are used in a yielding rock, a significantly high bolt density is required to effectively control the tunnel convergence.

4. Conclusion

The analytical model proposed in this paper can be used to predict the displacements and strains associated with deep circular tunnel openings. The dimensionless ratios R^*/a and U_a^*/a are functions of the bolt spacing (hence bolt density) and the bolt length. The influence of fully grouted bolts on tunnel wall stabilization is best evaluated on the basis of displacement control. The bolt spacing and length must be carefully selected in order to prevent excessive displacements in the vicinity of the tunnel wall. The proposed analytical solution provides a methodology for selecting an optimum reinforcement configuration. The theoretical model has been verified by laboratory simulation and analysis of case histories elsewhere (Indraratna and Kaiser, 1987).

5. References

Bawa, K.S., and Bumanis, A., 1972. Design Considerations for Underground Structures in Rock. Proc. North American Rapid Excavation and Tunnelling Conference, Chicago, pp. 393-417.

Freeman, T.J., 1978. The Behaviour of Fully Bonded Rock Bolts in the Kielder Experimental Tunnel. Tunnels and Tunnelling, Vol. 10, pp. 37-40.

Indraratna, B., and Kaiser, P.K., 1987. Control of Tunnel Convergence by Grouted

Bolts. Rapid Excavation & Tunnelling Conference, New Orleans, Lousiana, 20 p.

Indraratna, B., and Kaiser, P.K., 1989. Analytical Model for the Design of Grouted Rock Bolts. Int. Journal of Numerical and Analytical Methods in Geomechanics (in press).

Ito, Y., 1983. Design and Construction by NATM through Chogiezawa Fault Zone for Enasan Tunnel on Central Motorway (in Japanese). Tunnels & Underground, Vol.14, pp. 7-14.

John, M., 1976. Geotechnical Measurements in the Arlberg Tunnel and their Consequences on Construction (in German). Rock Mechanics, Supplementum 5, pp. 157-177.

Liu, B., and Huang, J. 1984. State of the Art of Rock Bolting in People's Republic of China. Int. Symp. on Rock Bolting, A.A. Balkema, pp. 285-294.

Sun Xueyi, 1984. Grouted Rock Bolts used in Underground Engineering in Soft Surrounding Rock or in Highly Stressed Regions. Int. Symp. on Rock Bolting, A.A., Balkema, pp. 93-100.

APPENDIX 1

List of Symbols

r = radial distance

K_0 = ratio of horizontal to vertical stress in situ

σ_0 = field stress for K_0=1 (hydrostatic loading)

σ_r = radial stress

σ_θ = tangential stress

σ_c = uni-axial compressive strength

σ_{cr} = $s.\sigma_c$ = post peak uniaxial compressive strength

s = post-peak strength reduction factor

σ_t = tensile strength

ϵ_f = failure strain in uniaxial compression test

ϵ_c = σ_c/E = critical strain

ϕ = angle of friction

c = cohesion intercept

α = dilation coefficient

E = Young's modulus

G = shear modulus

ν = Poisson's ratio

a = tunnel radius

L = length of bolts

d = diameter of bolts

S_T = circumferential (tangential) bolt spacing

S_L = longitudinal bolt spacing

z_n = distance from tunnel wall to neutral point on bolt

ρ = $a+z_n$ = radial distance to the neutral point on bolt

λ = friction factor for bolt-ground interaction

β = bolt density parameter

u_a = convergence of the unsupported tunnel

u_e = elastic portion of the total convergence

u_p = plastic portion of the total convergence

R^* = radius of the equivalent plastic zone

u_a^* = convergence of the reinforced tunnel

u_a^*/u_a = normalized convergence ratio

i = bolt effectiveness.

Solutions of the Equivalent Plastic zone Radius

Solutions of the Equivalent Plastic Zone Radius

Category (I): $R^* < \rho < (a+L)$ (minimal yielding)

$$\frac{R^*}{a} = \left(1 + \frac{1}{s} \left(\frac{m^* - 1}{m^* + 1} \right) \left(\frac{2\sigma_0}{\sigma_c^*} - 1 \right) \right)^{1/(m^*-1)}$$

Category (II): $\rho < R^* < (a+L)$ (major yielding)

$$\frac{R^*}{a} = \frac{\rho}{a} \left(\frac{1 + B_1}{1 + A_1} \right)^{1/(m'-1)}$$

where: $B_1 = \frac{1}{s} \left(\frac{m' - 1}{m' + 1} \right) \left(\frac{2\sigma_0}{\sigma_c'} - 1 \right)$

$$A_1 = \left(\frac{m' - 1}{m^* - 1} \right) \left(\frac{1 + \beta}{1 - \beta} \right) \{ (\rho/a)^{m^*-1} - 1 \}$$

Category (III): $R^* > (a+L)$ (excessive yielding)

$$\frac{R^*}{a} = [1 + (L/a)] \left(\frac{1 + B_1}{1 + A_2 + A_3} \right)^{1/(m-1)}$$

where: $B_1 = \frac{1}{s} \left(\frac{m-1}{m+1} \right) \left(\frac{2\sigma_0}{\sigma_c} - 1 \right)$

$$A_2 = \left(\frac{m-1}{m'-1} \right) \left([(a + L) / \rho]^{m'-1} - 1 \right)$$

$$A_3 = (1 + \beta) \left(\frac{m-1}{m^* - 1} \right) [(a+L)/\rho]^{m'-1} \{ (\rho/a)^{m^*-1} - 1 \}$$

5. Site investigation and selected topics

Geotechnical aspects of the construction of the Singapore MRT

J.N.Shirlaw
Moh and Associates, Taipei, Taiwan (Previously: Singapore MRTC)

D.F.Stewart
Stewart International, Bangkok, Thailand (Previously: Singapore MRTC)

ABSTRACT. The different methods of construction used for excavations and bored tunnels in the soft Singapore marine clay are briefly presented. Monitoring data from these deep excavations and two tunnel drives, particularly settlement and piezometer measurements, are summarised. Despite the use of diaphragm walls extending up to 54m below ground level for the excavations, and an exceptionally watertight precast tunnel lining, a significant proportion of the measured settlements were due to consolidation. Changes in stress due to excavation and tunnelling appear to have been a major factor in the porepressure changes recorded.

1 INTRODUCTION

Construction of the Singapore Mass Rapid Transit System (MRT) started in late 1983. The first section of line opened in 1987. Of the 42 stations, shown in Figure 1, 41 are operational (as of May 1990) with the remaining ones expected to become operational in the near future.

15 of the 42 stations are underground. These are largely the stations in the Central Business District (CBD). Connecting the underground stations are 11 route kilometers of twin bored tunnel and approximately 6 kilometers of cut and cover tunnel.

The work encountered a great variety of different ground conditions. Face conditions in the bored tunnels included:

. Massive, fresh quartzite
. Highly weathered and fractured shales and mudstones.
. Completely weathered granite
. Stiff clay containing quartzite boulders
. Open sands (beach or alluvial deposits)
. Soft marine clays
. Highly organic estuarine clays
. Mixed faces consisting of two or more of the materials given above

The open excavations for stations and cut and cover tunnels encountered a similarly wide variety of ground conditions. This variation meant that a number of different construction methods were used. Table 1 summarizes the major methods applied for open excavations and for bored tunnels.

As a general description of the works was given in Hulme, Potter and Shirlaw (1989), it is not proposed to give a similar overview here. Instead greater detail will be presented on some of the techniques and monitoring results from construction work in soft clay.

2 DEEP EXCAVATIONS IN SOFT CLAY

Singapore marine clay was encountered in many of the deep excavations for stations and cut and cover tunnels of the MRT system. Up to 45m depth of marine clay was found in some areas.

The clay is generally in two layers, a younger upper layer and an older, lower layer. In between the two marine clay members there is typically a thin, 2 to 5m thick, layer of alluvial sand or clay. The marine clay has a plasticity index that is between 30 and 80. An overconsolidated crust generally extends to a depth of about 10m below ground level. Below the crust the clay is typically lightly overconsolidated, except in coastal areas where extensive recent reclamation has resulted in clays which are still consolidating under the weight of the fill. More details of the nature and properties

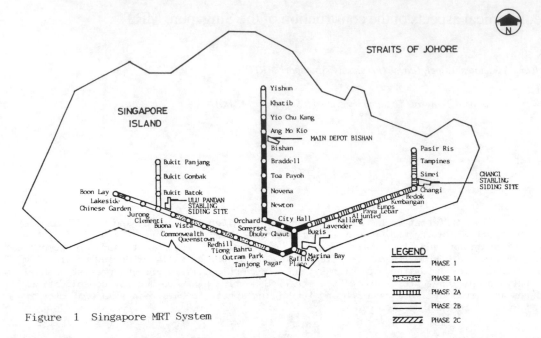

Figure 1 Singapore MRT System

LEGEND

———————	PHASE 1
▬▬▬▬	PHASE 1A
ⅢⅢⅢⅢ	PHASE 2A
══════	PHASE 2B
▨▨▨▨	PHASE 2C

of the Singapore marine clay are given in Tan and Lee (1977), Tan (1983).

A typical plot of measured shear strength, based on field vane testing, is presented as part of Figure 7. In these deep, soft clays excavations can result in large ground movements. For example deflections of up to 260 mm were measured at one sheetpile excavation in Singapore marine clay for an excavation depth of just 7.3m (Davies and Walsh, 1983). The large movements were caused by active pressures exceeding the ultimate passive resistance for a considerable depth below the final excavation level. In these circumstances no amount of strutting above excavation level can restrain the movement which

Table 1: Techniques used for open excavations and bored tunnelling.

		Open Excavations	Bored Tunnels
a.		Excavation Support	Tunnelling Method
		Cut Slopes Nailed Slopes Anchored Slopes King Pile & lagging Cellular Cofferdams Diaphragm Walls	Open (Greathead type) shields Drum digger Earth pressure Balance Shields N.A.T.M.
b.		Ancilliary Measures	Ancilliary Measures
		Jet Grouting Lime piles Underwater excavation	Compressed air Jet grouting Chemical grouting (sand lenses) only Lime piles

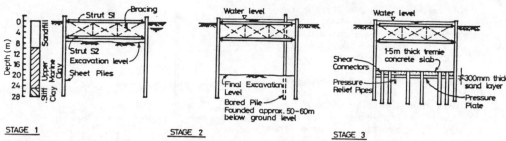

STAGE 1

1.1 Drive sheet piles
1.2 Excavate approx. 1.5m & install S1
1.3 Excavate approx. 6.5m & install S2 & bracing between S1 & S2

STAGE 2

2.1 Flood the cofferdam to the top level
2.2 Excavate under water using grabs, water jets & air lifting
2.3 Install bored piles using RCD method

STAGE 3

3.1 Place min. 300mm thick sand levelling layer
3.2 Install pressure relief pipes
3.3 Install pressure plates
3.4 Cast tremie concrete slab

Figure 2 Excavation Sequence, Marina Bay Station and Tunnels

occurs in the soil in the passive zone.

Excavations for the MRT system had to be, typically, 16m to 18m deep. For this depth of excavation the active pressure in the marine clay would exceed the ultimate passive pressure over the full depth of the marine clay. Where the marine clay extended for a significant depth below final excavation level, then it was a major design problem to limit deflections and consequent settlements to a reasonable level.

Four of the MRT station excavations had significant depths of marine clay below final excavation level. As shown in Table 2, two different strategies were adopted in

the design of the temporary works. In three cases, at Newton, Bugis and Lavender stations, a diaphragm wall was taken right through the marine clay and into weathered rock or hard old alluvium. The diaphragm walls were heavily reinforced to take the moments imposed by the net active pressure between the bottom strut and the hard stratum below the marine clay. Up to seven strut levels were required to restrain the diaphragm walls above the base of the excavation.

An alternative strategy was adopted at Marina Bay station, where the marine clay was the deepest encountered on the system, and the clay was still consolidating under 12m of hydraulic fill used to reclaim the

TABLE 2

Station	Max. Depth excavation	Typical Depth Soft Soils *	Retaining System Used	Special Measures
Newton	14.3 to 15m	16m	0.8m Diaphragm Walls	Jet grouting
Bugis	18.3m	35m	1.2/1.0m Diaphragm Walls	Lime Columns
Lavender	16.5m	23m	1.0m Diaphragm Walls	
Marina Bay	16.4m (+1.8m for tremie slab)	36m	Composite H pile/sheet pile	Underwater excavation

* Primainly marine clay, but also including fill, fluvial clays and sands.

509

area. As described by Clark & Prebaharan (1987), a composite H pile sheet pile retaining system was used which did not extend to a firm stratum. Despite the relatively high moment capacity of the composite wall, excavation was only possible to a depth of about 7 to 8m. At this excavation depth the maximum deflexion of the wall was approximately 100mm. As shown in the construction sequence (Figure 2), an initial, dry excavation was used to install two levels of struts, retaining the top of the wall. The excavation was then flooded. The water acted as a surcharge on the underlying soft clay, allowing excavation to a final depth of 18m below ground level. Once excavation was complete 1m diameter bored piles were installed. These were designed to prevent settlement of the station due to continuing consolidation of the soft marine clay, and were taken 12m into very dense old alluvium. After completion of the piling, a mass concrete base slab was poured using tremie pipes. The combination of tremie slab and piles (acting in tension) was designed to resist the base heave pressure as the cofferdam was dewatered.

The design of the anchored tremie slab required prediction of the base heave pressures to be resisted. A major consideration of the design was to ensure strain compatibility. The anchored, unreinforced tremie slab was a brittle system - it had little capacity for movement and failure would be catastrophic. The resistance provided by the shear strength of the soft clay was, in contrast, highly ductile, with large movements required to mobilize peak strength.

As described by Clark and Prebaharan, the base heave pressure to be resisted by the tremie slab was calculated from the equation.

Uplift pressure = $(q + \gamma H) - 0.7 N . Cu - \dfrac{0.8 d (h1 + h2) Cu}{L.h.}$

These terms represent:-
= Driving load - resistance from soil - adhesion along piles.

and are shown in Figure 3.

For the second term, the resistance from the shear strength of the soil, the 0.7 value was a factor introduced to provide strain compatibility. The value of 0.7 was based on the relationship between factor of safety and movement for deep excavations

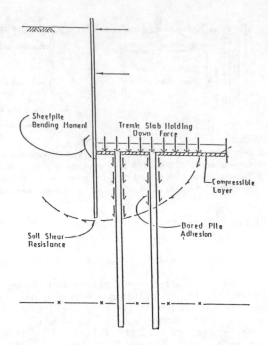

Figure 3 Forces Resisting Base Heave, Marina Bay (After Clark and Prebaharan, 1987)

shown in Clough, Hansen and Mana (1979).

The 'N' value used in the equation was based on the relationship between stability number and excavation geometry proposed by Bjerrum and Eide (1956). For the Marina Bay excavations 'N' values were typically between 5 and 7.

In order to be able to verify the calculated uplift pressures, total pressure cells were installed below the tremie slab for the first four cofferdams constructed using this method. Of the 10 pressure plates installed, 9 showed final pressures close to or lower than the predicted value after completion of dewatering. One plate recorded a final pressure 20% higher than the predicted value, but this pressure was still within the capacity of the tremie slab. These field measurements confirmed that the simple hand calculation methods used provided reasonable estimates of the uplift force on the tremie slab.

The other three excavations in deep soft clays, at Newton, Bugis and Lavender, relied on more convential methods for temporary support. Heavily reinforced diaphragm walls were used to transfer the

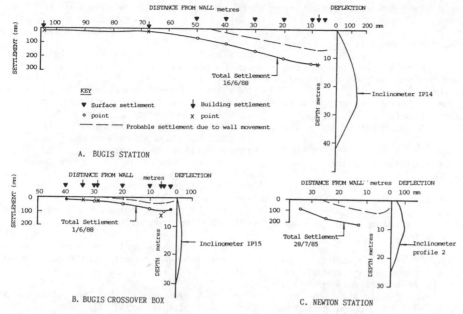

Figure 4   Measured Wall Deflexions and Settlements at Bugis and Newton Stations

out-of-balance forces in the marine clay down to hard strata. At both Newton and Bugis soil improvement was carried out to help limit deflections in the areas where the clay was deepest; jet grouting was used at Newton and lime piles at Bugis (Kado et al., 1987)

Design of the diaphragm walls was carried out using computer programs modeling the soil as elasto-plastic springs. Such programs are widely used. It is not proposed to discuss their application here, save to say that the measured lateral deflections of the diaphragm walls were reasonably well predicted.

What is of particular interest at the three stations is the magnitude of settlement measured. Unlike the station at Marina Bay, the three stations were all in urban areas, and therefore the magnitude of the settlement resulting from excavation, and its effect on the adjacent buildings, was of great interest and concern.

Figure 4 presents three profiles of wall deflection and measured settlement. Two profiles are shown at Bugis, one in the area of deepest marine clay in the centre of the station, the other for a section of the adjacent cross-over box where the total depth of marine clay was significantly smaller.

Nicholson (1987) has suggested a method of estimating settlements due to wall deflection in soft clays, a method based on that proposed by Milligan (1983), but modified to fit field observations. The suggested relationship is shown in Figure 5. The probable profile of settlement, assuming constant volume, and due only to

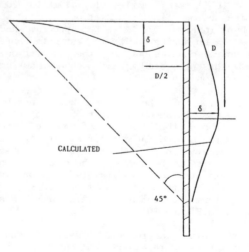

Figure 5   Relationship Between Deflexion and Settlement, as Proposed by Nicholson (1987)

wall deflection is also shown for the three cases in Figure 4.

For each of the three cases it is evident that:-
a) The volume of the settlement trough far exceeds the volume of the lateral movement of the wall.
b) The settlement trough extends well outside a theoretical zone of influence equal in width to the depth of the wall.

This suggests that a significant proportion of the settlement in these cases was caused by consolidation.

The interpreted consolidation settlements had a maximum value of approximately 100mm at Newton and Bugis stations and 70mm at the Bugis crossover box.

At Newton station deep wells were used to control the pore pressures in underlying fluvial sands. The resultant pore-pressure changes inside and outside the box are shown in Figure 6. With deep pumping inside the excavation, and a relatively small diaphragm wall penetration into weathered granite, the large pore pressure drops and consequent settlement can be easily understood.

Less obvious is the process by which a similar magnitude of consolidation settlement occurred at Bugis, where no deep dewatering was used. The only pumping at Bugis was from surface sumps to keep the base of the excavation as dry as possible. There was a 'plug' of low permeability (10^{-9} to 10^{-11}m/sec) marine clay 10m to 20m thick below final excavation level, as shown in Figure 7. A seepage program had been used during design to estimate pore pressure changes outside the excavation, and this predicted minimal change during excavation.

A plan of the station with nearby buildings is shown in Figure 8. The plan also shows the locations of some of the instruments used for monitoring the effects of excavation, although only those instruments to be discussed here are shown for clarity. One instrumented section is shown in Figure 9. The wall deflection and long term settlement in this area are shown in Figure 10.

At this instrumented section four piezometers were installed outside the wall:-

VP 3 B   In the old alluvium.

VP 5B and 5A   In the lower marine clay
VP 3A   At the base of the upper marine clay

VP 3 B started to register a significant drop in piezometric head very soon after the start of excavation; when the excavation level was only 5m this piezometer was already registing a pore pressure reduction of 2m head. The rapid reduction in pore pressures in the old alluvium continued as excavation proceeded (Figure 11). The reduction in pore pressure by 4th strut level (excavation depth 11 m) was nearly 5m head. This large drop in pore pressure in the old alluvium was of considerable concern due to the thick layer of compressible marine clay overlying the old alluvium. It was therefore decided to install a line of recharge wells between the piezometers and the diaphragm wall to minimize further head losses.

The recharge wells had an immediate effect on piezometer VP 3B, but as the nearest recharge well was only 8.5m away, this is hardly surprising. The measured Piezometric pressure in the old alluvium had recovered to about 3m head below the initial level by the time the station base slab had been completed. However the efficiency of the recharge wells gradually deteriorated (see Figure 11), until by the end of 1987 they were no longer effective. Once the recharge wells had ceased operating the recorded pressure in VP3B dropped again, to nearly 4m head below the initial level. This suggested that the general piezometric pressure in the old alluvium had in fact been significantly lower than that shown by the piezometer; and that the piezometer had been registering an artificially high level due to the nearby recharge well. It is possible that a similar distortion occurred with the piezometers in the marine clay, even though this soil was much less permeable than the underlying old alluvium, so the piezometer results from the period of recharge cannot be taken as representative of the general pore-pressure changes in the area. What can be taken as representative, however, are the measured pore-pressures before recharging started and after it had become ineffective. The distribution of measured pore pressures at 13 May 1987 (excavation at 11m depth) and 6 June 1988 (8 months after completion of the base slab) are shown in Figure 9. It can be seen that at 11m excavation depth there was a large loss of head in the old alluvium, but little change in the

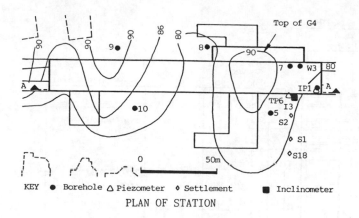

KEY ● Borehole △ Piezometer ◇ Settlement ■ Inclinometer

PLAN OF STATION

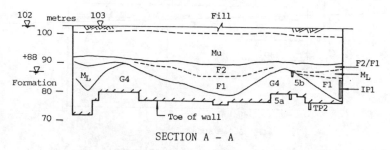

SECTION A - A

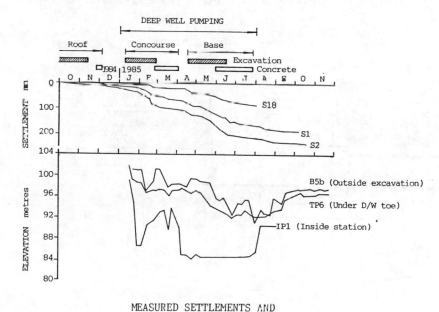

MEASURED SETTLEMENTS AND
PIEZOMETER READINGS

Figure 6 Pore Pressure Changes and Settlement at Newton Station
(Adapted from Nicholson (1987))

513

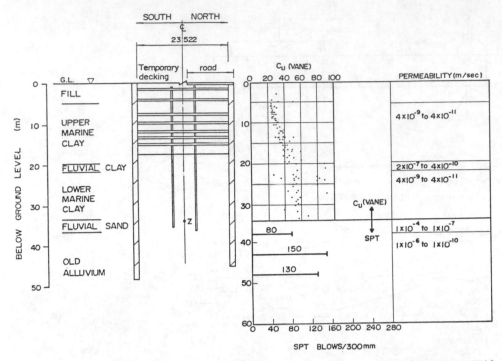

Figure 7 Soil Conditions at Bugis Station, Area of Inclinometer IP13

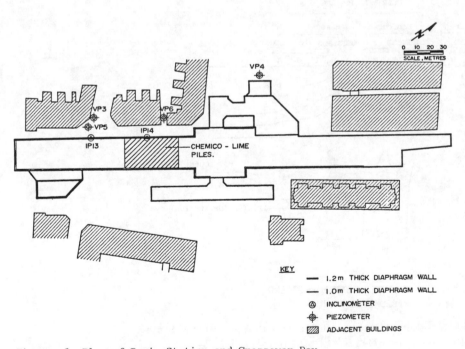

Figure 8 Plan of Bugis Station and Crossover Box

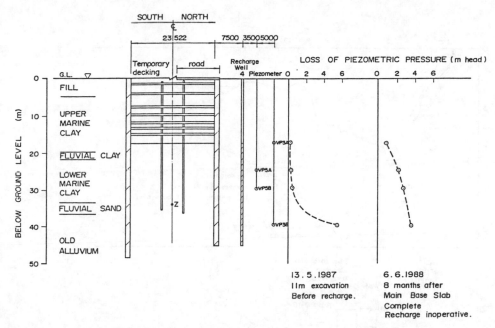

Figure 9 Changes in Pore Pressure Measured Outside the Excavation for Bugis Station

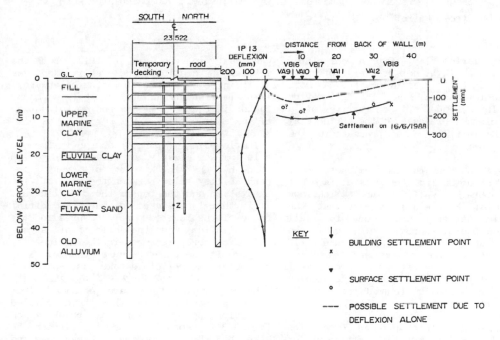

Figure 10 Wall Movement and Settlement Measured at IP13

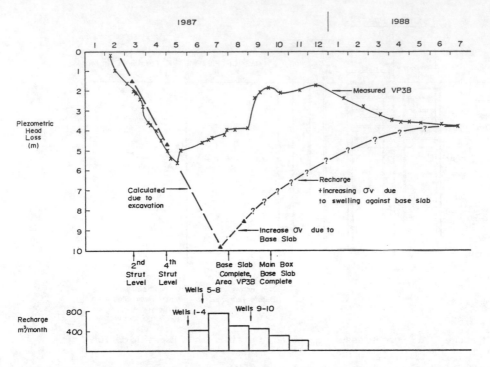

Figure 11 Measured Piezometric Head Loss in Old Alluvium, and Calculated Head
Loss Due To Stress Relief by Excavation

overlying marine clay. Over a year later there was a slight net recovery of pore pressure in the old alluvium, but a more significant loss of head in the marine clay. This sequence is suggestive of a process of underdrainage of the marine clay due to the initial drop in pressure in the old alluvium.

It was found that the maximum settlement occurred about 9m from the outside face of the diaphragm wall. The Nicholson relationship (Figure 5) suggests that the settlement at this point due to wall deflection should, approximately, equal the maximum wall deflection. Figure 12 compares the maximum deflection measured for inclinometer IP13 with the settlement recorded at point VB16. VB16 was located on a building on shallow foundations just 9m from the excavation.

It can be seen that up to an excavation depth of 11m (and a total settlement of about 80mm) the two values were following a very similar path. At this point, however, the settlement started to increase more quickly than the lateral deflection. Wall deflection stopped as soon as the base slab

was cast in the area of IP13, but the settlement continued inexorably , and was still increasing some 11 months afterwards. This is further indication that a major factor in the settlements was consolidation of the marine clay.

The lowering of pore pressures in the old alluvium was clearly not due to seepage up through the thick plug of very low permeability clay between the base of the excavation and the top of the old alluvium. It is therefore suggested that the initial cause of the lowered pore pressures in the old alluvium was the effect of stress relief on pore pressures in the marine clay plug. If the pore pressure change, Δu, due to stress relief is taken as: -

$$\Delta u = \Delta \sigma_v /2 + \Delta \sigma_h /2$$

Where $\Delta \sigma_v$, $\Delta \sigma_h$ are obtained from the Boussinesq formula, than the calculated reduction pore pressure at point z on Figure 9 is shown in Figure 11. It can be seen that the pore pressure change measured in VP3B was:

. Very close to that calculated for point z.

. Would have continued to a loss of pore

516

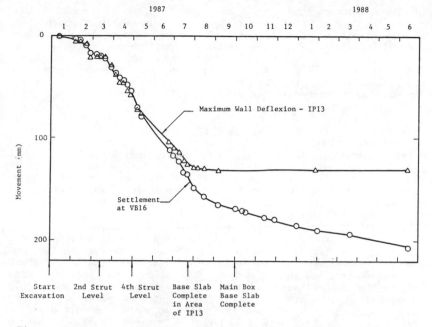

Figure 12   Development   of   Maximum   Wall   Movement   and Settlement at IP13

pressure of nearly 100kPa had recharge wells not been installed.

The loss of head in another piezometer in the old alluvium (VP4B), in an area where no recharging was carried out, is shown in Figure 13. The measured loss in this piezometer, outside an excavation for a vent shaft, exceeded 8m. Given the particular geometry of the vent shaft (see Figure 8), the loss of head at VP4B could be expected to be less than at VP3B. So the calculated maximum loss of nearly 10m at VP3B is not unreasonable.

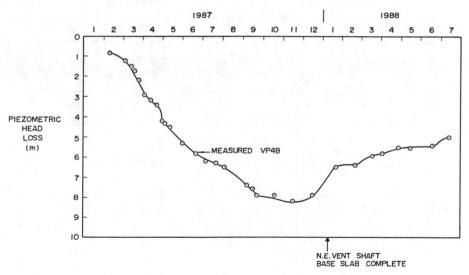

Figure 13   Measured   Piezometric Head Loss in Old   Alluvium, VP4B

The good agreement found between the calculated loss of head at the base of the marine clay below the excavation and the measured loss in the old alluvium over 27m away suggests that: -

. The old alluvium was permeable enough to allow very rapid transmission of reduced pore pressures despite the presence of the diaphragm wall.

. The old alluvium was not permeable enough to allow a significant volume of recharge into the general area of the excavation within the time scale of the excavation. It this context it is important to note that the old alluvium can reasonably be thought of as semi-infinite body, being at least 100m thick and forming nearby hills.

The fact that large consolidation settlements can occur due to excavations in soft clays has been noted elsewhere, and in itself is not surprising. The mechanisms by which such settlements occur have generally been taken as either: -

. Seepage through walls (particularly with sheet pile walls) or **through** permeable layers located close to the base of excavation.

. The dissipation of excess pore pressures induced by shearing of the clays outside the wall during deflection.

The mechanism suggested for the Bugis case, of changes in stress due to excavation inducing large changes in pore pressure far below the base of the excavation; and of those pressures than being transmitted laterally through more permeable layers, is one quite distinct from the two quoted above. Despite the fairly large movements of the diaphragm wall there was no evidence from the monitoring of excess pore pressures being generated.

3. TUNNELS IN SOFT CLAY

Nine of the Singapore MRT civil construction contracts included bored tunnels. Two contractors were each awarded two civil contracts including tunnels, so altogether seven contractors were involved in the tunnelling.

Each of the seven tunnelling contractors encountered the soft Singapore marine and estuarine clays at some point in their drives. The amount of marine clay encountered, however, varied considerably on each of the contracts. Some contractors were faced with soft clay for virtually the whole length of their drives, whereas others only encountered it in localized areas where old infilled channels cut across the line of the tunnel. The varying proportion of soft clay to length of drive and the particular preferences of the individual contractors resulted in a variety of different methods being used to tunnel through the soft clays. Three different types of shield machine were used, often in combination with compressed air and/or jet grouting.

TABLE 3: Progress rates and settlements for tunnels in soft clay.

Shield Type	Additional Measures	Average Progress Rate (m/wk)	Typical range of Settlements over twin tunnels (mm)
Greathead	Compressed air	15 (C104) 27 (C301)	30--120
Greathead	JGP and Compressed air	27	30--60
Greathead	JGP only	28	40--110
EPBS	None	40	70--140
Drum Digger	Compressed air	35	200--400

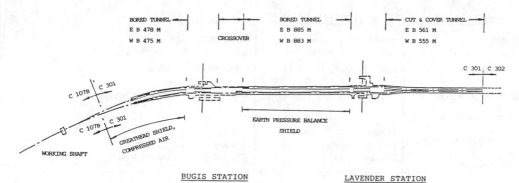

BUGIS STATION   LAVENDER STATION

Figure 14   Plan   of   Contract   301,   Showing   Stations   and   Tunnelling Methods

A general summary of the total settlements measured over the tunnels is soft clay and of the average progress rates achieved is presented in Table 3.

Of particular interest is the comparison of the surface settlements measured over EPB driven tunnels with those over conventional Greathead shields. The total settlements over the EPB shields slightly exceeded those over the Greathead/compressed air shields. All of the EPB measurements and a large proportion of the Greathead values were obtained on a single construction contact, number 301 (Figure 14). The tunnels approaching Bugis Station from the west were driven using the Greathead shields, and those from the east using EPBS. So the measurements were over tunnels at similar depths and separations, in similar ground conditions and by the same contractor.

A significant proportion of the settlement that occurred was due to consolidation of the marine clay. As shown in Figure 15 the consolidation settlements occurred:

. After the shields had passed under the measurement point or after the compressed air pressure was removed.
. Followed a straight line on a settlement/log time graph. This is clear from Figure 15 for the EPB tunnels but not so evident for the compressed air tunnels. However if only the settlement following depressurisation is studied, then it is much clearer (Figure 17).
. Was still continuing nearly a year after the completion of the tunnels.
. There was a delay of, typically, 3 to 7 weeks in the onset of the consolidation settlements. This occurred only in the contract 301

tunnels and was not noted in other tunnels in soft clay.

As shown in Figure 15, the lateral extent of the measured troughs of consolidation settlement was quite different over the compressed air and over the EPB shields. Over the compressed air shields the trough was relatively flat, with significant settlements being measured at 30m from the centre of the tunnels. The trough over the EPB shields followed a classical error function form, with a trough width parameter ("i") equal to approximately 0.5 depth to tunnel axis.

The only pore pressure measurements at these tunnels came from a BAT (Torstensson 1987) pore pressure probe jacked into the ground though the holes used for grouting around the concrete segments.

One observation made during installation of the BAT pore pressure probe was that there was a clear difference in the effectiveness of the grouting around the tunnel rings. In the EPB driven tunnels the ground was in direct contact with the ring, and the pore pressure probe could be driven out as soon as the grout plug was removed. For the tunnels driven using compressed air a thick skin of grout was found around the ring, and this had to be broken out with pneumatic tools before the pore pressure probe could be installed.

The pore pressure probe installed outside one of the EPB driven tunnels measured a large excess pore pressure of over 100kPa. This excess pore pressure clearly resulted from the method of shield driving. High face pressures of typically 1.2 to 1.4 times the total overburden pressure were used. In addition, the conventional ring grouting formed bulbs and fractures in the soft clay.

EXAMPLE 1	EXAMPLE 2	EXAMPLE 3
SINGLE TUNNEL 6.41 m OD DRUM DIGGER	TWIN TUNNELS 16 m c/c 5.91 m OD GREATHEAD SHIELDS	TWIN TUNNELS 16 m c/c 5.93 m OD EARTH PRESSURE BALANCE
DEPTH TO AXIS: 13 m	DEPTH TO AXIS: 15.3 m (E/B) : 16.0 m (W/B)	DEPTH TO AXIS: 18.9 m

AIR PRESSURE: 0.9 BARS	AIR PRESSURE: 1.3 BARS (E/B)
	              1.25 BARS(W/B)
		FACE PRESSURE (SHOVE): 3.5 BARS (E/B)
		                   : 3.8 BARS (W/B)

TEMPORARY LINING OF RIBS . WOODEN LAGGING PERMANENT LINING OF INSITU CONCRETE PLACED AFTER TUNNEL DEPRESSURISED	PRECAST CONCRETE LINING WITH HYDRO-SWELLING SEALING STRIP	(REST) : 2.5 BARS (E/B) : 2.3 BARS (W/B) PRECAST CONCRETE LINING WITH HYDRO-SWELLING SEALING STRIP

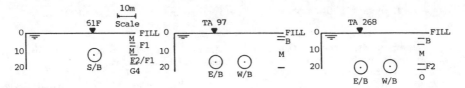

FIG 2(a) TUNNEL DETAILS

KEY

M = MARINE CLAY
F2= FLUVIAL CLAY
F1= FLUVIAL SAND
G4= WEATHERED GRANITE
O = OLD ALLUVIUM
B = BEACH SAND

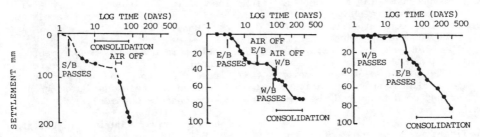

FIG 2(b) SETTLEMENT/LOG TIME FOR POINTS INDICATED IN FIG 2(a)

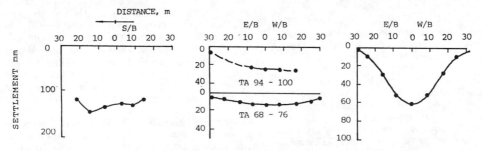

FIG 2(c) LATERAL SETTLEMENT TROUGH FOR PERIODS INDICATED IN FIG 2(b)

Figure 15 Three Different Forms of Settlement Trough Due To Consolidation Following Bored Tunnelling

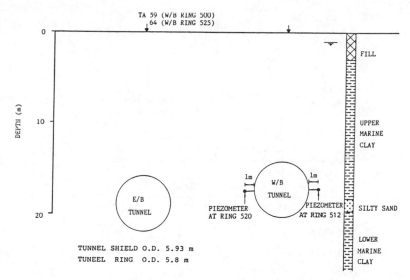

Figure 16 Tunnels, City Hall To Bugis, Showing Ground
Conditions and Piezometer Locations

Whereas the EPB shields were driven with a face pressure in excess of the insitu ground pressures, the compressed air shields were driven with a face pressure significantly less than the insitu ground pressure. However, there was still some evidence that the settlement was due to the dissipation of excess pore pressure rather than the result of a general lowering of pore pressures due to seepage into the tunnel. Figure 16 shows the location of a pore pressure probe driven into the central pillar between the two tunnels. On Figure 17, the recorded pore pressures in the central pillar between the two tunnels are plotted on the same log time scale as the settlements. It can be seen that the pore pressure initially dropped following depressurisation of the tunnel. It rose until it peaked some 44 days after the tunnel had returned to atmospheric pressure. During this period, the settlement recorded over the tunnels was very small in magnitude. The onset of more rapid settlement coincided with the excess pore pressures starting to dissipate.

For the Earth pressure balance shields the consolidation settlements were of error function form with a similar width to the initial settlements. The settlements can therefore be added directly to the initial ('ground loss') settlements, whether expressed as an absolute value or in terms of volume loss. A histogram of the total volume losses recorded 8 months after completion of the tunnels (Figure 18) shows that the volume losses recorded typically fall between the minimum and maximum values of the potential void around the rings at the tail of the shield.

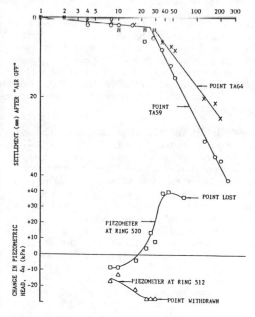

Figure 17 Measured Settlements and Pore Pressure Changes Plotted Again Log Time

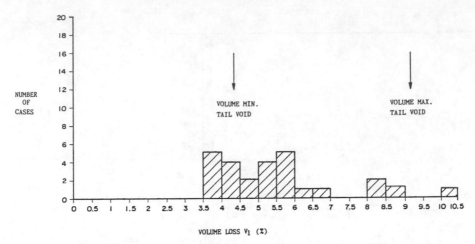

Figure 18 Histogram of Measured Volume Losses Over EPB Driven Tunnels

As discussed by Broms and Shirlaw (1989) this suggests that:-

a) The soft clay closed around the rings immediately, despite all efforts to grout around the rings as rapidly as possible.
b) Conventional ring grouting carried out after the soft clay had come down at the rings had little effect in reducing total settlements. This was in marked contrast to the tunnels with compressed air, where conventional ring grouting could be carried out before the clay came down on the rings, and movements were consequently reduced.
c) The outward displacements caused by an excess face pressure and grouting at the tail were matched by similar, but later inward movements due to dissipation of excess pore pressures generated by the outward displacement. The excess face pressure and tail grouting thus acted to delay the development of the final settlement but may not have significantly reduced their magnitudes.

In studying the settlements over the EPB driven tunnels, it is clear that the critical factor was not the good control of ground movements that could readily be obtained at the face, but the lack of any support at the tail of the shield. For clays volume losses are often plotted against the stability number (N) of the tunnel. If the volume losses over the EPB shields (including consolidation) are plotted against the stability number at the tail of the shield (Figure 19), it can be seen that:-

a) The settlements over the EPB shields were significantly lower than would be expected over an open shield at the same stability number.
b) The settlements over the open shields with compressed air (excluding consolidation settlements - see Hulme, Shirlaw and Hwang 1990) were significantly lower than those over the EPB shields, simply because of the support available at the tail from the compressed air.

The settlements over the EPB shields were acceptable because the shields were driven under a major road, and some buildings were near the edge of the settlement trough. Had the tunnels been driven under or close to buildings it is probable that the amount of ground movement measured would have been unacceptable.

The delay in the onset of consolidation settlements noted for both the EPB and Greathead/ compressed air driven tunnels was an unusual feature. Most previously published cases (Glossop and O'Reilly, 1982, Glossop and Farmer 1979, Morton and Dodds, 1979) record consolidation settlements commending immediately after the compressed air was removed, or when the tunnel passed under the monitoring station. One possible reason for the difference at the Contract 301 tunnels was the use of a hydro-swelling sealing strip together with precast concrete segments sealed on the outside face by a coal tar epoxy coating. The tunnels lined in this manner were exceptionally dry. It has been observed that tunnels generally act as drains (Ward

522

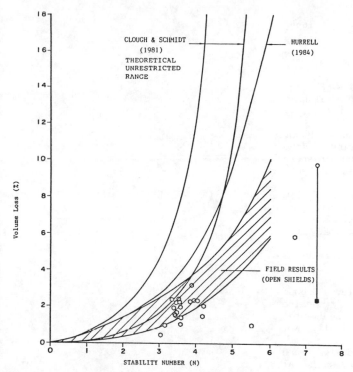

CIMPARISON OF SOME SUGGESTED RATIOS FOR N AND V1 WITH SOME FIELD
MEASURED VALUES IN SOFT CLAYS

O - SINGAPORE MARINE CLAY - OPEN SHIELD
●—O - SINGAPORE MARINE CLAY - EPB

Figure 19   Stability Number Against Volume Loss for Open and
            EPB Shield Driven Tunnels

and Pender 1981), so that normally the
dominant direction of flow would be towards
the tunnel. It is possible that the delay
in the start of consolidation settlements
in Singapore resulted from the
exceptionally watertight lining which
minimized the flow towards the tunnel.

4.   CONCLUSIONS

This paper has described aspects of both
station and bored tunnel construction in
soft clay in Singapore. The main
conclusions resulting from the review are:-

. In soft clay significant consolidation
  settlement can occur as a result of
  excavation even if relatively
  impermeable walls or linings are used.

. In order to be able to understand and
  predict these consolidation settlements,
  the effects on pore pressures of both
  stress changes and seepage need to be
  considered.

REFERENCES

Bjerrum, L. and O. Eide 1956,   Stability
  of struttled   excavations   in clay,
  Geotechnique, Vol. 6, 1: 32-47.
Clark, P.J., and N. Prebaharan   1987.
  Marina   Bay   Station,   Singapore.
  Excavation in soft   clay. Proc. 5# Int.
  Geot.   Seminar, Nanyang Tech. Inst.,
  Singapore; 95-108.
Clough, G.W, L.R. Hansen and A.I.   Mana
  1979. Prediction of   behavior   of
  supported excavations   under marginal
  stability   conditions, 3nd   Int.
  Conf.   on   Numerical   Methods   in
  Geomechanics, Aachen, Germany, Vol. III;
  1485-1502.
Davies,   R.V.   and N.M.   Walsh   1983.
  Geotechnical   aspects of soft   ground
  tunnelling.   Proc. Int. Sem. on constr.
  problems in   Soft soils, Nanyang Tech.
  Inst., Singapore.
Glossop,   N.H. and I.W.   Farmer   1979.
  Settlements associated with   the removal
  of   compressed   air   pressure   during
  tunnelling   in   alluvial   clay.

Geotechnique 29, No. 1, 67-72.

Glossop, N.H. and M.P. O'Reilly 1982. Settlement caused by tunnelling through soft maine silty clay. Tunnels and Tunnelling, Vol. 14, No. 9, 13-16.

Hulme, T.W., L.A.C. Potter and J.N. Shirlaw 1989. Singapore Mass Rapid Transit system; Construction. Proc. Inst. Civ. Engrs, Pt 1, 86; 709-770.

Hulme, T.W., J.N. Shirlaw and R.N. Hwang 1990. Settlements during the underground construction of the Singapore MRT. Proc. Xth Southeast Asian Geot. conf. Vol. 1; 521-526.

Kado, Y., T. Ishii, J.N. Shirlaw and K. Lim 1987. Chemico Lime pile soil improvement. 5th Int. Geot. Seminar, Nanyang Tech. Inst., Singapore; 207-218.

Milligan, G.W.E. 1983. Soil deformations near anchored sheetpile walls. Geotechnique, 1; 41-55.

Morton, J.D. and R.B. Dodds 1979. Ground subsidence associated with machine tunnelling in fluviodeltic sediments. Tunnels and Tunnelling Vol. 11, No. 8; 13-17.

Nicholson, D.P. 1987. The design and performance of the retaining walls at Newton Station. Proc. Mass Rapid Transit Conf. Singapore; 147-154.

Torstensson, B.A. and A.M. Petsonk 1986. A device for in-site measurement of hydranlie conductivity. Proc. 4th Int. Geot. Seminar, Nanyang Tech. Inst., Singapore; 157-162.

Ward, W.H. and M.J. Pender 1981. Tunnelling in soft ground. General Report. Proc. Xth. Int. Conf. Soil. Mech. and Found. Engng. Stockholm. Vol. 4; 261-276.

Developments in Geotechnical Aspects of Embankments, Excavations and Buried Structures, Balasubramaniam et al. (eds)
© 1991 Balkema, Rotterdam. ISBN 90 5410 019 2

The geotechnical aspects of Malaysia's North-South Expressway

R.A.Nicholls
Sir William Halcrow & Partners UK, seconded to Pengurusan Lebuhraya Berhad, Malaysia

ABSTRACT : It is just over two years since the concession agreement for the Expressway was initiated. The first section built under PLUS management is to be opened in December 1990. This paper highlights the important geotechnical aspects of the work to date and draws heavily on the recent Seminar on the same topic.

1   INTRODUCTION

During the late 1970's, the Government of Malaysia embarked upon an ambitious programme to construct the North-South Expressway. This work began in 1982 under the direction of the Malaysian Highway Authority (MHA). In 1986 it was decided to privatise the whole Expressway project. In July 1988 a concession agreement was awarded to Projek Lebuhraya Utara-Selatan Berhad (PLUS), a company incorporated by United Engineers (M) Bhd (UEM). In turn, the project management was awarded to Pengurusan Lebuhraya Berhad (PL).

The initial design was prepared by local consultants appointed by MHA. These designs were prepared on the basis that there would be on site adjustments and variations during construction. However, PLUS require the contractors to assume many of the risks associated with lump sum contracts.

The same local consultants were appointed by PLUS to optimise their designs under the project management of PL. The geotechnical aspects of this optimisation form the basis for this paper.

2   ROUTE GEOLOGY

The route of the Expressway follows the existing Highway 1 on the western side of Peninsular Malaysia. The underlying geology is shown on Figure 1. There are extensive stretches of Quaternary deposits, especially in the northern region centred around the Butterworth area, and in the southern region at Muar. These recent deposits are a feature of the west coast of the Peninsular, Raj & Singh (1990).

The important geotechnical problems centre around construction of embankments on the soft sediments and the stability of cuttings in the metasediments. The Expressway crosses almost the full extent of the Malaysian geological profile. Each geological unit has its own specific problems which have to be identified and solved.

3   THE CONCEPT OF A CENTRAL SOILS LABORATORY

Optimising designs of embankments on soft ground requires very high quality sampling and testing facilities. In 1987, whilst the site investigations were being planned, no such facilities existed in Malaysia. The concept of a central soils laboratory was introduced. This facility would undertake soft ground sampling, transportation of samples and subsequent testing. Soil Centralab Sdn Bhd (CSL) was formed in 1988 as a subsidiary of UEM.

The facilities are housed in a new building at Bangi, just south of Kuala Lumpur and within a short distance of the Expressway. The CSL has 40 consolidation machines, effective stress path testing equipment and the full range of soils testing facilities. In addition, their activities were extended to include a piezocone mounted on a crawler chassis, Marchetti dilatometer equipment and Geonor in situ shear vane equipment. The laboratory has its own piston sampling equipment and purpose made foam padded

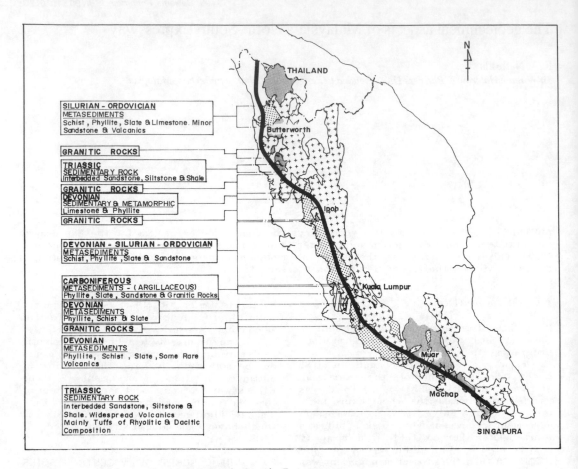

Figure 1   Geology along the North-South Expressway

sample transit containers.

The laboratory and field testing personnel were trained by Kiso-Jiban Consultants and with the exception of two senior technicians all members of staff had no previous experience of soil mechanics laboratories. The recruitment of staff without prior experience was a policy adopted from the outset and has proved very effective indeed.

4   SITE   INVESTIGATION: THE STATISTICS

The full length of the Expressway will be 867 km of which some 355 km are now operational. The remaining 512 km to be built were split into some 46 construction packages. The site investigation contracts, numbering some 75 contracts were split into four categories: soft ground conditions requiring piston sampling; drilling to investigate ground conditions where piled foundations for structures were needed; areas having cuttings where sampling and coring would prove difficult because of the rock types, and lastly mining ponds.

Some 35 site investigation contractors have been invited to tender for the work. CSL has been appointed to undertake the majority of the sampling works in the soft ground contracts. Currently, they employ two modified drilling rigs under contract to carry out the drilling works. A small group of contractors has specialised in drilling in the area of deep cuttings and these contractors are now experienced in appropriate methods of exploration.

At the present time, some 2,540 boreholes have been drilled with a total meterage of about 36.4 km. The site investigation contracts have been administered by PL on behalf of PLUS. The local consultants employed on a particular package have had the responsibility of supervising the site work. The scope of work for each contract has been prepared jointly between the local consultant and the Geotechnical Department of PL.

The site investigation work has been carried out to an exacting standard. PL (1989). This Standard Specification for Soils Investigation Contracts is now in its second addition. The document contains a section on the format for presentation of data to ensure a reasonably uniform presentation. This has been of great benefit to the geotechnical engineers when dealing with several contractors.

At the present time about 85% of the proposed site investigations have been completed at a cost of approximately Rgt. 15 million. This amounts to only 0.3% of the estimated construction costs, a minimal expenditure.

At the completion of the site investigations some 200 volumes of reports will have to be created. In addition to this storage of data the actual information is presented on A1 size drawings each covering approximately 600 m of Expressway. At present, 30% of this graphical archiving has been completed.

5   SITE INVESTIGATION: SOFT GROUND

In the initial phase of the soft ground investigation the major emphasis was placed upon soil profiling using carefully drilled boreholes and sampling techniques. As confidence has grown in the use of the piezocone, the emphasis has now changed in favour of using the piezocone as a profiling technique with the conventional boreholes as a means of confirming the piezocone derived profile and obtaining detailed design parameters.

5.1   Sampling

The importance of drilling boreholes in soft ground should not be overlooked in favour of the more exciting aspects of piston sampling techniques. Of prime importance is the advancing of a stable borehole with the bare minimum of disturbance for the soil especially in the vicinity of the sampling depth.

The preferred method is to use a single tube core barrel with a blade brazed across the face, which has tungsten carbide inserts. The blade itself shaves the clay as the barrel is rotated and the small clay shavings are transported to the surface by the bentonite drilling mud discharged at full diameter thus, minimising any jetting

tendency. The benefit of a core correl, as compared with a fish tail bit or tri-cone arrangement, is that it leaves relatively smooth walls which minimise the tendency for cave ins when the piston sampling equipment is lowered down the hole.

On the North-South Expressway project the following specification has been adopted for the stationary piston sampler, PL (1989):

Material	–	Stainless steel
Minimum diameter	–	75 mm
Nominal length	–	1000 mm
Wall thickness	–	1.5 mm to 2.0mm
Angle of cutting edge	–	6° to 10°
Clearance ratio	–	0% to 5%

In practice this specification has proved quite satisfactory in obtaining high quality samples in the soft silty clays.

The drilling machines used for the soft ground drilling works are crawler mounted and have been modified such that they have at least 1000 mm stroke on the hydraulic rams so that the piston sampler can be pushed to full depth in one continuous operation.

5.2   Transportation and Storage

Once the samples have been recovered, sealed and labelled they are placed in the padded transit container. The wooden boxes are standard size and the inner foam moulds can be changed dependent upon the sample tube size. These mouldings have sufficient thickness to ensure adequate cushioning during handling and transportation.

The sample containers are carried in a specially constructed suspended frame which can accommodate six containers. The frame is fitted permanently inside a 4 WD vehicle which itself has relatively soft suspension. The frame has scissor type suspension which is controlled by a substantial pneumatic shock absorber. The combination of foam padding, frame and vehicle suspension has minimised the sampling and handling disturbance that can take place when bringing samples back from remote locations.

Upon arrival at the central laboratory the samples are placed on shelf mounted rubber cradles to resist any movement. It is not necessary to move any sample tube to retrieve another tube for testing. The shelves are supported by stainless steel frames.

5.3   Testing

There are two objectives for obtaining undisturbed samples:

i.   detailed fabric description using samples split longitudinally, and

ii.   testing undisturbed specimens from a sample 800 mm in length.

At most sites one borehole with continuous piston sampling is undertaken so that each sample can be extruded and logged. This method gives an almost continuous detailed description of the soft ground profile. These split samples are then photographed. The detailed descriptions and photographs form part of the factual report.

The success of photographing full length split samples has encouraged the splitting of redundant specimens from full sample lengths and making up discontinuous photo-logs of boreholes. Engineers are finding the photo-logs very informative. The photographs are taken using large format negatives either 6 x 4.5 cm or 6 x 6 cm. The larger format negatives have yielded better definition than the more common 35 mm negative format.

The full length, 800 mm, samples are extruded using a motorised hydraulic extruder, which was designed and constructed locally. The hydraulic ram has a stroke greater than 1000mm and this allows a full sample to the extruded into a plastic former without any breaks in the sample. If specimens are to be tested the sample is chopped up into 100 mm long specimens, waxed and labelled.

One of the most important tests to be carried out on soft clays is the consolidation test. In the CSL there is the facility to test 40 specimens together at either 50.5 mm diameter or the larger 71.4 mm diameter. In general the testing follows the recommendation of BS 1377: 1975. However, modified loading procedures have been developed with the aim of speeding up the standard test and also to identify the pre-consolidation pressure with more confidence, Ho & Dobie (1990).

The conventional load increment method results in difficulties in assessing the pre-consolidation pressure when it falls midway between the load increment values. To avoid this problem the load is increased in uniform pressure increments of 5 kPa or 10 kPa until the pre-consolidation pressure is reached. Once the pre-consolidation is exceeded a load increment ratio of 0.5 is used for the 24 hour maintained loading.

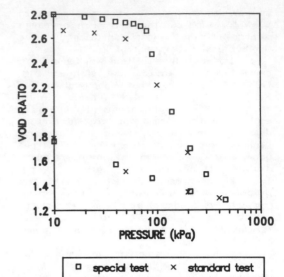

Figure 2   Comparison of void ratio versus log pressure results for standard and special procedures.

At pressures below the pre-consolidation pressure the load is increased as soon as primary consolidation is complete. Figure 2 shows the comparision between the standard BS method and the modified method. There is a better definition in the area of the pre-consolidation pressure and this can be defined within two to three hours.

The less glamorous tests have not been ignored at the CSL and standard testing such as Atterberg limits have been examined following different methods of material preparation, Rusli (1990). This latter investigation was instigated to assess the effects of air drying and oven drying. There was a greater variation on the plasticity chart for alluvial soils as compared with residual soils. It was found that oven drying samples prior to the test significantly lowered the liquid limit and the plasticity index, Figure 3. The investigation concluded that samples should preferably be tested in the as received condition and that factual laboratory reports should state which method of sample treatment was adopted.

As part of the classification testing for the soft ground, selected samples were subjected to clay mineralogy analysis, Rai & Ho (1990) and the measurement of the salinity of the extracted pore fluid, Nicholls & Ho (1990). The clay mineralogy study revealed the presence of montmorillonite

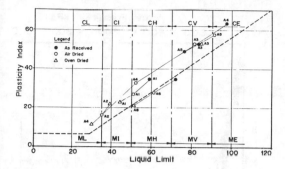

Figure 3 Variation of plasticity with different sample preparation for alluvial soils.

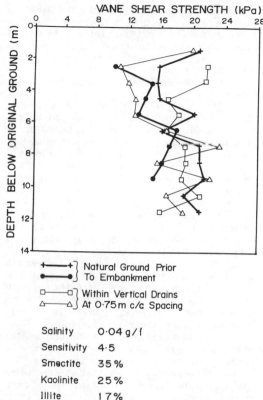

Legend (Figure 5):
+———+ } Natural Ground Prior To Embankment
●———● }
□———□ } Within Vertical Drains
△———△ } At 0·75m c/c Spacing

Salinity	0·04 g/l
Sensitivity	4·5
Smectite	35 %
Kaolinite	25 %
Illite	17 %

Figure 5 Shear strength between vertical drains

within the marine clays, especially in the northern region near Butterworth; towards the south kaolinite is the predominant clay mineral. Although the clay mineralogy study, together with scanning electron microscope studies, have revealed that the clays along the western lowland were deposited in a marine environment, the measurements of salinity on extracted pore fluid have revealed a very significant variation.

At the southern limit, in Johor Bahru, the salinity of the clay was the same as sea water (35 ~ 40 g/l), other clays sampled had salinity as low as 0.01 g/l. At the embankment trials near Butterworth the pore water salinity is approximately 13 g/l. Unlike the Norwegian quick clays the sensitivities of Malaysian marine clays lie predominantly between 3 and 5 whatever the value of salinity. There appears to be no correlation between in situ void ratio and salinity as might be expected in quick clay with very low values of salinity (Figure 4). The Malaysian clays are not particularly sensitive to disturbance. At a

site near Kuala Lumpur where vertical drains had been installed under an embankment, the measured shear vane strength between the drains at a spacing of 0.75 m c/c was very similar to the strength measured outside the embankment in the untreated material ground (Figure 5). At this location the clay was almost completely leached with a salinity of 0.04 g/l.

5.4 Dilatometer

The flat plate dilatometer was introduced in Europe in the late 1970s but has not been widely used in Malaysia. It was introduced into the site investigations for the Expressway to investigate whether the equipment could be an effective profiling tool. Its main advantage over more sophisticated and expensive equipment is that the dilatometer is robust and technically basic. It should prove more practical for general use than the piezocone for soil profiling.

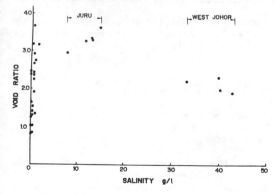

Figure 4 Void ratio versus salinity

The Marchetti dilatometer requires the dilation of a circular membrane into the ground for 1mm. The interpretation of various pressures to achieve this expansion can be used to assess the soil profile. The dilatometer is pushed into the ground and pressure measurements are taken at intervals of 0.2m.

Marchetti (1980) had proposed that three parameters could be used to profile the ground. This work was based upon Italian clays and some regional calibration for use in the Malaysian clays is required. The testing to date, Wong & Dobie (1990), has verified that the dilatometer can quite effectively predict the soil profile. The overconsolidation ratio proposed by Marchetti and others does not appear to be valid for the Malaysian clays and additional testing will be needed to yield satisfactory correlation for this region.

The undrained shear strength predicted using the Marchetti empirical factor slightly under predicts the in situ shear vane strength. The correlation needed for Malaysian clays is partly dependent upon the plasticity index.

An additional aspect of the dilatometer has been investigated and that is the closing pressure. This is the corrected residual gas pressure in the system when the membrane has returned to its original closed position. Campanella et al (1985) indicated that the residual pressure when adjusted for the membrane stiffness can be a valid indication of induced excess pore water. Comparisons were made between the corrected residual pressure, or closing pressure, and the transient excess pore water pressure measured with a piezocone. The results were quite encouraging. Figure 6.

The work with the dilatometer carried out by CSL shows that it can be used as a soil profiling tool quite effectively. As would be expected from any system requiring empirical factors additional testing in the Malaysian clays is needed before interpreted parameters can be used confidently. However, the work does indicate that the equipment has a very useful role to play in site investigations. With the dilatometer and compatible hydraulic penetration equipment mounted on a crawler chassis a very neat and efficient site investigation tool is made available.

5.5 Piezocone

The performance specification for this equipment required by the project was as follows:

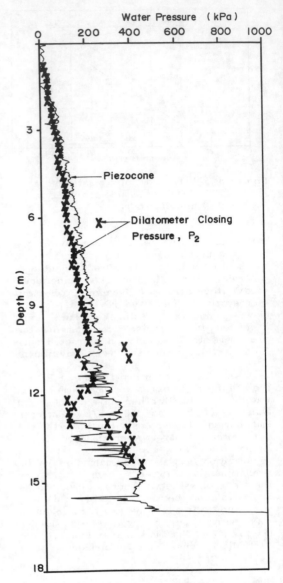

Figure 6   Dilatometer closing pressure versus piezocone measured water pressure

cone resistance sensitivity : 0.01 MPa

friction sleeve sensitivity : 0.1 kPa

pore water pressure sensitivity : 1 kPa

The equipment now used by CSL is supplied by Hogentogler and Co. Inc. of USA and was mounted on a crawler chassis to the design of the CSL engineers. The unit has proved a very effective piece of equipment. Examination of the above measurement sensitivities will reveal that very high quality transducers, data logging and management equipment is required.

Back to back tests with another piezocone having the same performance yielded some quite interesting results. At the same location earlier profiles did not agree at all. One of the piezocones had a temperature sensor and the variation in temperature during the test gave a clue to the problem.

Any transducer with surface mounted electrical resistance strain gauges will give a variation in electrical output when the ambient temperature changes. A manufacturer can use special metals in the transducer and indeed use special thermally stable gauges but when the transducer itself changes dimension as a result of the temperature the electrical output must change. Total temperature compensation is near impossible in a cylindrical load cell.

Examination of the data from the temperature probe showed that when the piezocone was being prepared on the ground surface its temperature rose quite considerably higher than the in situ ground temperature and the cone temperature would drop with penetration, Test 1 in Figure 7. As the load cell cooled it contracted and in consequence this would be read as increased cone resistance, Pauzi et al (1990).

The series of tests undertaken with two piezocone machines helped to understand the real need for temperature conditioning of the whole piezocone probe at a temperature similar to that of the ground. When this conditioning is carried out properly little trouble is experienced and profiles are then quite repeatable.

Once the aspect of temperature conditioning has been fully understood and a strict conditioning procedure followed the only other aspect to affect results is the effective saturation of the filters. The latter affects the measurements of the transient pore pressure. Experiments were performed by CSL to study the effect of bad or incomplete filter saturation. They were quite dramatic as shown on Figure 8, Wong et al (1990), where Test 1 was carried out with a poorly saturated filter and Test 2 used a fully saturated filter.

The piezocone is a very high technology piece of equipment and when operated under difficult field conditions the operators must follow a very stringent procedure and, what is more, they must understand why this is so otherwise there may be a tendency to short cut on the

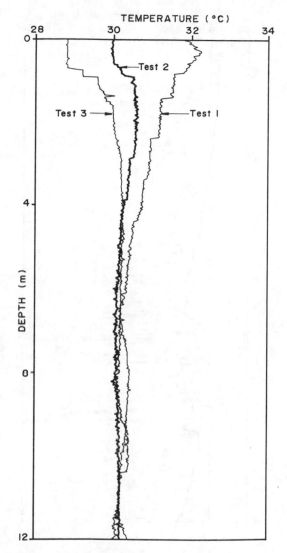

Figure 7   Temperature sensor profiles for a soft clay

essential pre-test preparation of the equipment. There is a definite learning curve in the operation of the piezocone to obtain the high quality repeatable profiles that are now being achieved.

The geotechnical engineers are now quite confident in using the piezocone to profile the soil. The addition of the pore pressure element has created a very important additional parameter which allows accurate soil profiling. Such is the confidence of the engineers using the piezocone data that now the initial investigations in areas of soft ground are

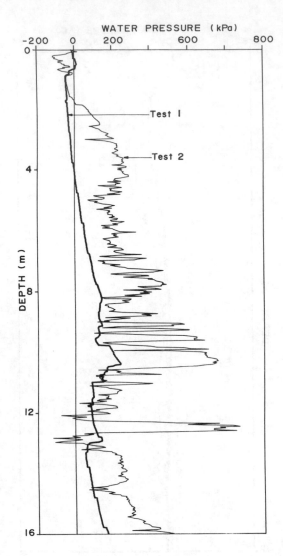

WATER PRESSURE (kPa)

Test 1

Test 2

Figure 8   Water pressure plots from adjacent locations showing effect of poor filter saturation

boreholes but obtaining a high density of testing at those borehole locations. Needless to say boreholes are located adjacent to piezocone soundings to confirm the interpretation.

The present indications are that in future at the required borehole locations, one of a pair of boreholes will be drilled to obtain continuous piston sampling and a subsequent photo-log of the profile; the second of the pair of boreholes will be for any in situ testing and sampling for subsequent laboratory testing. This seemingly expensive luxury can in fact be accommodated by an overall reduction in the number of boreholes that were customarily used to determine the soil profile.

Penetration profiling equipment is changing and the geotechnical engineer must change his ideas and use the new breed of equipment to his advantage.

6.   SITE INVESTIGATION : RESIDUAL SOILS

On the project the investigations in residual soils have been developed for the design of cut slopes, with ever increasing depths, and the empirical design methods have given way to a more fundamental approach. Aspects such as classifying weathering grades and identifying relict joints require material for geologists to study.

6.1   Mazier type samplers

In the majority of weathering profiles the conventional open drive sampler has a limited depth within which it can operate. It is common practice in some of the more extensive zones of cuttings to start boreholes with cable percussive methods and open drive samples. Such samples are an excellent method of obtaining bulk samples for laboratory testing, the tests being directed towards the use of cut material as embankment fill.

However, for the design of the cut slopes these drive samplers have limited use. As the soil profile becomes more competent some form of rotary drilled sampling system is required. Learning from the experience of the Hong Kong engineers the project adopted the Mazier type triple tube retractor core barrel as a means of sampling. The introduction of this retractor sampler was encouraging but problems with its use remained to be solved. Early specimens of the sampler had over size clearance between the cutting ring and the sampling tube, or inner core barrel.

carried out with the piezocone. Once these profiles have been studied the location of boreholes can be identified. The boreholes are now confirming soil profiles interpreted from the piezocone and used to obtain engineering parameters.

The emphasis of investigation in soft ground is changing on the project. The increased confidence in penetration soil profiling has allowed more effective location of boreholes. At these boreholes a whole suite of sampling and testing can be undertaken. The tendency is towards less

Various local modifications have been made to the Mazier type sampler used in Malaysia, locally made inner plastic liners have been made to suite the cutting shoes. This has largely reduced the problems of the surface of the samples dilating and soaking up water. Re-usable stainless steel tubes are also available. These are particularly efficient when using a manual sample extruder on site.

6.2 Foam flush drilling

One of the problems of drilling with the Mazier type system is that the field geologist does not have sufficient material to log, the system having been developed as a method of obtaining samples for laboratory use. The conventional rotary drilling system in residual soil can result in very little core recovery. This poor core recovery is normally attributed to the erosion of the in situ material during the drilling process, in addition, the erosion of the core that has passed within the cutting shoe.

Trials were carried out in the early phase of the project when drilling was being carried out in the difficult volcanic tuff. The face discharge bits were modified to reduce the jetting action but increase the upward flow behind the cutting edge. Bypass valves and monitoring gauges were added to the drilling rigs to attempt to control water flow to the minimum required. Greater control of the flush water and modified drill bits did improve core recovery, but the recovery ratio was still less than acceptable.

Again following the lead set by the engineers in Hong Kong the use of foam flush was introduced to the project. Even drilling in highly erodible formations the core recoveries have been quite spectacular, Chin and Tan (1990). Needless to say there has been a learning curve for each contractor undertaking a new system, and those contractors who have specialised in the cuttings investigations have been very enthusiastic and they are pleased with the results and the improved working conditions around the rigs.

The combination of high core recovery and the availability of the retractor sampling system has given the geologists as much material as possible to work with. For the first time the geologist can start to consider logging the ground profile before cuttings are opened up.

The availability of good core has resulted in improved logging methods and material classification system.

6.3 Material classification

The Expressway route passes through the granites, the sedimentary and the metasedimentary rocks all having different degrees of weathering. A classification of the materials has been attempted in the belief that once the material and the weathering profile have been systematically classified the work of design is made easier and more thorough, Singh et al (1990).

At the present time there is no universal classification system for the weathering of tropical rocks. The system adopted by the project was the result of an extensive literature study and much discussion, it is at best a compromise and accepted as such. One classification system has been adopted for the weathering classification of granites and volcanic rocks and another has been adopted for the weathering classification of the sedimentary rocks.

7. TRIALS

The construction of embankments over soft ground can be solved in a number of ways ranging from reasonable to very expensive solutions. The various sections of the Expressway have been awarded to local consultants and their designs for such situations reflect their experience or interest. Solutions to the problem have included, piled embankments and viaducts, dynamic replacement, soft ground replacement, acceleration of consolidation using vertical drains with and without surcharge affected by fill or vacuum and very low embankment only with preload.

As part of the optimisation of the design the local consultants were given a performance specification for the long term behaviour of the embankments. The solutions to achieve the requirements of the specification were numerous. The main areas of soft ground are the peat swamp to the south near Machap and the marine clays to the north between Changkat Jering and Sungai Muda in the Butterworth area.

As in all engineering solutions, the proposals need to be viable as well as practical. In the south between Machap and Sedenak sand and gravel are not available locally which precludes certain methods. In the north, in general, the marine clays are thick and the concept of soft ground removal is not practical.

7.1 Peat replacement at Machap

Between Machap and Sedenak the surface of the peat swamp is approximate +9m. During the initial design in the early 1980's this area was a neglected and largely untouched waterlogged swamp. During 1984 the Drainage and Irrigation Department started a programme of draining this swamp land and by 1988 the ground water table had been lowered to its present depth of 0.5m below ground level.

The original design had focused on piled viaducts over a considerable extent of the swamp. During the optimisation exercise and an additional site investigation it was concluded by the local consultant that a peat replacement method in certain restricted areas could be a viable alternative to a piled solution. The consultants proposals were to dewater the peat layer, some 6m deep, in stages and remove the peat. With a dewatered excavation residual soil would be placed as backfill and compacted in layers. Initial trials were undertaken to assess the permeability of the peat, a borehole pumping test was carried out as well as trench dewatering trials. These trials although not yielding totally conclusive results indicated that a dewatering method was practically possible.

A trial length of approximately 400m was chosen, access was not difficult and the trial area lay between two hills of residual soils. Prior to any drainage ditches being excavated the ground conditions were very difficult for excavators to move on. The site investigation crawler rigs were forced to move on timber matting.

Perimeter drainage ditches were excavated by hydraulic excavator and these were kept dewatered by intermittent pumping. The general lowering of the water table improved the ground and enabled the main 2m deep dewatering drains to be excavated without too much difficulty. These main drains were excavated in depth increments of 0.5m over distances of 20m, again with intermittent pumping.

The finally adopted method of excavation involved two stages of excavation. The first stage was achieved by removal of 3.5m of peat using hydraulic excavators. Following this stage a second team removed the remaining peat and underlying soft clay, this work commencing by advancing from the sloping residual soil. The second stage of excavation was kept dewatered by the manual construction of drainage ditches allowing the water to flow to the sumps of the main dewatering drains. As the second stage work advanced the equipment worked off the residual soil replacement fill. The details of this trial are reported by Toh et al (1990).

The trial has been successfully completed and the whole embankment is currently being monitored to measure any future settlement. The method of peat replacement has saved a considerable amount of money. It is now proposed to adopt this method during the main works over some 10km of peat swamp. Not all piled viaducts have been removed from the design, but where the peat replacement can work and the depth of compressible clay below the peat is not significant the replacement method is being proposed by the consultants.

7.2 Vertical drain trial, near Butterworth

In the northern region of the Peninsular, in the Butterworth area, the depth of compressible clay deposits can be up to 20m. The construction of embankments less than 2.5m high has not proved troublesome on the existing section of the Expressway. However historically wherever embankments have been higher than 2.5m, at bridge approach embankments the designers have adopted a piled approach.

As part of the optimisation process consideration has been given to using conventional soil mechanics in these transition embankments. There was a certain concern regarding designing with vertical drains to accelerate the consolidation process. The concern over the performance of vertical drains in the marine clay lead to the concept of the vertical drain trial embankment at Juru, just south of Butterworth.

The marine clay at the site is approximately 13m thick and in a uniform very soft to soft silty clay. Two embankments, 3.6m high have been constructed. Both embankments lie on a drainage blanket. The control embankment has the same instrumentation as the drained embankment. Vertical drains have been installed on a 1.2m square grid and the installation took place after all the instrumentation had been commissioned. Details of the trial are given by Hashimi et al (1990).

The laboratory values of C_v are approximately 0.3 m^2/year and based upon only a short period of pore pressure dissipation the back analysed field C_h values are in the range 0.6 to 0.8 m^2/year. These values are some 2 to 2.6 times the laboratory C_v values.

The monitored values of settlement

and lateral movement indicate the lateral movements beneath the toe areas of both embankments are quite similar, whereas the centreline settlement of the drained embankment is considerably greater than the control embankment. The early monitoring data indicate that the drains are accelerating the process of consolidation.

8. CONSTRUCTION ON MINING PONDS

A vast amount of alluvial tin mining has taken place between Kuala Lumpur and Ipoh. The proposed route of the Expressway passes through these mining lands.

The present soil profile is very disturbed and erratic, the base of the ponds which is often pinnacled limestone tends to be deeper towards the south. The nature of disturbed material depends upon the process of mining and since the filling of worked excavation is by hydraulic means the whole profile is very erratic.

The material creating the problem of settlement is the fine faction and its current state of consolidation. Excavations which are present day ponds are likely to contain under-consolidated fines.

Presently, the design proposals for constructing embankments on these ponds fall into two groups, Ting et al (1990) and Najib & Nicholls (1990) :

i. complete removal of soft material, and

ii. partial removal of soft material followed by ground improvement.

The work of constructing the Expressway embankments on the mining land has only just started and it is too premature to indicate which are the optimum ways of dealing with this problem.

Research is underway to attempt to solve the problems of consolidation of this man made fine material.

9. EROSION CONTROL

No account of the geotechnical aspects along the Expressway project would be complete without mention of the important issue of erosion control. Zaim (1990) has given an account of the current situation on the project and brief comments can be made for completeness.

PL identified the need for specialist input on erosion control. There is much experience in Malaysia with regard to erosion and its costly effects but there does not exist any one body collating this experience. Geostructures Consulting from UK were employed to undertake a specific study, the benefits of which would be incorporated into the design of the new work.

The result of the specialist input has been a revised specification on grassing and the treatment of slopes prior to grassing. However, it must be accepted that at the present time sufficient consideration is not given to erosion control by the engineering profession.

PLUS has sponsored a considerable research programme with the University of Malaya and aspects of slope stability and erosion control are being actively pursued. It is very encouraging that PLUS has initiated this work because PLUS as an organisation will be in existence for 30 years and the cost of erosion will be an important aspect. It is certain that the benefits from this combined research will be made available to the engineering profession in due course.

ACKNOWLEDGEMENT

The author wishes to thank the members of the PL Geotechnical Department for their strong support from the beginning of the project. Much reference has been made to the recent Seminar organised by PL and PLUS, and the contribution of the various authors is gratefully acknowledged.

REFERENCES

British Standards Institution (1975). Methods of test for soils for civil engineering purposes. BS 1377:1975. BSI. London.

Campanella, R.G., Robertson, P.K., Gillespie. D and Greig, J.(1985). Recent Developments in In-situ testing of soils. Proc. of 11th ICSMFE, San Francisco, Vol. 2 pp 849–854.

Chin, C.H. and Tan, S.L. (1990). Improved core recovery using foam flush. Seminar on Geotechnical Aspects of the North South Expressway. PLUS. Kuala Lumpur.

Hashimi, A., Khalit. Hi. O., How, K.T. and Chan, P.C. (1990). Vertical drain embankment trial at Sungai Juru. Ibid.

Ho, H.S. and Dobie, M.J.D. (1990). Consolidation testing : alternative loading procedure for soft clays. Ibid.

Marchetti, S. (1980). In-situ tests by flat dilatometer. Journal of the Geotechnical Engineering Division. A.S.C.E. Vol. 106. No. GT3. pp. 299–321.

Najib, Mhd and Nicholls. R.A. (1990).
Embankment construction over mining
ponds. Seminar on Geotechnical Aspects
of the North-South Expressway. PLUS.
Kuala Lumpur.

Nicholls, R.A. and Ho, H.S. (1990). Clay
salinity. Ibid.

Pauzi, I., Kassim, K. and Dobie. M.J.D. (1990).
Piezocone testing : effect of
temperature. Ibid.

PL (1989). Pengurusan Lebuhraya Berhad.
Standard Specification for Soils
Investigation Contracts. Second Edition
1989.

Raj, J.K. and Ho, H.S. (1990). Clay
mineralogy of Holocene Marine clays along
the North South Expressway. Seminar on
Geotechnical Aspects of the North-South
Expressway. PLUS. Kuala Lumpur.

Raj, J.K. and Singh, M. (1990).
Unconsolidated Holocene sediments along
the North-South Expressway. Ibid.

Ting, W.H., Oo,G.H., Ng,S.J. and Lau L.Y.
(1990). Geotechnical Aspects of Reclaiming
Mining Ponds Along An Expressway. Ibid.

Toh, C.T., Chua, S.K., Chee, S.K., Yeo, S.C.
and Chock. E.T.(1990). Peat replacement
trial at Machap. Ibid.

Wong J.T.F., Kassim, K. and Dobie. M.J.D.
(1990). Piezocone testing : saturation of
filter. Ibid.

Wong, J.T.F and Dobie, M.J.D. (1990).
Marchetti dilatometer : interpretation in
Malaysian alluvial clays. Ibid.

Developments in Geotechnical Aspects of Embankments, Excavations and Buried Structures, Balasubramaniam et al. (eds)
© *1991 Balkema, Rotterdam. ISBN 90 5410 019 2*

Pile-tip cap driving method for precast concrete pile

Tatsuo Ishikawa
Saga University, Japan

ABSTRACT: Precast concrete piles have been mainly driven into the ground by using diesel hammer. Unfortunately, to drive the piles with diesel hammer causes a lot of unavoidable and unplesant problems, such as noise, vibration and oil splash around the construction site. Pile-tip cap driving method for precast concrete pile is now developed. The tip of hollow precast concrete pile is coupled with steel cap. The cap is driven by dropping a hammer through its central longitudinal hollow as well as the pile itself is advanced by another hammer on the pile top. This pile-tip cap driving method has less noise and quake levels which are below the allowable values of the Japanese regulations for the specified construction work and no significant oil splashes. Based on basic energy equation, a dynamic formula for estimating the allowable bearing capacity of pile is proposed. A calculated value from this formula using last penetration shows good results close to the field measurements. Furthermore, finite element analysis on a model related to this method gives higher driving efficiency than the conventional pile driving method on pile top.

1 INTRODUCTION

Nowadays, diesel hammer is widely used to drive the precast concrete piles in engineering practice. Impacting the pile top by the diesel hammer produces satisfactory results with excellent work efficiency and reliable bearing capacity. Recently, pile driving with diesel hammer, creates social unpleasant problems, i.e. noise, vibration and oil splash around the construction site. With these reasons, the cement milk method has been developed. This method employs no impacting on the pile which causes lower pile capacity. Moreover, it produces pollution of underground water and increases the construction cost. With impacting the tip of hollow precast concrete pile, the impact efficiency is extremely improved. It also reduces the noise, quake of the ground and oil splash. This study aims not only to increase the bearing capacity of precast concrete pile by using pile-tip cap driving method but also to propose a dynamic formula for estimating the allowable pile capacity in order to compare the calculation results with those from field tests. It is found from this study that the noise and

vibration measured are reduced to be lower than that given by the Japanese regulations for the specified construction work by employing this method.

Finite element analysis on a model corresponding to pile-tip cap driving method shows a higher driving efficiency than the common pile driving method on pile top.

2 LOW ANNOYANCE DRIVING METHOD FOR PRECAST CONCRETE PILE

There are two basic driving methods of precast concrete piles with low annoyance to the surroundings. One is to directly drop the precast concrete pile into the prebored or preaugered hole and another is to strike on the top of the pile with hammer to drive it into the ground.

In engineering practice, there are many methods for pile installation. Cement milk method is one of the most applicable methods which is hereby briefly described. The hole is bored with a diameter a few larger than pile size by auger or other means down to the required depth. The hole walls are kept stable by using

bentonite suspension and cement milk. Then
the precast concrete pile is dropped into
the hole by its own weight as well as
applying pressure and light striking on
the pile. The bearing capacity of pile
consists of tip resistance and
circumferential friction resistance of
cemented material. Noise and quake is
considerably reduced. If the pile is not
impacted, large settlement may sometimes
occur due to the loose soil at pile tip.
However, high construction cost, long
working time period, pollution of
underground water, massive soil disposal
are inevitably also troublesome.

Another low annoyance striking method is
to drive the pile with noiseproofing
covered diesel hammer or oil pressure
hammer. Noise is really reduced with great
deal but not the quake.

3 PILE-TIP CAP DRIVING METHOD

This method requires the tip of hollow
precast concrete pile to be coupled
with steel cap which is driven into the
ground by dropping a weight hammer through
its central longitudinal hollow. Moreover,
the pile is also advanced by another
center holed hammer on the pile top. Not
only noise, quake, oil splash and
pollution of ground water can be reduced
but also the more reliable bearing
capacity of pile can be achieved, due to
impacting.

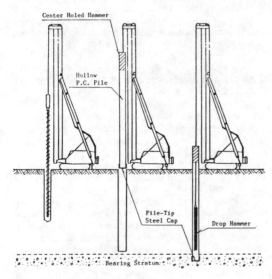

Fig. 1 Pile Installation by
Tip Cap Driving

3.1 Execution procedure of pile-tip cap driving method

Piles can be installed by employing pile-
tip cap driving method as follows (Fig.1):
  1. Dig out hole with diameter a few
larger than hollow precast concrete pile
by auger down to the required depth or
near to the bearing stratum.
  2. The steel cap is coupled to the tip
of hollow precast concrete pile and is
then dropped into this prebored hole.
  3. As the pile-tip steel cap is driven
by dropping weight hammer through
its central longitudinal hollow, the pile
is also driven by the center holed oil
pressure hammer on pile top.

4 BEARING CAPACITY OF PILE

4.1 Dynamic formula for bearing capacity of pile

Two basic formulas for estimating bearing
capacity of pile are static formula based
on engineering properties of soil and
dynamic formula derived from balance of
impacting energy of hammer together with
work done by pile driving.

To make use of dynamic formula for pile
capacity, many factors and variables are
needed which are not easy to evaluate.
This creates some troubles and
difficulties. However, it can still be
used in engineering practice for
estimating bearing capacity of pile and
gives reasonable good results.

The dynamic formula is based on the
balance of impact energy of hammer E_h,
work done RS by pile penetration of depth
S against resistance reaction R and energy
loss Z.

$$E_h = RS + Z \qquad (1)$$

Z can be treated theoretically and
practically which has been introduced in
many forms of dynamic formulas.

Let hammer weight be W_h and hammer speed
at driving be v_h, then the impact energy
will be

$$E_h = W_h \frac{v_h^2}{2g} \qquad (2)$$

Speed of hammer v_h and speed of pile v_p
before impact are changed after impact to
v_h' and v_p' respectively (Fig. 2).
With consideration of balance of
kinematic energy, the following formula
can be written.

$$\left(W_h\frac{v_h^2}{2g} + W_p\frac{v_p^2}{2g}\right)e_c = \left(W_h\frac{v_h'^2}{2g} + W_p\frac{v_p'^2}{2g}\right) \quad (3)$$

Where, e_c is efficiency coefficient with respect to energy loss of cushion at impact and Wp is weight of pile.

From the balance of momentum

$$\frac{W_h}{g}(v_h - v_h') = \frac{W_p}{g}(v_p - v_p') \quad (4)$$

Let ratio of relative speed before and after impact be defined as n, coefficient of restoration.

$$n = \frac{v_p - v_h}{v_h' - v_p'} \quad (5)$$

then

$$e_c = \frac{W_h + n^2 W_p}{W_h + W_p} \quad (6)$$

Effective energy of pile transferred from hammer is

$$E_h e_c = E_h \frac{W_h + n^2 W_p}{W_h + W_p} \quad (7)$$

E_h is kinematic energy of hammer at impact and is a little bit smaller than E_o which is theoretically calculated based on

Before Impact   **After Impact**

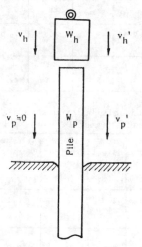

Fig. 2 Velocities of Pile and Hammer, before and after Impact

hammer head and explosion, due to friction loss of hammer driving. Let e be the efficiency corresponding to this loss, then

$$E_h = eE_o \quad (8)$$

and

$$E_h e_c = eE_o\frac{W_h + n^2 W_p}{W_h + W_p} \quad (9)$$

$E_h e_c$ (effective hammer energy) is the sum of works done by pile penetration, elastic deformation of pile and elastic deformation of soil.

Let R be resistance reaction, S be penetration depth, Cp be elastic displacement of pile, C be elastic deformation of soil, then

$$RS + m_1 R\frac{C_p}{2} + m_2 R\frac{C}{2} \quad (10)$$

Where, m_1 is modified coefficient for R which is not constant along the pile length due to the variation of circumferential friction, m_2 is modified coefficient for force acting on soil. From Eq.(9) and (10),

$$R\left(S + m_1\frac{C_p}{2} + m_2\frac{C}{2}\right) = eE_o\frac{W_h + n^2 W_p}{W_h + W_p} \quad (11)$$

Let E be Young's modulus of pile, A be cross area of pile, L be length of pile,

$$C_p = \frac{RL}{AE} \quad (12)$$

substituting (12) into (11)

$$R\left(S + m_1\frac{RL}{2AE} + m_2\frac{C}{2}\right) = eE_o\frac{W_h + n^2 W_p}{W_h + W_p} \quad (13)$$

Most of dynamic formulas for bearing capacity of pile are derived from the same basic equations, but with different modification or simplification.
Assume

$$e\frac{W_h + n^2 W_p}{W_h + W_p} = 1 \quad (14)$$

If energy loss is neglected

$$m_1 \frac{C_p}{2} + m_2 \frac{C}{2} = 0.02(m)$$ (15)

$$R = \frac{E_o}{S + 0.02}$$ (16)

When the safety factor of 5 is taken, the allowable bearing capacity R_a

$$R_a = \frac{1}{5} \frac{E_o}{S + 0.02} = \frac{E_o}{5S + 0.1}$$ (17)

Where, E_o (impact energy) is $2W_h H$ for diesel hammer according to results of vertical loading test.

Formula of the pile-tip cap driving method is derived from the energy balance in the same manner as other dynamic formulas. In pile-tip cap driving method, E_o equals $W_h H$ for drop hammer, elastic deformation of pile $C_p = 0$ and elastic deformation of soil $C = 0$ as pile tip is impacted (see Fig. 3).

The sum of $C_p + C$ is observed as rebound of pile at driving, which cannot be observed in this method.

From Eq.(11)

$$R\left(S + m_1 \frac{C_p}{2} + m_2 \frac{C}{2}\right) = RS$$ (18)

and assume

$$e \frac{W_h + n^2 W_p}{W_h + W_p} = 1$$ (14)

If the safety factor is taken as 5, then the allowable bearing capacity for pile-tip cap driving method is proposed to be

$$R_a = \frac{W_h \cdot H}{5S}$$ (19)

Allowable pile capacity R_a tf
Weight of drop hammer W_h tf
Height drop of hammer H m
Last (final) penetration S m

4.2 Result of field tests of pile driving

Results of the last penetration and the calculated allowable bearing capacity using the pile-tip cap driving method and the diesel hammer driving method under the same conditions and bearing layer are summarized in Tables 1 and 2. Test results show close relation between proposed dynamic formula and conventional diesel hammer driving formula. With impact energy equals one, the allowable bearing capacity from both methods are presented in Fig. 4. Because the formula for pile-tip cap driving method doesn't have a term

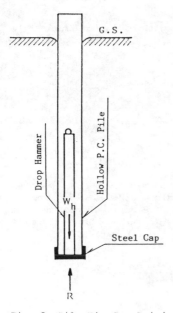

Fig. 3 Pile-Tip Cap Driving

Table 1 Allowable Bearing Capacity by Pile-Tip Driving Method

Location	Hammer Weight W (tf)	Drop Height H (m)	Final Penetration S(mm)	Allowable Bearing Capacity $Ra=\frac{WH}{5S}$(tf)	Reference $Ra=\frac{2WH}{5S+0.1}$(tf)
YAME FUKUOKA	0.7	2.5	7	50.0	26.0
HYOUGO SAGA	0.7	2.5	15	50.0	42.8

Table 2 Allowable Bearing Capacity by Diesel Hammer Driving Method

Location	Hammer Weight W (tf)	Drop Height H (m)	Final Penetration S(mm)	Allowable Bearing Capacity $Ra=\frac{2WH}{5S+0.1}$(tf)
YAME FUKUOKA	1.5	2.3	7	51.1
HYOUGO SAGA	1.5	2.2	5	52.8

for rebound of pile, so only a small depth of final penetration can increase the pile capacity sharply and rapidly. In fact, rebound of pile cannot be observed in the field.

The last penetration of pile employing pile-tip cap driving method as shown in Table 1 is measured by wire connected to steel cap extending to the ground surface. In construction practice, the last penetration should be measured and observed on the ground surface with rare eyes, like other conventional diesel hammer driving method.

Pile driving tests for both methods at Hyougo, Saga site show clearly the difference in driving efficiency. With the same impact energy for both methods, the last penetration using pile-tip cap driving method is three times greater than using diesel hammer driving method. This means that the penetration efficiency from hammer impact is quite high.

4.3 Vertical loading test results

Prestressed concrete pile with diameter 300 mm driven by the pile-tip cap driving method is tested by application of vertical loading. Load cycle and loading steps are summarized in Table 3. At turning point of load-settlement and residual-settlement curve, the yield load is found to be 90 tf. At elapsed time of thirty minutes of settlement-log time curve, settlement increases gradually at 105 tf and remarkably at the next load of 120 tf, the yield load is also estimated at 90 tf from this plot. So a half of that yield load should be long time allowable

bearing capacity of pile, i.e. Ra = 45 tf. From driving record, the pile is driven with hammer weight Wh = 1.4 tf, height drop H = 1.6 m, and the last penetration S is measured to be 11 mm. Then the allowable bearing capacity of pile, calculated from proposed dynamic formula (Eq.19), Ra is 41 tf which is close to 45 tf from vertical loading test.

5 NOISE

5.1 Evaluation of noise

Human ear can hear a wide range of acoustic pressure. Numerical evaluation needs a large number of figures. So logarithmic measurement is used and the acoustic pressure level is measured in dB. Even under the same acoustic pressure, human ears receive different strength if frequency is different. Measurement on acoustic pressure by noisemeter monitored by an acoustic sensor with specific frequency weight is defined as noise level. A measuring of noise with A characteristic is the most practical and acceptable to human feeling, so noise level is expressed as dB(A).

The allowable noise level according to regulations for the specified construction work at 30 m apart from the boundary of construction site should be less than 85 dB(A).

5.2 Noise test results

Noise test results by pile-tip cap driving method and diesel hammer driving method are shown in Fig.5. Measurement of noise level has been done according to JIS Z 8731 Measurement of Noise Level. Because

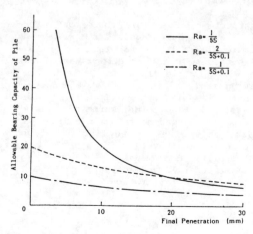

Fig. 4 Bearing Capacity and Penetration Curve

Table 3 Load Cycle and Step

Cycle	Load Step (tf)
1	0 - 15 - 30 - 0
2	0 - 30 - 45 - 60 - 0
3	0 - 30 - 60 - 75 - 90 - 60 - 0
4	0 - 30 - 60 - 90 - 105 - 120 - 60 - 0

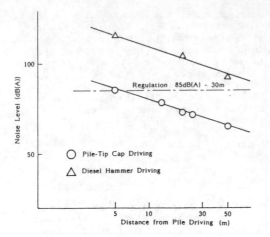

Fig. 5 Measurement of Noise with
        Distance

noise observed from pile driving give
peak values intermittently, so the average
of peak values is taken the measurement
value.

5.3 Discussion on noise

In general, the noise level by diesel
hammer driving method is 110-120 dB(A),
whereas by pile-tip cap driving method is
less around 30 dB(A). The measured noise
level at 30 m apart from site boundary is
70 dB(A), less than those given by the
regulations 80 dB(A). The noise level of
70 dB(A) generally equals to that produced
by traffic on the street. Impacting noise
from dropping hammer on steel cap in
underground can hardly be heard. Noise is
principally originated from impact of
center holed hammer on pile top. So it is
effective and reasonable to use pile-tip
cap driving method in order to avoid noise
problems.

6 VIBRATION

6.1 Evaluation of vibration

Evaluation of vibration created by pile
driving is essential and required like
noise. With different frequency, human can
not feel to have the same magnitude under
same acceleration. Vertical and horizontal
vibrations have different sensation levels.
Acceleration measured by vibrationmeter
monitored by quake sensor with specific
frequency weight is defined as vibration

level. Only vibration in vertical
direction is always measured and discussed
in engineering practice.

With regard to regulations for the
specified construction work the vibration
level at the boundary of construction site
should be less than 75 dB.

6.2 Vibration test results

Vibration test results by the pile-tip
cap driving method is shown in Fig.6.
Similar to noise, vibration on pile
driving gives the peak values
intermittently, measurement value is the
average of the peak values.

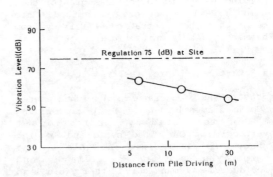

Fig. 6 Measurement of Vibration
        with Distance

6.3 Discussion of vibration

In general, the vibration level of
approximately 75 dB by diesel hammer
driving is observed at the boundary of
construction site. This corresponds to 8
gal in acceleration scale. With respect to
seismic intensity scale, this value lies
between weak to rather strong earthquake.
The vibration level of the pile-tip cap
driving is less than 75 dB at boundary of
construction site and 54 dB at 30 m apart
from pile driving point, which corresponds
to 0.8 gal. According to seismic intensity
scale, this gives no feeling or slight
earthquake sensitivity.

Generally, difference in impact energy
of hammer on pile and soil affects greatly
on the degree of vibration.

7 FINITE ELEMENT ANALYSIS

7.1 Analysis model

As the pile is impacted, how the impacting energy from pile body transfers to the soil as well as the stress developed on the pile body should be dynamically analysed in relation to the propagation of elastic wave. However, finite element method is one of the powerful analysis which employs static loading model. In this model, pile and soil are divided into mesh with isoparametric eight node element as shown in Fig.7. The model is treated as two dimensional elastic body with width of precast concrete pile of 30 cm, embedded length in soil of 10 m. The contact surface between pile and soil is replaced by 0.1 cm joint in the model (Black mesh in Fig.7). For diesel hammer driving , one ton force is loaded on pile top. Whereas, the same force is loaded on pile tip for pile-tip cap driving. Normal stress is taken as 0.33 kgf/cm^2, Young's modulus of precast concrete pile is taken as 3×10^5 kgf/cm^2, Poisson's ratio as 0.16. According to this model, soil is divided into four layers including bearing stratum with Young's modulus of 10, 30, 300 and 500 kgf/cm^2 respectively. Poisson's ratio is taken as 0.35 for all layers. Modulus of subgrade reaction, spring constant, of joint element k_n for the upper three layers are 10, 100 and 1,000 tf/m^3 respctively where k_n is 100 tf/m^3 for all three layers. Spring constant beneath pile tip in bearing stratum, k_s equals 100 tf/m^3 and k_n to 1 tf/m^3.

7.2 Analysis result and discussion

The distribution of vertical stress along the pile length is shown in Fig.8. For diesel hammer driving with loading on pile top, vertical stress reduces rapidly in third layer (E_3), about one third is found near pile tip. On the other hand, for pile-tip cap driving maximum vertical stress occurs at pile tip, which means that the impacting efficiency is quite high. Also, the vertical tensile stress in the latter case is found in finite element analysis around 1 m at pile tip. In this analysis, the pile-tip cap and pile is connected by spring which is far from reality. In this case, no tensile stress can occur in practice. Principal stress distribution for both methods are shown in Figs. 9 and 10. In case of using diesel hammer driving, the load on

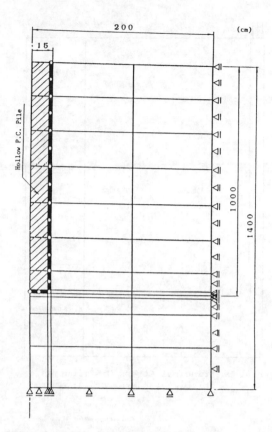

Fig. 7 Analysis Model for F.E.M.

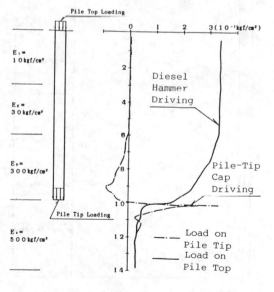

Fig. 8 Vertical Stress Distribution of Pile Central Longitudinal Section

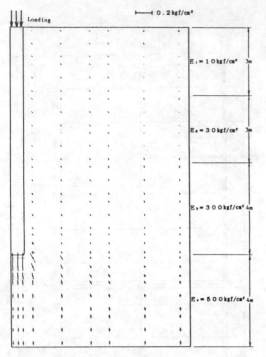

Fig. 9   Principal Stress Distribution
for Diesel hammer Driving

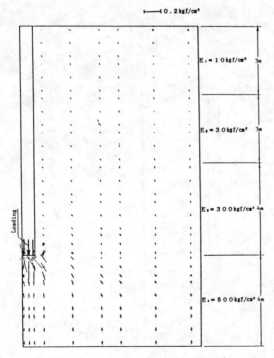

Fig. 10   Principal Stress Distribution
for Pile-Tip Cap driving

the pile top is distributed to the soil around and under the pile tip, which results into small principal stress. But for pile-tip cap driving, large compressive principal stress is distributed to the soil with magnitude approximately ten times larger.

8 CONCLUSION

Pile-tip cap driving method has less noise and less vibration levels which are below those given by regulations for the specified construction work. It also produces no significant oil splashes and no pollution to underground water.

The reliable bearing capacity of pile can be achieved due to impacting. Calculated values of proposed dynamic formula shows good results close to the field measurements.

Higher driving efficiency is also confirmed by finite element analysis.

ACKNOWLEGEMENT

The author wishes to thank to Seiji Matsushita, Shinei Foundation Construction Co., who has invented and developed the pile-tip cap driving method.

Developments in Geotechnical Aspects of Embankments, Excavations and Buried Structures, Balasubramaniam et al. (eds)
© *1991 Balkema, Rotterdam. ISBN 90 5410 019 2*

Stochastic three dimensional joint geometry modelling including a verification to an area in Stripa Mine, Sweden

Pinnaduwa H.S.W. Kulatilake & Deepa N. Wathugala
Department of Mining and Geological Engineering, University of Arizona, Tucson, Ariz., USA

ABSTRACT: At present, a three dimensional joint geometry modelling scheme which investigates statistical homogeneity, incorporates corrections for sampling biases and applications of stereological principles, and, also includes a formal verification procedure is not available in the literature. This paper shows development of 3D joint geometry models with aforementioned features. Joint data from Stripa mine were used in the investigation. Verifications performed so far has indicated the need to try out different schemes in modelling joint geometry parameters and to consider a number of combinations of such schemes in order to arrive at a realistic 3D joint geometry model which provides a good comparison with field data during verification. Further verification studies are recommended to check the validity of the suggested modelling schemes.

1 INTRODUCTION

Presence of discontinuities strongly affects the mechanical and hydraulic behaviour of rock masses. Discontinuities may be divided into major and minor. Faults, shear zones and dykes are categorized under major discontinuities and their geometry may be represented deterministically. On the other hand, due to statistical nature, geometry of minor discontinuities such as joint, bedding planes and fissures should be characterized statistically. This paper deals with minor discontinuities and henceforth they are referred to as "joints." Joint geometry network plays a vital role in civil, mining, and petroleum engineering projects associated with jointed rock. The need for a realistic representation of joint geometry in rock masses has been recognized.

Since joint geometry pattern can vary from one statistically homogeneous region to another, each statistically homogeneous region should be represented by a separate joint geometry model. Therefore, the first step in the procedure of joint geometry modelling in a rock mass should be the identification of statistically homogeneous regions. To model joint geometry in three dimensional

(3D) space, for a statistically homogeneous region, it is necessary to know the number of joint sets, and for each joint set, the intensity, spacing, location, orientation, shape and dimension distributions. These joint parameters are inherently statistical. Sample values of joint parameters provided by the field data usually subject to errors due to sampling biases and represent only one or two dimensional properties. Therefore, before inferring these parameters from sampling values, sampling biases should be corrected on field data. In addition, principles of stereology or geometrical probability (Kendall and Moran, 1963; Underwood, 1970; Santalo, 1976; Stoyan, Kendall and Mecke, 1987) need to be used in order to infer 3D parameters of the joint sets from 1D or 2D parameter values. A few three dimensional joint geometry modelling schemes have been suggested in the literature (Baecher et al. 1977; Veneziano, 1978; Dershowitz and Einstein, 1988; Billaux et al. 1989). However, at present, a 3D joint geometry modelling scheme which incorporates the aforementioned features and also includes a formal verification procedure is not available in the literature (Kulatilake 1988 & 89). This paper shows development of 3D joint geometry models which investigate statistical homogeneity, include corrections for sampling biases and applications of stereological principles to estimate 3D joint geometry parameters. Also the paper shows a procedure for verification of built up models using joint data from Stripa mine, Sweden.

2 DEVELOPMENT OF 3D JOINT GEOMETRY MODELS

2.1 Investigation of statistical homogeneity

Figure 1 shows the flow chart used to develop 3D stochastic joint geometry models. Orientation data from a thirty six meter long stretch of the ventilation drift, Stripa mine (Rouleau et al. 1981) were used to investigate the statistical homogeneity of the rock mass. Equal area polar plots and two statistical methods (Miller, 1983; Mahtab and Yegulalp, 1984) with new interpretations were used in this investigation. A ten meter long stretch from this thirty six meter stretch was identified as the largest statistically homogeneous region (Kulati-

Step 1: Obtain a statistically homogeneous region.

Step 2: Delineate joint sets.

Step 3: Apply correction for orientation bias for each joint set.

Step 4: Model the true orientation distribution for each joint set.

Step 5: Determine joint spacing distributions along scanlines taking into account the sampling bias on spacing.

Step 6: Infer joint spacing distribution along the mean pole direction (true spacings) for each joint set.

Step 7: Estimate the trace length distribution for each joint set taking into account the sampling biases.

Step 8: Infer joint size distribution for each joint set taking into account the sampling biases (joints are considered as circular discs).

Step 9: Estimate the mean joint center density in 3D (number/volume) for each joint set.

Step 10: Obtain a distribution for the random variable, " number of joint centers per chosen volume".

Step 11: Suggest a 3D stochastic joint geometry model by describing the joint geometry parameters:

(a) number of joint sets

(b) orientation distribution for each joint set

(c) spacing distribution for each joint set

(d) distribution for density in 3D for each joint set

(e) size (diameter when joint shape is considered as circular) distribution for each joint set.

Figure 1. Flow chart of the procedure used for development of stochastic joint geometry models.

lake et al. 1990a). Field joint data for this ten meter stretch is shown in Fig. 2.

2.2 Joint orientation modelling

Using the method given in Shanley and Mahtab (1976), orientation data in this region were delineated into four joint sets as shown in Fig. 3. For each joint set, high dispersion of the orientation data can be seen very clearly from Fig. 3. For such joint sets orientation bias correction may be significant. Joint data for each joint set come from the east and west walls and the floor of the drift. The procedure available for orientation bias correction for finite size joints intersecting finite size sampling domains (Kulatilake et al. 1990b) is directly applicable only for vertical sampling planes. Thus a general procedure applicable for sampling domains of any orientation was developed to correct for orientation bias (Wathugala et al. 1990). This procedure was applied to each joint set in order to correct for orientation bias. For example, Fig. 4 shows the effect of sampling bias correction on cluster 4. Table 1 provides the mean vectors and the spherical variances obtained for the four clusters.

Raw orientation data as well as orientation data corrected for sampling bias were subjected to chi-square goodness-of-fit tests to check the suitability of Bingham distribution (Bingham 1964) and the hemispherical normal distribution (Arnold 1941) in representing the statistical distribution of data. Results are given in Table 2. The P value given in Table 2 is the maximum significance level at which the adopted probability distribution is suitable to represent the statistical distribution of data. The P value should be at least 0.05 to accept the tried probability distribution to represent the data. Suitability of the tried probability distribution in representing the data increases with increasing P values. According to Table 2, only cluster 4 follows a Bingham distribution for raw data; neither of the two probability distributions satisfied the corrected data. Therefore, empirical distributions obtained for corrected data were chosen to represent the statistical distributions of joint set orientations. In Table 2, for each cluster, the difference between the chi-square values for raw and corrected data reflects the significance of sampling bias correction.

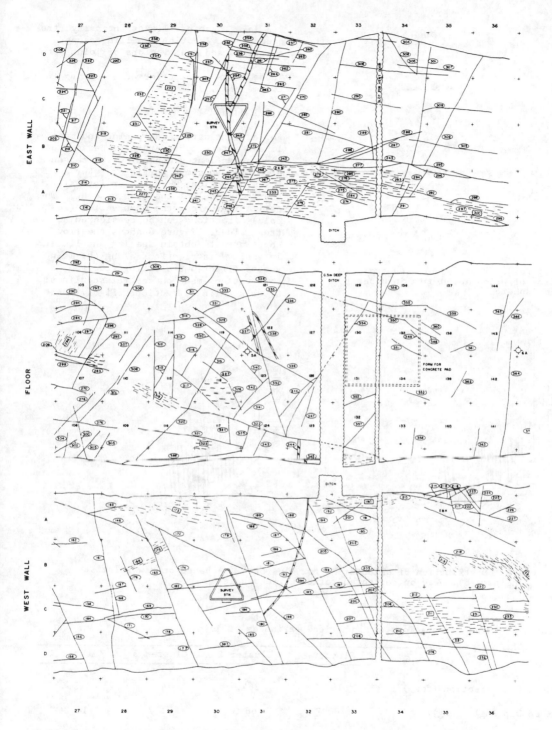

Figure 2. A typical joint map of a portion from ventilation drift in Stripa mine (after Rouleau et al. 1981).

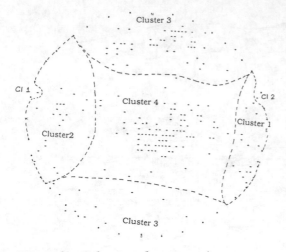

Figure 3. Polar equal area projection of orientation data in upper hemisphere.

2.3 Joint spacing, linear intensity (frequency) and location modelling along scanlines and mean vectors of joint sets

These tasks were performed according to two methods.

2.3.1 Method 1

For a joint set, mean spacing and mean linear intensity (average number per unit length) estimates depend on the chosen direction as shown in Fig. 5. For a fixed direction, mean linear intensity is the reciprocal of the mean spacing. The estimations along the mean vector (mean pole) directions can be considered as the true values. Figure 6 shows the flow chart used to obtain spacing and linear intensity distributions along the mean vector directions starting from the observations made on several scanlines. Seven scanline directions (Fig. 7) were

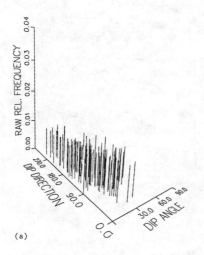

(a)

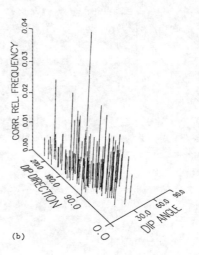

(b)

Figure 4. Distribution of orientation for cluster 4 (a) based on raw relative frequencies; (b) based on corrected relative frequencies.

Table 1. Mean Orientations and spherical variances of joint sets

	Cluster 1	Cluster 2	Cluster 3	Cluster 4
Mean Dip Direction (deg.)	102.2	264.7	14.9	113.4
Mean Dip Angle (deg)	80.7	66.3	71.7	14.8
Spherical variance	3.75×10^{-2}	5.69×10^{-2}	9.34×10^{-2}	7.90×10^{-2}

Table 2. Results of χ^2 goodness-of-fit tests on orientation data

Joint Set (# of data)		Hemispherical normal distribution			Bingham distribution		
		D.F.	χ^2 value	P value	D.F.	χ^2 value	P value
Cluster 1 (21)	observed data	4	24.76	<.005		Data	
	corrected data	4	15.59	<.005		insufficient	
Cluster 2 (24)	observed data	4	13.89	0.008		to run	
	corrected data	4	29.00	<.005		χ^2 test	
Cluster 3 (73)	observed data	6	41.87	<.005	6	24.93	<.001
	corrected data	6	42.20	<.005	6	33.35	<.001
Cluster 4 (114)	observed data	14	105.93	<.005	14	20.65	0.116
	corrected data	14	96.24	<.005	14	28.94	0.011

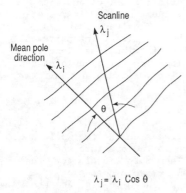

Figure 5. Influence of direction on mean linear intensity estimation.

chosen to analyze spacing and intensity in different directions to have a good coverage in 3D. Along each direction, several parallel scanlines were drawn either on the walls or on the floor of the drift, having joint traces coming from all four joint sets, to estimate spacing distribution as well as observed mean spacing. For each direction, observed spacing values were subjected to chi-square and Kolmogorov-Smirnov goodness-of-fit tests to check the suit-ability of exponential gamma, lognormal, normal, uniform and triangular distributions in representing the observed spacings. For all seven direc-tions the exponential distributions were found to be the best distributions to represent the distributions of observed spacing by satisfying the goodness-of-fit tests at very high significance levels. A typical fit obtained is shown in Fig. 8.

Chosen scanlines were of finite size. To obtain unbiased estimates of spacing and intensity, scanlines should be of infinite size. Sen and Kazi (1984)

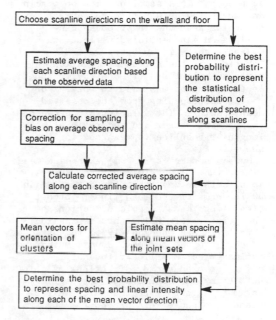

Figure 6. Flow chart used to esti-mate spacing and linear intensity distributions along the mean vector directions.

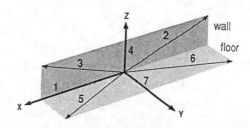

Figure 7. Chosen scanline directions for spacing and intensity analysis.

549

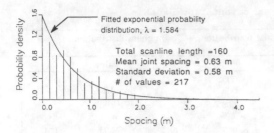

Figure 8. The exponential fit obtained for spacing along scanline direction 1.

pointed out this sampling bias on spacing and suggested an equation to correct for this bias when the observed spacing follows an exponential distribution. However, there is an error in their equation and the corrected equation is given in Kulatilake (1988 & '89). The latter equation was used to obtain corrected mean spacing from the observed mean spacing. The relationship between the possible error and the ratio of length of scanline to mean spacing is shown in Fig. 9.

Hudson and Priest (1983) expressed mean linear intensity resulting from several joint sets in any arbitrary direction in 3D in terms of the mean linear intensities along the mean pole directions of the joint sets. Using this expression, Karzulovic and Goodman (1985) suggested a procedure based on least square method to estimate the mean linear intensities along the mean vector directions of joint sets using the mean linear intensities estimated along several scanline directions. To apply this procedure, the number of scanlines should be greater than or equal to the number of joint sets. This procedure was used with aforementioned seven scanlines in order to estimate the mean linear intensities along the mean vectors of the four joint sets. The obtained 1D mean intensities along the mean vector directions are given in Table 3. The relative standard error obtained in the least square procedure was less than four percent with respect to any mean linear intensity estimated along scanline directions indicating quality estimations. These findings lead to the conclusion that joint spacing along mean vector directions follow exponential distributions with mean spacing values obtained in respective directions. Thus, according to statistical theory, linear intensity and linear location distributions along the mean vector direction for each joint set follow respectively Poisson and uniform distributions.

2.3.2 Method 2

First, joint traces on the two walls and the floor were sorted out into the four joint sets. Then for each cluster, the following analyses were conducted.

The flow chart shown in Fig. 6 was followed for each cluster until mean corrected spacings and mean linear intensities were found along the seven scanline directions. Then the relationship shown in Fig. 5 was used to estimate the corresponding mean linear intensities along the mean vector direction of the cluster. Finally, these values were averaged to obtain the final mean linear intensity estimate for each cluster. The

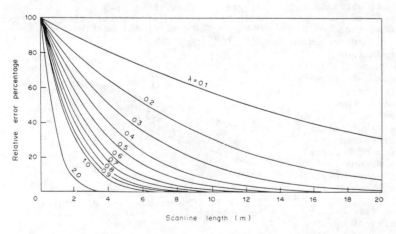

Figure 9. Relative error percentage in the estimated mean discontinuity spacing with scanline length.

Table 3. Mean linear intensities along the mean vectors of the joint sets.

	Cluster 1	Cluster 2	Cluster 3	Cluster 4
Method 1	0.064	0.509	1.488	1.801
Method 2	0.288	0.420	0.963	1.325

obtained values are given in Table 3. It is important to note the possibility of obtaining different estimates for 1D mean intensities based on different methods. However, when the angle between the mean vector direction and the scanline direction is greater than about seventy degrees, the accuracy of the estimation is questionable. Therefore, the estimations from such cases were not included in the averaging procedure. Again, the conclusions regarding the probability distribution types for spacing, intensity and location are same as for method 1.

2.4 Joint trace length and 3D joint size modelling

First, joint traces appearing on the two walls and the floor were sorted out into the four joint sets. Then for each cluster, the following modelling was performed; one on the wall data and the other based on the floor data. For example, Fig. 10 provides the subsequent modelling performed on the wall data for each joint cluster.

Due to finite size of the walls (10m x 4m) and the floor (10m x 4m), the observed traces are subject to censored and size biases. Since observed traces of length less than about 0.4m were neither mapped on the walls nor on the floor, it was assumed that the truncation limit for joint traces is 0.4m. Area sampling technique was used in sampling the joint traces on both the walls and on the floor. As for spacing, goodness-of-fit tests were carried out on the sampled joint traces. For all four joint clusters, gamma distribution was found to be the best distribution to represent observed trace length distribution on both walls and on the floor. A typical fit obtained is shown in Fig. 11. Note that the fits obtained for trace length distributions were not as great as the fits obtained for spacing distributions.

The method given in Kulatilake and Wu (1984a) was used to correct for censoring error and size bias, and to estimate mean trace length on an infinite 2D exposure using the observed trace data from a finite 2D exposure. The obtained results for the joint sets (Table 4) show the importance of the correction. An attempt was made to express the probability density function for the trace length distribution on 2D infinite exposure, f(l), from the probability density function obtained for observed traces on the 2D finite exposure, g(l). It was necessary to incorporate the estimated corrected mean trace length and correction factors in f(l) to account for the effect of size, censoring and the truncation biases on the probability distribution. For example, the following relationship was obtained between f(l) and g(l) for analysis based on the wall data

$$f(l) = \frac{Kg(l)[wh + \mu(w\sin\theta_A + h\cos\theta_A)]}{[wh + l(w\sin\theta_A + h\cos\theta_A)]} \qquad (1)$$

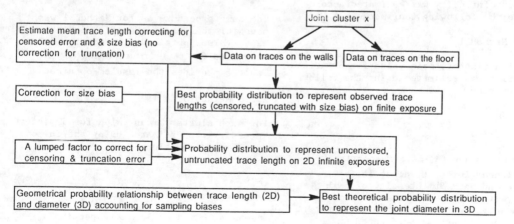

Figure 10. Flow chart for modelling joint size in 3D for each cluster.

Table 4. Observed and corrected mean trace lengths

	Cluster 1		Cluster 2		Cluster 3		Cluster 4	
	wall	floor	wall	floor	wall	floor	wall	floor
Observed $\mu(m)$	1.08	1.24	1.28	1.64	1.43	1.22	1.38	1.01
Corrected $\mu(m)$	0.53	0.69	1.18	2.53	1.38	0.97	0.79	0.61

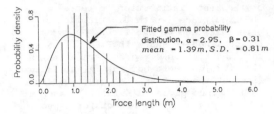

Figure 11. The gamma fit obtained for trace length distribution for cluster 4 (wall data).

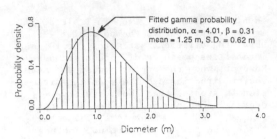

Figure 12. The gamma fit obtained for diameter distribution for cluster 4 (wall data).

where w and h are the width and the height of the wall, μ is the corrected mean trace length, θ_A is the average apparent dip of the wall data set,(1)is the trace length and K is a constant.

Assuming joints as finite circular discs, joint diameter distribution for a joint cluster was estimated from f(l) using the procedure given in Kulatilake and Wu (1986). Diameter distributions for all joint clusters were found to be best represented by Gamma distributions. Figure 12 shows the diameter distribution obtained for cluster 4 based on wall data.

2.5 Three dimensional joint intensity (number per unit volume) modelling

3D joint intensity was estimated according to the following four methods.

2.5.1 Method 1

For each cluster, mean value for 3D joint intensity was estimated using the following equation:

$$(\lambda v)_j = \frac{4(\lambda_l)_j}{\pi E(D^2)} \qquad (2)$$

where $(\lambda_l)_j$ and $(\lambda_v)_j$ are respectively the mean linear intensity along its mean vector and mean 3D intensity of j th joint set and $E(D^2)$ is the expected or the mean value of squared joint diameter. This equation was derived based on an equation presented by Oda (1982). For

each cluster, $E(D^2)$ can be estimated based on the diameter distribution obtained either from the wall data or from the floor data. Values obtained from Section 2.3.1 were used for $(\lambda_l)_j$. Table 5 provides the results obtained. Since linear intensity follows a Poisson distribution, it is reasonable to conclude that 3D intensity follows a Poisson distribution with corresponding $(\lambda_v)_j$. Mean 3D intensity for the rock mass was obtained through equation (3).

$$\text{mean } 3D \text{ joint intensity} = \sum_{j=1}^{4}(\lambda_V)_j \qquad (3)$$

2.5.2 Method 2

The same procedure as for Method 1 was used for this case. However, for this case, values obtained from Section 2.3.2 were used for $(\lambda_l)_j$ in computing $(\lambda_v)_j$. Table 5 provides the results obtained.

2.5.3 Method 3

For each cluster, mean value for 3D joint intensity was estimated using the following derived equation:

$$(\lambda_v)_j = \frac{(\lambda_a)_j}{E(D)E(|\sin v_1|)} \qquad (4)$$

where $(\lambda_a)_j$ and $(\lambda_v)_j$ are respectively, mean areal intensity (mean number of

552

Table 5. 3D mean joint intensities

	Method 1		Method 2		Method 3		Method 4	
	Wall	Floor	Wall	Floor	Wall	Floor	Wall	Floor
Cluster 1	.061	.042	.278	.190	.320	.166	.220	0.133
Cluster 2	.403	.568	.332	0.468	.344	.324	.360	.453
Cluster 3	.978	1.056	.633	0.683	.624	.461	.745	.733
Cluster 4	1.083	2.024	.796	1.266	.766	1.401	0.882	1.564
$\lambda_v = \sum$	2.525	3.690	2.039	2.607	2.054	2.352	2.207	2.883

joint centers per unit area) and mean 3D intensity of j the joint set, and, E(D) and $E(|\sin v_1|)$ are the expected or mean value of the diameter and $|\sin v_1|$ respectively. The angle v_1, is the angle between the joint plane and the sampling plane on which $(\lambda_a)_j$ was estimated. From a sampling domain, what we can directly estimate is the average number of joints per unit area. The procedure given in Kulatilake and Wu (1984b) was used to estimate $(\lambda_a)_j$ from the average number of joints per unit area. Table 5 provides the results obtained.

2.5.4 Method 4

Values of $(\lambda_v)_j$ obtained from the previous three methods were averaged to obtain these estimations. Results are given in Table 5.

Table 5 very clearly shows the possibility of obtaining different estimates for 3D mean intensities based on different methods.

3 JOINT SYSTEM MODELLING

Section two dealt with modelling of joint geometry parameters for the chosen statistically homogeneous region. Results obtained in Section two were used in this section to build joint geometry networks in 3D. For each joint cluster the following statistical models were used to generate joints in 3D: (1) Number of joints per certain volume is Poisson distributed with mean value $(\lambda_v)_j$ (2) Location of joint clusters in 3D is uniformly distributed (3) Orientation is

distributed according to the empirical distribution obtained for corrected data (4) Diameter is gamma distributed with the parameter values obtained.

4 JOINT GENERATION IN 3D, PREDICTION IN 2D, AND VERIFICATION

Only one of the several verification studies performed for the largest joint set (cluster #4) is discussed in this paper. The joints were generated in the volume shown in Fig. 13, according to the statistical model given in Section 3 using Monte Carlo simulation. In Fig. 13, D_{max} is the largest mean diameter of the four joint sets. In the chosen volume, the vertical plane EFGH of size (10m x 4m) (Fig. 13) was chosen to simulate the tunnel wall. For this case, $(\lambda_v)_j)$ obtained from Section 2.5.4 was

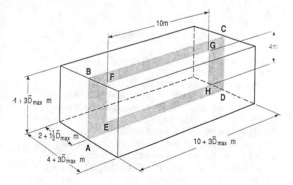

Figure 13. Volume and vertical section used in verification.

used. Diameter estimation obtained based on wall data was used. Joint traces appearing on EFGH were censored and truncated. A truncation limit of 0.4m was used. These joint traces were used to estimate the parameters given in Table 6. Several such simulations were repeated. Mean predictions were computed using the results obtained through thirty simulations. Table 6 shows a comparison between predictions and actual field data. In statistical sense, the agreement is quite good. Figure 14 provides a comparison between actual field data appeared on the wall and a prediction based on a simulation close to the mean prediction. Again, in statistical sense, the agreement is quite good.

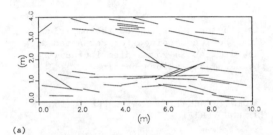

(a)

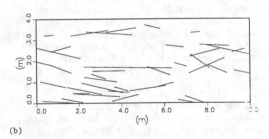

(b)

Figure 14. Comparison of joint traces on a wall (a) from field mapping (b) a prediction close to the mean prediction.

5 SUMMARY AND CONCLUSIONS

Development of 3D stochastic joint geometry models including corrections for sampling biases and applications of stereological techniques and checking validity of such built up models were addressed in the paper. Verification studies performed so far has indicated the need to try out different schemes in modelling joint geometry parameters and to consider a number of combinations of these schemes in order to come up with a realistic 3D joint geometry model which provides a good comparison with field

Table 6. A comparison between the field data and predictions on 2D from Monte Carlo simulation of the parameters in the model for joint cluster 4.

Parameter	Field value	Predicted range	Average prediction
R_0	0.01	0-0.02	0
R_1	0.18	0.07-0.33	0.17
R_2	0.81	0.67-0.93	0.83
mean ϕ	14.9	7.09-18.6	13.2
mean Θ	245.1	223.3-257.0	238.8
l_{obs} (m)	1.38	0.87-1.39	1.14
l_{corr} (m)	0.79	0.23-1.27	0.61
Number of joints	40	30-49	41

where
Θ = dip direction (deg.)
ϕ = dip angle (deg.)
l_{obs} = mean trace length on the finite window
l_{corr} = corrected mean trace length

Note: $R_0 = N_0/N$, $R_2 = N_2/N$ and $R_1 = N_1/N$
where
N_0 = expected number of joints with both ends censored intersecting the window
N_1 = expected number of joints with only one end observed intersecting the window
N_2 = expected number of joints with both ends observed intersecting the window
N = $N_0 + N_1 + N_2$

data during verification. The verification reported in the paper shows that the modelling scheme used for the verification has very good capability in producing 2D predictions which provide good agreement with 2D field data for the cluster studied. Further verification studies for other clusters of the same site as well as for different sites are recommended to check the validity of the suggested modelling schemes.

ACKNOWLEDGEMENTS

The authors are grateful to the Swedish Natural Science Research Council and the Arizona Mining and Mineral Resources Research Institute for providing partial financial support for this study.

REFERENCES

Arnold, K.J. 1941. On spherical probability distributions. Ph.D. dissertation, Massachusetts Institute of Technology, Cambridge, Massachusetts: 42.

Baecher, G.B., Lanney, N.A., and Einstein, H.H. 1977. Statistical description of rock properties and sampling. Proceedings of the 18th U.S. Symp. on Rock Mech. 5C1-8.

Billaux, D., Chiles, J.P., Hestir, K., and Long, J. 1989. Three dimensional statistical modelling of a fractured rock mass - an example from the Fanay-Augeres Mine. Int. J. Rock Mech. pp. 281-299.

Bingham, C. 1964. Distributions on the sphere and on the projective plane. Ph.D. dissertation, Yale University, New Haven, Connecticut: 93.

Dershowitz, W.S. and Einstein, H.H. 1988. Characterizing rock joint geometry with joint system models, Rock Mechanics and Rock Engineering, Vol. 21, pp. 21-51.

Hudson, J.A. and Priest, S.D. 1983. Discontinuity frequency in rock masses. Int. J. Rock Mech. Min. Sci. 20: 73-89.

Karzulovic, A. and Goodman, R.E. 1985. Determination of principal joint frequencies. Int. J. Rock Mech. Min. Sci. 22: 471-473.

Kendall, M.G., and Moran, P.A.P., 1963. Geometrical Probability, Griffin's monograph series, C. Griffin and Co., Ltd., London.

Kulatilake, P.H.S.W. 1988 & 1989. Stochastic joint geometry modelling: state-of-the-art. Proc. 29th U.S. Symp. on Rock Mech., Minneapolis, Minnesota: 215-229. Also invited papers at the conferences held in Switzerland, Sweden and Mexico.

Kulatilake, P.H.S.W., Wathugala, D.N., Poulton, M. and Stephansson, O. 1990a. Analysis of structural homogeneity of rock masses. Int. J. Engineering Geology (in press).

Kulatilake, P.H.S.W. and Wu, T.H. 1984a. Estimation of mean trace length of discontinuities. Rock Mech. and Rock Engineering. 17: 215-232.

Kulatilake, P.H.S.W., and Wu, T.H. 1984b. The density of discontinuity traces in sampling windows. Int. J. Rock Mech. Min. Sci. 21:345-347.

Kulatilake, P.H.S.W. and Wu, T.H. 1986. Relation between discontinuity size and trace length. Proc. 27th U.S. Symp. on Rock Mech., Tuscaloosa, Alabama:130-133.

Kulatilake, P.H.S.W. and Wu, T.H. and Watugala, D.N. 1990b. Probabilistic modelling of joint orientation. Int. J. Numerical and Analytical Methods in Geomechanics (in press).

Mahtab, M.A., and Yegulalp, T.M. 1984. A similarity test for grouping orientation data in rock mechanics. Proceedings of the 25th U.S. Symp. on Rock Mechanics, pp. 495-502.

Miller, S.M. 1983. A statistical method to evaluate homogeneity of structural populations. Mathematical Geology, Vol. 15, No. 2, pp. 317-328.

Oda, M. 1982. Fabric tensor for discontinuous geological materials. Soils and foundations. 22: 96-108.

Rouleau, A., Gale, J.E. and Baleshta, J. 1981. Fracture mapping in the ventilation drift at Stripa: procedures and results. Research report LBL-13071, Lawrence Berkeley Laboratory, Berkeley, California.

Santalo, L. 1976. Stochastic Geometry and Integral Calculus, Addison - Wesley Publ., Reading, MA.

Sen, Z. and Kazi, A. 1984. Discontinuity spacing and RQD estimates from finite length scanlines. Int. J. Rock Mech. Min. Sci. 21: 203-212.

Shanley, R.J. and Mahtab, M.A. 1976. Delineation and analysis of centers in orientation data. Math. Geology. 8: 9-23.

Stoyan, D., Kendall, W.S., and Mecke, J. 1987. Stochastic geometry and its applications, John Wiley Inc.

Terzaghi, R. 1965. Sources of error in joint surveys. Geotechnique. 15: 287-304.

Underwood, E.E. 1970. Quantitative stereology. Addison - Wesley, Mass.

Veneziano, D. 1978. Probabilistic model of joints in rock. Unpublished manuscript, M.I.T., Cambridge, MA.

Wathugala, D.N., Kulatilake, P.H.S.W., Wathugala, G.W. and Stephansson, O. 1990. A general procedure to correct sampling bias on joint orientation using a vector approach. Computers and Geotechnics (submitted for publication).

Developments in Geotechnical Aspects of Embankments, Excavations and Buried Structures, Balasubramaniam et al. (eds)
© 1991 Balkema, Rotterdam. ISBN 90 5410 019 2

Geotechnical characteristics of soft soils in Indonesia

J.S. Younger
Frank Graham International Ltd, UK (Presently: Institute of Technology, Bandung, Indonesia)

ABSTRACT: The properties and behaviour of some soft clay deposits in Indonesia are discussed. Quite significant areas of the country are covered in compressible materials of recent volcanic origin. Most have been deposited in a marine environment, but a notable exception is the Bandung lake material, salient features of which are also summarised. Swelling soils are also found and reference is made to one example in East Java. The importance of establishing suitable characteristic design parameters for these deposits is demonstrated.

1 INTRODUCTION

Indonesia has a significant amount of its land surface covered by Quaternary soft deposits. These soft deposits frequently exist in areas where developments have taken place or are projected. They are generally of recent volcanic origin and, with one or two exceptions, have been deposited in a marine environment. COX (1970) referred to the distribution of soft soils in Southeast Asia, and BRENNER et al. (1987) pointed out that many of those existing in Indonesia had significantly different characteristics as a result of their volcanic origin. A notable exception to the marine deposits is the deep recent volcanic mud deposit to the south of Bandung over which a major toll road has been constructed.

The recent volcanic nature of most of the soft material in Indonesia sometimes means the presence of less weathered clay minerals, such as smectites and sometimes allophanes and halloysites, for instance, in place of the more common illites and kaolinites which are found in older deposits. Identification of these minerals is important as a pointer to the likely behaviour of the deposits, which can be somewhat different from that found in more mature clay deposits.

When studying the manner in which a soft deposit is going to respond under loading a good understanding is required of the macro-structure of the deposit, and what drainage seams and lenses might be present (McGOWN et al., 1979; 1980) because, apart from the need to evaluate the total settlement that is likely to occur, the rate at which the settlement will take place also often requires to be assessed. Only when both the total and rate of settlement have been reasonably accurately analysed can rational decisions be taken on the type of ground improvement, if any, to be undertaken.

2 GEOTECHNICAL PRACTICE IN INDONESIA

Similar to other practice in the region, the wash-boring technique for borehole sinking is extensively used in Indonesia. The limitation of the method, which is not encouraged in standard practice in U.K. and other western countries, has been discussed elsewhere along with the requisite design parameters for soft ground engineering (COOK & YOUNGER, 1987; YOUNGER, 1988).

Results from standard laboratory practice, particularly with regard to the interpretation of rates of consolidation from conventional oedometer testing and analysis, often lead to an erroneous interpretation of in situ soft ground behaviour under load. Certain clay minerals are also sensitive to the method of preparation for classification testing, e.g. fersiallitic andosol soils (WESLEY,

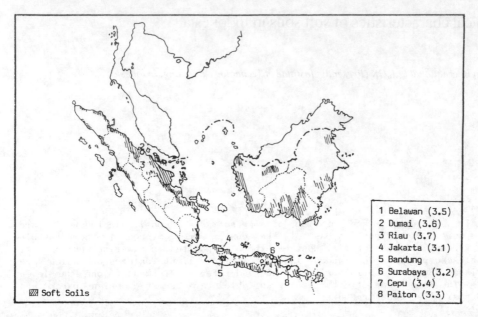

Fig. 1 Locations of Soft Soils

1973; COOK & YOUNGER, 1988; ANON, 1990) which include allophane or hallosite clay minerals. A feature of these soils is that they hold extraordinarily large amounts of water, in a Casagrande chart fall below the A-line, and exhibit higher permeabilities and shear strength than otherwise might be expected.

Some characteristics of Indonesian clays are now described, using a number of examples of marine clays, a swelling soil situation, and one freshwater clay deposit. The location of the soils discussed are shown in Figure 1.

3.0 MARINE DEPOSITS

Of the soft soils around the Indonesian archipelago which have been laid in marine conditions, those which have received most attention have been along the north Java coastline or on the east side of Sumatra as a result of developments in the two islands, which hold the majority of the population of Indonesia (60% in Java and 25% in Sumatra). Most of the clays are of recent origin and contain clay minerals which are less weathered than normally found in older environments. One consequence of this is a micro-structure giving a higher permeability and strength, as particularly referred to in one case, than might be considered likely on the

basis of the classification characteristics of the soils. Areas with swelling soils can also be found. Reference is made to examples in Java and Sumatra, and these indicate also the importance of careful site investigation.

3.1 Jakarta Area, West Java

The soft soils in the greater urban Jakarta area are compressible although there is a degree of overconsolidation, below which settlements are small. There is significant variation in thickness of the soft strata in the area and cohesionless layers are often encountered.

However, for highway embankment sections of some height, for example, disruptive long term settlements can take place, as can be observed at bridge abutments on the toll road section from the Soekarno-Hatta airport. All major building structures and concrete fly-overs, as part of the extensive road construction now taking place in the city, are supported on piles to sound supporting strata, usually at depths of 15 m or greater. As in most coastal lands, the soils to be found at the surface along the coast or along river outlets tend to be softer, since they are more recent or, in the latter case, have been laid in conditions of some turbidity.

558

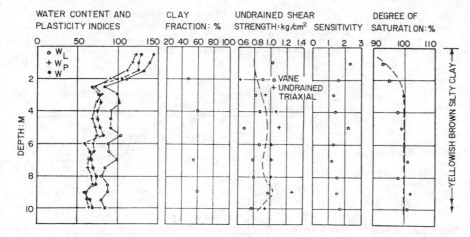

Fig.2 Engineering properties of tropical clay in Java (after MORI, 1982)

One profile of Jakarta materialis as shown in Figure 2 (MORI, 1982). Despite the relatively high natural moisture content, the soil is relatively insensitive and shear strengths are of the order of 80-100 kPa. A similar range of values was obtained for soil extracted from 3-6m depth at a location on the alignment of the outer ring road to the west of Jakarta. This soil was examined in the laboratory to check its reaction with lime. It was seen that lime is favourably reactive with the Jakarta soil tested (YOUNGER, 1988a). This would indicate that control of settlement using lime columns for lighter loads could be an option for ground improvement in Jakarta soil, provided the economics were favourable, although the use of a wide variety of piling options is now well-established for all types of structures.

MORRISON et al. (1984) presented a good summary of the geological conditions in the Jakarta area and described the characteristics of the coastal soft clays in the vicinity of the main harbour at Tanjung Priok. Lightly overconsolidated and normally consolidated clays, and beach sand along the shore line, are found above a stiff lahar formation clay. The two marine clays were found to be similar differing mainly in clay content and liquidity index, differences in the latter being ascribed to light overconsolidation by overlying beach sand. Figure 3 shows the plasticity characteristics of the clay, i.e. extremely highly plastic with smectite minerals present. A summary of consolidation characteristics is given in Figure 4. Continuous sampling indicated

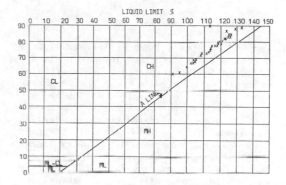

Fig. 3 Marine Clay Plasticity Chart

the lack of obvious drainage paths or silt seams; the clay appeared uniform, and slow rates of settlement were thus suggested, but in contradiction to recorded experience in the area. Because of this a second investigation with greater care in sampling and handling was undertaken, and the results showed a 10-100 times increase in laboratory determined values of permeability. Values (within a factor of 5-10) approached those obtained from constant head in situ permeability testing (WILKINSON,1968), where horizontal permeability is measured rather than vertical, as in the case of conventional laboratory testing. MORRISON et al. (1984) concluded that the C_v of the normally consolidated soft clays was very sensitive to the micro-fabric, which is easily disturbed in sampling and handling, and commented on the high dependence of C_v on void ratio. The considerable care taken in the invest-

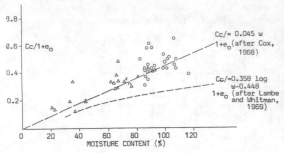

Fig. 4   Compression Index Summary
        Soft Marine Clays

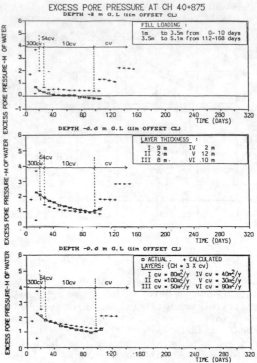

Fig. 5. Actual and Simulated Layered
Model Excess Pore Water Pressure at Ch.
40+875 for Given Depth at 11m Offset
from Centerline

igation confirmed that adequate rates of settlement could be achieved without recourse to providing assistance through means such as vertical drains.

3.2 Surabaya, East Java

Very deep strata (40m) of largely lightly overconsolidated soft marine materials are found in the Surabaya area, and affected the rate of construction of the Surabaya-Gempol Toll Highway, which was opened to traffic two years ago. The top 2-3m were a heavily overconsolidated crust, underlain by a sand layer. Void ratios in the main clay layers exceeded 3.0. Very large settlements (of the order of 2m) were found in areas of high fill. Surcharging was used generally as a means of accelerating settlement, although a residual settlement of up to 100mm has been observed for the general low embankment level which carries much of the highway. This settlement has been uniform and no disruption to line or level of the road profile has resulted because of this. Settlement at high fill areas behind bridge abutments has been greater with the usual bumps developing between the solidly founded concrete structures and unsupported approach fill.

An observation made on settlement monitoring during construction of the toll highway was that the rate was significantly faster than anticipated on the basis of conventional investigation evaluations, with C_v values being of the order of several hundred m^2/yr for loadings below a threshold embankment height of about 3.5m. This has also been found in a back analysis of early pore pressure dissipation behaviour (ADYAWATI, 1991), as shown in Figure 5. Once applied stress levels increased beyond the 3.5m

hold, settlements increased significantly, but with increased loading the rate of dissipation of pore pressures reduced very considerably, suggesting the passing from a lightly overconsolidated condition, calculated for the upper 15m, to a normally consolidated state (MOTT RENARDET, 1983). Subsequent increases in pore pressure were ascribed to 'destructu-ration' (MITCHELL, 1986).

3.3 Site near Paiton, East Java

A well documented case history in East Java was presented by BRENNER et al. (1987) for a near-shore deposit of soft volcanic silty clay. A significantly deep weathered crust was found as a result of dessication and chemical weathering, a typical situation along the north Java coast, in line with past large variations in sea level. Figures 6 and 7 reproduce recorded properties, and the plasticity charts and activity tests indicated the presence of smectite minerals. These

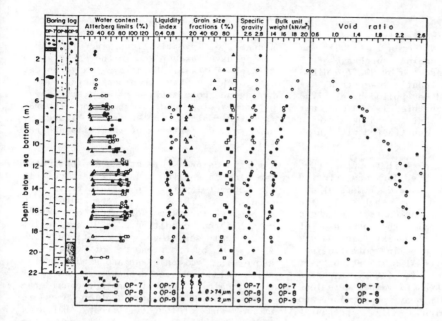

Fig. 6 Geotechnical profile in tidal shelf area

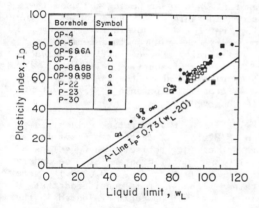

Fig. 7 Plasticity chart for
soft marine mud

along with kaolinite were identified by thermo-analytic methods, the kaolinite being an end product of halloysite weathering. Coefficients of consolidation of 6–11m²/yr were determined by conventional oedometer testing, and the values were considered high for the types of clay minerals found. Shear strengths were also high in spite of high liquid limits and moisture contents, and it was considered that the clay mineralogy was a principal factor responsible for these

unusual engineering properties, with emphasis placed on the importance of soil fabric (GONZALES DE VALLEJO et al. 1981, TERZAGHI, 1958), and the formations of aggregations in tropical volcanic soils.

3.1 Swelling Soil, East Java

Such soils, defined as black tropical soils (ANON., 1990), are found in several areas of Java, and have had a disruptive effect on the performance of highways. Smectite (montmorillonite) minerals are present causing large volume changes according to seasonal water table fluctuations.

Black tropical soils can be classified into non-expansive or potentially expansive on the basis of a series of simple index tests, which include Atterberg Limits, linear shrinkage, free swell and colloid content (ANON., 1990). Unlike non-swelling soils, for which plotting activity (SKEMPTON, 1953) in conjunction with clay fraction and PI has been found useful, montmorillonite soils cannot be classified in this manner, but there would appear to be a good correlation between LL and the exchangeable Na ions present. Full recommendations for reliable identification testing are presented in ANON. (1990).

A useful summary of the existing State-of-the-Art for testing swelling soils was presented by McELVANEY et al. (1990), who observed that the nominal surcharge of 7 kPa in testing, suggested by HOLTZ & GIBBS (1956), is significantly lower than the pressure that would be applied by a highway pavement structure let alone any intervening embankment loading, which would sensibly be constructed with non-swelling materials.

An example of swelling soil from the Cepu-Bojonegoro road in East Java was studied (SUPRIYONO, 1987). Liquid limits of 85-103% were recorded for the soil with points in the Casagrande chart lying above the A-line, confirming the presence of smectite minerals. The clay fraction was about 40%, and the swelling potential was thus very high on the classification chart proposed by SEED et al. (1962) as shown in Figure 8. Soaked CBR values were less than 1.8%, even with heavy compaction, and complementary high swell pressures developed on soaking.

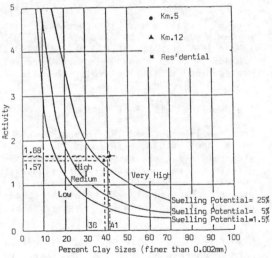

Fig. 8 Classification chart for swelling potential (after SEED et al., 1963)

These materials certainly indicate the beneficial application of pre-treatment. Post-construction trial attempts with polythene sheeting (CHEN, 1975) with geogrids have not prevented further seasonal damage to the pavement. Treatments using lime or cement as widely practised in Japan have been presented by AKAGI (1990), and would be worthy of consideration in this case.

3.5 Belawan Port, Sumatra

A number of investigations since the early 1980's have taken place in the near shore area of Belawan, Sumatra. A large number of boreholes have been sunk using wash-boring and more sophisticated techniques. A summary of ground conditions and laboratory test data are presented in Figure 9, and salient features of the soils are as follows.

The main compressible strata are the upper two clay layers (UC and MC), particularly the topmost layer which is in a very soft condition. These soils were found to be in a slightly overconsolidated to normally consolidated condition, a function of geological sea level fluctuations, as indicated by the presence of vegetation above the volcanic ash and sand layers, which also indicate a change in a past geological environment. Another consideration is quasi pre-consolidation pressure as a result of long term creep (BJERRUM, 1972). Recorded soil sensitivities were in the relatively insensitive to moderately sensitive range. However, it was also evident from examination of the sensitivities that when wash-boring was used, there was considerable disturbance caused to the soil, reflected also in low values in laboratory shear strength determinations. The plasticity characteristics indicated the presence of smectite minerals, with points plotting above the A-line, except where pronounced organic or silt contents were observed.

A wide range in void ratio values, e.g. from 0.87 to 3.12 in the top clay layer, were observed, reflected in C_c values of 0.5-1.0. Coefficients of secondary compression, C_α, of 0.01-0.1 were obtained although it is considered that C_α values were not properly evaluated in the conventional investigations carried out. Creep settlement became a significant feature in post construction sand-fill reclamation behaviour.

With regard to rates of primary consolidation, original design values of less than 5 m^2/yr were indicated, although BURDALL & NICHOLLS (1981) from high quality undisturbed samples indicated pre-critical compression values of C_v of 30-50 m^2/yr with post Pc' values dropping to 1-5 m^2/yr according to pore pressure dissipation measurements. Subsequent field performance showed field coefficients of consolidation to be of the order of 500 m^2/yr. Later, C_h values were assessed to be up to 2-5 times greater than corresponding C_v

ELEV (m)	SOIL DESCRIPTION	INDEX PROPERTIES	SHEAR STRENGTH	COMPRESSIBILITY
3,8to-6,1	**UPPER CLAY (UC)** silty clay, MH to CH grey, very soft small organic content and shell debris	W_n = 70 - 110 % LL = 80 - 120 % PI = 30 - 60 % clay = 20 - 30 % silt = 50 - 60 % sand = 15 - 20 % Gs = 2.65 Y_m = 13.5 - 17 kN/m³	C_u = 10 - 40 kN² q_c = 200 - 5600 kN/m² SPT = 1 - 10 blows/ft C' = 10 - 20 kN/m² ϕ' = 22° - 30° St = 1.5 - 4	e_o = 0.87 - 3.12 C_c = 0.5 - 1.0 C_v = 2.0 - 5 m²/yr P_p' = 1.0 - 2.0 Po' E_u = 20.000 25.000 kN/m² C_α = 0.01 - 01
14 to -20	**VOLCANIC ASH & SAND (VA)** volcanic ash and sand, fine to coarse silica overlies peat, decomposed wood	W_n = 40 - 60 % Dr = 45 - 60 % Gs = 2.65 D_{50} = 0.1 - 0.25mm Cu = 12 - 93 Y_m = 15 - 19 kN/m³	q_c = 800 - 4000 kN/m² SPT = 20 - 50 blows/ft ϕ' = 30° - 35°	e_o = 0.95 - 1.83 E = 10.000 - 20.000 kN/m² ν = 0.2 - 0.35
19 to -23	**MIDDLE CLAY (MC)** silty clay, MH to CH dark brown to grey, stiff, containing sand partings sheet debris	W_n = 55 - 80 % LL = 60 - 80 % PI = 20 - 40 % clay = 25 - 40 % silt = 50 - 60 % sand = 5 - 20 % Gs = 2.65 Y_m = 14 - 18 kN/m³	C_u = 50 - 70 kN/m² qc = 800 - 2000 kN/m² SPT = 10 - 30 blows/ft C' = 20 kN/m² ϕ' = 20° - 30° St = 1,5 - 3	e_o = 1,10 - 2,07 Cc = 0.25 - 0.80 Cv = 1.5 - 3 m²/yr Eu = 50.000 - 200.000 kN/m² C_α = 0.01 - 0.1
52 to -54	**LOWER SAND (LS)** fine sand, containing some silt	W_n = 40 - 60 % Dr = 50 - 60 % Gs = 2.65 D_{50} = 0.01 - 0.2mm Cu = 11 - 114 Y_m = 18 - 19 kN/m³	q_c = 4000 kN/m² SPT = 20 - 60 blows/ft ϕ' = 30° - 35°	e_o = 0.95 - 1.36 E = 30.000 - 60.000 kN/m² ν = 0.2 - 0.35
55 to -59	**LOWER CLAY (LC)** silty clay, CL to CH and ML to MH, grey mostly stiff containing fine sand partings	W_n = 40 - 65 % LL = 60 - 80 % PI = 30 - 50 % clay = 26 - 50 % silt = 40 - 70 % sand = 5 - 10 % Gs = 2.65 m = 16 - 18 kN/m³	C_u = 160 - 200 kN/m² q_c = no records SPT = 10 - 60 blows/ft C' = 10 - 20 kN/m²	e_o = 1.06 - 1,73 Cc = 0.3 - 0.70 Pp' = 1.0 - 2.0 Po' Eu = 150.000 - 225.000 kN/m² Cv = 1.0 - 3 m²/year
		W_n = Natural Moisture Content LL = Liquid Limit PI = Plasticity Index Gs = Specific Gravity Y_m = Moist Unit Weight Dr = Relative Density D_{50} = 50% finer than Diameter Cu = Uniformity Coeff.	C_u = Undrained Cohesion q_c = Cone Resistance C' = Drained Cohesion ϕ' = Drained Fiction Angle St = Sensitivity	e_o = Initial Void Ratio Cc = Compression Index Cv = Coeff. of Consolidation Po' = Preconsolidation Press Pn' = Effective Overburden Press Eu = Undrained Young's Mod. E = Drained Young's Mod Y = Poisson's Ratio C_α = Secondary Comp. Index

Legend (soil symbols): sand, silt, clay, peat, volcanic ash

Fig. 9   Summary of soil characteristics at Belawan

values. Subsequently, in situ permeability tests showed that in situ permeabilities were about 10-100 times those calculated from the original consolidation tests. This again illustrated the general fallacy of designing with rates of consolidation determined only on the basis of standard investigation and complementary oedometer test results (ROWE, 1972; MURRAY, 1971; SMITH, 1978; YOUNGER, 1988; JAYAPUTRA et al, 1990; YOUNGER et al., 1990).

Thixotropic behaviour was observed during pile driving.

3.6 Dumai, Riau, Sumatra

A development at Dumai, opposite Rupat Island and facing the Malacca Strait, involved a coastal and near shore investigation. The site, somewhat like that at Belawan, involved two upper compressible clay deposits. The deposits are Quaternary. The topmost layer which has an average thickness of about 10m was found to be a very soft dark grey clay of recent origin and to contain a significant percentage of organics. The immediately underlying layer, which has an approximately similar thickness, from the shore outwards, but reduces in thickness on shore, is a soft

Stratum	Soil Type	Grain Size Sandy Clay	W_n %	PI %	b kN/m³	q_u kPa	p'_c kPa	e	C_v m²/yr
1 (C1)	Very soft silty clay with organics (CH)	1- 9	50-70	70-120	60-90	12 -15.5	10-35 +	40- 87	1.7-3.5	1-2
2 (C2)	Soft silty clay with tuff inclusion (CL --> CH)	1-24	30-60	30- 75	20-65	14.7-18.6	30-80 +	140-600	0.9-2.1	2-3
3	Silty sand with sand	5-58	0-55	(20)	0-30	16.7				
4 (C3)	Stiff weathered clay (CH)	2	70-80	50- 60	70-90	16.7	150	250-350	1.3	Negl. Cons. Settle -ment
5 (C4)	Stiff to hard clay (CH)	1- 2	50-70	35- 50	70-80	17.0	200	350 +	0.9-1.2	

+ Stratum 1   5-10 kPa and
   Stratum 2   20-50 kPa in deep water

Fig. 10 Summary of soil testing results

clay with tuff inclinations, suggesting quite active volcanic activity during the period of deposition. A sand stratum, varying in thickness from a few metres to more than 6m separates these top soft clay layers from layers of stiff to hard clay to considerable depth. The topmost layer was found to have no effective resistance to penetration in SPT testing, with the second layer having N values of 1-4. Undisturbed soil sampling was achieved using thin-walled piston samples at 5m intervals. A summary of the soil properties determined in the laboratory is given in Figure 10.

Natural moisture contents were close to the plastic limit for the upper two layers, except near the top 3m or so of the second layer where the moisture content was significantly less, and the material less plastic. In general, points fell above the A-line as shown in Fig. 11.

The presence of tuff in the second layer is considered responsible for the higher bulk densities indicated. Low values in the top layer were related to the presence of organic matter.

In the uppermost layer the very high void ratios were found only in the top 2m. Below this the value of void ratio tended to decrease linearly from 2.1 to 1.7 with bulk densities increasing from 14.5 to 15.5 kN/m³, except where organics appeared. The top 4m of the second layer in shallow water showed lower void ratios, indicating overconsolidation from dessication following water level lowering in an earlier geological period. Further off-

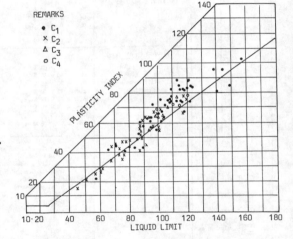

Fig. 11 Plasticity Characteristics at Dumai Site

shore this weathered section did not appear, and generally void ratios were of the order of 1.5, sometimes less, for this layer.

Undrained strength was found to increase with depth as indicated in Figure 12. Dessication was apparent for soil on land near the surface. Some lower values at depth were attributed to removal of overburden pressure by erosion. Sensitivities of 4-8 were measured for the uppermost clay layer and 2-5 for the second stratum. The sensitivities for the top layer are higher than indicated for the Belawan soils and point to the influential

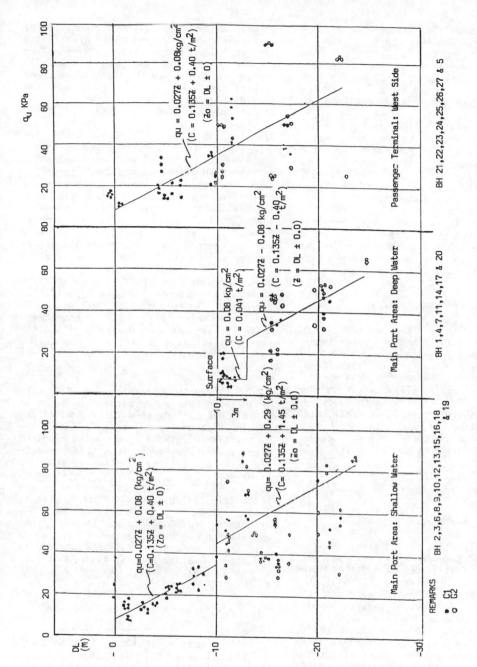

Fig. 12 Shear Strength with Depth Profiles at Dumai Site

presence of smectites, although no data on clay mineralogy was obtained.

Plots of p'c and p'o with depth showed all strata to be overconsolidated, with OCR ratios of 1.5-2.0 and 1.5-2.5 found for the two upper soft layers respectively.

For the rates of consolidation, the laboratory determined C_v values were quite low, but reference was made to the presence of interlocated sand seams, specifically for the second clay stratum, actual rates of consolidation were expected to be much higher than derived on the basis of the laboratory tests. m_v values for the top stratum, over the range of expected applied pressures, were found to be of the order of $5-15 \times 10^{-5}$ m^2/kN, while those in the second layer were about three times less.

3.7 East Riau Province, Sumatra

The low lying land in the eastern part of Riau, where oil extraction developments are in being, is largely covered by tropical jungle often contains significant depths of ombrogenous peat (SOFYAN, 1990) overlying soft marine clays having characteristics of the upper deposits described for Dumai above. The extremely high compressibility of the surface peat deposits has in some cases, caused problems of construction, even for very light loads.

4 FRESH WATER CLAY

4.1 Bandung, West Java

To the southeast of the city of Bandung there is a 50 km² deposit of soft volcanic mud, to a fairly uniform depth of over 30m. The highly compressible deposit has been the subject of considerable examination since a major toll highway to bypass the city of Bandung has been constructed for a length of 8km over the deposit. Characteristics of the materials and settlement behaviour have been presented in some detail (YOUNGER & SURATMAN, 1988; YOUNGER et al, 1989, 1990; JAYAPUTRA at el., 1990). The following summarises some of the salient features of the deposit.

The soil contains a mixture of clay minerals, mainly halloysite, allophane (although it is not possible to identify this in XRD and electron microscope testing), albite, cristobalite, minor quantities of montmorillonite (SIETRONICS,

1988), and also significant quantities of diatomaceus silica and some erratically distributed organics. There was also clear evidence of silt and sand seams in continuous boring logs, providing good lateral drainage, which would be expected considering the volcanic nature of the deposit (SETJADININGRAT, 1988).

Profiles of soil properties through the deposit at various locations indicated the presence of two buried weathered horizons, as shown in Figure 13 and for the test embankment site in Figure 14. The presence of allophanes and halloysites were also indicated in Atterberg Limit testing (WESLEY, 1973; ANON., 1990), as shown in Figure 15, wherein pre-drying of soil vastly changed the measured Atterberg values of the soil. The characteristics

of allophanes and halloysites indicate much higher permeabilities than would hold for other clay minerals. Combined with the open structures of diatomaceous silica, fast rates of drainage within the fine-grained mass itself were exhibited as discussed in YOUNGER et al. (1989, 1990).

Load-settlement profiles were also established using the parameters given in Table 1 and as indicated in Figures 13 and 14, as well as 'true' rates of consolidation (RIYANTO, 1988) these being compared with field measurements in a trial embankment section. Typical results are shown in Figure 16.

Stability calculations (BISHOP, 1958; HUANG, 1988) taking account of the drainage conditions found showed that embankments could be constructed up to 9m (including surcharge) fairly rapidly with a factor of safety against failure of 1.25 (YOUNGER et al., 1990a).

Table 1   Design C_c, C_R and C_α Values

Depth (m)	4.7m Section			6.2m Section		
	C_c	C_R	C_α	C_c	C_R	C_α
0.0- 3.0	0.50	0.05	0.005	0.50	0.05	0.005
3.0- 7.5	2.60	0.14	0.020	2.92	0.14	0.030
7.5-10.0	1.20	0.09	0.010	1.20	0.09	0.010
10.0-16.5	2.40	0.15	0.030	2.80	0.15	0.040
16.5-19.0	0.60	0.07	0.008	0.60	0.07	0.010
19.0-28.0	2.50	0.10	0.012	2.50	0.10	0.012
28.0-30.0	2.00	0.06	0.012	2.00	0.06	0.012

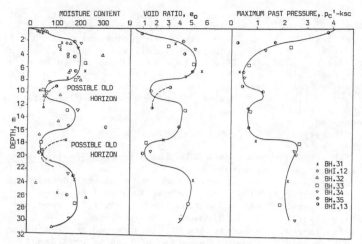

Fig. 13 Geotechnical properties of Bandung clay deposit along
bypass alignment (after YOUNGER, 1986)

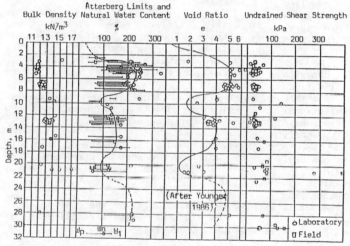

Fig. 14 Geotechnical properties of the Bandung clay deposit:
Details from Km. 37, Test Embankment Site

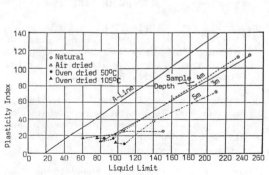

Fig. 15 Atterberg Limits and Influence of
Method of Preparation on These Properties

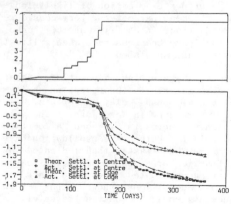

Fig. 16 Actual and Theoretical Settlement for 6.2m Embankment Section

5 CONCLUSIONS

Most of the soft clay deposits in Indonesia are coastal and many derive from geologically recent volcanic activity, an exception being the soft areas found in Kalimantan. The clays, with few exceptions, have been deposited in a marine environment, and are usually quite thick with layers sometimes separated by evidence of significant volcanic activity. Softest materials, from the surface down, are found near or offshore, and some of these layers are normally consolidated or lightly overconsolidated although there is frequent clear evidence of overconsolidation, and deeper strata and land surface strata are clearly overconsolidated. The main soft layers are highly compressible and have high void ratios. Frequently, a significant percentage of smectite minerals is present. Overconsolidation has resulted from significant changes in sea level as well as normal dessication processes. There is some evidence of delayed consolidation.

Under compression the main soft deposits have shown high rates of consolidation, higher than might be anticipated from the grain size distribution or conventional laboratory testing. Drainage seams have been clearly identified in properly conducted investigations, and the importance of soil fabric has been emphasised. It is also clear that the standard wash-boring method of sinking boreholes can cause considerable soil disturbance and lead to erroneous analysis of likely soil mass behaviour under applied load.

The more highly compressible strata have a significant long-term settlement, and it is advised that laboratory testing allows for the proper examination of likely long-term settlement parameters. Very considerable reductions in coefficients of consolidation have been observed when critical pressures are surpassed.

A deep fresh-water lake deposit of recent volcanic material, incorporating halloysite and allophane clay materials, was found at Bandung in West Java. The soil is highly compressible beyond a certain yield point. Rates of consolidation have been very much faster than indicated by clay fraction or standard oedometer tests. The overconsolidated crust and lack of undrained conditions showed that high embankments could be constructed safely. Certain areas contain swelling clay minerals and these have disrupted roads and structures. Proper identification of these soils, the presence of which can be expected from the geological history of the archipelago, will be a pre-requisite for future projects and use should be made of the guidelines for the identification and handling of all tropical residual soil types which have been published recently.

6 ACKNOWLEDGEMENTS

The data for the sites in Sumatra were provided by Dr. F.X. Toha, Institute of Technology Bandung. The work on Bandung clay and the swelling soil at Cepu was carried out on the ITB Master's programme in Highway Engineering and due acknowledgement is accorded to Professor Djuanda Suraatmadja, Head of Programme.

7 REFERENCES

Adyawati (1991). Pore pressure and Settlement Behaviour of Soft Marine Clays in Indonesia and Malaysia. MSc. Thesis, Inst. of Tech. Bandung - in course of preparation.

Anon (1990). Tropical Residual Soils. Geological Society of Engineering, Group Working Party Report, v.23, no. 1.

Akagi, T. (1990). Improvement of Clay Soil by Addition of Cement or Lime. Proc. 4th Int. Indonesian Soils Assoc. Conf., Bandung.

Bishop, A.W. (1958). The Use of Slip Circle in the Stability Analysis of Slopes, Geotechnique, v.5, pp.7-17.

Bjerrum, L. (1972). Engineering Geology of Norwegian Normally Consolidated Marine Clays as Related to Settlement of Buildings. 7th Rankine Lecture, Geotechnique, v. 17, pp. 18-118.

Brenner, R.P., Meier, R. & Udibowo, C. (1987). Geotechnical Properties of Volcanic Mud in a Near-Shore Area of East Java. Proc. 9th S.E. Asian Geotechnical Conf., Bangkok.

Burdall, A.C. & Nicholls, R.A. (1981). Belawan Port Reclamation. Proc. Symp. on Geotechnical Aspects of Coastal and Offshore Structures, Asian Inst. of Technology, Bangkok.

Chen, F.H. (1975). Foundations on Expansive Soils. Elsevier Scientific Publ. Co., Amsterdam.

Cox, J.B. (1970). Shear Strength Characteristics of the Recent Marine Clays in
S.E. Asia. Jour. SEASSE, v. 1, pp. 1-28.

Cook, J.R. & Younger, J.S. (1987). The
Engineering Geology of Road Projects in
North Borneo. Planning and Engineering
Geology, ed. Culshaw, Bell, Cripps &
O'Hara, Geol. Soc. Spec. Publ. No. 4,
pp. 419-428.

Cook, J.R. & Younger, J.S. (1988). Impli-
cation of Tropical Weathering in Road
Design & Construction. Proc. 2nd Pasca
Sarjana Workshop on Development in Hwy.&
Transportation Eng., v.2, Bandung.

Gonzales de Vallejo, L.T. Jimenez-Salas,
J.A. & Lequey Jimenez, S. (1981). Eng.
Geology of the Tropical Volcanic Soils
of La Laguna, Tenerife. Eng. Geology,
v. 17, pp. 1-17.

Holtz, W.G. & Gibbs, H.J. (1956). Engineering Properties of Expansive Clays,
Trans. ASCE, vol. 121, pp. 641-663.

Huang, Y.H. (1988). Stability Analysis of
Earth Slopes. Van Nostrand Reinhold Co.,
New York.

Jayaputra, A., Younger, J.S. & Liliawaty,
S. (1990). The Influence of Sample Size
and Disturbance on the Consolidation
Characteristics of Bandung Clay. Proc.
4th Int'l. Indonesian Soils Assoc.
Conf., Bandung.

McElvaney J., Supriyono & Nasution, S.
(1990). Laboratory Investigation of an
Expansive Clay from East Java. Proc. 4th
Int. Indonesian Soils Assoc. Conf.,
Bandung.

McGown, A., McNeil, N. & Gabr, A.W.
(1979). Predicted and Measured Behaviour of Clyde Alluvium Beneath Renfrew
Motorway Stage II Embankments. Proc.
7th Europe Conf. on Soil Mech. & Found.
Eng., Brighton, U.K., v. 3.

McGown, A., Marsland, A., Radwan, A.M. &
Gabr, A.W.A. (1980). Recording and
Interpreting Soil Macrofabric Data.
Geotechnique, v.30, no. 4.

Mitchell, J.K. (1986). Practical Problems
from Surprising Soil Behaviour. 20th
Karl Terzaghi Lecture (1984), Jour.
Geotechnical Eng. Div., ASCE, v. 112, n.
3. pp. 259-289.

Mori, H. (1982). Site Investigation and
Soil Sampling for Tropical Soils. ASCE

Geotech. Div. Specialty Conf., Hawaii.

Morrison, I.M., Wangsadinata, W. & Harris,
A.J. (1984). Tg. Priok Port Devel.
Investigations and Estimation of Rate of
Settlement of Land Reclamation. Proc.
of Symp. on Geot. Aspects of Mass and
Materials Transportation, Bangkok.

Mott Renardet (1983). Surabaya-Malang Hwy.
Project. Mid-term Review of Earthworks
and Ground Engineering, June, (unpubl.).

Murray, R.T. (1971). Emb'ments Constructed
on Soft Ground: Settlement Study at
Avonmouth. DoE, TRRL Report LR419,
Crowthorne, U.K.

Riyanto, J. (1988). Pore Water Pressure
Response of Bandung Clay Under Test
Embankment Loading for Padalarang-Cileu-
nyi Toll Highway Project. MSc Thesis,
Institute of Techn., Bandung, Indonesia.

Rowe, P.W. (1972). The Relevance of Soil
Fabric to Site Investigation Practice.
Geotechnique, 22, pp. 195-300.

Seed, H.B., Woodward, R.J. & Lundgren, R.
(1963). Predicting of Swelling Potential
for Compacted Clays. Trans. ASCE, v.128.

Setjadiningrat, C. (1988). The Properties
of Bandung Clay and Stability under
Emb'ment Loading. M.Sc. Thesis, Inst. of
Technology Bandung.

Sietronics. (1988). Report on Bandung Clay
Mineralogy,Canberra, Australia, (unpubl).

Skempton, A.W. (1953). The Colloidal
"Activity" of Clays, Proc. Int. Conf. on
S.M. and F.E., vol. 1, Zurich.

Smith, I.M. (1978). Computer Predictions
in Difficult Soil Conditions. Ch. 4.
Found. Eng. in Difficult Ground, ed.
F.G. Bell, Butterworth & Co, (Publ) Ltd.

Sofyan, I. (1990). Management of Subsurface Water in Peat. Proc. 19th Annual
Convention Indonesian Petroleum Assoc.

Supriyono. (1987). Strength and Volume
Change Characteristics of Cepu Clay.
M.Sc. Thesis, Inst. of Tech. Bandung.

Terzaghi, K. (1958). Design and Performance of the Sasamura Dam. Proc. Inst.
of Civil Eng., London, v.9, pp. 369-394.

Wesley, L.D. (1973). Some Basic Engineering Properties of Halloysites and Allophane Clays in Java, Indonesia. Geo-

technique, v. 23, pp. 471-494.

Wilkinson, W.B. (1968). Constant Head In Situ Permeability Tests in Clay Strata. Geotechnique, n. 18, pp. 172-194.

Younger, J.S. (1986). Construction over Soft Ground. Proc. Pasca Sarjana Workshop on Dev. in Hwy. & Transportation Engineering, Bandung, December.

Younger, J.S. & Suratman, I. (1988). Natural and Lime Stabilised Properties of Bandung Clay. Proc. 2nd. Int. Conf. on Geo-mechanics in Tropical Soils, Singapore, December.

Younger, J.S. (1988). Soft Ground Construction with Emphasis on Site Investigation and Use of Vertical Drainage. Proc. National Symp, HATTI, Jakarta.

Younger, J.S. (1988a). Ground Improvement Options for Highways in Indonesia. Proc. of Seminar on Ground Improvement Applic. to Indon. Soft Soils, August, Jakarta.

Younger, J.S., Riyanto, J. & Setjadiningrat, C. (1989). Bandung Clay: Characteristics and Response Under Trial Embankment Loadings. Proc.XII Int. Conf. on S.M. & F.E., Rio de Janeiro.

Younger, J.S., Javaputra, A. & Rachlan, A. (1990). Characteristics of Bandung Clay and Performance under Embankment Loading; A Review. Proc. 4th Int. Indonesian Soils Assoc. Conf, Bandung.

Younger, J.S., Setjadiningrat, C. & Nasution, S. (1990a). Embankment Stability on Bandung Clay. Proc. 10th S.E. Asia Geot. Conf., Taipei, Taiwan, R.O.C.

Developments in Geotechnical Aspects of Embankments, Excavations and Buried Structures, Balasubramaniam et al. (eds)
© 1991 Balkema, Rotterdam. ISBN 90 5410 019 2

Field and laboratory tests of tropical residual soils as materials for dam

Didiek Djarwadi
PT Bangun Cipta Kontraktor, Indonesia

ABSTRACT : Soil as engineering fill material in dam construction should be compacted to a dense state to obtain satisfactory engineering properties such as shear strength, compressibility and permeability.

This paper deals with laboratory and field compaction tests of tropical residual soil at Batam Island of Indonesia to achieve the required engineering properties as described in the Technical Specification.

Test results show that the compaction energy determined by 1 Modified Proctor Compaction Tests, according to ASTM D 1557, gives the most satisfactory engineering properties, and called ideal compaction.

Field compaction tests were carried out in order to obtain the minimum required degree of compaction at 95% of the laboratory compaction tests.

1 INTRODUCTION

Batam Island is located some 20 miles south-east of Singapore and facing the Singapore Strait. Considering its strategic location the Indonesian government developed Batam Island as an industrial, processing, and manufacturing area. Infrastructures such as highways, roads, airports, seaports and 5 dams were already constructed. Soils and geological investigations as well as field and laboratory tests have been intensively done in order to support the construction. Batam Island lies between longitudes 104° and 105° E and latitudes 1° N. The climate is typified by hot tropical conditions, with temperatures averaging at 31° C and high humidity. Yearly rainfall records indicated heavier periods in October to February. The geology of Batam Island is mainly dominated by sandstones at Central Batam Formation, metamorphic rocks at the western areas, and granite at eastern part as shown in Fig. 1 (LAPI ITB, 1985).

Previous investigation indicated that Batam Soils area related to Acrisols in the FAO Classification, or Yellowish Brown Ferallitic Soils according to French Classification. It was confirmed by distribution map of Red Tropical Soils in Southeast Asia (Morin and Todor, 1975) that Batam Soils belong to Tropical

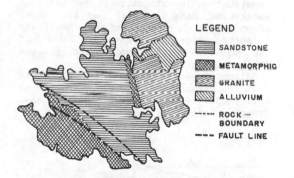

LEGEND

SANDSTONE
METAMORPHIC
GRANITE
ALLUVIUM
---- ROCK — BOUNDARY
--- FAULT LINE

Fig. 1 Geological formation in Batam Island (after LAPI ITB, 1985)

Residual Soils. In the last 5 years, two large earthfill dams were constructed, namely: Sei Ladi Dam and Mukakuning Dam. Field survey in order to identify and study the availability of embankment materials, exploration of borrow areas, laboratory tests, field compaction tests and construction of those 2 dams gave interesting results.

Sei Ladi Dam which was constructed in 1983 - 1985 is homogenous earthfill dam with 30 m in height, with a crest length of 1020 m, and mean upstream and downstream slopes of 3H/1V. Total embankment volume is 1,200,000 m3, while Mukakuning Dam which was constructed in 1985 - 1988 is a zoned earthfill dam with 30 m in height,a crest length of 380 m, and mean of upstream and downstream 3H/1V. Total embankment volume is 750,000 m3. The dams were built at Central Batam Formation in which sandstone mainly dominate in this region. In Sei ladi Dam, The sound rock with minor cracks and joints was found just 6 m below the swamp deposit. Grouting tests show that permeability of the foundations was very low, 10^{-6} cm/dt, such that grouting was not necessary. In order to extend the paths of water flow beneath the foundation, an impervious clay blanket with 50 m length and 2 m in thickness was constructed in front of the dam. Mukakuning Dam, on the other hand, lies on the valley which is along fault zones across the dam axis. Grouting tests indicated that the flow rate in the foundation was extremely high at 10 cm/dt especially on the fault zone.

Based on the geological conditions, a curtain grouting in 3 rows was constructed at 10 depth and extending to within 20 m in the fault zone. Fig.2 shows the typical cross section of the dams. Pre-construction investigation at Sei Ladi and Mukakuning dams indicated that the layer of soil available as embankment materials is not so thick. The maximum thickness of the deposits was 6 m, while the minimum thickness was 2.50 m.

Colour , texture, and depth of soils from the ground surface yielded different soil properties, although within the same group of soil.

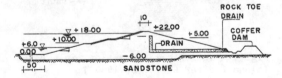

Fig.2a Typical cross section of Sei Ladi Dam

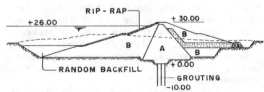

A = CORE

B = IMPERVIOUS FILL

Fig.2b Typical cross section of Muka kuning Dam

Material characteristics from pre-construction investigation shows that the material at the borrow area around Sei Ladi and Mukakuning Dams have good engineering properties as shown in Table I.

Table I. Material characteristics from pre-construction investigation

Parameter	Sei Ladi Dam	Mukakuning Dam	
		CH	CL
- Clay frac.(%)	40	50	35
- Specific gravity (kN/m3)	27.6	28.2	27.2
- Liquid limit (%)	35-58	50-66	30-42
- Plastic limit (%)	18-44	20-38	15-40
- Activity	0.70	0.85	0.67
- Natural moist. content (%)	40	48	35
- Optimum moist. content (%)	16-20	18-22	14-20
- Max dry density (kN/m3)	16.9-19	16.6-17.8	18.5-19.3
- Shear strength Friction angle	18°	16°	21°
Cohesian (t/m2)	6	6	3

2 SCOPE OF WORKS

In order to provide the embankment materials with satisfactory engineering properties, the soil investigation was carried out in an area of 50 ha surrounding the dams.Bulldozer was used to make an access in 20 m spacings. The scope of work to provide the suitable materials for dam were site recognition investigation, laboratory tests, and field embankment tests. The laboratory test was carried out in accordance with the ASTM Standards and/or AASHTO Standards, and the description of the test is shown in Table II.

2.1 SITE RECOGNITION INVESTIGATION

In this stage, hand boring or auger boring were used to take samples. Natural water content, colour, texture and description of the soil layers can be made. Boring was undertaken on the access at a spacing of 20 m such that each boring will represent the area of 400 m2. Laboratory

Table II. Description of the laboratory tests

Description of the best	Destination	
	ASTM	AASHTO
- Moisture content	D 2216	
- Specific gravity	D 854	T 100
- Liquid limit	D 423	T 89
- Plastic limit	D 424	T 90
- Gradation	C 136	T 27
- Hydrometer	D 422	T 88
- Thin walled tube samp.	D 1587	T 207
- Standard Proctor Comp.	D 698	
- Modified Proctor Comp.	D 1557	
- Direct Shear Test	D 3080	T 236
- Triaxial Test	D 2664	T 226
- Permeability Test	D 2434	T 215
- Consolidation Test	D 2435	T 216
- Field density test	D 1556	T 191
- Auger sampling	D 1453	

tests such as specific gravity, Atterberg limits, particle gradation and hydrometers were carried out in order to study the soil properties and availability as embankment materials. Soil distribution map and cross section can be made based on the test result identify the soil group.

The Unified Soil Classification System was adopted in identifying the soil. Based on the soil distribution map, test pits were plotted and samples in large quantities were taken by excavator. The number of test pits depends on the distribution of soils and preliminary laboratory test. Every test pit will represent 50.000 m3.

2.2 LABORATORY TESTS

In this stage samples taken from the test pits were tested. The test program includes compaction tests, permeability, direct shear, triaxial, and consolidation tests, while specific gravity, Atterberg limits and grain size analysis were determined from the trimming of compacted samples. Preliminary attempt in compaction was carried out in order to investigate the compaction characteristics due to the drying methods. Fig. 3 illustrates the effect of drying methods to the compaction values. Gidigasu (1974) presented similar result of compaction characteristic for lateritic soils from Ghana.

Drying methods of the soil from its in-situ moisture content has the effect of changing the soil properties. However, in cases where standard soil testing procedures in which the soil is air dried before testing cannot be used, drying may be preferable by the use of drying apparatus such that the temperature will not break up the soil properties.

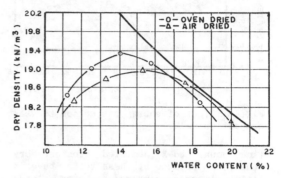

Fig. 3 Compaction characteristic of soil due to the drying method

2.2.1 COMPACTION TESTS

In 1933, Proctor described a laboratory compaction test which is now , called "Standard Proctor" compaction test. In the "modified" test the amount of compaction energy is over four and half times that of the Standard Proctor test by adopting bigger hammer, height of hammer drop, and number of layer at the size mould. Nowadays those tests are reference to ASTM D 698 and ASTM 1557, respectively.

Soil as an engineering material needs compaction in order to increase shear strength and decrease settlement and permeability. Compaction tests can determine the relationship between moisture contents and dry densities of the soil.

In this case, water plays an important role to produce the maximum dry density. A proper amount of mixing water in the compaction test will produce maximum compaction value. In Sei Ladi and Mukakuning dams, a series of compaction tests using energy equal to 1 Standard Proctor, 2 Standard Proctor, 1 Modified Proctor and 2 Modified Proctor were undertaken to investigate the level of compaction energy which can achieve the most satisfactory engineering properties.

Test results show that the use of more compaction energy tends to move the curves upward and to the left as shown in Fig. 4. The trend as shown in Fig.4 exists on most soils tested in the laboratory. A special phenomenon was found using energy compac-

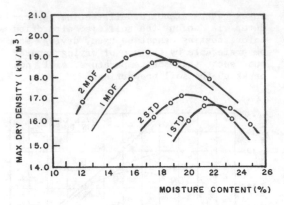

Fig.4 Effect of energy compaction to the compaction value

tion equal to 2 Modified Proctor. Dry densities produce by these tests have a little increase compared to using energy compaction equal to 1 Modified Proctor Test. More over, some test results indicated that dry densities produced by 2 Modified Proctor fall below the dry densities produced by 1 Modified Proctor Tests.

2.2.2 PERMEABILITY TEST

Permeability tests are performed to study the water flowing through the soil. In the construction of dams, it is important to control the level of water flowing through the dam body in order to retain the water in the reservoir and avoid instability of dam due to seepage. In this case compaction is very important, since it will produce the permeability value. Permeability depends on a number of factors such as the size and composition of soil grains, viscosity of fluid, void ratio , and degree of saturation. Wallace (1948) indicated that the increase in degree of saturation of soil causes an increase in permeability. There are at least 4 tests available to measure the permeability of soil : variable head test, constant head test, capillarity test, and the use of consolidation test's data to compute the permeability by mean of squeezing out of water from soil by application of a load.
In the construction of Sei Ladi and Muka-kuning Dams only 2 tests were adopted, i.e. constant head test referred to ASTM D 2434, and the computation of permeability from consolidation test's data.

2.2.3 DIRECT SHEAR TEST

Direct shear tests are performed to study the shear strength of soil in different loading mechanism with triaxial test. In direct shear test, the soil is stressed into failure by moving one part of the soil container relative to another, causing the soil to shear along the horizontal surface, while triaxial test consists of axially loading a soil specimen to induce failure. The loading mechanism of the tests is shown in Fig. 5.

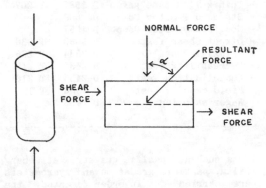

Fig. 5 Loading mechanism of Direct Shear and Triaxial Tests

In stability problems of embankment structures, the strength of the soil involved is required. The strength of cohesive soil is a combination of cohesion and internal friction of soil grains. Direct shear tests can be carried out under 3 condition such as consolidated undrained, unconsolidated -undrained, and drained tests. As already pointed out, the strength of soil depends on the internal friction of soil grains, therefore the test result will exhibit greater strength when tested in drained than when tested in undrained condition. By similar reason, consolidated specimen will exhibit greater strength than unconsolidated specimen.
With Reference to ASTM D 3080, Direct Shear tests in Sei Ladi and Mukakuning Dams were carried out in the consolidated undrained condition (Kezdi,1979). The results are then expressed in terms of normal stress-shear stress curves.

2.2.4 TRIAXIAL TEST

Triaxial tests are performed to study the shear strength of soils. In most testing programs, three samples in the same conditions are tested under different pressure. The final result of these tests can be plotted by three Mohr's circles determined from effective stress condition at failure. From the normal stress-shear stress relations, the cohesion and internal angle of friction which reflect the shear strength of soil, can be obtained in a certain condition.

Triaxial tests in Sei Ladi and Mukakuning Dams were carried out in the conditions of consolidated undrained on partly saturated and unconsolidated undrained on partly saturated tests (Bishop and Henkel,1962). The friction of soil mass depend on the pressure acting between the soil particles. Therefore, tests in consolidated condition will exhibit the greater strength than when tested in unconsolidated condition. Compaction gives "Strength" to the soil mass due to the interaction of soil particles associated with the alteration of pore water pressure in the soil mass. Based on this assumption, soil compacted with greater compaction energy will produce greater shear strength. In general, shear strength of soils can be presented by empirical equation as follows :

$$\tau = c + (\sigma - \mu) \tan \emptyset \qquad (1)$$

where τ is the shear strength of soil, c is the cohesion, σ is the normal stress, μ is the pore water pressure in the soil mass, and $\emptyset$ is the internal angle of friction of soil obtained from triaxial test.

2.2.5 CONSOLIDATION TEST

Consolidation tests are performed to determine the probable settlements of compacted soils under load. The rate and amount of settlements or consolidation of large soil masses, such as dam embankments, can be predicted from the results of laboratory consolidation tests on representative samples.

Consolidation process in the soil mass due to load increment, pushes the water to drain out from the soil mass. The rate of consolidation is directly related to the permeability of the soil since permeability controls the speed at which the pore water can escape. In this case, compaction of the soil can influence the rate of consolidation, since the level of compaction will also produce different permeability values. Compaction will produce the pre-consolidated value to the soil.

The two most important soil properties furnished by a consolidation test are the compression index (Cc) and coefficient of consolidation (Cv). Compression index indicated the compressibility of soil, while coefficient of consolidation indicated the rate of compression under a load increment. The coefficient of consolidation (Cv) can be computed using the following equations (Taylor,1948) as follows :

1) Square Root of time Fitting Method;

$$Cv = \frac{0.848 \ H^2}{t \ 90} \qquad (2)$$

2) Log Fitting Method;

$$Cv = \frac{0.197 \ H^2}{t \ 50} \qquad (3)$$

where H is the average thickness of drainage surface for the load increment, while t 90 and t 50 are the times at 90% and 50% compression , respectively. Compression index (Cc) can be found from the slope of void ratio vs log of pressure graph.

Based on several assumptions, Terzaghi (1925) developed the one dimensional consolidation theory to predict the settlements of structures in the field, which in general can be calculated using the formula;

$$\triangle H = \frac{Cc}{1 + eo} \ \log 10 \ \frac{Po + \triangle P}{Po} \ H \qquad (4)$$

where; $\triangle$ H is the consolidation settlement of layer H.
    H is the original thickness of soil.
    Cc is the compression index.
    eo is the initial void ratio.
    Po is the vertical soil pressure.
    $\triangle$ P is the change in effective vertical soil pressure at the center of the soil layer due to applied load.

2.3 FIELD COMPACTION TEST

Since the laboratory compaction behaviour may not correlate with the behaviour of compacted soils using roller in the field, field compaction tests were carried out in order to determine the type of roller, lifting thickness and number of passes.

The specification recommended a 15 ton pneumatic tyre roller, an 11-ton padfoot vibratory roller and an 11-ton smooth drum vibratory roller, travelling at speeds under 10 km/hr for construction of dams. In the preliminary attempts of field compaction tests on similar soil, compaction with 15-ton pneumatic tyre roller and 11-ton padfoot vibratory roller by 6, 8, and 12 passes, produce almost the same field densities as shown in Table III.

Table III. Field densities produce in preliminary test

Percentage of M D D

Number of passes	Vibratory padfoot roller	Pneumatic tyre roller
6	94.2	93.6
8	97.2	98.1
12	92.6	93.1

Based on the above results, the padfoot vibratory roller was utilized in the series of field compaction tests and in the construction of both Sei Ladi and Mukakuning dams. At series of field compaction tests were then carried out using an 11-ton padfoot vibratory roller by 4, 6, 8, 12 and 16 passes on 30 cm layer as specified. The compaction by 16 passes is more likely to study the effect of heavy energy of compaction to the change of soil properties.

The number of field compaction tests were dependent on the soil distribution in the borrow area, but every test was designated to represent at least 150,000 m3. Field compaction tests were carried out on levelled ground of 30 x 25 m2, divided into 5 strips of 6 x 25 m2. A spreading layers of 30 cm were then placed and compacted by 4, 6, 8, 12 and 16 passes, respectively. Compaction tests were stopped after reaching 4 layer, and samples were taken from the last layer at depth of 15 cm from the surface. Samples were taken by thin walled tube sampling according to the specification by ASTM D 1587. The number of samples taken for laboratory tests and test description are shown in Fig. 6 (Djarwadi,1988).

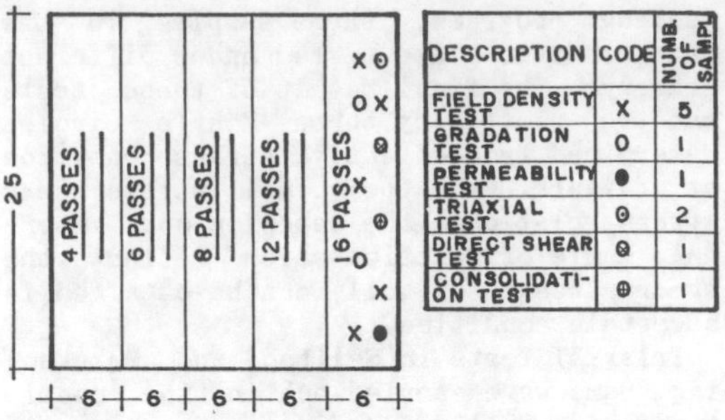

Fig. 6 Field compaction test aptent and description of tests

3 TEST RESULT

From borrow areas of Sei Ladi and Muka-kuning dams, 360 handbores and 42 test pits were explores, while 12 series of embankment tests were carried out. Test results from those activities can be presented as follows :

3.1 SAMPLING FROM THE FIELD

Test results from sampling (auger boring) consist of : elevation of ground, depth of boring, soil profile, soil group, texture, colour, and natural moisture content as shown Fig. 7.

DEPTH		SOIL GROUP	COLOUR	NATURAL WATER CONTENT	CLAY FRACTION
+20.12	0	ORGANIC	BLACK	—	—
	1	CLAY WITH LOW	YELLOWISH	38 %	42 %
	2	PLASTICITY (CL)	BROWN		
22.90	3	(CL)	BROWN	34 %	46 %
23.85	4	(CH)	GREY	36 %	54 %
24.95					
25.70	5	(CH)	GREY BLACK	28 %	62 %

Fig. 7 Typical sampling test results

3.2 LABORATORY TESTS

3.2.1 Atterberg limits

The location of Batam soils as embankment material in the Casagrande chart is shown in Fig. 8. It was confirmed that those soil fall within the approximate boundary of lateritic soils (Mitchell and Sitar,1975). The test result and interpretation indicated that A-line as described in Casagrande chart is not suited for Batam Soils. Test results indicated that in the high plasticity side "A-line" moved downward, while in the low plasticity side it moved upward. The "correction line" for Batam soils can be presented using empirical equation as follows. (Djarwadi 1983,1985).

$$IP = 0.68 (WL - 15) \qquad (5)$$

seems to be in agreement with the pre-construction investigation as shown in Fig. 10. Skempton (1953) indicated that Ia < 0.75 can be considered as a clay with low activity, and Ia > 1.25 as clay with high activity.

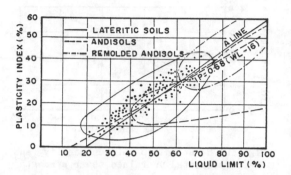

Fig.8   Position Batam soils in Casagrande chart and revision of A line

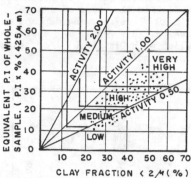

Fig.10   Potential expansiveness of Batam clay

3.2.2   Grain Size Analysis

Figure 9 shows the grain size distribution of Batam soils obtained from the borrow areas. The embankment materials, as described in the specification were clay with low and high plasticities. The clay fraction varies from 26% in the clay with low plasticity to 64% in the high plasticity. The test results also indicated that coarse fraction (sand and gravel) constitutes almost 10% in the clays.

3.2.4   Compaction test

Result of compaction tests by applying different compaction energy are shown in Fig. 11 and indicated that most of the compaction values using energy compaction equal to 2 Modified Proctor Compaction were less than those obtained by 1 Modified Proctor Compaction. This phenomenon are mostly found at the samples tested.

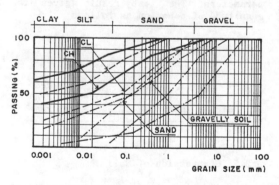

Fig.9   Grain size analysis of Batam soil

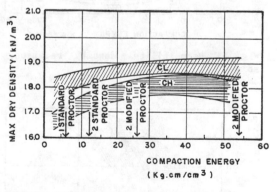

Fig.11   Compaction test result

3.2.3   Soil activities

The activity index of Batam clays fall between the range of 0.50 and 1.00, and

Futher investigation shows that one of the reasons maybe attributed to the degradation of soil particles as shown in Fig. 12.

3.2.5 Permeability

Permeability test results obtained for compacted soils from a series of compaction tests are shown in Fig. 13 and indicated that the permeability of compacted soils are related to their densities. The magnitude of permeability of compacted Batam clays falls in the range of 1×10^{-5} cm/sec to 1×10^{-8} cm/sec, and depends on its bulk density. This result is approximately similar with Ola (1980) for some African tropical soils.

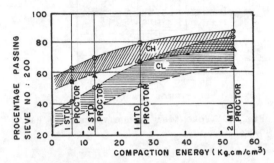

Fig.12 Degradation of soil particles due to compaction energy

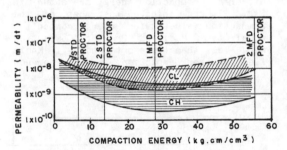

Fig. 13 Permeability - density relationship

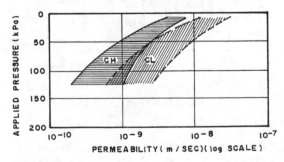

Fig. 14 Permeability - applied pressure relationship

The magnitude of permeability calculated from consolidation test data as shown in Fig. 14 gives an indication that permeability of soil are more or less equal to the test results refered to in ASTM D 2434.

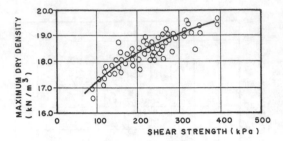

Fig. 15 Relationship between density and its consolidated undrained shear strength

3.2.6 Shear Strength of Soil

Three kinds of tests were carried out in the laboratory in order to obtain the shear strength of soil i.e. Triaxial CU test and U.U test, and Direct Shear CU tests. Two parameters influenced the magnitude of shear strength.

The shear strength of tropical soils is dependent on their density. Figure 15 shows the variation of shear strength with their densities obtained from Direct Shear CU test, and Fig. 16 shows the variation of internal angle of friction obtained from triaxial tests. It was difficult to express the relation of cohesion with other parameters. The test result gives the magnitude of cohesion as follows :

Test Method	Cohesion
Triaxial test	
- C.U	40 - 130 kPa
- U.U	90 - 210 kPa
Direct shear C.U	40 - 90 kPa

3.2.7 Compression of Soil

Consolidation test results indicated that the coefficient of consolidation ranges from 1×10^{-1} to 1×10^{-3} cm2/sec, and the compressibility of compacted tropical soil is less than would be expected of temperate zone clays with the same liquid limit. Figure 17 shows the relationship between liquid limit (WL) and

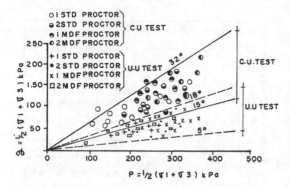

Fig. 16 Summary of triaxial test results

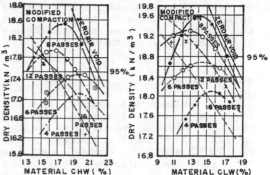

Fig. 18 Compaction test value in various moisture content

compression index (Cc) for samples from Sei Ladi and Mukakuning embankment materials. As can be observed from these figures, the average value of Cc for Batam clays for some value of WL appears to be somewhat smaller than that determined by Skempton (1944).

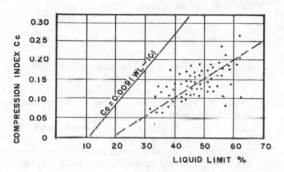

Fig. 17 Relationship between Compression Index and Liquid limit for Batam Soils

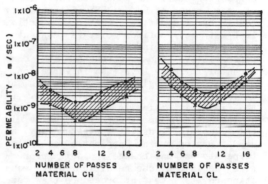

Fig. 19 Relationship between permeability and number of passes

3.3 Field Test Results

The series of laboratory test result indicated that compaction using compaction energy equal to 1 Modified Proctor Compaction gives the most satisfactory engineering properties among the others (Djarwadi, 1988). Field compaction tests indicated that compaction condition of more than 2% from optimum moisture content will result to compaction value less than 95% of laboratory tests as referred to ASTM D 1557, as shown in Fig. 18, while permeability test on compacted material gives the magnitude as shown in Fig 19.

Consolidation test result further indicated that the compression index ranges from 5×10^{-2} to 3×10^{-1}, while the coefficient of consolidation ranges from 2×10^{-1} cm2/sec to 2×10^{-3} cm2/sec.

Direct Shear Test on C.U condition yielded shear strengths of compacted material at the range of 250 kPa, while triaxial tests yielded strength parameters ranging from $14°$ to $26°$ for $\emptyset$, cohesion (c) ranging from 70 kPa to 120 kPa in the C.U conditions, and $\emptyset$ range of $5°$ to $12°$ and cohesion (c) range of 100 kPa to 180 kPa in the U.U conditions. The test evaluations on the series of field compaction tests indicated that the compaction using 11-ton padfoot roller with travelling speed at 10 km/hr on 30 cm layer by 8 passes, gives the most satisfactory result.

Comparing with the laboratory test, where the most satisfactory engineering properties was obtained by compaction using energy compaction equal to 1 Modified Proctor, the field compaction tests by 8 passes gives the field engineering properties at the range ± 5%, eventhough the laboratory compaction behavior may not correlate with the field compaction using machinery, and in the construction of Sei Ladi and Mukakuning Dams, the same compaction procedure were ultilized.

4 DISCUSSION

Test results from laboratory indicated that compaction using energy compaction equal to 1 Modified Proctor gives the most satisfactory engineering properties of embankment materials for Sei Ladi and Mukakuning Dams, while the field embankment tests using a padfoot roller with operation weight of 11 tons, travelling under 10 km/hr by 8 passes on a 30-cm lift thickness gave the optimum result.

The degree of compaction from field compaction tests as described above fall at the range of 96% to 98% of laboratory tests by 1 Modified Proctor Compaction test at their optimum moisture contents. Also the engineering properties such as permeability, shear strength, and compressibility only have ± 5% in variation.

Attention is given on the method of drying of samples for compaction test. In the field, air drying is appropriate in the drying of embankment materials, while in the laboratory, drying apparatus may be used. Drying using apparatus in high temperatures may change the soil properties as in previous results, while some standard soil testing procedure the temperature and duration of drying using apparatus should be carried out. The American Society of Testing & Materials (ASTM) recommended the use of drying apparatus with the temperature of samples not to exceed 60° C (140° F). From compaction test results using 1 Modified Proctor, the relation of optimum moisture content and their densities as shown in Fig. 20 can be presented using empirical equation as follows (Djarwadi, 1989).

$$MDD = 34.85 \ (\ 2.33 - OMC \ ) \qquad (6)$$

Some specifications require that the placement moisture content of embankment materials in dam construction shall be within ± 2% of the optimum moisture content determined by laboratory compaction test. Field compaction test results (Fig. 18) show that where the compaction carried out exceeds the upper limits, the degree of compaction falls belows 95% of the maximum compaction in the laboratory, while in the lower limit, compaction up to -4% from the optimum moisture content will produce a degree of compaction at least 95% of the maximum compaction value. This

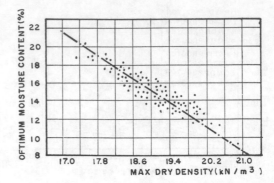

Fig. 20 Relationship between OMC and their densities

condition has the advantage, since in the tropical country during dry season, and temperature range 29° C to 34° C, usually the moisture content of embankment material drops below the optimum moisture content. Dewatering may be considered when the moisture content of embankment drop to 8% below the optimum moisture content.

5 CONCLUSION

From the test results described above, the following maybe pointed out :

- Tropical soils in its natural conditions are extremely heterogeneous materials which are difficult to sample and to test in a meaningful way.
- Tropical soils are widely used as embankment materials, apart from their heterogeneity, the soil properties may change as a result of the effect of drying, and heavy compaction energy that break the soil particles.
- Compaction using compaction energy equal to 1 Modified Proctor Compaction on Tropical soil at Batam gave the most satisfactory engineering properties.

ACKNOWLEDGEMENTS

The author sincerely thank Dr D.T. Bergado, Associate Professor, Geotechnical Division, Asian Institute of Technology, Bangkok, for his invaluable advice and in preparing and reviewing the final paper.

REFERENCE

AASHTO. Standard Specification for Transportation Materials and Method of Sampling and Testing, part II Method of Sampling and Testing, Washington DC: AASHTO.

ASTM. Annual book of ASTM Standards Vol 04/08 Soil an Rock, Building Stones. Philadelphia: ASTM.

Bishop,A.W and Henkel,D.J.1962. The measurement of soil properties in the triaxial test. London : Edward Arnold.

Djarwadi,D.1983. Report of soil Investigation for Embankment Material of Sei Ladi Dam Project (in Indonesian) Batam :unpublished.

Djarwadi,D.1985. Report of soil Investigation for Embankment Material of Mukakuning Dam Project (in Indonesian) Batam : unpublished.

Djarwadi,D.1988. Embankment test in Mukakuning Dam Project. Proc 2nd Intl Conf on Geomechanics in Tropical soil Vol I ; S'pore :381 - 395.

Djarwadi,D.1989. Geotechnical Properties of tropical clay at Batam Island. Proc 2nd Intl Conf on Geomechanics in tropical soils, Vol II, S'pore :545 - 587.

Gidigasu,M.D.1974. Degree of weathering in the Identification of laterite Materials for engineering purpose-A review, " engineering Geology", Vol 8 no. 3 : 253 - 266.

Kezdi,A.1979. Handbook of soil Mechanics, Vol 2. Soil Testing, Elsevier. Amsterdam.

Mitchell,J.K and Sitar,N.1982. Engineering Properties of tropical residual soils, Proc Conf on Engineering and Construction of Tropical and Residual Soils. ASCE : 30 -57.

Morin,W.J, Todor,P.C.1975. Laterite and Lateritic soils and other problems soils in tropic, An engineering Evaluation and Highway study for USAID, AID/CSD 3682, Lyon Associates Inc and Road Research Institute, Brazilian National Highway Department.

Ola,S.A.1980. Permeability of three compacted tropical soils. Quarterly Journal of Engineering Geology, London,Vol 13 : 13 - 95.

Proctor,R.R.1933. Design and construction of Rolled Earth Dams. Engineering News Record, Aug 31st, Sept 7 th, Sept 21 and Sept 28.

Skempton AW.1944. Notes on the Compressibility of clays. Quarterly Journal of Geology Society, London. Vol c : 119 - 135.

Skempton AW.1953. The Colloidal activity of clays. Proc 3rd I.C.S.M.F.E. Zurich. 1953.

Taylor,D.W.1948. Fundamental of soil Mechanics. John Wiley and Sons, New York.

Terzaghi, K.1925. Structure and volume of voids of soils, pp 10-12 and part of 13 of Erdbaumechanik of Bodenphysikalisher Grundlage, translated by Casagrande in "From Theory to Practice in Soil Mechanic, John Wiley and Sons, New York : 140 - 148.

Wallace,M.I.1948. Experimental Investigation of the Effect of degree of saturation on the permeability of sand. M Sc Thesis. Dept of Civil Eng. Massachusetts Institute of Technology.

Developments in Geotechnical Aspects of Embankments, Excavations and Buried Structures, Balasubramaniam et al. (eds)
© *1991 Balkema, Rotterdam. ISBN 90 5410 019 2*

Preliminary report on the sand liquefaction induced by the 16 July earthquake in Dagupan City (Pangasinan Province, Philippines)

G. Rantucci
Renardet, Consulting Engineers, Italy

ABSTRACT: During the 16 July earthquake in Central Luzon (Philippines) widespread sand liquefaction occurred in the Gulf of Lingayen in Pangasinan Province, with catastrophic consequences in Dagupan City. The paper describes the damage to structures in Dagupan and the evaluation of the liquefaction potential for the highly affected zone of Perez Blvd.

1 INTRODUCTION

A significant worldwide seismic activity followed by catastrophic consequences occurred during the period of October 1989 - September 1990 in different areas. In October, 1989 a 6.0 Magnitude earthquake struck Northern Thailand and the Burmese border area, a 4.5 Magnitude earthquake struck Northern Thailand and the Burmese border area, a 4.5 Magnitude quake of low intensity hit Tokyo and Eastern Japan, while a 6.9 quake occurred in Northern California, with the deadly collapsing of part of the San Francisco-Oakland Bay Bridge.

During 1990 most affected areas were in Middle East, Far East and China. An earthquake of Magnitude 6 jolted through the southern caucasian provinces of Russia followed in June by a quake of about Magnitude 6, which devastated the Caspian Sea provinces in Iran. On June 14, a 6.2 Magnitude quake hit the island of Panay in the Philippine Archipelago and again in the same country on July 16 1990, a catastrophic quake of 7.7 Magnitude affected the Central Luzon, followed by more than 1000 aftershocks during the next 45 days. During the period of August to September earthquakes were reported between the Philippines and Borneo (Magnitude 6) and in Mainland China.

The June and July earthquakes in the Philippines both were characterized by the extensive damage to structures due to the widespread occurrence of sand liquefaction.

2 THE 16 JULY EARTHQUAKE IN CENTRAL LUZON

The 16 July 1990 at 4.26 p.m. a catastrophic earthquake measuring 7.7 on the Richter scale (Intensity VIII, Rossi-Forel scale), with epicenter near the town of Cabanatuan and an approximate depth of 64 km jolted Central Luzon Island (Philippines) for about 45 sec., producing widespread damage to structures and more than 1,000 casualties.

Most affected places were the towns of Cabanatuan, Baguio, Dagupan, San Fernando in La Union and Manila. Coincidentally, Baguio, Dagupan and Manila, the three of them on the western side of Luzon, are located approximately a distance of 90 km from the epicenter.

The 16 July earthquake produced a widespread liquefaction of surficial sands in Pangasinan Province, with disastrous consequences in the area around Dagupan. It has not been possible yet to identify the entire zone affected by liquefaction, but reports from eyewitnesses and informations on collapsed structures lead to the conclusion that a large area, rather than the town of Dagupan only, was affected.

The aim of this paper is:
- to give an overview of earthquake induced damages in Dagupan;
- to describe the liquefaction phenomenon which affected this town and most probably part of the Pangasinan Province;
- to attempt an evaluation of the liquetion potential.

The study is based on the following data:

- boring logs made available by the Department of Public Works and Highways (DPWH)
- results of a field investigation executed by the company Foundation Specialist for the reconstruction of the collapsed Magsaysay Bridge
- data collected by the author after the quake, in various visits to Dagupan.

3 GEOMORPHOLOGY AND SOILS OF DAGUPAN CITY AND THEIR RELATIONSHIP WITH TECTONICS OF CENTRAL LUZON

The coastal town of Dagupan is located in the southeastern part of the Lingayen Gulf (northern coast of Luzon), not far from the famous touristic place of Hundred Islands. Most of Dagupan and surrounding countryside facing the Lingayen Gulf is flat.

From geological and surface features of Pangasinan it is concluded that the zone surrounding Dagupan was a lagoon, filled by loose sediments during the Quaternary. Dagupan City, in particular, was partly built on a reclaimed area with about topmost 1-2 m of sand transported from the nearby seaside quarry presently named Bonuan. Abandoned meanders of Calmay and Pantal Rivers, which are easily detected from aerophotos and maps, provide a useful indication on the river migrations in Dagupan Plain during the Quaternary.

Loose deltaic deposits characterize the town area with the top 10-18 m made of silty sand, overlying more than 40 m of clay with silty sand intercalations.

The bedrock, the depth of which is unknown but probably of the order of some hundreds of m, is presumably composed of intensely faulted intrusive and sedimentary rocks.

Tectonically speaking, the elongated N-S trending plain between Dagupan and Manila (Figs. 1 and 2) is crossed by a NS-SE system of faults responsible for the depression.

The 16 July quake, which induced the liquefaction of surface saturated sands in a large area of Pangasinan, was generated by block movements along the active Digdig Fault, located between Manila Trench and East Luzon Trench (Fig. 2). Also shown in Figure 2 are the major tectonic features of the Philippine Archipelago with the N-S trending subduction zones.

The damage was particularly severe in the City of Dagupan, between Perez Blvd. and Fernandez Ave. where most of the two to four storey buildings are located.

From the interpretation of boring logs it is concluded that loose sand, varying from 10 to 18 m in thickness, forms the topmost layer of Dagupan City. It is possible that this layer extends over a large part of the flat Quaternary sediments zone West, North-East and South of Dagupan, as can be inferred from the Geotectonic Sketch of Central Luzon (Fig. 1). The water table is located at 1.0 to 1.5 m depth in Dagupan Plain, as derived from boreholes.

Figure 3 shows the soil profile along the collapsed Magsaysay Bridge. This stratigraphic sequence can be considered representative of the highly damaged segment of Perez Blvd. located west of the Bridge. The reason for choosing the soil profile of Magsaysay Bridge is in the high concentration of damage on both sides of the bridge and in the availability of borehole data and laboratory tests executed after the quake.

Very limited borehole data related to a field investigation executed before the quake were made available for the zone near the concrete Quintos Bridge (not collapsed) in Fernandez Avenue.

4 PHYSICAL EVIDENCE OF SAND LIQUEFACTION AND DAMAGE DISTRIBUTION IN PANGASINAN AND TARLAC PROVINCES

The 16 July earthquake changed significantly the surface features of Dagupan City as described by the people in the locality.

Foundations with emission of gas and liquefied sand as well as sand boils were reported by eyewitnesses to have occurred after the beginning of the quake. The gas emission can be attributed to the presence of decomposed organic matter at various depths, as confirmed by boreholes.

Most spectacular aspects related to liquefaction in Perez Blvd. were the floating of the buried tank at the gasoline station, the tilting of many buildings with a three storey building tilted to about 15 degrees, the sinking of trucks parked on the roadside and the collapsing of Magsaysay Bridge. In general, road bulging, tilting of buildings and sinking of them often combined with tilting were the main element of the picture. Despite the widespread damaged, buildings were not significantly affected by cracks since the shaking due to the quake, by inducing liquefaction, minimized the effect of shear stresses on structures. It is well known in fact that shear stresses are not transmitted by liquids. Comparatively in Manila, where liquefaction did not occur, hundreds of buildings were affected by extensive crackings, concerning both the reinforced concrete frames and walls. The limited number of cracks, which were observed in

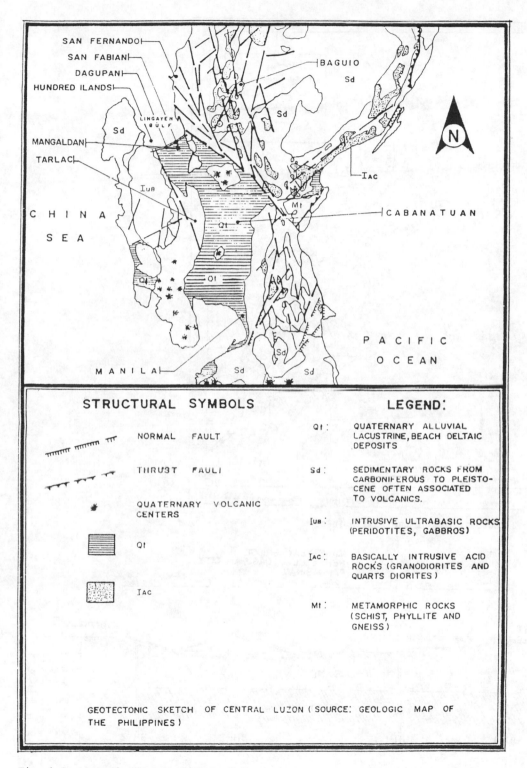

STRUCTURAL SYMBOLS

NORMAL FAULT

THRUST FAULT

QUATERNARY VOLCANIC
CENTERS

Qt

Iac

LEGEND:

Qt : QUATERNARY ALLUVIAL
LACUSTRINE, BEACH DELTAIC
DEPOSITS

Sd : SEDIMENTARY ROCKS FROM
CARBONIFEROUS TO PLEISTO-
CENE OFTEN ASSOCIATED
TO VOLCANICS.

Iub : INTRUSIVE ULTRABASIC ROCKS
(PERIDOTITES, GABBROS)

Iac : BASICALLY INTRUSIVE ACID
ROCKS (GRANODIORITES AND
QUARTS DIORITES)

Mt : METAMORPHIC ROCKS
(SCHIST, PHYLLITE AND
GNEISS)

GEOTECTONIC SKETCH OF CENTRAL LUZON (SOURCE: GEOLOGIC MAP OF
THE PHILIPPINES)

Fig. 1 Geotectonic sketch of Central Luzon.

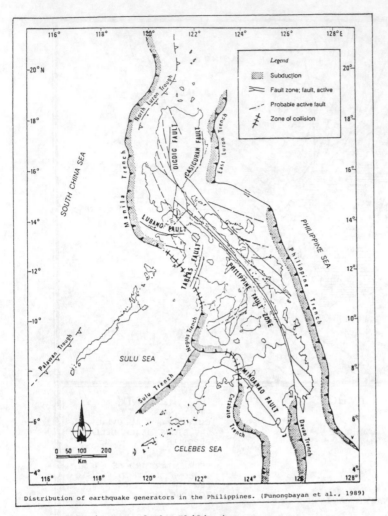

Distribution of earthquake generators in the Philippines. (Punongbayan et al., 1989)

Fig. 2 Tectonic map of the Philippines.

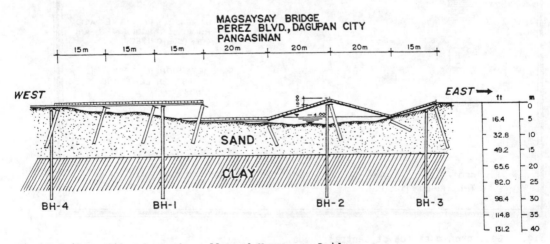

Fig. 3 Soil profile along the collapsed Magsaysay Bridge.

Dagupan buildings can be definitely attributed to stresses induced by the sinking and tilting of structures.

The area of Dagupan with highest damage concentration is located between Fernandez Ave. and Perez Blvd., where the seven-span Magsaysay concrete bridge collapsed. In the former most buildings underwent sinking only and the concrete bridge named Quintos, structurally similar to Magsaysay bridge and 450 m downstream, did not suffer any appreciable damage.

In general, the damage in Perez Blvd. including both bridge and building structures was higher than any other place within the affected zone. However, this does not mean that the liquefaction in Perez Blvd. was the most severe in the area.

Different behaviour of adjacent buildings was observed. Some of them were slightly affected or not affected at all by liquefaction, others were badly hit. Combined effects of density variations within the sand layer and different type of foundation adopted can be considered responsible for this case. Extensive liquefaction can explain also the significant damage which affected the road connecting Dagupan to San Fabian through Magdalgan (Fig. 1). The rigid pavement of this road was deformed into an undulated strip, with the seaside shoulder (both sides in some cases) sunk, for a length of several hundreds of m. The sinking of shoulders was estimated in the range of 30 cm to 100 cm below the pavement. Such behaviour, the occurrence of which was reported also in Niigata (Japan 1964), is due to fact that the embankment under the rigid pavement acted as a unique floating shield, thus preventing the dissipation of the accumulated pore water-pressure underneath. As a consequence of this local condition, pore pressure dissipation basically occurred along shoulder slopes, on both sides of the road, and shoulders appears as being vertically cut from the embankment.

The coastal road connecting directly Dagupan to San Fabian suffered the most severe damage; in some portions it was affected by multiple, wide (50-90 cm) and long longitudinal cracks (10-30 m), while in other zones the road embankment was split into tilted blocks with portions of concrete pavement still on top.

In the countryside near Dagupan City a number of Barangays (small villages) often surrounded by water have been reported to have undergone severe sinking. Most probably liquefaction occurred in a vast countryside zone with an effectiveness at least comparable to that of Dagupan. Barangays, in fact, are small communities based on agriculture; therefore, loads transmitted by houses are usually very low.

Intense liquefaction with sand boils, ground fissuring and subsidence were observed in the towns of Tarlac, Gerona and Paniqui (source "Observer", July 1990 Special Issue, published by the Department of Science and Technology of the Philippines) in the province of Tarlac.

Ground fissuring along Mac Arthur Highway in Tarlac, most probably due to liquefaction, was reported.

Some 40 km East of Dagupan the 650 m long truss bridge (Plaridel) in Carmen (Rosales), over the Agno River, collapsed. There is no evidence, however, of liquefaction in the area, although river bed sediments are composed of loose to medium-dense fine sands.

In general, after the 16 July earthquake large areas in Dagupan City and countrysides were inundated by water most certainly from the pore pressure dissipation occurred after the quake.

5 THE DAMAGE INDUCED IN DAGUPAN CITY

The structural damage was assessed by inspecting buildings, bridges, road platforms and pavements, electrical and sewerage city network.

Generally speaking, it was found that surroundings of Perez Blvd. Fernandez St. and some minor zones were catastrophically hit, but remaining areas were marginally affected by the quake. This difference can be partly explained by saying that lightly affected areas had probably higher relative density surface soils or building foundations had been properly designed and constructed or both aspects. Definitely, the highest concentration of light to heavy buildings occurs in Dagupan city center. In areas with scattered light one-storey concrete houses or wooden one-two storey houses, liquefaction was not so evident since surface disuniform loading was minimal.

In the highly affected areas where buildings basically underwent sinking and tilting with bulging of streets, the phenomenon of sand liquefaction triggered by the quake can be considered the major cause of foundation failure.

Most of buildings in Dagupan are located in the city proper and a majority of them are two to three storeys, a limited number reaching four to five storeys.

A preliminary classification of earthquake affected services and damaged structures, based on settlement and tilting of building foundations, is given below:

Fig. 4 Contiguous buildings with quasi-uniform severe sinking in Fernandez Avenue.

Fig. 6 Contiguous and isolated buildings in Perez Blvd. with sinking and limited appreciable tilting.

Fig. 5 Prudential Bank building, at the corner of a street in Fernandez Avenue. The structurte underwent sinking for more than 1 m and some tilting as well.

5.1 Buildings

a) Isolated structures or contiguous building which underwent quasi-uniform sinking are concentrated in Fernandez Avenue. Their appreciable sinking

range varies between 20 and 40 cm, occasionally reaching 80-100 cm (Figs. 4 and 5). Some cases of significant sinking with appreciable tilting occurred in buildings at the corner of roads intersecting Fernandez Ave. In these cases the lack of lateral support was an essential factor for the tilting.

b) Isolated and contiguous buildings with sinking and limited tilting (< 5 degrees) are shown in Figs. 5 and 6. Structures under this class are many

Fig. 7 Isolated buildings in Perez Blvd., with more than 1.5 m sinking and about 10 degrees tilting. The two pictures are in sequence.

Fig. 8 Isolated building in Perez Blvd. with 15 tilting and severe foundation settlement. Further rotation of the structure was prevented by the close-by building.

in Perez Blvd. and their sinking range varies between 10 and 100 cm.

c) Buildings with significant nonuniform sinking and severe tilting; this class is represented by a number of cases along Perez Blvd., (Figs. 7). While their sinking range is 40 to 150 cm, tilting mostly varies from 5 to 10 degrees.

In the spectacular case of Fig. 8 the tilting was estimated in about 15 degrees, but the entire rotation and collapsing of the structure was definitely prevented by the close-by building.

d) Already collapsed buildings. This class of structures, although represented in Dagupan, is not the most frequent one, if compared to damages due to sinking and tilting.

5.2 Roads and pavements

Roads in Dagupan City were catastrophically affected by the sand liquefaction and the consequent bulging they underwent was the major cause for the entire destruction of the rigid pavement. The well of city streets was due to the lightweight of road structure in comparison to the heavier load of buildings.

Concrete pavements of houses and courtyards were disrupted almost everywhere, with local occurrences of sand boils and foundations.

Figure 9 shows the damaged pavement of the Gasoline Station in Perez Blvd. at which floating of the buried oil tank occurred.

5.3 Bridges

The Magsaysay Bridge in Perez Blvd. collapsed during the quake (Figs. 3, 10, 11). Four out of seven bridge spans moved down the river bed, the remaining three being significantly affected by the

Fig. 9 View of the gasoline station in Perez Blvd. at which floating of the buried oil tank occurred.

Fig. 11 Collapsed Magsaysay Bridge in Perez Blvd.; view from the left bank.

Fig. 10 Collapsed Magsaysay Bridge in Perez Blvd.; view from the right bank.

quake, since the supporting piers underwent tilting. Lateral displacement of soils towards the center of the channel, due to liquefaction, was responsible for tilting of piers from right and left banks towards the center of the river (Fig. 3).

At this bridge location, cofferdams established on the left bank for the widening of the bridge underwent sinking and tilting despite their estimated length of 15 m and their embedment of about 10 m. A useful information about the depth at which liquefaction propagated during the

earthquake will be obtained by verifying the exact length of these sheet piles. Their present depth of embedment most probably coincides with the depth of the liquefaction zone. Comparatively, Quintos Bridge, about 450 m downstream in Fernandez St. and structurally similar to Magsaysay bridge, did not suffer any damage.

5.4 Utilities

Together with building foundations failure and bulging of roads in between, electric poles were tilted and part of the electricity network in the city area was disrupted. Road bulging automatically involved existing sewerage lines.

It has not been possible yet to make an inventory of all building and of the damage they suffered in order to derive a correlation between damage, foundation type and eventually SPT N values. Such an investigation would not only provide a clear picture of surficial distribution of damage, but also guide in assessing the relationship between soil behaviour and foundation type.

5.5 Evaluation of liquefaction potential

The simplified procedure proposed by Seed and Idriss (1971) with some modifications by Seed (1979) has been adopted for the evaluation of liquefaction potential.

The bedrock under Dagupan area is a structural depression filled by deltaic sediments, basically composed of alterna-

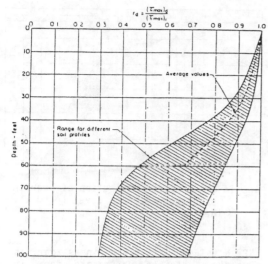

RANGE OF VALUES OF r_d FOR DIFFERENT SOIL PROFILES

Fig. 12 Plot of r_d versus depth (Seed and Idriss, 1971).

ting layers of sand and clay.

The propagation of seismic waves during an earthquake is a complex matter. Rocks forming the basement under Dagupan are intensely faulted which introduce a tectonic complexity aspect.

Simplifying the problem, however, during the 16 July quake the shear wave propagating from the epicenter, by travelling through and along the bedrock, but following the highest velocity travel path, reached the rock basement under Dagupan and then propagated upward. The horizontal component of this wave induced the liquefaction of surface sands forming the topmost layer of the town.

During earthquakes, under the effect of ground motion, saturated sands can undergo a significant increase of pore-water pressure. When pore-water pressure build-up equals the overburden pressure (confining pressure), effective stress becomes zero, the sand completely loosing its shear strength and thus undergoing liquefaction. Laboratory tests performed by Seed and Lee (1966) confirmed this theory. Loose saturated sands were tested under cyclic loading conditions and "during the ninth stress cycle the pore pressure suddenly increased to a value equal to the externally applied confining pressure and the sample developed large strains, which in the tenth cycle exceeded the 20%, in fact, the soil had liquefied, the effective confining pressure was reduced to zero, and over a wide range of strains the soil

could be observed to be in a fluid condition" (Seed & Lee, 1966). Four main factors are recognised to influence the condition causing liquefaction in saturated sands with a relative density lower than 60%.

- void ratio (relative density)
- confining pressure
- number of cycles
- magnitude of cyclic stress

Following the procedure proposed by H.B. Seed and I.M. Idriss (1971), horizontal shearing stresses developed by an earthquake at the depth "h", in a soil with the unit weight of "Γ", are expressed by the equation,

$$\tau_h = f = ma = h\,\Gamma/g\,a \qquad (1)$$

where "g" is the acceleration due to gravity and "a" the acceleration induced by the quake at the site.

To account for the deformability of soil sediments Eq. (1) is more appropriately expressed by,

$$\tau_{max} = a_{max}\,r_d\,(\Gamma\,h)/g \qquad (2)$$

where "r_d" is the reduction factor, estimated from Fig. 12, varying as function of depth. The maximum shear strength τ_{max}, however, is not considered a suitable value either, since shear stresses during a quake are typically irregular.

By appropriate weighting of individual stress cycles and based on laboratory test data, Seed and Idris (1971) propose Eq. (2) in the form

$$\tau_{av} = 0.65\,\tau_{max} = 0.65\,a_{max}\,r_d\,(\Gamma\,h)/g \qquad (3)$$

where the factor 0.65 accounts for the reduction of maximum stress to average stress.

The number of significant stress cycles "N_c" is evaluated from Table 1.

Stress ratios causing a peak cyclic pore pressure ratio of 100 % under multidirectional soil shaking in the field are given by the equation

Table 1

Earthquake Magnitude	No. Equiv. Stress Cycles, N_c
7	10
7.5	20
8	30

by Seed and Idriss (1971)

591

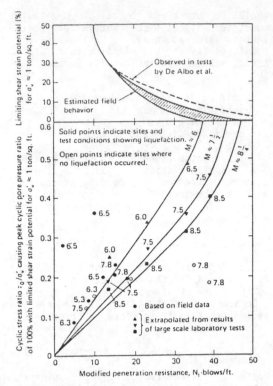

Fig. 13 Correlation between field liquefaction behaviour and modified penetration resistance N_1 (Seed, 1979).

$$(\tau_h/\sigma_o')_{1-field} = (\tau_o/\sigma_v') \qquad (4)$$

where (τ_o/σ_o') is derived from Fig. 13 (Seed, 1979), based on data from large-scale shear tests, σ_o' being the effective overburden pressure.

when $\tau_{av} = \tau_h$ in Eq. (3) and Eq. (4), for a given number of cycles, it means that the minimal value of shear stresses required to cause liquefaction was actually generated by the earthquake and then

$$0.65\ r_d\ \Gamma h\ a_{max}/g = (\tau_o/\sigma_v')\ \sigma_o' \qquad (5)$$

The zone of the collapsed Magsaysay bridge and the segment of Perez St. on the western side of the bridge were selected for the evaluation of the liquefaction potential.

The reason for choosing this area is basically due to the fact that the highest damage occurred there, and most of field investigations data are concentrated along Perez Blvd.

From the analysis of available data the following conclusions were derived:

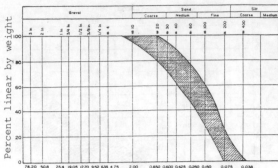

Fig. 14 Grain size distribution of Dagupan sand from ground level to 50 ft. depth.

- SPT N values, from field investigations executed after the earthquake, show in general good correlation with the water content of the the sand at the N test level. Although the correspondence was not so good when comparing data from different boreholes, the trend was consistent since for high N values, low water contents were found and vice versa. From in situ water content determinations at different depths, where N values were available, a plot of N versus water content was derived for a number of boreholes. It was found that there is a linear relation between them and that the best fitting lines for different bore-holes are almost parallel to one another. The separation of these lines most probably reflects different execution technique by various SPT operators and/or equipments with different characteristics.

- At two boreholes in Magsaysay Bridge, determinations of specific gravity of sand were performed at different depths. The values are distributed in the range 2.66 -2.75 g/cu cm, and the average value of 2.71 g/cu cm was adopted for calculation purposes.

- From all available determinations of water content it was found that they are distributed in the range 18% to 36% for different N values at various depths. Using the average specific gravity of 2.71 g/cu cm and water content, in situ unit weights were determined and a broad correlation between N values, water content and unit weight was derived. From such correlation data in Table 2, concerning Magsaysay Bridge, were obtained.

- Figure 14 shows the grain size distribution of Dagupan sand, in the area between Perez Blvd. and Fernandez Av., from ground surface to about 50 ft depth

(15 m). The gradation range was derived from samples at the Magsaysay Bridge and near the Quintos Bridge in Fernandez Av. Dagupan City surface soil can be defined as a mix of medium and fine sand with some silt and a $D_{50} = 0.2$ mm.

From above considerations the following qualitative correlation was established for the Magsaysay Bridge sequence:

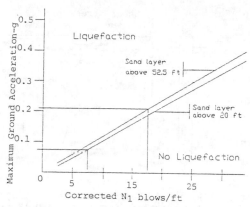

Fig. 15 Plot of ground acceleration versus N_1 values for liquefaction zones extending from 2 ft down to 20 and 52.5 ft depths.

Table 2

Depth ft	N	N_{lav}	Rel. Density D %	Unit Weight d	(pcf) d
0-5					115
0-15	2-3	4	41	119	
15-35	4-10	7.5	46	121	
35-52.5	15-30	18	60	128	

where N_{lav} stands for average corrected N, based on the equation $N_1 = N \ C_N$. Unit weights d_{wet} and d_{ps} represent in situ densities for submerged and partially saturated conditions, respectively. The correction factor C_N is determined from the relation proposed by Peck, Hanson and Thorburn (1979)

$$C_N = 0.77 \ \log_{10} 20/p$$

where p is the effective overburden pressure at the SPT elevation.

For the evaluation of the liquefaction potential, first a_{max}/g versus (τ_0 / σ_v') for different N_1 values is derived from Eq. 5 (and Fig. 13) at 52.5 ft and 20 ft depths (Fig. 16), for water table at 5 ft depth and soil properties listed in Table 2. The reason for choosing these depths is that 20 ft (minimum river bed elevation) and 52.5 ft (maximum depth of the sand layer) can be considered, respectively, as minimum and maximum depth for lower limit values of the liquefaction zone.

Figure 15 shows the plot of ground accelerations a_{max}/g versur N_1, for 52.5 and 20 ft. For $N_1 = 18$ (and a Relative Density of 60%) at $a_{max}/g = 0.21$ the entire sand layer (16.5 m thick) can undergo liquefaction; for $N_1 = 7.5$ (and a Relative Density of 46%) at $a_{max}/g = 0.08$ liquefaction can extend down 20 ft only.

In the area below each plot there is no liquefaction while above it liquefaction is expected to occur.

To verify liquefaction potential and lower limit of the liquefaction zone within the range 20 to 52.5 ft, the equivalent average cyclic stresses required for lique-

Table 3

Depth ft	r_d		τ_h psf	a_{max}/g	τ_{av} psf	a_{max}/g	τ_{av} psf
5	0.995	0.65	575	0.14	52	0.18	67
10	0.98	0.65	1170	0.14	105	0.18	134
15	0.97	0.65	1765	0.14	156	0.18	200
20	0.95	0.65	2370	0.14	205	0.18	263
25	0.93	0.65	2975	0.14	252	0.18	324
30	0.90	0.65	3580	0.14	293	0.18	377
35	0.88	0.65	4185	0.14	335	0.18	431
40	0.82	0.65	4825	0.14	360	0.18	463
45	0.80	0.65	5645	0.14	411	0.18	528
50	0.75	0.65	6105	0.14	417	0.18	536
52.5	0.725	0.65	6425	0.14	424	0.18	545

Table 4

Depth ft	N_{lav}	τ_h/σ_0'	σ_0' psf	τ_h psf
5		0.05	575	29
10	4	0.05	858	43
15		0.05	1140	57
20		0.10	1421	142
25	7.5	0.10	1705	171
30		0.10	1988	199
35		0.10	2270	227
40		0.18	2598	468
45	18	0.18	2925	527
50		0.18	3253	586
52.5		0.18	3416	615

faction and stresses induced by the quake are computed by Eqs. (2) and (4), at different levels, for $a_{max}/g = 0.14$ and 0.18.

From Figure 13, (Seed, 1979), which shows

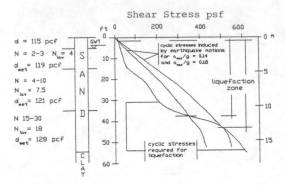

Fig. 16 Plot of shear stresses causing liquefaction (for 0.14 and 0.18 g) and shear stresses required for liquefaction.

the correlation between field liquefaction behaviour of sands (for level ground conditions) and modified penetration resistance, (τ_0 / σ_v') is derived. Figure 13, which applies to earthquakes with magnitudes from 6 to 8.25, for stress ratios required to cause peak cyclic pore-pressure ratios of 100% and limited strain potential, was obtained by combining all reliable data from,
- comparison on field liquefaction correlation with correlation based on liquefaction criteria adopted in China (Seed, 1979);
- tests by De Alba et al. on large-scale shear tests on the Monterrey N. 0 sand deposited by pluvial compaction;
- tests at the Waterways Experiment Station (WES) by Bieganousky and Marcuson (1977).

Table 3 and 4 show calculations of τ_{av} from Eq. (2) and τ_h from (3), respectively, at different depths for $a_{max}/g = 0.14$ and $a_{max}/g = 0.18$.

Figure 16, with the plot of τ_{av} and τ_h versus depth, shows that the liquefaction zone for Magsaysay Bridge conditions can extend from about 2 to 38 and 44 ft for different ground accelerations. The acceleration $a_{max}/g = 0.14$, however, has been sufficient to produce a liquefaction zone extending from 2 to 38 ft, with 36 ft (11 m) of thickness. This liquefaction zone, generated by the acceleration of 0.14 g, by propagating for about 18 ft (6 m) below the minimum elevation of the river bed, seems to be most consistent with tilting and sinking of bridge piers (Fig. 3). In addition to that, a liquefaction zone extending down about 40 ft quite well combines with the tilting and sinking of

the sheet piled cofferdam located on the right bank of the river as described in paragraph 5.

The section of Perez Blvd., West of the collapsed bridge, is characterized by a 56 ft (17 m) thick sand layer, with clay lenses at the depth of 39 to 48 ft (12-15 m). The sand forming topmost 40 ft of this road section is slightly denser than top sand layer in Magsaysay Bridge. Assuming that liquefaction in this zone affected the top 40 ft (12 m) of sand only, a value of about $a_{max}/g = 0.15$ would have been presumably sufficient for it.

Since top 40 ft of Perez Blvd. are mainly composed of loose to medium dense sand, the range $a_{max}/g = 0.14 - 0.15$ can be considered as the minimum value required for the liquefaction in this area.

6 CONCLUSIVE REMARKS

For medium level accelerations, 0.14-0.18 g, there is still a potential for liquefaction along Perez Blvd. and most likely in other areas of Dagupan and surroundings. For a better evaluation of liquefaction potential in Dagupan City, extensive field investigations and appropriate laboratory testing are required in order to identify location and size of dangerous areas and establish a local calibration for the N SPT. To minimize field investigation costs it would be sufficient to establish a reliable correlation between modified N_1 values and relative densities, identifying as well the modified N_1 value above which liquefaction is not likely to occurr.

An essential aspect is also the aerial extent in the northern Pangasinan and Tarlac Provinces of zones with loose sand at ground level and potential for liquefaction.

Liquefaction occurrence and sand boils were, in fact, reported in the town of Tarlac, which is located about 60 km South-east of Dagupan.

The problem of identifying liquefaction prone areas, although not so immediate as the reconstruction of Dagupan City, definitely has important economical implications both in terms of reconstruction of collapsed structures and future development of the region.

ACKNOWLEDGEMENTS

The author wishes to express his sincere thanks to Undersecretary Edmundo V. Mir (Dept. of Public Works and Highways) for the help received in collecting data from

Local Authorities and from the private company Foundation Specialists Inc. The author also expressed his sincere appreciation to Dr. D.T. Bergado and Mrs. Uraiwan Singchinsuk of the Division of Geotechnical and Transportation Engineering, Asian Institute of Technology, Bangkok, Thailand, for their valuable assistance in the preparation of the final manuscripts of this paper.

REFERENCES

Seed, H.B. and Lee, K.L. (1966). Liquefaction of saturated sands during cyclic loading. Journal of Soil Mechanics and Foundation Division. ASCE Vol. 92 No. SM6 Proc. Paper 4972 No. 1966, pp. 105-134.

Koizumi, Y. (1966). Change in density of sand subsoil caused by the Niigata earthquake. Soil and Foundation. Tokyo, Japan. Vol. 8 No. 2 1pp. 38-44.

Seed, H.B. and Idriss, I.M. (1967). Analysis of soil liquefaction: Niigata earthquake. Journal of Soil Mechanhics and Foundation Division. ASCE, Vol. 93, Proc. Paper 4223 May 1967, No. SM3, pp. 83-108.

Lee, K.L. and Seed, H.B. (1967). Cyuclic stresses causing liquefaction of sand. Journ. of Soil Mech. & Found. Div. ASCE Vol. 93, No. SM1, Proc. Paper 5058, Jan. 1967, pp. 47-70.

Peacock, W.N. and Seed, H.B. (1968). Sand liquefaction under cyclic loading simple shear conditions. Journal of Soil Mechanics and Foundation Division. ASCE Vol. 94, No. SM3, Proc. Paper 5957, May 1968, pp. 689-708.

Seed, H.B. and Idriss, I.M. (1971). Simplified procedure for evaluating soil liquefaction potential. Journal of Soil Mechanics and Foundation Division. ASCE Vol. 97, No. SM9, pp. 1249-1273.

De Alba P., Seed, H.B., and Chan, C.K. (1976). Sand liquefaction in large scale simple shear test. Journal of the Geotechnical Engineering Division. ASCE Vol. 102, No. GT9, Proc. Paper 12403, Sept. 1976, pp. 909-927.

Bieganousky, W.A. and W.F. Marcuson III (1977). Liquefaction Potential of Dams and Foundations. Report 2, Laboratory Standard Penetration Test on Platte River Sand and Standard Concrete Sand. WES Report n. 72-6 Viksburg, Miss. March 1977.

Seed, H.B. (1979). Soil liquefaction and cyclic mobility evaluation for level ground during earthquakes. Journ. Soil Mech. & Found. Div. ASCE Vol 105, No. GT2, Feb. 1979, pp. 201-255.

Punongbayan, R.S. (1987). Disaster Preparedness Systems for Natural Hazards in the Philippines: An Assessment. Dept. of Science and Technology, Philippine Instit. of Volcanology.